To : Paul Green

Study hard, work hard and make Ken happy.

Mahi Saleh

April 22, 2015

Principles of
Metal Casting

About the Authors

Mahi Sahoo, Ph.D., was a research scientist at CANMET Materials Technology Laboratory. He has received many awards, including the Gold Medal and also Scientific Merit Award from the American Foundry Society for contributions to non-ferrous casting technology. He is Fellow of ASM Int and CIM. Dr. Sahoo has published more than 250 peer-reviewed papers and numerous book chapters, and has delivered research presentations around the world.

Sam Sahu, Ph. D., was manager of Metallurgy and Technical Services at Waukesha Foundry. He has received the Scientific Merit Award from The American Foundry Society. Dr. Sahu hold five patents and has published 26 peer-reviewed scientific papers and one book chapter. He has 35 years of production foundry experience.

About SME

SME connects all those who are passionate about making things that improve our world. As a nonprofit organization, SME has served practitioners, companies, educators, government and communities across the manufacturing spectrum for more than 80 years. Through its strategic areas of events, media, membership, training and development, and the SME Education Foundation, SME shares knowledge to advance manufacturing. At SME, we are making the future. Together.

Principles of Metal Casting

Third Edition

Mahi Sahoo

Sudhari (Sam) Sahu

New York Chicago San Francisco
Athens London Madrid
Mexico City Milan New Delhi
Singapore Sydney Toronto

Cataloging-in-Publication Data is on file with the Library of Congress.

McGraw-Hill Education books are available at special quantity discounts to use as premiums and sales promotions, or for use in corporate training programs. To contact a representative please visit the Contact Us page at www.mhprofessional.com.

Principles of Metal Casting, Third Edition

1 2 3 4 5 6 7 8 9 0   QVS/QVS   19 18 17 16 15 14

ISBN 978-0-07-178975-2

MHID 0-07-178975-8

The pages within this book were printed on acid-free paper.

Sponsoring Editor Judy Bass	**Proofread** Suzanne Rapcavage
Acquisitions Coordinator Amy Stonebraker	**Indexer** Robert Swanson
Editorial Supervisor David E. Fogarty	**Production Supervisor** Lynn M. Messina
Project Manager Charu Khanna, MPS Limited	**Composition** MPS Limited
Copy Editor Surendra Shivam	**Art Director, Cover** Jeff Weeks

This book is dedicated to Late Prof. Carl R. Loper whose motivation and guidelines made this book possible.

Contents

Contributors

Andrew Adams *Senior Product Applications Manager, FOSECO, Cleveland, OH* (CHAPTER 5)

Al Alagarsamy *President, Alagarsamy Consulting, Spring, Texas* (CHAPTER 11)

Christof Heisser *President, MAGMA Foundry Technologies, Inc., Schaumburg, IL* (CHAPTER 6)

John R. McIntyre *President, Anderson Global, Muskegon Heights, MI* (CHAPTER 2)

Gregory V. Miskinis *Director of Research and Process Development, Waupaca Foundry, Inc., Waupaca, WI* (CHAPTER 13)

Raymond Monroe *Executive Vice President, Steel Founders' Society of America, Crystal Lake, IL* (CHAPTER 12)

Alfred Spada *Publisher/Editor, American Foundry Society, Schaumburg, IL* (CHAPTER 1)

Mahi Sahoo *President, Suraja Consulting Inc., Formerly Senior Research Scientist and Program Manager (Casting Technology) at CANMET Materials Technology Laboratory, Ottawa, Ontario, Canada* (CHAPTERS 1, 2, 4, 7–10, 14)

Sudhari "Sam" Sahu *President, Creative Technical Solutions, Inc., Glendale, WI, Formerly Manager of Metallurgy and Technical Services, Waukesha Foundry, Waukesha, WI.* (CHAPTER 12)

Laurence V. Whiting *Formerly Senior Research Scientist at CANMET Materials Technology Laboratory, Ottawa, Ontario, Canada* (CHAPTERS 3 AND 14)

Preface

This book, *Principles of Metal Casting*, first published in 1955, was later modified in 1967 (second edition). The authors were Richard W. Heine, Carl R. Loper, Jr., and Philip C. Rosenthal. Since then, casting technology has developed significantly in all areas such as melting, casting processes, sand systems, alloy development, heat treatment, processing technologies, etc. The 1967 publication has been a good resource for undergraduate and graduate students as well as personnel in metal casting facilities. Following consultation with Tom Prucha, vice president of Technical Services of the American Foundry Society (AFS) and several well-known foundry metallurgists, it was decided to consolidate the contents of the second edition and add new chapters on solidification modeling, casting defects, and zinc and zinc alloys, but include the latest technology in all areas. Similar to the second edition, the third edition is aimed primarily at students of metallurgy and foundry science and technology as well as practicing foundry engineers and research metallurgists.

While the second edition contained 25 chapters, the third edition has condensed the number of chapters down to 14. The following changes have been incorporated in the third edition:

1. The five chapters comprising molding processes and materials, molding process equipment and mechanization, molding sands, cores, and core materials have been condensed to one chapter called Casting Processes.

2. The two chapters on aluminum and magnesium foundry practices, and aluminum and magnesium casting alloys have been separated into two chapters, one on aluminum and aluminum alloys and the other on magnesium and magnesium alloys.

3. The two chapters on copper-alloy foundry practice and copper-base casting alloys have been covered in one chapter (Copper and Copper Alloys).

4. The three chapters on steel castings, steel melting in the foundry, and metallurgy of cast steels come under one chapter called Steel Castings.

5. The six chapters on the family of cast irons, melting of cast irons, gray-iron foundry practice, metallurgy of gray iron, ductile iron, and malleable iron are condensed into one chapter as Cast Irons.

6. The chapter on casting design considerations has been eliminated. Three new chapters—namely casting process simulation, zinc and zinc alloys, and casting defects—have been added. Thus, the third edition contains 14 chapters.

Of the three authors of the 1967 edition, Profs. Heine and Rosenthal had passed away when the remaining author, Prof. Carl Loper decided to rewrite the book. He had preliminary discussions with Sudhari (Sam) Sahu and Mahi Sahoo to join him to contribute to several chapters, and the outline of the third edition was developed. Since McGraw-Hill had published the second edition in partnership with AFS, it was decided to follow the same format for the third edition. Unfortunately, Prof. Loper also passed away in 2011. Since then, Sahoo and Sahu looked after the publication of the book and contacted several prominent people from the industry to contribute to some chapters. They are (1) Al Spada, AFS, Chapter 1 (Introduction), (2) John R. McIntyre, Anderson Global, Chapter 2 (Patterns), (3) Laurence V. Whiting, retired from CANMET Materials Technologies Laboratory, Chapter 3 (Casting Processes) and Chapter 14 (Casting Defects), (4) Andy Adams, Foseco, Chapter 5 (Gating and Risering of Castings), (5) Christof Heisser, Magma Foundry Technologies Inc., Chapter 6 (Casting Process Simulation), (6) Al Alagarsamy, Alagarsamy Consulting, Chapter 11 (Cast Irons), (7) Raymond Monroe, Steel Founders' Society of America, Chapter 12 (Steel Castings), and (8) Gregory Miskinis, Waupaca Foundry, Chapter 13 (Cleaning and Inspection). These contributors have many years of industrial experience and are well known in their respective fields of casting technology. We are thankful to them for their efforts and contributions.

Some publications from organizations such as ASM International, AFS, SFSA, Ductile Iron Society, etc., have been a useful source of information on various fields of casting.

We sincerely hope that this book will meet the expectations of undergraduate and graduate students, research scientists, and those in the metal casting industry.

Mahi Sahoo
Sudhari (Sam) Sahu

Acknowledgments

We have received excellent assistance from many societies, foundries and other industries, publishing companies, and individuals. They have provided photographs, figures, and tables for the book. We are grateful to all of them. They have been mentioned below.

Our special thanks to Laura Moreno, director, Special Publications, AFS and her staff for innumerable assistance in providing many references, figures, and photographs from the AFS library, and helping us to get reprint permissions from many publishers, societies, foundries, and product and equipment manufacturers. Thanks are also extended to the staff of Modern Casting at AFS for providing data and photographs for Chapters 1, 3, and 11. AFS is also a partner with McGraw-Hill in this publication.

In addition to AFS, we would like to express our gratitude to the following:

1. ASM International (Ms. Sue Sellers, Product and Project Administrator)
2. Copper Development Association, New York (Mr. James Michel)
3. North American Die Casting Association (Mr. Steve Udvardy)
4. Steel Founders' Society of America (SFSA)
5. TMS, The Minerals, Metals and Materials Society
6. ECK Industries, Manitowoc, WI (Mr. David Weiss)
7. PIAD Precision Casting, Greensburg, PA
8. Prof. Ravi Ravindran and his research group at Ryerson University
9. Prof. Sumanth Shankar at McMaster University
10. Drs. Yemi Fasoyinu, Kumar Sadayappan, and Wojciech Kasprzak of CANMET Metals, Hamilton
11. Mr. David Ashe of Natural Resources Canada
12. Anderson Global, Muskegon Heights, MI
13. Magma Foundry Technologies Inc., Schaumburg, IL
14. Foseco, Cleveland, OH
15. Waupaca Foundry, Waupaca, WI
16. Springer Science + Business media
17. Maney Publishing
18. Metallurgical Society of Canadian Inst. of Mining, Metallurgy and Petroleum (CIM)
19. Society of Automotive Engineers (SAE)
20. American Society for Testing and Materials (ASTM)
21. International Standards Organization (ISO)
22. International Lead Zinc Research Organization (ILZRO) (Dr. Frank Goodwin)
23. Dietert Foundry Testing Equipment Inc.

24. American Colloid
25. Didion International Inc.
26. DISA Industries, Inc.
27. Gudgeon
28. Palmer Manufacturing & Supply, Inc.
29. U.S. Pipe
30. Vulcan Engineering Inc.
31. Custom Aluminum, Cambridge, Ontario
32. Dr. Roberto Boeri, National University of Mar Del Plata, Conitet, Argentina
33. Castwell Products, Skokie, IL
34. Dr. George Vander Voort, Vander Voort Consulting LLC
35. De Pere Foundry Inc.
36. Capture 3D, Inc.
37. Inspec Tech Corp
38. Ms. Montserrat Tallec for publishing and graphics for most of the chapters
39. Rob Blair (SFSA) for assistance in artwork and graphics for the steel chapters
40. Dr. Robert D. Foley, University of Alabama at Birmingham
41. Society for Manufacturing Engineers

Mahi Sahoo
Sudhari (Sam) Sahu

CHAPTER 1

Introduction

Mahi Sahoo
Suraja Consulting Inc.

Alfred Spada
American Foundry Society

A casting is a metal object produced by solidifying molten metal in a mold. The shape of the object is determined by the shape of the mold cavity. The casting process, also known as founding, involves melting metal and pouring it into the mold cavity, which is close to the final dimensions of the finished form. Many types of complex objects ranging in size from a few grams to thousands of kilograms are produced in a metal casting facility. Castings are produced by various casting processes such as sand, permanent mold, investment, and lost foam. While all metals can be cast, the most predominant are iron, steel, aluminum, copper, magnesium, and zinc-based alloys. The science of treating the molten metal and designing the molds for smooth flow of molten metal into the mold cavity to minimize air entrapment are essential parts of casting technology to produce premium quality cast components.

Cast products are different from wrought products. The former are obtained when molten metal solidifies in a desired form. By contrast, wrought products start as cast ingots that are then thermomechanically shaped by different processes such as forging, rolling, and extrusion. In terms of value and volume, metal casting ranks second only to sheet steel in the metal producing industry.

Metal castings are used in more than 90% of all manufactured goods and find a wide range of applications in various sectors such as transportation (automotive, railway, naval, aerospace), mining, forestry, power generation, petrochemical, construction machinery, sporting goods, household appliances, and farm equipment. Typical cast components in major alloys are shown in Fig. 1.1.

Advantages of Casting Process

Certain advantages are inherent in the metal casting processes. These may form the basis for choosing casting as a process to be preferred over other shaping processes. Some of the reasons for the success of the casting process are as follows [1, 2]:

- The most intricate of shapes, both external and internal, may be cast. As a result, many other manufacturing operations such as machining, forging, and welding may be minimized or eliminated.

1

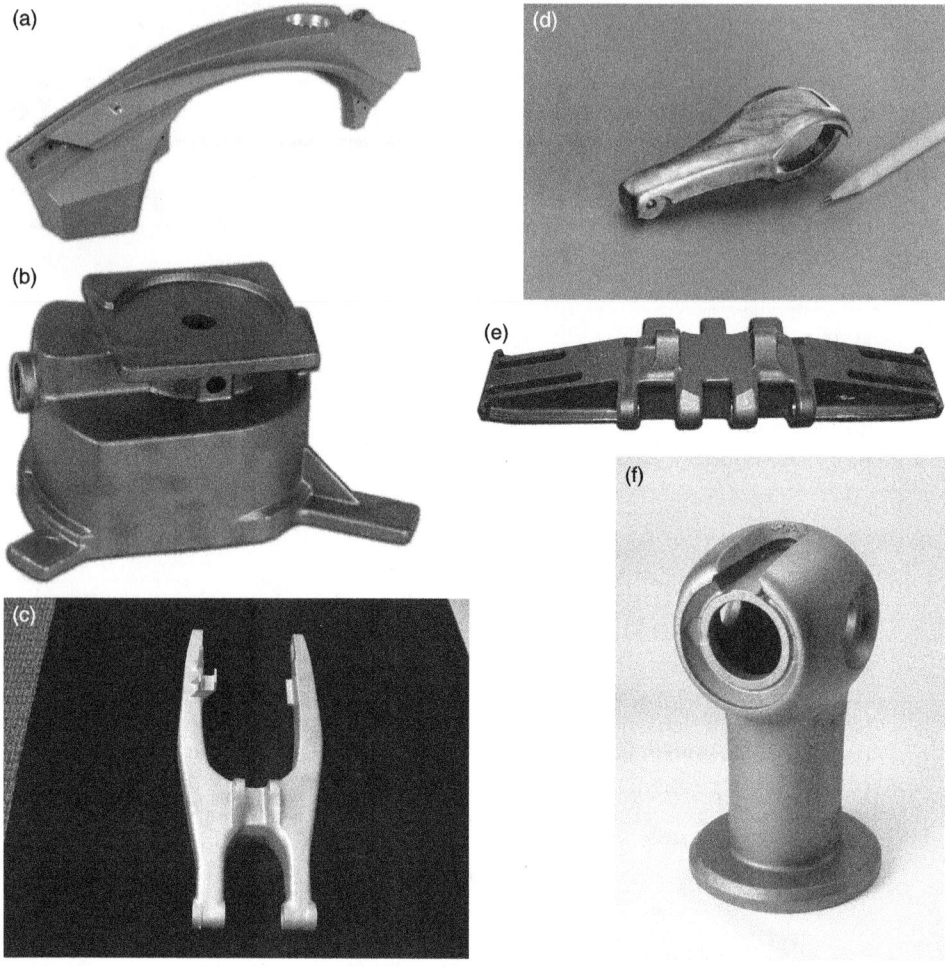

FIGURE 1.1   Typical castings in major alloys. (a) This motorcycle frame component was produced via the nobake sand casting process in 356 aluminum with T6 treatment temper. (b) The bronze alloy used for this dental suction pump was selected for its high strength, mechanical properties, and wear resistance. (c) Produced for a racing motorcycle, this one-piece magnesium casting replaced a three-piece aluminum part. The component is 33% lighter than the original, which impacts the overall performance of the bike. (d) This miniature zinc casket arm weighs less than 6 oz. (e) This NASA component for the space shuttle crawler transporter, produced with modified 4320 steel alloy via V-process casting, met reduced surface hardness requirements while maintaining high material strength. (f) This ductile iron green sand casting is the main structural element of the Spartan hydrant, enclosing and protecting its working parts.

- Because of their metallurgical nature, some metals can only be cast to shape since they cannot be hot-worked into bars, rods, plates, or other shapes from ingot form as a preliminary to other processing. A good example of casting is the family of cast irons which are low cost, extremely useful, and exceed the total of other metals in tonnage cast.

- Casting is a simplified manufacturing process. An object cast as a single piece often would otherwise require multiple manufacturing steps (stamping and welding, for example) to be produced any other way.
- Casting can be a low-cost, high-volume production process, where large numbers of a given component may be produced rapidly. Typical examples are plumbing parts and automotive components such as engine blocks, manifolds, brake calipers, steering knuckles, and control arms.
- Extremely large, heavy metal objects such as pump housings, valves, and hydroelectric plant parts which could weigh up to 200 tons may be cast. These components would be difficult or economically impossible to produce otherwise.
- Some engineering properties such as machinability, bearing, and strength are obtained more favorably in cast metals. In addition, more uniform properties from a directional standpoint can be expected, which is not generally true for wrought products.
- Casting technology has progressed significantly, allowing products to be cast with very thin cross sections, often referred to as "thin-wall-casting"; such capabilities allow designers to reduce the casting weight that is often assumed necessary for production.
- One has to consider the economic advantages of the casting process. In the aerospace industry, some components are still being machined out of forged or rolled pieces despite the fact such pieces can be cast more economically to meet the design criteria, especially with respect to strength and toughness.

In some cases, the casting process may give way to other methods of metal processing. For example, machining produces smooth surfaces and dimensional accuracy not obtainable in any other way; forging aids in developing the ultimate tensile strength and toughness in steel; welding provides a convenient method of joining or fabricating wrought or cast parts into more complex structures; and stamping produces lightweight steel metal parts. Thus the engineer may select from a number of metal-processing methods, singularly or in combination, which is most suited to the needs of his or her work.

The Foundry Industry

The world production of castings for different metals is shown in Table 1.1 for the years 2009 through 2011. It is evident that the world economic recession (2008-09) affected the production of castings. However, as the economy continues to improve, production of castings will increase and the global forecast is for 102 million tons of casting by 2015. China is the largest producer of castings followed by the United States and India. The top 10 countries in the world producing ferrous and nonferrous castings are listed in Table 1.2.

According to U.S. Department of Commerce statistics, metal casting remains 1 of the 10 largest industries when rated on a value-added basis. In the United States, the foundry industry is a $23 billion industry, employing more than 200,000 people

Metal	2009	2010	2011
Gray iron	37,749	43,258	45,870
Ductile iron	29,404	23,451	24,782
Malleable iron	1,013	–	–
Steel	9,070	10,215	10,342
Copper alloys	1,488	1,652	1,799
Aluminum alloys	9,477	10,879	11,319
Magnesium alloys	149	196	181
Zinc alloys	470	528	505
Total	80,895	91,673	98,593

Source: From Spada [3].

Note: Global forecast is for 102 million tons by 2015.

TABLE 1.1 World Production of Castings during 2009 to 2011 (in metric tons)

Country	Castings (in million metric tons)	Number of Casting Plants
China	41.3	30,000
USA	10.33	2,040
India	9.99	4,500
Germany	5.47	612
Japan	4.76	1,612
Russia	4.2	1,350
Brazil	3.34	1,325
Korea	2.34	890
Italy	2.21	1,111
France	2.05	441

Source: From Spada [3].

TABLE 1.2 Global Production of Castings in 2011

(see Table 1.3). As shown below, the 2012 shipments came to $34 billion based on GDP, housing starts, auto, railcar and truck production, construction activity, end-user and supplier interviews. Although the number of foundries is gradually decreasing, it is still a vital part of the manufacturing sector, producing 10.33 million tons of castings for different sectors, as mentioned earlier. At present 77% of the demand for castings in the United States is met by U.S. metalcasters. The lack of capacity in a number of foundries and high cost of production have led to imports from other countries such as China, India, Brazil, Mexico, and other Asian and European countries as shown in Fig. 1.2. In 2009, the United States imported 2.1 million tons of castings.

2,001 metal casting facilities

More than 700 ferrous and 1,300 nonferrous foundries

Employs more than 200,000

80% are small business (less than 100 employees)

2012 shipments: $34 billion

In 1955 and 1991 there were 6,150 and 3,200 plants, respectively

Second in production in the world

Lost approximately 150 plants since the beginning of recession

Source: From Spada [3].

TABLE 1.3   Health of U.S. Metal Casting Industry in 2013

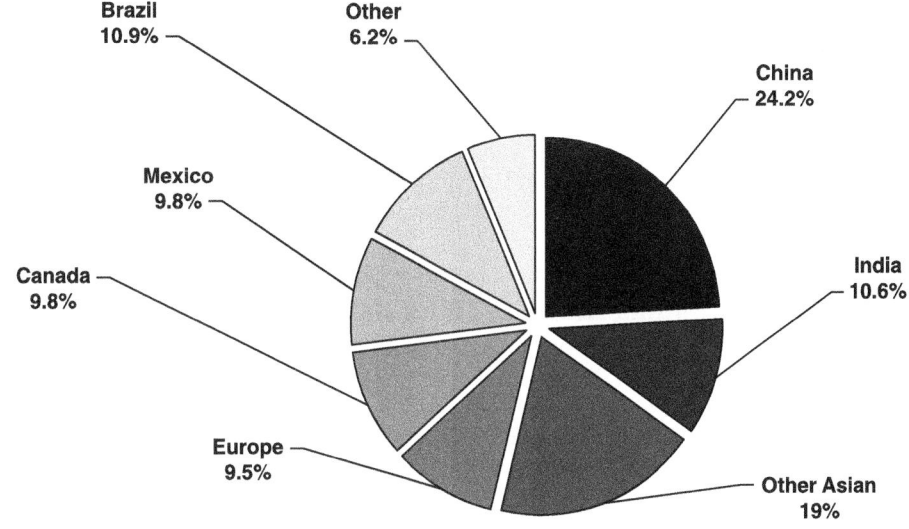

FIGURE 1.2   Import of castings to the United States from other countries. (From Spada [3].)

Future of the Casting Industry

Worldwide, the aluminum, steel, iron, and copper alloy casting industry is a mature industrial sector, with an established infrastructure, significant metal sources and scrap stream, a large alloy selection and multiple choices for casting process selection. This is not the case for the magnesium casting industry. Magnesium does play an important role in the transportation industry and other industrial applications. Being a structural material with low density and a very high strength to weight ratio, magnesium will reduce vehicle weight, and in turn, reduce greenhouse gas emissions and increase fuel economy. However, at present, it lacks a stronger infrastructure and technical support, including more casting process options, new alloy development with better creep properties at more than 150°C for automotive engine applications, environmentally friendly cover gas for melt protection, and scrap processing technologies. These technical barriers have been discussed in a recent publication on magnesium by the American Foundry Society [4].

The metal casting industry has to be more proactive in adapting new technologies to be competitive. Some of these include

- Enhanced adoption of simulation modeling to optimize casting design
- Vacuum- and pressure-assisted casting process to enhance casting properties
- Automated pouring designed for job shops to improve productivity and quality
- Ablation process for sand casting to achieve sound castings with mechanical properties similar to those permanent mold castings
- Application of rapid prototyping for tool-less casting applications
- Affordable automated grinding systems to improve productivity and reduce cost
- Metal-matrix composite casting including nanoparticle casting

Another important area for the growth of the foundry industry is part consolidation and conversion to castings. Some examples of castings converted from other manufacturing processes are shown in Fig. 1.3.

Education and Training in the Foundry Industry

Casting technology encompasses many branches of science and engineering including physics, chemistry, metallurgy, mechanical engineering, chemical engineering, and computational modeling. Chemical reactions involved in melting of metals and treating of liquid metals, formation of crystal structures, thermodynamic principles as applied to the determination of phase diagrams, design of gating and risering systems, and principles behind grain refinement and heat treatment all deal with some branch of science and engineering. Although there is still a lot of art and craft involved in the casting process, a good understanding of casting technology for the foundry engineer and technologist, mold maker, and pattern maker, would be useful to produce premium quality castings by minimizing casting defects. Application of solidification modeling in the design of gating and risering systems would eliminate the guesswork and contribute to the production of sound castings.

The engineer who designs a casting must have accurate information about the properties of the cast metals to be used. Handbook data may not be useful in the design of components that would lead to low mechanical properties. To provide a foundation for foundry work, course work in the principles of metal casting finds a place in the educational preparation of student engineers. In addition, training offered by casting institutes or societies must incorporate developments in all aspects of metal casting such as alloy development, melting, melt treatment, sand, mold design, and gating and risering design.

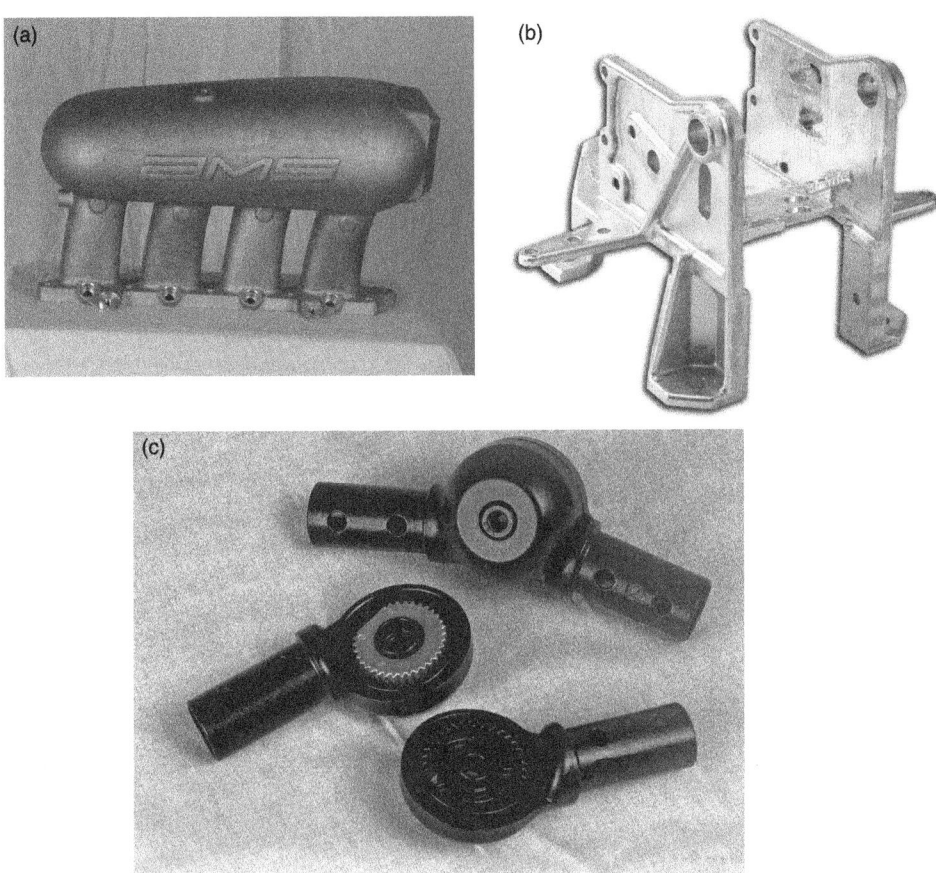

FIGURE 1.3 Examples of part consolidation and conversion. (a) This four-cyclinder intake manifold was cast in A356 aluminum via the green sand casting process. The casting was converted from a sheet metal weldment by Eagle Aluminum Cast Products, Muskegon, MI. The manifold is designed for the high-performance automotive aftermarket, specifically the Mitsubishi Evolution. (b) This steering/linkage assembly was cast in brass via the permanent mold casting process. The casting was converted from a multipiece weldment by Piad Precision Casting Corp., Greensburg, PA. The assembly is used in a pallet truck application and delivered the customer a 10% weight reduction over the weldment. (c) This locking hinge assembly was cast in 4140 steel via the investment casting process. The casting was converted from an aluminum fabrication by Signicast Investment Castings, Hartford, WI. The locking hinge fits on a wheelchair and is more durable than its aluminum counterparts. (From Spada [3].)

References

1. R. W. Heine, C. R. Loper, Jr., and P. C. Rosenthal, *Principles of Metal Casting*, McGraw-Hill Book Company, New York, 1967.

2. *ASM Handbook,* Vol. 15, *Casting,* 2008, "History and Trends of Metal Casting," pp. 3–15.

3. A. Spada, *Modern Casting: Annual Census of Casting Production,* 47th Census of World Casting Production, December 2013, pp. 18–23.

4. Mahi Sahoo (principal author), *Technology for Magnesium Castings: Design, Products and Applications,* American Foundry Society, Schaumburg, IL, 2011, pp. 1–7.

CHAPTER 2

Patterns

John McIntyre
Anderson Global

Mahi Sahoo
Suraja Consulting Inc.

Introduction

Patterns are the casting manufacturers' mold-shaping tool. The mold cavity, and therefore ultimately the casting, is made from the pattern. Even if only one casting is desired it is necessary to have a pattern, but a great many castings may be made from a single pattern. Obtaining suitable pattern equipment is thus the first step in making castings.

The exceptions to needing a pattern are some rapid prototyping methods where (a) the mold shape is milled from a block of sand using a computer numerically controlled (CNC) machine and (b) the mold shape is printed or laser sintered from grains of sand that are spread, a layer at a time, on a machine that hardens each layer using the 3D CAD (computer-aided design) model of the mold, as explained later under "Rapid Prototyping."

In the case of a lost foam pattern or a lost wax pattern, aluminum molds are made first and the disposable patterns formed in them.

Patternmaking

Patternmaking is divided between that which is done within foundries and that which is done by separate businesses called *pattern shops, foundry tooling manufacturers,* or *mold manufacturers.* Foundries often have pattern departments. Some foundries have both wood and metal-pattern facilities. However, most pattern departments in foundries are more concerned with repair, maintenance, and minor changes than with producing new patterns. The vast majority of patterns are made by pattern shops, foundry tooling manufacturers, or mold manufacturers which are independent of the foundry and operate as separate businesses [1, 2].

Patternmaking, the art and science of making patterns which will produce the desired casting dimensions, is not within the scope of this book. Certain principles that are applied to patterns, however, should be of common knowledge to all who may be concerned with castings.

There are two ways of getting around patternmaking and still getting a casting: (1) using an old casting as a pattern, and (2) using robots and CNC machines to cut a mold out of a block of chemically bonded sand.

CAD CAM

The contoured complex shape of castings, and therefore foundry tooling, has been made by wood patternmakers since the beginning of the casting industry. They would take 2D drawings and would saw, sand, and carve the casting shape for the tool. Metal tooling was then traced using "duplicators," which are machines that trace the wood pattern with a stylus and a cutting tool would cut the metal tool.

Since around 1980, CAD and CAM (computer-aided manufacturing) have been used more and more each year. Since around 2000, the majority of all castings have been defined with CAD and the tooling made with CAM.

Figure 2.1 shows a CAD model of an exhaust elbow with translucent core prints and risers. It should be noted that casting process simulation tools can be used to provide patternmakers and tool shops with the design of the entire mold including gating and riser systems as shown in Fig. 2.2*a* and *b*.

CNC Machined Foundry Tooling

After the CAD models are complete, CAM programs are made on a computer using a variety of CAM software. The software defines the "path" the cutting tool will take when cutting the foundry tooling.

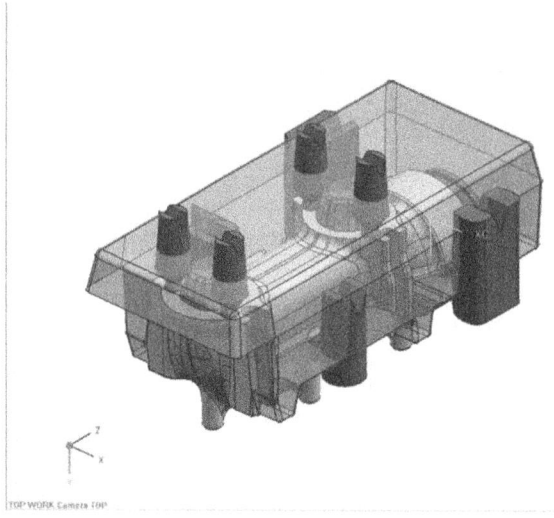

FIGURE 2.1   CAD model of an exhaust elbow with translucent core prints and risers. (From Anderson Global [3].)

(a)

(b)

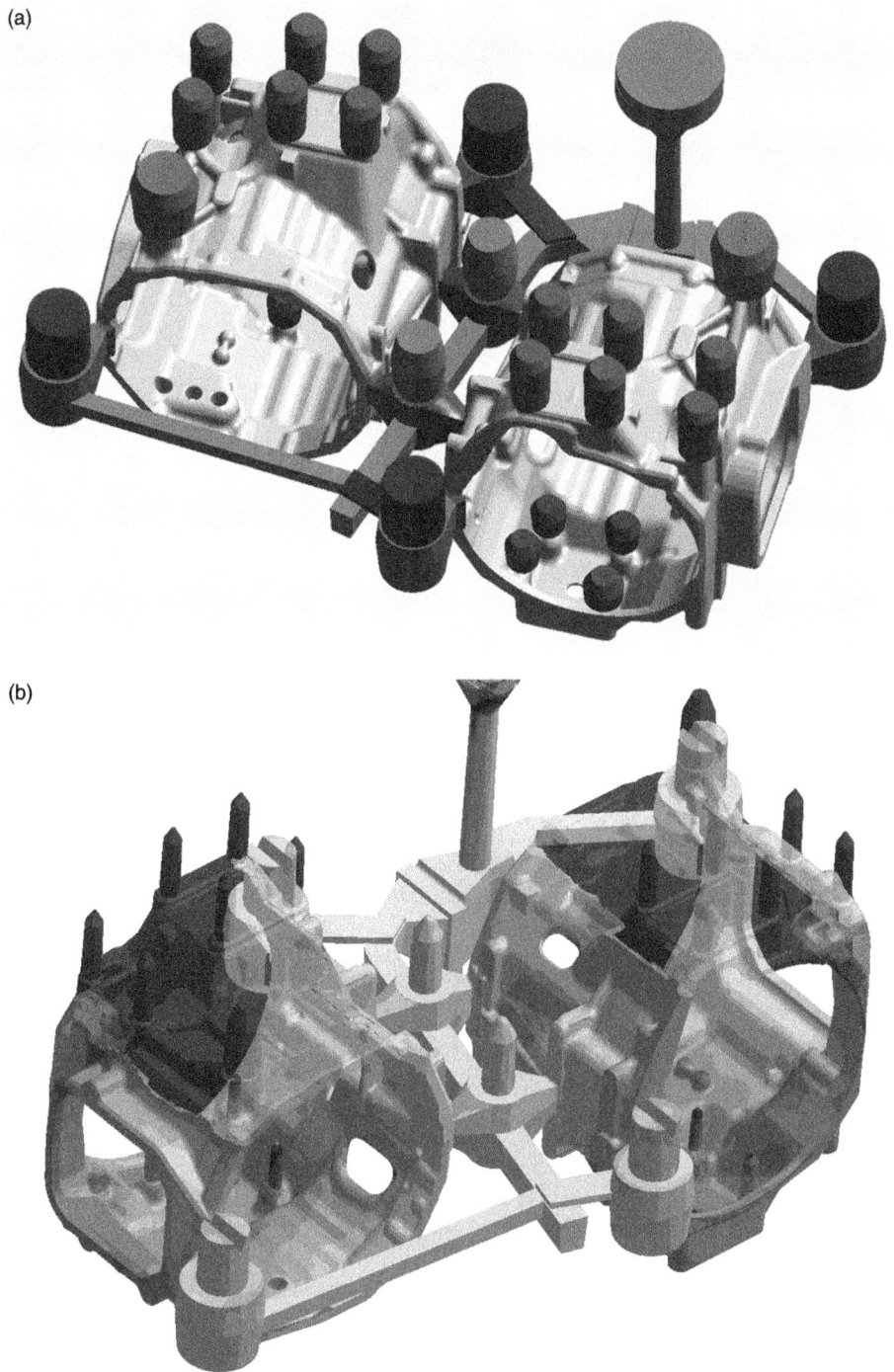

Figure 2.2a and b   Two examples of patternmaking showing entire mold with gating and riser systems. (From Heisser [4].)

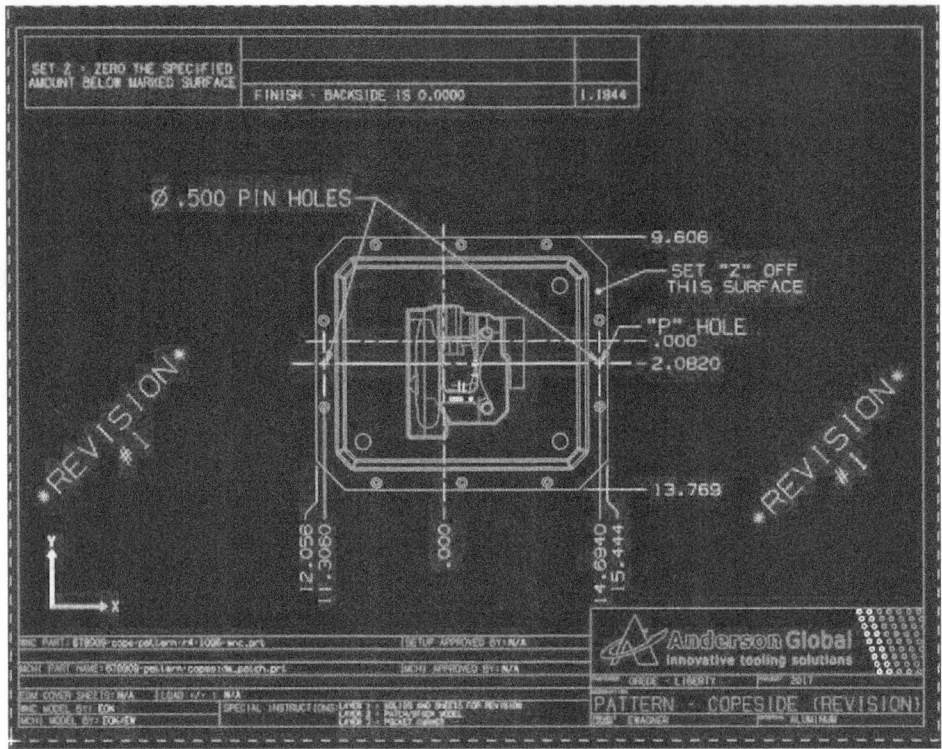

Figure 2.3 A CAM setup sheet (with CAD drawing) that aids the machinist in setting up the CAM tool path to the tool to be machined. (From Anderson Global [3].)

Figure 2.3 shows a CAM setup sheet, which aids the machinist in setting up the CAM tool path to the tool to be machined. A typical CNC machining facility is shown in Fig. 2.4.

Pattern Materials

Pattern materials have been described in detail in [5–7]. Wood is the most common pattern material. Patternmakers use two types of pattern lumber; soft and hard woods. The soft woods are primarily used in master patterns and for short-run production patterns. Hard wood is used for long-run production patterns where wear is a major factor. Typical soft woods are Sugar Pine, Northern White Pine, Idaho Pine, and Spanish Cedar in North America. Mahogany is the best known and best suited of all hard woods for pattern use. It should be pointed out that the type of wood used can vary from country to country depending on availability.

Plaster and gypsum cements are used for patternmaking in the automotive, aircraft, and plastic industries.

Plastic materials for pattern construction are available in two forms: solid shapes for machining and castable liquids. Plastics are also used as glues and adhesives, coating lacquers and enamels, and seals and gaskets. Plastics in the foundry industry are also used in the replacement or reproduction of worn out wood or metal patterns.

FIGURE 2.4   View of a typical CNC machining facility. These machines are 5-axis machining centers, which reduce machining time for foundry tooling components. (From Anderson Global [3].)

Magnesium alloys are mainly used for core box work. Its weight being 35% that of aluminum, core boxes are easier to handle. It is free machining, takes mirror-like surfaces, and can be machined at high speed.

Pattern Coatings and Release Agents

Two types of coatings are used for the patterns: paints and metallic coatings. The paint-type coatings are usually used on wood or plastic patterns. Nitrocellulose base lacquers, epoxies, and polyurethane coating are the most commonly used paintings, but the type used can vary from foundry to foundry depending on the sand, binder, release agents, and so forth. Metallic coatings are of three types: electroplated, those produced by chemical reduction, and metal sprayed. Hard chrome plating (mostly for ferrous metals), electroplated nickel, and electroplated copper belong to the first group. Electroless nickel (90 to 92% Ni with the balance being P) is an example of a chemically reduced coating for pattern and core boxes. Metal spraying or metalizing can be done to apply a thin coating on wood, epoxy, or metal substrate. Atomized liquid metal can be sprayed by arc spray or plasma spray. Details on pattern coatings and release agents can be found in References [5] to [7].

Rapid Prototyping

Traditionally sand castings involve significant cost and lead time of several weeks or months. This is overcome by rapid prototyping (RP), which is a technique that can quickly construct a scale model of a part using the three-dimensional CAD data.

RP process and CNC machining contribute to automated patternmaking. It is not uncommon to hear the term RP, rapid manufacturing, rapid casting, and rapid tooling to describe the acquisition of components within a shortened time frame. RP has also been referred to as solid-form manufacturing, computer-automated manufacturing, and layered manufacturing. It allows designers to quickly create tangible prototypes of their design, rather than just two-dimensional pictures. Prototypes can be used for design testing. Some advantages of RP include making components faster and less expensively, decreasing development time, increasing effective communication, decreasing costly mistakes, minimizing sustainable engineering changes, and extending product life time by adding necessary features and eliminating redundant features early in the design. References [8] to [17] deal with rapid prototyping.

Steps involved in RP are the following:

1. Construction of the CAD model and conversion to STL format.
2. Processing the STL file in the RP machine to create sliced layers of the model.
3. Creation of the first layer of the physical model. The model is then lowered by the thickness of the next layer, and the process is repeated until completion of the model.
4. The model and any supports are removed. The surface of the model is then finished and cleaned.

A large number of competing technologies are available in the marketplace for RP. These are

1. Selective laser sintering (SLS) creates a sand mold directly by sintering layers of coated shell sand [12]. The process is slow and the mold is expendable. Hence, it is not a viable option for quantities higher than 1.
2. Laminated object manufacturing (LOM) utilizes paper layers to create an object of the desired shape and size that could be used directly as a pattern for metal casting.
3. Stereolithography (SLA) and fused deposition modeling (FDM) create patterns for investment casting or models that can be indirectly used for other molding processes.
4. Three-dimensional printing (3DP) can create objects via a process similar to inkjet printing which are subsequently infiltrated to increase their strength.
5. Multi-jet modeling (MJM) builds parts in layers from the ground up, in extremely fine details using a very durable, acrylic-based resin.
6. Direct metal laser sintering (DMLS) is an additive RP technology, also referred to as SLS or selective laser melting (SLM). The technology consists of fusing metal powder into a solid part by melting it locally using the focused laser beam. Parts are built up additively layer by layer, typically using layers 20 µm thick. The process allows for highly complex geometries to be created directly from the 3D CAD data automatically, in hours without any tooling.

RP has been widely used for years by the defense and aerospace industries, but other industries such as pump, medical, and automotive are starting to take advantage of its benefits.

Figure 2.5 shows a printed core package that will be able to make a V-8 block without tooling. The printed cores are suitable for most casting processes to prove process or make prototypes. The process is too slow and expensive to use in production.

Figure 2.5 Printed core package for a V-8 block. (From Anderson Global [3].)

Types of Patterns

Several types of patterns are used in foundries. Depending on the casting requirements, the pattern may conform to one of the following types:

- Single or loose patterns
- Gated patterns (loose)
- Match-plate patterns
- Cope and drag patterns
- Vertically parted patterns
- Lost foam patterns
- Lost wax patterns

Each of the pattern types has characteristic uses.

Loose Patterns

Loose patterns are single copies of the casting but incorporate the various allowances and core prints necessary for producing the casting. They generally are of wood or modeling board construction but may be made of metal, plaster, plastics, wax, or any other suitable material. Relatively few castings are made from any one loose pattern since hand molding is practiced and the process is slow and costly. The parting surface may be hand-formed. Gating systems are hand-cut in the sand. Drawing the pattern from the

FIGURE 2.6 Example of a loose pattern. (From Anderson Global [3].)

sand, after rapping it to loosen it from the sand, is also done by hand. Consequently, casting dimensions vary. A loose pattern is shown in Fig. 2.6. Such a pattern might be used for producing prototype castings, to make tooling castings, and to make fixture castings.

Gated Patterns

Gated loose patterns are an improvement on ungated loose patterns. The gating system is actually a part of the pattern and eliminates hand-cutting the gates. More rapid molding of small quantities of castings results with this type of pattern.

Match-Plate Patterns

Large-quantity production of small castings requires match-plate patterns or more specialized types of pattern equipment. Figure 2.7a and b shows the gating system attached to the match plate. The cope and drag portions of the pattern are mounted on opposite sides of a wood or metal plate conforming to the parting line. Match plates are also integrally cast in which cast pattern and plate are cast as one piece in sand or plaster molds. Match plates with more accurate patterns are fully machined from aluminum or other metal. Figure 2.8 shows a metal match-plate pattern.

Gating systems are almost always attached to the plate. Match plates are generally used with some type of molding machine, in order to obtain maximum speed of molding. The improved production rate possible with these patterns serves to compensate for their increased cost. Plates also increase the dimensional accuracy of the casting. A limitation of the match-plate pattern arises in the weight of mold and flask, which can be handled by the molder when using manual squeezer molding machines. Automatic match-plate molding machines can handle large and heavy molds.

Cope and Drag Pattern Plates

Cope and drag pattern plates are shown in Fig. 2.9. Cope and drag plates consist of the cope and drag parts of the pattern mounted on separate plates. The cope and drag halves of the mold may thus be made separately by workers on different molding machines. The molding of medium and large castings on a molding machine is greatly facilitated by this type of pattern equipment. Separate cope and drag plates are more costly, but this type of pattern equipment is usually necessary in high-speed mechanized or automated molding. Separate pattern plates require accurate alignment of the two mold halves by means of guide and locating pins and bushings in flasks in order that the upper and lower parts of the casting may match.

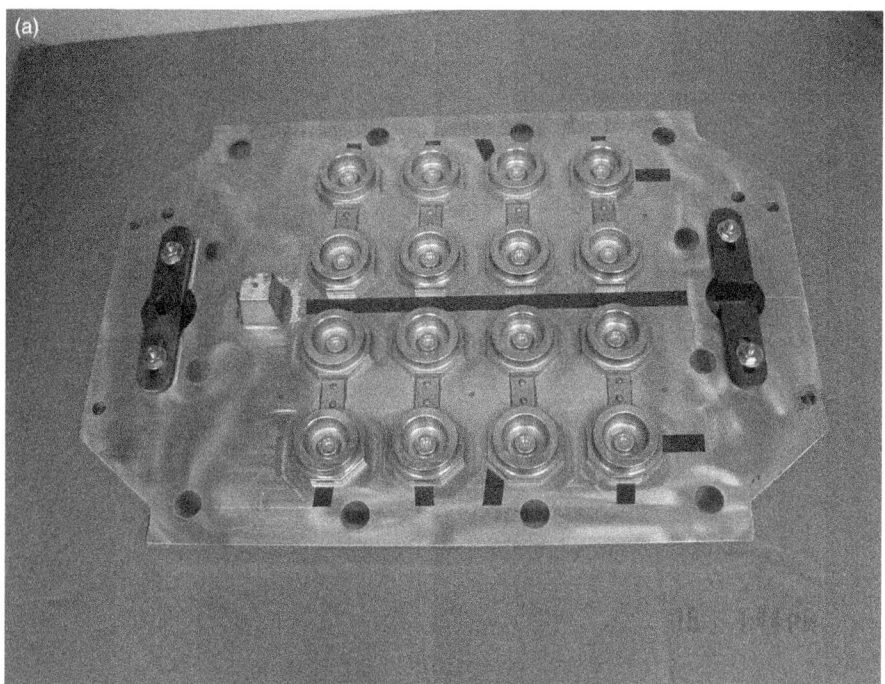

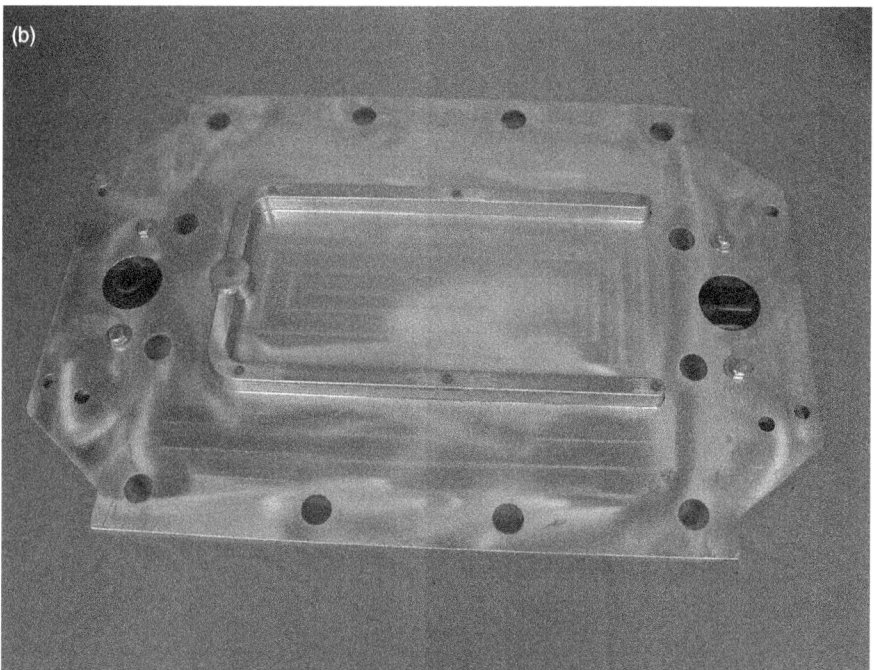

Figure 2.7   (*a*) Gating system attached to the match plate, drag part. (*b*) Gating system attached to the match plate, cope part. (From Ducharme et al. [18].)

Figure 2.8   Metal match-plate pattern. (From Anderson Global [3].)

Figure 2.9   Cope and drag pattern plates. (From Anderson Global [3].)

Vertically Parted Patterns

Vertically parted patterns create a green sand mold that is a continuous line of sand that has a ram cavity on one side of the sand and a door cavity on the opposite. This process is excellent at core-less, high production castings and can accept cored work, but is limited in sizes and weights of cores.

Lost Foam Patterns

Lost foam patterns are polystyrene replicas of the casting to be cast. They are made in aluminum molds which have poly beads blown into them and steam injected in to expand the beads and form the patterns. This process is good for castings that require a better surface finish and have areas that would be difficult to core out with sand cores. Atypical foam pattern is shown in Fig. 3.77. Chapter 3 describes the lost foam casting process.

Lost Wax Patterns

The investment casting process has been described in chapter 3 (Fig. 3.74). Lost wax patterns are wax replicas of the casting and made in aluminum molds in which the wax is injected into. After the wax pattern is made, multiple patterns are assembled onto a wax gating tree. Once the wax patterns are assembled to the gating tree, multiple dips into ceramic slurry create a ceramic shell that is fired in an oven, which melts the wax. The metal is poured into the ceramic shell to create the castings. This process is used when extreme accuracy is required for a near net or net shaped casting. Titanium turbine blades for jet engines are one of the many types of castings made in this process.

Follow Board

Loose patterns having an irregular parting line are difficult to mold without a follow board, or *match*. The pattern match serves to support the loose pattern during molding of the drag half of the mold and also establishes the parting surface when the match is removed.

Master Pattern

A master pattern, often made of wood or modeling board, is used as an original for casting aluminum pressure cast match plates or pressure cast patterns to be mounted to a flat parted match plate. Several pressure cast patterns may be cast from the master and mounted on a flat parted match plate or pattern plate after they have the mounting surface machined. A master pattern incorporates certain dimensional allowances, discussed in the following section.

Pattern Allowances

Although the pattern is used to produce a casting of the desired dimensions, it is not dimensionally identical with the casting. For metallurgical and mechanical reasons, a number of allowances must be made on the pattern if the casting is to be dimensionally correct.

Shrinkage Allowance

Shrinkage allowance on patterns is a correction for solidification shrinkage of the metal and its contraction during cooling to room temperature. The total contraction is volumetric, but the correction of it is usually expressed linearly. Pattern shrinkage allowance is the amount of pattern that must be made larger than the casting to provide for total contraction. It may vary from a negligible amount of up to $\frac{5}{8}$ in/ft (15.9 mm/30 cm), depending on the metal and the nature of the casting. Typical shrinkage allowances are given in Table 2.1. The linear allowances in Table 2.1 are representative of castings in sand molds.

Casting Alloys	Pattern Dimension		Type of Construction	Section Thickness		Contraction	
	inch	mm		inch	mm	in/ft	mm/m
Gray cast iron	Up to 24	Up to 61	Open construction	–		$\frac{1}{8}$	10.5
	25–48	63.5–122	Open construction	–		$\frac{1}{10}$	8.2
	Over 48	Over 122	Open construction	–		$\frac{1}{12}$	6.9
	Up to 24	Up to 61	Cored construction	–		$\frac{1}{8}$	10.5
	25–36	63.5–91.4	Cored construction	–		$\frac{1}{10}$	8.2
Cast steel	Up to 24	Up to 61	Open construction	–		$\frac{1}{4}$	21.0
	25–72	63.5–183	Open construction	–		$\frac{3}{16}$	15.7
	Over 72	Over 183	Open construction	–		$\frac{5}{32}$	13.1
	Up to 18	Up to 45.7	Cored construction	–		$\frac{1}{4}$	21.0
	19–48	48.3–122	Cored construction	–		$\frac{3}{16}$	15.7
	49–66	124.5–167.6	Cored construction	–		$\frac{5}{32}$	13.1
	Over 66	Over 167.6	Cored construction	–		$\frac{1}{8}$	10.5
Malleable cast iron				$\frac{1}{16}$	1.6	$\frac{11}{64}$	14.4
				$\frac{1}{8}$	3.2	$\frac{5}{32}$	13.1
				$\frac{3}{16}$	4.8	$\frac{19}{128}$	12.5
				$\frac{1}{4}$	6.4	$\frac{9}{64}$	11.8
				$\frac{3}{8}$	9.5	$\frac{1}{8}$	10.5
				$\frac{1}{2}$	12.7	$\frac{7}{64}$	9.2
				$\frac{5}{8}$	15.9	$\frac{3}{32}$	7.9
				$\frac{3}{4}$	19.1	$\frac{5}{64}$	6.6
				$\frac{7}{8}$	22.2	$\frac{3}{64}$	3.9
				1	25.4	$\frac{1}{32}$	2.6
Aluminum	Up to 48	Up to 122	Open construction	–		$\frac{5}{32}$	13.1
	49–72	124.5–183	Open construction	–		$\frac{9}{64}$	11.9
	Over 72	Over 183	Open construction	–		$\frac{1}{8}$	10.5
	Up to 24	Up to 61	Cored construction	–		$\frac{5}{32}$	13.1
	25–48	63.5–122	Cored construction	–		$\frac{1}{8} - \frac{1}{16}$	10.5–5.2
	Over 48	Over 122	Cored construction	–		$\frac{9}{64} - \frac{1}{8}$	11.8–10.5
Magnesium	Up to 48	Up to 122	Open construction	–		$\frac{11}{16}$	57.4
	Over 48	Over 122	Open construction	–		$\frac{5}{32}$	13.1
	Up to 24	Up to 61	Cored construction	–		$\frac{5}{32}$	13.1
	Over 24	Over 61	Cored construction	–		$\frac{5}{32} - \frac{1}{8}$	13.1–10.5
Brass	–			–		$\frac{3}{16}$	15.7
Bronze	–			–		$\frac{1}{8} - \frac{1}{4}$	10.5–21

Source: From Heine et al. [1].

TABLE 2.1 Pattern-Shrinkage Allowances (Before Specifying, Consult the Patternmaker and Foundryman)

Casting Alloys	Pattern Size		Bore		Finish	
	inch	cm	inch	mm	inch	mm
Cast iron	Up to 12	30.5	$\frac{1}{8}$	3.2	$\frac{3}{32}$	2.4
	13–24	33–61	$\frac{3}{16}$	4.8	$\frac{1}{8}$	3.2
	25–42	63.5–106.7	$\frac{1}{4}$	6.4	$\frac{3}{16}$	4.8
	43–60	109.2–152.4	$\frac{5}{16}$	7.9	$\frac{1}{4}$	6.4
	61–80	155–203.2	$\frac{3}{8}$	9.5	$\frac{5}{16}$	7.9
	81–120	205.7–304.8	$\frac{7}{16}$	11.1	$\frac{3}{8}$	9.5
	Over to 120	309.8	Special instructions	Special instructions	Special instructions	Special instructions
Cast steel	Up to 12	30.5	$\frac{3}{16}$	4.8	$\frac{1}{8}$	3.2
	13–24	33–61	$\frac{1}{4}$	6.4	$\frac{3}{16}$	4.8
	25–42	63.5–106.7	$\frac{5}{16}$	7.9	$\frac{5}{16}$	7.9
	43–60	109.2–152.4	$\frac{3}{8}$	9.5	$\frac{3}{8}$	9.5
	61–80	155–203.2	$\frac{1}{2}$	12.7	$\frac{7}{16}$	11.1
	81–120	205.7–304.8	$\frac{5}{8}$	15.9	$\frac{1}{2}$	12.7
	Over to 120	309.8	Special instructions	Special instructions	Special instructions	Special instructions
Malleable iron	Up to 6	Up to 15.2	$\frac{1}{16}$	1.6	$\frac{1}{16}$	1.6
	7–9	22.9	$\frac{3}{32}$	2.4	$\frac{1}{16}$	1.6
	10–12	22.9–30.5	$\frac{3}{32}$	2.4	$\frac{3}{32}$	2.4
	13–24	30.5–61	$\frac{5}{32}$	4.0	$\frac{1}{8}$	3.2
	25–35	61–88.9	$\frac{3}{16}$	4.8	$\frac{3}{16}$	4.8
	Over 36	Over 91.4	Special instructions	Special instructions	Special instructions	Special instructions
Brass, Bronze, and Aluminum-alloy casting	Up to 12	Up to 30.5	$\frac{3}{32}$	2.4	$\frac{1}{16}$	1.6
	13–24	33–61	$\frac{3}{16}$	4.8	$\frac{1}{8}$	3.2
	25–36	63–91.4	$\frac{3}{16}$	4.8	$\frac{5}{32}$	4.0
	Over 36	Over 91.4	Special instructions	Special instructions	Special instructions	Special instructions

Source: From Heine et al. [1].

TABLE 2.2   Guide to Pattern Machine–Finish Allowances unless Otherwise Specified

However, special conditions prevail with some metals. White iron, for example, shrinks about $\frac{1}{4}$ in/ft (6.4 mm/30 cm) when cast, but during annealing it grows about $\frac{1}{8}$ in/ft (3.2 mm/30 cm), resulting in a net shrinkage of $\frac{1}{8}$ in/ft (3.2 mm/30 cm). Some grades of ductile iron have practically no shrinkage. The amount of shrinkage is usually determined by the casting source and based on similar types of castings in the particular alloy.

Sometimes double allowances are made if a pattern is first made in wood or modeling board and then in some other metal, as in making master patterns for pressure cast match plates or pressure cast loose patterns. For example, an aluminum pressure cast match plate or pressure cast aluminum pattern made from a wood master pattern may require a total allowance of $\frac{9}{32}$ in/ft (7.2 mm/30 cm) on the wood pattern if a gray-iron casting is to be made. The total allowance on the original wood pattern will then provide for shrinkage of the aluminum pressure cast plate and of gray-iron castings made from the aluminum pressure cast plate.

Machine Finish Allowance

Machine finish allowance is the amount the dimensions on a casting are made oversize to provide stock for machining. Typical finish allowances are presented in Table 2.2. It can be seen that these allowances are influenced by the metal, the casting design, and the method of casting and cleaning. The values in Table 2.2 are for castings made in a conventional sand mold. Other casting processes permit different finish allowances to be used. In general, machine finish allowance may be a minimum if the surfaces to be machined are entirely in the drag half of the mold, since dimensional variation and other defects are usually least prevalent there. Smaller castings can have less machine stock than those values given in Table 2.2.

Pattern Draft

Draft is the taper allowed on vertical faces of a pattern to permit its removal from the sand or other molding medium without tearing the mold-cavity surfaces. A taper of 0.06 in/ft (1.5 mm/30 cm) is common for vertical walls on patterns drawn by hand. Machine-drawn patterns require about 1° taper. In some cases, even vertical walls 6 to 9 in (152 to 229 mm) deep may be drawn by machine if the pattern is very smooth and clean and the drawing equipment is properly aligned. In the case of pockets or deep cavities in the pattern, considerably more draft is necessary to avoid tearing the mold during withdrawal of the pattern.

Size Tolerance

The variation that may be permitted on a given casting dimension is called its tolerance, and is equal to the difference between the minimum and the maximum limits for any specified dimension. Typical values for heavy castings which require maximum tolerance are given in Table 2.3.

Metal	Tolerance	
	inch	mm
Gray cast iron	$\frac{1}{16}$	1.6
Malleable iron	$\frac{3}{32}$	2.4
Cast steel	$\frac{5}{32}$	4.0
Aluminum alloys	$\frac{5}{64}$	2.0
Magnesium alloys	$\frac{11}{64}$	4.4
Brass	$\frac{3}{32}$	2.4
Bronze	$\frac{1}{8}$	3.2

Source: From Heine et al. [1].

TABLE 2.3 Typical Tolerance for Casting Weighing 455 kg (1000 lb) or More

The values in Table 2.3 are approximately maximum values. A common rule states that size tolerance should be at least half the shrinkage allowance. However, where there is considerable experience with a casting, and cooperation between the foundry and the casting purchaser exists, much closer tolerance may be established. Where such conditions prevail, tolerance of only a few thousandths of an inch (25.4 mm) may be maintained with some casting processes.

Smaller castings and castings made in permanent molds or precision sand can be made much tighter than indicated in Table 2.3. Tolerances of as little as +/−0.01 in (0.25 mm) can be held in many cases.

Distortion Allowance

Certain objects, such as large flat plates and dome- or U-shaped castings, sometimes distort when reproduced from a straight or perfect pattern. In such cases, the pattern may be intentionally distorted, or "faked." The distorted pattern then produces a casting of the proper shape and size.

Example of Allowances

Examples of other allowances would be: loading metal walls to allow for internal core movement and distortion, core print to mold clearance allowance, core print to core print clearance allowance for multiple core stackups, core wash allowance for cores that need to be dipped in ceramic slurry to resist burn-on, and mold coating allowance for permanent molds that always have a ceramic coating sprayed and baked on to form a barrier between the molten aluminum and the steel or iron mold.

Functions of Patterns

The main purpose of a pattern is its use in molding. However, to produce a casting successfully and render it suitable for further processing, the pattern may be required to perform other functions besides producing a mold cavity. These are briefly considered as follows.

Molding the Gating System

Good gating practice for castings generally requires that the system of channels and feeding reservoirs (gates and risers) for introducing metal into the mold cavity be attached to the pattern. The gating system may then obtain the benefits of machine molding.

Establishing the Parting Line

On a flat pattern plate, the parting surface is a simple plane. Many castings, however, require curved parting surfaces (Fig. 2.10). Because of the contoured shape of the castings, the parting lines need to follow this and come back to a flat plane at the perimeter.

Making Core Prints

When a casting requires cores, provision is made on the pattern for core prints. Core prints are portions of the pattern and mold cavity which serve to anchor the core in proper position in the mold. The core print is added to the pattern, but does not appear on the casting because it's blocked off by the core. Core prints are illustrated in Fig. 2.11.

Establishing Locating Points

The foundry, pattern shop, or machine shop employs locating points or surfaces on the casting to check the casting dimensions. Machining operations may also use the locating points in establishing the position of machined surfaces relative to the balance of the casting.

Figure 2.10   Curved parting surfaces. (From Anderson Global [3].)

Minimizing Casting Defects Attributable to the Pattern

Properly constructed, clean, and smooth-surfaced patterns are a necessity in making good castings. Patterns with rough, nicked surfaces and undercuts, loosely mounted, and in a generally poor condition contribute substantially to defective castings containing sand inclusions and other imperfections.

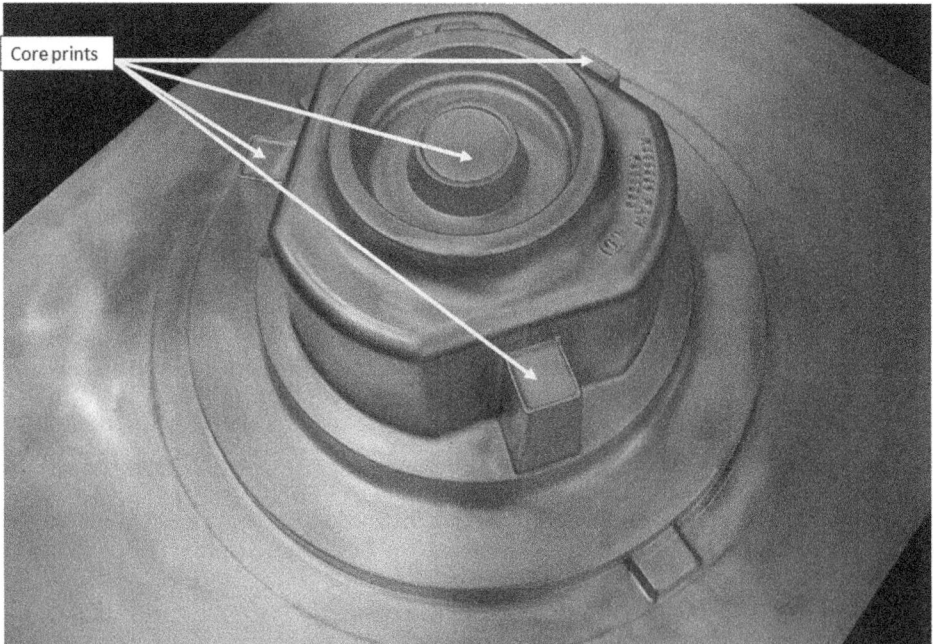

Figure 2.11   Core prints. (From Anderson Global [3].)

Providing for Ram-Up Cores

Sometimes a part of a mold cavity is made with cores, which are positioned by the pattern before the molding sand is rammed. The ram-up core then is held by the sand that has been packed around it. This is usually done only for prototypes or very low-volume castings.

Providing Economy in Molding

The pattern should be constructed to achieve all possible savings in cost of the casting. Here such items may be considered as the number of castings in the mold, the proper size of the pattern plate to fit available molding equipment, method of molding, and other factors.

Core Boxes

Core boxes, although not referred to as patterns, are an essential part of the pattern equipment for a casting requiring cores. Core boxes are constructed of wood or metal. The simplest type of box is the dump box, illustrated in Fig. 2.12. The top of the box is flat, and the core is removed by placing a plate over the box and inverting it.

A split box is a two-piece box, which can have a flat or contoured parting surface. A simple gang core box and accompanying pattern are shown in Fig. 2.13a and b. A gang box permits making several cores in the same box simultaneously.

Some core boxes require provisions for electrical or gas heating if they are to be used for shell core making, hot-box core making, or inorganic core making. Cores blown into the core boxes by machine need vents in the core boxes to relieve the compressed air

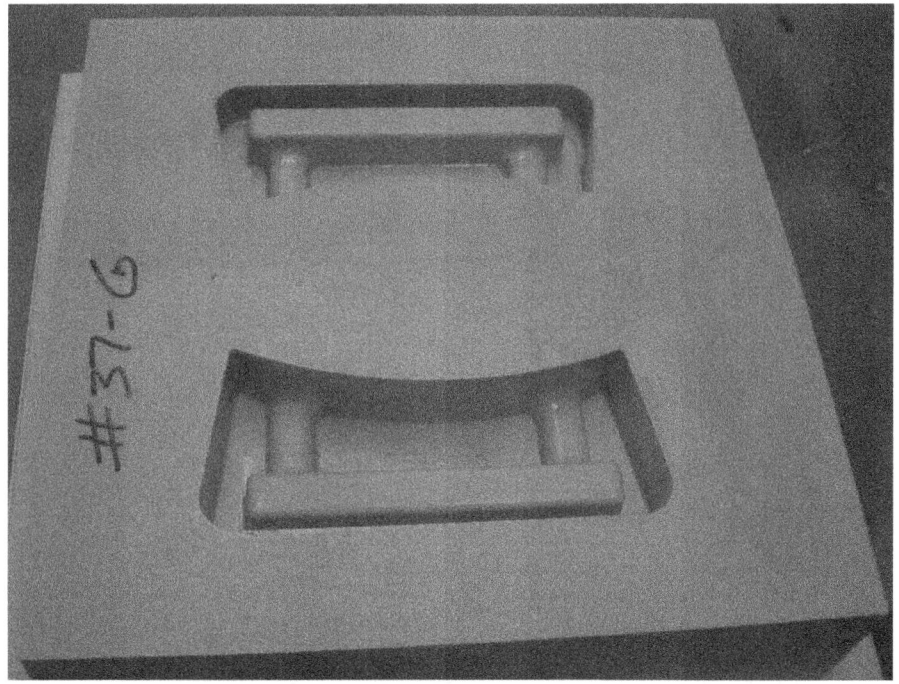

Figure 2.12 Dump box. (From Anderson Global [3].)

Figure 2.13 (*a*) Simple vertically parted split "ganged" core box. (*b*) Pattern for the gang core box. (From Anderson Global [3].)

pressure used to blow the sand into the box. Core boxes can be blown vertically through openings on the top of the box or horizontally through "blow tubes." If using blow tubes, tamp pins are used to shape the core where the tube leaves a core stub before curing. The main blown core processes are amine cold box, SO_2 cold box, furan hot box, and shell or an inorganic resin.

The importance of good pattern equipment cannot be overemphasized. Patterns that take into account the problems of molding and coremaking, proper gating and risering, ease of cleaning, and further processing promote quality in castings. As was pointed out earlier, the subject is one for detailed treatment, beyond the scope of this text. Further information on construction and principles of patterns may be obtained from some of the references listed at the end of this chapter.

References

1. R. W. Heine, C. R. Loper, Jr., and P. C. Rosenthal, "Principles of Metal Casting," McGraw-Hill Book Company, New York, 1967, chapter on "Patterns," pp. 8–22.

2. *ASM Handbook*, Vol. 15, *Casting*, 2008, "Patterns and Patternmaking," pp. 488–496.

3. Anderson Global, Muskegon Heights, Michigan, USA.

4. C. Heisser, MAGMA Foundry Technologies, Inc., Schaumburg, IL, USA.

5. American Foundry Society, *Patternmaker's Manual, 1986*, 2d ed., Reprinted in 1994, 2006, 2009, and 2011.

6. American Foundry Society, *Pattern Making Guide, Text and Illustrations*, 2d ed., E. Hamilton, Des Plaines, IL, 1990.

7. Patternmaking, Cast Metals Technology Series, American Foundry Society, 1981.

8. R. Gustafson, E. Guinn, and D. Tait, "Rapid Prototyping for Pattern and Foundry Tooling," *Modern Casting*, 85: 48–50, Feb. 1995.

9. M. Frank, F. Peters, X. Luo, and S. Oberbroeckling, "Rapid Patternmaking," *Steel Founders' Society of America, National Technical and Operating Conference*, Chicago, IL, Paper 4.6, Dec. 11–13, 2008.

10. T. Hahn, N. Demarczyk, D. Gantner, and T. Becker, "Comparison between Rapid Moldmaking Technologies," *Transactions of the American Foundry Society*, Paper 05-007, Vol. 113, 299–309, 2005.

11. S. Singamneni, O. Diegel, D. Singh, and N. McKenna, "Rapid Casting of Light Metals: An Experimental Investigation Using Taguchi Methods," *International Journal of Metalcasting*, 5(3): 25–36, Sept. 2011.

12. Y. Tang, J. Y. H. Fuh, H. T. Lo, Y. S. Wong, and L. Lu, "Direct Laser Sintering of Silica Sand," *Materials & Design*, 24(8): 623–629, 2003.

13. http://www.efunda.com/processes/rapid_prototyping/intro.cfm. eFunda is an engineering online resource for technical information (accessed Dec. 27, 2013).

14. "Rapid Prototyping," http://en.wikipedia.org/wiki/Rapid_prototyping. Wikipedia, The Free Dictionary (accessed Dec. 27, 2013).

15. "Direct Metal Laser Sintering (DMLS)," http://en.wikipedia.org/wiki/Direct_metal_laser_sintering. Wikipedia, The Free Dictionary (accessed Dec. 27, 2013).

16. J. Shah, "Rapid Prototyping Technologies Enhance Military Part," *Metal Casting Design and Purchasing*, American Foundry Society, Sept./Oct. 2010, pp. 20–21.

17. "Capitalize on Rapid Manufacturing, Metal Casting Design and Purchasing," Vol. 13, No. 6, pp. 41–43, Nov./Dec. 2011.

18. S. Ducharme, M. Sahoo, and K. Sadayappan (eds.), *Casting Copper-Base Alloys*, American Foundry Society, Schaumburg, IL, 2007, p. 178.

CHAPTER 3

Casting Processes

Laurence V. Whiting
Formerly with CANMET Materials Technology Laboratory
Natural Resources Canada

Introduction

Many casting methods have been developed to produce metal shapes. Although some are uniquely suited to produce specific kinds of castings; foundries frequently use more than one casting process for historical and economic reasons. Factors such as casting size, shape, complexity, quantity required, surface finish, and dimensional accuracy, together with alloy requirements, must all be taken into account in the selection of the most economical casting process.

Good castings require good molds with careful attention to detail. Unfortunately, most foundry operations involve both the human element and an element of "what can go wrong will go wrong." The Canada Centre for Mining, Metallurgy and Energy Technology (CANMET), after visiting 100 Canadian foundries, determined that sand and molding were responsible for 80% of the defective castings, while melting and metal composition accounted for 20% [1]. Although the better automotive foundries operate with 0.7% defective castings, other shops may produce an excessive number and variety of defective castings. As a result a large number of terms are used to describe defects. Because molding processes involve so many operations, it is only possible to optimize costs once the interplay of cause and effect on all possible defects is understood. It can be a mistake to try to control a single defect. Although the first part of this chapter only alludes to defects, later sections make wholesale reference to various defects. The reader is referred to the sections on defects later in this chapter which discuss greensand molding and cores, and to Chapter 14.

Because of the importance of the mold, casting processes are often described by the materials and methods employed during molding. The term *molding process* refers to the method of making the mold and the materials used. Whereas *casting process* conveys a broader meaning, often including the molding process, the method of introducing the metal into the mold cavity, or all the processes used in making the casting. It is perhaps convenient to further divide the casting process for liquid metals into two groups according to whether (1) the molds can be reused after the castings are ejected or (2) the molds consisting of refractory aggregates must be destroyed to recover the castings. The majority of molds consist of two or more pieces that form a cavity into which liquid metal is poured to produce one or more connected castings.

Reusable metal or ceramic molds are either not wetted by liquid metal or have coatings that survive the thermal attack of liquid metal. Processes that use the mold more than once are

1. Gravity permanent molding uses a mold made of metal or graphite. Simple metal shapes form cores that are later extracted. Liquid metal enters the mold cavity under the force of gravity. Semipermanent molding is a variant that uses complex-shaped cores made of chemically bonded sand. In low-pressure permanent molding, the metal is forced into the die from below.

2. High-pressure die casting forces liquid or semisolid metal (squeeze casting) into a metal mold under high pressure.

3. Centrifugal casting runs a measured volume of liquid metal into a spinning mold.

4. Continuous casting produces sheet and simple shapes (rods, rail) in a continuous manner from a port in the bottom or side of a furnace or a trough.

Casting rates in reusable molds are limited by the mold temperature, the amount of heat that has to be extracted into the mold before the castings are strong enough to be ejected, and the time it takes to cool the mold so the next casting can be poured. Many firms using reusable molds do not regard themselves as foundries, although the operations employed are similar.

Processes that use the mold only once require the use of patterns and share a number of common features:

1. The use of a pattern and perhaps one or more core boxes

2. Some type of aggregate mixture made up of a granular refractory and a binder

3. A means of shaping the aggregate mixture around the pattern

4. The aggregate is hardened or a bond is formed while in contact with the pattern

5. Withdrawal of the pattern from the hardened aggregate mold

Examples are shown in Fig. 3.1. The two halves of the mold and core pieces are assembled to make a complete mold, and then the metal is poured into the mold cavity.

The following molding processes produce expendable single use molds, and differ in the method of forming the mold, the granular refractory, and the method of bonding it.

1. Greensand or clay-bonded sand molding in which sand, clay, and water are intimately mixed to produce a pliable but not pasty solid which is then molded. Older variants are
   a. Natural sand and loam molding have been largely supplanted by synthetic mixes.
   b. Dry sand molding where the greensand mold is fired or dried.

2. Chemically bonded sands employ polymeric or chemical glues to hold the sand grains in place. These processes are frequently known by the names of the binder.
   a. Shell molding or Croning process—a high-strength binder used as a thin shell of a mold without backing for small rangy steel castings.
   b. Furan Nobake
   c. Alkyd Urethane Nobake
   d. Phenolic Urethane Nobake
   e. Furan Warm Box
   f. Gas-cured binder systems

MAKING A GREEN SAND MOLD

Iron casting to be produced in the subsequent illustrations of molding.

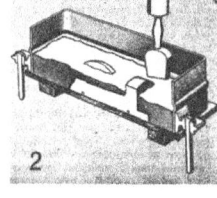

1. Cross-section of the first step in making a green sand mold. Bottom half of the pattern is on the mold board and surrounded by the bottom or drag half of the flask.

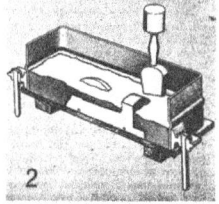

2. Molding sand is rammed around the pattern in multiple steps to provide uniform density.

3. After the bottom half of the mold is filled, it is rolled upright and the top half of the pattern and flask are put in place to complete the mold.

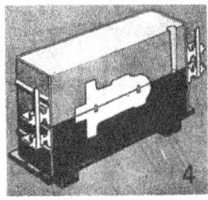

4. Section through the completed mold with pattern still in place and the sprue hole formed for entrance of molten metal.

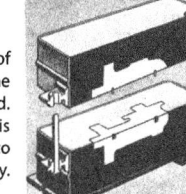

5. Cope and drag halves of mold are separated so that the pattern can be removed. The gate channel is then cut from the sprue to the mold cavity.

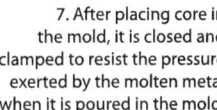

6. The core of bonded sand is made separately to form the internal passages of the casting.

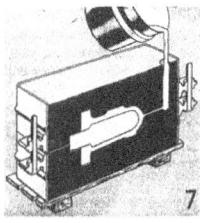

7. After placing core in the mold, it is closed and clamped to resist the pressure exerted by the molten metal when it is poured in the mold.

FIGURE 3.1 Basic steps in making a greensand casting of a valve body. (From Metal Casters Reference & Guide [2].)

3. Inorganic binder systems
   a. CO_2 process
   b. Phosphate self-setting binder
   c. Portland cement
4. Plaster molding uses a ceramic (originally Plaster of Paris) to make a mold.
5. Ceramic molding uses two ceramic slabs as in the gravity permanent mold process which are formed, cured, and then used for casting.
6. Lost wax or investment casting uses a wax replica to which a thick ceramic coating is applied. The wax is removed, the ceramic fired and the casting poured.
7. Ablation casting—water is used to reduce mold wall thicknesses during solidification to control the size of crystal structures formed during solidification (specifically dendrite arm spacing) and thereby improving the mechanical properties.
8. Loose sand processes
   a. The V-process wraps loose sand in a plastic film and air is removed; the force of atmospheric pressure hardens the sand and makes it rigid.
   b. Evaporative pattern process uses a foam replica immersed in loose sand.

9. There are processes (vacuum casting) in which metal is sucked or forced up into the mold from a reservoir. The Cosworth process uses an electromagnetic pump.

10. Slush casting, also known as thixocasting and rheocasting.

Foundries need to determine their market niche, review available process constraints and advantages, and then make their equipment purchases to comply with their niche. Equipment dictates not only production rates but also casting tolerances, surface finish, quality control procedures, labor (amount and skills), energy, and costs. Once selected, it is expensive to compete in a market with a different level of excellence. Production rates in the core room, melting, and molding departments must mesh. For example, if a molding line is running with casting cavities that require 5 tons/h, and the melt department has only 4 tons/h available, then the molding line must juggle jobs and their patterns, otherwise, the line will be idle 20% of the time. Processing restrictions need to be identified and solutions found. A molding machine must be linked with a pouring deck or pouring carousel with adequate mold storage between them to buffer some of the vagaries of usage rates. Molding rates are steady, but pouring operations stop to replenish ladles, therefore intermediate areas (pouring decks and carousels) soon fill up with unpoured molds. Additional mold storage is required for castings to solidify and their strength reaches levels sufficient for shakeout. Older mains (50 or 60 Hz) frequency electric induction melting furnaces were intended as heel melters, that is, the furnace can only be powered at maximum kilowatt when full, and only 20 to 30% of capacity is available for tapping and then ingots or scrap immediately charged to replenish bath levels. Medium frequency induction furnaces are usually emptied, i.e., batch melting. Operating a mains frequency furnace as a batch melter decreases potential production by 20%. Foundry management needs to be vigilant as to their equipment constraints, otherwise new, less productive procedures get established.

If on the other hand, the market niche is automotive castings, then the need to produce perhaps a million castings will be central in selecting the casting process. The process will need to be highly mechanized, produce defect-free castings with reproducible mechanical properties, surface finish, minimal finishing, machining, and heat treatment costs. Cost differentials of the order of pennies per casting multiply quickly when a million castings are required. So other factors such as labor, scrap rate, casting yield, energy, inspection rates and costs, waste streams and their costs can be as important as the capital cost of the process. Table 3.1 addresses particulars of some of the casting processes discussed in this chapter. The higher operating costs for certain processes may, in some cases, be outweighed by their greater precision. With greater precision, certain finishing or machining operations may possibly be eliminated, thus actually reducing the net cost of the casting. Besides the direct costs of sand and binder, the variable costs—labor, floor space, downtime, maintenance, utilities, cleaning, casting yield, and overall sand consumption—need to be considered. Figure 3.2 shows the number of U.S. foundries using the various casting processes in 2013 [3].

The various casting processes will now be explained in more detail. However, not all foundry alloys can be cast by all methods. Several of these casting processes were developed for specific alloys and products, and although other alloys could be cast, the additional costs are unwarranted.

Casting Process	Weight Range, kg (lb)	Economical Quantity	Thin Section, mm (in)	Cast Hole Dimensions, mm (in)	Surface Finish, μm (μin)	Dimensional Accuracy, mm (in)
Greensand	0.1–100 (0.2–220)	1–100,000	>5 (0.2)	>6 (0.25)	6–25 (250–1,000)	–
CO_2/silicate	0.1–100,000 (0.2–220,000)	1–1,000	>5 (0.2)	>6 (0.25)	6–25 (250–1,000)	–
Cold box	0.1–100,000 (0.2–220,000)	1–1,000	>5 (0.2)	>6 (0.25)	6–25 (250–1,000)	–
Shell mold	0.1–50 (0.2–110)	1,000+	>2 (0.1)	>3 (0.1)	3–12.5 (125–500)	±0.5–100 (±0.02–4)
Ceramic	0.1–5,000 (0.2–10,000)	50+	>1.5 (0.06)	–	0.8–3.2 (30–125)	±0.4–100 (±0.02–4)
Investment	0.1–50 (0.2–110)	1,000+	>1.5 (0.06)	>1 (0.05)	0.8–3.2 (30–125)	±0.5–100 (±0.02–4)
Gravity	0.1–100 (0.2–220)	1,000+	>5 (0.2)	>6 (0.25)	3–6 (125–250)	±0.25–100 (±0.01–4)
Low pressure		1,000+				±0.5–100 (±0.02–4)

Source: American Foundry Society [4].

TABLE 3.1 Process Capabilities of Some Casting Methods

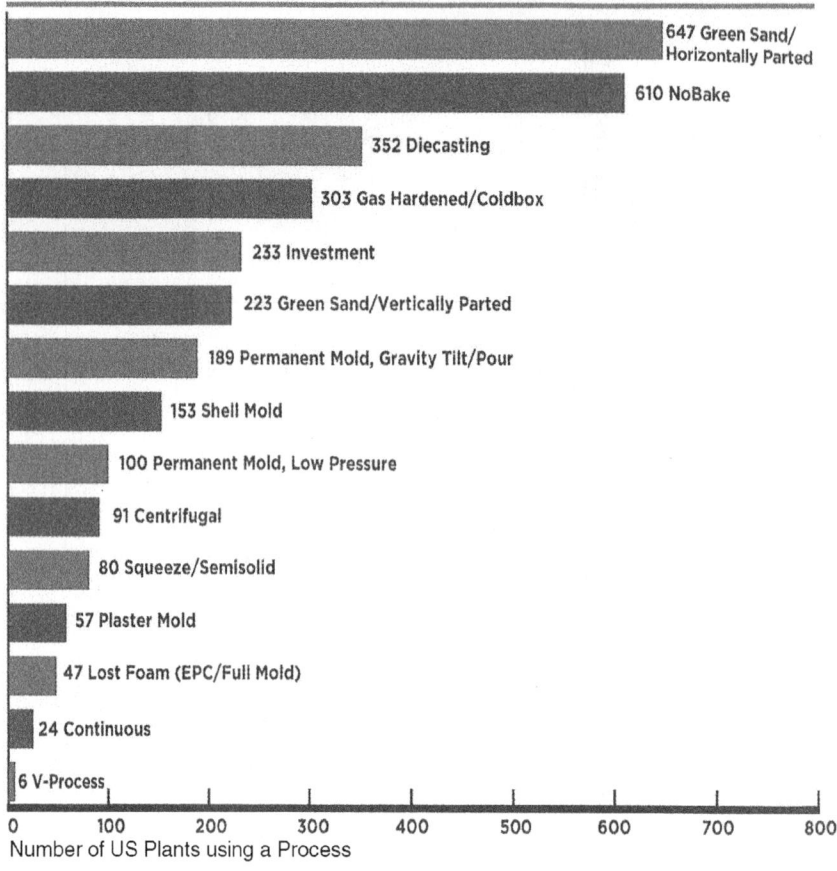

FIGURE 3.2 Molding methods used in U.S. casting facilities. (Courtesy of Metal Casting Design & Purchasing [3].)

Foundry Equipment for Reusable Molds
Gravity Permanent Molding

Gravity permanent molding is also known as gravity die casting and die casting in other countries, whereas die casting is known as pressure die casting elsewhere. Gravity permanent molding uses a mold made of metal, graphite, or ceramic. The system for getting liquid metal into the casting (otherwise called the gating system, which contains pouring cup, down sprue, runner, and ingates, see Chapter 5), along with feeders (relief risers) and overflow are all machined or cast into the die (Fig. 3.3). Two die halves are clamped together usually pneumatically by backing plates which help operate a series of metal ejector pins and simple-shaped metal cores. Liquid metal enters the gating system and the mold cavity under the force of gravity. A variant of this is a semipermanent molding which uses complex-shaped cores made of chemically bonded sand (Fig. 3.4). The outer metal mold is reusable whereas the sand cores are sacrificed as their shape precludes simple extraction. Nevertheless, the cores must withstand the heat without distorting; although high binder levels do not collapse readily (see later in the chapter in the discussion of cores). Typical examples are domestic brass water faucets and cast iron water fittings.

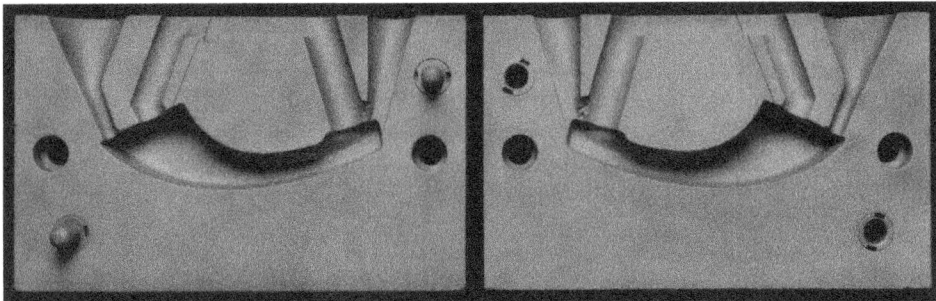

Figure 3.3 Gravity permanent mold for a bathroom faucet showing both dies, core prints, rigging, and locator pins. This single cavity die is poured with the die at left starting with the pouring cup on the left lowered, and ending with the mold horizontal to reduce turbulence during filling. (From Dion et al. [5].)

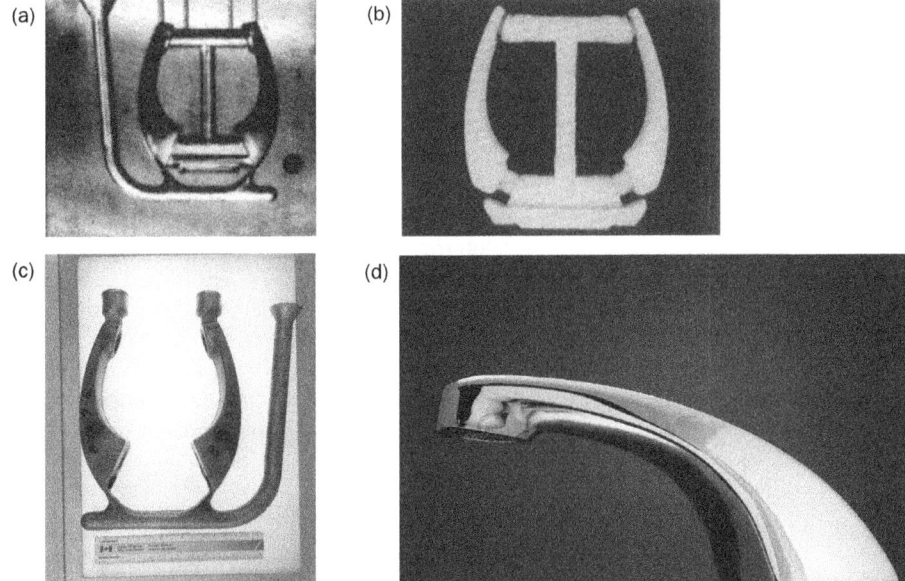

Figure 3.4 Gravity permanent mold with two cavities showing (a) a die, (b) the core, and (c) the unfettled casting. (d) The polished and plated casting in silicon brass. (From Dion et al. [5].)

Mold Materials

The earliest reusable dies were stone molds, used in the Bronze Age for casting weapons, axe blades, spear heads, etc. These days, cast metal or metal blanks are machined into dies. Blanks, in a variety of sizes, can be purchased on the open market, but must be free of casting defects, particularly porosity and shrinkage. Prior to the advent of CNC machining, the costs for machining dies and the consequent errors were extremely high, but foundries can now purchase dies from specialty die makers at relatively low prices. Repeat orders and renewals lower the cost, as the setup costs (machine manipulations) are a one-time expenditure, especially if modifications are required during the development phase.

The most common materials for permanent molds were cast irons, normally a high carbon equivalent gray iron. Eaton Corporation, a principal U.S. producer, preferred an iron consisting of 3.45 to 3.65% carbon (C), 2.45 to 2.65% silicon (Si), 0.6 to 1.0% manganese (Mn), up to 0.35% phosphorus, and no more than 0.15% sulfur. Eastern Europe, which tended to do more permanent molding, favored a lower-silicon-content cast iron with consequently higher carbon, 3.5 to 3.7% C, 1.8 to 2.0% Si, and 0.6 to 0.8% Mn. In Poland, similar compositions of cast irons were used, but with the addition of 0.8 to 1.2% copper. Such materials are preferred for ingot molds and are also used for gravity permanent molding [6]. Sound cast iron blanks are readily cast with some features and contoured backs. Cast iron is easily machined and provides excellent cooling rates for aluminum. However, the trend has been toward H-13 tool steel.

Pouring temperatures and consequently mold temperatures are far higher for the casting of gray iron than copper or aluminum castings. Molds undergo high temperature creep, distortion due to higher temperatures at the mold interface (mold halves curve away from each other), and thermal fatigue due to the frequent expansion and contraction, particularly at grain boundaries. Cast iron molds fail by oxidation and subsequent cracking of the layer around graphite grains. The oxides have a larger structure, and the consequent expansion causes microcracks which get larger and more visible as the number of mold cycles (shots) increases. High-pearlitic structures are preferred over ferritic carbides, although pearlite slowly decomposes with the precipitation of graphite (particularly near the mold surface) and undergoes dimensional changes. As the degree of cracking worsens, so too does the surface finish of the castings produced in the mold. Die life depends on the material being cast (different pouring temperatures and amount of heat being transferred), the size of the casting, the complexity of the casting (die creep occurs at constrained angular sections) as well as cycle times. Shorter cycle times generally mean higher and more uneven mold temperatures, even when the mold is dipped into a water bath for applying washes. This leads to a significant reduction in die life and is a major cause of dispute if the customer owns the dies.[1] In production, molds tend to be hotter in the region of the gating system, as this area sees the hottest metal although a rounded thick section is used to lower the surface area to volume and thereby reduce heat transfer, while at the same time reducing turbulence. Typical mold life for cast iron molds used for permanent mold casting of metals is shown in Table 3.2.

Metal	Pouring Temperature		Mold Operating Temperature*		Approximate Mold Life
	°F	°C	°F	°C	No. of Shots*
Cast iron	2,300–2,700	1,260–1,480	600–800	315–430	5,000–20,000
Aluminum base	1,300–1,400	700–760	650–800	340–430	Up to 100,000
Copper base	1,900–2,100	1,040–1,150	250–500	120–260	5,000–20,000
Magnesium base	1,200–1,300	650–700	300–600	150–315	20,000–100,000
Zinc base	730–800	390–430	400–500	200–260	>100,000

*Operating temperatures will be higher near sprues and where backing hardware is close.
Source: Sahoo [7].

TABLE 3.2   Cast Iron Mold Conditions for Permanent Mold Casting of Various Metals

---

[1]When the customer owns the dies, a contract will usually specify the expected die life, and the foundry's cost of failing to meet that life expectancy.

Some tool steels, refractory metals, and superalloys suffer less soldering and heat checking and give longer service life as molds, core, and die inserts. While this comes at a higher cost, the longer life and improved surface finish is justified if production numbers warrant. Many of these alloys require annealing prior to machining followed by another heat treatment to restore properties. Some alloys, such as H-13, work harden during extensive machining and may need additional softening. Depending on casting variables, H-13 can produce 250,000 castings and with repeated repairs, several million. The surface wear tends to be predictable and wear resistance of H-13 is superior to steel or cast iron. Table 3.3 gives the nominal composition of typical die insert materials, while Table 3.4 shows the recommended heat treatments. Because the various die components are exposed to different thermal and mechanical conditions, Table 3.5 gives some suggestions for die components.

Copper beryllium or nickel beryllium dies can also be used for casting copper. Copper beryllium blanks are more expensive than steel or cast iron, but their physical properties make them more cost effective when the volume of castings is high enough. Copper beryllium has high thermal conductivity (10 times that of tool steel), elevated temperature ductility, and good machinability. Copper beryllium dies for casting copper alloys are good for 30,000 to 50,000 shots depending on the shape, size, and complexity of the casting. Dies can be reworked once or twice, where their better machinability makes rework costs lower in comparison to tool steels. The mold life of copper beryllium dies can be extended to 150,000 shots (copper alloys), far more than other die materials [8].

Graphite molds can be made from a mix of 70% graphite grains of about 85 AFS GFN and 30% binders composed of pitch, carbonaceous cement, starch, and water. The material is molded by squeezing it around a pattern at 60 to 120 psi (4.2 to 8.5 kg/cm^2). It is then dried and fired in a reducing atmosphere at 1,800 to 2,000°F (1,000 to 1,100°C)

Material	C	Mn	Si	Cr	Mo	V	W	Co	Fe
Tool steels									
H-13	0.40	0.30	1.0	5.0	1.35	1.0	–	–	Rem.
H-19	0.40	–	–	4.3	0.40	2.1	4.3	4.3	Rem.
H-21	0.40	–				0.25	9.0		Rem.
Material	C	Ni	Al	Cr	Mo	Ti	W	Co	Fe
Refractory metals									
Rene 41*	0.1	Rem.	1.5	19.0	10.0	3.0	–	11.0	3.0
Anviloy 1150†	–	4.0	–	–	4.0	–	Rem.	–	2.0
Mo-TZM	0.12Zr				Rem.	0.5			
Material	Ni	Cr	Al	W	Co	Mo	C	Fe	Mn
Superalloys									
Inconel 617	Rem.	22.0	1.15	–	12.5	9.0	0.1	–	–
Haynes 230	Rem.	22.0	0.3	14	5.0	2.0	0.1	3.0	0.5

*Proprietary name of General Electric Co., USA.
†Proprietary name of P. R. Mallory Co., USA.

Source: From Sahoo [7].

TABLE 3.3 Composition of Some Die Materials

Material	Anneal Temp., Anneal Time, Cooling	Stress Relief Temp., Stress Relief Time, Cooling	Hard. Temp., Hard. Time, Cooling	Temper Temp., Temper Time,* Cooling
H-13	1550–1625°F $\frac{1}{4}$ h/in 50°F/h	1200°F 2 h at temp. Furn. cool	1795–1975°F 45–15 min. Air quench	1000–1200°F 1–2 h* Air
H-19	1600–1650 g $\frac{1}{4}$ h/in 50°F/h	1200°F 2 h at temp. Furn. cool	2100–2200°F 15 min. Air quench	1000–1300°F 1–2 h* Air
H-21	1600–1650°F 2 h at temp. 50°F/h	1200°F 2 h at temp. Furn. cool	2150–2200°F 15 min. Air quench	800–1300°F 1–2 h* Air
ORO45	1560°F 2 h at temp. 50°F/h	1200°F 2 h at temp. Furn. Cool	1830–1975°F 15–45 min. Air quench	1000–1200°F 1–2 h* Air
Rene 41			1950–1975°F 4 h Air	1400°F 16 h Air
Anviloy 1150	1850°F 1 h/in of thickness Vacuum or dry hydrogen atmosphere			
Mo-TZM	2000–2250°F (See manual for best temp, time, and cooling.) Vacuum or dry hydrogen atmosphere			

*Double or triple temper is desirable. Consult manufacturer's tables and graphs in order to achieve the final desired hardness.

Source: Sahoo [7].

TABLE 3.4 Recommended Heat Treatment for Die Materials

to form a solid mold or core piece. After assembling the mold, reactive metals such as titanium alloys can be poured under vacuum to prevent oxidation of the metal.

Permanent graphite molds are also made by machining the mold cavity into solid blocks of graphite. These molds can then be used in the permanent mold casting process. However, as graphite begins to oxidize above 750°F (400°C), the mold will soon start to deteriorate. A mold coating of ethyl silicate, which when heated deposits silica, will increase the mold life. Graphite mold liners are used in centrifugal casting of brass and bronze bearing, sleeves, etc. Graphite molds are used to cast steel railroad wheels using controlled pressure pouring and mechanical handling of the graphite mold [9]. Little machining is required to the as-cast surface.

Aluminum alloys and ceramics such as silicon carbide have also been used as permanent molds.

Use	Steel	Use	Steel
Ejector pins	H-13, H-13[N]	Movable holder block	P-20
Ejector plate	1020	Stationary holder block	P-20
Ejector plate retainer	1030	Angle pins	A-2
Pillars and rails	1020	Slides	A-2
Bottom clamp plate	1020		
Die inserts and cores	H-13, H-21, H-19, QRO-45*, Mo-TZM, Anviloy 1150†, Rene 41‡, superalloys	Runner blocks and risers	H-21, H-19, QRO-45[1] TM special§, superalloys

Notes: N = nitrided.
*Proprietary name of Bofors Aktiebolaget, Sweden.
†Proprietary name of P. R. Mallory.
‡Proprietary name of General Electric Company.
§Proprietary name of Teledyne Vasco.
Source: Sahoo [7].

TABLE 3.5 Die-Set Construction Materials

Hot dies are coated with insulating alumina washes, conducting graphite washes, or adjacent sections may be separately coated to adjust properties and aid directional solidification. Test bar molds receive an insulating coating on the grips and a conducting coating on the gauge. Acetylene black or soot from an acetylene torch is used to protect molds during the casting of gray iron and to help strip the casting (lubricates). The acetylene black has to be applied before each shot. Insulating coatings are applied to a hot surface (typically 400°F or 200°C) by spraying. The die halves are usually blasted with glass beads or fine sand prior to coating. Coatings not only protect the mold from soldering but aid in releasing the casting. Coatings can also improve surface finish. Instruments are available to measure coating thickness on steel and cast iron dies. Although some coatings last for perhaps a hundred shots, they need to be inspected frequently to remove oxides and ensure continuity. Recipes for some spray coatings are shown in Table 3.6.

Permanent Mold Casting Machines

As molten metal freezes quickly in contact with the relatively cold mold, splashing and turbulence must be minimized to avoid surface defects, i.e., laps and wrinkles. The gating systems are designed to get the metal into the mold cavity as smoothly as possible, usually from the bottom. To even the flow, the mold is often started in a tilted position then turned upright. While the operation can be done manually, larger molds are difficult to handle and open especially when hot. Therefore, casting machines now do most of the operations.

Gravity permanent mold machines can be programmed to manipulate the mold through a series of operations. When used for copper alloys, the hot dies are immersed face down in a tank of graphite wash to cool the mold and coat the die surface with graphite wash. The machine then presents the die halves to the operator to insert cores and metal insets. The machine then closes the mold and sets it in the correct angle for pouring. The operator then ladles the correct amount of molten metal into a cup and the machine rotates the mold to pour the metal into the dies, ending in the vertical orientation. After

Composition % by Weight (remainder water)

Coating No.	Insulators					Lubricants				
	Sodium Silicate	Whiting	Fireclay	Metal Oxide	Diatomaceous Earth	Soapstone*	Talc*	Mica*	Graphite	Boric Acid
1	2	–	4	–	–	–	–	–	1	–
2	8	–	4	–	–	–	–	–	–	–
3	–	7	–	–	–	–	–	–	–	7
4†	12	9	–	–	–	–	–	–	–	–
5	5	11	–	2	–	–	5	–	–	–
6	9	–	4	–	–	14	–	–	–	–
7	11	–	–	17	–	–	–	–	–	–
8	–	–	4	–	–	23	–	5	–	–
9	7	–	1	–	–	23	–	2	–	–
10	23	–	–	–	–	–	20	–	–	–
11	30	–	–	–	5	–	10	–	–	–
12	18	–	–	–	41	–	–	–	–	–
13	8	–	–	60	–	–	–	–	–	–
14	7	–	–	–	–	–	62	–	–	–
15	20	53	–	–	–	–	–	–	–	–

*Also serves as an insulator.

†Plus silicon carbide 2% by weight for water resistance.

Source: Sahoo [7].

Table 3.6 Compositions of Some Coatings for Permanent Molds

an appropriate time the mold is opened and the casting is ejected. The cycle is repeated. For aluminum alloys, the cycle time can be of the order of 210 to 360 seconds for steady-state mold temperature conditions, of which 90 to 120 seconds are taken for the metal to solidify and be strong enough to be ejected. Mold temperatures should be around 625°F (330°C) for aluminum; although temperatures can exceed 840°F (450°C) around the sprue. Molds are preheated before starting to reduce internal stresses and speed steady-state operation. Molds for casting copper alloys will have surface temperatures of 285°F (140°C) after immersion in graphite wash. Residence times for iron castings need to be only seconds, with castings ejected red hot, otherwise mold life is reduced and cycle times lengthened. To increase productivity, several gravity permanent molding machines are either rotated on a carousel while the molds cool, or an operator can service a couple of machines besides a bail-out resistance furnace. For castings with cores, it is helpful if the castings are put in a barrel rather than tossed on the floor or a shallow container, as the extra retained heat will aid in collapsing the cores.

In the low-pressure casting method, the mold is positioned over a stalk projecting through the lid of a sealed furnace (Fig. 3.5). Molten metal is forced up the stalk into the mold in a controlled manner using programmable pressure transducers to control both metal flow rate and pressure in the mold (as long as bath level is monitored). The metal in the stalk acts as a riser to feed shrinkage in the casting, and after a suitable interval air pressure is released, and excess metal returns to the furnace. This operation is suitable for readily oxidizable alloys as the metal in the stalk is exposed to less air. The machine can be programmed to brush the mold to remove dross, etc. before coating is reapplied. Low pressure is used for brass spigots and automotive aluminum wheels. Figure 3.6 shows a multicavity die half, and the castings produced from it.

In vacuum casting, the mold is contained in a vacuum chamber with a nozzle connected below. The nozzle is covered with an aluminum foil that seals the vacuum. When the nozzle dips into the melt, the foil melts and the vacuum draws metal up into the mold. The method tends to be labor intensive and is also applicable to sand molds.

Advantages of permanent molding are

1. Near-net shape castings that only require removal of gates and fins manually or with trim dies and presses.
2. Intricate castings with complex shapes and reentrant angles[2] as small as $\frac{3}{16}$ in (5 mm).
3. Excellent surface finish, 150 to 200 μin RMS (4 to 5 μm), although 70 μin (2 μm) are attainable, dense, chill cast surfaces, free of near-surface defects which can be polished and plated.
4. Excellent uniformity and tolerance control as there are few process variables:
   a. ±0.010 in (±0.25 mm) across the parting line for 1-in sections
   b. In the same mold half ±0.008 in (0.2 mm) up to 1 in
   c. Part to part variation ±0.002 in (0.5 mm)
5. Rapid solidification produces sound castings with fine grain-size and high mechanical properties for potentially lower weight designs. Rapid freezing avoids massive shrinkage. Microshrinkage can be avoided by alloy selection and controlling the direction of solidification.

---

[2]Reentrant angle is the inside curvature of a corner, which during casting becomes a hot spot as heat accumulates in the adjacent shape mold media from two directions and leads to shrinkage in the last metal to freeze.

Figure 3.5 Low-pressure casting machine showing mold positioned over the furnace and a casting at right. (Courtesy Premier Aluminum.)

6. Pressure tight castings suitable for hydraulic fittings, valve bodies, and pump housings

7. Few environmental issues

Limitations of gravity permanent molding are

1. Size limited to 1 oz to 50 lb (30 g to 23 kg) for copper, 20 lb for iron

2. Long production runs required to amortize the cost of the die

High-Pressure Die Casting

In the high-pressure die-casting process, molten metal is injected under high pressure into a cavity in a reusable steel die. Pressures of 1 to 100 ksi (70 to 7000 kg/cm^2) are used. The casting is ejected at the end of the cycle and the process is repeated. Modern

(a)

(b)

FIGURE 3.6 Model BP 155 low-pressure casting machine for copper alloys: The hot dies are immersed in tanks containing a graphite wash, (a) a multicavity die half is presented for core placement before the mold is closed and moved over a stalk projecting from the furnace for filling, and (b) returned showing the PM castings produced. (Courtesy IMR USA).

automated die-casting machines can mass produce small parts. Advances have gradually increased the casting sizes. For example, magnesium castings up to 6 lb can be produced, but rangy thin-walled parts (e.g., dashboards) cause dies to become too large and heavy for the casting machines. The advantage of high-pressure die casting is that parts can be cast to close tolerances and require minimal machining other than removal of flash and gates.

H. Doehler invented die casting in 1905 [10]. As it was first used for low-melting point metals, particularly zinc, tin, and lead alloys, the injection mechanism was immersed in the metal bath because the ingress of tin and zinc (galvanizing) into the steel components is slow at zinc casting temperatures. Die casting reached a peak after World War II, but has made a dramatic recovery since 1997 in the production of magnesium automotive castings. By 2006, die casting of magnesium parts alone had reached 300,000 tons [11]. As magnesium is normally melted in steel pots, both "hot chamber" and "cold chamber" die-casting machines can be used. In the cold-chamber machine, a measured quantity of molten metal is ladled into an injection cylinder and the "shot" injected immediately. Contact between steel injection components and the molten metal needs to be minimized when casting higher-melting-point metals, i.e., alloys of aluminum and copper. Steps of the cold-chamber process are shown schematically in Fig. 3.7.

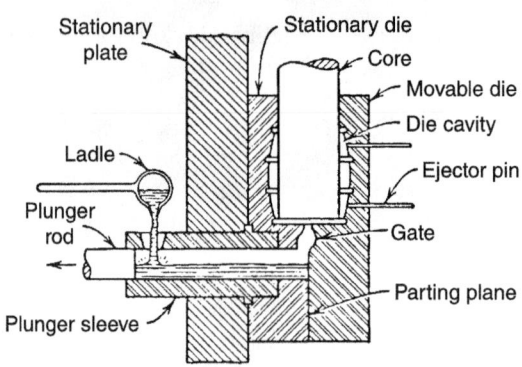

A. Metal poured into sleeve

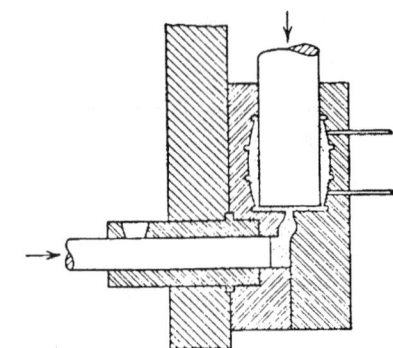

B. Plunger forces metal into die

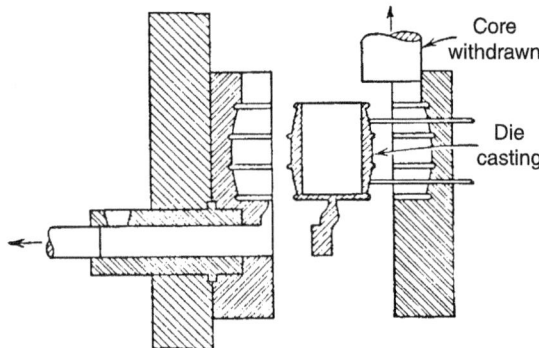

C. Die separated and casting ejected

Figure 3.7   Steps of the cold-chamber die-casting process. (Casting Copper-Base Alloys [7].)

Die-Casting Machines

Die-casting machines consist essentially of (1) a high-pressure hydraulic system, with valves, cylinders, accumulators, and electrical servo-controls; (2) a "shot-end" which receives molten metal in the shot sleeve and a water-cooled plunger which forces it into the die at high pressure; (3) a stationary and a moving platen on which the die halves are mounted; and (4) a mechanism to lock the two halves of the die together and prevent them from being forced apart under the high metal-injection pressure. Important secondary systems include die heating or cooling, die lubrication, core actuators, and casting ejection. Parting lines may be vertical or horizontal. Tie bars or frames become massive for strength and rigidity.

Machines are rated in terms of their die-clamping force, in tons. The clamping force, the injection pressure, and the projected area of the maximum die cavity limit the size of castings which can be produced. For magnesium alloys, hot-chamber machines are typically 400 tons, while for copper-base alloys, machines can range from 250 to 2,500 tons. The larger the machines are, of course, the more expensive they are and they operate on a slower cycle. Therefore, the smallest machine capable of producing the casting will be the most economical. Figure 3.8 shows the schematic of a hot-chamber die-casting machine and Fig. 3.9 shows a magnesium casting.

Electric motors larger than 250 hp are built with copper rotor structures by expensive fabrication techniques. Squirrel cages of smaller rotors are usually die-cast aluminum alloys, but high conductivity copper grades would be more efficient. H-13 tool steel fails quickly when subjected to copper at temperatures exceeding 2,000°F (1,100°C) from the rapid onset of heat checking, a fatigue phenomenon caused by temperature gradients. Computer simulations indicate the die surfaces have a temperature-induced cyclical strain exceeding 1.9%. Copper rotors could be die cast in Inconel 617 and Haynes 230 dies (see Table 3.3). Practical procedures, including heated die inserts and insulation, were developed to produce the rotor shown in Fig. 3.10.

Advantages of die casting are

1. Production rates of 150 to 250 cycles per hour, with rates of 500 possible
2. Near-net shape castings that only require removal of gates and fins
3. Excellent surface finish, free of near-surface defects, that can be polished directly
4. Uniformity and close tolerance control of ±0.001 to 0.003 in (0.025 to 0.076 mm)
5. Thin sections down to 0.015 in (0.4 mm) in small zinc castings or 0.08 in (2 mm) in magnesium due to the high pressures involved
6. Sound castings with high mechanical properties with fine grain size due to the rapid solidification under high pressure

Die casting is applicable to a comparatively small number of suitable alloys. Other issues are high maintenance costs, expensive tooling, and angles on the casting and cores must be tapered in order for the casting to be ejected from the dies (which can produce underlying porosity that can be opened up by heat treatment, however, technology has improved and parts can be heat treated). When there is a need for reverse tapers or lack of space for core/slide mechanisms, loose pieces or inserts can be manually inserted, and recovered once the casting is ejected.

Squeeze casting builds on die casting experience. Two types of machines are commercially available: vertical and horizontal-vertical squeeze casting [7, 12, p. 993]. Degassed

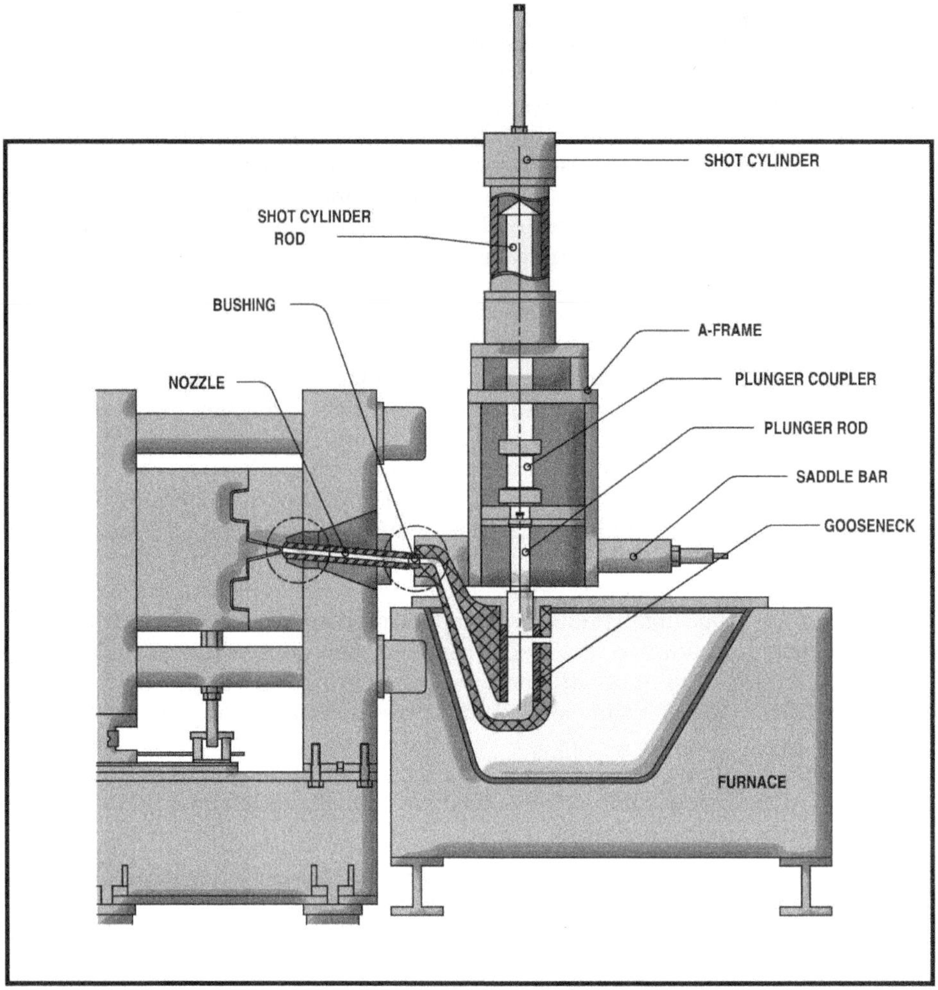

SHOT CYLINDER

SHOT CYLINDER ROD

BUSHING

NOZZLE

A-FRAME

PLUNGER COUPLER

PLUNGER ROD

SADDLE BAR

GOOSENECK

FURNACE

FIGURE 3.8 Hot-chamber die-casting machine. (Courtesy of NADCA.)

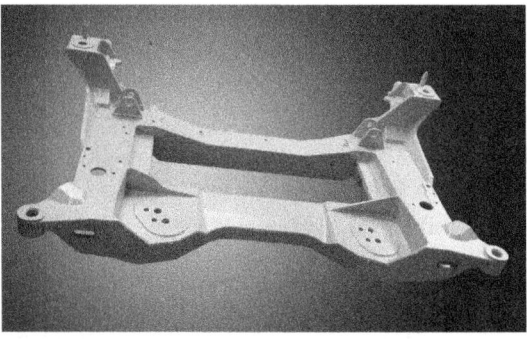

FIGURE 3.9 A die-cast magnesium engine cradle (as purchased from a General Motors dealer).

FIGURE 3.10   A copper rotor casting made by die casting. (Courtesy Copper Development Association.)

Al-Si alloys are ladled into a vertical shot sleeve and very slowly forced into the die cavity. The metal solidifies while the pressure of 10 to 15 ksi (70 to 100 MPa) is applied.

Semisolid casting uses the phenomenon that at 50% solid a metal alloy at rest behaves as a solid, but under stress flows as a liquid—it is thixotropic. More importantly, Al-Si alloys in a slurry do not have the dendritic primary phase structure. In addition, a semisolid slurry has already released much of the sensible heat (heating) and heat of fusion. This lowers the heat load and extends the die life. Machines with clamping forces from 75 to 1,600 mt are available, which are capable of producing parts from $\frac{1}{2}$ oz to 20 lb (15 g to 9 kg).

Centrifugal Casting

Centrifugal casting refers to the forces used to distribute the metal in the mold rather than a specific molding process. However, as molds for centrifugal casting are usually specially designed, it is considered a process. Centrifugal casting can be divided into three categories: true centrifugal casting, semicentrifugal casting, and centrifuging.

The essential feature of centrifugal casting is the introduction of molten metal into a mold which is rotated. The centrifugal force not only shapes but feeds the casting as the molten metal is thrown by the force of gravity into the detail of the mold. Although sand-lined molds were once used, coated metal molds are now preferred. Shorter castings may be cast with the spinning axis vertical.

The process has the following inherent advantages:

1. The process leads naturally to progressive solidification.
2. Dross and nonmetallics, being less dense, stay on the inside surface, where they may be removed by machining.

3. The extremely high casting pressures promote soundness by providing feeding metal, and the pressure effectively increases gas solubility, thereby less gas bubbles are formed and gas is either trapped or liberated after freezing.

4. The permanent chill-molds produce fine-grained castings of exceptional soundness and uniformity.

5. The high pouring rates make it possible to take advantage of lower pouring temperatures, which minimize gas absorption from furnace atmospheres and tend to decrease grain size.

6. Tubular sections can be cast without a core thus avoiding disposal of hazardous wastes.

The quality of any centrifugal casting depends on the extent to which it is possible to utilize all of these factors.

True centrifugal casting is the most widely used centrifugal process. Molds may be rotated on any axis—horizontal, vertical, or at any angle. The essential features are that there are no sprues or runners, the metal being forced directly against the periphery of the spinning mold, and the castings are essentially tubular. No central core is used, so the thickness of the tube and the inside diameter are determined by the volume of metal poured. Figure 3.11 shows various methods, and Fig. 3.12 shows a ductile iron casting immediately after pouring. Typical examples of this class of product are the thousands of tons of spun cast iron and ductile iron pipes produced for water and sanitation services, stainless steel pipes for the chemical industry, and the large-diameter, copper-alloy, hollow suction rolls used on papermaking machines.

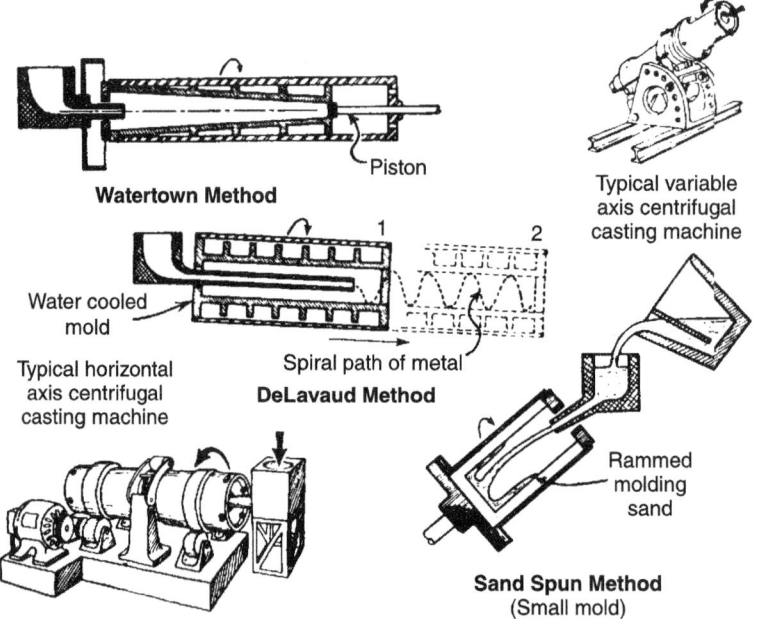

Figure 3.11 Variations of true centrifugal casting process. (From Metalcaster's Reference & Guide [2].)

Figure 3.12   20-ft ductile iron pipe immediately after pouring on a horizontal centrifugal casting machine. (Courtesy U.S. Pipe and Modern Casting.)

Such rolls may be 3 ft in diameter by 40 ft long (1 × 10 m), and castings weighing over 20 tons have been made.

Castings made on *horizontal-axis casting* machines have a uniform bore, although rims, enlarged ends, etc., can be made by enlarging the external alignment at the appropriate location (mold surface).

The first metal poured into the spinning mold tends to lie against the bottom surface until it picks up the rotational speed of the mold, at which point it rapidly spreads over the entire mold surface. The initial heat extraction rates are very high, and it is therefore important that initial pouring rates into the mold are high enough to ensure that cold shuts and similar defects are not formed. Mold temperature is also important. Metal is normally poured against the downward-spinning side of the mold to assist acceleration of the metal, and much of the technology of horizontal casting, particularly of long tubes, has been concerned with metal delivery into the mold. Thus, some form of pouring spout into the mold is normally used and, in long castings, the spout into the mold may be retracted as pouring proceeds. Alternatively, the pouring spout may be in the form of a weir, with metal overflowing along its whole length.

If the rotational speed is too low, metal rests on the bottom of the spinning mold too long and is carried only part way up the mold wall, falling back to cause "raining" (which generates laps and entrapped oxides). The critical velocity at which metal is "picked up" and forms a complete tube increases with the volume of liquid, i.e., increases with tube thickness. In practice this is usually not a problem, as rotational speeds sufficient to generate forces of 50 to 100 G[3] are normally employed, there being no demonstrated advantage in exceeding this.

---

[3]G is the acceleration due to the force of gravity.

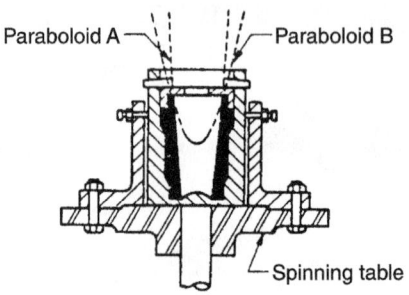

Figure 3.13 Vertical centrifugal casting method showing effect of rotational speed on the shape of the inner surface. (From Casting Copper-Base Alloys [7].)

Castings can also be produced by pouring a given weight of metal into a mold rotating about a *vertical axis*. The metal is accelerated by and distributed around the inside mold surface. Dross, slag, and other nonmetallics are centrifuged to the inside. Unlike horizontal casting, it is not possible to obtain a uniform bore. This is because the earth's gravitational force acts in a vertical direction, whereas the centrifugal force caused by mold rotation acts horizontally. The resultant vector force on the molten metal varies from top to bottom, causing the internal section to assume a parabolic shape. Thus, the lower end of the casting will always have a somewhat thicker wall than the upper end. The effect of spinning speed can be seen in Fig. 3.13. Although high spinning speeds still have a parabolic shape (A), the walls are more nearly vertical, compared with (B), where lower spinning speeds cause walls to be more inclined. This behavior can be utilized to advantage in the production of tubular parts requiring tapered inside diameters. The vertical-axis centrifugal casting process, because of the above mentioned condition, is not well-suited for the production of long, pipe-like shapes. Its greatest application is for relatively short cylinders and rings.

Centrifuging

Centrifuging is probably the least important of the three types of centrifugal casting. A typical example uses the investment casting "Christmas trees" of assembled lost-wax patterns. Metal is poured down a central sprue while the mold is spinning and centrifugal force is used to fill runners, gates, and feed the cavities. Castings are relatively small, and spinning speeds are adjusted to give a relatively low force of about 10 G in molds with small diameters. An alternate version of the centrifuging process using more conventional sand molds is shown in Fig. 3.14; this can, of course, be applied to simple cope-and-drag molds as well as stacks of molds. The objective is to fill the mold cavities quickly and uniformly, and utilize the centrifugal pressure to achieve the best reproduction of the mold detail and enhance feeding. In larger castings and sprues, turbulent flow becomes a problem but can be mitigated to some extent by using curved "spiral taper" runners whose contours correspond to the movement and acceleration of metal in the runner caused by the applied centrifugal force.

Centrifuging differs from the other two processes in that one or two sand or investment molds are mounted on their sides and spun from the ends of arms (Fig. 3.15c). Liquid metal is poured down the central axis and quickly finds its way into the mold cavities. Another variant uses a ladle with a single mold mounted above, and slung on

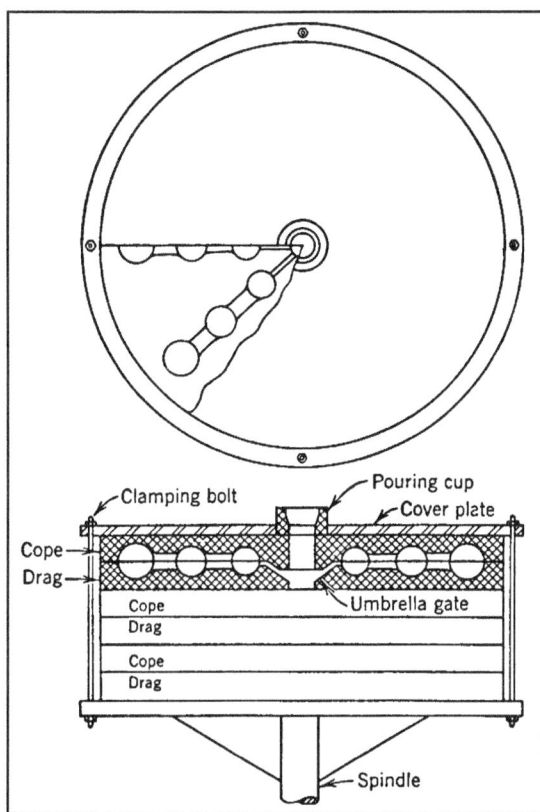

Figure 3.14 Centrifuging process showing a dry sand stack of three molds. (From Casting Copper-Base Alloys [7].)

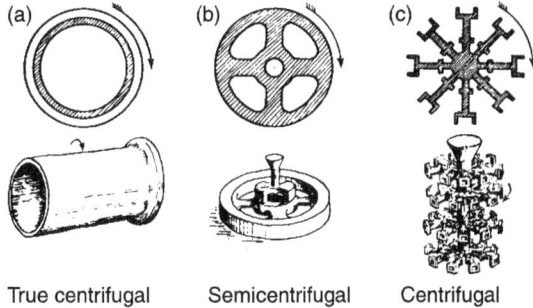

Figure 3.15 (a) True centrifugal casting, (b) semicentrifugal, and (c) centrifugal casting using an Xmas tree. (From Metalcaster's Reference & Guide [2].).

the end of one arm with a counterweight on the other. The ladle is filled with liquid metal, and once the apparatus reaches the assigned rotational velocity, a lever tips ladle and mold so that liquid flows from one to the other. The apparatus is usually enclosed for safety.

Semicentrifugal Casting

Semicentrifugal casting is another centrifuging method where a circular mold is spun around its central axis as metal is poured into a central sprue which later becomes the casting's feeder. The centrifugal force is usually low (<10 G). However, the casting is always symmetrical and concentric with the mold, and all contours are formed by (sand) mold or cores. Again, the objective is to achieve rapid filling, enhanced feeding, and perhaps to induce directional solidification from the periphery to the center. A typical example is shown in Fig. 3.15*b*. Castings can be made with or without a central core. Other examples of this type of centrifugal casting are large gear wheels, where the spokes act as runners to the outer rim. The central sprue acts as the riser, and, as will be discussed later, the centrifugal force causes the open metal surface to take the form of a parabola. As the riser is "piped" even before solidification starts, it is therefore necessary to make the riser considerably higher than would be the case with static casting [12, p. 665].

The centrifugal force at any point in a true centrifugal casting is represented by

$$F = (W\, 4\pi^2 N^2 R)/(G \times 3600 \times 12) \tag{3.1}$$

where   F = centrifugal force in pounds
   W = weight in pounds
   N = number of revolutions per minute
   R = radius (distance of that point from the axis of rotation in inches)
   G = acceleration due to gravity

The forces on a solid disc are different from those of a spinning hollow tube due to hydrostatic pressure in the thin annulus. The two forces, centrifugal acceleration and metal head, have different effects on the structure and soundness of the cast parts, and may be manipulated when choosing the operating conditions of the casting machine. Thus, the internal pressure or metal head generated by centrifugal casting is a hydrostatic pressure, and it is this which provides the feed metal to ensure a sound casting. Solidification must proceed from the outer periphery toward the center, otherwise shrinkage will not be fed. Centrifugal acceleration, or "G" force, affects the cleanliness of the casting by assisting the migrations of inclusions and dross to the inner surface. In addition, any density differences between solidifying constituents will be magnified. This applies particularly to alloys containing lead which, because of its greater density, tends to segregate toward the outer sections of the casting (such as centrifugally cast babbitt liners in large bronze bearings). Steels with more than 0.4% carbon may segregate. "Banded" structures are commonly encountered, particularly if the machine vibrates.

Mold Materials

Although any mold material can be used for centrifugal casting, cast iron is normally used for high production runs and where rapid heat extraction is required. Such molds are usually dressed with wash, or even with dry sand. Nonmetallic mold materials ranging from ceramic for investment, shell, or oil-baked core sands are widely used depending on cost, complexity, dimensional stability, chill rates, and collapsibility, to ensure freedom from hot tearing. Greensand is too soft. Molds may be cooled externally to increase production rates and induce directional solidification.

Continuous Casting

Continuous casting is used to produce simple shapes, i.e., rods, steel rails, billets, or slabs. Billets are used as blanks for machining of bushings and gears as well as for extrusion of wire, rods, and slightly more complex shapes, such as aluminum door casings. These operations are seldom performed in foundries, nevertheless, continuous casting offers diversification and significant returns. Continuous casting was used for 95.6% of the world's 2012 production of 1,500 million tonnes of steel [13], 90% of the 46 million tonnes of aluminum, and 90% of the 24 million tonnes of copper [14]. In comparison, the world's 2010 production of steel, aluminum, and copper-shaped castings was 10, 13, and 1.8 million tonnes, respectively [15].

Continuous casting was first used in 1830 to cast lead pipes, and in 1930 came the revolutionary use of a continuously cooled graphite mold to cast copper. Brass plates were cast in 1935. It was in 1956 to 1961 that the first curved strand continuous caster was built in Eisenwerke, Switzerland. It can be said that continuous casting was really achieved when a Japanese plant produced 176,000 tons of steel bloom in 1982 before having to stop to change molds [12, p. 923]. Curved copper plate molds are used to cast and turn the strand from vertical to horizontal and resolve logistical and safety issues by lowering the building height.

The metal is transferred either from a tundish (Fig. 3.16), or directly through a nozzle in the bottom, in which case, the process is called *closed head*. In some cases, the nozzle is cemented into the side of the furnace, which allows the furnace to be tilted slightly from melting only and maintenance, to a melting and casting position. Both methods use a dummy bar or starter rod which fits into the nozzle (mold) and temporarily prevents liquid metal from flowing out. It is then withdrawn, often a bit at a time. Cooling water is fed into the mold and around the starter rod (and later the continuous casting). The cooling water allows the solidifying casting sufficient time to form a skin, which allows more liquid to feed into the center (unless a core is used). In this manner, a rod or sheet is withdrawn from the mold by pinch rolls. Extraction rates will depend on casting temperature, delay time between extractions, roll speed which depends on the section thickness being cast (and cooling rate), and the characteristics of the alloy. Steel rods are usually lubricated by dripping oil around the side of the mold; however, the oxides on nonferrous metals reduce the friction to a tolerable level. Once the casting passes through the withdrawing rollers, the dummy bar is severed. Multiple strands are often cast simultaneously in steel. Rods are cut into convenient lengths, while sheets are temporarily coiled. They will be later rolled to final specified thicknesses and to remove imperfections.

Figure 3.17 shows examples of what can be accomplished when shaped molds and suspended graphite cores are employed. The improved mechanical properties of metals produced by continuous castings can be seen by comparing the values of some copper alloys in ASTM B505 with the same alloy sand cast in ASTM B584. In addition to mechanical and metallurgical advantages, the process yields can be as high as 98%. However, the serious disadvantage is the close proximity of molten metal and water, which due to the 1000-fold volume increase requires special walls to any building.

Single Use Molds

Greensand Systems

Greensand molding is the most common process used for casting small- to medium-size castings. The process requires a tempered sand to be compacted about the pattern (usually in a flask to contain the force) by one or more mechanical means such as jolting, squeezing, ramming, vibrating, slinging, or blowing. The plasticity of the bonding clay

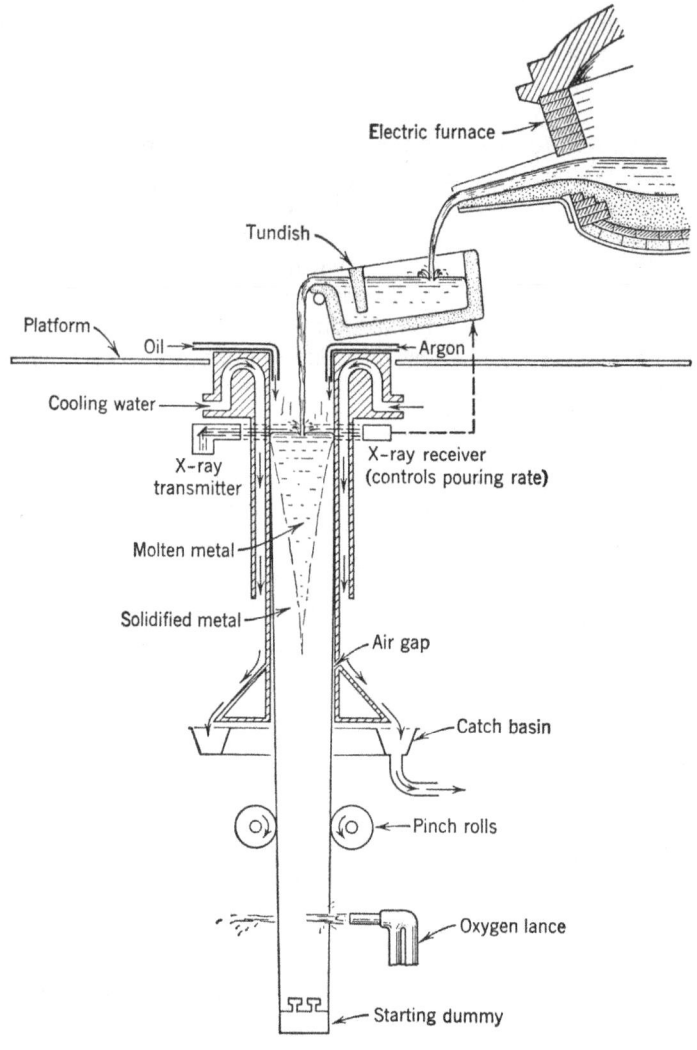

Continuous Casting (Steel)

FIGURE 3.16   Continuous casting of steel. (From Metalcaster's Reference & Guide [2].)

is sufficient to produce a mold which has enough rigidity to hold its shape during pattern removal (sometimes the flask is also removed), core placement, pouring, and solidification. A simplified version of the process is illustrated in Fig. 3.1.

The relative ease with which the molding aggregate can be prepared (by mulling or mixing), transported, and stored contributes to the economy and versatility of the

Figure 3.17 Sections of cast shapes produced by vertical closed head continuous casting. Hollow shapes produced using an uncooled graphite mandrel suspended in the mold. (Artwork courtesy *ASM Metals Handbook*, Vol. 15, *Casting*.)

process. More importantly, the sand mixture can be reused many times before it must be discarded. In recycling the material, however, it is necessary to compensate for chemical, physical, and thermal changes caused by exposure to the molten metal and to control the effects of contamination by foreign and degraded matter. The quality of castings made in greensand molds is highly dependent on the properties of the molding aggregate as well as those of the mold. These parameters can easily change from hour to hour or from day to day unless close control is maintained.

Sand System Equipment

Immediately after mold shakeout and removal of the casting, preferably with contained cores and some adhering sand, the recovery and reuse of the sand starts with the following basic steps:

1. Shakeout is followed by a vibratory or rotary screen.
2. Sand is conveyed back to the mulling area.
3. Sand is transferred to storage.
4. Sand is dropped from the silo in such a manner as to re-blend the sand.
5. Continuous or batch mulling with additions of new clay and new sand, water, etc.
6. Sand is discharged from muller.
7. Sand is transferred to storage hoppers at molding stations.
8. Sand is reused in molding.

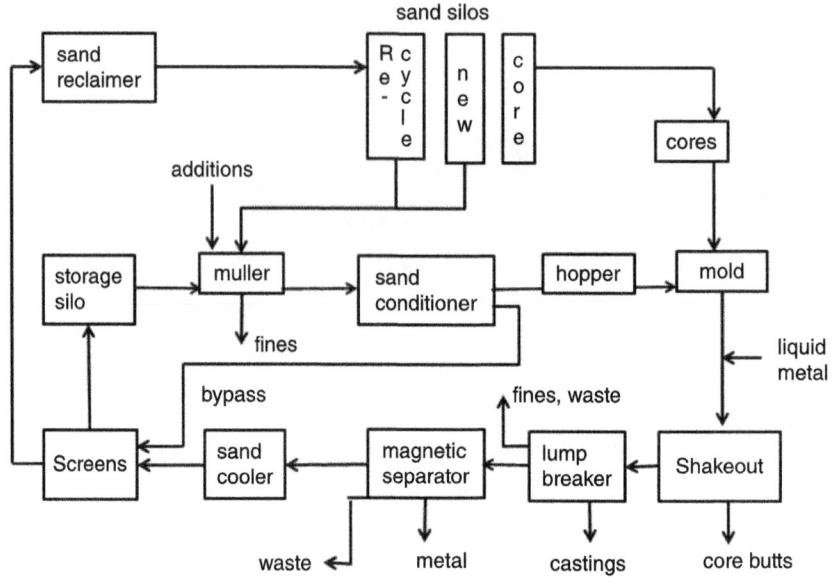

FIGURE 3.18   Flow sheet of sand circulation in a mechanized foundry.

A simplified flow sheet of sand circulation is shown in Fig. 3.18. The degree of mechanization in the sand system will depend on the volume of sand handled (4 or 5 tons of sand for each ton of ferrous casting poured). Conveyors, elevators, or fork-lifts move sand and molds between stations. A Battelle study conducted in 1975 for U.S. Bureau of Mines estimated that mold and core making consumed 13% of the energy in a gray iron (cupola) foundry and 10% in a steel (arc) foundry [16]. The following is a brief description of some of the equipment to fulfill the previous eight steps.

1. A suitable time after pouring (some castings need controlled cooling), molds are moved to the shakeout, which in its simplest form may consist of oscillating bars to separate sand from castings. Larger foundries convey the castings via a shaker pan conveyor, consisting of a vibratory oscillating deck. Castings and sand are then separated in a rotary drum fitted with flutes or ribs to tumble the castings. Light thin-walled castings should not be tumbled. Some rotary shakeouts are used as sand coolers with additional air to take heat away and even water additions. Carbon monoxide, dust, and noxious organic compounds liberated from the sand during shakeout should be exhausted from the foundry. The amount of carbon monoxide liberated from molds during shakeout depends on the pouring temperature, casting thickness, residence time in the mold, and the amount and nature of the organics in the sand (chemically bonded core sand, seacoal, or cereal). Carbon monoxide levels can exceed health and safety limits of 25 ppm over 8 hours and require removal of air and dust from the shakeout area by the appropriate exhaust system.

   Sand which falls through on shakeout is transferred to a coarse vibratory or rotary screen for removal of large contaminant particles such as core butts, wire, hard sand lumps, metal splash, and flash. Magnetic separation in the form of a simple magnetized drum at the end of a conveyor belt may be combined with a cross-belt magnetic separator to remove sand-sized particles such as chromite sand and silica-fayalite reaction products.

Castings may be further tumbled with old risers or by flights in a rotating drum to vibrate castings to remove surface sand, break down cores, and snap off risers and gating systems. Sand and peel from the casting surfaces, and core butts are devoid of live clay, and contain damaged sand, fines, etc., which should not be returned to the sand system. Figure 3.19 shows an example of a shakeout drum. Ferrous castings usually have a black scale on the surface unless sufficient seacoal was used. To remove this scale, castings are usually shot blasted by sand, glass beads, or steel shot in a variety of equipment. Duration and severity of shot blasting will depend on the strength of the casting alloy and customer requirements for surface finish. Machines are available to process castings batch-wise (in stop-start operations) or continuously, either in rotary tumblers, on tables, or hanging from rails. Tumblers are usually more effective at removing cores, unless blast can be easily directed into the core cavity. Because the clay on the sand-sized material from these last two operations has been destroyed; it should not be returned directly to the greensand system. Castings are then sent to the cleaning room where operations continue to further improve the appearance of the casting and carefully remove additional metal. It is not uncommon to find half the foundry staff cleaning castings, which begs the

Figure 3.19 Castings exiting from a shakeout drum. (Courtesy of Didion International Inc.)

question whether this is normal for the type of casting or due to poor molds or planning. Cleaning and fettling operations are described in Chapter 13.

2. If the sand circulates rapidly and the returning sand temperature exceeds 120°F (50°C), it needs to be cooled before going to storage [17]. A number of actions can be used to cool the sand. Lumps must be broken up, usually by an over-the-conveyor-belt lumpbreaker mounted after the magnetic separator. Surge hoppers may be required on larger systems to even variations. When sand exceeds 120°F (50°C) most of the time, specialized equipment is needed to cool and blend the sand. Volatilization of water is the most effective way and can be accomplished by controlled water additions, mixing sand with water, and contact with air. The effects of hot sand on greensand properties will be addressed later.

   The cheapest solutions. Spray water on the sand, which also suppresses dust and helps water penetrate clay layers. However, sensors should be used to sense sand on the belt, and shut the water off when there is no sand. Metal chains or fingers can be suspended over the conveyor to riffle the sand and help steam escape. Longer conveyors allow more time to vent heat.

   Larger sand systems. Water additions can be made using a programmable controller that meters water additions after sensing the sand flow, belt weigh scales, and sand temperature (infrared). Such systems can be retrofitted using an over-the-belt system to thoroughly mix the sand, but as retention time depends on belt speed, temperature reduction is limited. Sand coolers cool the sand by adding moisture, fluidize the sand so that blowing air carries some of the moisture and heat away, homogenize the sand, and subsequently improve the efficiency of mulling operations. Sand coolers may be rotating drums with large flights and air blowing, fluidized bed oscillators, continuous rotating plough coolers, or fluidized bed vertical coolers (Fig. 3.20). Some of these methods require considerable floor space. Air blowing can remove excessive fines, and these methods require dust collection. Sand leaving the cooler should have a temperature of 100 to 110°F (38 to 43°C) with 1 to 1.5% moisture. In some cases, dry additions can be made, but moisture should be 0.5% higher [18].

3. Sand is often transferred to storage adjacent to the mulling station using bucket elevators. The shaft for the bucket elevator up to the silo should be open, to vent

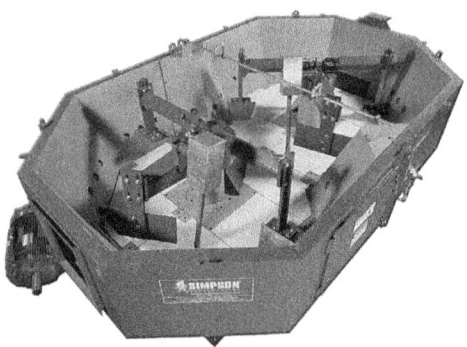

Figure 3.20   A sand cooler. (Courtesy of Simpson Technologies.)

hot air and continue cooling hot sand. Pipes at 45° angles can be set into the side of the silo to vent steam. Air can be forced into silos to prevent caking. The foundry should have at least a 10-hour supply of sand, otherwise sand temperatures will be excessive during any second shift. If sand is discharged into storage directly from a conveyor, the sand should enter without impacting the sides of the silo or dropping too far onto a heap. This avoids excessive sand segregation caused by larger grains of sand, core butts, and metallic and clay balls from rolling down the sides of a pile. Sand impacting the walls can cause fines to stick to the sides. Plastic liners are sometimes used to reduce sticking and friction. Hence, it is important that the sand system should be full at all times. Inadequate sand supplies are also more susceptible to greater swings as different casting jobs change the clay burn-out, core sand additions, muller additions, etc. Sand silos need to be inspected at least once a week late in a shift. Low sand levels promote sand segregation from heaping and piping. Pipes form over exits and cause *first in, first out* issues.

4. The bottom of a sand silo is generally reinforced by hard facing and live bottoms consisting of flutes or plates to direct sand from as much of the silo's cross section as possible. This reduces the sand segregation that occurs when sand heaps (coarser material rolls down). A movie by Martin Marietta [19] showed that when two sands of different narrow screen-size were dyed then blended and subject to many typical foundry operations that the color changes indicated that segregation had occurred. Materials of different sizes and densities are particularly susceptible to separation, which is why drug companies pelletize or encapsulate drugs and other diluents.

5. Methods used for sand mulling depend on the sand/clay to be bonded, the amount required, and the end use. The best mulling facilities produce considerably more green strength (greensand, green bricks, or green strength—uncured) than inadequate mixing. Consequently, more clay must be used when less efficient mulling methods are used. Mullers do more than mix; they develop the strength and plasticity of the clay/sand/water bond by working or kneading the sand. It takes time and energy to get the water into the clay and spread the clay over the sand grains. The muller is a key ingredient when designing a sand system. There are three basic kinds of mullers: vertical wheel, horizontal wheel, and high-shear types. Factors that need to be considered are the tonnage required for molding (including spillage, type of molding machines, strike off, and safety margin), sand properties, sand-to-metal ratio (including core sand), grain fineness number, grain shape, type, and form of binders, and whether a preblend can be added in the cooler (saving 15 to 25% of effort) or the muller. Mulling efficiency, sand-to-metal ratio, and amount of new and core sand will affect cycle times. Also some core sands (i.e., phenolic urethanes) are more difficult to coat.

   Mix mullers are batch mullers with one or two chambers with wheels that roll over the sand (Fig. 3.21). Two plows move the sand into and out of the path of the spring-loaded wheels that knead the clay and sand under fairly high pressure. Electric power consumption increases significantly during a mulling cycle, increasing until the end of the cycle. It is the work the wheels do that spreads the clay and temper water over the sand grains, and work the water into the clay. Some of the water goes toward cooling the sand. The duration of

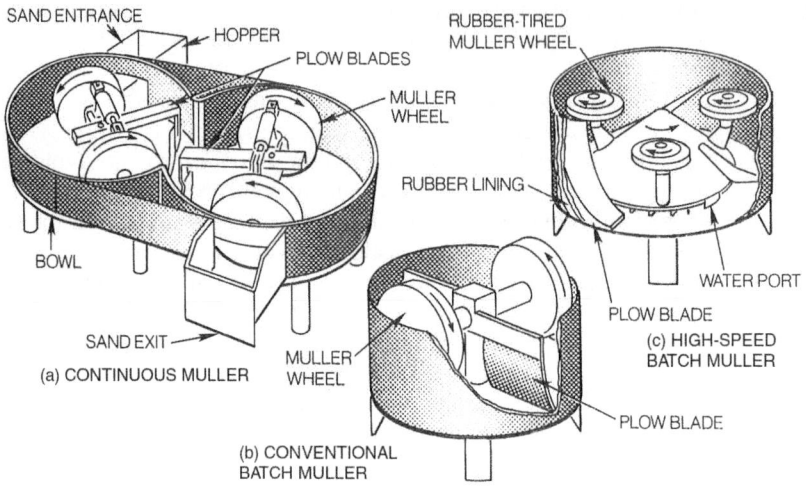

FIGURE 3.21   Various muller designs, (a) continuous, (b) conventional batch, and (c) high-speed. (Photo is used with the permission of Penton Media Inc.)

the cycle is dependent on the type and amount of clay bond, sand grain size, and the water content. The batch gradually becomes plastic. Over the years, wheels have become wider to work more sand with each pass. Many foundries fit the wheels with rubber tires to simplify maintenance and slow the wear on the wheels, which tend to wear on one side unless wheels can be reversed. The plows are faced with hard material. Wear and height of plows and wheels need to be checked and adjusted periodically. Modern mullers are built with the intention that all areas that wear should be easily replaced. Larger mullers are usually fitted with load cells to adjust additions, screw feeders for clay, sand, and other additives, and control mechanisms to monitor either moisture or compactibility. Return sand, water, clay, and new sand additions are metered into the muller. In the early 1980s mechanical compactibility testers were used to extract sand samples and to control sand properties. The Hartley moisture tester arrived by the mid-1980s. Modern versions determine moisture levels and compactibility, and log the time, batch size, moisture, and compactibility. Mullers should not be overfilled. The cycle for a simple mix muller on recycled sand is about 10 minutes, a short period to fill, 8 minutes for mulling, and 1 to 2 minutes to empty. Although shorter cycles increase production, sand properties and muller efficiency suffer. Larger mullers are usually enclosed under an exhaust hood. However, the exhaust should be away from the clay addition port and not too strong as to pull out too much of the clay [the bag-house filter needs to be checked by a methylene blue (MB) clay test periodically].

Higher-speed horizontal wheel mullers are better able to rebound bentonite bonded sands. Batch and continuous mullers are available. These are better able to process hundreds of tons per hour than the slower vertical wheel mullers. High-speed mullers employ centrifugal force to provide the pressure for proper mulling (Fig. 3.21c). Manganese steel plows sweep the sand into the path of lightweight rubber-tired wheels which provide the kneading action to mull the sand on the side of the bowl. Residence time for the sand should be of the order of 90 seconds.

Eirich speed mullers use one or more high-speed rotating plows to throw the sand at the side of the bowl. The bowl of the mixer may also rotate to create counter flow. A stationary plow scrapes sand from the wall. The sand is mulled while airborne and if large volumes of air are introduced, it can be beneficial for rapid cooling of sands in iron foundries. However, fines and sand temperature must be monitored. Overmulling can cause hot sand.

Figure 3.22 shows the average power input into the motors on the three major types of mullers. The vertical wheel continuous mixer operates at approximately 95% of its full-load motor output continuously. The two batch machines spend time charging and discharging sand during their cycles. As a result the horizontal wheel machine operates at 83% of root mean-square motor load even though it exceeds 100% for short periods, and the high shear mixer 71%, as it is designed to operate at 90% of rated power input. After processing a variety of sands of different sand-to-metal ratios, retention per cycle times, and grain fineness, Smith was able to produce an approximate mixer size for these variables (Fig. 3.23) [20].

6. After mulling, sand is conveyed quickly to hoppers in the molding area. Sand may be further conditioned by belt aerators (Fig. 3.24), which develop little green strength, but fluff up the sand, break clods, and generally develop good flowability. Many fit over a conveyor belt and beat the sand with a series of wire brushes or blades. Aerators produce consistent sand with fewer lumps. Aerators

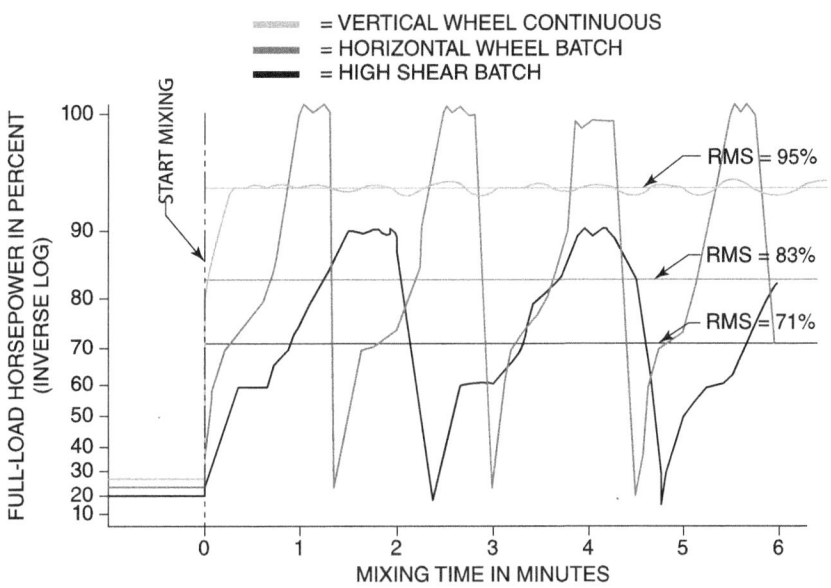

FIGURE 3.22 Muller motor load in terms of rated output and an overlay of average power input for each of the three types of mixers. (From Smith, courtesy of Modern Casting [20].)

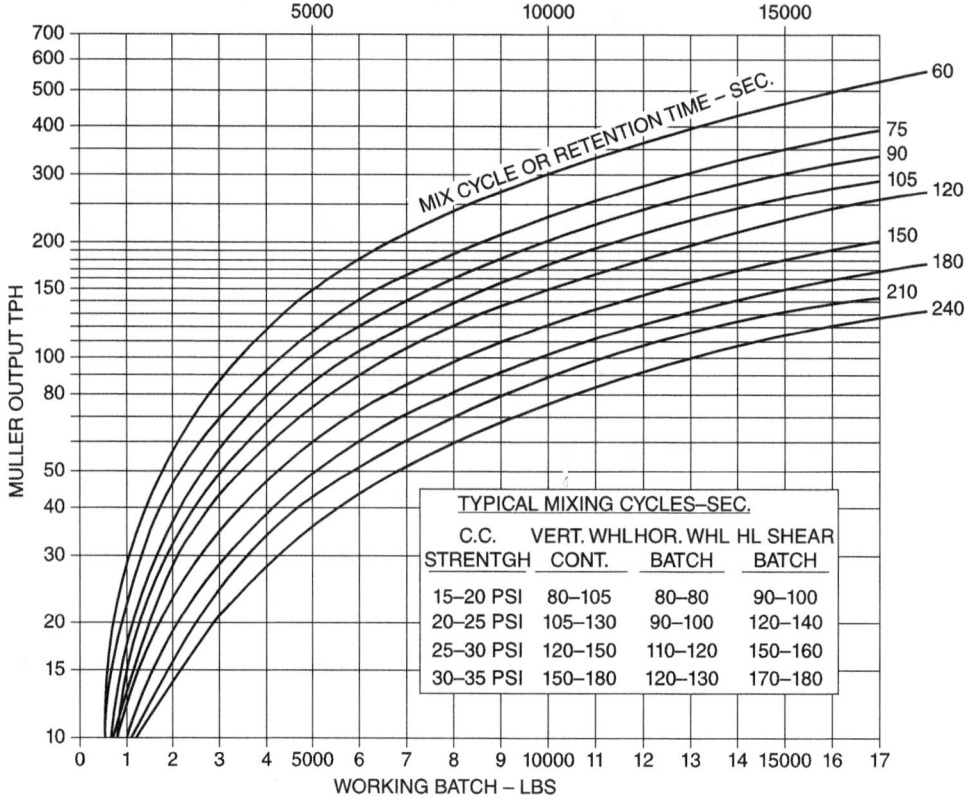

FIGURE 3.23 Sand tonnage output of a mixer as a function of the batch size and cycle time. (From Smith, courtesy of Modern Casting [20].)

need to be placed near the molding machine hopper, as each change in belt direction compacts the sand. Aerators should reduce the bulk density below 60% and there should be less than 15% over $\frac{3}{4}$ to 1 in in size. Moisture losses from long, slow belts can be appreciable from warm sands. In nonmechanized molding lines, facing sand is screened which not only aerates the sand but removes metal, core buts, and clay balls, while the riddling action makes the sand more moldable. The detritus is thrown on the floor; unfortunately, from there it is later returned to the system sand.

7. *Molding machines.* Various types of molding machines have been developed for different applications; however, not all of these are equally efficient in producing quality molds. Each machine utilizes one or more basic means of mechanical sand compaction. Some procedures are more amenable to automation and high operating speeds. It is important to understand the behavior of greensand as these different compaction forces are applied. With this knowledge, a more intelligent selection can be made in matching a molding system to the type of parts to be cast.

Figure 3.24 Belt aerator before installation over sand belt between muller and molding machine. (Courtesy Vulcan Engineering.)

The principal objective of any molding technique should be the production of a mold whose surface will neither wash nor buckle when exposed to sudden contact with hot metal, and a mold rigid enough to retain its shape and dimensions while permitting the rapid escape of gases. This requires the compaction of a suitably formulated sand to produce high, controlled levels of mold density across the entire mold. Two things are necessary when forming a mold with greensand: (1) the sand mass as a whole must move to its final resting place within the flask, and (2) the sand mass must be compacted, or densified, by moving the individual grains closer together. As the work done to compact the sand increases, there is a limiting bulk density between 100 and 115 lb/ft^3 (1600 and 1840 kg/m^3) at which the sand grain to sand grain contact limits further compaction. There will be 60 to 65% solids, the rest voids (Table 3.7). As compaction occurs, the mechanical properties of the sand change as shown in Fig. 3.25. As the sand density increases, the green compressive strength, green shear strength and tensile strength, and mold hardness increase. More work has to be done to overcome the strength of the sand and further increase density. The high frictional forces between the sand and the adjacent surfaces (flask walls and pattern), and between the grains themselves, create a natural resistance against movement or compaction. The machine and pattern should not move or distort while the force is applied. Lastly, the sand may spring back when the force is released. The response of the sand mass to outside forces depends upon the nature of

Material	Bulk Density		Solids, %	Voids, %
	lb/ft³	g/cc		
Silica sand (no clay)	100–115	1.6–1.84	60–70	40–30
Dry sand with clay, 60–75 AFS GFN, 9–11% AFS Clay, 4–8% LOI	90–105	1.44–1.68	54.4–63.5	46–36.5
As above + tempered with water, mulled, riddled	50–65	0.8–1.04	30.2–39.2	69.8–60.8
Above sand compacted to mold hardness 70–85	65–85	1.04–1.36	39.2–51.3	60.8–48.7
Above sand compacted to mold hardness 90–95	90–100	1.44–1.60	54.4–63.5	45.6–36.5

Note: True density of solid silica is 165.5 lb/ft³ or 2.64 g/cc.
Source: Heine [21].

TABLE 3.7   Bulk Densities of Compacted Molding Sands

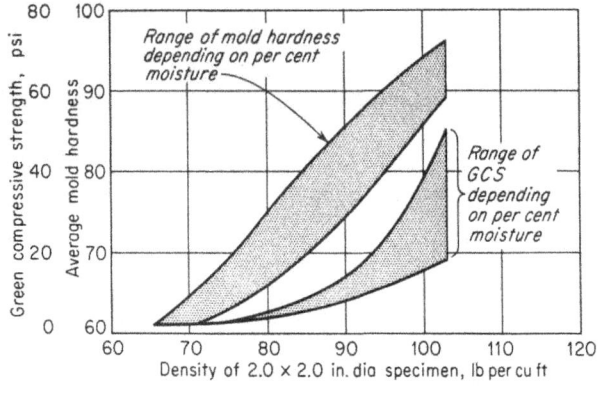

FIGURE 3.25   Change in sand properties with compaction. (From Heine [21].)

the force as well as its intensity. It was found that ramming at higher speeds results in greater compaction than the same force at lower speed. Comparisons of the relative effectiveness of ramming, jolting, and squeezing are given in Fig. 3.26.

Ramming, either manually or pneumatically, consists of hammering the sand mass with blows applied to small areas. Although sands can be uniformly and densely rammed into small spaces in this manner, the overall level of mold compaction tends to be erratic (Fig. 3.26a).

Jolting utilizes the inertia of the sand mass itself, causing it to settle and compact against horizontal surfaces of the pattern and parting surface. The work table with flask, pattern, and sand are lifted by a pneumatically operated piston, and then allowed to fall against the base of the machine under the influence of gravity. The power of

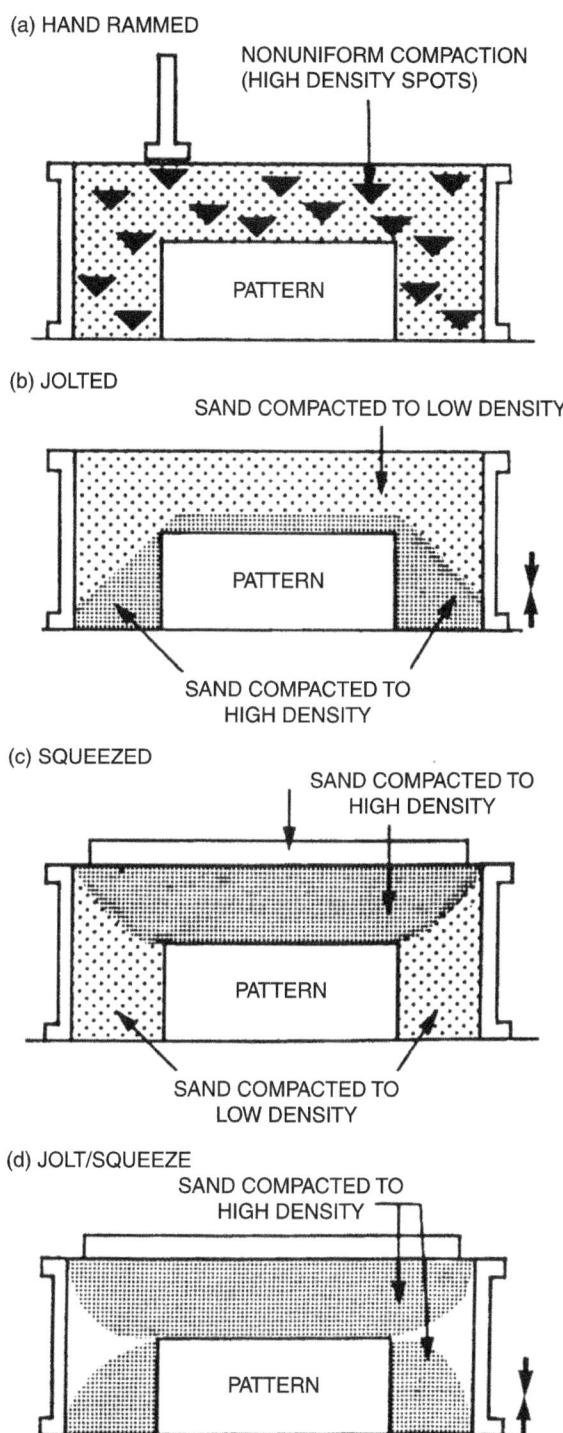

Figure 3.26 Compaction of molding sands using various molding methods. (From Sahoo [7].)

jolting results from the conversion of the momentum of the falling sand to work when the mass is stopped instantaneously by the base:

$$\text{Power of jolting} = MV/A \tag{3.2}$$

where M is the weight of sand, V is the velocity of jolt or $\sqrt{2gh}$, A the jolted area, and h the stroke of the jolt. The power of jolting from this formula is essentially independent of the flask area, and dependent on the jolt stroke, a machine characteristic. Although the number of times the jolting is performed gradually increases sand compaction, after about 20 strokes the sand asymptotically approaches maximum hardness and density, therefore in general 20 or fewer jolts are given. The capacity of the machine is dependent on the size of the cylinder doing the lifting. As the upper portions are much less affected (Fig. 3.26b), some jolt-type equipment places heavy weights on top of the sand mass, to increase compaction in the upper layers. Noise levels were high and the impacts were hard on the machinery and patterns.

Squeezing of the sand in a flask produces a compaction pattern which is the reverse of that produced by jolting. The degree of compaction is greatest in contact with the squeeze head, diminishing as the distance from the squeeze head increases. With relatively high flasks and shallow or flat-back patterns, squeezing alone may result in relatively soft mold faces even when using high squeeze pressures. With higher patterns, flat squeeze heads create high-density compaction against the upper pattern surfaces and relatively low compaction against the pattern and flask sidewalls or parting surface (Fig. 3.26c). The limiting effects of these high-density sand columns can be offset in some measure by use of a profiled squeeze head or by use of multiple pressure feet with hydraulic equalization (as with high-pressure squeeze machines). For simple squeeze machines using air pressure introduced into a pneumatic cylinder the maximum squeezing force is

$$\text{MF} = P \times \tfrac{1}{4}\,\pi d^2 - W \tag{3.3}$$

where MF is molding force, d is the diameter of the cylinder, P is air pressure in squeeze cylinder fed from the foundry's compressed air, which as the molding machines are often at the ends of air lines, is no more than 115 psi (8.1 kg/cm^2) and variable, and W is the weight of flask, sand, and accessories on the molding machine table. This means that flask size is generally limited by the size of the cylinder. Thus, a 20-in (50-cm) diameter cylinder can only effectively squeeze a 16×20 in (40×50 cm) flask. The machines cannot adequately compact the sand in larger flasks.

Jolt/squeeze machines combine the compaction characteristics of both jolting and squeezing (Fig. 3.26d), thus taking advantage of both forces. Areas adjacent to flask sidewalls still tend to be less dense, however, unless the flask is relatively shallow. Deep pockets of sand inadequately compacted by jolting cannot be markedly improved by the application of subsequent squeeze pressure, however high.

Molding pressures of 20 to 50 psi (1.4 to 3.5 kg/cm^2) were typical in small squeeze machines. Pressures approaching 150 psi (10.5 kg/cm^2) are required to produce the maximum sand bulk density. Mold forces were approaching 1 million tons by the mid-1980s. Patterns had to be made of metal, and both the cope and drag had to be squeezed simultaneously (with the pattern between them). However, for sand subjected to high squeeze pressure to achieve high mold hardness it also needs high green compression strength and high clay content. Sands also need excellent flowability in order to achieve uniform bulk density.

Bulk density can be evaluated in the lab with the standard AFS 2-in cylinder, but can be difficult to measure on the foundry floor. Sanders and Clem [22] developed the

"One-to-Ten Ram" test to correlate sand properties to mold properties (Fig. 3.27). While some permeability testers are transportable, the mold "B" or "C" hardness gauge is a handheld gauge that measures movement of a ball when applied with a fixed force against a flat surface (Fig. 3.28). This gauge coupled with the "One-to-Ten Ram" test allows rapid assessment of potential problems. For example, a mold hardness of less than 85 will likely result in mold wall movement and roughness in iron castings with an

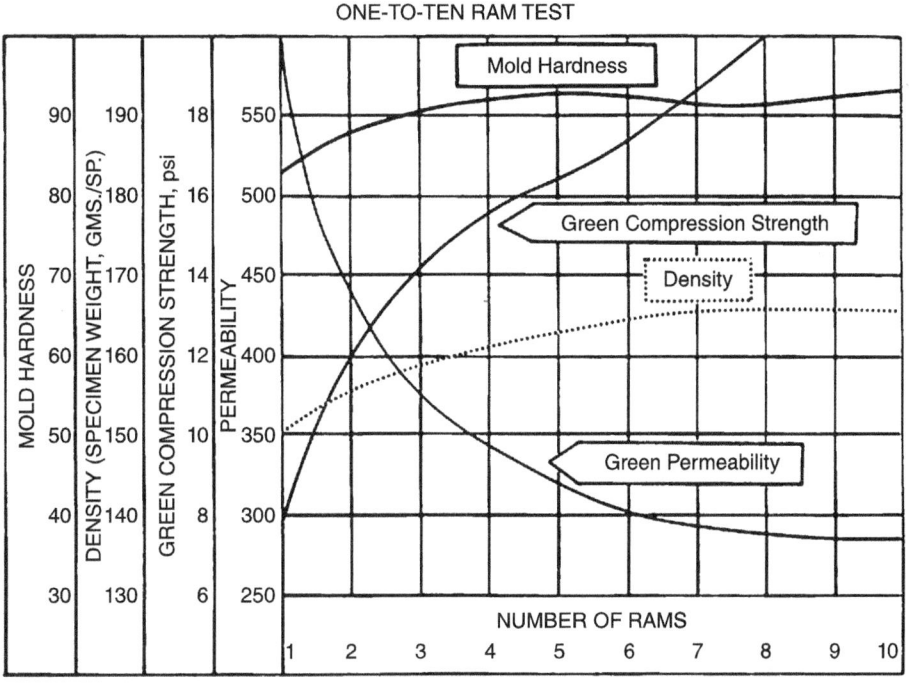

FIGURE 3.27 The "One-to-Ten Ram" test correlates sand properties with mold properties. (From Sanders [23].)

FIGURE 3.28 Mold "B" hardness tester. (From Sanders [23, p. 360].)

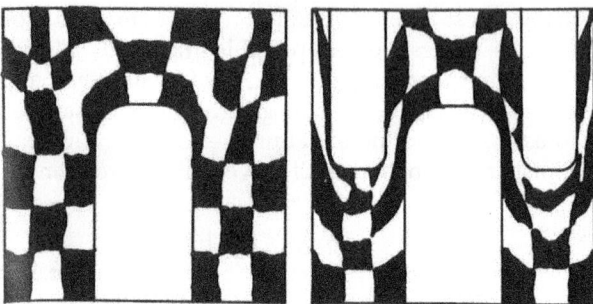

Figure 3.29 Effect of squeeze pressure by flat and contoured boards on mold wall hardness for a deep draw. (From Heine [21].)

appreciable metallostatic head. Many older molding techniques produced poor mold hardness with patterns requiring a deep draw (the pattern extends high in the flask resulting in a significant vertical wall in the mold), see Fig. 3.29.

Venkoba Rao and Roshan [24] used a high-pressure molding machine to study mold density. Figure 3.30 shows that squeeze pressures of 430 to 570 psi (30 to 40 kg/cm^2) are required to achieve bulk densities of 1.7 g/cc. Figure 3.31 shows squeeze pressures of only 140 psi (10 kg/cm^2) are required to produce a mold hardness in excess of 90 on the

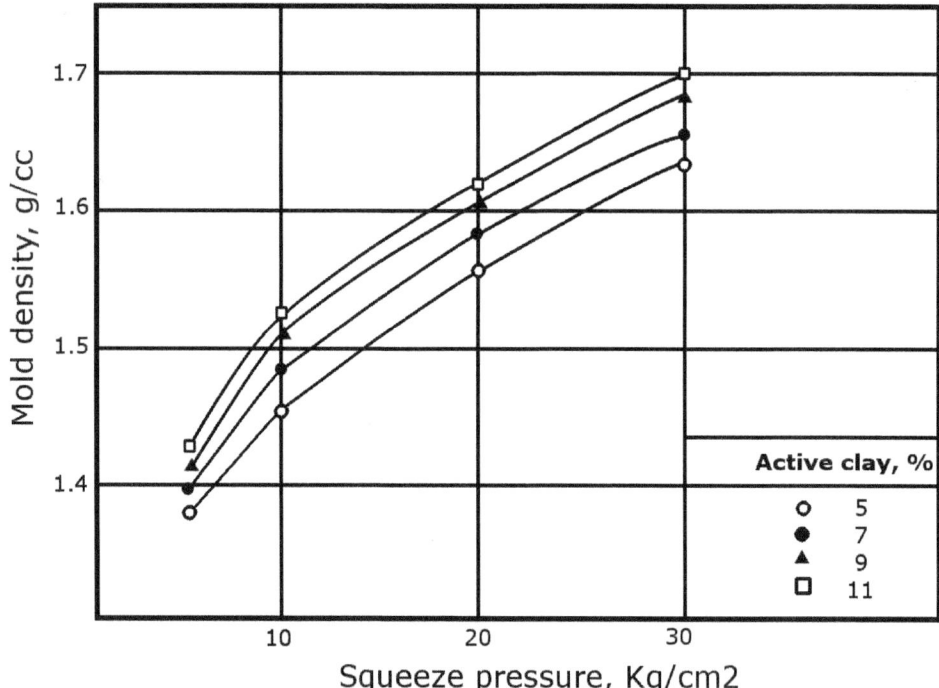

Figure 3.30 Effect of active clay on bulk density with squeeze pressure. (From Venkoba Rao and Roshan [24].)

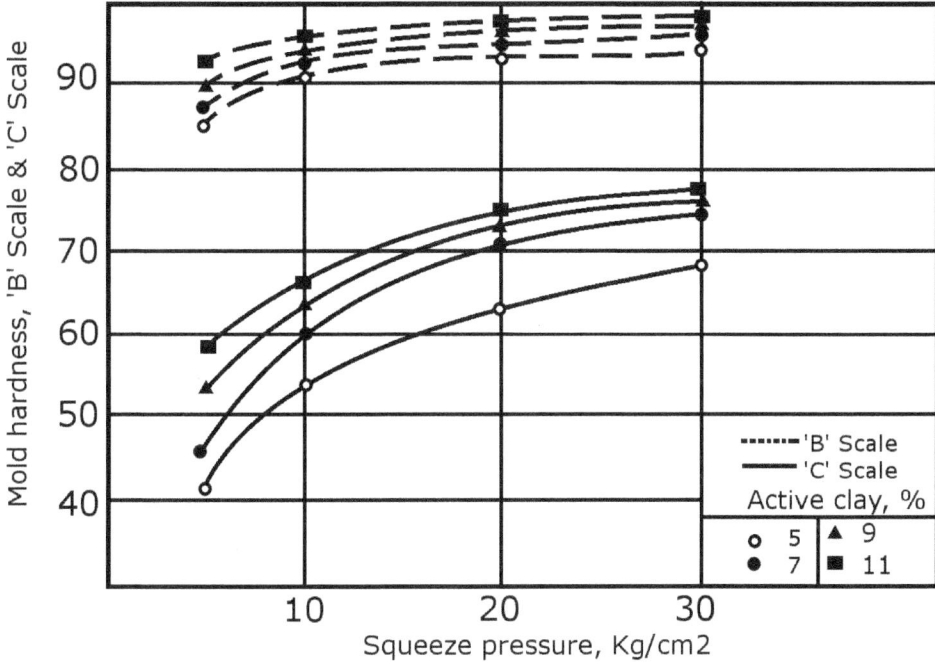

FIGURE 3.31 Effect of active clay on mold hardness with squeeze pressure. (From Venkoba Rao and Roshan [24].)

top of the mold, but Fig. 3.32 shows they fall dramatically at intermediate levels and are lowest on the parting line. Mold hardness on a sidewall is worse, particularly if the pattern is too close to the sidewall of the flask (Fig. 3.33). Density increases with mold hardness, clay and moisture contents, and coarser sand. Mold hardness only measures the surface layer of a mold and not underlying areas.

Recent developments of molding machines provide greater automation for molding and pattern changing, and largely eliminate much of the experienced labor required in jolt squeeze methods. As mentioned above, speed in filling the flask is important in achieving high compaction. Manufacturers have developed new methods of filling the flask using compressed air to fluidize the sand which is then blown or shot, or sucked by vacuum into the flask followed by compaction by high pressure hydraulic squeeze [25]. The first part of the operation is similar to core blowing only the equipment is larger. Impulse or explosive action on the full flasks offers the ultimate compaction speed for producing molds of very high density. This method is also known as air pressure wave or gas pressure wave when initiated by the explosive force of controlled combustion of pressurized gas/air mixtures (Fig. 3.34). The profile of sand properties is represented in Fig. 3.35. To date, such equipment has found only limited application in North America; however, its developers claim a uniquely uniform and efficient compaction behavior. This is ascribed both to the uniformity of pressure distribution within the sand mass in promoting movement and to a thixotropic behavior of the clay bonding material, inducing high flowability under the effect of shock pressure. These molding methods offer high dimensional accuracy, consistency, high production rates, lower

a. Test mold

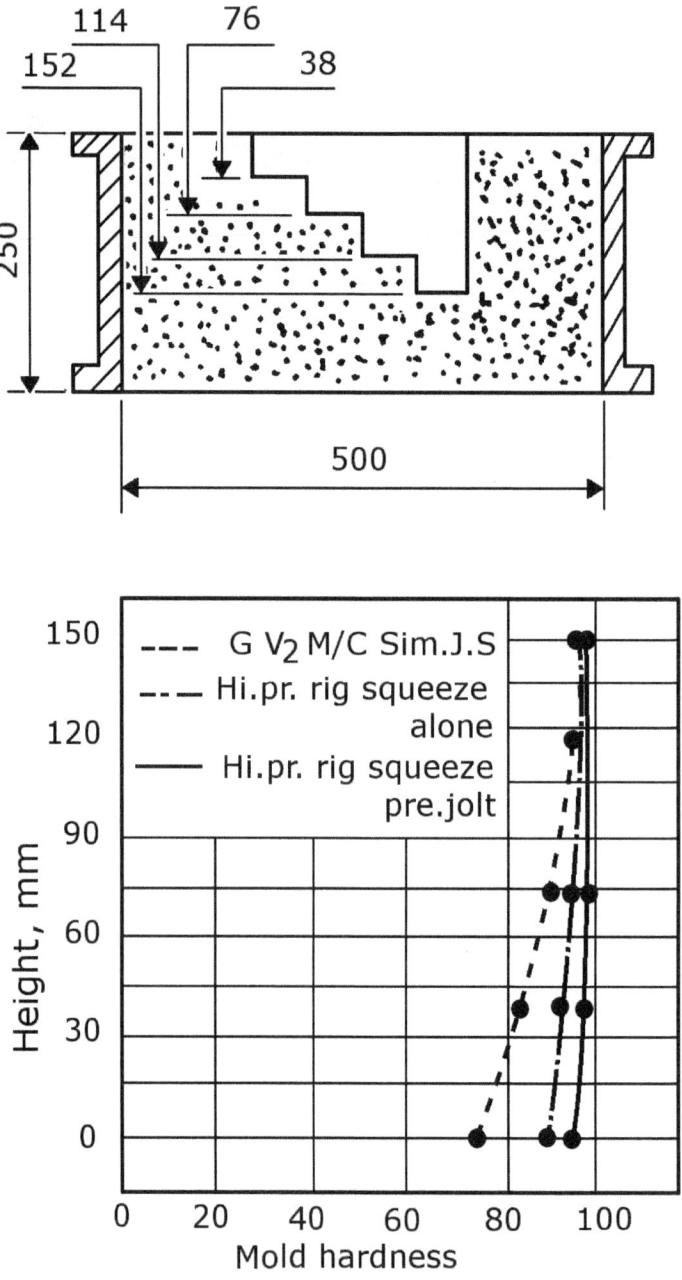

Figure 3.32 Effect of pattern height on mold hardness at parting line. (From Venkoba Rao and Roshan [24].)

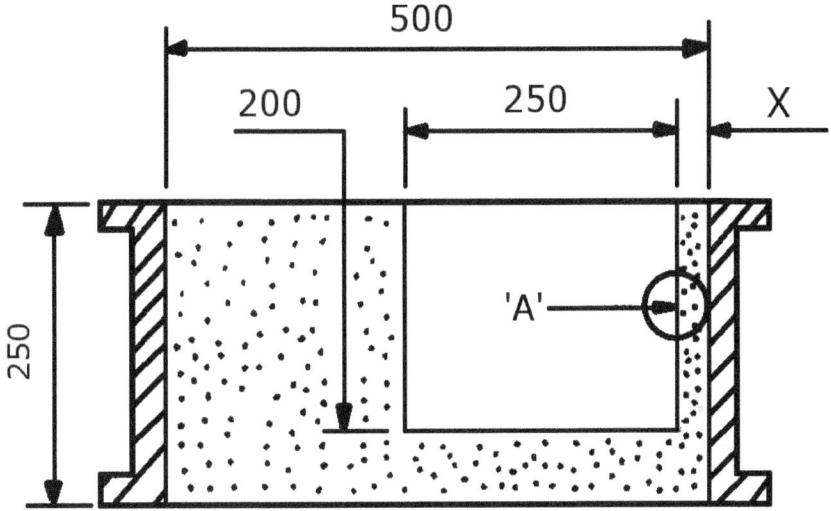

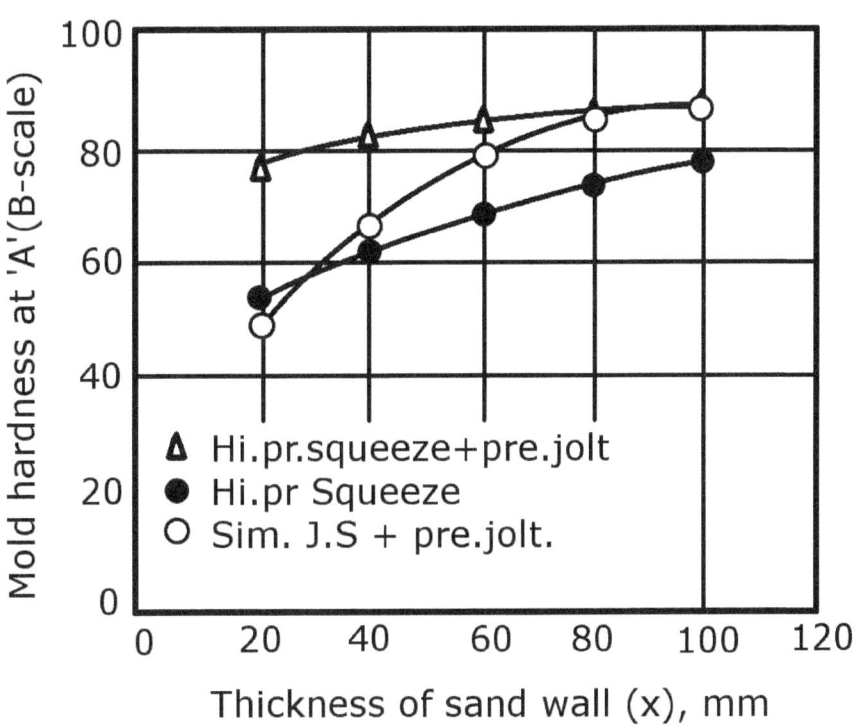

FIGURE 3.33   Effect of sidewall thickness on mold hardness. (From Venkoba Rao and Roshan [24].)

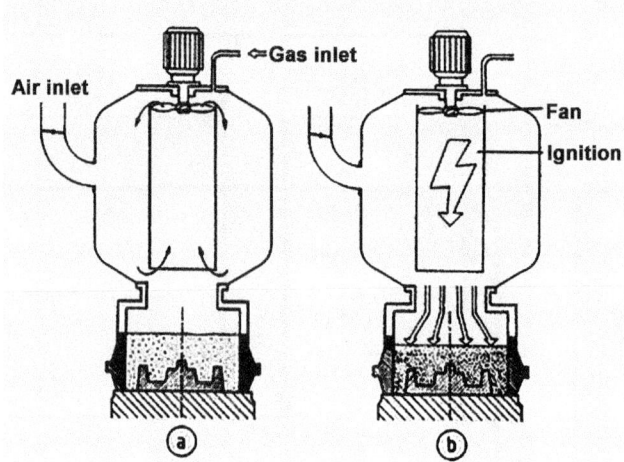

Figure 3.34   Mold compaction by combustion with natural gas: (*a*) air and gas mixed, (*b*) impact from combustion. (From Schaarschmidt [25].)

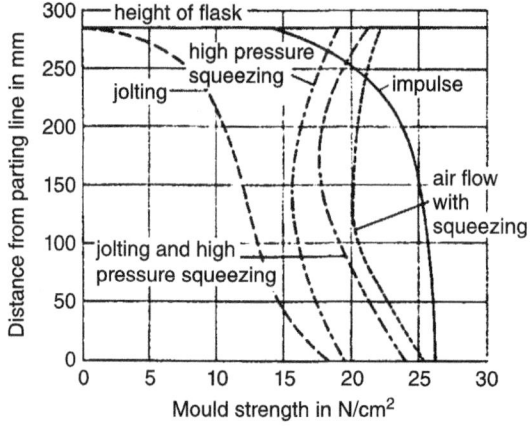

Figure 3.35   Distribution of mold strength within the mold half with compaction by current molding processes. (From Schaarschmidt [25].)

cost, and near net shape castings, all in demand by the automotive industry. Greensands must have high strength and low compactibility (otherwise the permeability is too low with high-pressure molding). High-pressure molding solves the issues of swells and distortion of mold walls by control of sand properties.

Sand slinging, when done properly, will produce molds of high, uniform density. In this process, small clumps of sand are thrown against the pattern in rapid succession, each slightly overlapping the preceding clump. Compaction is by impact, the degree of which is controlled by the velocity of an impeller, which hurls the sand by centrifugal force. Sand is conveyed by belt into the slinger head, which contains a rotor with blades that pick up the sand as it comes off the conveyor and propel it against the pattern. A rotor operating at up to 1800 rpm can eject sand at 10,000 fpm (3000 m/min). Slower

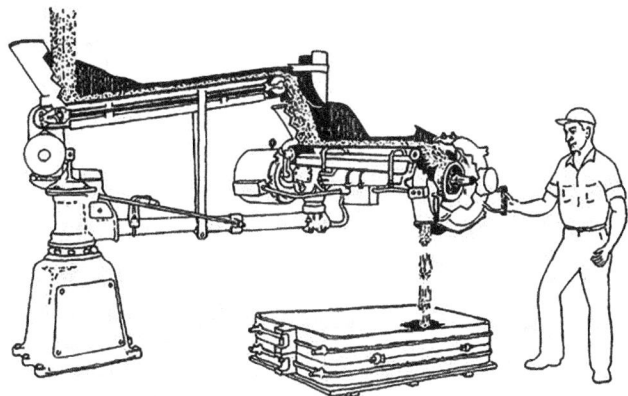

Filling A Large Flask With A Hand Slinger

FIGURE 3.36 Small sand slinger. (From Metalcaster's Reference & Guide [2].)

speeds of 1200 rpm are used at first to avoid excessive wear of the pattern. Backing sand is applied at higher velocities.

Several types of sand slingers are available. Stationary slingers (Fig. 3.36) have the operator standing or seated next to the head, where he can control the sand stream as molds pass on tracks beneath. Motive slingers carry their conditioned sand in tanks, and operate from rails from where they can produce medium and larger castings for floor or pit work.

To aid in stripping the pattern, either a zircon flour or liquid parting agent (particularly for automatic molding) is used to lubricate and reduce friction while stripping the pattern to avoid damaging the mold. Excess liquid parting agent can pool in pockets and reentrant angles where it can damage the mold as well as cause gas defects. It is critical that the pattern be withdrawn with the correct alignment. Rapping and vibration are also used to strip the pattern from the mold, particularly with chemical binders where the binder can also bond with the pattern. Excessive sidewall movement is to be avoided as it will deform the mold.

In summary, it should be pointed out that the true economy of a greensand molding system depends upon the effectiveness of (1) a well-formulated system sand, (2) a well-designed and operated sand recycling system, (3) a systematic testing program for monitoring the properties of the system sand and those of its ingredients, and (4) selection of molding and handling equipment adapted to the foundry's product mix.

Flasks

Most greensand molding lines use metal flasks to contain the sand and permit handling the molds in order to get them into position for pouring. Molding flasks may be classified as follows:

1. Removable flasks
   a. Snap flasks
   b. Pop-off flasks
   c. Slip flasks
2. Tight or permanent flasks

Removable flasks are used for match-plate molding and cope and drag molding of small to moderate size. They are used because only a single flask is required per machine setup. After the mold is made, the flask is removed and replaced with a jacket so the mold can be weighted and poured. A *slip flask* has the side walls tapered 4° to allow the mold to be removed. *Pop-off* flasks have corrugations on the tapered sides to hold the sand in place. Pop-off flasks have expandable sides, so they can be removed after the mold is completed (Fig. 3.37). Snap flasks are hinged on one corner so they can be opened.

Removable flasks are subject to warpage if dropped or mishandled and do not provide the most rigid support of the mold. Flask sizes are usually described by their width and length at the parting line [thus a 16 to 32 flask is 16 in wide and 32 in long (40 × 80 cm) at the parting line] and by their depth. These dimensions are usually marked on the side as flasks are often stacked. The drag height is usually limited to 2 in (5 cm) of sand at the bottom and also at the sides. The cope height determines the height of the sprue, and therefore the metallostatic pressure applied by the molten metal. High cope heights favor elimination of gases and promote feeding, but also may cause mold cavity enlargement (mold wall movement). *Upsets* are small flasks used to extend the cope or drag. They are usually metal and are bolted on to either flask. Removable flasks are typically used for floor or jolt-squeeze molding and allow the mold to be vented more easily.

Tight, rigid, or *permanent flasks* remain around the mold until after the casting is poured and shaken out. Tight flasks are made of steel and can be further reinforced; they are more rigid and resistant to warpage (Fig. 3.38). Alignment of cope and drag is ensured by the use of pins and bushings. The cope can be clamped to the drag, so weights may not be required. Tight or permanent molds are generally used in highly mechanized molding lines. Molds should not deflect under sideways forces from high-pressure molding.

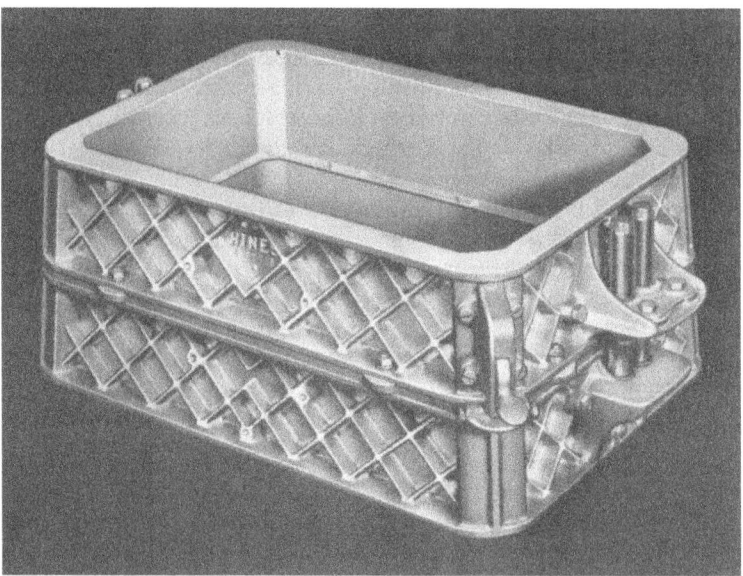

Figure 3.37 Pop-off flask with double guide pins.

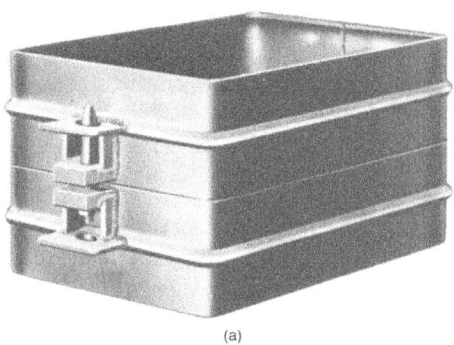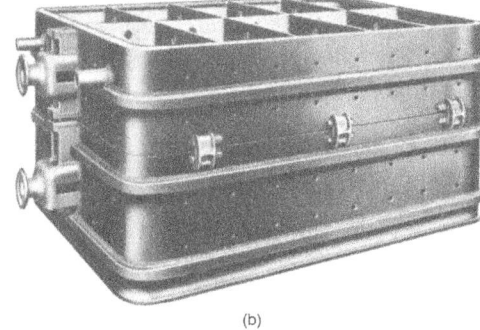

(a)   (b)

FIGURE 3.38   Tight or permanent flasks: (a) common type, (b) reinforced.

Bottom boards are required to contain the sand prior to molding and squeezing the sand in the flask during molding. As the bottom boards fit inside the flask, they are generally ¼ in smaller, as seen in Figs. 3.1 and 3.27. In floor molding, wood is used but metal boards are used in automatic molding lines and larger floor molds as wood distorts under higher pressures and contact with damp sand. Only one is required as cope and drag are the same size. These boards stay with the mold during pouring. For rollover machines, the board engages the flask. In some molding machines, the board may be part of the squeeze head. In such instances, contoured squeeze heads achieve more uniform mold hardness in the mold.

Flaskless molding is also used in greensand molding with vertical parting lines. Vertical parting lines avoid the great variability of mold hardness. Normally mold hardness falls quickly as the distance from the parting line increases. With vertical parting, two patterns squeeze a block of sand between them. Both faces benefit from the squeezing pressure. Control of compactibility is critical if the patterns are not to be damaged by contact. If stops are used to protect the patterns, then the sand could be too weak if the compactibility were too high for sand compression to be accomplished. This can be particularly critical if the molding machine pushes 100 yd (90 m) of molds (just sand, no jackets or retainers). With the exception of the first 10 yd (9 m) most molds contain hot metal. Ten to twelve percent of mostly western bentonite is required to raise the hot strength so the molds maintain their shape and not disintegrate until the shakeout station is reached. A molding machine is shown in Fig. 3.39.

Molds with a horizontal parting line need a weight on the cope to keep the cope down and avoid run-outs of metal at the parting line. The metallostatic pressure at the parting line will try to buoy up the cope unless sufficient force is used to hold the cope down. The force, F_c, on the cope can be calculated from the following relationship:

$$F_c = P_c \times A_c \qquad (3.4)$$

where A_c is the estimated mold cavity area at the cope parting line, and P_c is the metallostatic pressure at the cope parting surface, which can be calculated by $P_c = w \times h$, where w is the weight of a cubic inch of metal and h is the effective height of metal above the parting line.

In the event that the casting is all in the drag, then sprue height is the effective height of the metal head above the cope parting line. For safety, the height can be the

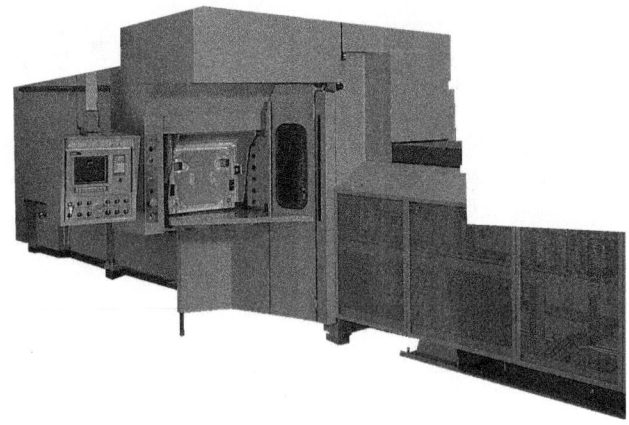

FIGURE 3.39 Molding machine for flaskless molds with a vertical parting line. (Courtesy Disa Industries Inc.)

sprue height h_c for the cope. For ferrous castings, w is 0.26 lb/in^3. Therefore the force pushing up on the cope is

$$F_c = 0.26 \times A_c \times h_c \qquad \text{or} \qquad (F_c = 0.008 \times A_c \times h_c \text{ in the metric units}) \qquad (3.5)$$

The weight of the sand in the mold can be subtracted from the force required. However, a factor of 1.5 to 2.0 is often used to overcome the dynamic-pressure effect. Weights must be available in a variety of sizes and shapes to place on each mold prior to pouring. In practice, weights can be moved from one mold to another as pouring progresses. Many lines incorporate the manipulation of weights as part of the line operations. However, in the latter case, weights gradually become lighter as new work requires new positions for pouring sprues, and new holes are made. Flat weights covering a greater area are preferred, as tall weights can crush the mold prior to pouring. Tight flasks can be clamped together.

Multiple parting lines are used in floor molding for complicated castings, or when horn gates, bottom gates, and swirl gates are used to introduce liquid metal into the bottom of the casting cavity in order to avoid cascading metals prone to oxidation or hydrogen pick-up through the casting cavity.

Molding Materials

Sand

The word "sand" describes mineral particles of 1/400 to 1/12 of an inch (0.06 to 2 mm) in diameter[4] although the foundry term generally means the refractory minerals silica, olivine, zircon, chromite, etc. Properties of some typical foundry sands are shown in Table 3.8. Low density synthetic mullites, with low heat conductivity, are being developed for lost foam casting of magnesium and for chemically bonded molds for thin-walled casting of steel.

Sand and mineral aggregates are usually sized. While a number of standard sieves are in use, foundry sands in North America are usually screened using U.S.

---

[4]Sieve sizes minus 8, plus 270 mesh.

	Silica	Olivine	Chromite	Zircon
Composition	>95% SiO_2	80% forsterite Mg_2SiO_4	$FeCr_2O_4$	$ZrSiO_4$
Impurities	Clay limonite glaucomite	Serpentine 7–10% fayalite talc, silica	Mg, Fe, Al spinels	2–15% silica
Origin	United States	Washington, North Carolina, Norway	South Africa, Russia, Zimbabwe	Florida, Australia, India, South Africa
Mineral Properties				
Specific gravity	2.65–2.67	3.27–3.37	4.3–4.5	4.5–4.7
Mohs hardness	6–6.5	6.5–7.0 variable	5.5–7.0 variable	7–7.5
Fusion point	1,713°C 3,110°F	1,205–1,900°C 3,200–3,600°F	1,760–2,000°C 3,200–3,600°F	2,200–2,500°C 3,700–4,000°F
Result on heating	Phase change	Hydrous Mg silicates	Hydrous iron oxide	
Thermal expansion in/in	0.018	0.0083	0.0045	0.0037
Maximum thermal expansion at	550°C 1,100°F	420°C 790°F		
High temperature reaction	Acidic	Basic	Basic-neutral	Slightly acidic
Sand Properties				
Color	White, gray, yellow, brown	Green, olivine, yellow-brown	Black	White, grayish or orange
Shape	Rounded	Angular	Angular-subangular	Subangular to rounded
Grain size AFS GFN	27–120	30–240		65–120
AFS screen dist'n	2–4 screens	2–4 screens	53–90	2 screens
Bulk density lb/ft^3	100–108	98–103	4 screens	152–183
pH	6.4–7.4	8.2–8.7	156–165	5–7
Acid demand at pH7	0–35	9–16	7–10	0–0.5
Clay	0–1.8%		2–4	
Loss on ignition, %	0–0.8	0.4–0.7	Gain of 0.4–0.5	0.05–0.10

Source: Garner [26].

TABLE 3.8 Characteristics of Foundry Sands

Standard Screens. The mesh and wire sizes of other sets vary and therefore so does the size of particles that can pass through the holes. The important parameters of the U.S. Standard Screens are shown in Table 3.9. Table 3.10 shows the calculation of the ASF Grain Fineness Number from the screen distribution. The foundry industry prefers to chart the distribution as the percentage retained on each screen so the number over 10% retained can be counted. Many other industries prefer the cumulative distribution (Fig. 3.40).

U.S. Series Equivalent Number	Tyler Screen Scale, meshes/in	Openings, mm	Openings, in[¶]	Mesh Openings, μm	Permissible Variation in Average Opening, %±	Diameter of Wire, in
4	4	4.099	0.187	4760	3	0.065
6	6	3.35	0.132	3350	3	0.036
8*	8*	2.36	0.0937	2360	3	0.035
12	10	1.70	0.0661	1700	3	0.032
16*	14*	1.18	0.0469	1180	3	0.025
20	20	0.850	0.0331	850	5	0.0172
30	28	0.600	0.0234	600	5	0.0125
40	35	0.425	0.0165	425	5	0.0122
50	48	0.300	0.0117	300	5	0.0092
70	65	0.212	0.0083	212	5	0.0072
100	100	0.150	0.0059	150	6	0.0042
140	150	0.106	0.0041	106	6	0.0026
200	200	0.075	0.0029	75	7	0.0021
270	270	0.053	0.0021	53	7	0.0016
400*	400*	0.037	0.0015	37		

*Not used for sands.
[¶]Most differ by a factor of $\sqrt{2}$ or 1.414.
Source: ASTM E11-70 [2, 27, 28].

TABLE 3.9   Screen Scale Sieves

USA Sieve Series Number	Amount of 50 g Sample Retained on Sieve		Multiplier	Product
	Grams	Percent		
6	0	0	3	0
12	0	0	5	0
20	0.02	0.044	10	0.53
30	0.01	0.022	20	0.44
40	4.88	10.707	30	321.19
50	15.96	35.015	40	1400.61
70	15.94	34.972	50	1746.57
100	6.30	13.822	70	967.53
140	1.50	3.291	100	329.09
200	0.49	1.075	140	150.50
270	0.26	0.570	200	114.09
Pan	0.22	0.483	300	144.80
Total	45.58	100		5177.36

The sample size was 50 g, the AFS clay content 4.42 g, and the amount of sand screened was 45.58 g.
AFS GFN = total product/total percent retained = 5177.36/100 = 51.8

Source: Metalcaster's Reference & Guide [2], *AFS Mold and Core Test Handbook* [28].

TABLE 3.10   Typical Calculation of AFS Grain Fineness Number

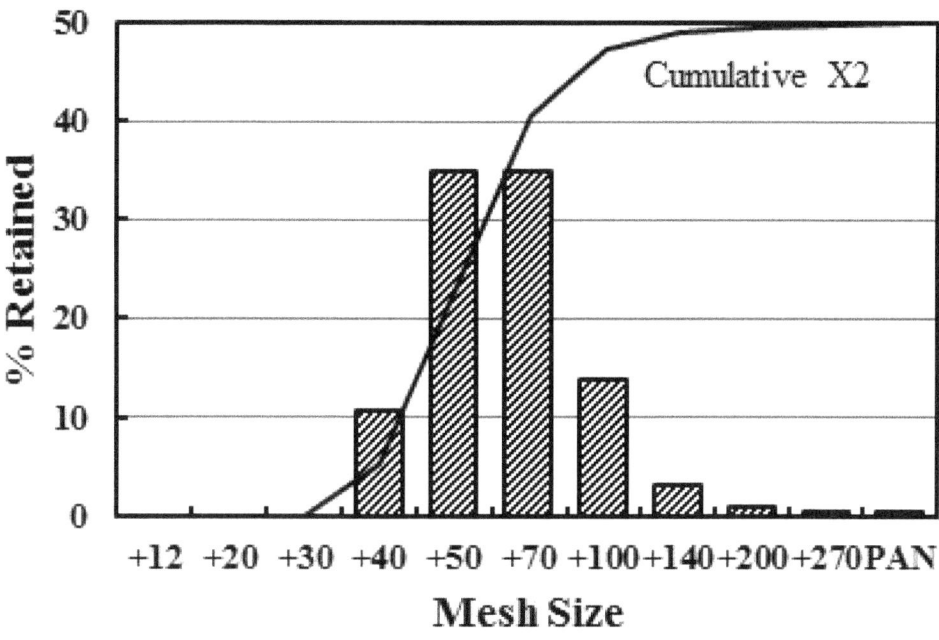

FIGURE 3.40   The screen distribution of the sand in Table 3.10.

Silica

Refractory sands have been used by the foundry industry for hundreds of years. The greensand molding described by Biringuccio in 1540 is more akin to dry sand molding [29]. Contemporary greensand molding became common by the seventeenth century. Early foundrymen were able to recognize suitable sand deposits from which they were able to make molds. Sands from seashores contain too much calcite, from sea shells, which react violently to heat and with many metals. Geologists know that rock is weathered and fractured by rain, wind, and cold. The particles are then washed away and transported downstream, to be deposited when the currents are no longer strong enough to carry the particles. Grains are further fractured, polished, picked up, and transported still further down river eventually forming sand bars in estuaries and lake inlets. The constant tumbling produces rounded sand grains. Glaciers tend to produce more angular particles. On the other hand, deserts or coastal sand dunes tend to frost the surface from constant jostling by winds. Early foundries used natural sand deposits that contained a wide range of particle sizes, lots of fines, and a mix of clays (up to 20% total fines). The natural sands, known by their place of origin, just required a controlled amount of water to make them pliable for molding. Most of these smaller deposits have been depleted, and the foundry industry now uses synthetic mixes of refractory sand and clays to produce molding mixes with carefully controlled properties. Large silica deposits are found in the central United States around what was once a large inland sea. Glaciers later pushed these deposits southward, notably into Wisconsin, Illinois, Indiana, Oklahoma, and Arkansas. The Michigan Bank sands were transported by wind to their present location [26].

The main users of mined high-purity quartz sand (99.8% silica, SiO_2) were once glass making and smelting, whereas now silica sands are used for fracking of shale (to

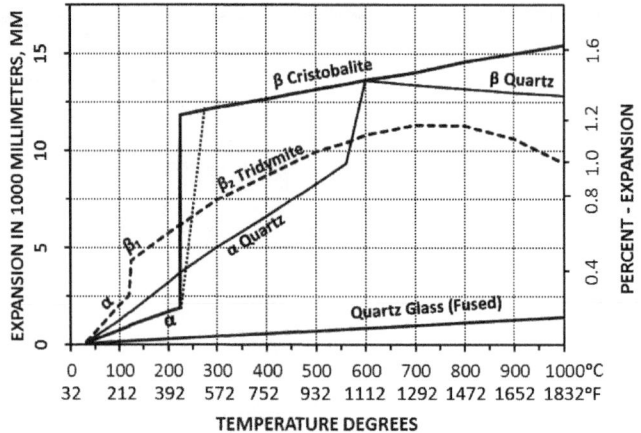

FIGURE 3.41   Thermal expansion of silica materials. (From Sanders [23].)

keep cracks in oil and gas wells open). Foundries are less important. Figure 3.41 shows the thermal expansion of some silica materials. The NA automotive industry prefers lower purity bank and lake sands (94% SiO_2), as the lower quartz content lessens the impact of quartz expansion. Sand deposits are usually processed by first removing brush and top soil, the sand layer is mined, screened by a coarse rotary screen to remove roots, then mixed with water, screened by a scalping screen to remove root hairs and organics, then deslimed to remove clay and silt (<270 mesh). More water is added and the sand stockpiled on the ground or fed onto banks of screens to further split sand into different sized fractions, followed by desliming, drying in rotary dryers, and then to storage bins.[5] Blending the various grades to make specific foundry products is done just before sending a batch to the customer, by blower truck, railroad car, canvas super-sacks, or paper bags. Constant testing is performed and piling of the final product is avoided, otherwise the sand will separate. Table 3.11 shows the size distribution of a number of commercial foundry silica sands.

Most naturally occurring silica sands are seldom pure, even after refinement. The impurities may seriously affect not only the color of the sand, but other physical and chemical properties, such as melting point, pH and acid demand, weight change on heating, and high-temperature reactivity particularly with molten metal and slag.

Olivine

Another foundry sand is olivine or magnesium-iron orthosilicate, $(Mg,Fe)_2SiO_4$, which occurs as a continuous solid solution of two minerals, forsterite (Mg_2SiO_4), and fayalite (Fe_2SiO_4). Olivine is obtained from rocks containing at least 90% forsterite, less than 10% fayalite, and inclusions of magnetite, spinels, and apatite $[Ca_5(PO_4.CO_3)_3]$. Suitable mountainous deposits of olivine exist in North Carolina, Washington, and in neighboring British Columbia. The rock is crushed, water added followed by gravity separation of mineral impurities (mainly serpentine and talc), dewatered, dried and then screened to produce various grades (Table 3.12). Crushing produces angular grains with

---

[5]Sand with more than 0.25% moisture will not flow through the gates of foundry equipment.

Mesh	Nominal Size, mm	1	2	3	4	5	6
20	1.18	0	0	0	0	0	0
30	0.850	Trace	1.5	0	0	2.5	0.5
40	0.600	30	12	0.1	0	10	1
50	0.425	52	31	9	1.5	29	3
70	0.30	13	32	35	5	39	10
100	0.212	4	15	33	25	17.5	49
140	0.150	1	6	18	40	2.0	26
200	0.106	0–0.05	2	4	22	0	9
270	0.053	0	0.5	1	5.0	0	1
Pan		0	0	Trace	2.0	0	0.5
AFS GFN		40	53	70	106	50	83
Surface area		122	176	250	390	184	340
%SiO_2		99.8	99.7	99.5	99.5	94.8	94.8
Shape		Round	Round	Round	Round	Sub-round	Sub-round

TABLE 3.11 Physical Analyses of Some Silica Sands

Mesh	Nominal Size, mm	Washington Olivine	N. Carolina Olivine	Chromite 50/60	Chromite 70/80	Zircon
20	1.18			0.5	Trace	0
30	0.850	Trace	Trace	2.5	0.6	0.5
40	0.600	0.4	1.7	16.2	3.5	1
50	0.425	26.1	26.1	32	12.2	3
70	0.30	35.4	38.3	24	26.4	10
100	0.212	26.4	26.3	15.2	28.0	49
140	0.150	6.4	5.3	6.5	22.0	26
200	0.106	2.7	1.5	2.8	6.2	9
270	0.053	1.6	0.5	0.1	1.0	1
Pan		0.8	0.3	Trace	Trace	0.5
AFS GFN		6.3	58	52	72	80
Surface area		248	227	133	200	210
SiO_2						94.8
Shape		Angular	Angular	Angular subangular	Angular	Elliptical to subangular

TABLE 3.12 Physical Analyses of Some Olivine, Chromite, and Zircon Sands in Percent

fractures. As a result, the mulling of new olivine sands produce large amounts of fines, which must be extracted during shakeout or extra bond added in compensation. Recycling knocks off sharp edges, so although the amount of fines produced gradually decreases, there is no easy way to remove fines from greensand without also removing clay. Excessive force on the grains increases fracturing. Heat transforms the crystal structure to hydrous magnesium silicates (serpentine). Sands with different chemical composition will have different physical properties. Because of the presence of trace amounts of nickel (it is substituted in the spinel structure and cannot be leached) some West Coast states restrict the use of olivine.

A major advantage of olivine is that it contains no free silica, so foundry workers do not get silicosis unless silica core sands are used. Olivine is more refractory than quartz, with one-third the thermal expansion (Fig. 3.42). Olivine sand molds have higher thermal conductivity, so cooling rates will be greater than silica sand molds, although some chemical binders can have a bigger effect. Olivine can be bonded with clays, sodium silicate, and organic chemical binders other than those that are acid catalyzed because the highly basic olivine consumes the acid catalyst. Olivine sand systems require more

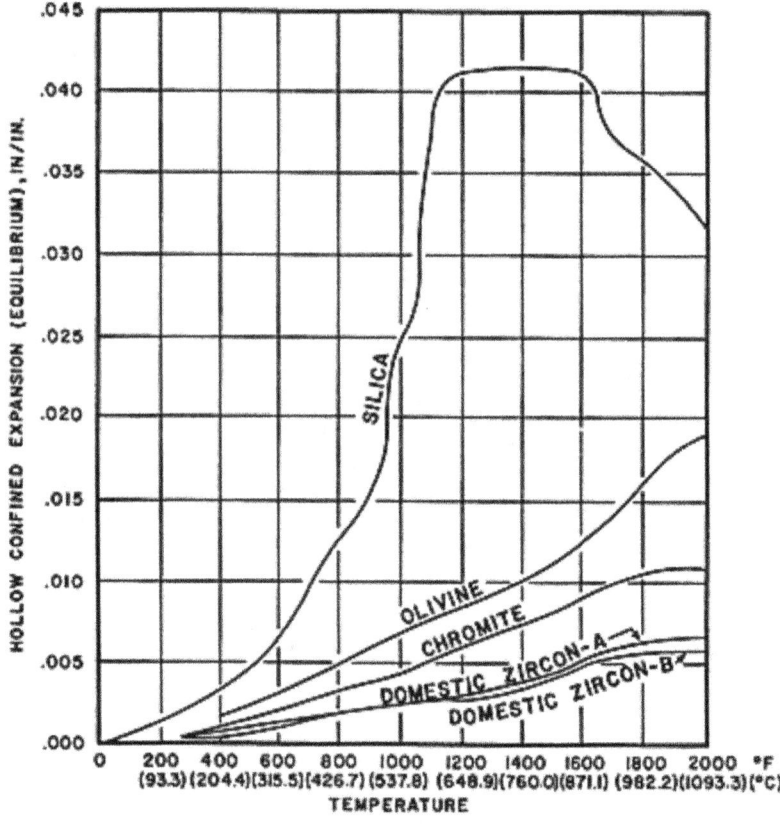

Figure 3.42 Thermal expansion of nonsilica aggregates as compared with silica sand. (From Garner [26].)

energy to process than silica sands because of the slightly higher bulk density. Contamination of olivine system sands by silica (shell core sand) lowers the sintering point of the system sand and causes the casting surface finish to deteriorate.

Green Diamond sand is another magnesium orthosilicate product that has been melted, then granulated by quenching in water. The company claims the product has all the advantages of natural olivine, and avoids the fines and dust problems, so that less new sand additions are required. The product has lower bulk density than natural olivine; as a result Green Diamond sand produces greensand with higher permeability than the equivalent natural olivine so that pouring temperatures can be lowered [30].

Zircon

Zircon (zirconium silicate) is a by-product of mining deep sand deposits for another titanium-bearing heavy sand. Hence, the availability of zircon depends on the demand for titanium. Deposits are mined in South Africa, Australia, India, and northern Florida. Australian deposits contain about 15% zircon, whereas deposits in Florida are only 1.5 to 4% and the zircon is too fine and only suitable for mold washes. Zircon obtained by a costly method of electrostatic separation can be 99.5% pure. Occasionally, zircon sands obtained using oil-flotation become available. Although the purity is high, the sand grains have an oil coating which obstructs the application of chemical binders.

Zircon is the most heat-stable of all available foundry sands and has a thermal expansion coefficient less than one-eighth that of silica (see Fig. 3.42). However, it is the combination of high thermal conductivity and density (both of which are nearly twice that of silica) which makes zircon most attractive. These two properties give zircon a heat absorption rate nearly four times that of quartz. Zircon sand inserts are used as local chills to achieve directional solidification. Whether used as a mold or core aggregate or as an ingredient of a wash coating, zircon is virtually unwetted by most molten metals, thus resulting in excellent casting surface finish. Zircon has the added advantage of chemical stability and is compatible with all known foundry binders.

Chromite

Chromite is one of a series of compounds with the spinel structure, four of these minerals—chromite $FeCr_2O_4$, picrochromite $MgCr_2O_4$, hercynite $FeAl_2O_4$, and spinel $MgAl_2O_4$—form a continuous series, with the physical properties dependent on the relative amount of each component (from 0 to 100%). Chromite is always associated with rocks containing Mg, Ca, and Fe, and low in silica. However, chromite has the highest melting point and crystallizes out first, and if it has a higher specific gravity than the molten rock it can be found in bands or strata. In some areas chromite is mined, crushed in rod mills to release other minerals (mainly talc and serpentine), and washed. Nevertheless, chromite grains vary from deposit to deposit. Chromite, when heated in oxygen, gains weight (0.5 to 0.6%) due to the exothermic reaction of ferrous to ferric iron starting at 698°F (370°C), peaking at 788 to 806°F (420 to 430°C), and ending at 1022°F (550°C). Chromite sand is more chilling than silica, is not easily wetted by molten metals, and is highly refractory, however, is not rounded. It sometimes contains hydrous impurities which cause a variable basicity and high acid demand, and make it difficult to use with some resin binder systems. The hydrous impurities can cause pinholing and blows when the water penetrates the castings.

Chromite–4.5% western bentonite mixes produce a reducing atmosphere with high CO_2 content due to reaction of carbon from cast iron reacting with water in the bentonite. On the other hand, seacoal must be added to silica sands when casting iron, where it produces a reducing atmosphere with high CO and low CO_2 contents. Studies have shown that mechanical attrition or scrubbing of chromite sand grains to remove chemically bonded surface coatings from degraded bentonites is effective. High-intensity magnetic separation (cross-belt separator) of chromite grains from silica grains, followed by dust removal is necessary to recycle chromite. Contamination by more than 3% silica causes serious chemical reactions at steel casting temperatures. The author remembers a case where it took a day to remove a 1-ft (30-cm) diameter core from a steel casting, even then the cavity had severe burn-on. A 50:50 mix of silica and chromite was used to save money. For the same reason chromite dust cannot be used as a coating.

Several other aggregates are used occasionally, such as mullite, which is stable up to 3300°F (1800°C). Mullite is rare in nature and is made by calcining aluminum silicates—andalusite, kyanite, and sillimanite—then processing the clinker to separate and round the grains. ASK produced a very low density insulating aggregate for thin-walled steel castings, but many companies seem to be using ceramic molding (hot molds) with mullite.

Mineral Properties

The mineral properties of the four sands are shown in Table 3.8. The mineral density or specific gravity differs from the bulk density of the sand because the packed sand contains 35 to 40% voids. Next, the melting point of the refractory material should be greater than the temperature of the metal being poured. In contrast, phase and compositional changes affect the thermal expansion which not only affects the mold during casting but also causes fracturing of the sand grains to produce fines. These fines alter system sand properties, usually to their detriment. Silica undergoes a phase change from α- to β-quartz at about 1112°F (600°C) with a lattice expansion, and later to tridymite. The silica slowly reverts to α-quartz on cooling. This behavior is shown in Fig. 3.41, which indicates the thermal expansion of quartz to be 1.4% from ambient to normal casting temperatures. This characteristic of quartz causes expansion of mold surfaces when exposed to thermal shock and, unless precautionary measures are taken, can result in casting surface defects such as rat tails, buckles, and scabs. Other disadvantages of silica sands are their relatively low chilling power and respiratory problems caused by the insolubility of airborne dust particles deposited in the lungs.

During iron and steel casting, the higher pouring temperatures increase the reactivity of the sand system and make reactions between liquid metal and slag with the sand more likely. For example, the presence of one reaction product in particular, fayalite or Fe_2SiO_4, turns the silica sand black and its effects are readily visible under a microscope.

Sand color and shape can be examined at a magnification of 25 to 40 times. Even pocket microscopes are suitable for such examinations. The microscope will show whether the grain shape is simple, compound, or fractured, and whether colored impurities are compound or separate. Clay particles, which are present in many natural sands, will predominate in the −200 and −270 mesh fractions.

The size distribution of the sand is also important. If all the grains were round and of similar size, then the mold would have the maximum permeability and the minimum bulk density and strength. This is because the grains would pack in layers with adjacent layers covering the gaps (ideally cubic or hexagonal close packed structures with 12 contacts), the bulk density would be lower and percent voids higher (26% but could be as high as 45%). The best mixtures are three and four screen mixes, that is, sand whose distribution shows more than 10% retained on each of a series of three or four screens, as shown in Fig. 3.40. Such sands would fill more of the gaps between grains giving higher bulk density, lower percent voids, and permeability, but more points of contact between grains. Great care should be exercised with three or four screen sands as any piling or drop causes larger grains to roll and spread to the outside of hoppers and conveyors. Once sands segregate, they will not spontaneously converge.

Sand shape is an indicator of durability and how it will be used in the foundry. Figure 3.43 shows the various sand shapes. Compound or twinned grains, as well as fractured grains are generally undesirable as they disintegrate and round off during mulling and thermal recycling. Round grains are the most desirable as they do not pack densely and the spaces between the grains allow mold gases to pass through the sand, i.e., they are more permeable to gases. Similarly, round grained sands are preferred for core sands, because of their higher permeability (gases are more likely to vent out core

Original Magnification 20X
Angular Sand Grains

Original Magnification 20X
Rounded Sand Grains

Original Magnification 20X
Subangular Sand Grains

Original Magnification 40X
Compound Sand Grains

FIGURE 3.43   Sand grain shapes. (From *AFS Mold and Core Test Handbook* [28].)

prints than bleed out through the casting), and to obtain an easier shakeout. Angular sands can be desirable for molds because the "keying" action of the projections between other grains gives a stronger mold. North America has abundant supplies of round grained sands in the region southwest of Lake Michigan and some of the other great lakes. On the other hand, Japanese foundries use a mix of angular sand obtained from crushed local silica rock with round grained sands imported from Sarawak and Australia to reduce the cost of silica.

Sand shape is also an indication of the surface area, and thus binder requirement to fully envelop or coat the sand grains. Round grains, such as those in Ottawa silica, are still a long way from being perfectly round, and have a surface area of between 1.29 and 1.38 times the theoretical surface of true spheres. The coarse grades tend to be more rounded than the finer fractions. Subangular grains are 1.50 times, and angular grains 1.65 times the index of surface area of the theoretical round grain. Sanders and Doelman [23, 31] extensively reviewed the methods for calculating the surface area of sand grains of each screen in the U.S. Standard Sieve series. It involves calculating the number of grains per gram from the average sized particle on each screen. The surface area of spherical particles on each screen can be calculated, and the correction for shape applied. If the surface area factors shown in Table 3.13 are applied for each screen, then the total surface area in $cm^2/100$ g of sand can be calculated. Corrections for angularity might have to be made based on an inspection of the sand on each screen.

The surface area is important as foundry sands must be coated with water, clay, or chemical binders. Only the water and binder in the bridges between adjacent sand grains contribute to bond strength. Figure 3.44 shows just how much temper water is required to coat successively smaller silica sand grains. Excess binder and temper water, along with restricted passages for gas emissions can cause scabs and other casting defects. Surface area also dictates the minimum binder levels required to coat different sands.

Retained on U.S. Sieve Sizes	Silica	Olivine	Chromite	Zircon
12	×12	×12	×8	×6
20	×24	×24	×16	×12
30	×40	×40	×30	×24
40	×60	×60	×40	×35
50	×90	×85	×60	×50
70	×130	×120	×85	×70
100	×190	×170	×120	×100
140	×270	×240	×170	×140
200	×400	×350	×250	×200
270	×600	×500	×350	×300
400	×900	×800	×600	×500

Source: From Sanders [23], Modern Casting [27], Sanders and Doelman [31].

TABLE 3.13 Surface Area Factors

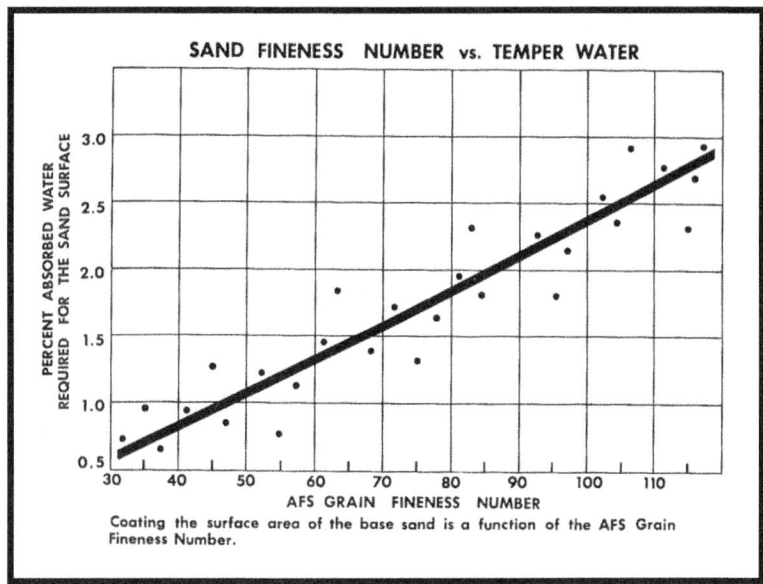

SAND FINENESS NUMBER vs. TEMPER WATER

Coating the surface area of the base sand is a function of the AFS Grain Fineness Number.

FIGURE **3.44** Coating the sand surface area is a function of the AFS GFN. (From Sanders [23, p. 423].)

Acid demand and sand pH have implications. For example, hydrous impurities, beside causing blowing and pinholes, have high acid demand and are detrimental to some organic binders.

Weight changes on heating can also indicate the presence of impurities. Loss of weight on ignition is probably one of the cheapest pseudo-chemical tests. For example, sand deposits and the effectiveness of sand cleaning operations can vary and the weight changes reflect this.

Clay

Clays are earthy or stony aggregates with innumerable small flat fragments of the order of 1/5000 to 1/50,000 of an inch in diameter (10^{-4} to 10^{-5} cm). They are composed chiefly of oxides of aluminum and silicon in various ratios, in the form of platelets of hydrous aluminum silicates. Other minerals are also present and play an important role in clay properties and structures. Clays have the unique property of absorbing moisture into their structures in such a manner as to form intermediate water layers between their platelets, resulting in adhesive, plastic masses. There are numerous types of clay found in nature and these vary significantly in properties, composition, and structure (Table 3.14). From a foundry standpoint, only two groups of clays are of interest: kaolinites and montmorillonites, which can be used to bond sand grains together.

Kaolinite clays, when uncontaminated, are characterized as being highly refractory minerals. The kaolinite group of clays contains three types—fireclay ($Al_2O_3.2SiO_2.2H_2O$), ball clay, and kaolin (china) clay—of which only fireclay is important to foundries. The structure of fireclay is two layers, one of silicon-oxygen tetrahedrals in a hexagonal configuration, and the other of aluminum with six hydroxyl groups also in a hexagonal configuration. Fireclays are the bonding material in naturally bonded sands, and

Amorphous	Crystalline
Allophane group	Two-layer type: one alumina and one silica sheet Equidimensional—Kaolinite group Elongated Halloysite group
	Three layer type: two layers of Si-O tetrahedron and one dioctahedral or trioctahedral layers Expanding lattice montmorillonites and vermiculites Nonexpanding lattice—illite group (clay-like shales)
	Regular mixed layer types: chlorite group
	Chain type structures Attapulgite (Fuller's earth type) Epiolite Palygorskite

Source: From Sanders [23, p. 551].

TABLE 3.14 Classification of Clays

also occur as free deposits. These clays exhibit lower green bonding strengths than the montmorillonite clays. About 12% by weight of fireclay is required to attain the same green compression strength as would occur if 4.5% of a montmorillonite clay were used in various sand/clay mixtures and brought to the proper temper by the addition of water and thorough mulling. Similarly, sands bonded with fireclay require about 5% by weight of water as compared with 3% for sands with montmorillonite clay. In many other respects, however, fireclay properties compare favorably with the montmorillonites (Table 3.15). Typical chemical analyses of various clays are shown in Table 3.16.

Montmorillonite, ($Al_2O_3.4SiO_2.4$ to $7H_2O$), is the primary mineral occurring in bentonite clays. Better commercial grades contain more than 80% montmorillonite. The remaining material varies with the source, but consists of feldspar, quartz, gypsum, and calcium carbonate and assorted minerals. The two important montmorillonites are southern bentonite and western bentonite, which are used extensively for bonding of sands for greensand molding. These two bentonites have significantly different properties, which need to be understood before selecting them. The montmorillonite structure consists of three layers, two silicon-oxygen sheets and one aluminum-hydroxyl sheet between the other two (Fig. 3.45). The central sheet is never fully aluminum hydroxyl, as some of the aluminum ions are substituted by magnesium or iron. Divalent magnesium creates an imbalance which is satisfied by loosely attached ions of sodium, calcium, magnesium, or hydrogen.

Southern bentonite contains calcium in its crystalline structure, in the form of bivalent ions (Ca^{++}). For this reason, Southern bentonite is commonly called calcium bentonite. Southern bentonite does not swell appreciably when moistened, and absorbs only slightly more water than does fireclay; it absorbs and loses such water with relative ease. By comparison, western bentonite contains monovalent sodium ions (Na^+) and is thus commonly referred to as sodium bentonite. Western bentonite absorbs water slowly, but does so with a high degree of volumetric expansion (swelling) accompanied by the formation of a thixotropic gel. While the United States has abundant deposits of both bentonites, Europe has only calcium bentonite, so they replace some of the calcium ions by sodium ions. Unfortunately this process leaves the clay vulnerable

Leading types of clay bonds	Southern Bentonite	Western Bentonite	Fireclay
Mineral classification	Montmorillonite	Montmorillonite	Kaolinite
Mined in	Mississippi-Alabama Strip	Wyoming and South Dakota Strip	Ohio-Pennsylvania Illinois-Alabama Strip, drift, shaft
Method of mining Average grain size	81% finer than 0.5 microns	70% finer than 0.2 microns	70% finer than 20 microns

Note: A micron is about 1/25000 of an inch.

Relative Value of Various Type of Clays

		MONTMORILLONITE	
	Kaolinite	Calcium (S)	Sodium (W)
Green compression strength	3	1	2
Dry compression strength	2	3	1
Permeability	3	2	1
Least tempering water required	3	1	2
Greatest durability	2	3	1
Highest flowability	2	1	3
Best collapsibility	2	1	3
Sand toughness	2	3	1
Least contraction at 2500°F (1370°C)	1	2	3
Fewest lumps in shakeout sand	2	1	3
Least intensive mixing effort required	3	1	2

1 = best, 2 = second best, 3 = worst of the three.

Source: Sanders [23, p. 562].

TABLE 3.15 Comparison of Bonding Agents and Clay Types

to contamination by other ions, which due to the salinity of Europe's rivers, becomes a problem when system sands are recycled many times. Highly soluble salts such as sodium sulfate, sodium carbonate, sodium chloride, calcium chloride, and potassium carbonate are detrimental to bentonite-bonded sand. Ion exchange resins can be used to desalinate the water.

Both types of bentonite have high sintering temperatures and are about equally effective in developing high green compression strength and permeability when used to bond greensand. As seen in Table 3.14, however, other characteristics of the two bentonites differ significantly.

Sands bonded with southern bentonite require less energy to mull, since this clay absorbs water more readily. Such sands are also more flowable (less stiff), and are thus easier to squeeze and ram into molds. Since these sands possess lower dry and hot

	National Western	Western	No. 200 Southern	Kaolinite
SiO_2	55.4	58–64	56–59	49.8
Al_2O_3	20.1	18–27	18–21	37.6
Fe_2O_3	3.7	2.5–3.6	5.4–9.0	2.4
CaO	0.5	0.1–1.0	1.2–3.5	0.7
MgO	2.8	2.5–3.2	3.0–3.3	0.2
Na_2O	0.6	1.5–2.7	0.3–0.5	Trace
Bound water	5.5	4.5–9.0	5.5–6.5	7.8
Moisture at 200°F (93°C)	8.0			

Table 3.16 Typical Analyses of Clay Minerals

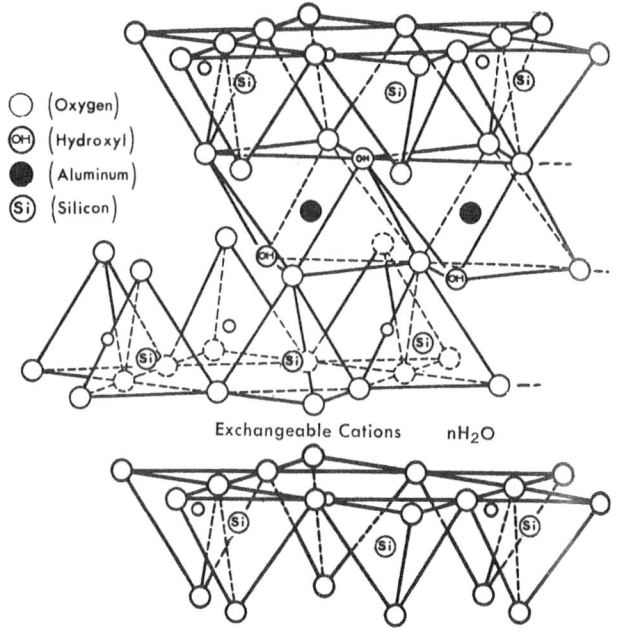

Figure 3.45 Structure of montmorillonite clay. (From Sanders [23, p. 548].)

strength, they are more collapsible at elevated temperatures. Hence, there is a lesser tendency toward formation of hot tears in castings of wide-freezing-range alloys, and fewer lumps of hot sand are formed at shakeout. Unfortunately, these same characteristics contribute to the possibility of mold erosion during pouring, especially with heavier castings—the results of which are usually sand inclusions in the castings. Sands bonded with southern bentonite clay tend to dry out rapidly when left exposed to air. This includes the sharp edges in open molds which are allowed to stand too long before closure. Another disadvantage of southern bentonite clay is its susceptibility to the formation of sand expansion defects. Many authorities ascribe this behavior to the poor wet-tensile strength.

On the other hand, western bentonite has excellent dry and hot strength. When used to bond greensand, it offers the maximum resistance to the formation of sand expansion defects. Conversely, all those factors listed above in favor of southern bentonite, will be disadvantages for western bentonite.

No single clay has optimum characteristics in all respects as a bonding material for greensand mixtures. Clays vary from deposit to deposit, and even within a deposit, therefore new greensand properties containing the same levels of clay or clay blends (strength, shear, tensile) vary widely. Sand property requirements differ from one foundry to another, depending upon many factors. It is possible to optimize sand properties for given conditions by utilizing the different clays either singly or in combination. Once it is determined that a particular ratio of specific clays best suits foundry requirements for general production conditions, it is often possible to purchase such a blended premix from a supplier. These blends may contain replacement clays, cereals, or seacoal, etc. (the blends differ from the working levels, because less durable additives need to be replaced in greater proportions, i.e., southern bentonite).

One way to show the durability of specific clays is by differential thermal analysis (DTA) [22, 32]. Two small samples, one of the clay and a control are heated side by side in a metal block. Thermocouples measure the differential between the two samples. If an exothermic reaction occurs in the clay then temperature in the clay exceeds that in the control. Conversely, if an endothermic reaction occurs then temperature in the clay lags that in the control. Figure 3.46 shows the DTA traces from three typical clays. Clays from different regions show slightly different responses. Figure 3.46 shows that the two montmorillonites start to lose physically held water above 212°F (100°C). Then gradually the chemically held water is expelled from all three clays. When this is complete, a sudden exothermic reaction occurs. In the case of fireclay, there is a structural change to primary mullite at 1360°F (680°C), followed by a second alteration to secondary mullite at about 2200°F (1200°C). In the case of the two montmorillonites, the structure is lost between 1550 and 1650°F (850 and 900°C). The new structure is mullite, cristobalite, or cordierite depending on iron, calcium, and magnesium concentrations.

In a greensand mold, southern and western bentonites lose their effectiveness when the sand reaches 840 and 1075°F (450 and 580°C), respectively [33]. Hofmann [32] claims that southern bentonite in molds degrades at 850°F (454°C) and western bentonite at 1100 to 1200°F (590 to 650°C). Research at CANMET has shown that when 2-in-(5-cm-) thick plates were poured at typical pouring temperatures for steel, cast iron, brass, and aluminum, that 50% of western bentonite survived at 0.9, 0.8, and 0.55 in, respectively, away from the plates when steel, iron, and copper were poured and essentially very little western was damaged by aluminum. More than 50% was destroyed inside these distances and less outside. For southern bentonite, 50% was destroyed at 1.3, 1.0, 0.6 in, and immediately adjacent to the plate, respectively [34]. On the other hand, DTA analyses are conducted over relatively short time periods with the temperature increasing constantly, whereas the plate castings in greensand molds were left until cool in order to salvage sand at various distances from the casting. Therefore, the clay was left to degrade over a longer time. Although it is desirable to clean the last of the sand from castings, that sand contains damaged clay, fractured sand grains, etc., and is detrimental to good sand control.

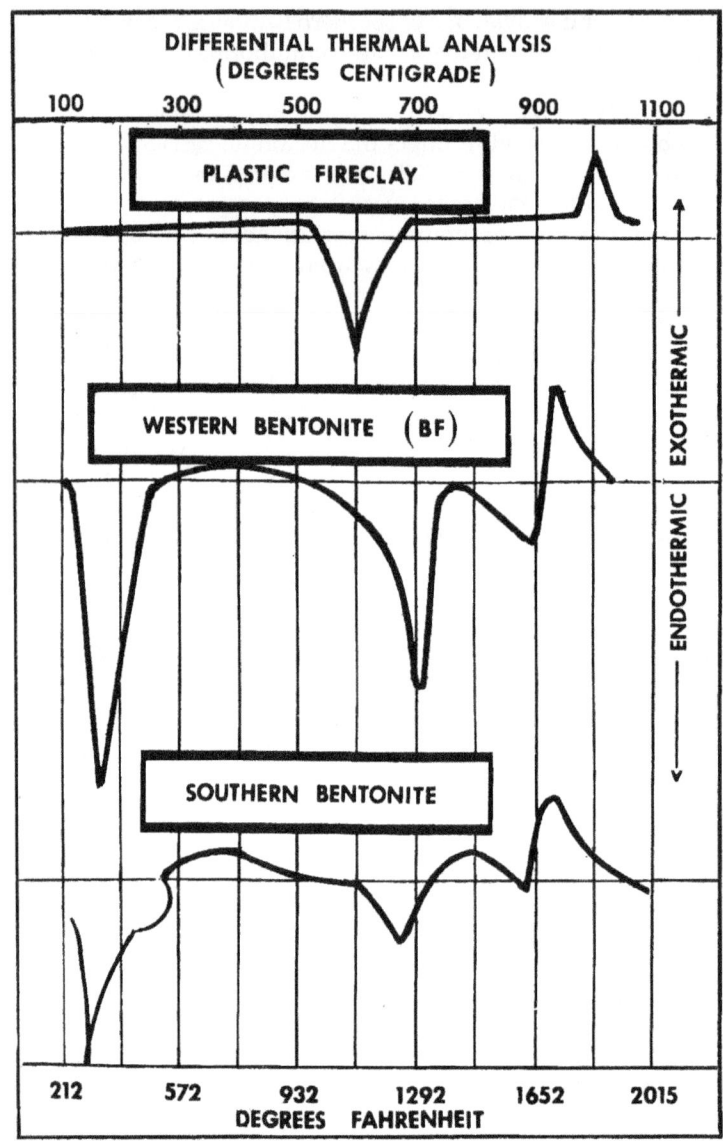

FIGURE 3.46 Differential thermal analyses (DTA) of three typical clays. (From Sanders and Doelman, Part II [33].)

Over time, dead clay builds up on the surface of recycled system sands [35]. Castings show surface roughness. Hofmann discovered sand density of a quartz sand can fall from 2.65 to 2.35 g/cm³. Microscopic examination showed grains were enveloped in many layers of burnt bonding clay. The higher the pouring temperature and the larger the castings, the greater the proportion of sand affected and the thicker the dead clay layer and the more serious the problem, called oolization. These same materials occur in sedimentary rocks: oolitic limestone and iron ores. While the weight of the oolitic layer can be determined (Fig. 3.47), the procedure is not recommended. Hydrofluoric acid, used to dissolve the layers, also attacks bones. Antidotes must be available and applied immediately.

Water

Water is added in amounts from 1.5 to 8% to activate the clay in the sand and to develop plasticity and strength. Water in molding sands is often referred to as *tempering water*.

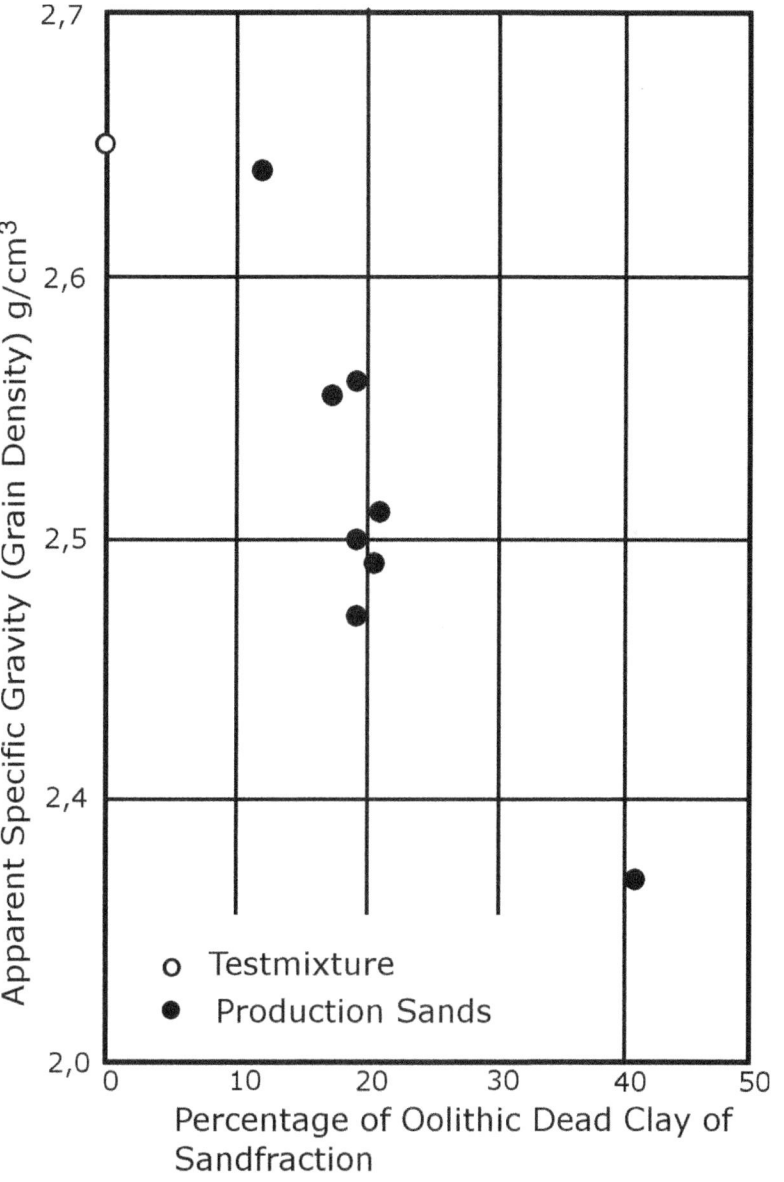

FIGURE 3.47 Oolitic clay on sand grain surfaces. (From Hofmann [35].)

Some of this water is adsorbed into the clay and becomes chemically bonded with the clay and helps develop strength. The rest of the water is physically bonded to dead clay on the surface of the sand grains and other particles in the sand. The more fines in the sand the greater the amount of water is required to physically wet the surface of these particles. Other ingredients added to modify sand properties not only require surface water but will actually take up water. Some of this water is necessary to lubricate the slippage of sand grains during compaction of the molding sand, but more water actually weakens the bond. The biggest problem with water in molding sand is that when it is exposed to heat, it turns to steam with a 1000-fold increase in volume. Just as metal is trying to replace the air in the mold, mold venting is hindered by this steam along with gases from other materials breaking down. Needless to say, control of water and materials that take up water becomes critical. Excess water is largely responsible for the following defects: (1) scabs, buckles, and rat-tails; (2) rough finish, burn-on, burn-in, and metal penetration; (3) blows, pinholes, gas, porosity; as well as (4) sand stuck to the pattern, cracked, or broken molds.

The total temper water in a synthetic system sand depends on what is in the system sand. Water is adsorbed by (1) the surface area of the sand grains, (2) by the clay, and (3) by other additives and residues. Water requirements depend on the fineness and surface area of the sand (see Fig. 3.44). A silica system sand of AFS GFN 50 requires 1.0% water just to wet the sand grains, whereas a sand of AFS GFN 100 needs 2.4% water. Table 3.17 shows the water requirements for additional sand additives. Hence, if there is 6% clay in the sand, then add 3.0% additional water, and so on for each additive. For chemically bonded sands, there is a minimal binder volume required to at least wet the sand, otherwise distribution is uneven and bonding becomes erratic.

The addition of water or organic matter to montmorillonite clays has been studied using X-ray methods. The size of crystal structure of a montmorillonite cell is a-axis 5.2 Å, b-axis 8.9 Å and thickness, or c-axis 9.4 Å.[6] Water absorbs onto the silicon-oxygen layers, causing the gap between the two layers in Fig. 3.45 to expand. The height of the blocks in Fig. 3.45 is 9.2 to 9.6Å (c dimension), whereas the basal spacing between the two silicon-oxygen layers in adjacent cells along the c-axis varies from 9.6 to 21.4 Å (d spacing) as more water is absorbed (Table 3.18). In clay crystals, the structure is

Additive	Extra Water in % for Each 1% of Additive
Fireclay	0.35–0.5
Bentonite	0.5*
AFS Clay (10 μm)	0.25–0.33
Seacoal or grain cellulose	0.06–0.10
Corn flour	0.15–0.22
Wood flour	0.20–0.275

*Correct total for compactability.

Source: Sanders [23].

TABLE 3.17   Temper Water Requirements for System Sand Additives

---

[6]The angstrom, symbol Å, is used for atomic distances and is 10^{-10} m or 0.1 nm.

Condition	d Spacing in Å
Anhydrous	9.5
Single water layer	12.3
Double water layer	15.4
Triple water layer	18.4
Ethylene glycol	17.0
Single organic layer	23.3
Multiple organic layer	Around 60

Source: Sanders [23, p. 551].

TABLE 3.18   Effect of Absorbed Materials on the Basal Spacing in Montmorillonite

repeated again and again in all directions. However, water allows clays to cleave along these water layers into flakes with the silicon-oxygen layers exposed. Cell structure in the *a*- and *b*-axes remains essentially continuous in the flakes. The expansion of these *d* spacings is what has been called *swelling in bentonitic clays*. Drying removes water and the flakes contract approximately back to their original spacing. Kaolinite flakes do not attract waters layers, their spacing is fixed and they do not swell.

In the 200 mesh U.S. Standard Sieve commercial material, clay particles are of the order of 1/5000 to 1/50,000 of an inch in diameter (10^{-4} to 10^{-5} cm), hence there could be up to 50,000 stacked platelets, or flakes in each. The internal atoms of trivalent aluminum cations can be substituted by divalent cations, resulting in a small negative charge present on the upper and lower basal oxygen planes of the crystal. This negative charge is balanced by Na, Ca, or a small amount of Mg. The addition of water can transport other ions into these water layers. The effect of these foreign ions can affect the properties.

Oil/Clay-Bonded Molding Sand

When casting plaques with larger letters, the presence of excess moisture or the rapid drying of sharp edges around letters can make moisture control in greensand systems quite problematic. The need for good sand properties under difficult conditions led to the development of waterless binders. These binders are made plastic, or tempered by oil rather than water. Some oils are particularly beneficial, imparting higher vaporization temperatures, lower rates of gas evolution, and reduced chilling effect than aqueous clay-bonded sand. The binder clay, known as *bentone*, is obtained by treating western bentonite with organic salts to convert it to an oil-swelling material rather than a water-swelling clay. Such modified western bentonites have relatively low dry- and hot-compression strengths. As with southern bentonite, these oil/clay-bonded sands are not used to produce heavy section castings.

Sand for copper-base alloy casting production is usually a washed-and-dried silica sand of AFS GFN 100. Approximately 5% bentone can be tempered with 2% coastal oil (SAE 40). Once mulled, the green compression strength is typically 8 to 12 psi (0.6 to 0.85 kg/cm^2), dry compression strengths more than 50 psi (3.5 kg/cm^2) and hot compression strength more than 100 psi (7.0 kg/cm^2). Oil-bonded sands must be mulled, as mixing does not coat the grains properly. Proper mulling develops the maximum bond strength and thoroughly distributes the plasticized binder. The more normal procedure is to add the clay to the sand and then the oil, although some add the oil to the sand before adding the clay.

As the heat-affected zone in molds progresses outward from the casting with time, it is critical to shake out the mold as soon as the casting solidifies and is strong enough to withstand shakeout. Otherwise the oil is pyrolyzed, and the bentone damaged and loses its capacity to be replasticized. When the sand is recycled, the amount of additives required will depend on the depth of the heat-affected sand, which depends on pouring temperature, casting section thickness, and the delay before shakeout. Premature shakeout may cause the oil to ignite from contact with the hot castings, accelerating losses, and creating smoke and fumes. Sand testing is required to determine the residual oil to ensure it does not fall below minimum requirements.

Other Molding Sand Additives

Cereals in the form of finely ground corn flour, or gelatinized and ground starch from corn, may be added up to 2% to greensand to improve green and dry strength and more importantly improve collapsibility by burning out to counteract the expansion of silica sand. Similarly, wood flour or other cellulose materials (cob flour, cereal hulls, and carbonized cellulose) are added in amounts between 0.5 and 2% to improve collapsibility, flowability, and to counteract silica expansion. Cellulose products such as wood flour and wheat husks have replaced starches for high-pressure molding.

Seacoal, finely divided soft coal or coal dust, is used in ferrous molding sands to improve surface finish and ease cleaning of castings. The name, seacoal, either originated from the fines that collected in the holds of ships used to transport coal or, according to Sanders, to a 1620 patent by Lord Dudley for smelting and casting iron [23, p. 696]. Seacoal prevents sand burn-on by forming a reducing atmosphere in the mold cavity which deposits carbon on the sand, and limits the ingress of atmospheric oxygen necessary for the formation of fayalite. It is the volatile fraction of the coal that is critical to surface finish and mold atmosphere. The higher temperature form is less useful in this respect. Both this residue and the ash are detrimental to sand properties as they require a disproportionate amount of temper water. Typical amounts are 2 to 5% seacoal of which volatiles should be between 1.5 and 3%. While 5% is the suggested maximum, there seems to be a trap at 7 to 8% at which it seems impossible to lower system sands without surface finish and burn-on becoming problems (possibly because the higher water requirement of ash affecting the water gas reaction). A number of substitutes have been used: Gilsonite (a highly volatile coal found in Utah and Colorado) and asphalt or ground pitch. Pitch additions improve green strength but lower hot compression. More recently, graphite and iron oxide have been used in an attempt to reduce pollution during pouring and shakeout. Typical analyses are given in Table 3.19. The selection of seacoal or substitute depends on a number of factors:

1. Grain fineness of base sand and its distribution
2. Type and amount of bonding material used
3. Permeability of molding sand desired
4. Thickness of casting to be produced
5. Desired surface finish of the casting
6. Pouring temperature and pouring time
7. Compaction of sand surrounding the casting cavity
8. Length of time molds stand after pouring

	Seacoal	Gilsonite	Pitch	Graphite + Iron Oxide	
				Graphite	**Oxide**
Volatiles %	36–40	75–80	50–60	0.4	0
Total carbon %	80–85	85	92	99.6	0
Ash %	3–5	0.5	0.25	0.1–0.3	100*
Level in sand %	3–5	0.2–0.5	50:50 with seacoal		

Note: *Will depend on the oxide and temperature.

Source: Metalcaster's Reference & Guide [2], Sanders [23].

TABLE **3.19** Properties of Seacoal and Seacoal Substitutes in Percent

Most researchers believe that seacoal or coke in the molding sand volatilizes and burns, thereby consuming the oxygen in the mold cavity to provide a reducing atmosphere. The volatile fraction of the seacoal distills off and immediately consumes some of the oxygen in the mold cavity. Heat is absorbed by the mold from the molten cast iron, which as the sand reaches the coking temperature gives off hydrogen, methane, ethane, tars, light oils, and others. This faction gives a sooting action (also known as lustrous carbon) at the mold/metal interface which not only allows the metal to lie evenly on the mold surface but prevents metal penetration and metal/silica reactions. Gases and fines in the mold produce back pressure, which reduce metal penetration.

Seacoal should have a similar size and distribution as the system sand, although coarser grades are recommended for heavier castings. Seacoal additions increase green compression strength, dry compression strength, mold hardness, deformation, but decrease hot strength, permeability, sand flowability, volume changes in molding sand, and mold wall movement.

Green Shell Carb is a commercial carbon additive which has replaced charcoal. It is added at the 3% level to improve the casting finish of cast irons.

Iron oxide in the form of hematite, Fe_3O_4, (also sold as Klean Surf) in amounts of 0.5 to 2% is added to core sands for iron casting to prevent metal penetration, burn-on and pinholes. Iron oxide forms a eutectic with silica at 2150°F (1178°C) which improves hot compression strength and hot plasticity by forming a glaze just below the surface of the sand mold or core. The iron oxide also tends to improve the surface finish.

Silica flour is finely ground silica. A number of grades are available based on 95% passing 100, 140, or even 200 mesh standard sieves. The addition of fine particles to coarse sand will increase the density (99.8 lb/ft³ or 1600 kg/m³) of foundry sand, by partially filling the 39.5% voids that remain. It has been added in the belief that it prevents penetration and metal burn-on by blocking the penetration of liquid metal into the sand interstices. Up to 1% silica flour reduces permeability and increases temper water requirements, mold toughness, handling, and mold thermal conductivity. It improves high temperature mold properties for steel foundries. However, the increase in expansion defects must be offset by greater additions of cereal, wood flour, and clay. Finer sand would have been a better solution. Silica flour's use has declined due to concern over silicosis.

Additives for magnesium

Because most magnesium alloys catch fire just below the melting point, many additives have been used to suppress the vigorous oxidation. The oxide skin is not protective above

500°C and as the reactions with air and water are highly exothermic, large magnesium castings that catch fire should be left to burn and not disturbed until cool. Greensand was used extensively during the War years, and Elliott reported that large wheel casings for the B-36 bomber were cast in greensand [36]. Moisture levels should be kept below 2 to 3%, and ethylene glycol can be substituted for water, or preferably the oil/clay mixes (sold as petrobond). Additions of 3 to as high as 8% sulfur have been used (larger cross sections require more sulfur). Others have used coarse angular sands with 2% sulfur with either 2% boric acid (H_3BO_3) or potassium borofluoride (KBF_4) [23]. Ammonium silicofluoride [$(NH_4)_2SiF_6$ or bararite], which is cheaper, has also been used. These additions work when volatile to protect the surface. However, the sands are difficult to mold and dry rapidly becoming very friable. Coatings containing these same additives or Teflon have been used [37]. Molds are often flushed with sulfur hexafluoride, but as SiF_6 has been assigned a CO_2 equivalence of 300 for its effect on the ozone layer, its use is being curtailed. Tilch reported developing a coating containing inhibitors for chemically bonded molds, and the author can confirm that coatings work with phenolic urethane bonded molds [38].

Theories of Clay-Sand Bonding

Clay and water are responsible for the characteristics of soils, ordinary muds, and the ceramic muds used for pottery. Consequently, foundry specialists were able to learn from bonding studies conducted in other fields and industries. There are two early theories of clay bonding in foundry sands. This section summarizes the review by Smiernow and colleagues [39].

Grim developed his "wedge-block" theory in 1945 using an optical microscope (150×) [40]. Grim was aware of the rapid swelling of montmorillonite clays once water reaches two per layer and that maximum green strength occurs when there are three water layers per cell. Nevertheless, based on optical evidence he assumed that the green strength of clay-bonded sand is mostly determined by the structure of the bridges of clay binder linking the grains in compacted sand. Grim assumed that montmorillonites easily break down into fine elementary platelets during sand preparation [39]. When there are no other aggregates and coarse bentonite particles, all the clay is broken down and deposited on other flakes in a uniform arrangement (Fig. 3.48b). The contacts between the grains appear as wedge-shaped blocks. Sands bonded with other clays (kaolin or illites) appear different, and the coating on the sand grains is irregular and uneven (Fig. 3.48a). Other clays, unlike montmorillonites, are not readily broken down into very fine particles during mulling with sand and water. The envelopes around sand grains consisting of these clays contain larger clay crystals and aggregates dispersed in a matrix of very fine particles.

(a) 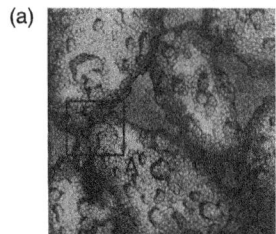   (b)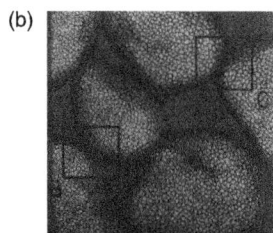

FIGURE 3.48   (a) Schematic of kaoline-bonded sand. (b) Schematic of montmorillonite-bonded sand. (Grim and Cuthbert [40]; Copyright © University of Illinois at Urbana-Champaign. Used with permission.)

Grim assumed that the strength of bentonite-bonded sands depends on the extent that the clay breaks down during sand preparation, the uniformity of the envelope around the sand grains, and the likely strength of the binder bridges [39]. Bentonite-bonded sands have superior strength because all the binder is uniform elementary platelets which build into strong binder bridges free from zones of weakness. Whereas, sands prepared using other clays have zones of weakness due to presence of coarse aggregates. While Wyoming bentonite (Na) completely breaks down after only a short period of milling (predates muller development), the strength was largely due to the uniformity, and further milling had little effect. However, other clays should get progressively stronger with milling as the coarse particles become smaller and more uniform.

Grim's hypothesis was able to explain some phenomena, but there were contradictions. While Na bentonite-sand-water mixes quickly develop strength on mulling within a few minutes, contrary to Grim's theory all clays including Na bentonite continue to develop strength. While all foundry clays benefit from extra mulling, the gap in strength persists, and with synthetic sands actually increases. The development of the electron microscope following Grim's work showed the structure of binder bridges with illite-type clays did indeed show irregular contact in the bridges between sand grains. However, a Na-activated German Ca bentonite—synthetic sand mix—when compressed showed a similar structure in the bridges. Smooth round bridges were only found in sand from the moisture condensation zone after pouring.

Previous work had shown kaoline flakes to be hexagonal with a and b dimensions about 3000 Å but may be up to 40000 Å, and 500 Å thick (but up to 2000). Bentonites are about 20 Å thick, and nearly square about 10 to 100 times wider. Western (Na) bentonite appears film-like, whereas southern (Ca) bentonites are irregular aggregates but decrease in size when dispersed in water [41].

Boenisch [49], on the other hand, explains the differences in bonding as being due to the specific surface area of the clay, the type and strength of water bonds on the clay surface and the hydration envelopes of the adsorbed cations at room temperature and at 212°F (100°C). Clay bridges are not an issue. Boenisch was more concerned with what happens in the mold once the casting has been poured and moisture boils away from the mold wall back into the body of the mold. This advancing condensation zone is responsible for scabs and expansion defects if the bond fails.

Boenisch studied the strength of synthetic sands at different temperatures. Figure 3.49 shows the properties of a sand containing approximately 4% German southern bentonite. The green strength is greatest at 2.3% moisture, although this effect is lost above 80°C (176°F). Another weak maximum occurs in warm sand at 7% moisture. An extrapolated line of the warm strength curve at 100°C (212°F) cuts the abscissa at 2.3% moisture. This implies that high green strength of temper point sand falls to zero at the boiling point of water. The dotted wet strength line peaks at 4.5% moisture.

Next Boenisch added soda in varying amounts to the Ca bentonite followed by a short period of mulling [49]. Figure 3.50 shows the green and wet strengths as functions of sand moisture for various soda levels. At 7% moisture, the decrease in properties is greatest in the overtempered moisture region, particularly when no soda has been added (Ca bentonite); but at 75 mequiv.[7] per 100 g of bentonite, the reduction in the overtempered region is minimal. Further additions of soda above the ion exchange

---

[7] A milliequivalent is 1/1000 of an equivalent weight or 1/1000 the gram molecular weight divided by the charge.

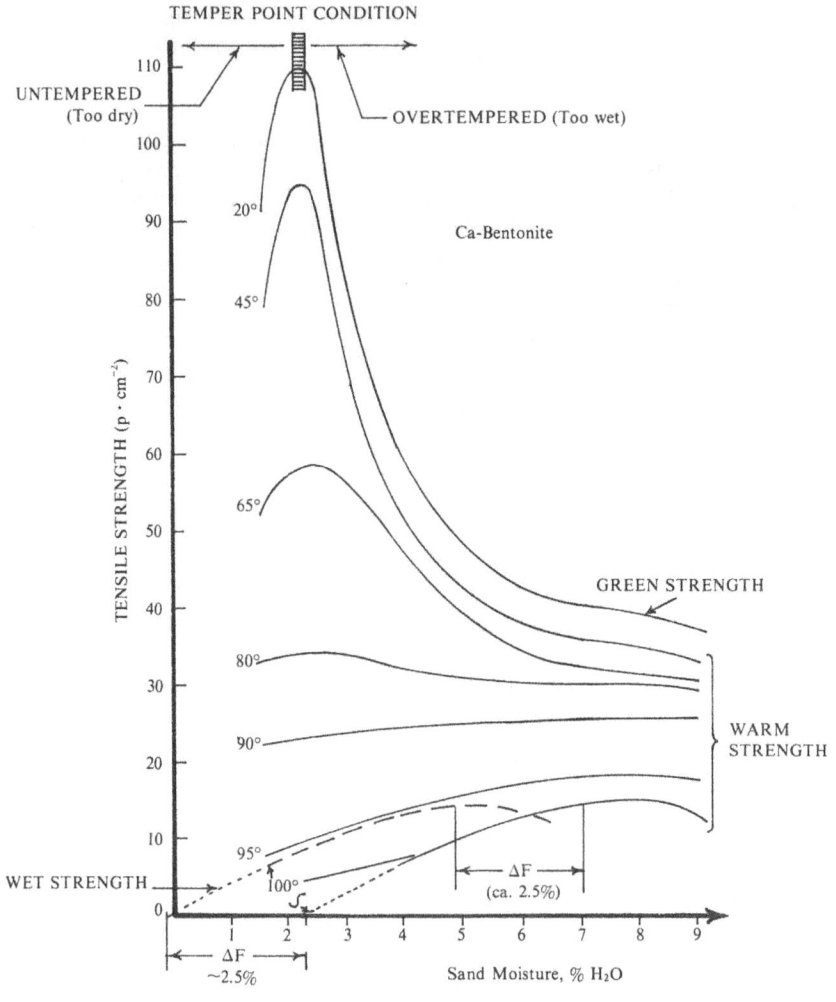

FIGURE 3.49 Properties of a sand containing approximately 4% German southern bentonite. (From Smiernow et al. [39].)

capacity (75 mequiv. per 100 g in this case and enough to convert the clay to Na bentonite) decrease strength again. Boenisch calls the "water sensitivity" the decrease in strength between the lower and higher mold moisture contents. The difference in value of 3% and 7% moisture was taken from the curves in Fig. 3.50a and plotted in Fig. 3.51 versus the amount of added $Na+$ ions. The curve for water sensitivity has a minimum at 75 mequiv., therefore, Na-bentonite has the lowest water sensitivity.

Water adsorbs on the silicon-oxygen layers on the clay flakes, they expand and cleave into flakes with the silicon-oxygen layers exposed. There is a small negative charge present on the upper and lower basal oxygen planes of the crystal. In addition, there are adsorbed cations on the surface of the clay flakes. The number of cations is a function of the specific surface and is greatest in bentonites. The clay flakes and the cations are surrounded by electric force fields which decrease with increasing distance

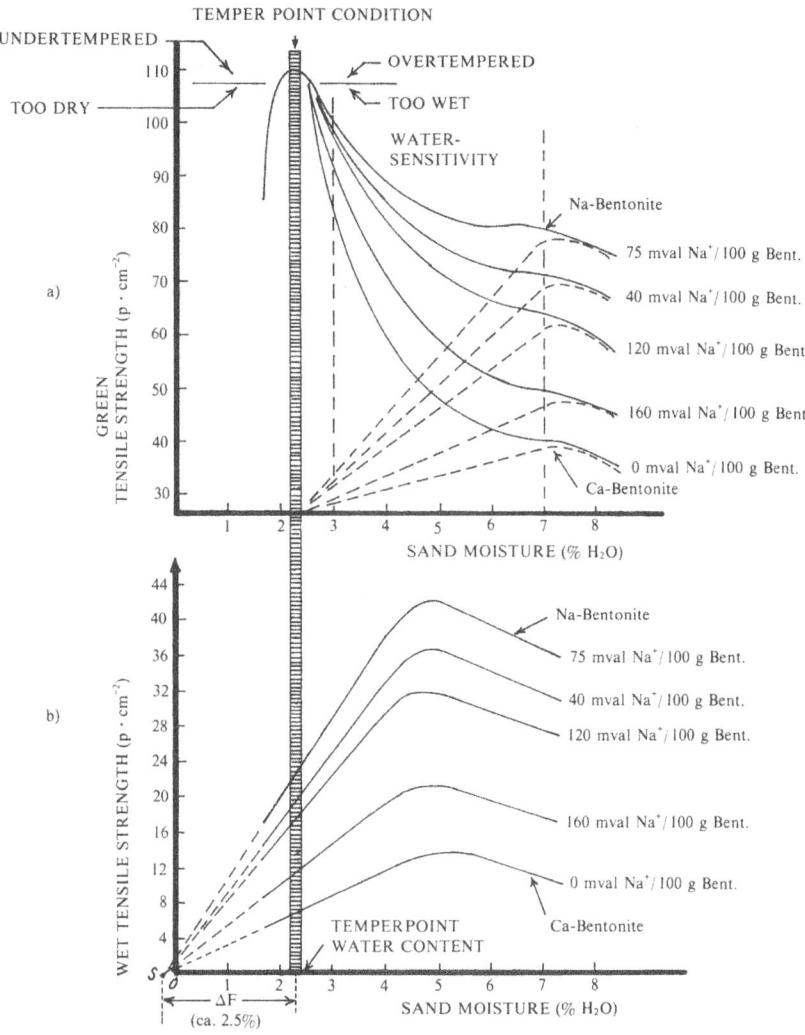

FIGURE 3.50 Green and wet strengths as function of sand moisture for various additions of soda. (From Smiernow et al. [39].)

from the source. Water dipoles close to the clay flakes are bonded and orientated more strongly than those further away.

Boenisch assumed that two different kinds of bonding were operating, albeit at different moisture levels [49]. Greensand strengths below the temper point water content (clay-water ratio of 10:4) should be conditioned by electrostatic van der Waals attractive forces between a continuous network of clay flakes or clay and sand particles, as well as their linkage through an orientated water network. These hinder the slip of particles against each other when external forces operate on the mold and thus affect the sand strength. Relatively lower strengths of overtempered sands could come about by the linkage of particles through ion hydration shells when the distance between particles caused by swelling becomes too large and electrostatic attraction is no longer possible.

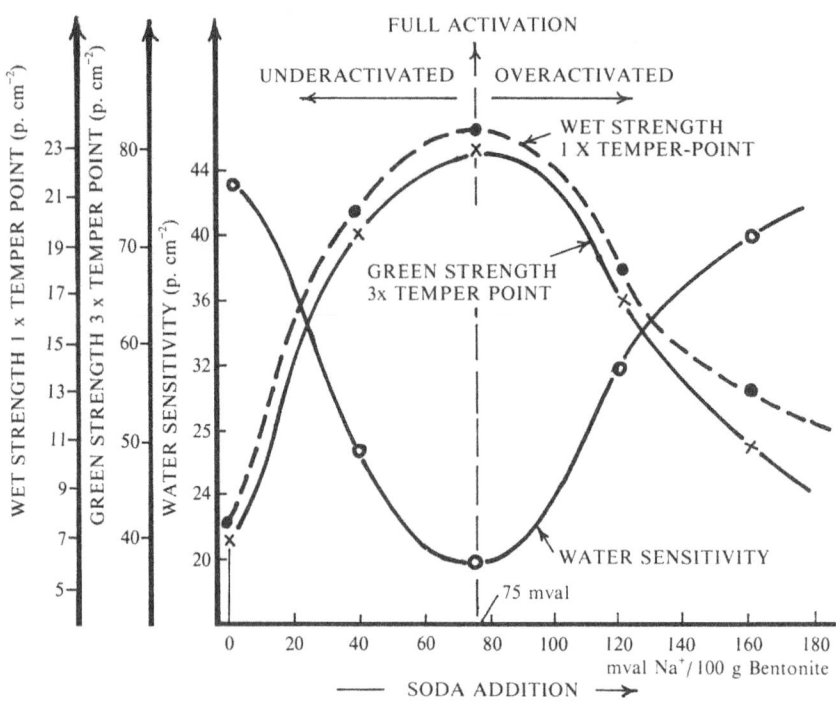

FIGURE 3.51   The difference in value of 3% and 7% moisture was taken from the curves in Fig. 3.50a and plotted versus the amount of added Na⁺ ions. (From Smiernow et al. [39].)

The bonding between particles through layer water should be designated as surface bonding and that through hydration water as bridge bonding. Surface bonding acts across the surface of clay particles where water is more rigidly held along with few adsorbed cations. Surface bonding operates at low temperatures and low moisture levels. Surface bonding is highest at the temper point (clay-water ratio 10:4). However, increased moisture (toward 10:8 clay-water ratio or twice the temper point water) weakens surface bonding and it falls to zero by 3 times the temper water (clay-water ratio 10:12). Surface bonding ceases at temperatures of 212°F (100°C), (Table 3.20). At this temperature, the layer water simmers. Bridge bonding is zero at the temper point and reaches its highest value at three times the temper point (water-clay ratio 10:12). Bridge bonding decreases at still higher moisture levels. Bridge bonding reaches full

	Temper Point	2 × Temper Point	3 × Temper Point	4 × Temper Point
Room temperature	Surface—strong Bridge—none	Surface—weak Bridge—weak	Surface—none Bridge—strong	Surface—none Bridge—weak
212°F (100°C)	Surface—none Bridge—none	Surface—none Bridge—weak	Surface—none Bridge—strong	Surface—none Bridge—weak

Source: Smiernow et al. [39].

TABLE 3.20   Effect of Temper Point and Temperature on Surface and Bridge Bonding

strength immediately after a swelling jump, but only as each flake undergoes the swelling jump. Hydration water is more strongly bonded to cations, especially if the cation is small and has a high charge. Schwiete and colleagues noted hydration water with Ca ions simmers only at 300°F (150°C). In practice, humid molding sands will lose water by boiling in the water layer [42].

Defects Originating from Greensand Molds

The following is a brief description of some of the sand-related casting defects and some of the solutions for solving them. The cause, identification, and solutions for defects are discussed in greater detail in Chapter 14. Nevertheless, the terminology should be useful when reading the next section on system sands.

Veining or finning. This defect occurs as projections in the form of irregular veins which are usually raised from the casting surface, either singly or in networks. In greensand molding, these projections often appear at fillets in conjunction with scabs or swells. Remedies may include regulating sand composition and sealing mold cracks, taking measures to avoid scabs, increasing mold rigidity and reducing metallostatic pressure.

Swells. When the metal pressure exceeds the strength of the sand, the casting swells and is oversized. Swells also show as massive and irregular projections on the surfaces of castings, generally rough and extended. Swells may be avoided by either (1) increasing the ramming density in areas of the mold subject to the defect, increasing the binder content of the sand; (2) modifying the location of the casting in the mold together with the gating and risering system to reduce the pressure of liquid metal; or (3) substituting dried or chemically bonded sand for greensand.

Erosion, cut, or wash. When the flow of liquid metal is too high or the bond destroyed, sand will wash away. These defects are irregular, rough projections on the surfaces of a casting, usually near ingates, in line with a gate, or extending along the course of metal flow. Sand inclusions often occur in other regions of the casting. The defect can be reduced by (1) increasing the proportion of western bentonite to increase the hot strength; (2) increasing dry compression strength of sand; (3) harder, more uniform ramming; (4) modifying the ingates to spread the entry of liquid metal over as wide an area as possible; or (5) using ceramic sprues, runners, gates, and/or strainer cores or protect them with resistant coatings.

Crush. When parts of the mold are damaged during withdrawal of the pattern, placing cores, or closing the mold, sections of the mold may end up misaligned. Crush shows as projections on the vertical or oblique surfaces of the casting. Crush will show on several surfaces, and as pieces of the mold can break off, will be combined with sand inclusions. Crush can be prevented by (1) improving the design of the casting, if possible, to provide greater draft or increased wall thickness; (2) maintenance to flask hardware, particularly alignment pins; (3) use of chaplets, etc., to support cores; and (4) exercising care when closing.

Blowholes and pinholes. Holes near the surfaces of a casting may only show during subsequent machining operations. By then their cause becomes a job for a detective. They can be caused by outgassing of wet sand, improperly cured binders, gassy metal, normal decomposition of core binders, lack of proper venting, or poor refractoriness of the system sand reacting with the metal. The holes are smooth-walled cavities, essentially spherical, sometimes contacting the external casting surface (and therefore elongated). The largest cavities (blowholes) are often isolated, while the smallest (pinholes) appear in groups. The inside walls can be shiny or oxidized depending on their cause.

Gas bubbles rising up toward the cope can be trapped by early formation of a metal skin. The best preventative action is sand control. Cavities that occur deeper inside the casting (gas, shrinkage, and even some misruns) are not usually sand-related defects but can complicate defects analysis. Gas and shrinkage often occur together in the late stages of solidification. Any analysis must estimate their relative importance from the degree that gas rounds the spaces between dendrites projecting into a cavity.

Blow. A casting with a blow shows a smooth depression on surfaces, or gas pockets just under the cast surface or in heavy sections above a cored cavity. The depressions or holes will be discolored, sometimes black, gold-colored, or reddish brown. The causes are wet or overslicked sand (a repair with a spatula can bring water to the surface, making it slick), sand with low permeability, insufficient drying of core washes, improperly cured or mixed chemical binders, local variability of permeability, and lack of venting in cores. Hard and uneven ramming of sand adjacent to the pattern may also cause blows. Venting the mold at the parting line and above high portions of the casting in the cope will help to reduce the tendency to blow.

Burn-on. Burn-on is a strongly adherent crust of sand on the casting surface that is not removed by normal shot-blast cleaning, but requires grinding. Burn-on may be prevented by (1) obtaining good rammed density particularly in hollows, e.g., between risers and casting walls; (2) using mold and core washes when excessive head pressure or sand coarseness exists; or (3) decreasing the pouring temperature. In some cases, the metallostatic head on fixed-height molding machines can be decreased by increasing the size of the pouring cup and pouring short.

Metal penetration. This defect consists of an irregular projection, consisting of an intimate mixture of sand and metal which adhere strongly to the casting. Penetration often occurs where the sand is the hottest (cores, concave sections) and poorly compacted. The remedies for metal penetration are the same as for the prevention of burn-on.

Scabs and expansion scabs. Low-strength wet condensation zones can expand when exposed to the heat of liquid metal going into the mold cavity. Liquid metal infiltrates the space behind the crust forming a thin layer of metal which comprises the scab. If infiltration does not occur, the defect will instead be a buckle. Scabs form on the upper horizontal mold surface (cope scab), or on the lower horizontal mold surface (drag scab). The incidence of scabs is reduced by (1) increasing the clay level, especially western (sodium) bentonite; (2) reducing the moisture level of the sand; or (3) reducing stresses on the mold surfaces by gating which achieves rapid filling of the mold.

Rat tail. Rat tails are shallow, groove-like defects up to 5 mm (0.2 in) deep that usually occur only on horizontal drag surfaces of a casting. Rat tails most often extend from gate locations, delineating the metal flow pattern in the mold. For sands with a strong tendency toward this type of defect, rat tails may also be accompanied by more severe sand-expansion defects such as buckles and scabs. The remedies for avoiding rat tails are the same as for scabs.

Greensand Testing

The main objective in controlling recycled system sand is to maintain sand characteristics and properties within acceptable working limits. Many foundries endorse the practice of sand testing but seem to attach little significance to the actual test results. This is usually because sand properties are not correlated with casting quality. Both good and bad castings have resulted from "tested" sands on different occasions.

Variations in casting size, shape, and design as well as variations in gating and molding practice impose different demands on sand behavior. Sand technology has not yet reached the point where we can calculate particular sand properties. However, such information can be determined by observation and test experience. The working limits for any given sand property will always include some degree of compromise, reflecting the range necessary to compensate for variations in product mix. No one property can be optimized without compromising other properties due to the many interrelationships. Nevertheless, controllers coupled to electronic sand testing go a long way in reaching a compromise.

Establishing acceptable working limits for the properties of a general-purpose system sand requires correlating the properties with outcomes for the entire range of castings produced. A single sand formulation cannot be expected to perform over too wide a range of casting sizes and section thicknesses. Such a practice would seriously compromise production and would result in increased scrap rates and poor economies.

Practical foundry sand technology should focus on six to eight tests that are important tools for system sand control. Larger samples collected in the foundry should normally be reduced with the aid of a splitter or quartering, but not for tests 2 to 6, as any loss of moisture could change the results. The methodologies of the following sand tests are outlined in the *AFS Mold and Core Test Handbook* [28]. These tests are

1. Sand temperature
2. Moisture content
3. Compactibility
4. Specimen weight
5. Permeability
6. Green compression strength
7. Methylene blue clay
8. Screen distribution

Sand temperature. System sands should be sampled carefully and stored in a closed container. Sand temperature should be taken immediately as many problems can be caused by hot molding sands; see below. It is convenient to install thermocouples in batch hoppers. For each 10 F degrees above 110, an additional 1.6 lb of water is required per ton of mulled molding sand. In addition, sand properties are temperature dependent above 110°F (43°C), so sand lab results will not be representative.

Moisture content. Not all of the water added to molding sands is absorbed by the clay bond; some is absorbed onto various inorganic and organic additives or contaminants present in the sand. Whereas the compactibility test is an accurate indicator of sand temper, wide differences in total moisture may be encountered at a constant level of compactibility. For this reason, the moisture test should be made as a supplement to the compactibility test. When extra moisture is required to reach a given sand temper, it indicates there has been a build-up of moisture-absorbing contaminants. Conversely, unusually low values of total moisture may indicate an excess of new, unbonded sand in the system. Unusual fluctuations in total moisture signal the need for additional tests (such as tests for green compression strength, permeability, and loss-on-ignition) in order to evaluate the cause. Several automatic moisture controllers are available to monitor and control moisture levels in system sands. Clay absorbs only a limited amount of water. Excess moisture acts as a lubricant making the sand more plastic, moldable but weaker. Moisture levels for ferrous foundries using predominantly 6 to 7% western bentonite, seacoal, and normal compactibility will be 4 to 4.5%. Nonferrous foundries with 6% southern bentonite and finer sands typically have system sands with 3% moisture.

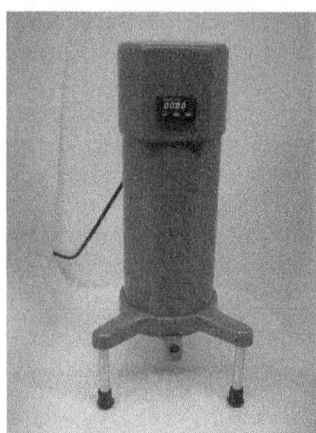

Figure 3.52   Greensand moisture tester with blower motor at top and sample pan location at bottom (Courtesy of Dietert Foundry Testing Equipment Inc.)

The moisture test apparatus shown in Fig. 3.52 passes 265°F (129°C) air through system sand supported on a screen for 5 minutes. People using this older equipment use exactly 50.00 g so percent moisture can be estimated without the aid of a calculator. In the latter case, it is critical that moisture and green strength samples should not be weighed in air-conditioned rooms. The older equipment tends to lose parts and the volume and temperature of the air passed through the sample deviates from specification. Physically bonded water is lost at lower temperatures and the chemically bonded water at higher temperature, but the loss is continuous and seamless, hence it is important that time and temperature be controlled. The new generation of equipment avoids these issues.

Compactibility. Compactibility of a tempered sand is measured as the percentage decrease in height of a loosely-packed sample when subjected to a controlled compaction force. The test is highly sensitive to moisture changes and can be used to supplement or monitor moisture determinations (there is usually a linear relationship between compactibility and moisture for system sands with less than 7% clay), see Fig. 3.53. The molder's "feel test," which relates to compactibility, indicates the degree of sand temper (clay/water ratio).

Compactibility is measured by filling a 4.5- by 2-in-diameter (11.4- by 5-cm) cylinder level with the top and then a 10.5 lb weight (plus 3.5 lb rammer assembly) is dropped 3 times a distance of 2 in (4.76 kg a distance of 5 cm) onto the sand. The sand rammer (Fig. 3.54) is mounted on a large concrete block or often a steel post braced on a structural member. In the latter case, a reaction base should be used to reduce bounce back. A practiced motion of a crank is used to lift and drop the weight without throwing or catching the weight. The tube is highly polished and must be lubricated each time with parting oil. Failure to lubricate leads to an error in compactibility (about 4 points), sand density, and green compression strength measurements. As tubes wear, fiction reduces compaction; hence, a new tube should be kept hidden away for comparison purposes. Compactibility measures volume reduction. The densities of materials have no effect.

Refinements to the compactibility test enable closer control of sand properties and have made quicker and more accurate testing possible on the production floor. As compactibility indicates the degree of temper, many foundries installed compactibility test

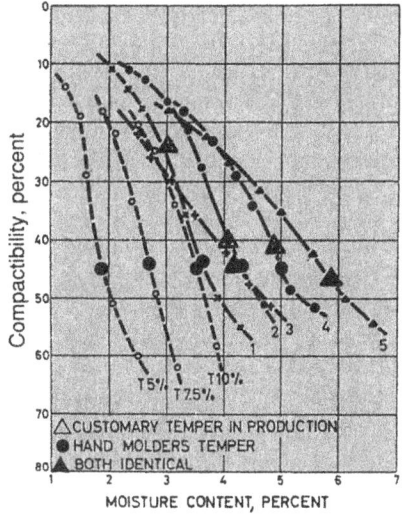

Figure 3.53 Compactability versus moisture for a number of system sands. (From Rich and Krysiak, courtesy Modern Casting [43].)

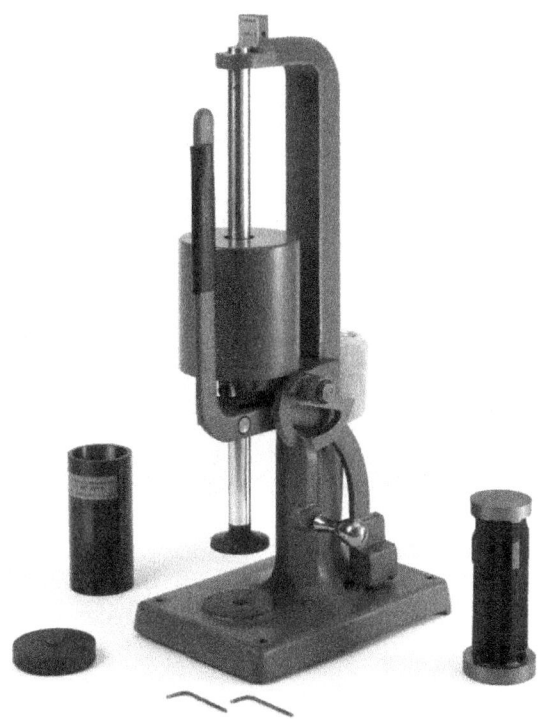

Figure 3.54 Sand rammer with compactability scale and specimen tube. (Courtesy of Modern Casting & Dietert Equipment.)

equipment on the muller deck for use by the operator as a means of controlling moisture content. In addition to the laboratory rammer-type compactibility tester, automatic units became available in the 1980s that continuously sampled and tested the sand in the muller, adding the necessary amounts of water at the proper times in order to reach the designated value for compactibility. These systems worked for both batch-type and continuous mullers. Foundries now employ additional moisture sensors before the muller.

Specimen weight. A simple and convenient test for evaluating changes in a system sand involves the measurement of the weight of sand required to form a standard 2- by 2-in-diameter AFS rammed specimen. In this case, the tube is not filled, but only the right weight of sand is used to make a 2-in-high specimen within ±1/16 in (1.5 mm). Specimens outside this range will give erroneous permeabilities and either will not fit the jaws for the green strength test or produce zeroing errors. Specimen weight, normally from 160 to 145 g, should be relatively constant, and changes are cause for concern as they affect density and packing. The accumulation of dead clay, fines, and ash (materials on USA Sieve Nos. 200, 270, and pan) in the sand causes a gradual reduction in sample weight. If, however, the specimen weight shows an increasing trend it is an indication that excessive amounts of new or unbonded sand are entering the system sand.

Bulk density is also of concern when people have to handle molds but can also be used as a simple quality control test of binder stickiness and packing. Problems can be easily detected by occasionally weighing molds from known patterns in flasks tagged with colored paint.

Permeability. The permeability of a compacted or cured sand specimen is a measure of the ease with which gas can flow through its mass, and is directly related to the venting qualities of the sand when used as a mold or core. The fineness, shape, and distribution of the sand grains, type and amount of bond material, degree of tempering, moisture level, and density of compaction are all factors influencing permeability. Increased permeability may indicate coarser grain size or lower mold density, factors that cause burn-in and penetration (rougher casting finish). Reduced permeability reflects conditions which may cause gas-related defects (blowholes, pinholes, and so on). Under conditions of reasonably good control, the permeability of a system sand should not fluctuate widely; however, sustained production of unusually heavy or unusually light castings may cause changes in this permeability. The sample used for specimen weight is placed in the apparatus shown in Fig. 3.55. It measures the time required for a given volume of gas to flow through the 2×2 in sample. Permeability should be 125 to 175 for ferrous and more than 50 for nonferrous foundries. Although the frequency of permeability testing will vary from one foundry to another, hourly testing is good practice. Unfortunately, permeability on a 3-ram test only assesses the fully rammed condition; permeability in the mold can only be assessed using the mold B hardness tester; see Fig. 3.29. Lab results vary from 90 to 95, but hardnesses on molds vary from 60 to 95 depending on location, molding method, pattern, etc.

Green compression strength. Although many strength tests were developed for greensands, green compression strength (GCS) is one of the few practiced today.[8] The Dietert Universal Sand Tester triple beam balance for GCS and tensile strength is shown in Fig. 3.56. Electronic machines are now available. The GCS test is used to determine

---

[8]Other tests were tensile, dry compression, dry tensile, shear, and hot compression strength which measured strength at molten steel temperatures to determine how molds would withstand heavy section castings. The equipment was bulky and the test lengthy, so those labs that had them found better use for the space.

FIGURE 3.55 Instrument for determining permeability. Sample tube fits at top (Courtesy of Dietert Foundry Testing Equipment Inc.)

FIGURE 3.56 The universal sand strength machine is a triple beam balance with 3 scales and sample testing locations for different attachments (Courtesy of Dietert Foundry Testing Equipment Inc.)

the effects of the clay bond in system sands, and is often the primary means for establishing the clay addition to be made. There are several factors affecting green compression strength in addition to clay content. They include the degree of mulling, moisture level, compactibility, and the kind and amount of additives. GCS can be used in conjunction with the compactibility and moisture to evaluate the active clay level and muller efficiency using the Weninger diagram for system sand with less than 7% clay; see next section. Active clay should be compared with the MB clay every two sand cycles.

Methylene blue clay. The MB clay test provides a quick and accurate method for determining the level of active clay (montmorillonite) in clay-bonded sand. This measurement involves determining the "base-exchange" capacity of the bentonite present in the sand, and titrating with methylene blue dye to replace the exchangeable ions found only in the structures of active clay materials. The test does not measure the presence of dead burned clay, inert fines, or silt as would be included in the "total clay" value found by use of the older AFS clay-wash/decantation procedure. The MB test is useful for monitoring the active clay level in system sands as well as for checking incoming bentonite shipments to the foundry.

The MB clay test is performed by boiling 5 g of dried system sand with 50 ml of 2% tetrasodium pyrophosphate solution for 10 minutes in a 250 ml Erlenmeyer flask or by ultrasonic vibration with 5 g of 220 mesh silicon carbide in a stainless steel beaker. Then 80 to 90% of the required amount of methylene blue solution (0.01 mequiv./mL or 3.739 g/L) is added followed by 2 minutes of mechanical agitation with an ASTM soil disc driven by a 1550-rpm motor. A drop of the supernatant liquid is transferred onto a #50 hardened filter paper. Methylene blue is absorbed by montmorillonite, and any residual dye is free to leach into the paper around the original dark blue sandy spot. However, its progress is dependent on MB concentration and the amount of water (a single drop). The sandy spot is about $\frac{1}{4}$ in; the water stain, about $\frac{5}{8}$; and the end-point about mid-way (6, 15, and 10–11 mm). Reference papers should be kept for comparison to maintain a uniform-sized blue-green halo. Another mL of methylene blue is added, and another droplet compared. The procedure is repeated until the end-point is reached. The titre is compared against the titre of 0.3 g of reference dry clay and 4.7 g of foundry new sand (6% clay).

The following issues of the MB clay test should be considered:

1. Clays are difficult to dry (some call for 3 hours at 230 to 248°F (110 to 120°C) in thin layers whereas Graham and Praski (44) said no more than 5 minutes). Thicker layers of fine clays (200 mesh) do not transmit heat nor allow gases to permeate. Clays are hydroscopic and will pick up 4 to 12% moisture. Blends of clays have been used.

2. Six percent dry clay typically requires 30 mL of MB solution, whereas if the same dried clay and sand are mulled (larger sample) and then titrated with methylene blue, the titre will be 34 to 36 mL. We found 6 mL per percent clay worked 95% of the time at 30 foundries. Others have also used 6 mL [45].

3. Solutions of reagent grade methylene blue are reproducible when weighed accurately, transferred to a volumetric flask, dissolved and made to the mark at typical ambient temperature. The solution needs to be stored in the dark, as it is sensitive to UV light.

4. Although the AFS test specifies #50, more than 50% of filters on the market work with some reservations.

5. The titration should last 10 minutes from start to finish for reproducible results, as the halo is fugitive.

6. The boiling method usually gives higher results on system sands than the ultrasonic method. Graham and Praski [44] concluded dye was not absorbed at working sites and did not produce any bonding power.

Screen size. Screen size is determined by passing a weighed quantity of dried sand through a stack of ten U.S. Series screens shaken for a period of 20 minutes. Most foundries use 50 g, although often the residue from the AFS clay test[9] offers a sample free of clay, dead clay, and 10-μm fines. The weight of sand retained on each screen is multiplied by a factor equivalent to the surface area (metric only), and the total divided by the weight to

---

[9]There are two tests for fines the one mentioned here and an earlier method that was based on the settling rate (1 in/min) of finely suspended particles in which a portion of the supernatant liquid is decanted, and more water is mixed until the supernatant liquid is clear, which has been largely supplanted by washing the 50 g sand sample on a 20-μm screen.

calculate the grain fineness number (AFS GFN). Table 3.10 shows the methodology. The plots and methodology differ from other industries. Figure 3.40 shows the bell-shaped curve familiar to foundrymen, and the cumulative distribution used for aggregates.

Several types of sieve shakers are available. Some rotary shakers work for coarse steel foundry sands. The stack must be securely fastened to limit rotation and erosion of the screens (brass fines in the pan). For nonferrous system sands, additional rapping is necessary to get particles through the finer screens. As these sands contain orders of magnitude more particles, it may be necessary to increase the time allowed with less vigorous sieve shakers or still finer materials. Results with angular sands depend on the type of shaker used. RO-TAP (a brand of rotating and tapping) shakers are large, and a bit noisy, but can be enclosed and are necessary for finer sands and additives (Fig. 3.57).

Screen size is particularly important for brass foundries, as surface finish dramatically affects cleaning, polishing, and plating costs. Changes in screen size can be rapid or long term, and should be avoided and controlled. Factors that should be addressed are: timing, quantity, and frequency of sand disposal, belts that sling sand into hoppers, etc., and cause loss of fines and sand segregation, maintaining an appropriate mix of light and heavily cored jobs, new sand additions, screen size of new and core sands, clay additions and clay burn-out from heavy section castings, and excessive metal splash. Coarser screens (6 to 12 mesh) need to be inspected for clay balls, core butts, and metallic balls. The latter are the sparks that are seen running across the mold during pouring, and in brass foundries constitute 0.7 to 1.5% by weight (iodomethane density

FIGURE 3.57   W.S. TYLER RO-TAP test sieve shaker with stack of screens. (W.S. TYLER® and RO-TAP® are registered trademarks of Haver Tyler Corporation.)

separation). Their number tend to increase, unless the dry system sand is screened weekly (20 mesh).

Foundries may find it necessary to conduct other tests, but on less frequent schedules. These tests may include AFS Fines, Loss on Ignition (1800°F or 982°C), and Volatiles [weight losses at several lower temperatures to differentiate cereals from seacoal 900°F (482°C)]. Rowell flowability is a quick test and puts a number to part of the molder's feel with respect to edge retention.

System Sand Properties

Molding sands which are repeatedly cycled and reused in a foundry are called *system sands*. The properties and composition of a system sand is partly trial and error and partly by design, as requirements vary from one foundry to another. The matter is further complicated by the deterioration of the clays during each cycle by exposure to the molten metal and excessive heat. In addition, debris and contaminants mix with the sand, and the temperature of the sand gradually increases during the day with the consequent change in sand properties. The goal for any system sand is reaching "steady-state" conditions. The sand is cleaned every cycle and appropriate additions of new material are made to either dilute or replenish materials formed or lost during each casting cycle. Lee [17] showed that ferrous foundries with sand to metal ratios approaching 3:1, had the additional problem of combating high sand temperatures. The objective is to maintain the characteristics and properties of the system sand within acceptable limits.

System sands should have appropriate grain fineness for the surface finish required and no greater than absolutely necessary. However, operations which alter the selected fineness are new sand additions, unbonded core sands entering during shakeout, fines generated by thermal and mechanical stresses to the sand grains, and still more fines from the thermal degradation of clay and other additives. Modern molding machines, particularly those employing high density and high pressure, can utilize coarser sands without sacrificing surface finish.

System sands in brass and bronze foundries are typically 95 to 100 AFS GFN. Coarser sands cause a noticeable deterioration in the casting's surface finish. Figure 3.58*b* shows

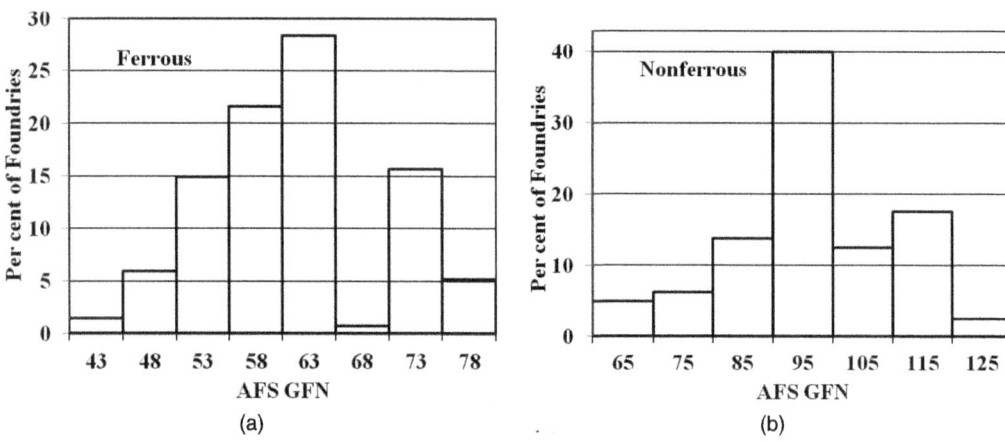

FIGURE 3.58 AFS grain fineness of numerous (*a*) ferrous and (*b*) nonferrous foundries.

data that CANMET collected at some 40 nonferrous foundries.[10] As core sands are typically 70 AFS GFN, in order to limit binder requirements and core blow defects, new sand additions tend to be about 105 AFS GFN. Many foundries purchased 120, 130, or even 180 in one case, but this made little difference to the final size other than to increase AFS clay and lead to five-screen system sands with poor permeability. In some ways, it would be advantageous to choose the same base sand for both molding and coremaking, as this avoids changes to the grain size or distribution when the system sand becomes diluted by broken cores. The ideal base sand is a four-screen grain distribution (more than 10% retained on four screens) as these sands require the least amounts of clay bond and temper water to develop optimum properties. Otherwise, new and core sands need to be three-screen sands to limit the system sand to four screens. In practice, the difference in the size of new sand and the system sand was often 20 points or larger. Figure 3.58*a* shows from the data collected by CANMET for some 68 smaller iron foundries, where wear to sand grains coupled with core sand addition decreased the size of system sands (higher AFS GFN). Both ferrous and nonferrous foundry grain sizes show that new sand additions are inadequate to offset the ingress of core sand. System sands for aluminum foundries tend to be about 70 to 80 and steel foundries 45 AFS GFN. However, by the mid-1990s Canadian steel foundries were using chemically bonded sands, and only one used dry sand molding for special tasks. In the case of aluminum foundries, procuring a shiny finish not only requires the right grain size, but also adequate mold hardness (>80) and metallostatic head (at least 2.5 in or 6 cm).

A clay-sand mix must be tempered; that is the clay and other additives require water to wet and activate the clay. When the clay bond is activated, the mix will reach a low bulk density. Mixes with insufficient or excess moisture are less desirable. Unfortunately, system sands change with repeated cycling (sand size, cores, new sand, bond, the proportion of fines, other additives, unpoured molds, and temperature), so the moisture requirements change. Therefore, controlling moisture at the muller is not particularly effective with synthetic system sands, especially if the results of one batch are used for the next. Figure 3.59 shows moisture levels at ferrous and nonferrous foundries. Foundries operating at the lower end probably have too little clay, whereas those at the upper end have large amounts of additives, fines, and dead clays, particularly ferrous foundries, where the additives are used to control mold atmosphere and sand expansion defects. In addition, nonferrous foundries use more southern bentonite, which requires much less water than western.

The concept of compactibility and its measurement were developed by Hofmann in 1969 [46]. Coincidently Dietert developed the original MB clay test [47]. Compactibility relates riddled bulk densities between 50 and 70 lb/ft^3 to mold densities between 95 and 100 lb/ft^3 (3-ram sample weights 152 to 165 g) for compactibilities in the range 30 to 50%. Sands with compactibilities less than 30% or greater than 60% are difficult to mull, as plows simply push masses of sand around without working it. During the period 1965 to 1980, many researchers were determining regression lines between the various properties, particularly strength, moisture, and clay content in synthetic system sand.

---

[10]Ten nonferrous and one ferrous foundry used olivine sand during CANMET's 1995 to 1997 visits. Properties were distributed evenly throughout the system sand properties: AFS GFN, moisture, compactibility, strength, permeability, MB, and AFS clay shown here and in the following sections; although AFS clay tended to be a couple of percent higher. Sand densities were 187 down to 171, indicating either contamination by silica core sands (density 156) or the variability of density with olivine grain size.

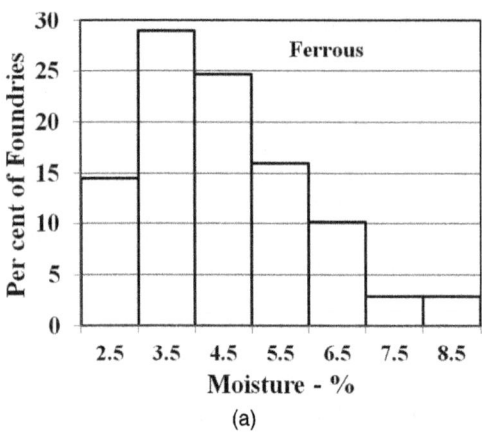 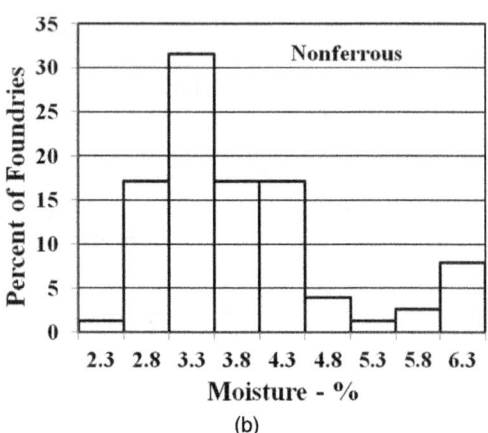

Figure 3.59 Moisture levels in (a) ferrous and (b) nonferrous foundries.

There was a concerted effort to predict the physical properties of the sand from the ingredients in the sand: fineness, clay type and content, fines, seacoal, and particularly water-clay ratio. The programs were also expected to formulate the additions required to maintain sand consistency.

Compactibility control implies that the moisture of the sand in the mixer is adjusted so that the compactibility remains on target. Even if the composition of the sand changes, the target compactibility will still be met as the moisture will be adjusted for the particular batch of sand. Although water is the variable being adjusted, it is compactibility that is controlled. Figure 3.54 shows the relationship between compactibility and water for five–system sands and three silica-southern bentonite (5, 7.5, and 10%) mixes [43]. The type of molding dictates the preferred compactibility, but 45 to 50% is a good compromise. High-pressure molding tends to use compactibilities of 30 to 40%—lower and the sand will be friable and subject to cuts and washes.

In the late 1960s, mulling produced variable properties with commercial western bentonites which were less than 85% pure. Green strength sometimes greatly exceeded those previously reported. The effect was greatest at similar water-clay ratios, reminiscent of the expansion of clay flakes predicted in the clay-water bonding theories. Also, sands of different fineness would obtain similar green strengths at the same clay-water ratios. Western bentonites were frequently added as slurries rather than powders, as this accelerated the coating of sand grains. Wenninger and Volkmar studied fully mulled sands with both Ca and Na bentonites and found [48]:

> maximum dry and hot strengths occur at water to clay ratios of 0.6:1.00, or four layers of water
> minimum bulk density occurs at water to clay ratios of 0.45:1.00, or three layers of water maximum green strength occurs at water to clay ratios of 0.30:1.00, or two layers

Single layers of water were so rigidly attached that they lacked the cohesion to bond the sand. Wenninger and Volkmar concluded that for the maximum 3-ram strength, two layers of water in the clay were essential, but less than three layers. The new sand test, compactibility, which is related to bulk density, is very sensitive to the clay size going from three layers to two layers: a contraction of 15% (see Table 3.18). The authors tested fully mulled sands at constant density and 3 rams and developed a series of nomograms. As bentonite levels were increased properties continued to increase up to the surface saturation level for the sand size as more water-clay entities became engaged. A variety of size sands were used, including those with seacoal. The authors constructed the diagram shown in Fig. 3.60, but stated they had removed the curvature [48].

Available clay is the clay percentage that is activated according to the compression strength and moisture in the Wenninger diagram (see Fig. 3.60). Available clay may not be all the clay in the foundry sand. The MB clay may be different as this is the amount determined in water suspension. For example, clay balls do not contribute to green strength.

The authors found that only about 50% of the available clay was actually being used to bond the sand grains, the other half was not only ineffective but absorbed water and contributed to the poor mold conditions. The working bond is the clay level at the intersection of green compression strength and compactibility. The authors defined muller efficiency as

$$\% \text{ muller efficiency} = \text{working bond} \times 100 / \text{available clay} \qquad (3.6)$$

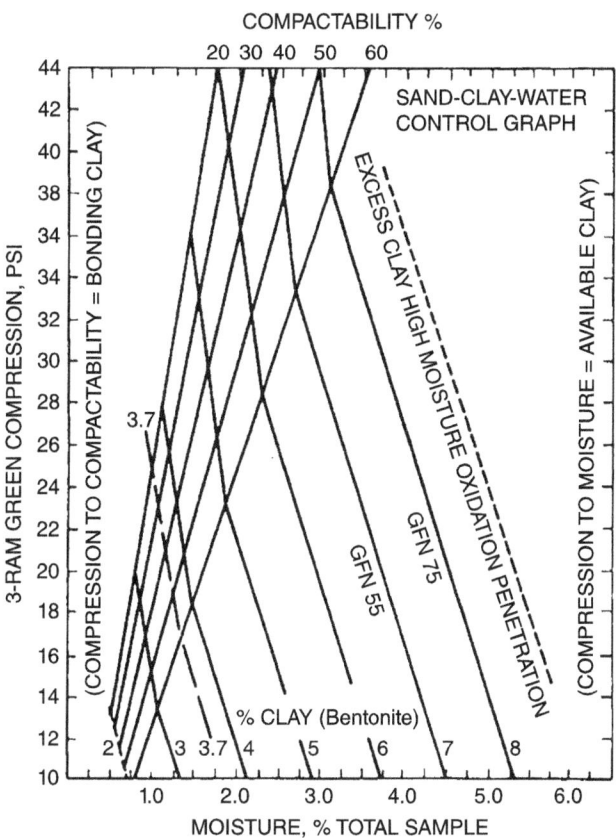

FIGURE 3.60 The Wenninger diagram: A graph for evaluating efficiencies of available bentonite within foundry system sands. (From Wenninger and Volkmar [48].)

Early mullers were shown to be not particularly effective at distributing the clay. As the researchers used fully mulled sands, they used this numerical evaluation of muller efficiency to convince manufacturers to make design changes [48].

Shih and colleagues published a series of papers extending the 4 to 8% clay levels of Wenninger and Volkmar to 4 to 15%. The higher levels are typical of those used in high-pressure molding, where greater hot strength and improved casting dimensional tolerances are required [45]. Attempts to find a relationship between clay and compactibility (moisture) showed a distinct break around 7% clay (Fig. 3.61). The various regression lines published formed a basis for the many computer programs written to predict sand properties and for statistical quality control (SQC). Maxima in the green strength/moisture diagrams occurred at 7% WesternB, 11% SouthernB, 6 + 6% WB + SB, 9 + 3% WB + seacoal, 12 + 4 SB + seacoal and 6 + 6 + 3.5% WB + SB + seacoal (Figs. 3.62 and 3.63). Silica grain fineness or olivine had little effect. They found that for foundries mulling to a constant compactibility there were two options, either a shorter mulling time with less water or a longer mulling time with more water. However, these alternatives result in different clay/water ratios and different strengths and bulk densities. The shorter cycle

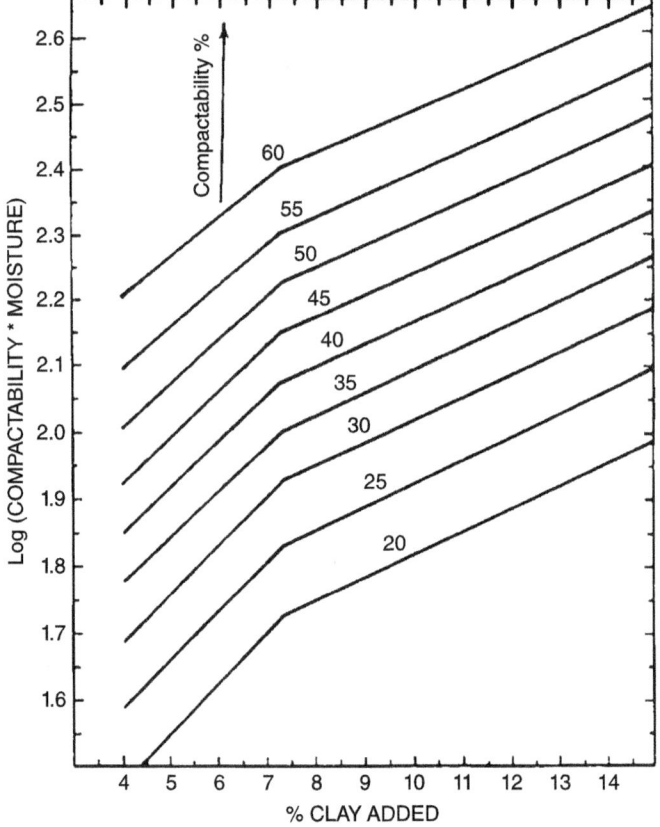

Figure 3.61 Relationship of compactability and moisture to the percent of Na bentonite. (From Shih et al. [45].)

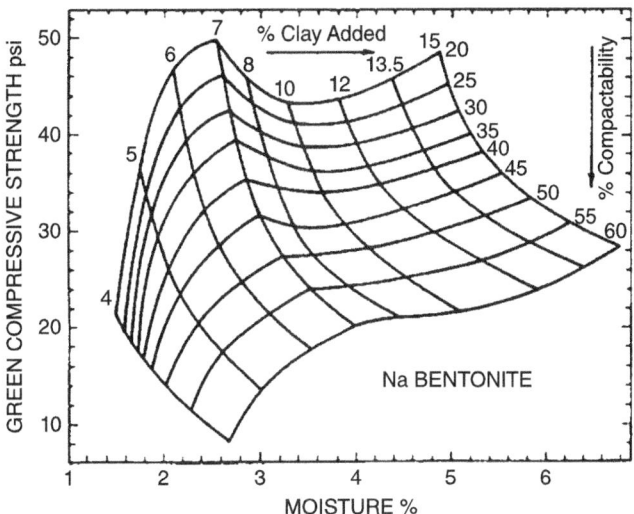

Figure 3.62   Relationship of GCS, compactability and moisture for the indicated percentage of Na bentonite. (From Shih et al. [45].)

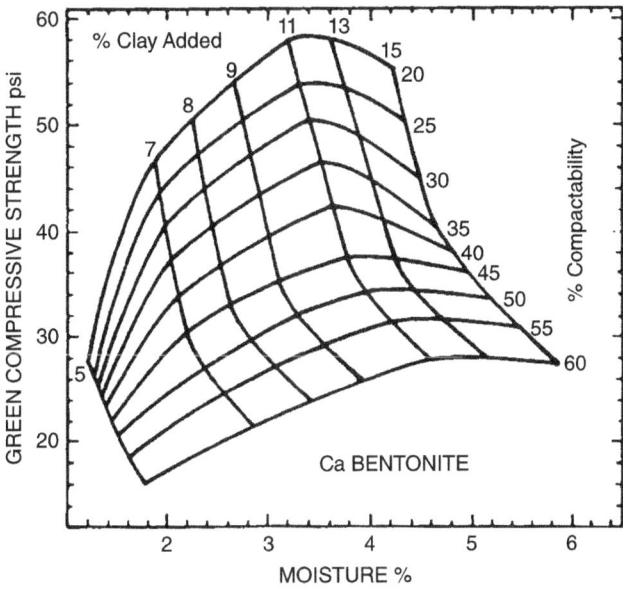

Figure 3.63   Relationship of GCS, compactability and moisture for the indicated percentage of Ca bentonite. (From Shih et al. [45].)

produces sand with a lower moisture content that is less fully processed. The Wenninger diagram is a good approximation for foundries with less than 8% clay, although foundries should always perform a weekly MB clay test to confirm. Above this level additional tests will be required.

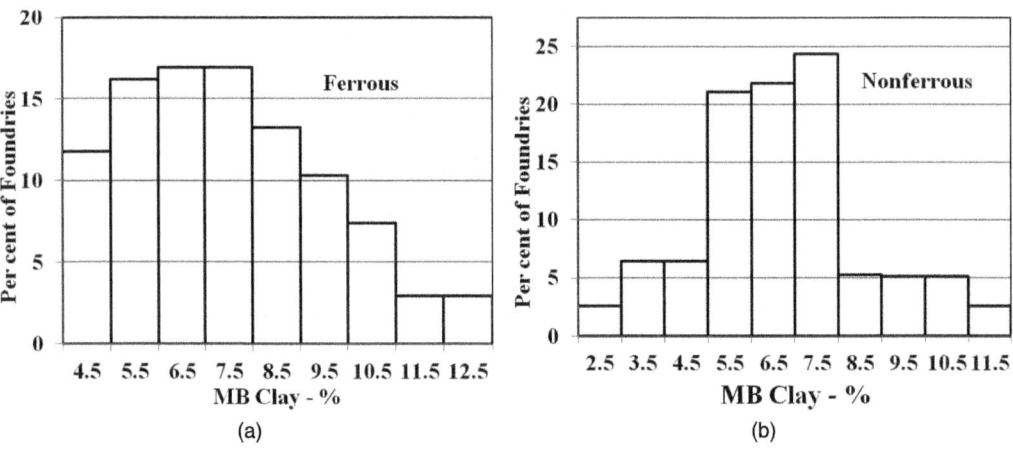

FIGURE 3.64 MB clay levels in (a) ferrous and (b) nonferrous foundries.

Canadian ferrous foundries used between 5½ and 9% largely western bentonite clay, whereas the nonferrous foundries used between 5½ and 7½% mostly southern bentonite (Fig. 3.64). AFS fines, which include live and dead clay, were 8 to 14% in both ferrous and nonferrous foundries, although a number had more fines. As a result, permeability, which should be in the range 140 to 175 for ferrous, had a few foundries with sand that was too open, but many were closed and lacked the ability to pass gas. The finer nonferrous foundry sands should have permeabilities in the range 50 to 100, but a large number had less (Fig. 3.65). Many packed their molds to a low bulk density and were blithely unaware of their problem; they just traded gas defects for poor surface finish. Mold "B" hardness on 3-ram lab specimens were 90 to 93, but considerably less in the molds. Compactibilities employed in ferrous foundries covered the entire range from 30 to 52, reflecting the different molding methods, whereas nonferrous foundries used 45 to 52% compactibility. As a result green strength, which reflects mulling efficiency, as well as moisture and compactibility, was generally poor in ferrous foundries and only slightly better in nonferrous foundries.

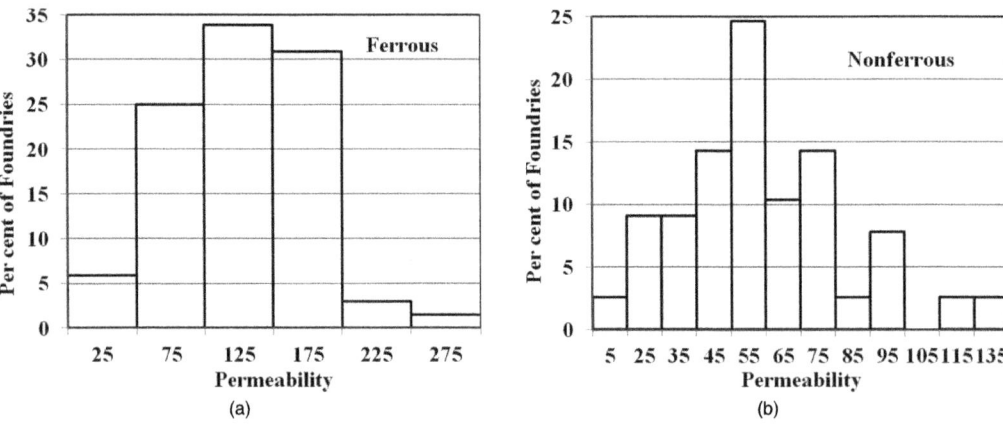

FIGURE 3.65 Sand permeability levels in (a) ferrous and (b) nonferrous foundries.

Boenisch published a number of papers in the mid-1980s on European foundry sands with more than 8% clay for use in wave and high-compression molding [49]. As part of this work, he developed a new sand test, the Deformation Limit, a numerical quantity that describes the distance in mm that a compacted sand mass can be deformed before cohesion is lost. The test uses shear stress, as other tests proved difficult to interpret. The test responds to most greensand parameters and components but especially to moisture. The deformation limit on mulled new sand and systems sands are vastly different due to the plasticity of system sands. Whirl mixers are about 10 times faster than conventional mix mullers in blending system sands, as new clay additions tend toward sand grains that are already coated, rather than to new or core sands.

Boenisch found that below 6% bentonite, molding sands cannot be improved by increasing the moisture level. Clay levels between 8 and 10% (i.e., excess clay over that required to coat the sand) are necessary to improve the deformation limit. Deformation limits of 0.2 to 0.4 mm at shear strengths of 4 to 8 N/cm^2 (equivalent to 3 to 10 rams) are desirable. Deformation limits outside these ranges produce different types of defects. Boenisch concluded that increased bentonite usage was the price paid for inadequate mulling—cycle times too short and moisture levels too low to integrate the considerable quantities of core and new sand being added to system sands used for wave compression molding.

The type of clay selected for bonding washed-and-dried base sand will be determined not only by the size, shape, and thickness of the casting to be produced but also by the equipment available for sand preparation, handling, molding, shakeout, and storage. General characteristics required of any system sand include the following:

1. Flowability adequate to fill the flask by gravity
2. Compactibility suitable for attaining the desired mold hardness and deformation
3. Adequate green compression strength to prevent mold damage
4. Dry- and hot-compression strengths sufficient to avoid chronic mold erosion problems
5. Freedom from excessive hard sand lumps at shakeout
6. Ability to consistently produce quality castings

Foundries producing small- or medium-size castings of light section thickness nonferrous castings often find it possible to take advantage of the better mulling, ramming, and shakeout properties afforded by the use of a southern (calcium) bentonite bonding clay. Conversely, producers of large, heavy-section iron castings tend to select western (sodium) bentonite because of its superior dry- and hot-strength and its good resistance to sand expansion defects. In general practice, foundries usually bond their system sands with a blend of different clays in order to optimize system sand properties for the vast range of conditions encountered in production. As pointed out earlier, southern bentonite degrades at lower temperatures than western, therefore when the system sand contains both clays, additions of southern on remixing will have to be higher in order to maintain the proportion in the rebounded system sand. The MB clay test measures the total montmorillonite clay and cannot identify the type.

System sands containing fireclay or southern bentonite as the main binders can have better resistance to sand expansion defects by the addition of small, controlled amounts of organic additives—usually a form of cellulose (wood flour, cob flour, rice hulls, and so on). These materials should be of uniform fineness and low in ash content. They burn out of the sand at low temperatures and, by a process not fully understood or agreed upon by authorities, prevent the cracking and spalling behavior of mold surfaces which are associated with the occurrence of rat tails, buckles, and scabs. The effectiveness for such cellulose additives are 1 to 2% by weight. It is a common mistake to use heavier additions of these additives to correct persistent sand expansion problems; other corrective means should be explored instead. Excessive cellulose additions require additional temper water to develop proper green compression strength. This plus the gas-forming tendency of the additive, combine to aggravate sand expansion problems (by the formation of wet-condensation layers) and cause expansion defects. High levels of cellulose make the sand friable, brittle, increase the frequency of blowholes and, in extreme cases, cause sooty pockmarks on casting surfaces.

It is difficult to monitor the working level of cellulose in a system sand. Although there are tests to evaluate the overall level of organic materials present (LOI and volatiles at various intermediate temperatures), such tests cannot distinguish between desirable additives and organics arising from unwanted core binders.

Hot sand. Greensand molds absorb a portion of the superheat and heat of solidification from the castings. During shakeout, hot castings and steam carry some of the heat away. However, if the sand is reused again, the sand acquires still more heat. In 1980, AFS Committee A80-2 reported a relationship between return sand temperatures on a conveyor belt as a function of sand metal ratio (Fig. 3.66) [17]. Sand temperature is highly variable without extensive blending, and measurements are erratic. Figure 3.66 shows just a 10 to 20 C rise above ambient until the sand to metal ratio reaches 10:1. However, sand temperatures from larger castings measured on the return sand belt increase significantly, rising 90°F (50 C) for a sand metal ratio of 8:1. Water additions into a muller are not adsorbed until the temperature falls below 160°F (71°C). Hot sand in the mixer steams and condensation causes problems: sand sticks in hoppers and onto patterns, mold surfaces dry out and become friable. Greensand is not stable at temperatures above 120°F (49°C) [50].

Although it was realized that hot sands cause numerous casting defects and were a major cost, attempts in the 1950s, 1960s, and early 1970s to study the issues by mulling hot sand and testing it at temperature produced results that were erratic and not easily

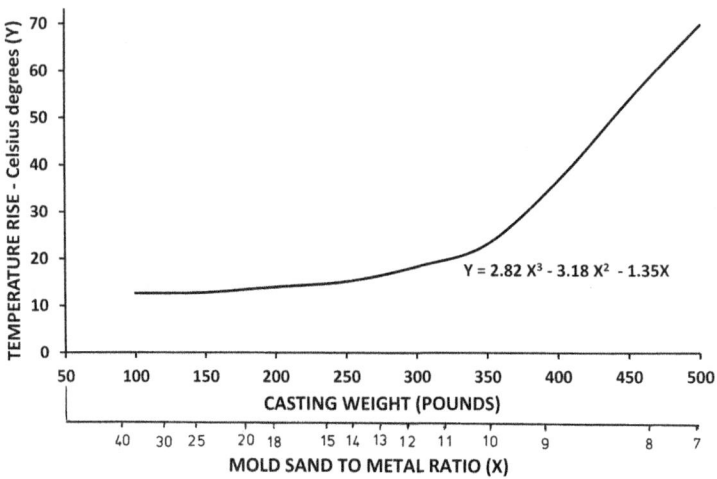

FIGURE 3.66   The effect of metal/sand ratio on sand temperature. (From Lee [17].)

reproduced. Schumacker and colleagues showed that bentonite slurries are gels above 160°F (71°C), but not gels when cooler [50, 51]. Above 140°F (60°C), evaporation of water is too rapid for stable and reproducible sand properties. At these temperatures, water layers in the clay flakes are not stable and mulling to produce clay-sand-moisture bonding is not effective. Many studies represented sand properties as changing linearly above 120°F (49°C) (i.e., lines were drawn through the scatter of data points as it was difficult to work above 50°C. Even today one occasionally sees data from very large mullers on short cycles, where the sand retains enough water so that sands tested at lower temperatures produce reasonable regression lines. Schumacker and Heine [51] concluded:

1. The hot water layer in the clay flakes has low viscosity and is not held as rigidly, so the more open sand-clay structure dries out and the edges of the mold collapse.

2. Hot sands are up to 6% lighter than cold sands. Sand dimensions may not be kept.

3. Green and dry compressive strength are lower in hot sands.

4. Permeability is lower, because the condensation zone either acts as a barrier or causes back pressure.

5. Compactibility is lower at the same moisture level.

6. The surfaces and edges [within 1/8 in (3 mm) or so] of molds lose 22% of the water within 1 minute and 44% within 2 minutes of stripping from the pattern causing the surface to be friable and crumble when exposed to flowing metal.

Unless extensive steps are made to cool the sand before it enters the muller, hot sand will cause innumerable problems, casting defects and costs. Hot sands continue to cause problems in small- to medium-sized foundries in spite of the available knowledge.

Greensand Molding

Used system sand has a great deal of impurities, ranging from slag (or dross from non-ferrous casting), metal balls, sand that has reacted, core sand of a different size, core lumps, dead clay, floor sweepings, metal flash, the odd casting, ear plugs, etc. Larger items are removed by sieves, although it is not uncommon to find foreign material in the sand system. The biggest issues involve: "How many pounds of clay and other additives are required to replenish what has been burned out and how many fines have come in?" The answer depends on the size of the castings being made and the sand durability. Sand durability refers to the system sands, ability to endure punishment and keeping that quality. The following factors affect sand durability:

1. The sand to metal ratio (typically 3:1 for iron, more for nonferrous).

2. The amount and size of the core sand that will dilute the system sand.

3. The amount of new sand and its size required to make good losses and dilute core sand.

4. An estimate of fines produced versus lost to the exhaust systems at shakeout, muller, etc.

5. Loss of clay and additives to the exhaust in the muller when make-up additions are made.

6. An estimate of the amount of sand discarded on shakeout and sand adhering to castings after shakeout.

7. Heat liberated and absorbed by sand, which depends on metal pouring temperature, type of metal, and the time hot castings remain in the mold during solidification and before shakeout.

8. The amount of carbonaceous material in the sand. The ignition temperature, burning rate, and an uncertain amount and condition of what survives.

9. The greater the rammed or squeezed density, the more rapidly heat will be transferred to the mold, and the greater the destruction of system sand materials.

10. The amount and type of organic matter in the system sand. Organics can ferment, burn out, or affect the clay bond.

11. Problems in the sand system: hoppers and conveyors can cause piping, funneling, bridging, or hang-ups that affect sand distribution. Hot sand will not cool in hoppers and causes numerous problems.

12. A build-up of fines from angular sands, seacoal, and heat degradation of the sand affect durability.

13. Type and source of clay which also affects its durability.

14. Uniformity of muller batch size affects the effective addition of bond and additives.

15. Amount of tramp materials, build-up of magnetic particles, fayalite Fe_2SiO_4 on the surface of sand grains, metal particles [flash and splash (have personally measured 8% brass)], etc., which affect durability and wetting.

16. The amount of the surface layer that burns onto the casting and usually stays on the casting until sandblasting. This layer contains dead clay, fayalite, and damaged sand, which produce fines and require excessive temper water and causes loss of strength and permeability.

Greensand molding has a number of advantages and limitations. The advantages are

1. Greensand offers great flexibility for production processes. Mechanical equipment is available that performs many molding and allied operations in a highly mechanized manner to produce molds approaching 500 an hour for both small and medium-sized castings.

2. Greensand can be reused many times when reconditioned with just water, clay, and a small addition of new silica.

3. Greensand offers a simple easy route from pattern to a mold ready to pour.

4. Greensand molding is generally the least costly method of molding.

The limitations of greensand molding are

1. Some casting designs require the use of other casting processes. Long thin projections or rangy castings in a greensand mold cavity tend to suffer washes as sand is washed away by the molten metal or may not be moldable. Cooling fins on air-cooled engine cylinder blocks and heads are examples of castings that should not be made in greensand as greater strength is required and because of difficulties in packing and stripping cavities.

2. Certain metals and some castings are subject to defects if poured into molds containing moisture. Some delicate cores made of some chemically bonded sands are subject to premature failure when subjected to moisture or the high humidity of a mold cavity for prolonged periods.

3. More intricate castings are better produced by other casting methods.

4. The dimensional accuracy and surface finish of greensand castings may not be adequate. A dimensional variation of ±1/64 in on small castings or ±1/16 to ±3/32 in on larger castings may be required. Although variations of less than this can be achieved, equipment maintenance becomes an ongoing issue. The recent requirement of near net shape has dictated machining tolerances with thicknesses less than the casting skin: the layer of oxides, fayalite, silica, and burnt clay on the surface of castings. Machining would normally undercut this layer to reduce tool wear and costs.

5. Larger castings require greater mold strength and resistance to erosion than available in greensand.

Dry Sand

Dry sand molds are actually made with molding sand in the greensand condition. However, the sand mixture is altered to achieve strength and other properties after the mold is dried. Dry sand molding is performed in the same manner as greensand molding, but on considerably larger castings. Dry sand mold cavities are coated or sprayed with a mixture, that when dried gives greater hardness or refractory properties to the mold. The entire mold is dried in an oven using circulating hot air at 300 to 650°F (150 to 340°C). As the drying operation is time consuming, some foundries use a skin-dried mold. In this process the surface of the mold is dried to a depth of $\frac{1}{4}$ to 1 in (6 to 25 mm) using torches, radiant heating lamps, or electric heaters. Skin-dried molds must be poured soon after drying, otherwise moisture will migrate to the mold surface. Castings between 1 and more than 100 tons were made in this manner in large pits. Steel railway frogs (switches) are still made in dry sand but in large molding flasks. Large castings generally have poorer surface finish and tolerances of about $\pm\frac{1}{4}$ in (±6 mm), unless special requirements dictate closer control. Mold construction requires special handling equipment to assemble the mold, and cores with gagger bars (heavy steel rods) inserted into the casting to give lifting points. Further difficulties are experienced during gating, pouring, feeding, and cleaning of large castings. Considerable floor space, heavy equipment, and time are required to handle large castings.

Loam Molding

Loam molding is still used in bell foundries. Step-by-step descriptions for the casting of bells, guns, cannon balls, and statues were given in Biringuccio's *De la Pirotechnia,* published in 1540 [29]. The molding media, which is still in use today, is a mix of sand, clay, animal hair, manure, and water. Substitutions by silty material or organic debris were also practiced, with wool-cloth clippings being a particular favorite of Biringuccio. The earliest manuscript on casting bells was written around 1100 by the monk Theophilus [29]. The molding media is shaped to form the central core for the bell and the outer mold. The various portions are oven dried. The molds are assembled often in pits from the various pieces, and frequently with multiple parting lines. Reassembly of the baked forms can be a problem if distortion occurs during baking. Producing consistent wall thickness is always a tricky problem with cone-like cores inside another cone. As bell alloys (Cu-Sn) are particularly prone to cracking, the loam media forms collapsible cores that shrink as the organics burn out and reduce the tendency to hot tear when the solidifying casting is constrained by a central core. The organics function in the manner of cellulose, wood flour, or green shell carb in greensand where they also give good surface rendition and close tolerances. In the case of the earlier Cu-Pb cannons, the central bore was an iron rod coated with clay, which is held in place by chaplets and core prints. Iron canons produced after 1774 were bored.

Molding Aids

A number of molding aids that can improve the quality of castings may be incorporated in the mold depending on their compatibility with the molding machinery. Most require additional operations at the molding station that delay closing the mold. Whether they decrease productivity will depend on the severity of the casting defects they alleviate; the foundry needs to do a cost benefit analysis.

Thick ($\frac{1}{2}$ in or 1.2 cm) ceramic filters are used in the gating system to remove oxides from ductile iron, aluminum, and copper alloys. Filters to remove dross from the magnesium treatments for ductile iron need to be quite large as they soon block. As the area necessary for both filtration and flow needs to be adequate, filters need a special compartment in the runner system to rigidly hold the filter in place. The sprue base is generally too small and the pouring basin lacks the metallostatic head. Metal needs access to upstream side, but not flow around the filter (on later inspection, both sides will be covered but not the edges). A number of filter compartments are shown in Chapter 5 (Figs. 5.20 and 5.21). The slightest movement of the filter will result in the metal flowing around the filter. Filters from different sources often vary in thickness and allow this to happen. Filters should be checked to ensure metal actually flowed through them. Filters are recommended for removing the hydrogen from aluminum (as they are associated with oxides) and copper alloys, where porosity in heavier (and hotter) sections of castings can be a problem. Filters also improve mechanical properties in copper alloys. In the case of aluminum, the holes in the filters must be large enough for the metallostatic head to overcome the surface tension (of the meniscus) in the capillaries, otherwise metal will either flow around, through only a small section of the filter, or not at all. However, this author made several studies of filters used for aluminum alloys, and the only oxides on the filters were silica [52]. Particle size was typical of the system sand, indicating poor molding techniques or excessive flow through the gating system. Rapid melting, temperature control, no contact with flue gases, and clean scrap are all critical in controlling hydrogen pickup.

Skim gates of tin plate (0.05 to 0.2 in thick) or fiber are also used at the bottom of the sprue (between the drag and cope), in the runner, or below a pouring basin to act as a choke to keep the sprue full. This ensures slag floats to the top of the pouring basin rather than entering the mold.

Pouring basins solve a number of issues. They reduce the amount of splash, particularly the bright red droplets running over the mold in brass and iron foundries. They vary from just a simple depression around the sprue to long runners with complex dams to get metal quietly from the edge of the mold to a central sprue when the pouring ladle cannot reach the sprue without a long drop. Basins show clearly when the sprue below is full. When the basin is kept full during pouring, there is less chance of air being aspirated into the mold cavity. They increase the cope height and can increase the metallostatic head when coupled with appropriate riser sleeves. However, molded basins must be either dammed around the base or glued to the top mold surface, otherwise metal will escape through any gap. Generally, with most automatic molding machines, all that can be done is to increase the diameter at the opening. Care is required near the end of pouring with a raised pouring basin, because if metal overflows a riser, that riser will be ineffective, as any flash acts as a cooling fin freezing the riser prematurely.

Breaker cores or rings. Shell sand cores placed in the gating or casting ingate to thin a section so that the casting breaks cleanly from the rigging either during mold shakeout or during the early stages of separating sand from the castings. These have a sharp notch which does not significantly affect flow.

Feeding aids—riser or feeder sleeves, blind risers, popoffs. Riser sleeves are available in a variety of diameters and heights. Sleeves are insulating, so they improve the efficiency of risers and improve casting yield. Sleeves are put on the pattern prior to ramming up the sand. Exothermic sleeves are also available for iron and steel castings. However, as they can contaminate remelt, risers may have limited recyclability.[11] Exothermic sleeves take about 2.0 to 2.6 times longer to freeze than cylindrical castings of the same size in greensand. Since they can be smaller, the casting yield is greatly improved and savings are considerably greater for thicker castings [53]. To facilitate the removal of risers, a popoff core is used at the base. The popoff not only provides a sharp notch which facilitates sawing off the riser, but reduces the chance of sink around the riser caused by a hot region of sand.

Risers are usually placed in the cope section of the mold, however, it is sometimes necessary to place risers in other areas of the casting. When a riser's feeding distance is too small to feed from top to bottom of a casting, shrink bobs (or blind risers, covered in Chapter 5) are placed at one or more parting lines off to the side of the casting. These risers by necessity must be closed, otherwise metal will run out of the mold. Because risers must feed metal to the casting, it is necessary for the metal to stay liquid and for the meniscus to shrink away from the top surface. An inverted cone is provided to start the meniscus and to allow air to ingress into the cavity and release the vacuum for metal to feed. A number of companies provide ready-made shrink bobs.

Vents

Both mold and cores need to be vented to get trapped air and mold gases out of the casting cavity. If open to the casting cavity, vents need to be approximately 1/16 in (1.3 mm) in diameter to prevent metal from entering and acting as chills. Molds also benefit from vents at the parting line, about 1 in (2.5 cm) from the casting running parallel to the casting, and then away to the edge of the mold. These vents collect gases and lead them away from the casting.

Cores should be vented whenever the casting design calls for a core that is nearly completely surrounded by metal. It is important to supplement the permeability of core sand which might vent the products of binder decomposition with additional passages that lead from the interior of the core through core prints and hence to the outside of the mold. Otherwise, the casting could contain blow holes, particularly if the metal freezes before the gas rises to the upper surface and out through the sand. The latter is made difficult by the metal surface skinning over and restricting escape of any gas bubbles. Core prints should have a vent from the core print to the edge of the flask. As Biringuccio said when he wrote *Book VI* in 1540 "the more vents you make on your molds, the surer you will be of a good result in your casting" [29].

---

[11] The thermite reaction uses iron oxide and +12 mesh aluminum. Chapter 5 covers risers.

Most chemically bonded cores, with the exception of larger shell cores, usually have the vents made after the resin has cured. Because of the strength of the bond this requires drilling the core usually by hand, a labor-intensive operation. Damage to pencil cores can be high, and many foundries employ people with greater dexterity for such tasks. Larger shell cores are made hollow, as excess coated sand is poured out during the blowing process. Core oil-bonded cores are pricked with a rod 1/8 in indiameter or larger, and as the sand usually has sufficient green strength, inspection is all that is required before baking. Smaller oil-bonded cores have a wire embedded in them. Others can have the wire rammed into the core during manufacture. The wire is then withdrawn before it is removed from the core box. Longer cores require guides and are labor intensive. Many cores are made in two sections then bonded together and cleaned up. It is normal to use a hacksaw blade to cut a groove in the core prior to affixing the two halves.

Chills are steel, cast iron, or zircon sand inserts to control freezing in castings when risers are not practical. Chills are usually about an inch thick, but larger castings may benefit from bigger chills, and in rare cases iron-shaped castings are used. Metal chills should be precoated and be warm and dry when placed in the mold, otherwise condensation and moisture could cause blows. Alternatively, a dry wash coating may be applied.

Chemically Bonded Molds and Cores

The following sections describe the processes used to form molds and cores from chemically bonded sand. While the processes seem quite simple to an observer, a rudimentary understanding of the underlying chemical processes is required to realize why certain operations are critical to producing good molds.

In the case of greensand, clay and water hold the sand grains in place with weak forces within the clay flakes. Chemically bonded binders employ polymeric and chemical glues to surround and hold the sand grains in place. With few exceptions, none bond with the silica in the sand. These chemical processes are known by either the names of the binder (organic or carbon-based molecular binders such as phenolic urethane, furan, epoxy) or an inorganic binder such as sodium silicate (also known as CO_2, as the latter is used to cure the binder) or by what happens in the molding box after chemically bonded sand is blown in and then cured (cold box, warm box, hot box, or gas-cured). Production rates are limited by the curing time (seconds to hours), the available number of patterns and core boxes, and whether the binders stick to the pattern which then requires extensive cleaning.

The processes have a number of features in common. All involve applying a mix of chemicals and sometimes a catalyst to sand grains, packing or blowing the mix into a core box or over a pattern, curing the mix, stripping the mold or core, and assembling the mold and cores to form the finished mold, which will be later destroyed by the heat from the casting. Whereas greensand involves a simpler physical attraction to bond sand grains together when contact is close, chemical binders enmesh the sand grains in a matrix of resins of high molecular weights. Moving greensand around only makes new physical attractions as long as the moisture level remains within the optimum range. Moving chemically bonded sand around once reaction has been initiated breaks bonds, reduces polymer molecular weight from shortened polymer chains, and decreases the strength of the overall three-dimensional matrix. Although some reactions start in the mixer, others start upon addition of a reagent to the core box. An understanding of the chemistry involved in the various binder processes is helpful in getting the best return from the use of effort and resources. Binders constitute 0.8 to 2.5% the weight of sand, and almost without exception, they are destroyed by the casting process. Some binder chemicals are products from oil refining and sensitive to the vagaries of international oil pricing, others suffer production shortages in the chemical industry. Both can be costly but they offer castings with better surface finish, dimensional accuracy, and reproducibility than greensand.

Major changes in mold and core making occurred after World War II and again after the oil embargos of 1973 and 1979. These two events caused a shift in foundry thinking toward energy efficiency. Then in the 1980s it was environmental concerns and the Clean Air Act of 1990. Legislation affected what could be done on the foundry floor over concerns about foundry air and disposal and emissions of volatile organic compounds (cause of atmospheric pollution). Smaller flexible operations quickly adopted energy efficient operations, whereas large capital-intensive operations could not change immediately. Table 3.21 shows the year that new significant binders first appeared in North America, along with cure temperature and typical chemical mechanism. Figure 3.67 shows the shift in binder usage as each system was adopted and replaced labor-intensive processes that frequently required experienced workers for good results [54, 55].

Commercial Introduction	Process	Cure Temperature	Cure Mechanism
1950	Core oil	Warm 205	Neutral/oxidation
1950*	Shell, liquid, and flake	Hot box	Neutral/hexa addition
1952	Silicate/CO_2	Gas cold box	Dehydration + acid
1953	Air set oils		
1958	Phenolic acid catalyzed	Nobake	Acidic
1958	Furan acid catalyzed	Nobake	Acidic
1960	Furan	Nobake	
1962	Phenolic	Hotbox	
1965	Oil urethane	Nobake	Urethane + oxidation
1967	Phenolic/CO_2	Gas cold box	Acidic
1968	Phenolc/urethane/amine	Gas cold box	Increas. basic with cat.
1968	Silicate/ester catalyzed	Nobake	Basic/saponif./dehydr.
1970	Phenolic urethane	Nobake	Basic
1977	Furan/SO_2	Gas cold box	Acidic
1978	Polyol urethane	Nobake	
1978	Warm box	Warm box	Acidic
1982	FRC/SO_2 acrylic epoxy	Gas	Slightly acidic
1983	Epoxy/SO_2	Gas	Acidic
1984	Phenolic ester	Nobake	Basic/K salt residue
1985	Phenolic ester	Cold box	
1992	Alumina phosphate	Nobake	pH starts at 2 then up to 5

*U.S. introduction.

Source: *Foundry Management & Technology* [54], Carey [55].

TABLE 3.21 Chemically Bonded Core and Mold Binders

Because these binder systems use chemical reactions to enmesh the sand grains, many factors will affect them. They include the nature of the sand surfaces, what's on the surface (impurities, especially water), how much surface, the temperatures of sand, pattern and chemicals, and the type of chemical reaction. A rudimentary understanding of how these factors influence mold properties is important.

Sand. Sand quality, size, and shape are more critical for chemically bonded binders than for greensand. Size distribution, grain shape, and surface area dictate the amount of binder required to coat the sand. The factors were shown in Tables 3.8 to 3.11. Angular grains require 25% more resin than rounded grains. More resin means greater cost, more gas evolved, poorer sand flowability, and less permeable cores and molds. Silica content is important, as any impurities tend to be on the finer grains (minus 140 mesh). The Acid Demand Value Test (ADV) is an acid-base titration that reacts with acidic or basic minerals on the sand grains, and indicates the potential to consume critical binder reagents (usually acid catalysts) and affect the binder curing, strength, and other properties [28]. Sand reclamation can cause an accumulation of basic materials, sodium from the CO_2 process, potassium from phenolic esters, and calcium carbonate or lime from thermally reclaimed lake sands or mined olivine. The constant handling of sand, particularly pneumatic transport, can cause fines to accumulate. Fines smaller than 0.004 in (0.1 mm or 140 mesh) should be about 1% by weight, just enough to block metal penetration from the force of the metallostatic head, particularly for ferrous castings. If the sand is to be reclaimed, then the foundry becomes its own sand supplier. Sand quality becomes even more critical. Issues that need to be addressed are: compressive strength and grain friability, and surface condition, particularly after transport or mechanical reclamation, and how the sand survives transportation without knocking off corners and retaining

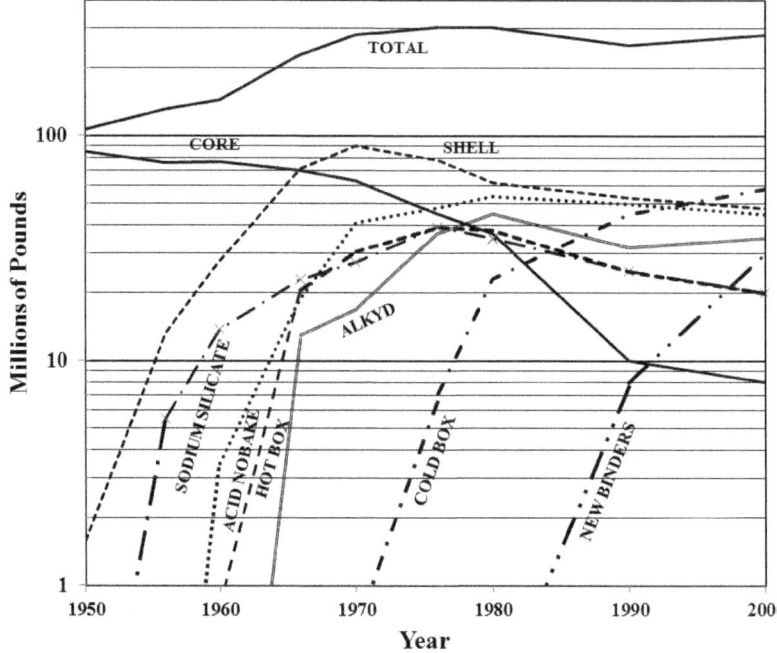

Figure 3.67   Chemical binder sales in United States by type. (Composite of Carey articles [54, 55].)

the shards. Often large numbers of cleavage plains can be seen in new sand under polarized light, which under the duress of thermal shock during casting result in cleavage.

Moisture. Water can enter the sand system from many sources, resin, binder, reclamation, sand, cold pattern, high humidity, condensation, or even rain. Moisture interferes with the resin/catalyst reactions by diluting the acid catalyst, etc., and binder ratios may need to be adjusted with the seasons or prolonged rainy weather. Humidity dramatically affects tensile strengths and storage life of cores. The water content of air can exceed 5% during the summer.

The 10°C rule of thumb (or 18°F). This rule is said to apply to the rate of chemical reactions. It states that for every 10 C rise or drop (or 18°F degrees) in temperature, the rate of reaction doubles or halves, respectively. The Arrhenius equation relates the rate constant for any reaction with temperature to be

$$k = A\, e^{-Ea/RT} \tag{3.7}$$

where the rate k is the total number of collisions between reacting molecules which equals a constant A times the fraction of collisions $\exp(-Ea/RT)$ whose energy exceeds a threshold "activation energy" Ea at temperature T (in Kelvin). R is the universal gas constant.

If the temperature rises from T_1 to T_2 and the rate constant from k_1 to k_2, then the ratio is

$$\frac{k_2}{k_1} = \frac{A\exp\left(-\dfrac{Ea}{RT_2}\right)}{A\exp\left(-\dfrac{Ea}{RT_1}\right)} = \exp\left(-\frac{Ea}{R}\left[\frac{1}{T_2} - \frac{1}{T_1}\right]\right) \tag{3.8}$$

For many organic reactions the activation energy is about 50 kJ/mol, and if the temperature increases from 23 to 33°C (or 295 to 305°K), then the reaction rate doubles. This is a rule of thumb that applies only to organic reactions at ambient temperatures with activation energies around 50 kJ/mol (and by assumption chemical nobake resins). Although the Arrhenius equation applies to other reactions at various temperatures; this rule of thumb is not applicable.

It is therefore important that anything contacting the curing sand be temperature controlled to between 70 and 80°F (21 to 27°C), otherwise cure rates, work, and strip times will not be controlled, and the molding sequence becomes unsynchronized and out of control. Therefore, in northern climes the sand silo should be indoors, sand and binders should be warmed, and reclaimed sand cooled, patterns and binder fresh from storage should be acclimatized before use in the nobake or cold box processes. At lower temperatures, resins become quite viscous and 85% phosphoric acid can precipitate and become less concentrated. It is difficult to coat sand with viscous liquids as the resin will not spread. Cold tooling can cause condensation.

Other factors that need to be considered are: higher curing temperatures and heat drying of coatings may lower the strength of the cores; the foundry needs to conduct quality control on all its supplies, not just the sand and binder but additives, coatings, and their carrier. Has the supplier changed the source, size, or composition?

When a foundry is faced with the decision to change or start using a chemical binder system, there are several factors they need to consider. The relative properties of mold and core binder systems are given in Table 3.22. Additional factors include

ORGANIC

	Binder Level BOS, %	Relative Tensile Strength (1)	Rate of Gas Evolution (1)	Hot Distortion (1)	Ease of Shake-out (2)	Effect of Humidity and Storage (2)	Curing Speed (1)	Strip Time (3)	Resistance to Overcure (2)	Optimum Curing Temperature, °C (°F)	Rebonded Reclaimed Sand (2)	Flowability (2)	Pouring Smoke (1)	Metals to Be Avoided
THERMOSETTING														
SHELL Dry blend and hot coat	1–4.5	H	H	M	F	G	H	0.5–6 m	G	260 (550)	G	E	M	
HOT BOX Furan and phenolic	1.5–2	H	H	M	G	G	H	0.5–2 m	F	230 (450)	P	F	M	Steel
WARM BOX	1–1.5	H	L		G	G	H	0.5–1 m	P	175 (350)	P	G	L	
CORE OIL	3.5 (4)	M	H	H	F	G	L	1 h/in	P	205 (400)	E	F	M	
SELF-SETTING														
FURAN NOBAKES														
High nitrogen furan—acid	0.8–1.5	M	M	L	G	G	M	5–45 m	F	27 (80)	P	G	M	Steel
Medium nitrogen furan—acid	0.8–1.5	M-	M	L	G	G	M	5–45 m	F	27 (80)	P	G	M	
Low nitrogen furan—acid	0.8–1.5	M-	M	L	G	G	M	5–45 m	F	27 (80)	P	G	M	
URETHANES														
Alkyd-organometallic	1.0–1.5	M	H	L	F	G	L	10–90 m	E	27 (80)	F	F	H	(5)
Phenolic—pyridine	1.0–1.5	M	H	M	G	F	H	2–20 m	E	27 (80)	P	G	M	
PHENOLIC ACID NOBAKE	0.8–1.	M-H	M-	var	G	G	M+	2–45 m	G+	27 (80)	F-	G	M-	
VAPOR-CURED (COLD BOX)														
Phenolic urethane—amine	0.8–1.5	M	H	M	G	F	H	20–60 s	E	24 (75)	P	G	M	
Phenolic/ester cured	1–2	M	L	H	G	G	H	20–0 s	E	24 (75)	F	G+	M	
Epoxy—SO_2	0.6–1.2	M+	M	M	G	G	H	20–60 s	E	24 (75)	P	G	M	(5)

INORGANIC													
SELF-SETTING													
Sodium silicate—ester cured		L	L	M	P	P	F	15–120 m	F	24 (75)	F	P	L
Cement—hydraulic cured	8	VL	L	L	P	G		15–120 m		24 (75)	F	F	VL
Phosphate—oxide cured	2.5–4	M	L	L	G-F	P	F-G	10–60 m	g-g	24 (75)	F	F-	VL
VAPOR CURED													
Sodium silicate—CO₂	3–5	L	L	L	P	P	P	1/2–2 m	P	24 (75)	F	P	VL

Notes:

(1) H = high, M = medium, L = low, N = none

(2) E = excellent, G = good, F = fair, P = poor

(3) Rapid strip times require special mixing equipment

(4) Typically 1% oil, 1.5% water and 1% cereal

(5) Iron oxide required for steel

Source: Chemically Bonded Cores & Molds, AFS [65].

TABLE 3.22 Comparison of the Properties of Various Mold and Core Binder Systems

131

production rate, capital outlay, space, operating cost, quality level required by existing or potential customers, emissions and disposal, neighbors, existing equipment, skill of labor force, etc.

Chances are most existing foundries already use at least one chemical binder system for core making. As many organic compounds start to decompose above 750°F (400°C), the air in foundries is tainted. Carbon monoxide is easy to measure, and if present, then it can be assumed that other pyrolysis products are present. Carbon monoxide levels are high in pouring and shakeout areas. The author has measured up to 150 ppm carbon monoxide at head height above greensand molds in ferrous foundries and half that in the shakeout area. Foundries usually operate with doors open, which overwhelms the exhaust system. In foundries with chemically bonded molds it is not uncommon to find very high carbon monoxide levels in storage or other unventilated areas. Exhausts should be above the roof level and not in line with intakes. Some jurisdictions specify 8 hour exposure levels of 10 or 20 ppm carbon monoxide. Table 3.23 gives major emissions from chemical binders. Other noxious chemicals are also released, and worse could be inhaled on dust particles. An AFS report lists some of these. Waste foundry sands may contain many noxious chemicals and government regulations deal with disposal and storage. Phenols are notorious for tainting ground water, due to their taste.

Handling Machinery for Chemically Bonded Sand
Equipment for Nobake and Cold Box Binder Systems

The sand silo should be of an appropriate size and inside the building in northern climes to minimize condensation and cold sand. The silo's exit design should address sand segregation issues. A sand heater should be positioned right before the mixer, whereas sand coolers are placed after the reclaimer. Temperature control avoids swings in work/strip times. Bypasses should always be considered.

Mixers. Older batch mixers only work for chemical binders with long working lives (slow cure), although recent faster models with multiple mixing chambers have extended the range. The original continuous mixers left material in the mixing trough when turned off. On restart, the first material out had to be rejected. Any mixed sand not exposed to liquid metal can be difficult to recycle. Modern high-intensity continuous mixers eject all bonded sand (Fig. 3.68). Some binders require three pumps although two will do if catalysts can be premixed with one of the resins. Always consult the instructions before mixing, as resin monomers and catalyst will react violently when concentrated. Dry additive feeders can also be fitted to the mixer. Gear-driven metering pumps are accurate to within 2%, and many can be programmed for different blends. Positive displacement pumps are less reliable but can be improved with surge chambers. Some mixers have adjustable pitch blades to improve mixing at low binder levels. The mixer production rate should match the expected flask size, as patterns must be covered and compacted well before the work time has expired, thereby avoiding erosion defects.

Compaction tables fitted with vibrators can be used to compact the sand against the pattern. Mold and core density and strength are improved, thereby reducing mold wall movement and metal penetration. However, final properties are dependent on the binder and its level. Sand mixes should not be moved, molded, or compacted after their working life. For simple molds or cores, simply moving the pattern across the flow of sand may be sufficient, otherwise vibrating for 3 to 5 seconds, hand tucking, or gentle ramming can be used to compact the sand. More complex shapes with deep pockets, loose pieces, and undercuts may need combinations, although loose pieces that move

Table 3.23 is shown below.

Mold and Core Binder System	Work Area	CO	CH_2O	HCN	NH_3	MDI	SO_2	H_2S	Phenol	C_6H_6	Toluene	Frufuryl Alcohol	CH_3OH
Limit		50	3	10	50	0.02	5	20	5	10	200	50	200
Furan Nobake Phos. acid cat.	MM/C	O	√(1)	O	O	O	O	O	O	O	O	√	O
	P/S	X	O	√(2)	√(2)	O	O	O	O	O	O	O	O
Furan Nobake Sulfonic acid cat.	MM/C	O	√(1)	O	O	O	O	(3)	O	(5)	(4)	O	(6)
	P/S	X	O	√(2)	√(2)	O	X	√	O	√(5)	√(4)	O	O
Phenolic nobake Sulfonic acid cat.	MM/C	O	X	O	O	O	O	(3)	X	(5)	(4)	O	(6)
	P/S	X	O	O	O	O	O	√	√	√(5)	√(7)	O	O
Alkyd urethane	MM/C	O	O	O	O	√	O	O	O	O	O	O	O
	P/S	X	O	√	O	√	O	O	O	√	√(7)	O	O
Phenolic urethane	MM/C	O	√	O	O	√	O	O	√	O	O(7)	O	O
	P/S	X	O	√	√	√	O	O	√	√	√(7)	O	O
Sodium silicate, Ester. cement	MM/C	O	O	O	O	O	O	O	O	O	O	O	O
	P/S	X	(8)	O	O	O	O	O	(9)	(9)	(9)	O	O
Phosphate	MM/C	O	O	O	O	O	O	O	O	O	O	O	O
	P/S	√	O	O	O	O	O	O	O	O	O	O	O

O = absent

√ = normally present

X = present in sufficient concentrations to be a health hazard

Assuming normal ventilation, optimized binder usage, and proper handling of binder system

MM/C = mixing, molding and coremaking

P/S = pouring and shakeout

Notes:

(a) Volatiles at MM/C refer to sand at 80°F.

(b) For phenolic modified furans, consult the entries for each individual system-furan nobake or phenolic nobake.

(c) Controlling CO below 50 ppm results in other volatiles being below their respective PELs or below their detectible limits.

Source: Chemically Bonded Cores & Molds, AFS [65, Table 8].

(1) O if binder is formaldehyde free.
(2) O if binder is nitrogen (urea) free.
(3) Possibly present if reclaimed sand is used that was originally catalyzed with sulfonic acid.
(4) Possibly present if toluene sulfonic acid (TSA) is catalyst used.
(5) Possibly present if benzene sulfonic acid (TSA) is catalyst used.
(6) Possibly present if methyl alcohol is used as solvent for sulfonic acid.
(7) Plus other aromatic hydrocarbons.
(8) Formaldehyde and other aldehydes could be present depending on the ester hardener or sand additives used with the system.
(9) Possibly present depending on specific sand additive used with the system.
(10) Powdered hardener may produce nuisance dust.

Table 3.23   Airborne Contaminants at Two Work Stations

133

Figure 3.68   (*a*) Continuous mixer M2000XLD and (*b*) adjustable blades in the mixing chamber M500XL. Resins are delivered via the hoses into the mixing chamber along with fresh sand. (Courtesy Palmer Manufacturing & Supply.)

once the sand reaches their level can shift and either cause lack of compaction or gaps. Compaction should cease at this point. Portable vibrators can be used on larger molds or cores, but the boxes must have somewhere solid to clamp a vibrator and transmit the forces. Compaction and good mold density are critical in reducing defects, binder levels, and costs. The alternative is higher binder levels and gas defects.

Patterns can be of most materials: wood, epoxy, plastics, aluminum, iron, etc. Although plastics are cheaper, sand causes them to wear badly, and certain binder/plastic combinations may cause the plastic to soften, as could cleaning solvents (consult plastic manufacturer). Aluminum will react with weak acid catalysts, and iron with stronger and booster acid catalysts, which erode the pattern and retard the cure and strip time. Two to three degrees of draft are required to strip patterns. Release or parting agents might be used, depending on compatibility with pattern materials and binder system. It is best to consult the manufacturer's instructions as liquid agents can impair polymerization. Patterns must be cleaned if a build-up of release agents and binder material occurs, otherwise sticking problems get compounded.

Stripping and removal of patterns. Each foundry seems to have its own method of separating patterns from molds and cores. Among the most common are rollover/draw units, chains and hoist, or pin lifts. Appropriate draft and clean patterns are critical as well as drawing perpendicular to the vertical parts of the flask, box, or pattern to avoid breakage. Rougher methods might require some adjustment to the strip or draw strength.

The layout design for the location of mixer, compaction table(s), rollover or stripping areas along with handling systems for sand, empty flasks, molds, cores, etc., is critical. Think of the old time and motion studies and add up the time and motions required to complete each operation, including cleaning boxes and patterns, and then see how the operations can be orchestrated perhaps with additional stations. Also consider the flow of raw materials. The time to fill and compact the mold/core box should be greater than the work time. The strip time will dictate the delay before stripping and indicate the catalyst and other reaction parameters (Fig. 3.69). Straight roller conveyors can move molds and flasks but do not organize events. For this reason, most foundries adopted the roller conveyor loop as flasks and patterns move in one direction, movement is minimized, and there is a better chance that strip times will be observed. The

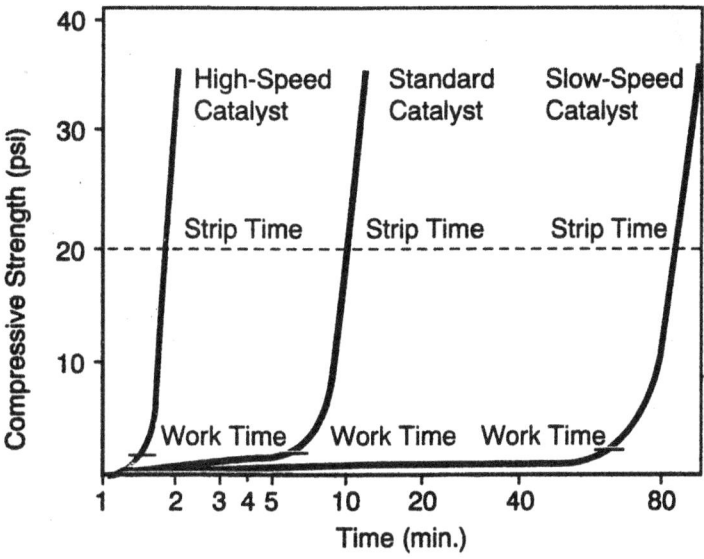

FIGURE 3.69   Onset of hardness in urethane-bonded sands, showing work and strip times for various catalysts. (From Carey [54, 56]. Used with permission of Penton Media Inc.)

multistation turntable goes one step further by controlling the duration of the operations. The turntable rotates and is indexed to (1) position a box under the mixer for filling and vibrating, (2) leveling, (3) surface compaction and automatic strike-off, (4–8) curing, (9) rollover and strip, (10) inspection and cleaning, (11) pattern cleaning, and (12) core setting, and back to the mixer for refilling. For very large molds, the mixer and its sand supply have to be moveable.

Equipment for Gas-Cured and Thermosetting Chemical Binders

Mixers. All types of mixers can be used for gas-cured and warm box binders. However, mixers with sand control and metering pumps are quicker and relatively fool proof. Mixing chambers might need cleaning depending on the bench life of the uncured sand and frequency of operation. Bench life of uncured binders is the period the mixed sands can be stored before the chemicals set or no longer function properly.

Core/mold blowers (Fig. 3.70) can accommodate many of the warm box, hot box, or gas-cured binder systems depending on their features. The features may include

1. A sand chamber to store sand and act as a pressure vessel to fluidize the bonded sand with compressed air and blow it into a core box after first passing through a blowplate. The pressure is then released and the chamber opened for another sand charge.

2. Core boxes can be of plastic for cold box binders, but cast iron for hot box binders to accommodate the heating chambers attached to each plate and resist the flames of the gas burners. Strengthening ribs can be used. The core boxes can be vertical or horizontally positioned. Core boxes need an array of special vents, blocked with wire screens to retain the sand. The location of these vents in rangy cores (i.e., water channels in engine blocks) can be critical if permeability is to be consistent, otherwise cores can vent a stream of gas through the solidifying metal where the core's permeability is low rather than through the core prints.

3. A squeeze table to squeeze the core box against the blowplate. In some machines, additional tooling enables the core to be gassed, i.e., a gas manifold is needed between the core box and the blowplate, and can be used to remove noxious gases to a scrubber, or even eject the core.

4. Gas or electric resistance heating for hot or warm core box. In this case, the blowplate will need to be water cooled on larger core boxes to stop sand from curing in the passages.

5. Programmable controller to time functions and control movements in the core blower.

A vaporizer will be required, as most amines are liquids. Sulfur dioxide and amine gases will need to be first purged from the core with additional air while still in the core box, and the air scrubbed by alkaline or acid washes, respectively.

Cores for ferrous castings usually require coatings to protect the core from erosion, penetration, and wetting by the metal, and improve the casting finish. Coatings are applied once the binder is cured, although some can be coated sooner (see the section on cores). Once the coating has penetrated a couple of grains into the core, the solvent needs to be completely removed, not a simple task once the impermeable coating is in place. Whereas alcohols can be burned or torched and chlorinated solvents are volatilized, water-based coatings present a more difficult case. Coating dryers may be batch or continuous

FIGURE 3.70   Core blower. (Courtesy Palmer Manufacturing & Supply.)

with cores on a conveyor or hung from a rail. Molds are usually torched. Although almost any heat source can raise the temperature of sand to 400°F (200°C), the aim is to get the heat in and away from the cores quickly with adequate air circulation to remove moisture.

Chemical storage facilities should minimize contamination from moisture, and be temperature controlled. Heat accelerates the decomposition and lowers the shelf life of organic chemicals. Barrels should be stored with expiry dates clearly visible. A first-in, first-out policy is best, as life-expired resins may not cure. Freezing can separate mixtures. Acid catalysts should be stored separately from the resin monomer, as in concentrated form they react violently on contact. Acids are corrosive to metals and skin. Resins should not be handled. Spills should be cleaned according to instructions on the MSDS Safety Sheets that are stored nearby. Some jurisdictions require that hazardous chemicals be stored in separate secure buildings with a good fire rating.

Sand Testing for Chemically Bonded Sands

The methodologies of the following sand tests are outlined in the *AFS Mold and Core Test Handbook* along with many references to original research [28].

Sand temperature. Although this was discussed under greensand testing, due to its importance perhaps it should be done more frequently.

AFS grain size. This was discussed under greensand testing.

Acid demand value (ADV). Basic ingredients in the sand can have a critical effect on some organic binder systems, although the inorganic binders (ethyl silicates and polymeric phosphate) are less sensitive. Acid demand measures the basic components in 50 g of sand by first consuming them with acid and determining the equivalent amount of base to bring the sand mix to neutral, pH 7, or another end-point depending on the ionization of the acid catalyst to be used. ADV is the milliliters of 0.1N hydrochloric acid that reacted with the sand. The sand pH (another test) is not an indication of the acid requirement. For example, clay is basic but insoluble; therefore, clay has little effect on the pH of water and sand, but has a large effect on acid demand.

Foundry sands with high ADV can consume acid catalysts in furan and phenolic resin nobakes, and disrupt the polymerization reaction. The effect of ADV will not be predictable from sand to sand because different impurities react at different rates. In the case of phenolic resin-isocyanate polymer binders, foundry sands with high ADV lower the bench life, as alkaline materials in the sand will speed the reaction between the two ingredients. Bond developed prior to molding is destroyed during molding and so reduced strength occurs. Recycled sands can contain metal, metal oxides, lime, or limestone depending on the method of reclamation, all serious consumers of acid catalysts.

Tensile strength. Tensile strength indicates whether molds and cores can withstand handling as well as metallostatic pressure. Because tensile strength is measured on special dog-bone specimens, they may not be representative of the actual mold conditions due to a different method of compaction. Dog-bones should weigh 100 to 103 g, otherwise properties will be lower and not reproducible, and unfortunately not representative of molds. The machine shown in Fig. 3.57 is fitted with a special jig to pull the dog-bones. Many years ago, the opinion was that mold tensiles should be around 200 psi; however, this could be wasteful of binder, and many satisfactory molds for steel were of the order of 70 psi. In the case of cores, handling is very important, but so too is collapsibility. Shakeout can be difficult if the mold or core retain their strength [61, 62]. These are factors of section size and metal poured. For this reason, binder levels in cores for nonferrous castings should be as low as possible (<1%). Air-set sodium silicate and shell are slow to collapse. In addition, organic binders are pyrolysed (thermal decomposition in the absence of oxygen) not to water and carbon monoxide, but the original monomers, and other gaseous decomposition products. Levels should also be low to avoid toxicity issues.

Hardness. At one time scratch hardness was used, but this has been replaced by an impact penetration gauge which puts a number (not tensile but an estimate of resistance) to the degree of cure below the surface. As many binders develop a skin, this tool informs the user if the mold is too weak to strip. Measurements on the top surface could be meaningless if it takes a long time to fill the flask. Figure 3.69 shows the onset of strength for various urethane formulations. Knowledge of work and strip times is critical, otherwise chemical bonding and strength will be lost if the sand is moved after the work time, and the early plasticity of the part will no longer enable easy separation of the pattern and mold once the strip time is exceeded. Factors that affect the work and strip times are binder levels, mixer calibration, sand and pattern temperatures, sand impurities, mold density, sand moisture, atmospheric humidity, nature of the acid catalyst, age, and water levels of the binder chemicals. Regular permeability and hardness testing of actual molds will reveal whether work and strip times are being observed, as the alternatives are higher binder costs, scrap, stripping labor, pattern damage, and wear.

Mold density and permeability. Generally, the more resin-coated sand packed into a mold or core, the higher the density and the stronger it is. Round grains pack better than angular. Densities greater than 1.55 g/cc solve handling and penetration defects, but can lead to permeability and expansion issues. Densities of less than 1.4 g/cc will cause defects. Short of weighing tagged flasks with the same pattern at regular intervals, mold densities are difficult to track. Solid cores, on the other hand, should not vary in weight. Nevertheless, many molding and core blowing problems can only be discovered by taking a permeability tester out on the foundry floor and measuring the permeability across flat surfaces. Areas of low permeability (poor location of vents in cores) in drags and cores will leave tracks of gas bubbles through the casting from areas of low permeability. The measured permeability of properly applied mold and core coatings will be 1 or 2. Otherwise, the measured permeability should be compared with dense dog-bones. It is critical that as many vents as possible be incorporated into molds and cores. You can never have enough vents.

Calibration and cleanliness of continuous mixers. Sand and binder weights per unit of time should be checked regularly to ensure blockages, wear and drift have not occurred. This may involve disconnecting lines unless additional valves are installed for sampling. In the former case, the nozzles inside the mixer need to be checked for any blockage. Generally, the binder entry and sand exit are all that should be cleaned at this time. The sand coating in the chamber becomes the effective lining for the mixing chamber and finishes blending and distributing the binder on the sand grains, and should not be disturbed. Calibration and cleaning is cost effective as control of setup and strip times is critical.

Hot distortion. The British Cast Iron Research Association developed a Hot Distortion Tester which measures the distortion of a $1 \times 4 \times \frac{1}{8}$ or $\frac{1}{4}$-in-thick ($2.5 \times 10 \times 0.3$ or 0.6 cm) wafer of bonded sand [63]. The wafer is heated by a gas-fired Meeker burner (a broad Bunsen burner), which simulates the effect of metal entering the mold cavity and then staying in contact with the core. The wafer first expands on the hot side then sags as it fails. The unit plotted the time/distortion. Wafers of shell sand usually survive from $3\frac{1}{2}$ to $4\frac{1}{2}$ minutes depending on the binder level. Other binders last a much shorter period. Some actually fail immediately. Cores made of such sand would sag before the metal surrounded the core and it would then try to float. Molds suffer run-outs unless securely boxed. Although this test can be simulated easily, it may only be of value when considering a new binder system or certain types of casting defects.

Organic Binders

In this chapter, the word *organic* is used in the traditional scientific manner to mean carbon based. Whereas inorganic refers to compounds made up of elements predominantly from the rest of the periodic table. Unlike the food and retail businesses, organic foundry binders have environmental issues as they release undesirable chemicals such as phenols, benzene, other aromatics, formaldehyde, carbon monoxide, etc., both in gaseous form and in the sand (see EPA listings). Inorganics have much less of an environmental footprint, foundry air is less hazardous and molding aggregates can generally be discarded without further treatment.

Shell Hot Box

The Croning or shell process was developed in Germany during World War II. Originally, the materials were mixed, then dumped onto a pattern and heated to melt and set the binder. As the ingredients were dry, there were numerous problems, the most serious being separation of the ingredients [54, 57]. Special plants hot mix sand (at 300°F) with particles of phenolic novolak resin and a lubricant calcium stearate. During a wet mull, the resin melts and coats the sand grains. An aqueous solution of hexamethylene-tetramine (hexa, a compound of ammonia and formaldehyde) is thoroughly mixed with the sand. The water cools the sand and the coating device and vibratory screens break the lumps. Copious amounts of air circulate to cool the sand on long conveyors until it is free-flowing. Most foundries buy coated shell sand from commercial sand coaters. Shell sand is available in bags with binder levels varying by 0.5% increments from 2 to 4.5% resin. The bags should be stored in dry, cool areas and not stacked too high, otherwise lumps can form. The advantages are

1. Nonflammable.
2. Free flowing sand which is easy to pour in and out of core boxes.
3. Free of volatile organic compounds (VOCs).
4. Easier to handle than the liquid counterparts, with no measuring of ingredients.
5. Quickly formed into dense, strong cores or molds with less pressure than other binders.
6. High hot strength and resistance to decomposition and distortion.
7. Excellent dimensional tolerances.
8. Although the binder levels are higher than other binder systems, the level of gas defects is lower as the cores can be hollow, and the use of a coarser sand improves the permeability so that core gases vent through core prints rather than through the casting.
9. Low nitrogen grades are available for steel castings.
10. Coatings are rarely required, so the cost of refractory, application, drying, labor, and maintenance of all the equipment are avoided.
11. Less sand is required to make thin-walled molds and hollow cores, and the uncured sand that is drained from the curing core is reusable.

Leo Baekeland in New York was the first in 1907 to study the elimination reaction of phenol and formaldehyde to produce polyoxybenzylmethylenglycolanhydride, Eq. (3.9), which became Bakelite, an early plastic, and the precursor of modern novalac resins.

$$C_6H_5OH + CH_2O = H_2O +$$ (3.9)

novalac

When the coated sand first contacts the hot pattern [below 500°F (260°C) for larger cores and 525 to 600°F (275 to 315°C) for small solid cores], a number of physical transitions and chemical reactions occur. First, the novalac resin begins to melt, causing the viscosity of the coating to drop, as the hexa decomposes to ammonia and formaldehyde. The resin begins to polymerize and the phenol in the resin reacts with the formaldehyde liberating a small amount of water. The binder flows between the sand grains and bridges lock them together. As the phenol-formaldehyde reaction continues the resin becomes thermosetting and the whole mass is rigid. The color of the sand is now golden yellow, and indicates a strong cured mold. If heat is still applied, the resin cures further, the color turns brown and the sand is weaker. Further heating the sand decomposes the binder, and it turns black. At this point, the surface is burnt and carbonized. Castings in contact with this core will have penetration defects on their surfaces, and/ or core breakage can occur. The best color is dark gold or light brown. Many overcook in the belief less gas is generated later, but less binder is more effective.

The sequence for the foundry process are first, the coated sand is either dumped by gravity or blown from an adjacent storage hopper into contact with a heated pattern plate or core box. The curing reaction takes place from the heated surface inward, and is normally allowed to progress to the point where a crust or shell of desired thickness is formed. At this point, the remaining uncured sand is drained back into the storage hopper usually by inverting the core box and a large section of the core blower. The cured shell mold or core is stripped or ejected. For best results, the process should be highly instrumented with the liberal use of sensors and a programmable controller. Die release agents can be applied every 10 applications. Many foundries block the ends of cores so if metal gets into the core print, it does not fill the core - a run-in defect. A schematic representation of the shell molding process is given in Fig. 3.71. The most authoritative reference on all aspects of the shell process is the AFS publication, *Shell Process Foundry Practice* [64].

The phenol-formaldehyde shell process has found greater application for coremaking than for molding. Its initial attraction as a molding method lay in its capability to hold closer tolerances than greensand molds, while providing improved surface finish. From the outset, its disadvantages were lower productivity, expensive pattern equipment, and higher energy costs. The advent of automated, high-density greensand molding served to swing the balance further away from shell molding as a viable economic process. Nevertheless, steel foundries still employ the process for small rangy castings where shell's high hot strength is an advantage. In addition, the draft on side walls is of the order of $\frac{1}{4}$ degree, so minimal machining is required so casting such as steel toothed bevel gears and truck fifth wheels can be cast in shell molds.

Shell coremaking has been quite another story. Its initial attraction in the 1950s was due to the fact that it offered a means of eliminating the traditional core oils, with their

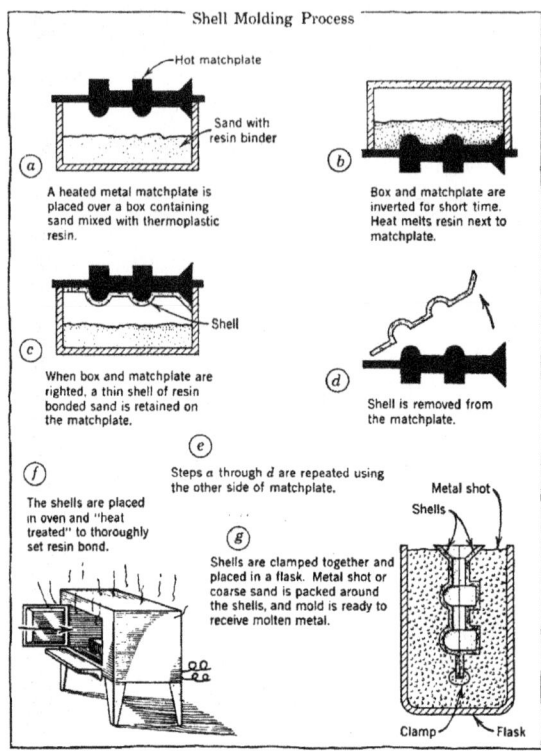

FIGURE 3.71   Forming a shell mold. (From Sahoo [7].)

core blowers, core driers, reinforcing wires, baking ovens, and skilled coremaking labor. The popularity of shell coremaking accelerated rapidly from 1950 onward, peaking in about 1970. Although the Shell Process has declined in popularity, it has benefited from new products such as the low odor–low hexa type. This decline has resulted from at least three factors: (1) the rapid escalation of energy costs beginning in the 1970s, curtailed supplies of natural gas for heating the boxes; (2) environmental regulations dealing with solid waste disposal and the identification of shell sand wastes as sources of free phenol; and (3) the introduction of other fast-cure coremaking binders competitive with the shell process. On the other hand, although cost per ton of coated sand is high, actual consumption per part and sand disposal can be significantly lower, coatings are seldom required, so drying costs and application costs are avoided, finish is excellent so cleaning room costs are lower. Reclaimed sand is not generally used, but the Canadian Foundry Association's Sand Reclamation Committee found in 1991 that shell resin is generally insensitive to surface impurities.

Furan Nobake

The acid catalyzed nobake furan binder system was introduced in 1958 [54, 58]. It consists of two main ingredients: furfuryl alcohol in monomeric form or a low molecular weight polymeric form and an acid to initiate or further polymerization. In 1960, the binder formulation was modified with an acid salt catalyst to use as a hot box core process. Furfuryl alcohol or 2-furylmethanol ($C_5H_6O_2$) has the $-CH_2OH$ group on the number 2 position of furan. It is made from its aldehyde, a by-product in the sulfuric acid treatment of waste vegetable materials (corn husks, rice husks, sugar cane) to obtain the sugar, xylose. From time to time, there are shortages of furfuryl alcohol, and foundries are forced to switch to another binder until supplies return to normal. Although furfuryl alcohol polymerizes easily, extenders such as urea, formaldehyde, water, phenol or phenolic derivatives, and proprietary chemicals are added to lower the cost. They also increase the viscosity, which does not help when coating sand. Certain ingredients in these extenders are of concern; urea adds nitrogen; phosphoric acid the phosphorus and formaldehyde the free formaldehyde, although modifiers can be added to "scavenge" formaldehyde. Although there are many grades, they are usually classified as low, medium, and high (Table 3.24). Water delays both the cure and the final tensile strength which are reduced approximately in proportion to the water content. The polymerization is accompanied by liberation of more water and might initially produce the polymer shown in Eq. (3.10), but time and UV light cause cross-linking and other changes. It is likely the coordinated double bonds are responsible for the sand's color changes (green to black to brown).

$$(3.10)$$

Furfuryl alcohol and a polymer formed with a strong acid catalyst.

Because the binder produces water as a product of the reaction, molds and cores cure from the outside in, i.e., surfaces exposed to air cure first. Molders often use a large nail to feel the depth that the binder has cured, i.e., they are concerned about what they call the "deep set." Consequently, stronger acids are used to optimize cure rate and deep set properties. Acids in the order of increasing reactivity are: 75% phosphoric, 85% phosphoric, toluene sulfonic, xylene sulfonic, and benzene sulfonic. Five to ten percent sulfuric can be added to phosphoric acid to boost its reactivity. Phosphoric acid may become viscous or even crystallize in cold weather, if not warmed. The sulfonic acids are usually diluted by methanol which is flammable.

Although composition and grade selected for the binder will be dictated by the metal to be poured (Table 3.25), the cure time required, the complexity and thickness of the casting, whether and what type of reclamation is to be used. A typical binder level

Grade	% Nitrogen	% Water
Low	0–3	0–5
Medium	2–8	5–15
High	5–11	10–30

Source: AFS [65].

TABLE 3.24 Grades of Furan Nobake Binders

Metal	Class	Suggested Amounts	
		% Nitrogen	% Water
Gray iron	20, 30	5–11	8–15
Gray iron	40, 50, 60	0–4	0–10
Ductile iron	All	0–5	0–8
Steel	All	0	0
Nonferrous*	–	4–11	8–15

*Furan can turn the surface of aluminum castings green.
Source: AFS [65].

TABLE 3.25   Furan Grade for Various Metals

for silica molds is 0.8 to 1.5%, and will depend on the type, shape, and purity of the sand. Cores are usually made with a higher binder level. A good starting binder level is 1.0 resin (1% based on sand), 0.30 catalyst (30% based on binder), and 100 of sand—units are pounds, kilograms, etc.

Catalyst and resin should be stored separately. They should not be mixed in the pure state; only on sand (slow cure time) with the sand first coated with catalyst in a batch mixer for 2 minutes, then the resin added. Reversing the process results in rapid curing and bond failure where the acid meets the sand. In continuous mixers, the catalyst is added before the resin. If there is a need to dilute the catalyst, the catalyst should be added to water. Addition of water to acid (catalyst) is highly exothermic, and a steam explosion will probably occur, throwing acid around. Plastic containers are better temporary storage as acids attack metals. Transport of acid containers is governed by safety rules in most jurisdictions. The materials are all water soluble, so small spills, tools, or even hands can be easily cleaned. Even so, contact with the chemicals and bonded sand should be avoided.

Advantages:

1. High strength for the binder level, high hot strength, but good shakeout due to collapsibility.
2. Excellent flowability.
3. Lower smoke and odor during the casting process than other solvent-based binders.
4. Can be used with all sands but olivine, and any pattern material although furfuryl alcohol may dissolve shellacs, varnish, or paint.
5. Cores can be stored pending need, as the bond has good resistance to humidity.
6. Wide selection of binder formulations to balance cost and performance.
7. Good selection of acids and combinations to control catalytic activity.
8. Low temperature variants available to control curing and coating.
9. Color change indicates rapid cure and warns of insufficient work time.

Disadvantages:

1. Chemicals are irritating to skin and eyes. These acid catalysts are corrosive.
2. Work/strip times very sensitive to temperature.

3. Extenders lower cost but reduce properties.

4. Nitrogen from urea and sulfur from acids complicate the sand reclamation process.

5. Mechanical reclamation causes an acidic sand surface (high ADV).

6. Some formulations contain phenol which can leach into ground water.

Alkyd Urethane Nobake

The urethane nobake and urethane cold box binders are the second and third highest sellers in North America after furan [54, 56]. It was discovered that the isocycanate reacted with core oil, causing the metallic dryer to perform better by forming a urethane. Companies such as ASK (previously Ashland Oil) developed and patented a series of sand binders based on MDI [see Eq. 3.11)] starting in the early 1950s called *air-set* and *cold-set* binders. The drawback with these early binders was a prolonged drying at elevated temperatures. However, from them came better formulations. The current binders have three parts: two co-reactants, one an oil urethane, the other an MDI isocyanate, and a catalyst. Just as furfuryl alcohol was able to polymerize with itself, so can MDI. Isomers of MDI can also polymerize to form polymers of the type $-[-C_6H_4NCO-CH_2-]_n-$. However, resins or vegetable oils containing active phenols $-[-C_6H_3OH-]-$ can react with the MDI polymer in the presence of an amine catalyst to cross link and form a solid urethane Eq (3.11), but not to the degree as for the next binder system.

1-isocyanato-4-[(4-isocyanatophe-nyl) methyl] benzene or MDI

polymeric MDI

$$(3.11)$$

The alkyd urethane nobake binder (also known as oil-urethane or polyester–urethane) consists of three parts. Part A is an oil-type resin or modified vegetable oil with phenyl hydroxyl groups which will form a urethane group with the isocycanate and a catalyst. Many such resins are available commercially. Part A is typically 1.0 to 1.5% BOS (based on sand) for an AFS GFN 55 round grain silica sand. Part B, the catalyst, consists of several amines and a metallic drying agent to facilitate the oxidation reaction, stage two. At one time, lead was the metallic drying agent, but it has largely been replaced by less noxious metals. Strip time can be adjusted from 10 minutes to 4 hours depending on the amine selected and its concentration. Part B should be between 1 to 10% of the resin. Most manufacturers preblend the catalyst with part A, as some pumps are difficult to adjust with sufficient accuracy. Part C is a polyisocyanate, which because of the isocyante group contains 10% by weight nitrogen. Part C should be added at 18 to 20% of Part A. Too much Part C causes gas defects [65]. As Part C reacts with water, its container should be mildly pressured with nitrogen, or the liquid extracted for the process replaced by air that has passed through silica gel to remove atmospheric moisture. Water in the sand (>0.2%) or in Part A will retard the cure and result in loss of strength. Even without a heating stage, small molds and cores can be poured 20 minutes after stripping, larger molds may require 4 to

24 hours. Most washes are compatible once the mold/core has cured, but coating solvents should be evaporated. High atmospheric moisture will increase strip time. Half a percent moisture in the sand can quadruple cure time and reduce tensile strength by a half.

Alkyd urethane binders cure in three stages [54]. The first stage determines the work and strip times and involves a reaction of the hydroxyl group on a phenolic resin with the isocyanate (-NCO) group to form a weak urethane bond. This leaves the sand strong enough for stripping. In the second stage, unsaturated (double) bonds in alkyd resin react with oxygen. If the cure is insufficient, the hot strength of the mold or core is poor. The cure can be improved by heating, with the third stage occurring at 400°F, which is the cross linking of the various polymers.

The alkyd urethane system being less sensitive to conditions has the following advantages:

1. Suitable for ferrous and nonferrous metals.
2. Can be mixed in conventional mullers and continuous mixers.
3. The mix is flowable but mild compaction is necessary to achieve good strength.
4. The plasticity of the sand at stage one permits patterns with minimum draft.
5. Strip time is less sensitive to sand type, temperature, moisture, additives, and impurities than other nobake or cold box binders.
6. The temperature and nature of the pattern material affects stripping time, but less so than with other binders.
7. Shakeout characteristics are good and sand is reclaimable.

Disadvantages are:

1. The time between stripping and pouring is longer than other binder systems due to the oxygen cure.
2. Large molds need to be stored several days for the oxygen cure to complete.

Phenolic Urethane Nobake

Although announced in 1970, consumption has grown from 2.7 million pounds in 1970 to 300 million pounds worldwide by 2001 [66]. This binder also has three parts: two co-reactants, polybenzylic-ether-phenolic resin and polyphenyl polyisocyanate (polymeric MDI isocyanate), and an amine catalyst. This time, it's basically the same functional groups but now on two polymers that react together in the presence of a catalyst (usually a pyridine derivative, benzene with one carbon in the ring replaced by nitrogen) to cross link the two polymers and form a solid three-dimensional urethane matrix with no by-products. The reaction in Eq. (3.12) is faster than in the alkyd urethane system, and the sand remains free flowing almost to the end of the working life. The work time is about 75% of the strip time (30 seconds to 30 minutes) but as the reaction is so fast, the sand becomes very rigid and does not yield easily for stripping. Patterns need to be inspected frequently and well maintained. Molds can be poured when the reaction is 80% complete, about 4 hours. As there are no by-products, the sand cures evenly at the same rate in thick and thin sections. Cross-linked chains tend to be short because the binder cures so rapidly, so the cores are sensitive to ambient humidity when stored. It is better to use high-speed mixers in order to take advantage of the short working life [54, 56].

$$\left[\underset{\text{Solvent}}{\bigotimes\!-\!OH}\right] \quad \left[\underset{\text{Solvent}}{OCN\!-\!\bigotimes}\right] \quad \begin{array}{c}\text{Amine}\\ \text{Catalyst}\end{array} \longrightarrow \left[\bigotimes\!-\!O\!-\!\overset{\overset{\displaystyle O}{\|}}{C}\!-\!\overset{\overset{\displaystyle H}{|}}{N}\!-\!\bigotimes\right] \quad (3.12)$$

Total binder is typically 0.8 to 1.75% (based on sand or BOS). Part I is the polyphenolic resin and Part II the polymeric MDI. Part II is hydrophilic and needs to be protected from moisture. Although the binder was designed to operate with equal Parts I and II, many foundries use more Part I, often from 52.5 to 60%, as Part II contains most of the nitrogen. According to Ashland, Part II tends to cause lustrous carbon defects in ferrous and nickel castings [56]. The defect appears as wrinkled skin often following flow marks. It is caused by high quantities of carbonaceous decomposition products. These alloys normally need a reducing atmosphere, so this amounts to too much of a good thing. Naro recommends trying shorter, deeper ingates and rapid mold filling with nonturbulent flow, failing these higher pouring temperatures, 2% red Fe_2O_3 with an additional 0.1 to 0.2% binder, or binder solvents made of biodiesel [66]. Red Fe_2O_3 is said to release oxygen into the mold cavity to form carbon monoxide rather than lustrous carbon. Hematite also improves casting surface finish, provides chill, and reduces gas and sand expansion defects.

The amine catalyst, Part III, is selected to fit work/strip times into foundry operations. Fine-tuning is accomplished by adjusting the concentration. Part III is normally between 2 and 9% of the Part I resin. If the addition falls outside this range, a different amine should be used. As the amount of catalyst is very small, the delivery pump should be accurate to ±1%. Improved mixing is achieved by introducing the catalyst into the Part I delivery line just ahead of the discharge into the mixer. The nitrogen content of the binder is between 3.0 and 3.8%, or 0.04% of the sand, at the upper end of the range of low-grade furans.

New formulations and amine catalysts with smaller environmental footprints are constantly being announced and marketed. There are also similar binders, the phenol-free polyether-polyol-urethane and a special low VOC version both used for nonferrous castings. The polyether has as Part I a polyol resin, Part II an MDI isocyanate, and Part III an amine catalyst. The hydroxyl groups are distributed on nonaromatic chains rather than phenol. Both Part I and Part II are hydrophilic and must be protected from moisture during storage and use.

The phenolic urethane nobake being less sensitive to conditions has the following advantages:

1. Suitable for ferrous and nonferrous metals.
2. The mix is flowable and workable almost to the end of the working time but mild compaction is necessary to achieve good strength. The delayed cure means there is no loss of strength even if the full working life is used.
3. There is greater productivity and pattern turnaround as a result of the short period between work and strip times, and the periods are easily adjusted.
4. Phenolic urethane can be used with almost any sand, however, as the catalyst is alkaline low, ADV are preferred, but adjustments to the binder can be made even for mechanically reclaimed sands with high ADV.

5. Aqueous coatings can be applied immediately after stripping as long as they are oven-dried immediately; coatings that light-off should be delayed 10 minutes [56].

6. Sand is easily reclaimable by mechanical means as the coating is relatively brittle. Thermal reclamation requires one million BTU/ton of sand to remove the resin and its residues from the sand. The binder supplies much of the heat.

Disadvantages are:

1. Sand temperature should be controlled within 80 to 90°F (27 to 32°C), and moisture less than 0.2% in winter and 0.1% in summer, otherwise the reaction is delayed.

2. Stripping time is critical, and the mold or core must be stripped while it is still a little plastic, otherwise breakage and sticking occur. Patterns need to be kept clean and release agents should be used. A paint mitten over a rubber glove works well to remove sand and apply release agent. Wooden patterns should be impregnated with the urethane release agent to stabilize the wood and not painted or lacquered as all three parts will dissolve them.

Furan Warm Box

The furan warm box process uses the same coating and production equipment, the same procedures and techniques as the furan and urethane hot box processes (not covered). These hot box processes use ammonium salts rather than the acids in the nobakes. The salt decomposes once exposed to core box temperatures of 400 to 500°F (205 to 260°C) into ammonia and the same acids as in the cold box processes [54, 59, 65]. The additional temperature increases the rate 50,000 times, enough to finish the reaction as the sand approaches the core box temperature, 20 to 40 seconds. Although a fraction of dissolved salt in the mixed sand can initiate the same reactions as the various nobake processes described earlier, the reaction is slow, but proceeds enough to shorten bench life, particularly if the hopper above the core blower is not shielded or cooled. The hot box processes use 1.5 to 2.0% resin (BOS).

The chemistry of the furan warm box process is quite different. The resin component contains very little water, 70% furfuryl alcohol, and may be modified with small amounts of urea-formaldehyde or phenolformaldehyde. Urea is the source of the nitrogen, which can be up to 2.5%, but free formaldehyde is not present. The catalyst is usually a copper salt of toluene sulfonic acid and some copper chloride dissolved in water and methanol. Just as in the hot box process the copper sulfonic acid breaks down into the acid, but at a lower temperature than the ammonium salt. There are several sulfonic acids that can be used to vary the cure speed, bench life, through-cure, and resistance to humidity during storage. Bench life is typically about 8 hours. The mixed sand should not be handled, especially the catalyst. The off gases being low in phenolformaldehyde result in improved air quality. The reaction between the resin and catalyst is quite violent, so pure materials should not be mixed. The best temperatures for the core boxes are between 360 and 400°F (180 and 205°C). The resin has low viscosity, and 1.0 to 1.5% (BOS) is used depending on the sand type and condition, core configuration, metal to be poured, and shakeout collapsibility required.

The warm box process uses new sands with low moisture and low ADV, although the latter can be compensated for by increasing the amount of catalyst. Mined natural olivine with its high calcite content should not be used. Any type of mixing system can be used, although blending may be a little more difficult without pumps to add the ingredients slowly. The furan warm box is a pleasure to use.

Advantages:

1. The lower resin content means lower gas emissions and less chance of lustrous carbon. Consequently iron oxide powder feeds on the continuous mixer are not required.

2. The excellent sand flowability allows for a significant reduction in blow pressure (compared to hot box) and potentially larger blow tubes eliminates riffling of sand into blow box cavities. The excellent flowability means less tool wear and high bulk densities that do not require coatings.

3. Although resistance to high relative humidity is excellent, strength, though higher, still falls off starting at about 60% relative humidity (Fig. 3.72).

4. Lower core box temperatures reduce energy consumption, improve operating environment, and cause less distortion and warpage in tooling.

5. No free formaldehyde and no formaldehyde emissions from core boxes and spent sands meet present EPA standards.

6. Short cycle time with no post curing.

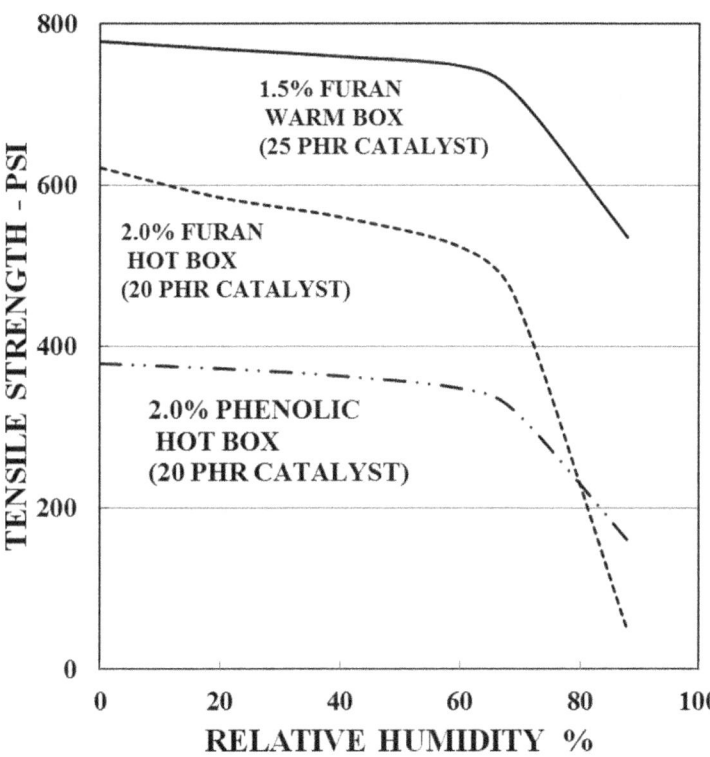

FIGURE 3.72   Resistance to humidity of several binders. (From Carey [54, 59].)

The major disadvantage is the relatively high cost, 2 to 3 times for the resin and 3 to 7 for the catalyst compared to the hot box. However, if the Shell Process produces a solid core because of the design, then warm box may be justified.

Gas-Cured Binder Systems

Over the years several resins have been developed that allow cores and molds to be used immediately after production and cleaning. For example, a gaseous amine can be used with the original phenolic urethane resins rather than a Part III that contains a solution of an amine. The advantage is that any weight of sand can be cured almost instantly, with excellent dimensional accuracy and stability of molds or cores and lower labor and energy costs. The disadvantages are that the amine has to be heated slightly to vaporize it, mixed with air, and as the amine has a strong odor, the core or mold has to be flushed to remove excess amine and the gases chemically cleaned. The process requires a vaporizer and scrubber system. Core blowers or shooters are the most common production method with PUCB (phenolic urethane cold box) sand mixes. Core boxes can be top blow, edge blow, vertically or horizontally split, and filled at pressures of 35 to 45 psi [54, 60, 65].

Although almost any material can be used for construction of core boxes, iron holds up the best as it resists release agents, cleaning solvents, scratching, gouging, and erosion from the sand. A variety of different vents can be used, but the preferred type is a wire screen mesh of 25 to 30 mesh to retain the sand, while passing the most gas. The combined area of all exit vents should be 70% of those of the input blow tubes. Strategic placement of vents is necessary for the catalyst gases to pass through all sections, and most importantly to achieve a uniform cured sand permeability.

The CO_2 process has been modified in 1968, with the addition of an ester to the water glass. This binder allows additives to improve shelf life, and minimizes casting defects while offering superior surface finish. Carbon dioxide only needs good ventilation and no nearby pits to collect in. A CO_2-cured alkaline phenolic cold box binder is also available that can be hand-cured or blown.

There have been a number of SO_2 processes, the first, or Sapic Process in 1975, had a number of issues [60, 65]. It was later modified to consist of an epoxy resin and a peroxide. The current system uses two resins, an epoxy (as in the adhesive) and an acrylic (the ingredient in a shatterproof clear plastic that can substitute for glass). This binder system has the highest strength, the longest mixed bench life and the best casting properties. The two resins can be mixed in different proportions to tailor core making, handling, and casting properties. Part A is an epoxy resin with an organic hydroperoxide. Part B contains acrylic resin, epoxy resin, additives, and solvents if necessary. Upon exposure to SO_2, the acrylic cures rapidly by a free radical mechanism with the SO_2 acting as a catalyst. This gives the system early handling strength. The epoxy cures as it absorbs the SO_2, and the two react with the SO_2 to become a solid mass. The epoxy reacts slowly enough to get a good release and ultimately high hot strength. Binder levels are from 0.5 to 2%. High hot strength formulations are used for iron and steel which avoid veining and erosion type defects. Aluminum foundries require a formulation with better shakeout characteristics. Generally, SO_2/N_2 mixtures are used as this lowers the amount of SO_2 used and facilitates purging the SO_2 left in the core. The flushed gases have to be scrubbed. The used sand can be reclaimed mechanically, although many prefer thermal reclamation as it lowers the LOI, but requires removal of SO_2 from off-gases of reclaimers.

Table 3.26 shows a brief comparison of the performance of some of the organic binder systems. This concludes the section on organic binders. Only some of the more

Process	Binder Level	Tensile in psi after 5 Minutes	Cure Time 20 lb or 10 kg Core	Bench life at 90°F (32°C)	Shelf Life of Resins	Moisture Resistance
CO_2-cured alkaline phenolic	3.0%	65	75 s	1–2 h	6 mo	Good
Phenolic urethane	1.3%	150	5–10 s	1–2 h	>1 y	Good
Methyl formate alkaline phenolic	1.75%	94	10–15 s	2–3 h	3 mo	Good
Acrylic-epoxy SO_2	1.1%	140	5–10 s	Weeks	>1 y	Good
Silicate-CO_2	3.5%	43	75 s	1–2 h	>1 y	Poor

Source: Carey [54], Carey and Sturtz [60].

TABLE 3.26 Performance Properties of Common Cold Box Systems

popular systems have been covered due to space limitations. However, most of the techniques and operating issues that occur with some systems occur in others. There are still foundries using core oils for a few special jobs, however, the reader is referred to the Second Edition of Dietert or Carey [54, 60, 65 (AFS), 68].

Inorganic Binders

The principal types of inorganic binders (other than clay) are silicates, phosphates, and cement.

CO_2 Process

Although the CO_2 process was discovered in 1898, silicates have only been used in foundries since 1947, and originally as a heat-cured binder [60, 65]. The cold-box CO_2-silicate (gas-cured) system was introduced in 1968. This nobake silicate binder uses organic esters, an aqueous solution of sodium silicate, and CO_2 gas. There are no environmental or waste disposal issues. The binder is odorless, nonflammable, and can be used with all sand aggregates to make molds and cores. No noxious gases and minimal emissions are released during pouring, cooling, or at shakeout and although a considerable volume of gases are liberated, the main ingredient is steam.

Chemical companies supply liquid sodium silicates with varying proportions of silica and sodium oxide with a wide range of viscosities because properties vary with the silica (SiO_2): soda(Na_2O) ratio. Typical viscosities are between 48 and 52 Baumé (Be) and silica soda ratios (SiO_2:Na_2O) between 2.0 and 2.8 to 1. Ratios of 1.9 to 3.2 are used in other applications. Sodium silicate formulations with a wide range of chemical and physical properties can be made by controlling the ratio and water content. The curing process involves two mechanisms [Eq. (3.13)]. In the first reaction, sodium metasilicate, Na_2SiO_3, reacts with CO_2 gas to form sodium carbonate and silica gel. At this stage, the system has reached 20 to 40% of its ultimate strength. The second mechanism is the dehydration of the silica gel to form a glass-like bond [69].

$$Na_2SiO_3 + 2H_2O + CO_2 \rightarrow Na_2CO_3 + Si(OH)_4 \rightarrow SiO_2 \qquad (3.13)$$

Sodium silicate-sand mixes can be made by batch or continuous mixers. Continuous mixers need only a single pump to meter the correct amount of binder. Overmixing reduces the bench life of the sand mixture. Mixes have a bench life of up to a week (longer in a plastic bag), although higher silicate ratios shorten the life. Hoppers of mixed sand should be covered with plastic sheets or damp sacking to prevent premature hardening (crusting).

The binder level for cores and molds varies from 3 to 6% depending on the type of sand, grain fineness, and sand purity. Additional factors are the nature of the metal, the pouring temperature, and the required erosion resistance. A washed-and-dried, rounded, silica sand grain (AFS GFN 55) requires approximately 2.5 to 3% of binder. Finer sands (AFS GFN 120-140) require from 1.5 to 3% more binder. It is recommended that test cores with an average amount of binder be used when determining the amount of binder for a specific job. Initial tensile strength of cores gassed for 5 seconds with CO_2 vary from 37 to 45 psi (2.6 to 3.2 kg/cm^2) depending on the binder level. The strength increases to a maximum of 100 to 200 psi (7 to 14 kg/cm^2) after 24 hours at room temperature, partly from some dehydration of unreacted silicates and the continued gelling of the silicate. However, high humidity in the first 24 hours can reduce dehydration and strength development. Formulations containing organic additives such as sugars, starches, or carbohydrates are particularly susceptible to high humidity; and a short heating cycle is needed. When a core is hardened by carefully controlled gassing and further hardened by dehydration and polymerization during the subsequent 24 hours, good strengths are maintained over a long period of storage. It is always better to date sodium-silicate-bonded/CO_2-hardened cores, so that the oldest cores are used first.

A CO_2 system should supply approximately 1 to 2 lb of carbon dioxide (8.5 to 17 ft^3) for every 100 lb of silicate-coated sand (1 to 2 kg per 100 kg of sand or 0.53 to 1.1 m^3). In addition, the process requires a CO_2 regulator, flow meter, and timer for success and uniformity. Probes, hoods, or cups are used for gassing. Gassing with CO_2 must be done carefully as overgassing and undergassing adversely affect properties. Fiberglass tubing and lanced vents are used to distribute CO_2 to all parts of the mold. Nevertheless, there is always a tendency to overgas some parts of the core or mold to ensure that all parts are cured. The gassing operation is more critical with the higher ratio binder, as shorter gassing times are needed, they are easily overgassed, and shelf life shortened. Overgassing produces bicarbonate which is evident as white crystals on the mold surface. The following general rules apply to all gassing techniques:

1. Gassing times needed for strength development are proportional to the binder level.
2. Longer gassing times at moderate flow and pressure are superior to shorter times at higher pressures for strength development.
3. Lower silicate ratios require more CO_2 and the cores have poorer collapsibility.
4. CO_2 pressure is 20 to 60 psi (1.4 to 4.2 kg/cm^2); but 30 to 40 psig (2 to 3 kg/cm^2) works best.
5. The hardening rate is temperature-sensitive and very slow below 50°F (10°C).

Small cores and molds can be successfully produced by combining core blowers or core shooters with programmed gassing stations operating on predetermined cycles. Higher production rates are possible with two core boxes; while one box is being gassed, the other can be filled. Carbon dioxide can be injected through (1) a hollow pattern or double-walled core box, (2) a hood covering the box, or (3) a mandrel into a core. Larger cores can be cured with $\frac{3}{16}$ in (5 mm) diameter lance pipes. Holes are punched with a rod approximately 6 in (15 cm) apart. The lance is then successively inserted into each hole and the gas applied for 10 to 15 seconds. The gas will permeate and cure a section of sand about 3 in (7.5 cm) around the hole. Large cores and molds are easily gassed with a full gasket designed to cover the flask or box.

Core boxes and patterns can be wood, metal, or plastic. With the exception of wood, they should be washed and metal core box tooling can be soaked overnight. This soaking treatment simplifies cleaning blow vents. Slotted screen vents are very effective when using sodium silicate core-box tooling, but are easily damaged during removal of the hardened and encrusted sand mixture.

A Japanese variant of the process uses less binder. The packed flask is moved over a rubber pad, a chamber lowered on to it, and the chamber evacuated before introducing the CO_2 gas into the chamber. This avoids overgassing, considerably reduces the gassing time and the need for ducts and holes. The Japanese process not only uses 40% less binder, but the sand can be mechanically reclaimed for reuse in the process. The conventional process introduces higher levels of sodium, which would poison the cure with rebonded sand [70].

Sometimes additives are necessary and vital to the success of the casting. Additives such as sugars are used to improve shakeout or collapsibility. They are sold with the silicate binder, but can be added separately to a sand mixture. However, reducing the binder level or using a higher ratio silicate can improve shakeout. The use of additives in the silicate mold and core mixtures will not remedy poor formulations or faulty procedures. Additives seriously lower bond strengths, but additives are recommended to

1. Improve collapsibility
2. Control expansion
3. Prevent burn-on/burn-in and metal penetration
4. Promote peel of castings and improving casting finish
5. Improve flowability of the sand mixture
6. Create the proper mold atmosphere
7. Prevent excessive drying of the sand mixture
8. Prevent sticking of sand mixture to patterns and core boxes

Iron oxide, plumbago, and polymers are added to silicate-bonded sands for copper-base alloys.

Phosphate Self-Setting Binder

Phosphate binder systems consist of liquid mixtures of monoaluminum phosphate and orthophosphoric acid which are hardened by reacting with an addition of magnesium oxide (modified with zinc in Part A or B). This inorganic binder was developed by Ashland (known as ASK since 1993) as an alternative to sodium silicate and the demand for more environmentally friendly systems [65, 71].

Recommended binder levels are 2.5 to 3.0% for molds and 3.5 to 4.0% for cores. The amount of powdered Part B oxide is from 18 to 35% of the binder mix, depending on sand temperature and impurities, mixing efficiency, and the desired work life. The powder should be added by positive-type conveyors. If a continuous mixer is used the blade angles should be adjusted to improve mixing time (reduced to 2/3 of rated capacity). Varying the Part B from 18 to 35% [of a total 3.5% binder (BOS)] causes the work time to go from 18 to 2 minutes and the strip time from 65 to 8 minutes at 75°F (24°C). The reaction is strongly exothermic, so sand should always be present. As long as the powder is kept dry, the only storage issue is the acid in Part A. The binder will bond any sand. However, there is a reaction with zircon, so shakeout will be more difficult. As mined and crushed olivine contains calcite, more binder will be required. Sand can be reclaimed by mechanical methods.

Portland Cement Molds

Cement has been used as a foundry binder since the turn of the nineteenth century. In the 1930s the Randupson cement-sand process was developed in France. This binder material is used in pits primarily to cast steel ingot molds and by the ship-building industry. It is also used to cast bronze propellers up to 20 tons. The advantages of cement-sand are that it resists erosion from molten metal, has high permeability and evolves only a small amount of gas. The cement does not cause surface contamination of steels and is inexpensive, but has limited tensile strength. Coarse sands (AFS GFN 40-46) are used. Compression strength can reach 160 psi (11 kg/cm^2) within 12 hours, 200 psi (14 kg/cm^2) after 24 hours. Additives such as molasses, calcium chloride, and carbon dioxide are used to increase setting time, whereas polyoxyethylene improves fluidity. Samples lose strength rapidly above 370°C. If accurate surface contours and overlapping blades are required, then the pattern will have the upper side of one blade and the lower side of the adjacent blade to successively make sections of concrete which must be removed to free the pattern and replaced to assemble the mold. The Portland cement requires a mold life of the order of 5 to 7 days to contain and feed a solidifying bronze casting. The advantages are very low cost, the binder is not affected by sands, the energy requirements are minimal other than the energy originally required to make cement, and it has no disposal issues. The disadvantages are the slow cure, poor shakeout, high cement and water additions, poor stripping (rigid cure), and difficult reclamation. Smaller propellers can be similarly made in chemically bonded sand molds where excessive hot strength and mold life are not required.

Commercial Portland cement is a mix of lime, silica, alumina or hematite, and gypsum. Tricalcim silicate {$3CaO.SiO_2$} and dicalcium silicate {$2CaO.SiO_2$} in Portland cement hydrate on addition of water into a gel of calcium silicate (tobormorite gel) and lime:

$$2\{3CaO.SiO_2\} + 6H_2O \rightarrow \{3CaO.2SiO_2 + 3H_2O\} + 3Ca(OH)_2 \qquad (3.14)$$

The compounds of CaO, Al$_2$O$_3$, and Fe$_2$O$_3$ with silica undergo similar hydrations. The calcium hydroxide reacts over the next week to form insoluble CaCO$_3$ [72].

Plaster Mold Casting

Low-melting-point nonferrous alloys (i.e., bronze sculptures) can be cast in plaster molds consisting of gypsum and other modifiers. The molds have an excellent surface finish, good detail, and better dimensional accuracy than sand castings.

Plaster of Paris (CaSO$_4$. ½ H$_2$O) or gypsum (CaSO$_4$.2H$_2$O) is blended with water to form a slurry, which is poured onto a pattern where it solidifies as a decahydrate. Plaster formulations are usually mixed with large amounts of water, between 120 and 180 parts water by weight to 100 parts of the dry mix (Fig. 3.73). The slurry thus formed is poured over the pattern (sometimes with vibration to lessen air bubbles) and allowed to set (15 to 20 minutes for metal casting plaster, less if contamination occurs). The mold and pattern are then separated (often with the aid of compressed air), and the mold heated to remove free and chemically combined water. Temperatures between 400 and 1400°F (200 to 760°C) are used for periods that depend on air circulation in the furnace, the

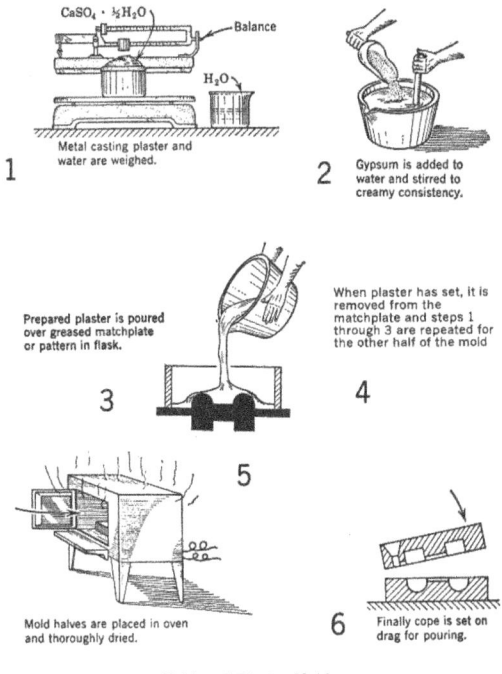

Making A Plaster Mold

FIGURE **3.73**   Basic production steps to make a plaster mold. (From Metalcaster's Reference & Guide [2].)

weight, and cross sections of the mold. Dehydration is necessary to completely remove moisture, as the presence of any residual amounts in contact with liquid nonferrous metals could cause gas-related defects. Different formulations are available, because unadulterated calcium sulfate is too friable after dehydration. The higher temperatures used for dehydration cause mold shrinkages between 1 and 2%. Similarly, the mold strength is lowered. The molds have relatively good strength, dense, but impermeable. Vacuum or pressure assistance can be employed to assist filling the mold cavity. The plaster is relatively insulating, so thin sections can be made. Fibrous or refractory materials (i.e., sand) are usually blended with the plaster. A common formula for casting molds consists of 70 to 80% plaster of Paris and 20 to 30% other minerals.

A process used in plaster molding for commercial production of copper-base alloy castings is the Capaco process, which utilizes conventional, reusable patterns. This process is highly mechanized, automatic, and continuous in operation, resulting in closely timed operations. The plaster formulation consists of approximately an 80/20 weight ratio of gypsum to fibrous mineral. This process requires a high weight ratio of water/plaster; generally, 140 to 160 parts water to 100 parts plaster. The high water content serves to improve mold permeability after dehydration.

Frothing agents may be used with vigorous stirring to promote the encapsulation of air to produce a foamy suspension of air and plaster with water; which when it solidifies produces a permeable, cellular structure. Molds containing 50% by volume air bubbles are possible. The AFS permeability can be raised from 1–2 to 15–30 units. However, as strength is decreased, flexible rubber patterns may be required to extract complex shapes. The rubber molds slip into a plaster form for support.

The proprietary Antioch Process uses a molding compound formulation: 50% sand, 40% gypsum, 8% talc plus other ingredients. The molds must be specially treated by autoclaving under steam pressure to control the crystalline structure and properties of the gypsum cement. The granular structure provides the permeability. Molds are denser than the foamed plaster, but weaker than conventional plaster molds. Although the final strength is lower, the initial cure has sufficient strength that molds can be stripped from permanent patterns, particularly if they can be laid flat.

Some of the newer plaster formulations can speed up the investment casting process (see the following section). After applying a mold wash to expendable wax or plastic trees, they can be inverted in a simple container, and the plaster slurry added. Vents (to overcome the lack of permeability), risers, ingates, sprues, and pouring cup should be part of the tree. The molds are then heated to burn out the wax or plastic, with the furnace programmed for burn out and extraction temperatures. In this way, copper alloys can be poured into a hot mold to improve filling thin sections.

Advantages of plaster mold casting are:

1. Accurate reproduction of fine surface details and smoothness.

2. Superior dimensional accuracy within 1 mm with larger castings to 0.4 mm with small castings, double across a parting line. Savings in machining are often more than sufficient to compensate for the higher cost of mold materials and processing of plaster molds.

3. Large-scale production of small detailed castings is possible. Multiple cavity molds result in more efficient production line operation.

4. High refractory value of plaster means smaller gating systems and lower pouring temperatures.

5. Uses wood, plastic, rubber, or metal patterns for molds and cores. Rubber molds can also be made as rapid-prototype models. Plaster casting is frequently used as a precursor to die casting until suitable dies are fabricated. Easy to use for prototyping.

Disadvantages are:

1. Molds and cores must be baked, which can cause expansion and cracking.

2. Mold strength is reduced after baking.

3. Can only be used for low melting point nonferrous alloys.

Investment Casting

The investment casting process is unique because of the way the mold cavity is created. Investment casting uses dispensable patterns which are completely surrounded or invested—that is, with no parting lines. The pattern is coated with a slurry of molding aggregate which hardens by chemical reaction. Patterns are most often made of fusible materials such as waxes, plastics, or fusible alloys (at one time, even frozen mercury was used) which are removed from the mold by melting-out. As wax has been used for centuries, the process has come to be known as lost-wax, or La cire-perdue.

The investment casting process has been used for centuries for making statues and fine jewelry. Fireclay washes invest the wax patterns produced from the original art.

Natural plant and animal specimens have also been used as patterns, in which removal is accomplished by total ignition and extraction of residual ash. Ancient artifacts of this type, cast in gold and silver as well as in bronze, have been found in the archeological remains of many early cultures. Prior to 1940, investment casting was used for the commercial production of statuary, jewelry, and dental hardware (crowns, bridges, and inlays). The molds for small, intricate castings were frequently centrifuged, or spun, at low speeds to assist the filling of the mold cavity.

Investment casting was industrialized during World War II to produce intricate, precision parts requiring minimal machining. Since then, the technology of the process, materials, and its automation, have improved to the point that investment casting is now an important commercial process, especially for the casting of refractory alloys.

There are two variants of the process: (1) investment flask casting, where a solid mold is enclosed in a flask; and (2) investment shell casting, where a pattern is enclosed in a shell of molding aggregate. In the latter, the shell must be strong enough to permit removing the pattern material and then casting. This method saves mold materials but is only used for smaller parts. A flow chart is shown in Fig. 3.74.

The steps for both variants of the investment casting process are

1. Production of heat-disposable special waxes or plastic patterns.
2. Assembly of patterns, the pouring cup and gating system into a tree.
3. "Investment," or coating the pattern assembly with ceramic to form a monolithic mold.

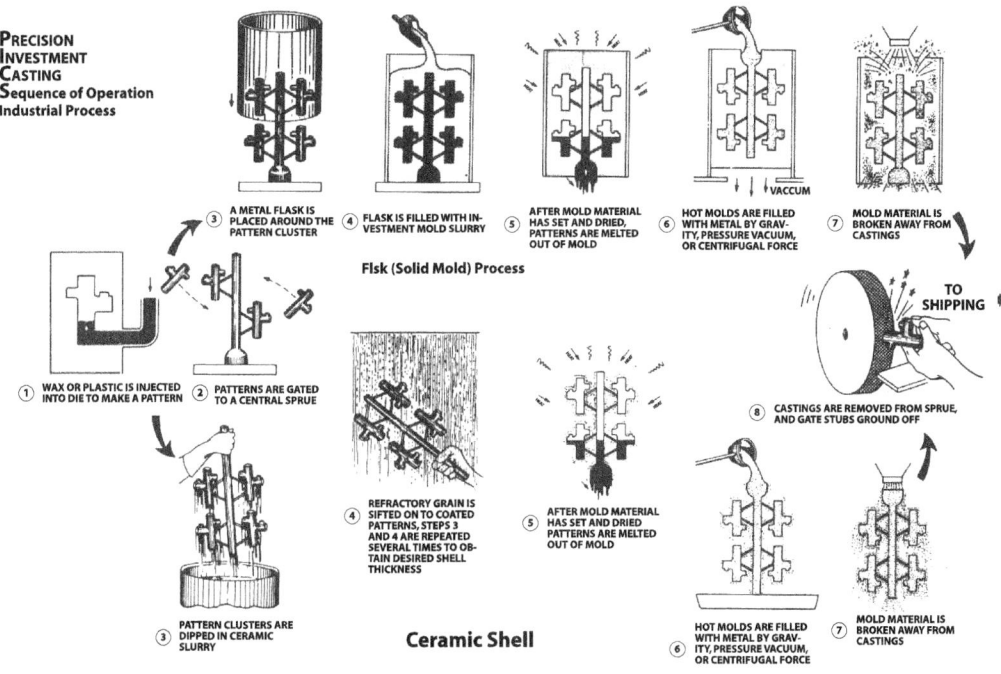

Figure 3.74 Basic production steps for investment casting methods. (From Metalcaster's Reference & Guide [2].)

4. Melting and removal of the wax.

5. Firing of the ceramic mold to burn out the last traces of the pattern material, developing the high-temperature bond, and preheating the mold to the final temperature for pouring which is often about 800°F (or 500°C) below the pouring temperature and can be critical to the success of the process.

6. Pouring by gravity, vacuum, centrifugal, etc.

7. Knockout, removal of rigging, finishing, and inspection.

Although patterns can be hand poured, more satisfactory finishes are produced by hot injection with less surface bubbles into relatively complex, expensive steel dies. The injection machines vary in their degree of mechanization, the simplest manually operated and the most complicated fully automated. Similar to die casting and permanent mold casting, the cost of tooling is a critical factor when selecting the process, as this cost must be amortized against production within a reasonable and economic time period.

The dies for the expendable pattern must have three shrinkage allowances for (1) the anticipated dimensional changes of the wax, (2) drying and firing the mold, and (3) the casting. Patterns are cleaned, inspected, and then joined to gates, risers, vents, runners, sprue, and pouring cup using heated tools resembling electric soldering irons. Cores are of a soluble wax or ceramic materials. The "tree" or "cluster" thus produced may be comprised of dozens, or even hundreds of patterns, depending on their size and complexity. These assemblies are ready for investment.

In the investment flask casting process, the pattern assembly is placed in an inverted position on a pedestal and surrounded by a flask which is sealed against the pedestal. As with ceramic molding, the pattern is often coated with a fine refractory wash and dried. Slurry is then poured slowly into the flask to invest the pattern cluster, with vibration or low vacuum applied to dislodge and remove any air bubbles. The mold aggregate is then left to harden chemically.

In the investment shell process, the tree is first dipped into the slurry to form a thin, continuous bubble-free coating. The wet assembly is uniformly covered with a dry, granular refractory layer (stuccoing with silica, zircon, or aluminum silicate sands) either by dusting or fluidized bed. The assembly is alternately dipped and dusted until the desired shell thickness is obtained. This process is frequently automated; see robot in the EPS Process. The ceramic shell is then allowed to harden and dry, although intermediate drying steps are often required. The thickness depends on the casting weight and the metal to be poured. The shell-building technique should minimize mold shrinkage and cracking.

Once the coating is dry, the expendable patterns are removed by "dewaxing," a term applied whether the patterns are wax, plastic, or more recently foam. The entire assembly is heated and wax drained from the mold. The shell is then heated to vaporize and combust any residual material and cure the refractory coating. Rapid expansion of the pattern material during the early stages of dewaxing can cause cracking of the shell-type mold which, at this stage, is still relatively weak. The temperature is ramped up slowly to minimize problems, or dewaxing is preferably done under pressure in an autoclave (sometimes with steam). Molds are slowly cured at incandescent temperatures.

Molds are cooled to a lower temperature for pouring. Many metals benefit from a mold temperature around 800°F (500°C) below their pouring temperature. Once solid the ceramic is broken away and gating removed. Vapor-blasting with fine abrasive effectively removes adherent ceramic without destroying casting finish and accuracy.

Advantages of investment casting are:

1. Intricate, complex thin-walled castings, excellent smooth surface finish (40 to 125 μin).
2. Linear tolerance of ±0.005 in (±0.13 mm) for dimensions in the first inch, plus ±0.005 in for each additional inch. Better tolerances are possible, depending on alloy and the part's configuration.
3. Minimal machining requirements.
4. Ability to cast superalloys, including titanium.
5. Recently 3D CAD models are used to "print" three-dimensional photo-polymers, thermo-polymers, or other materials. These prototypes are attached to a gating system.

Disadvantages of investment casting are:

1. Long lead time.
2. Very high pattern costs and mechanized equipment.

Ceramic Molding

There were originally two proprietary techniques developed to make expendable ceramic molds:

1. The Unicast process is a true ceramic molding process.
2. The Shaw process employs ethyl silicate and calcined clay slurries.

Unicast Process

The process uses calcined clays fired at high temperature to release oxides. These oxides are blended with water to form a slurry, sometimes with ammonium or calcium phosphate to bond the refractory oxides. The slurry is then poured over the pattern. However, the preferred method is a dry processing method in which molds are made with a clay containing between 4 and 9% moisture in dies under a pressure of 1 to 10 ton/in^2. The molds are then stripped and fired between 1660 and 2400°F (900 to 1320°C). This variant requires more expensive dies and equipment than the Shaw process.

Ethylsilicate Slurry or Shaw Process

The Shaw process, a U.K. variant, blends coarse and fine sillimanite, hydrolyzed ethyl silicate, and a liquid catalyst or gelling agent into a slurry. Sillimanite is one of the clays, which is calcined for this process. The defining characteristic of the Shaw process is the gelling agent that is added to the slurry within a minute of pouring the slurry over a pattern. The amount controls the gelling time. In a matter of minutes, the flexible mold can be stripped from the pattern. The molds, flexibility aids in the stripping from the pattern. The molds are immediately flamed with a low-temperature gas flame. Flaming causes a network of craze-cracks, which gradually progresses throughout the mold. This produces a mold with excellent gas permeability that is partially immune to thermal shock. Assembled molds are fired at red heat for about $\frac{1}{2}$ h/in of mold thickness. This further strengthens the bond and produces an inert, gas-free, erosion-resistant, collapsible mold, which can be poured hot or cold.

This method may be used for castings of aluminum, magnesium, brasses and bronzes, cast iron, carbon steels, low and high alloy steels, and superalloys.

Most pattern materials (wooden, metal, plaster and plastic) can be used in the Shaw process. Aluminum and brass patterns produce castings that are the most accurate and smooth. As the molding materials are poured over the pattern rather than being rammed against it, there is negligible wear on patterns. In this respect, the Shaw process is analogous to the Capaco process for plaster mold casting, which was described earlier. Low pattern costs permit design changes to be made quickly and very inexpensively. Conventional patterns with parting lines reduce the dimensional accuracy across the parting line.

The advantageous features of the Shaw process include the following:

1. Very thin sections can be cast, as molds tolerate high mold and pouring temperatures.
2. Castings with excellent surface finishes 80 to 125 μin.
3. Dimensional tolerances of ±0.003 to ±0.015 in (±0.08 to 0.4 mm) depending on section size and configuration.
4. Faithful rendition of detail with no draft requirements; design complexity far exceeding that of conventional sand castings but not exceeding die casting.
5. Low pattern costs, significant for low production numbers.
6. Simplicity of pattern construction and molding permit short lead times.
7. Adaptable to casting sizes up to 100 lb (45 kg) or higher.
8. Due to the inert and erosion-resistant properties of the molding materials, castings are unusually clean and free from gas and slag defects—assuming best melting techniques.
9. Minimal machining required.
10. Could be used for prototype work.

While the Shaw process has advantages, it cannot compete with investment casting for long runs for castings less than 5 lb (typical for investment castings). Casting tolerances are inferior to investment castings, particularly across parting lines. The following designs tend to be undesirable for production with the Shaw process:

1. Thin, flat, rangy components
2. Designs necessitating large print areas
3. Designs requiring deep mold pockets

The nature of the Shaw process is such that high material and labor costs are incurred. The use of exothermic materials helps to reduce these costs and improves casting quality. An evaluation of the advantages and limitations of the process indicates a definite place for it in the precision casting field.

Ablation Casting

During ablation casting the mold wall is removed (ablated) and coolant applied to the surface of the casting. An aggregate is bound with a binder that is soluble in the cooling media that is also used to extract the latent heat of solidification, a sort of ceramic molding process with a casting variation. In this manner, the crystal structure (size, dendrite arm spacing, orientation, etc.) of the metal casting is controlled during solidification and thereby mechanical properties. The molds are filled with molten alloy by gravity or counter-gravity methods as quiescently as possible. The coolant/ablation media is then directed at the mold in such a manner so that the casting solidifies from the impact point upward (as coolant runs down). A feeder can be positioned at the top to feed the casting, and should be the last to freeze. Magnesium and aluminum castings produced by this process have sound, fine-grained structures with excellent to superior properties.

As the ablated mold cannot support the casting, some other support mechanism is required. Production rates will be limited by the availability of the equipment. Additional information on the process is given in Chapter 7 on aluminum (see Fig. 7.38), and a casting is shown in Fig. 7.39. The proponents of the process claim the process to be environmentally friendly with no emissions. However, depending on the binder, large volumes of aqueous solutions may require extensive treatment, not to mention problems that large volumes of steam would create in a typical foundry environment (humidity, heat, and condensation in the foundry or baghouse). Drying and resizing of aggregate may present further issues.

Loose Sand Processes

Vacuum or V-Process

The V-process was invented in Japan in 1971 and was only available for license. The equipment consists of a rudimentary sand system with transportation, film heater, vibrating table, pattern carrier, and flasks, although automated systems are available for increased production rates. The pattern is perforated with small holes at regular intervals and mounted on the reinforced pattern carrier which acts as a vacuum chamber and is supported by the vibration table. Thin plastic film (0.003 to 0.008 in thick) is heated until it becomes elastic. The softened film is lowered over the pattern just as a partial vacuum is applied, and the film takes the exact contours of the pattern. A flask is placed over the pattern and loose sand is poured in with vibration. The sand surface is struck off and a cold plastic film placed on top. Vacuum is applied to the flask and atmospheric pressure squeezes and makes the sand rigid. Vacuum to the back of the pattern is released and the mold stripped. The process is repeated for the second half of the mold. The gating system is part of the pattern, and once the two halves are assembled, pouring cups, etc., can be added above the top film with care and holes patched immediately. Vacuum must be maintained to both halves the entire time, and only released after pouring and metal solidification, whereupon all the sand falls into a grate to be returned to a sand silo. Handling very large molds with long flexible vacuum tubes attached can be difficult. The vacuum system with fail safe valves can usually only handle one mold at a time, so production rates are low and limited by the available sizes of the pattern carrier and flasks. Typical castings are steel valve bodies, cast iron bathtubs, ship anchors, mill liners, and large frames.

Advantages of the V-process are:

1. Most metals except magnesium and titanium can be cast with thin and thick sections.

2. Patterns are not worn from sand contact, zero degree draft as plastic lubricates stripping.

3. Castings with fewer imperfections, good metal fluidity, no gas defects and less scrap.

4. Good dimensional accuracy, with tolerances of ± 0.10 in (0.25 mm) for the first 6 in (152 mm) and ± 0.002 in/in (0.05 mm) thereafter. Add ± 0.020 in (0.05 mm) across the parting line. Cross sections as small as ⅛ in (2.3 mm) can be achieved. Minimal fettling and machining.

5. Fine surface finish of 125 to 150 RMS.

6. Next to no odor or fume, little waste, no binders, or sand testing, low cost.

Disadvantages are:

1. V-process requires plated pattern equipment.

2. Low production rates.

Evaporative Pattern Process

A casting method in which a low-density foam pattern was embedded in greensand was patented by H.F. Schroyer in 1956 [73]. This process is now referred to as the *full mold process* and nobakes are used instead of greensand. Others have invested foam in the lost wax process in the same manner as wax or plastic patterns. Foam can be embedded in ceramic molds. Others attribute the lost foam process to Prof. M.C. Flemmings in 1964 for the use of loose sand. The lost wax and ceramic processes burn out the foam, while in the other two, the heat from the advancing molten metal evaporates the foam. In one way or another, these methods use the fact that foams evaporate when heated, i.e., the foam is lost. As the term *lost foam* has been used for several processes, the more accurate name will be used. Special thin coatings are used to fill the cracks between the expanded foam cells, otherwise metal enters the cracks, they weaken the casting and they spoil the appearance. The first major commercial use was by Ford Motors around 1980. Following an uproar over the frivolous use of freon (a greenhouse gas) to expand Styrofoam coffee cups, Ford abandoned the process even though low-density polystyrene was (and is) expanded using heat and vacuum or steam. The surfaces of early castings showed the foam structure due to lack of penetration of grains by the wash, but the problem was soon eliminated.

The original foam and still the most popular is expanded polystyrene (EPS), which can be used for aluminum, although new formulations are available. Copolymers of polystyrene and polymethylmethacrylate are used for copper and some ferrous alloys to reduce carbon defects [74, 75]. Ductile iron is particularly susceptible to elephant skin due to carbon pick-up from EPS. EPS foam when exposed to heat first collapses, then melts at 320°F (160°C), starts to volatilize at 525 to 572°F (275 to 300°C), ending at 860 to 932°F (460 to 500°C) with the production of 230 to 760 mL of gas per gram of EPS at 1380 to 2370°F (750 to 1300°C) [76]. Because the thermal degradation of the foam when first contacted by an advancing metal front is complicated, a number of casting defects can occur. Foam characteristics affect interactions at the metal front, the mold filling, and consequently casting quality.

Polystyrene as produced by chemical manufacturers consists of small granules which are then expanded in hot molds, usually by intermediaries who supply foundries with the foam patterns. Foam is not used immediately as it tends to shrink slightly for up to a week before stabilizing. The final foam density is above 1.1 to 1.6 lb/ft³ (18 to 26 g/L). Foundries assemble the patterns from the various components including the gating system using a hot wire for cutting or a hot glue gun. A key issue in the casting finish is the initial polystyrene granule size, the section thickness (number of grains thick), the amount of expansion, and the bead fusion level, or the gap between the beads is such that when a coating is applied does not bridge the gap, but can leave fins (when the coating penetrates the beads) or dimples. Too high a fusion level results in a smooth surface which causes too thin a coating as excess runs off easily and leaves too thin a layer (Figs. 3.75 and 3.76). Whereas many of the coatings used for cores are fine grained refractories, such as zircon flour, that produce impermeable coatings to both metal and gas, such is not the case with coatings for EPS patterns. EPS are usually dipped in a water-based refractory emulsion, and then dried before use. The coating forms a mechanical barrier between the metal and sand, improves casting finish, and allows styrene degradation products to pass through from mold cavity into the backing sand. The commercial coatings for aluminum casting are mica-based refractories for medium permeability and silica-based for low permeability. Coatings for iron have greater permeability and are based on alumino-silicate kaolin clay. Hollow fiber sprues are sometimes used to reduce the gas volume swept into the casting cavity (Fig. 3.77).

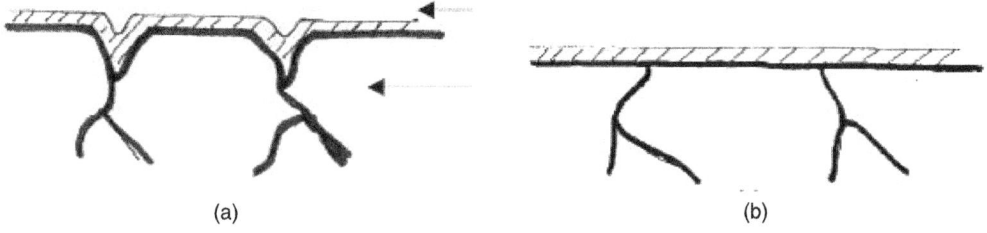

(a)   (b)

Figure 3.75   Schematic representation of foam with (a) low and (b) high bead fusion which cause coating and defective surfaces in EPP castings. (From Sahoo [74] and Bichler and Ravindran [77].)

Figure 3.76   A robot dipping an evaporative foam pattern in a wash tank. Similar robots are used for dipping investment trees. (Courtesy BRP Spruce Pine and Modern Casting.)

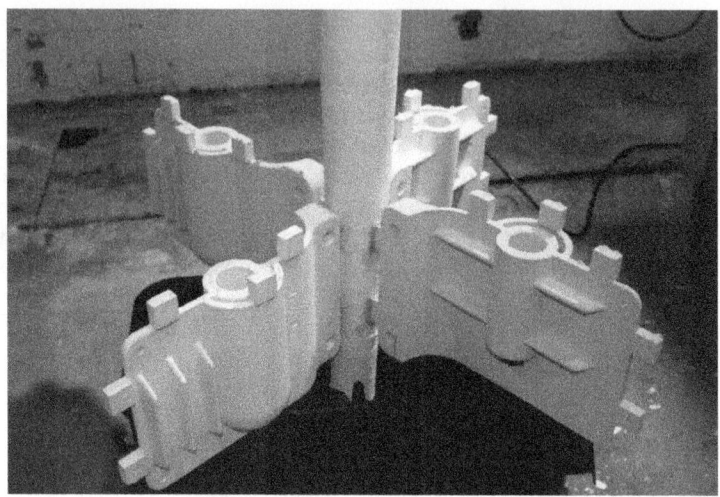

FIGURE 3.77   An example of coated foam patterns glued onto a hollow foam sprue prior to immersion in sand. The casting is a conglomerate of difficult to cast components. (From Fasoyinu et al. [78].)

The finished pattern is held in a box while the backing sand is poured around and allowed to penetrate cavities using vibratory tables. Coarse silica sand is satisfactory for most industrial metals, but as magnesium has one of the lowest heat emissions per unit volume on solidification, backing sands of low conductivity and low thermal expansion such as synthetic mullite are superior. Both silica and mullite require inhibitors for magnesium, but mullite, being less reactive, requires only potassium fluoroborate rather than a combination of fluoroborate and sulphur.

Advantages of the evaporative pattern process are:

1. Complex internal passages can be produced that are difficult or cannot be produced by other methods.
2. Linear tolerances of ±0.005 in (0.13 mm) and 60 to 250 RMS finishes.
3. Cleaning costs are low; shakeout only means rotating the casting for sand to pour out.
4. The evaporative pattern process uses loose sand which readily collapses rather than rigid cores. Good for alloys prone to hot tearing under constraint, which occurs when solidifying metal, at its weakest, forms around a rigid core.
5. Reduced machining costs.
6. U.S. Department of Energy reckoned energy savings of 27% over sand casting; setup costs are less than for die casting; and die life is 3 to 4 times longer.
7. Many complex parts can be combined into a single casting, thereby eliminating multiple machining, fitting, and assembly operations.

One disadvantage is that although the sand can be reused, foam degradation products, particularly phenolics, gradually build up in the loose sand, and occasional treatment in a fluidized bed thermal reclaimer is needed.

Cores and Molding Aids

Core binders and equipment for making cores was covered in the section on chemically bonded sands. This section covers issues specific to cores, namely casting defects, core washes, core supports, core setting and core collapsibility or knockout.

Cores as a Source of Casting Defects

Poor or defective cores should not be delivered to the molding line. Therefore, cores should be inspected before delivery, otherwise the foundry will continue to expend more work on the defect. Defects that can be caught by inspection in the core room provided there is enough light[12] are:

1. *Off-gauge or off-size* cores.
2. *Core sticke*r. Core sand that adheres to the core box will leave a blemish on the core. Rough pitted surfaces can reduce permeability and allow core gases to escape into the casting.
3. *Inaccurate core assemblies.* One or more cores are glued together to make an even more complex core which can be incorrectly assembled.
4. *Fins.* All fins and projections on a core will show up as cavities in the final casting, unless they are cleaned off in the core room.
5. *Missing vents.* Shell cores should be hollow, or cores that are drilled out to vent gases through core prints may be incomplete, causing gases to blow through the casting rather than the core print.
6. *Cracked cores.* Broken or cracked cores will produce fins when liquid metal penetrates the crack. Knocked off corners can fill with metal, and cause problems if core prints fill with metal which later flows into a central core vent.

Even though good, clean, dimensionally accurate cores are delivered to the molding line, cores may still be the cause of casting defects. The core sand/binder mixture, the coating, and the need to position the core correctly and hold it in the mold before closing the flask can cause problems. When the metal is poured, thermal effects cause the core binders to decompose. Most metals, with the exception of aluminum and magnesium, are considerably denser than cores and the core will be buoyant and try to shift from its original position. Some of the casting defects attributable to cores from these sources are:

1. *Blow.* As cores give off gases, they may cause blows or streams of gas bubbles that are trapped in the casting if the metal freezes too quickly. Underbaked shell cores, damp cores (storage issues), improperly bonded sand, and improperly dried coatings are also causes of additional gas. Improper location of vents used during core blowing can cause uneven permeability over the length of a core, and allow gases to vent from the core rather than through vents in the mold.
2. *Dirt.* Loose or easily eroded sand may result in dirt defects on castings.
3. *Core raise.* This defect occurs when the core floats and moves up close to the cope. Weak cores or ones that are not properly supported with chaplets may be

---

[12]Lighting levels in core rooms are often inadequate to inspect for defects.

the cause of this defect. Undersized cores not held tightly by the core prints may also rise. If the hot strength is inadequate, then the core can distort before the solidifying metal holds it in place.

4. *Core shift.* A core can shift horizontally if it is not held by chaplets or prints or if it is not centered when originally set. The evidence of a shift is that one casting wall is thinner and the other thicker than intended.

5. *Cracked core, fins.* A core can either be already cracked or crack because of buoyancy effects. A stringer core or better support with prints or chaplets is indicated. Cracks or fins can seal off internal passages.

6. *Metal penetration.* This defect is especially troublesome in larger castings where cores can be heated to the melting point of the metal before the metal freezes. Metal can then seep into the core, the resultant mass of metal and refractory is difficult to dislodge. Soft ramming and sand with a low sintering point seem to aggravate the defect.

7. *Core wash-out.* A core can wash (erode) when the surface collapses too quickly. Refractory coatings or more binder may help. Additives that increase hot strength such as silica flour may help resist washing.

8. *Hot tears.* A core which does not allow the casting to contract may cause the solidifying casting to rupture or tear. Too high a hot strength and excessive life in the hot distortion test for the core sand is a contributing cause. Examples of casting materials susceptible to hot tears are malleable iron, some high alloy steels, copper 85-5-5-5, Al-Cu, zinc die cast alloys, etc.

9. *Veins.* Veins on castings are fin-shaped extensions inside the core cavities. It seems that cracks in the sand core develop as the core is heated, and are filled with metal. Addition of iron oxide improves the hot strength. The problem is most often encountered in ferrous castings when a thin core is surrounded by substantial amounts of metal.

10. *Scabs and buckles.* Sand expansion defects of the type encountered in molding sands can also occur in cores.

11. *Crush.* A crush is usually a defect in the mold which is caused during core setting or closing the mold. Sand from the mold is displaced by the core. If the core is oversized, when the mold is closed, the core prints will be enlarged by the core, and this can flake sand off the walls. The defect appears as a depression on the surface of the casting where the mold wall was crushed into the mold cavity. The problem is caused by lack of proper care when closing the mold, or lose or worn guide pins.

12. *Fissures.* Fissures appear as rough, grainy-looking masses attached directly to the cored surface of the casting. They appear in areas where core sand has collapsed and been pushed aside by metal.

Core defects are related to partly mechanical problems and partly problems with core-sand properties. Sand is particularly erosive, and is hard on moving parts. Maintenance and inspection of wear locations of machinery needs to be done regularly. Control of core sand properties and checking of programmable controllers should be almost a daily task. Dirt, blockages, and wear can easily change flow rates. Sand temperature is an important and often overlooked variable.

Core/Mold Coatings

The manufacture of coatings is a relatively new industry. Coatings are applied to molds and cores to improve the surface finish and reduce the casting defects that occur on their surfaces. To accomplish these tasks the coating must have a higher melting point than the sand, and form an impenetrable barrier between sand grains to metal ingress. The coating must adhere to sand, penetrate the surface sand matrix, resist metal erosion, lower gas evolution, and enhance peel.

A coating may have five components to maintain its consistency for application, they are (1) a refractory, (2) a carrier, (3) an emulsifier, (4) a binder, and (5) a chemical modifier.

To the foundryman, the refractory is the most important. The size, shape, and size distribution, and the chemical inertness are critical. The smaller the grains, the longer it will remain in suspension. It must also match the space between sand grains on the mold or core surface, too small and capillary action will take them too far into the matrix; too large and they will not penetrate or stay in suspension. Very fine particles cause the surface to crack as it dries. Refractories used in coatings are classified as high—magnesite, chromite, zircon, mullite; medium—olivine, alumina silica, in combination with graphite; low—silica, talc, mica, clay, coke, and combinations. The low refractory coatings are used for casting aluminum and thin section copper and cast iron. The medium refractory coatings are for heavy section copper and gray and ductile iron. The high refractory coatings are used for carbon and alloy steels. If the expansion characteristics are too different from the base sand, the coating could spall, and another component will be needed to alter the expansion characteristic. The refractory should be inert, for example, an acid refractory such as silica or alumina should not be used with a basic liquid such as manganese steel. Nevertheless, special coatings have been developed to change the microstructure of the solidifying casting. For example, tellurium can be added to a coating for gray iron to cause the casting surface to be white iron for wear resistance. Carbon can be added to change the mold atmosphere.

The carrier is normally a liquid: water, alcohols, or chlorinated hydrocarbons. Most chlorinated solvents have been phased out. 1,1,1 Trichloroethane was frequently used in early recipes and as a cleaning agent in the metals industry, but was banned under the Montreal Protocol in 1996 for its effect on the ozone layer (its isomer 1,1,2 trichloroethane is a carcinogen). Chlorinated solvents were used as sprays and dried quickly. Alcohols, usually isopropanol, are still used where volatile organic compounds (VOCs) are not restricted, as they can be set on fire to burn slowly and rid the mold/core of residual liquid. Water has a number of serious drawbacks, it is not easily dried, spoilage, wetting, penetration, the pH is variable, and the pH of emulsions may even change and affect the emulsifying agent's characteristics.

A common emulsifying agent in water-based coatings is bentonite (bentone for organic carrier-based coatings). The bentonites swell providing a gel-like network to contain the refractory. An organic binder is used to harden the coating once it has been applied. They help resist erosion. The fifth ingredients are (1) modifiers to adjust the penetration of the coating into the first couple of layers of sand grains—these are usually wetting agents to break down air bubbles caused by agitation in dip tanks, (2) preservatives to prevent biological breakdown, and (3) stabilizers to maintain the emulsion and retard separation.

Coatings come either as premixed liquids, dry powders, or in paste form. It is up to the customer to mix the binders with the correct carrier and keep the viscosity and specific gravity in the specified range for the application method within the correct Baumé. Specific gravity can be used as it can be measured quickly with a hydrometer (a simple float for battery acid, radiator fluid, or wine making, cost $40). On the other hand, viscosity is a measure of resistance to flow and for liquids can be measured cheaply using a Zahn cup, which requires the time it takes for the cup to empty to be converted to viscosity (centistokes) using an equation. Viscosity of a fluid can be extremely sensitive to temperature, whereas specific gravity is not. While the two are not equivalent, returning to the same value for the same carrier system will work (provided the viscosity is measured at the same temperature). As evaporation can be a problem, the emulsions should be monitored and controlled. To avoid contamination by impurities, pure solvents or distilled water should be added as required. The coating can be applied by dipping, spray, or if absolutely necessary, brushed. Some binder systems can be coated immediately, while others need to cure. The coating needs to penetrate two layers of sand grains. Hence, water-based coatings need to be oven dried or heated by microwave to eliminate the water. Cores are often hung on hooks to facilitate air drying using forced hot air. Organic solvents may actually soften some of the polymers if contact is prolonged. Cores made by warm, hot box, or gas-cured systems can often be coated immediately. The delay or whether baking is required as measured by scratch hardness for nobake cores and molds is given in Table 3.27.

Chaplets

The supports for cores are called chaplets, which come in a variety of shapes and sizes. Perforated chaplets (flat or curved) are made in perforated tin sheet (for cast iron), aluminum, copper, or steel sheets for the corresponding casting. Several sizes of perforation holes are available to facilitate metal flow and reduce the mass. Other forms of

Chemical Binder	Water	Water + Wetting	IPA	Naphtha	Trichlor	Trich IPA	CH$_2$Cl$_2$	CH$_3$OH
Furan nobake	At strip	1 h	15 min		At strip	At strip	At strip	15 min.
Phenolic-acid cat.					15 min	15 min	15 min	
Alkyd urethane	15 min		At strip	1 h	At strip	At strip	At strip	At strip
Phenolic/ amine cat.	Bake		1 h	At strip	At strip	At strip	At strip	15 min
Phenolic ester/gas	Bake	Bake	1 h	At strip	At strip	At strip	At strip	1 h
Sodium silicate			1 h	At strip	At strip	At strip	At strip	At strip

Notes: Water with wetting agent, IPA is isopropyl alcohol, Trich is 1,1,1 tricloroethane, CH$_2$Cl$_2$ is methylene chloride, CH$_3$OH is methanol.

No effect when the coating is applied 1 minute after stripping; 15-minute, 1-hour, or 5-minute baking at 250°F (120°C).

No entry - expect some effect after baking, and when using water or water with wetting agent expect moderate to significant effect, even when postponing casting for an hour after treatment.

Source: AFS [65].

TABLE 3.27   Effect of Coating Carrier on Cured Binder Systems

chaplets are available—double head or riveted plates, angle stem, bridge, radiator, stem, cast chaplets, etc. Some are intended to hold cores in place, others to support, and some to hold cores down once the cores' buoyancy becomes a problem. Figure 3.78 shows a variety of chaplets. Tubular and threaded inserts are also made to form holes in thin flanges or threaded holes in the casting, i.e., not to support cores. It is important that chaplets bond completely with the liquid metal before it freezes and no blows or voids form. Cleanliness and storage of chaplets is a major problem.

Larger cores or long rangy cores in the mold cavity cannot always be supported by core prints. Sand cores tend to float in all metals except aluminum and magnesium alloys. Long thin cores tend to float more readily than short chunky ones. According to Archimedes' Principle, the force acting on the core is equal to the weight of metal displaced, for example, a 1 cubic foot sand core with a specific gravity of 100 lb/ft^3 immersed in liquid iron or steel of 450 lb/ft^3, has a force on it of 100-450 or 350 lb upwards. Table 3.28 gives the ratio of buoyant force to core weight. Once the core is surrounded by liquid metal, the depth has no effect on the force. More force is required above the core to keep it down than to support it. In fact, the chaplet below a core only has to support the core itself. A core print in a greensand mold can support a load of 5 psi (0.35 kg/cm^2) without giving way. Oil-bonded cores can support 75 psi (5.3 kg/cm^2) without a chaplet cutting into them, and chemically-bonded cores even more. Table 3.29 shows some loads that chaplets can sustain in liquid iron. A chaplet with a ¼ inch stem would normally have a 1 or 1.25 inch head, and support a force of 80 lb (36 kg). This force would not damage a core, but would severely damage a greensand mold. For this

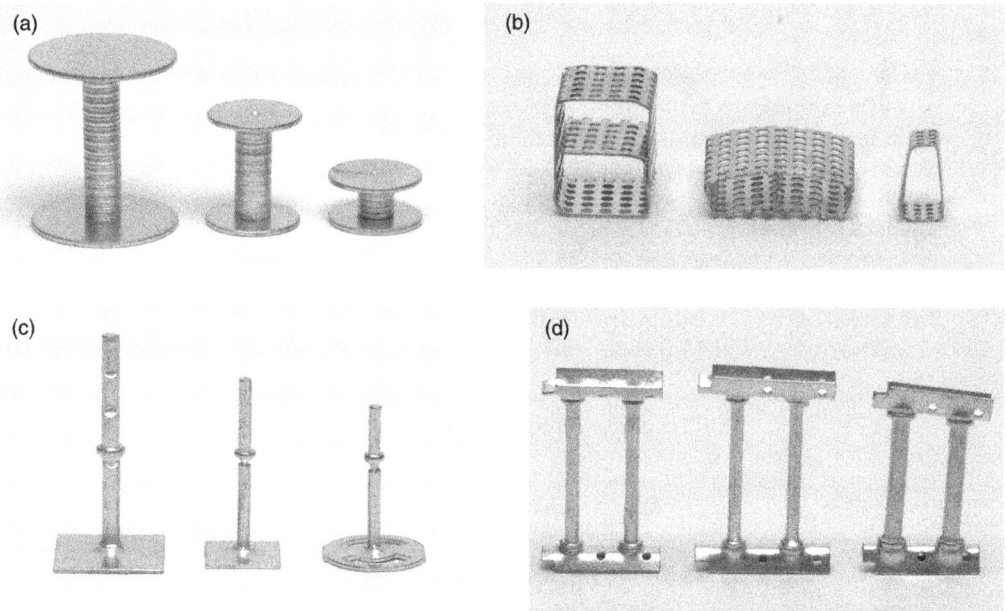

FIGURE 3.78 Examples of chaplets. (a) Double head chaplets, tin coated mild steel, (b) perforated chaplets, tin coated steel sheet, (c) radiator chaplets, and (d) separator chaplets. (Courtesy Smith & Richardson Inc.)

Metal	Ratio*
Magnesium	0.13
Aluminum	0.66
Brass	4.25
Copper	4.50
Cast iron	3.50
Steel	3.90

*Approximate as composition affects density.

TABLE 3.28 Ratio of Buoyant Force to Weight of Core

Stem Diameter, in	Double Head			Stem Chaplet		
	Head (square)	Safe Load, lb	Head (square)	Thin Metal	Thick Metal
0.067	0.375	5			
0.125	0.5	20	0.5	20	10
0.193	0.75	45	0.75	45	22
0.25	1.0	80	0.88	80	40
0.375	1.5	180	1.25	180	90
0.50	2.0	320	1.5	320	160
0.625	2.5	500	1.75	500	250

Source: Dietert [68].

TABLE 3.29 Load in Pounds Safely Sustained by Chaplets in Molten Cast Iron

reason, the chaplet should be supported by a 4 x 4 inch pad to resist the force, which must be rammed up in the mold. Perforated chaplets should be positioned with the side with greatest area against greensand, and may not require pads. The larger heads of double headed chaplets are more likely to trap gas bubbles.

To determine the chaplet size, one has to determine the core's volume surrounded by metal, assuming a density of 100 lb/ft^3 or 0.06 lb/in^3. The lift is then found by multiplying this volume by the ratio in Table 3.28. Core prints or areas where the core cuts through the mold, top or bottom, are assumed to support 5 psi of contact area. Remaining forces have to be exerted by chaplets.

Weight supported by core print (lb) = Area of core prints (in^2) × Support supplied (5 lb/in^2)

$$\text{Lift on core} - \text{Support supplied} = \text{Load on chaplet}$$

$$\text{Chaplet area needed} = \frac{\text{load on chaplet (lb)}}{5 \text{ lb/in}^2 \text{ (load carrying capacity per in}^2 \text{ chaplet in greensand)}}$$

These three equations give the size of the upper chaplet. A smaller chaplet below the core supports the weight of the core, prior to filling the mold. Custom-made cores are available, provided their run-size is cost effective.

Dietert's manual on core practice lists issues with chaplets, and recommends cores and castings be designed to avoid chaplets, as they cause leaks and blows, and leave unsightly marks. If they have to be used they should be as small and light as possible [68].

Core Setting

Core setting is the operation of setting cores in the molds. Cores must be of the correct size, and positioned properly with respect to the mold cavity so that cored out cavities are in their required location and orientation in the casting. Good core setting involves advanced planning so that cores are correctly positioned and firmly held when the metal is poured. Instructions, drawings, and even jigs should be provided for complicated castings.

Small cores can be placed by hand. Larger cores may require hoist or crane service. Sometimes a number of cores are assembled and set in at one time. This can require an assembly fixture, such as those required for automotive-motor-block cores. When a number of cores are assembled, dimensional errors are additive and some kind of fixture or gauge is often necessary. Provided molds and cores are always the same, then automatic core setting equipment can be programmed and used. Oversized cores cannot be made to fit into core prints by pressing them more firmly into today's harder molds, so dimensional accuracy is again an issue.

Core Knockout

Core knockout depends as much on the thought put into the process in advance as the effort for the speedy removal of cores from the castings with the least amount of atmospheric pollution, energy, and labor. Dietert listed several approaches to the efficient removal of cores [68].

1. *Design of casting.* The designer or pattern maker should provide ample sized openings from which the cores can be removed from the casting. The expense of core knockout is increased considerably by small convoluted openings. Many times cores crack into lumps but do not change in diameter, so larger openings certainly help.

2. *Selection of the core sand.* The foundry can also select the sand to maximize the degree to which the core degrades. Coarser sands conduct heat better than finer sands. Thus coarser sands heat up faster and the bond will degrade faster and more thoroughly leading to easier knockout. Olivine and heavier sands also conduct more heat into the sand. In the case of permanent mold casting, the castings cannot be left in the mold, so all that can be done to retain the heat as long as possible by minimizing air flow, is put them in a drum or oven. The latter is particularly advantageous if the castings are going to be heat treated, as long as the furnace atmosphere is not an issue.

3. *Selection of the binder.* The more binder, the stronger the cores and generally the higher the hot strength so the longer it takes for the core to collapse [61, 62]. Not all chemically bonded binders degrade at the same rate, some degrade so quickly and are too weak to be used for cast iron. Those designed for aluminum casting have the best collapsibility for aluminum. Shell cores, with a durability of 4 minutes in the "hot distortion test" would never reach a high enough temperature to collapse in an aluminum casting.

4. *Time casting is left in the mold.* The longer the mold is left intact, the more heat passes into the core and the further cores collapse. Although longer cooling reduces the stress in the casting, a partial heat treatment may reduce the development of properties later. Pyrolysis of organic binders starts at 400°C and is well underway at 650°C. The decomposing organic polymer binders will liberate polymer monomers and smaller fragments. For example, 1000 g of

sand core (bulk density 1.64 g/cc) takes up 600 cc with 240 cc of that being voids, if there is 1% binder and it decomposes to an average molecular weight of 100, and could occupy 7 L at 700°C. The gas doesn't come out all at once because pyrolysis is slow, but air (1/5 oxygen) molecules cannot enter the core as they would face the 7 L plus a molecule of water and one of carbon monoxide from reaction of the oxygen with any carbonaceous material. The residual carbonaceous volume would not be affected, as there would be one less –CH2– in the chain, and if the Ideal Gas Law were to apply, PV = nRT, n and hence V are unchanged.

5. *Selection of knockout equipment.* Hand tools used in knocking out cores include pointed bars, hammers, mallets, and pneumatic chisels. Mechanical tools include vibrators, bumpers, hydraulic jets, jolters, and tumbling barrels with flutes that toss the castings. Mechanical vibrators are connected to metal frames which are vibrated rapidly by an electric motor connected by an eccentric shaft. About 1,000 to 3,000 hammer blows per minute are used, often directly on the riser. However, because of the noise and vibration effects to the operator, simple rigs have evolved which clamp the casting and then vibrate it. High-frequency drive machines have been tried, but they can be self-destructive and noisy. Hydraulic units using water under 2.5 to 15 kpsi (17 to 105 MPa) have also been used to remove core oil cores. However, water recovery and erosion of equipment are a bit of a problem. Normal shakeout units and shot blast are not particularly helpful unless the core has completely collapsed. Loading the castings into an oven and trying to burn out the core is also used, however, air supply is a problem and it affects the heat treatment unless this procedure is used in conjunction with the heat treatment.[13]

Sand Reclamation

Sand reclamation has been successfully practiced for many years for the following reasons:

1. Rising costs for disposal of chemically bonded sands used in molds and cores, particularly those containing aromatic compounds (phenols, styrene, etc.), which are used once as opposed to 10 times or more for greensand

2. Rising costs of new sand and higher freight rates for delivery

3. Increasing environmental restrictions on the mining of new sands

4. Increasing environmental restrictions on the permissibility of solid waste disposal and its effects on the availability and cost of using approved landfill sites

Strong economic incentives exist for the reclamation and reuse of foundry sands, both as a means to offset higher material costs and to avoid the sometimes exorbitant expenses of waste sand disposal [79].

Although the reuse of green molding sand does rejuvenate and restore the molding aggregate to proper working condition, it does not fall within the usual concept of

---

[13]The author once tried oxygen injection on a couple of small brass valves. While cleanout was instantaneous, it cannot be recommended. Perhaps 50% air, 50% oxygen could be tried.

reclamation. Reclamation is normally applied to sand systems in which most, if not substantially all, of the sand binder material is removed from the grains. While this does not preclude reclamation of clay-bonded sand systems, reclamation is more frequently applied to chemically bonded sands. Nevertheless, as much chemically bonded core sand joins the greensand, the only way for many foundries to recover an adequate amount of sand is to reclaim greensand.

There are basically three commercial systems for sand reclamation: dry, wet, and thermal methods. Some systems may contain more than one of these basic methods, depending on foundry requirements as dictated by the binder systems in use. Figure 3.79 is a flow chart for the reclamation of greensand and chemically bonded sands, showing procedures that are common to most systems. Greensand has to be dried to complete removal of metallics and must be calcined at a higher temperature than organic binders. CO_2 and some binders respond well to just mechanical or dry reclamation. For a sand laundry operating on sands from many foundries, additional silos to segregate sand from different foundries would be required as well as a classifier after the mechanical reclaimers to control grain fineness.

Dry reclamation. Once cores and molds are mechanically reduced to small lumps (by crushing, grinding, abrasive blasting, etc.), the sand is subjected to a scrubbing operation which is designed to abrade away the adherent layers of binder on the sand grains. Scrubbing is accomplished by either mechanical or pneumatic forces which create sand-to-metal and sand-grain to sand-grain impingements. Both scrubbing techniques work efficiently with chemically bonded resins, although the pneumatic system is somewhat more effective. Both scrubbing techniques require the sand be reduced to near-grain

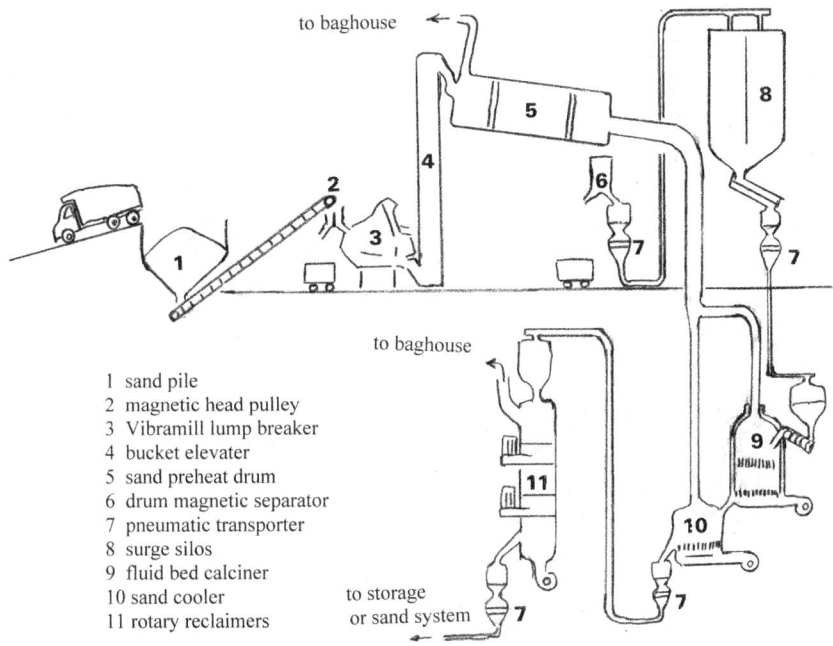

1 sand pile
2 magnetic head pulley
3 Vibramill lump breaker
4 bucket elevater
5 sand preheat drum
6 drum magnetic separator
7 pneumatic transporter
8 surge silos
9 fluid bed calciner
10 sand cooler
11 rotary reclaimers

Figure 3.79 Flow chart for clay-bonded and chemically bonded sand reclamation. (Redrawn from Whiting and Merlin [79].)

size for best efficiency and all metallic particles be removed to reduce wear on scrubbers. The systems must include dust collection equipment to remove fines, particularly since the fines contain a high proportion of the removed binder.

Figure 3.80 shows a commercially successful mechanical system for scrubbing the surface of sand grains by mainly sand-to-sand contact rather than the equipment [80]. Sand can be recycled through several times in small plants or units can be operated in serial or parallel in larger systems.

Wet reclamation. The thoroughly crushed shakeout sand is subjected to mechanical scrubbing in the presence of water during wet reclamation. As the solvent action of the water is important here; wet reclamation is unsuitable for sands bonded with insoluble organic resins. Alphaset resin is water soluble. Clay-bonded molding sands respond well to this, as the residual clay goes into suspension readily and is washed away. In spite of the relatively high water requirements of wet reclamation and the increasingly severe EPA regulations pertaining to waste water disposal, wet reclamation has been receiving renewed attention recently due to growing interest in the use of inorganic binders such as silicates. Silicates respond best to wet reclamation because of the need to reduce residual amounts of Na_2O to low levels. The use of electrolytic polymers to precipitate salts below their saturation levels makes the treatment and recirculation of the system water supply possible.

Thermal reclamation. Thermal reclamation removes organic binder materials from thoroughly crushed sand by heating to temperatures of 1400 to 1600°F (760 to 871°C). This

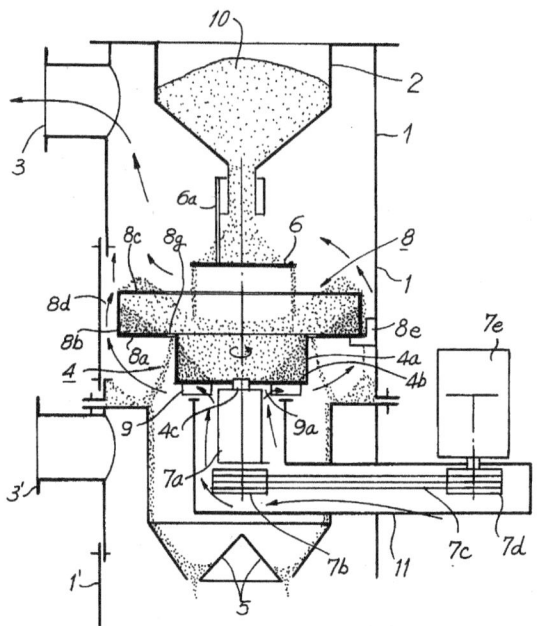

FIGURE 3.80 Nippon Chuzo rotary reclaimer. Sand is scrubbed in a centrifugal apparatus consisting of a (1) funnel to direct a measured quantity of (10) sand at 5 tonnes/h into a (4) rotating drum mounted on a (7) rotating shaft. Air flowing into the chamber removes dust out the (3) side of the unit, while reclaimed sand exits through the (5) bottom. (From U.S. Patent [80].)

FIGURE 3.81   Four tons per hour fluidized bed sand reclaimer. (Courtesy of Gudgeon Thermfire International Inc.)

may be done either in a rotary kiln, a multiple-hearth furnace, or in a fluidized bed system (Fig. 3.81). Thermal reclamation often yields sand, which is superior to the new sand entering the system. This is because the grains are not only free of residual binder but are also smoother and more rounded. Likewise, quartz grains become thermally stabilized by the treatment. Thermal reclamation is not a new procedure, having been used for reprocessing shell (phenolic resin) sands in a number of installations. Although the most expensive process to operate, thermal reclamation is the most efficient method for handling organic binders. Nevertheless, their operators can suddenly find they are in the incinerator business and need to install expensive afterburners and air recycling systems to reduce emissions. As of August 2013, it seems this might just be the case in the United States, as fairly tight regulations exist on the emissions from calciners and sand cooler/dryers as well as record keeping and monitoring of emissions [81]. In 1991, German foundries had to recirculate a large proportion of the air back through the bed to reduce emissions.

Biringuccio [29] took great pains to tell his readers in 1540 that "whoever wishes to exercise this art well and with certainty must do everything with exactness. You must always jealously and fearfully assure yourself of this as far as possible before coming to the actual casting, closing every crack with clay, and making strong … you should make certain with opportune remedies, for if you do otherwise, the deception is always paid for." And he went on.

We are still telling everyone in the molding department—
ATTENTION TO DETAIL.

References

1. R. K. Buhr, R. D. Warda, L. V. Whiting, K. G. Davis, and M. Sahoo, "The State of Foundry Technology as Measured by a Mobile Laboratory," *AFS Transactions*, 96: 171–182, 1988.

2. Metal Casters Reference & Guide, 2d ed., AFS, 1989.

3. MCDP Staff, "The Domestic Picture," *Metal Casting Design & Purchasing*, Vol. 15, No. 1, Jan/Feb 2013, pp 42–43.

4. American Foundry Society, Schaumburg, IL, www.afsinc.org/.

5. J. L. Dion, M. Sahoo, D. Cousineau, and C. Bibby, "Production of a Faucet in Low-Lead Copper-Base Alloys by Permanent-Mold Casting," *AFS Transactions*, 102: 559–564, 1994.

6. K. G. Davis and J. G. Magny, "Some Experiments on Castability and Gating in the Permanent Mold Casting of Copper-Base Alloys," *AFS Transactions*, 98: 41–48, 1990.

7. Casting Copper-Base Alloys, Editors: S. Ducharme, M. Sahoo and K. Sadayappan, AFS, Schaumburg, 2007.

8. "Copper Alloys for Permanent Mold Casting," Copper Development Association, 1998.

9. J. A. Rassenfoss, "Mold Materials for Ferrous Casting," *AFS Cast Metals Journal*, 2(4): 48–61, Dec. 1977.

10. H. H. Doehler, "Casting Apparatus," U.S. Patent No. 856772A, Oct. 18, 1906.

11. International Magnesium Association, Wauconda, IL, www.intlmag.org/.

12. *ASM Handbook*, Vol. 15, *Casting*, ASM International, 2008.

13. Worldsteel Association, Worldsteel's Statistical Yearbook 2013, Brussels and Beijing.

14. Copper Development Association, New York, NY, www.copper.org/.

15. Modern Casting, "46th Census of World Casting Production," Dec. 2012, pp. 26–29.

16. Battelle Columbus Laboratories, Energy Use in Metallurgical and Nonmetallic Mineral Processing (Phase 4) Energy Data and Flowsheets, High Priority Commodities), for U. S. Bureau of Mines PB245-759, Jan. 1975, pp. 87–102.

17. R. S. Lee, "Sand System Functions: Requirements and Specifications," Committee 80A-2 Report, AFS 80-47: 209–226, 1980.

18. J. H. Morgan, "The Cooling of Sand and Castings," *Foundry Trade Journal*, 154(3265): 777–787, Jun. 16, 1983.

19. Martin Marietta Corporation. 1978. Sand segregation. [S.l.]: Martin Marietta Industrial Sands Division.

20. R. J. Smith, "Selecting a Muller for a Green Sand Operation," Modern Casting, 80(6): 66–68, Jun. 1990.

21. R. W. Heine, T. J. Bosworth, J. J. Parker, E. H. King, and J. S. Schumacher, "Sand Movement and Compaction in Greensand Molding," *AFS Transactions*, Vol. 67: p. 47, 1959.

22. C. A. Sanders and A. G. Clem, "How the One-to-Ten Ram Test Measures Sand and Mold Properties," *AFS Transactions*, 60: 21, 1952.

23. C. A. Sanders, *Foundry Sand Practice*, American Colloid Co. Skokie, IL.

24. T. S. Venkoba Rao and H. Md. Roshan, "Studies on Dimensional Accuracy and Consistency in High Pressure Molded Castings," *AFS Transactions*, 96: 37–46, 1988.

25. E. Schaarschmidt, "Modern Green Sand Moulding Systems," *Casting Plant & Technology*, 4: 12–23, 1985.

26. T. E. Garner, Jr., "Mineralogy of Foundry Sands and Its Effect on Performance and Properties," *AFS Transactions*, 85: 399–415, 1977.

27. AFS Mold & Core Aggregate Committee Report, "Determining Sieve Analysis of Foundry Sands," *Modern Casting*, Nov. 1982, pp. 31–33.

28. *AFS Mold and Core Test Handbook*, 2d ed., AFS 2001/12.

29. V. Biringuccio, "De la Pirotechnia," published in 1540, translated by C. S. Smith and M. T. Gnudi, Dover Publications, 1990.

30. S. Wetzel, "Replacing Olivine Sand," *Modern Casting*, 102: 20–24, Dec. 2012.

31. C. A. Sanders and R. L. Doelman, "A Review of Sand Surface Area Relationships," *AFS Transactions*, 76: 85–91, 1968.

32. F. Hofmann, "Investigations in the Effect of Heat in the Bonding Properties of Various Bentonites," *AFS Transactions*, 93: 377–384, 1985 and 66: 305–311, 1958.

33. C. A. Sanders and R. L. Doelman, "The Durability of Bonding Clays, Parts I to X," *AFS Transactions*, 1967–1970.

34. L. V. Whiting, R. D. Warda, E. F. Darke, and R. J. Lacroix, "The Methylene Blue Test and Clay Durability," MTL 91-76(OP-J), p. 36; NRCAN, Government of Canada, 1991.

35. F. Hofmann, "Property Changes and Conditioning of Repeatedly Circulating Foundry System Sands," *AFS Transactions*, 75: 338–352, 1967.

36. H. E. Elliott and J. G. Mezoff, "A New Gating Technique for Magnesium Alloy Castings," *AFS Transactions*, 55: 241–251, 1947.

37. J. D. Hanawalt and M. O. Kada, "Oxidation Inhibiting Techniques for Magnesium Sand Casting," *AFS Transactions*, 80: 87–90, 1972.

38. W. Tilch, Casting Plant + Technology International, 1/2004, p. 11.

39. G. A. Smiernow, E. I. Doheny, and J. G. Kay, "Binding Mechanisms in Sand Aggregates," *AFS Transactions*, 88: 659–682, 1980.

40. R. E. Grim and F. L. Cuthbert, "The Bonding Action of Clays," Engineering Experiment Station of the University of Illinois, Bulletin 357, Urbana, IL, 1945.

41. E. A. Hauser, *Colloidal Phenomena*, McGraw-Hill Book Co., New York, 1939.

42. H. E. Schwiete and G. Ziegler, Ch. Kliesch Thermochemische Untersuchungen über die Dehydatation des Montmorillonits (Koln: Forschungsbar. D. Wirtschafts u. Verkehrsministeriuma Nordrh.-Westf. No. 545), 1958.

43. C. Rich and M. B. Krysiak, "Monitoring Sand Temper through True Compactibility," *Modern Casting*: 26–28, Mar. 1993.

44. A. L. Graham and R. M. Praski, "Methylene Blue Testing of System Sands and Preblended Additives," *AFS Transactions*, 86: 315–322, 1978.

45. T. S. Shih, R. A. Green, and R. W. Heine, "Evaluation of Green Sand Properties and Clay Behaviour at 7 to 15% Bentonite Levels: Parts I to IV," *AFS Transactions*, 92: 467–474, 1984; 93: 689–698, 1985; p. 14, 1986; 1987.

46. F. Hofmann, H. W. Dietert, and A. L. Graham, "Compactibility Testing—A New Approach in Sand Research," *AFS Transactions*, 77: 134, 1969.

47. H. W. Dietert, J. S. Schumacher, and A. L. Graham, "A New Methylene Blue Procedure for Determination of Clay in System Sand," *AFS Transactions*, 78: 208, 1970.

48. C. E. Wenninger and A. P. Volkmar, "A New Control Tool: A Graph for Evaluating Efficiencies of Available Bentonite within Foundry System Sands," *AFS Transactions*, 78: 17–24, 1970 and A. P. Volkmar, "Twenty-Five Years of Green Sand Control," *AFS Transactions*, 104: 1269–1274, 1996.

49. D. Boenisch and N. Ruhland, "New Concepts of Green Sand Technology," *Foundry Management & Technology*, Feb. 1988, pp. 21–27, Mar. 1988, pp. 38, 44–49.

50. J. S. Schumacher, R. A. Green, G. D. Hansen, D. A. Heinz, H. J. Gallaway, and R. W. Heine, "Why Does Hot Sand Cause Problems," *AFS Transactions*, Part I 82: 183, 1974; Part II; 441–446, 1975; Part III: 385–388, 1976.

51. J. S. Schumacher and R. W. Heine, "The Problem of Hot Molding Sands—1958 Revisited," *AFS Transactions*, 91: 879–888, 1983.

52. Unpublished confidential Government of Canada reports.

53. H. S. Murthy, K. S. S. Murthy, and M. R. Seshadri, "Exothermic Sleeves: Effect of Feeding of Nodular Iron Castings" *AFS Transactions*, 101: 9–15, 1993.

54. P. R. Carey (ed.), "Sand Binder Systems," *Foundry Management & Technology*, otherwise Parts 1 to 13 were published from Feb. 1995 to Apr. 1996.

55. P. R. Carey, "Practical Aspects of Resin Binder Processes," *Foundry Management & Technology*, Part 1, Feb. 1995, Jan. 1976, or Feb. 1989.

56. P. R. Carey and L. Hause, "Sand Binder Systems, Parts III and IV Urethane Binders," *Foundry Management & Technology*, Part III, 123(5): 27–31, May 1995; Part IV, 123(6): 25–91, Jun. 1995.

57. P. R. Carey, L. Roubitchek, and J. Green, "Sand Binder Systems, Part VI—The Shell Process," *Foundry Management & Technology*, Aug. 1995.

58. P. R. Carey and M. Lott, "Sand Binder Systems, Part V—Furan No-Bake," *Foundry Management & Technology*, 123(8): 58–75, Aug. 1995.

59. P. R. Carey, "Sand Binder Systems Part XII—Hot Box, Warm Box & Core Oil," *Foundry Management & Technology*, 124(3): 31–37, Mar. 1996.

60. P. R. Carey and G. Sturtz, "Sand Binder Systems, Part X—Silicate CO_2, Silicate No-Bake CO_2 and Alkaline Phenolic," *Foundry Management & Technology*, Mar. 1996, Part IX Acrylic Epoxy/SO2, Dec. 1995 and Jan. 1996.

61. L. V. Whiting, P. D. Newcombe, and D. Cousineau, "Core Collapsibility for PM Casting of Cu-Based Alloys: A Final Report," *AFS Transactions*, 106: 313–321, 1998.

62. C. Henry, R. Showman, and G. Wandtke, "Process Variables Affecting Al Casting Shakeout of Coldbox Cores," *AFS Transactions*, 107: 633–642, 1999.

63. A. D. Morgan and E. W. Fasham, "The BCIRA Hot Distortion Tester for Quality Control of Chemically Bonded Sands," *AFS Transactions*, 83: 73–80, 1975.

64. *Shell Process Foundry Practice*, 2d ed., AFS 1973 (later versions exist).

65. "Chemically Bonded Cores & Molds," AFS, 2000.

66. R. L. Naro, "Battling the Elusive Lustrous Carbon Defect," *Modern Casting*, 93: 34, 35, May 2003.

67. R. W. Heine, C. R. Loper, and P. C. Rosenthal, *Principles of Metal Casting*, 2d ed., McGraw-Hill, New York, 1967.

68. H. W. Dietert, *Foundry Core Practice*, 3d ed., AFS 1966, GM 6612.

69. K. Srimagesh, "Chemistry of Sodium Silicate as a Sand Binder," *AFS International Cast Metal J.*, 4(1): 50–63, Mar. 1979.

70. K. Kobayashi, "Recent Development of Vacuum-Gas Hardened Moulding Process (the VRH Process)," *Transactions of the Japan Foundrymen's Society*, 5: 11–13, 1886 and "Vacuum Replacement Hardening Process," *Foundry Trade Journal*, V166A: 456, 457, Jul. 24, 1992.

71. R. A. Bambauer, H. J. Langer, and S. C. Akey, "Inorganic Foundry Binder Systems and their Uses," US Patent 5,382,289, Jan. 17, 1995.

72. J. F. Wallace and J. P. Hrusovsky, "Properties of a Cement-Sand System Using Very High Early Cement Binder," *AFS Transactions*, 79: 269–278, 1971.

73. H. F. Shroyer, "Cavityless Casting Mold and Method of Making Same," U.S. Patent 2,830,343, Apr. 15, 1958.

74. M. Sahoo (ed.), "Technology for Magnesium Castings: Design, Products & Applications," AFS, 2011.

75. F. Sonnenburg, "Lost Foam Casting Made Simple," AFS Lost Foam Division 11, AFS, 2008.

76. Z.-T. Fan and S. Ji, "Low Pressure Lost Foam for Casting Magnesium Alloys," *Science & Technology*, 1: 727–734, 2005.

77. L. Bichler and C. Ravindran, "Effect of Vacuum and Selected LFC Process Parameters on Mold Filling of AZ91E Castings: Part II Casting Porosity and Density," *AFS Transactions*, Paper 06-039, 2006.

78. Y. Fasoyinu, P. D. Newcombe, and M. Sahoo, "Lost Foam Casting of Magnesium Alloys AZ91D and AM50," *AFS Transactions*, Paper 06-079, 2006.

79. L. V. Whiting and P. Merlin, "Central Sand Laundry Economics," *AFS Transactions*, 100: 1049–1055, 1992.

80. K. Kondo, "Method of and Apparatus for Reclaiming Molding Sand," U.S. Patent 4,436,138, Mar. 13, 1984.

81. D. Oman and B. Esch, "XI Recent Developments in Regulations and Enforcement of Air Requirements for Thermal Sand Relaimers and Sand Coolers/Dryers," *AFS 25th Environmental Health and Safety Conference*, Pittsburgh, PA, Aug. 13–14, 2013.

CHAPTER 4

Solidification of Metals and Alloys

Mahi Sahoo
Suraja Consulting Inc.

Introduction

The cast structure developed during solidification determines the mechanical properties of castings. The majority of castings are used in the as-cast condition. Hence, metallurgical control of solidification is necessary to achieve the required structure in cast metals.

Solidification occurs by the nucleation of minute grains or crystals, which then grow under the influence of the crystallographic and thermal conditions that prevail. The size and character of these grains are controlled by the alloy and by the cooling rate. Growth ceases when all the available liquid metal has solidified [1].

Other changes are also taking place during the freezing process. Heat is being extracted from the molten metal as soon as the metal enters the mold. This heat is often referred to as *superheat*, since it represents that which must be removed before solidification can begin. The latent heat of fusion is also evolved. This must be transferred to the surrounding mold before complete solidification can be achieved. Finally, the solid metal transfers heat to the mold, and then to the atmosphere as it cools to room temperature.

During the three stages of cooling (i.e., liquid, liquid-solid, and solid) shrinkage is also occurring. Thus the metal contracts as it loses superheat, as it transforms to the solid, and as the solid cools to room temperature. The shrinkage is due to changes in density in the three stages of cooling. A schematic of the density change is shown in Fig. 4.1 [2].

There are, therefore, three major points for consideration when a casting solidifies:

1. Growth of the solid grains

2. Heat evolution and transfer

3. Dimensional changes

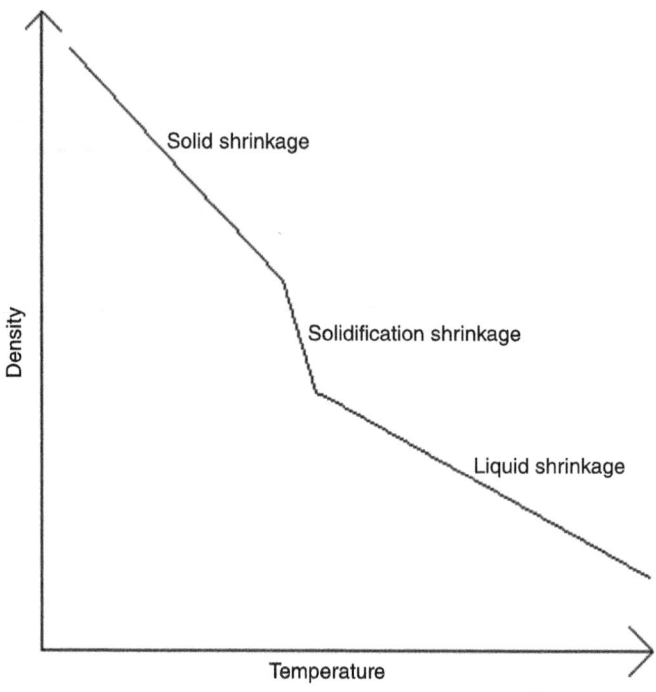

Figure 4.1   Density change with temperature in an alloy. (From *Basic Principles of Gating and Risering* [2].)

Additional variables are present which add to the complexity of the process. These include the effect of the mold material and its thickness, the mold geometry, the metal thickness, and such metal properties as its thermal conductivity and solidification temperature range, thermal gradient, heat transfer from the mold to the atmosphere, control of grain size by the use of inoculants or vibration, and others. Since these variables operate through their effect on the solidification process, major attention is given to the process itself. The effect of the variables is considered secondarily, or can be deduced from the discussion.

The relationship of other properties, such as fluidity, hot tearing tendencies, hot isostatic pressing, casting under pressure, etc., to the solidification process is also covered in this chapter.

If attention is confined for the time being to binary alloy systems, it is apparent immediately that solidification proceeds:

- At constant temperature (pure metals and eutectic alloys)
- Over a temperature range (solid solutions)
- By a combination of solidification over a temperature range followed by constant-temperature freezing (proeutectic-plus eutectic-type freezing)

Metalcasters are primarily concerned with the solidification of alloys. Before discussing the solidification of alloys, it would be wise to deal with fundamental aspects of solidification as described below.

Distribution Coefficient and Constitutional Undercooling

Solidification of an alloy produces nonuniform distribution of solutes in the solid, which can be explained in terms of distribution coefficient (k_0) with reference to the phase diagram. The equilibrium distribution coefficient is defined as

$$k_0 = C_S/C_L \qquad (4.1)$$

where C_S and C_L are the equilibrium solute concentrations in the solid and liquid, respectively [3]. k_0 is less than 1 when the melting point is lowered by the addition of the solute (Fig. 4.2). Alternatively, when the melting point is raised following the solute addition, k_0 is greater than 1 (Fig. 4.3). This results in a difference in composition between the solid-liquid interface and the bulk liquid. Irrespective of the distribution coefficient, the liquidus temperature of the liquid in contact with the interface is lower than that of the liquid at a greater distance from the interface. The concentration and temperature profile ahead of the solid-liquid interface is shown in Fig. 4.4 for k_0 less than 1 [1]. Because the solid is lower in solute concentration than the liquid at T_1 (Fig. 4.4a) a concentration gradient develops immediately ahead of the interface as shown in Fig. 4.4b. This concentration gradient will affect the liquidus temperature as shown in Fig. 4.4c. If this concentration gradient is not as steep as the temperature gradient (line OH in Fig. 4.4c), the interface will advance uniformly. If the temperature gradient is in some intermediate position as shown by line OI in Fig. 4.4c, the liquid, for a certain distance ahead of the interface, is at a temperature lower than the equilibrium freezing temperature. This liquid (shaded area) is undercooled resulting from the presence of a solute-enriched boundary layer at the interface. Hence, this is called constitutional undercooling. If the temperature rises sharply at the interface, constitutional undercooling will be promoted. The instability created by this condition promotes the growth of spikes into the liquid. Continued growth of these spikes into the liquid in a direction opposite the heat flow, and in lateral direction as well, results in a typical treelike, or dendritic structure as shown in Fig. 4.26. If the cooling rate eventually

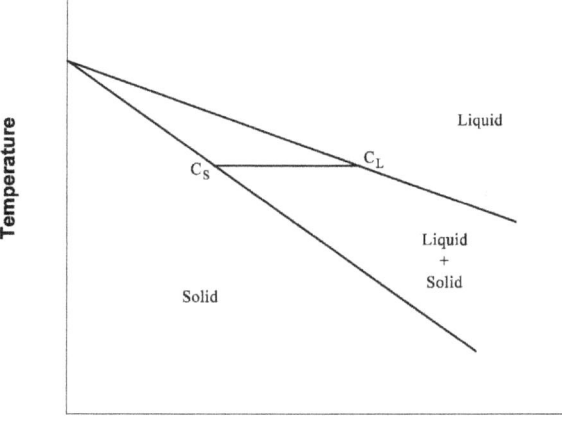

FIGURE 4.2   Equilibrium distribution coefficient for $k_0 < 1$. (From Gruzleski [3].)

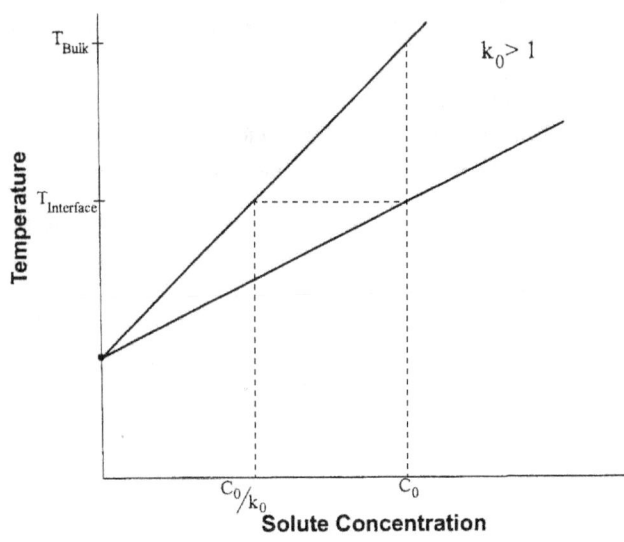

FIGURE 4.3   Equilibrium distribution coefficient for $k_0 > 1$. (From Gruzleski [3].)

results in a temperature gradient indicated by line OL in Fig. 4.4c, the supercooling differential ΔT may then be low enough to promote random nucleation and equiaxed grains. The maximum amount of supercooling will be the freezing range of the alloy at zero temperature gradient.

If the temperature changes slowly with distance in front of the interface, the undercooling may be avoided. Growth rate (R) and temperature gradient (G_L) in the liquid are two important conditions in the determination of constitutional undercooling, which can be expressed as Eq. (4.2) [4]

$$\frac{G}{R} = \frac{mC_0}{D} \cdot \frac{1 - k_0}{k_0} \tag{4.2}$$

where m is the slope of the liquidus and D is the diffusion coefficient of the solute in the liquid. Low gradients will promote constitutional undercooling while high gradients will eliminate it.

Cell formation is a result of the instability produced by constitutional undercooling. The interface is featureless except a few grains boundaries at zero constitutional undercooling. At very small amounts of the undercooling, one-dimensional solute-rich cells, which resemble pipes, form at the interface. With more constitutional undercooling, caused by high freezing rate or low temperature gradient, a two-dimensional cellular structure develops. This is illustrated in Fig. 4.5, which shows a linear transition from planar to cellular growth when the solute concentration C_0 is plotted against the ratio of temperature gradient to growth rate [3].

Nucleation

Nucleation is the first step when a liquid transforms to a solid on cooling in a mold cavity. The initial solid to form is called *nucleus*.

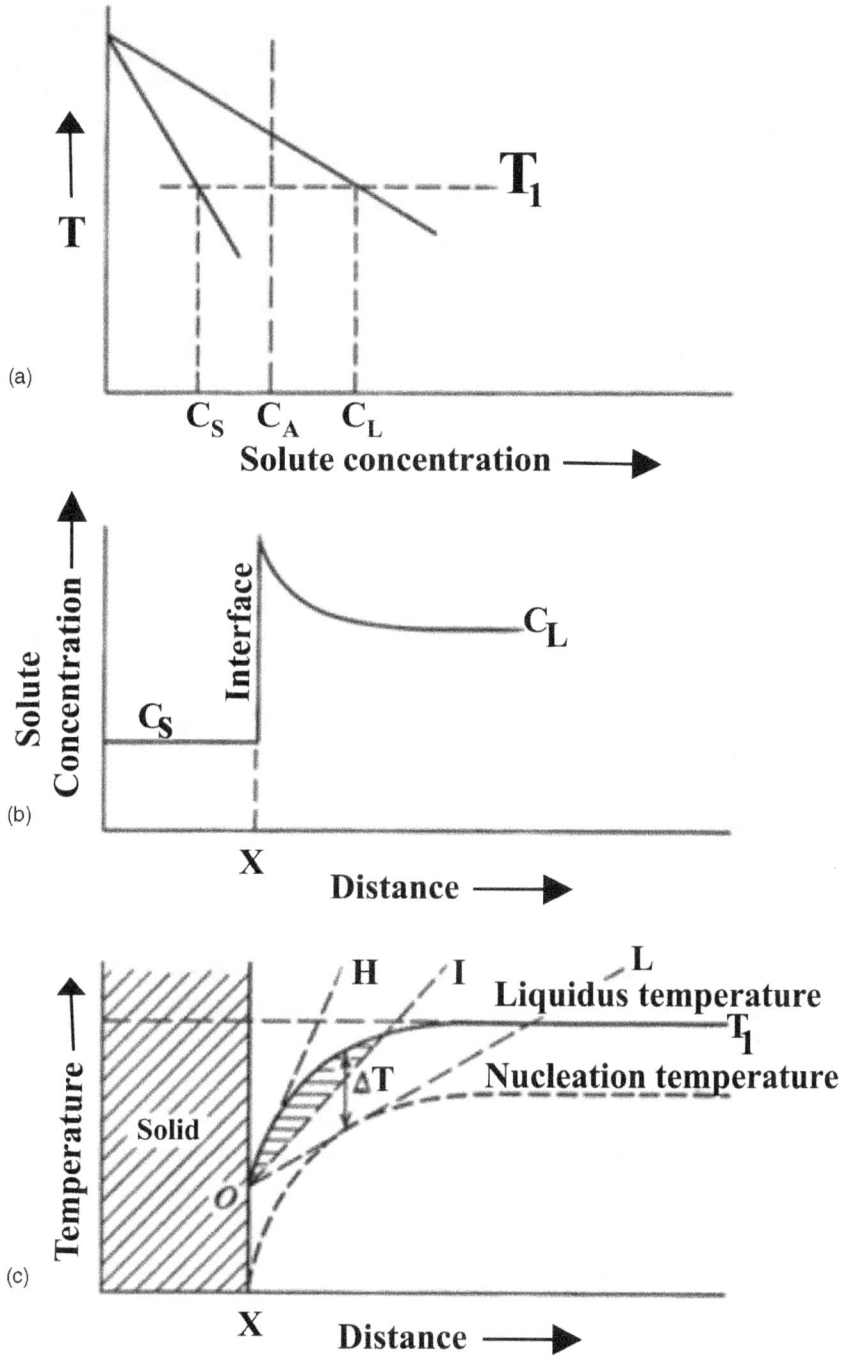

FIGURE 4.4   Schematic illustration of (a) the solute concentration in solid, C_S, and in liquid, C_L, at temperature T_1; (b) the composition gradient existing in the liquid at the solid-liquid interface; (c) the effect of this gradient on the liquid as temperature in the vicinity of the interface. (From Heine et al. [1].)

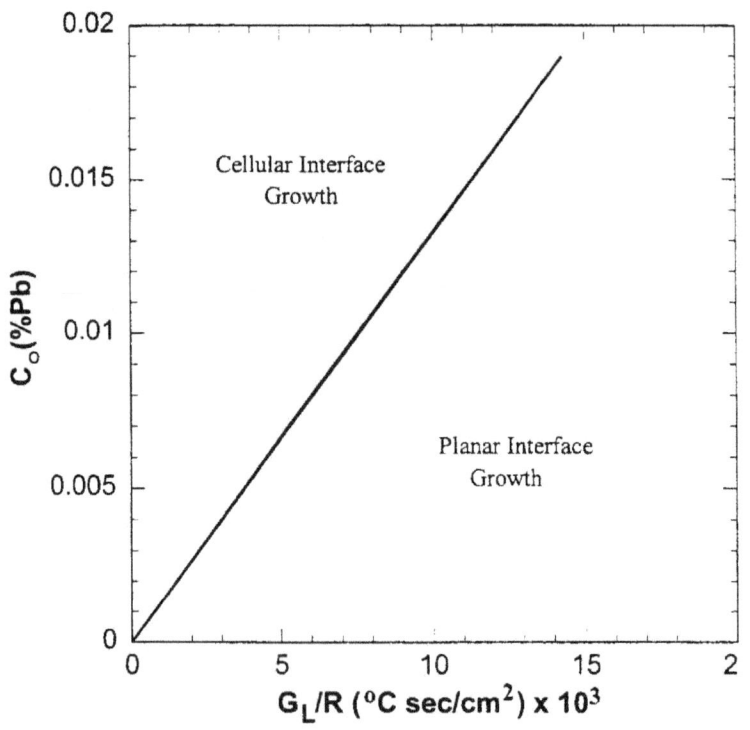

FIGURE 4.5   Linear transition from a planar to a cellular interface when C_0 is plotted versus (G_L/R). (From Gruzleski [3].)

There are two types of nucleation: homogeneous and heterogeneous. In the homogeneous nucleation process, the nucleus starts from the bulk of the liquid itself. Homogeneous nucleation rarely occurs in the normal casting process. Heterogeneous nucleation means the development of new crystals on the surface of an existing particle or a foreign substrate. Typical foreign substrates in commercial castings are inclusions (oxides, nitrides, and carbides), mold-wall, melt surface, and special additives such as grain refiners. For heterogeneous nucleation, it is assumed that a critical radius is formed on the substrate in the shape of a spherical cap, as illustrated in Fig. 4.6 [3]. The stability of the nucleus is governed by the critical angle and relative magnitude of the surface energies acting at the point of nucleus, substrate, and liquid contact. This is expressed as

$$\gamma_{LS} = \gamma_{SN} + \gamma_{NL} \cos \theta \qquad (4.3)$$

where θ is the contact angle; γ_{LS}, γ_{SN}, and γ_{NL} are the surface energies between liquid and substrate, substrate and nucleus, and nucleus and liquid, respectively. Equation (4.3) can be expressed as

$$\cos \theta = \gamma_{LS} - \gamma_{SN}/\gamma_{NL} = M \qquad (4.4)$$

where M is a measure of the tendency of the nucleus to spread on the substrate surface, also called *wetting*. Complete wetting occurs when θ approaches $0°$. Partial

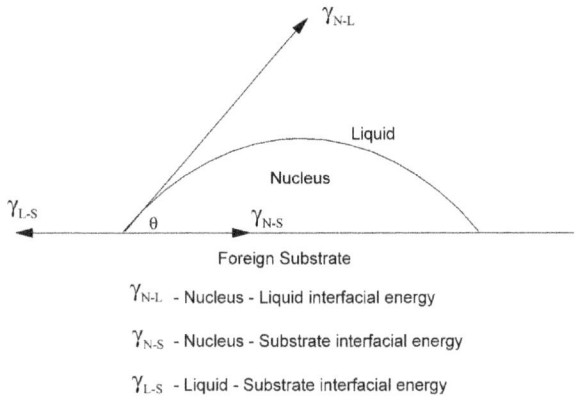

FIGURE 4.6   Contact angle and the interplay of the three surface energies. (From Gruzleski [3].)

wetting occurs when θ increases, until $\theta = 180°$, when there is no wetting and is equivalent to homogeneous nucleation. The contact angle is affected by the surface energies in the above equation, which is in turn related to atomic mismatch across the interface or the crystallographic or epitaxial relationship with each other. Complete wetting depends on the relative values of the interfacial energies as shown in Eqs. (4.3) and (4.4).

Cooling Curves

A convenient way to study the solidification process is to perform thermal analysis by plotting the cooling curve by means of a thermocouple situated in the center of a small crucible of liquid (Fig. 4.7a). A typical equilibrium cooling curve for a pure metal and eutectic alloys is shown in Fig. 4.7b [3], where T_E is the equilibrium freezing temperature. The real cooling curve is shown in Fig. 4.7c, where T_N is the nucleation temperature, and T_G is the growth temperature. It may be noted that T_N is not necessarily T_{min}, but considering the difficulty involved in determining T_N, T_{min} is taken as T_N. T_{min} is the temperature at which latent heat evolution exceeds heat loss to the surroundings. As shown, there is an undercooling ($\Delta T_N = T_E - T_N$) to cause the formation of the final solid. The latent heat evolution follows nucleation and the liquid heats up to T_G at which most of the freezing occurs. The reheating to T_G is called *recalescence*. $\Delta T_G = T_E - T_G$ is the undercooling needed for growth.

Schematic of a cooling curve for an alloy is shown in Fig. 4.8. It shows the undercooling, freezing range, and primary and secondary stages of solidification. The microstructure is determined in the primary stage of solidification and the microstructure develops in a cellular or dendritic way. The primary growth process determines the number, size, and morphology of crystals formed from the melt, which affect the properties of the cast products [5].

Lateral growth of dendrites, development of segregation patterns in both macro and micro scales, growth of secondary phases from the melt, development of pores (due to shrinkage or gas evolution), hot tearing, etc., develop in the secondary stage of solidification.

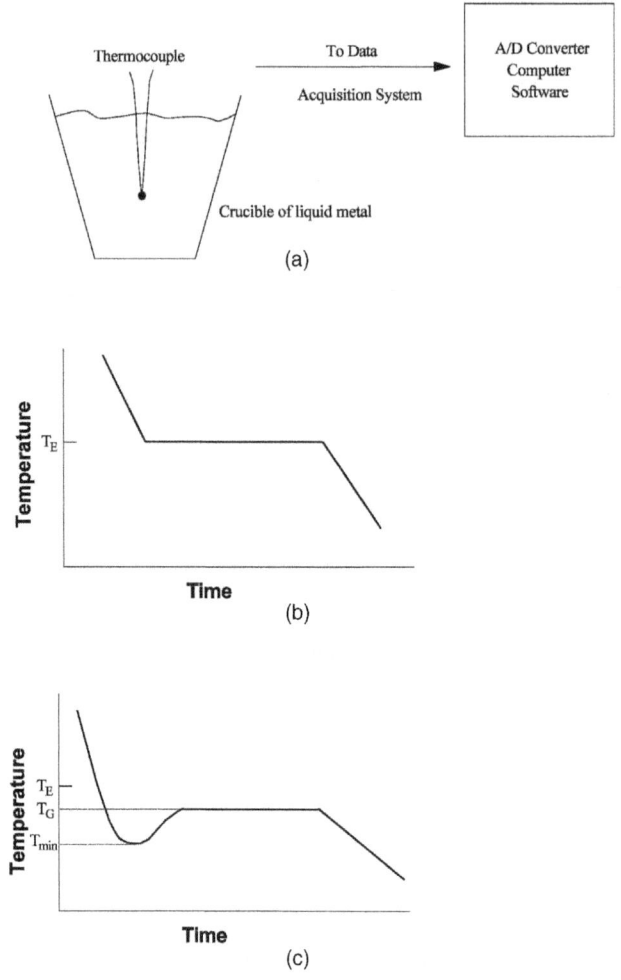

Figure 4.7 Illustrating (a) schematic of a thermal analysis setup; (b) equilibrium cooling curve for a pure metal; (c) real cooling curve for a pure metal. (From Gruzleski [3].)

When molten metal is poured into a cold mold, the liquid coming into contact with the mold wall may be undercooled to a temperature below the critical nucleation temperature (point 2 in Fig. 4.9) and the solid phase starts to grow inward [5]. After a short time when the surface crystals have formed a continuous layer covering the inner walls (point 3), the heat transfer through the walls is balanced by the latent heat released by the growing crystals, the temperature will adjust to a steady-state condition. For a short period of time (point 4), heat is conducted toward the undercooled center of the melt, but soon the thermocouple records a nearly constant temperature (point 5) during the time it takes for the growing crystals to penetrate the entire volume of the sample (point 6). Under conditions prevailing during this stage the crystals grow into the liquid in a very small temperature gradient and the plateau temperature recorded in the center of the sample is very close to the actual growth temperature of the crystals T_G.

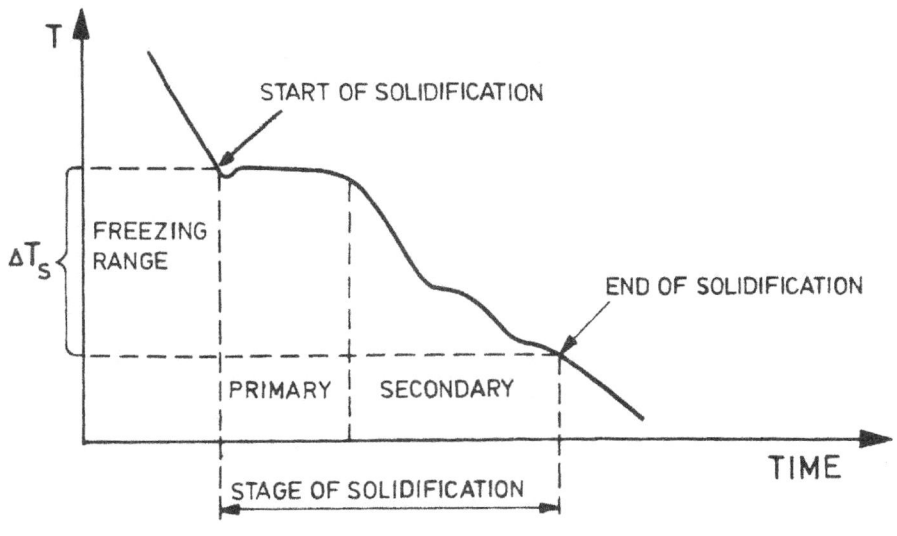

FIGURE 4.8   Schematic cooling curve for an alloy. (From Backerud et al. [5].)

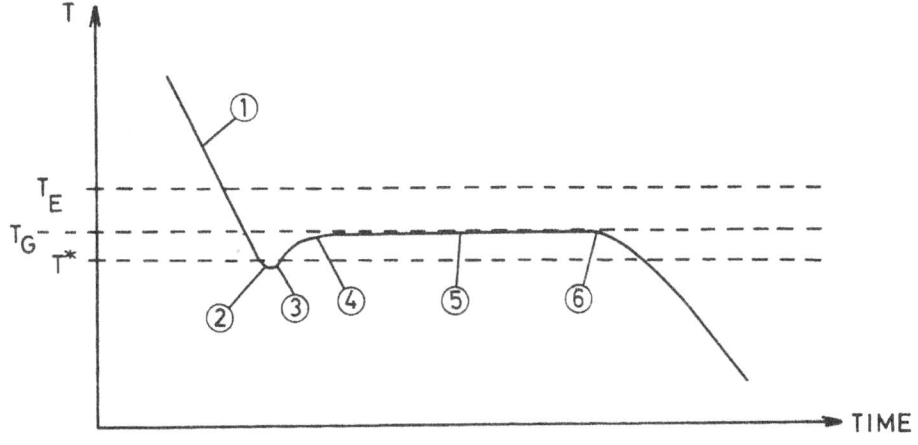

FIGURE 4.9   Cooling curve for a Cu—9.5% Mn alloy. (From Backerud et al. [5].)

Commercial Alloys Based on Binary Systems

Different commercial alloys based on binary systems that can be expected are the following [5]:

- Systems with complete solubility in both liquid and solid phases with $k_0 > 1$; e.g., Cu-Ni (Fig. 8.15). In this case, at low Ni concentrations, k_0 is 4.7. The first solid formed is enriched in Ni. There is a depletion of solute in front of the growing crystal, as well as in the area between cells and dendrite arms. Cellular dendritic growth occurs at a Ni content of only 0.01% and it persists up to 0.1% Ni. At concentrations more than 0.5% Ni, the structure is completely dendritic with correspondingly smaller changes in segregation distance. This change in morphology is accompanied by a corresponding increase in the undercooling. With increasing Ni contents, the solidification range also increases leading to porosity and heavy macrosegregation.

- Systems with complete solubility in both solid and liquid phases with a minimum in the melting point and $k_0 < 1$; e.g., Cu-Mn [6]. In this case, k_0 is approximately 0.4 for low concentrations of Mn. Because of comparatively small differences in solute solubility between the solid and liquid phases at a given temperature, less pronounced segregation and coarse cellular structure can be expected. Undercooling increases with Mn content and a change in structure from cellular to dendritic morphology starts at about 0.5% Mn with fully developed dendritic structure at 1% Mn.

- Eutectic systems with high solubility in the solid phase and $k_0 < 1$; e.g., Cu-Ag, Al-Mg (Fig. 7.6), Al-Cu (Fig. 7.2). Similar to the Cu-Mn system, a cellular structure develops at low solute concentrations followed by a fully dendritic structure at high solute concentrations due to increased undercooling. Because of increased freezing ranges at high-solute contents, porosity and heavy macrosegregation can be expected.

- Eutectic systems with low solubility in the solid phase and $k_0 \ll 1$; e.g., Cu-Bi. A very low k_0 value leads to pronounced build-up of solute in front of the growing solid phase and a dendritic growth pattern at a high degree of undercooling.

- Peritectic systems with $k_0 < 1$; e.g., Cu-Zn (Fig. 8.3). The k_0 for the Cu-Zn system is close to unity (~0.7) and due to very low solute differences between the liquid and solid phases, cellular-dendritic growth is observed at low Zn concentrations (~1%) and regular dendritic structure at about 5% Zn.

- Peritectic systems with $k_0 > 1$; e.g., Cu-Co. k_0 for this system is close to unity (~1.5) and hence, the growth undercooling is also low. The cellular structure changes to cellular-dendritic and then to fully dendritic with increasing solute concentrations.

- Monotectic systems with $k_0 \ll 1$; e.g., Cu-Pb (Fig. 8.9). The copper-rich part of this system is similar to the Cu-Bi system and similar structures such as cellular, cellular-dendritic, and dendritic are observed with gradual increase in Pb contents. Low undercooling has been measured at low Pb levels. A large proportion of lead is found as spherical particles in solidified Cu-Pb samples (Fig. 8.10).

Freezing of a Pure Metal

Skin Effects; Solidification in a Mold

The freezing (or melting) temperatures of pure metals have been established very accurately, so accurately, in fact, that these points serve as a means for standardizing thermocouples. When a pure metal is allowed to freeze in a mold, that portion of the liquid which first reaches the freezing temperature begins to solidify. This usually occurs next to the mold wall, where heat extraction is greatest. The chilling action of the mold wall results in the formation of a thin "skin," or shell, of solid metal surrounding the liquid.

With sufficient extraction of heat through this thin wall of metal, the liquid begins to freeze onto it and the wall increases in thickness, growing progressively inward to the center, as determined by the existing temperature gradient. The interface between the liquid and solid is relatively smooth because the metal is freezing at constant temperature. Actually, there is a mild change in the character of the interface as the front advances, as described later. The liquid metal near the mold wall is supercooled and solidifies as small equiaxed grains. These grains are formed and grow by attachment of atoms from the liquid onto foreign particles such as oxide inclusions. The mold surface also provides numerous nucleation sites.

It has been found that the thickness of the skin frozen in any given time can be expressed by the function shown in Eq. (4.5) [1].

$$D = k\sqrt{t - c} \qquad (4.5)$$

where k, c = constants
t = time
D = thickness

The magnitude of the constant k is determined by the size of the casting and how fast heat can be extracted by the mold. The constant c is determined largely by the degree of superheat. During the time the first skin of fine, equiaxed grains is forming, latent heat of fusion is released which tends to reduce the amount of undercooling (or amount of thermal gradient) in the remaining liquid. The result is a tendency to stop further nucleation. Growth, however, continues on some of the grains already formed. This growth is controlled by the heat transfer from the casting into the mold. Since growth is dependent on crystal orientation, only those grains which are favorably oriented continue to grow and the less favorably oriented grains are pinched off. This leads to the formation of columnar grains with an outer skin of fine equiaxed grains as shown in Fig. 4.10. The columnar zone can extend up to the center of the casting in case of pure metals. However, in alloys the columnar zone is followed by another zone of randomly oriented equiaxed grains caused by the heat transfer out of the last solidifying areas. It is also possible to get completely equiaxed structure depending on the pouring temperature (superheat), type of mold, and if any grain refiner used. These possibilities are shown schematically in Fig. 4.11.

Freezing of Alloys
Mode of Solidification

The mode of solidification depends on the freezing range of the alloys and the section thickness. Alloys can be divided into three groups based on the solidification range. Group I alloys have a short freezing range of less than 50°C; some examples being aluminum bronzes, high-strength yellow brasses, Al-Zn alloy 713, ductile iron, and A27 (Grade 60-30) steel. Group II alloys have a medium solidification range of 50 to 110°C; some examples being copper-nickel alloys, Al-Mg alloy A535, high-Si-Mo iron, and A743 (Grade CA40) steel. The long freezing range alloys such as red brasses, semi-red brasses, silicon brasses, Al-Cu alloys (201 or 206 type), hypoeutectic gray iron and 440C steel with a solidification range of more than 80°C belong to Group III.

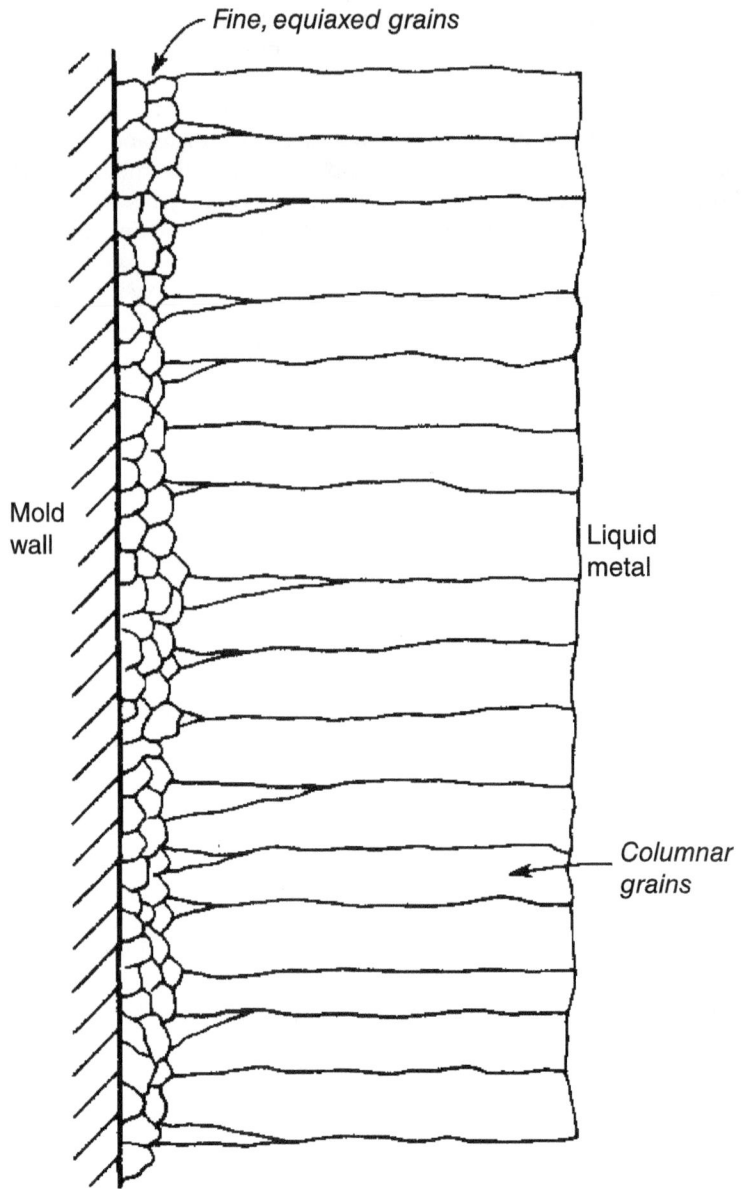

Figure 4.10 Schematic of the solidification front of a pure metal. (From Heine et al. [1].)

The solidification of alloys differs from pure metals in three ways:

1. Solidification of alloys usually occurs over a temperature range.
2. The composition of the solid which first forms and separates from the liquid is different than that of the liquid.
3. There may be more than one solid phase separating from the liquid.

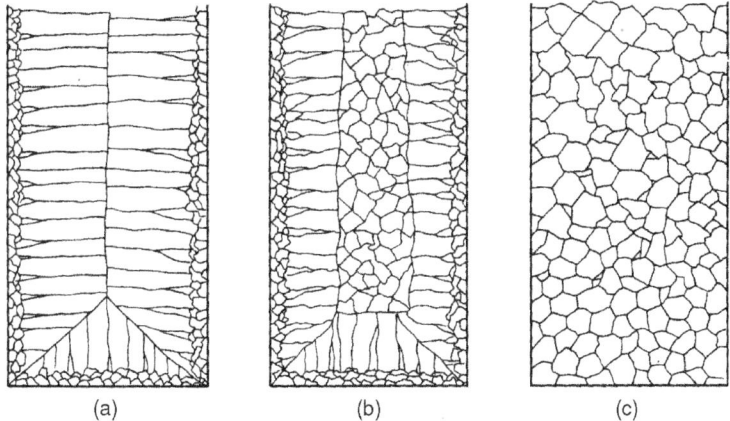

FIGURE 4.11 Schematic of the cast structure of ingots showing (a) columnar grains with an outer layer of fine, equiaxed structure; (b) outer equiaxed followed by columnar and central equiaxed structures; and (c) completely equiaxed structure. (From Heine et al. [1].)

Solidification of short and medium freezing range alloys is similar to what has been described for pure metals and eutectics. Their mode of freezing is depicted schematically in Fig. 4.12.

The mechanism of freezing of the long freezing range alloys is different from that of the short and medium freezing range alloys. Solidification starts with the formation of

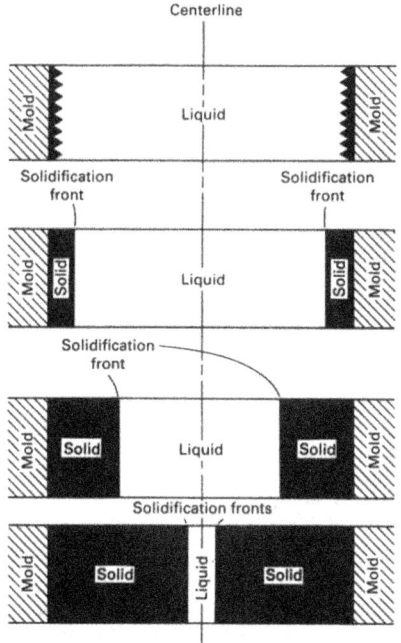

FIGURE 4.12 Mode of solidification in a short or medium freezing range alloy showing a planar interface. (From *Basic Principles of Gating and Risering* [2, p. 94].)

numerous crystallites which are invariably poorer in alloying elements than the liquid from which they formed. As a result of rejection of alloying elements, the liquid gets enriched in the alloying elements. This leads to lowering of the freezing point of this liquid and hence, crystal growth cannot continue temporarily.

Heat extraction by the mold continues which lowers the temperature of this liquid and the liquid in the unaffected region beyond it nearer to the interior of the casting. This decrease in temperature causes a second batch of crystallites to nucleate just outside the enriched area as indicated in the area marked "b" in Fig. 4.13 [7]. In turn, growth of this second batch of crystallites is restricted and a third batch of crystallites quickly forms toward the interior of the casting (area marked "c"), just beyond the enriched region which now surrounds the second batch of crystallites. This process is repeated until the small crystallites have been nucleated through the casting as shown schematically in

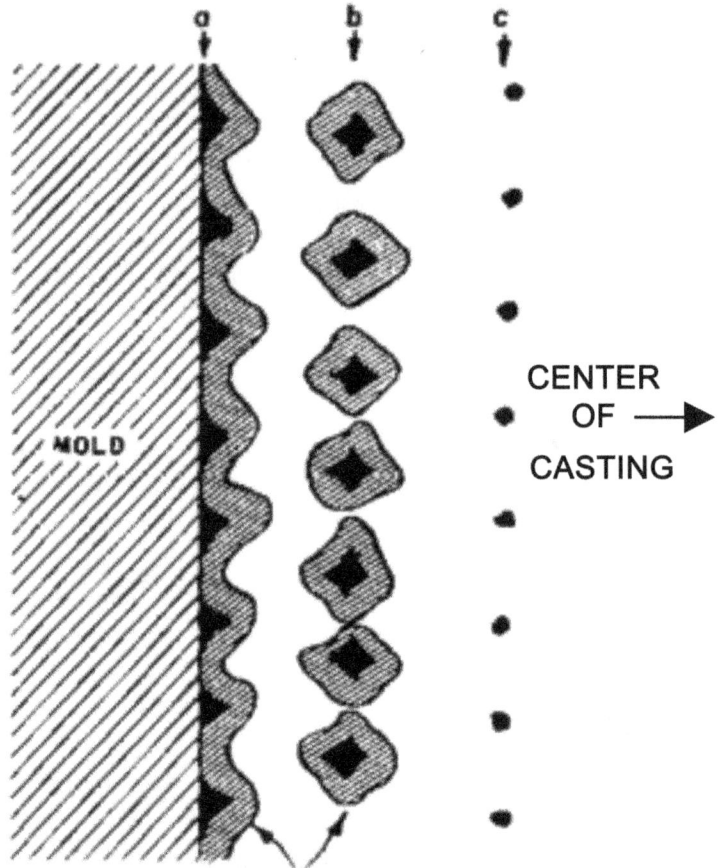

CENTER
OF →
CASTING

LOW FREEZING POINT LIQUID, ENRICHED
IN SOLUTE ELEMENT, WHICH RETARDS
GROWTH OF CRYSTALLITES

FIGURE 4.13 Schematic representation of the restriction of crystal growth in the freezing of a long freezing range alloy. (From Ducharme et al. [7, p. 185].)

Centerline

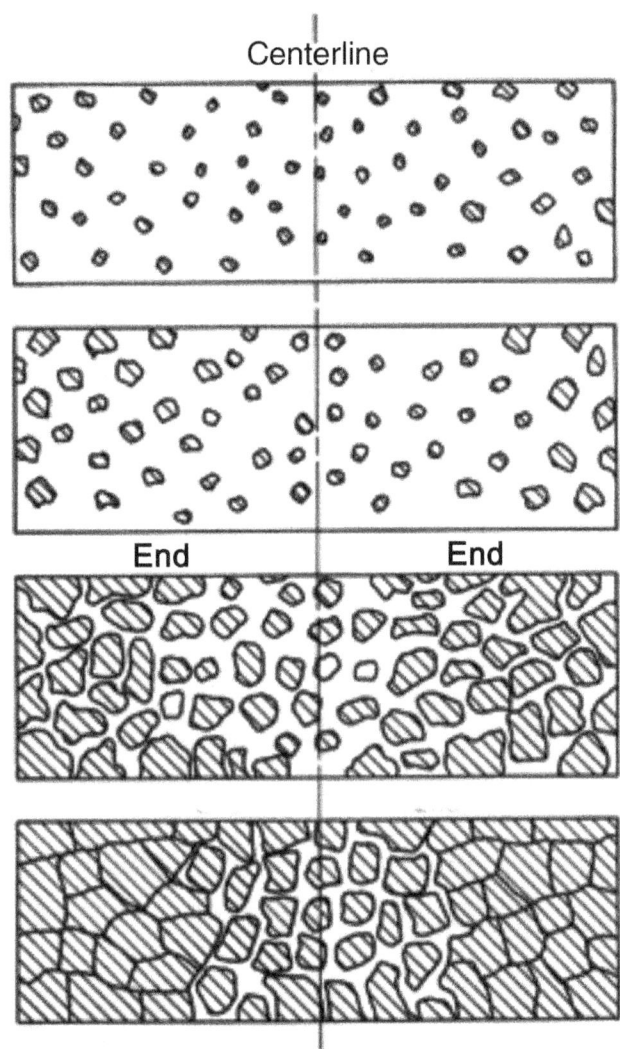

End   End

FIGURE 4.14 Mode of solidification in a long freezing range alloy showing equiaxed structure. (From *Basic Principles of Gating and Risering* [2, p. 94].)

Fig. 4.14. These crystallites start growing, although growth of those close to the mold wall is slightly faster. In actual castings a mix of macrostructures representing the skin forming short freezing range alloys and the long freezing range alloys can be expected depending on the alloy content, rate of solidification, and thermal gradients. With steeper thermal gradients (high superheat), the tendency for dendritic growth is greater. This dendritic growth can continue to the center of the section to get a fully columnar casting. Frequently, the growth of these dendrites stop altogether at a certain stage depending on the melt superheat (thermal gradient) and the central region of the casting freezes in mushy manner to get some equiaxed grains. This is shown in Fig. 4.15.

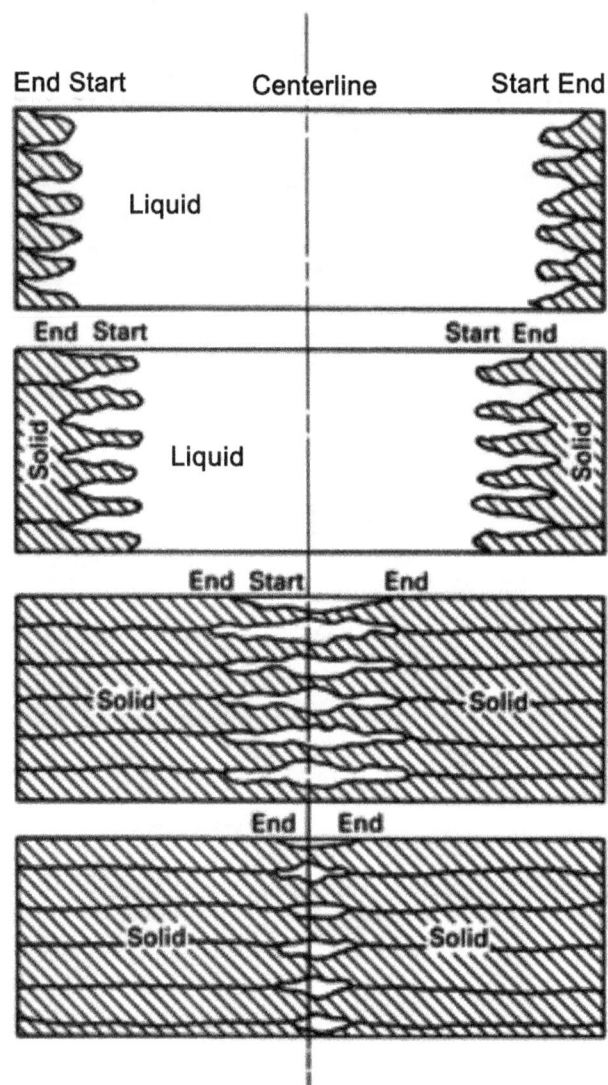

FIGURE 4.15 Mode of solidification in a long freezing range alloy showing dendritic structure. (From *Basic Principles of Gating and Risering* [2, p. 94].)

Directional Solidification

It is necessary to maintain open channels for the flow of the liquid metal during the entire period of solidification. Directional solidification, where solidification begins at a point farthest from the source of feed metal and moves uniformly in the direction toward the source (riser or runner) helps to promote this open channel. This is easily obtained in the case of short freezing range alloys as illustrated in Fig. 4.16, marked "A" [8]. For long freezing range alloys with medium section thickness (13 to 38 mm),

THE METALLOGRAPHY OF GUNMETAL ALLOYS

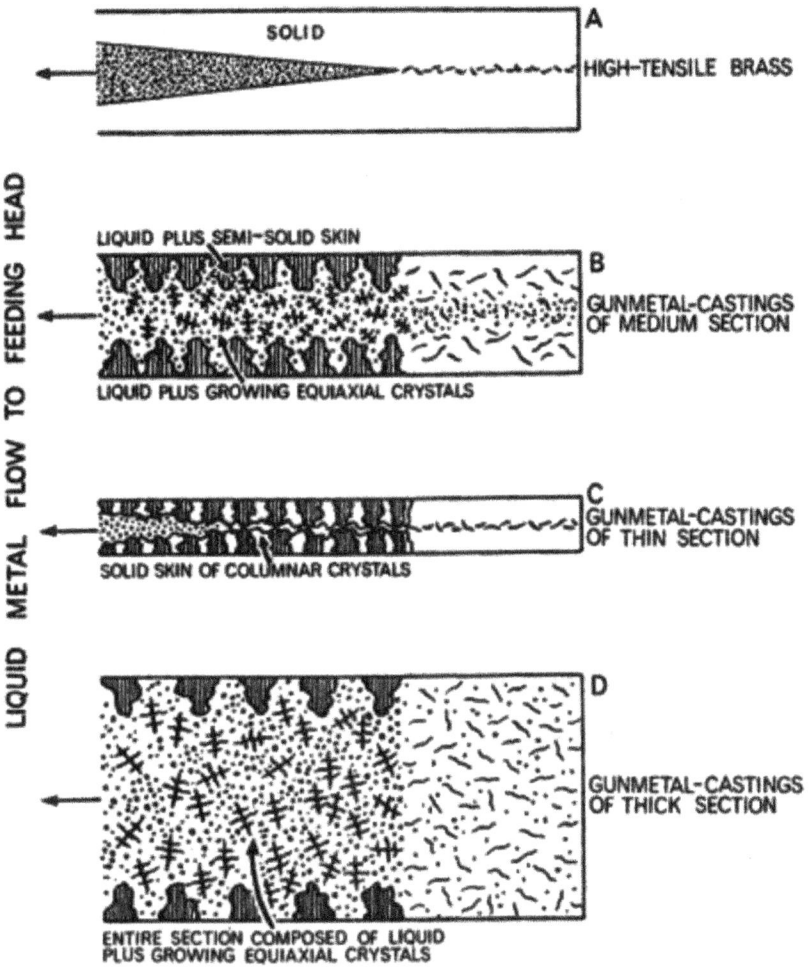

Figure 4.16   Mode of solidification in short and long freezing range alloys with different section thicknesses. (From Hudson and Hudson [8].)

the solidification is nonprogressive and the flow of feed metal at the centerline of the casting is obstructed by the formation of equiaxial crystals and in addition, lateral feeding is required at the surface of the casting during the late stages of solidification (Fig. 4.16). When molten metal is poured, solidification starts at the metal-mold interface and columnar crystals grow from the interface toward the center of the casting. The equiaxial crystals and the liquid metal form a "mushy" zone with restricted movement within the partially solidified shell of columnar crystals. "Mass feeding" occurs in the first stage of solidification by movement of the mushy mass of residual liquid and equiaxial crystals. This is facilitated by the proper design of gating and risering systems.

The crystals grow gradually within the residual liquid metal and when they touch each other, a skeleton of crystal framework is formed. Molten metal will continue to solidify in this skeleton and the solidification shrinkage is compensated by "channel feeding" which is extremely difficult to achieve in the second stage of solidification leading to the formation of interdendritic porosity. The evolution of gas during solidification also contributes to such interdendritic shrinkage cavities.

Solidification characteristics of thick section (>38 mm) castings are depicted in Fig. 4.16, marked "D" for sand castings. The grain structure consists of mostly equiaxed grains with possibly small columnar crystals extending from the mold surface. The length of the columnar grains depends on the melt superheat. It is recommended to use metal chills in the case of long freezing range alloys to increase the temperature gradient and promote progressive solidification leading to reduced interdendritic shrinkage porosity. End-chills are sometimes inadequate to develop directional solidification and wedge chills placed against the face of the casting are more effective to provide the necessary chilling. This is illustrated in Fig. 4.17 [7].

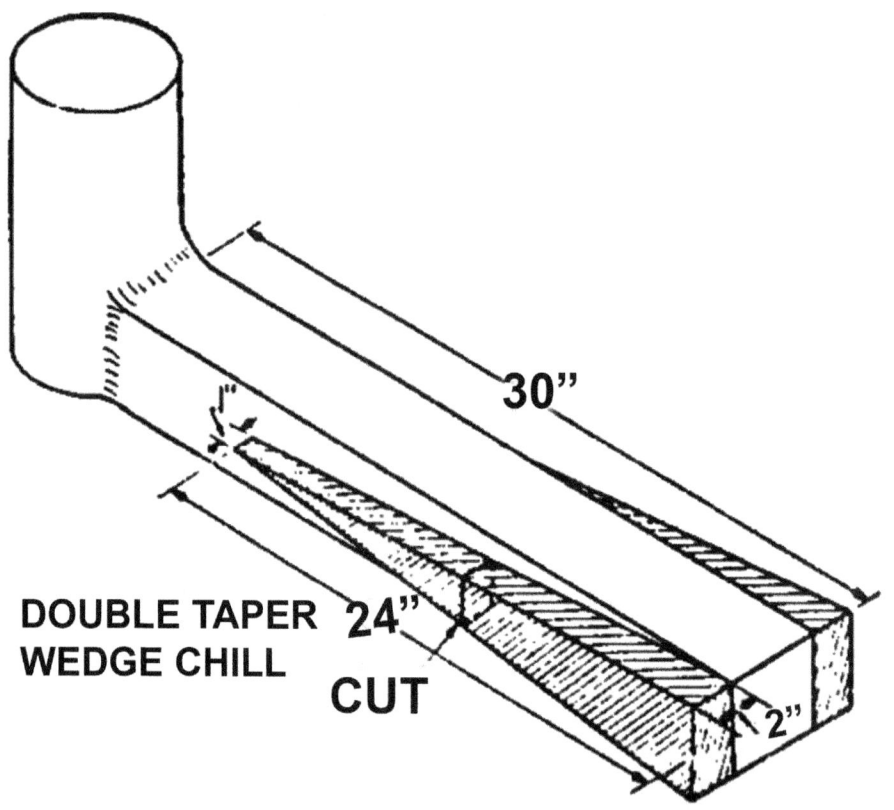

DOUBLE TAPER WEDGE CHILL 30" 24" CUT 2"

FIGURE 4.17 Arrangement of wedge chills in a long freezing range alloy such as the gun metal (Cu-Sn-Pb alloy) to promote directional solidification toward the riser. (From Ducharme et al. [7].)

Factors which affect the solidification characteristics of commercial castings are the following:

- *Effect of mold material.* High thermal conductivity and heat capacity mold materials produce a steeper thermal gradient and promote progressive solidification.
- *Effect of freezing range.* Short freezing ranges promote progressive solidification and long freezing ranges promote a low degree of progressive solidification.
- *Effect of thermal conductivity of solidifying metal.* Low thermal conductivity produces a steep temperature gradient leading to progressive solidification.
- *Effect of solidification temperature.* A high solidification temperature results in a steep temperature gradient leading to progressive solidification.

Other Forms of Solidification

Many commercial alloys undergo eutectic or peritectic reactions. Of these, eutectic solidification is the most significant to many commercial Al-, Cu-, Mg-, Zn-base alloys, and cast irons. Advantages of eutectic alloys include lower melting points and hence, lower processing temperatures and better fluidity and hence, thin sections can be filled easily. Commercial alloys contain several constituents. To understand their microstructures, one can focus on simple binary alloys to start with.

Eutectic Solidification

Binary eutectic alloys display a variety of microstructures. Their morphologies can be reduced to an orderly arrangement by various distinguished characterizing parameters, namely the entropy of solution (ΔS) of each phase and the volume fraction V_f of the minor constituent. If ΔS is small for both phases, regular rod or lamellar structures (nonfaceted) form which tend to be insensitive to changes in freezing rate R during solidification. However, should ΔS for the minor constituent exceed some critical value (5.5 calK^{-1}mol^{-1}), then the phases tend to facet. This causes the resulting microstructures to be markedly dependent on the rate at which it froze and the particular V_f value for the system. This is illustrated in Fig. 4.18 [9, 6]. In the faceted/nonfaceted region, the microstructure changes from broken lamellar (Region 3) to anomalous irregular flake to irregular to finally quasiregular with increasing V_f. Ledeburite (white iron of the Fe$_3$C-Fe system) is a typical example of the quasiregular structure.

Unidirectional experiments have been performed to show the growth rate dependence of the various eutectic morphologies [10–13]. Regular eutectics exhibiting lamellar or rodlike microstructures show very little growth rate dependence because of the absence of facets. Typical microstructures of the Cd-Zn eutectic system are given in Fig. 4.19 [13], showing the lamellar structures. However, a wide range of structures can be produced in anomalous eutectics depending on the volume fraction of the faceting phase and the particular eutectic. In the case of the Zn-Ge eutectic system, directionally solidified between 2.7 and 4800 mm·h^{-1}, the faceted Ge phase grew as branched dendrites (Fig. 4.20a) [12] to branched/unbranched plates (Fig. 4.20c) [12] to fibrous form (Fig. 4.20d) [12]—similar to what has been observed in the Al-Si eutectic system. It should be noted that the fibrous form of the faceted Ge or Si phase can be obtained by Na or Sr modification as shown in Fig. 4.21 [12].

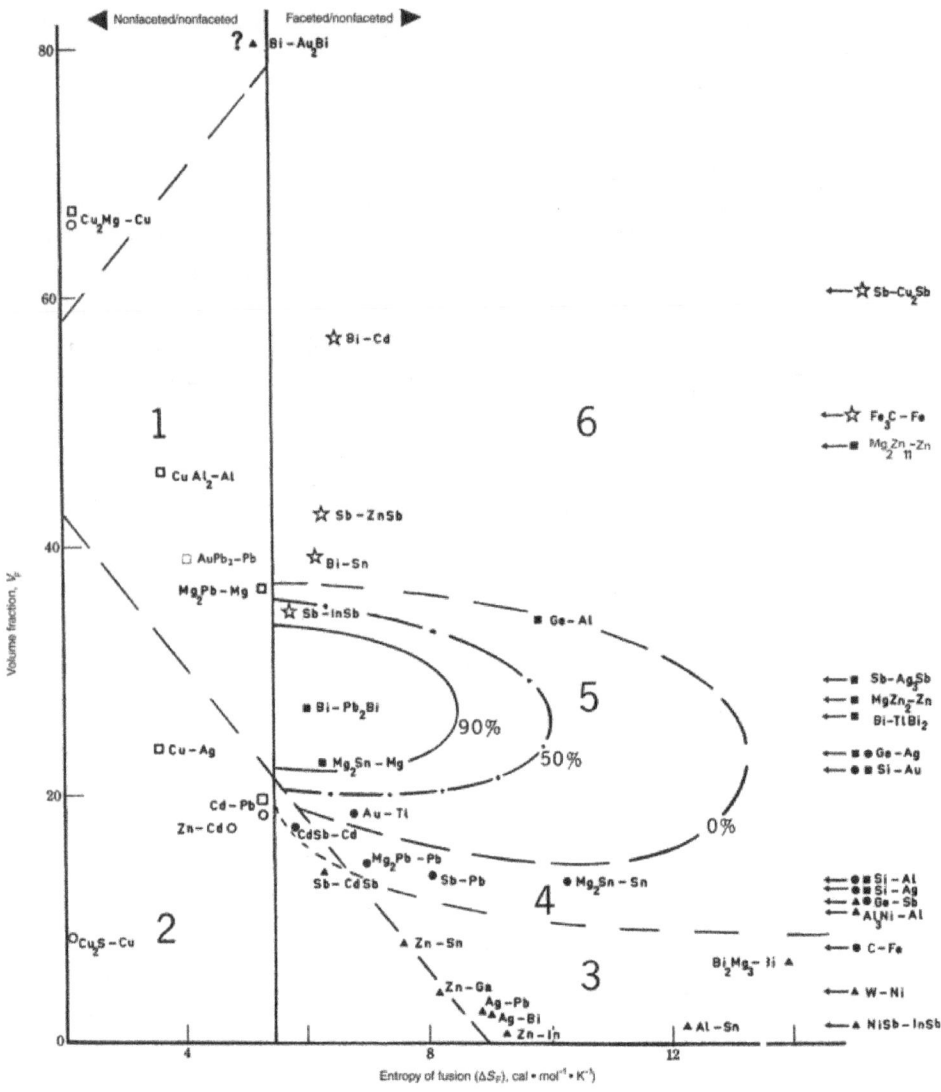

Figure 4.18   Classification of binary eutectic systems based on entropy and volume fraction of the minor phase. (From Croker et al. [9], figure supplied by ASM International.)

Directionally Solidified Eutectic Composites

Unidirectional solidification of eutectics with lamellar or rodlike structures would lead to alignment of the rods/lamellae along the growth direction and they perform as composites. Their tensile properties are equivalent to fiber-reinforced composites. A significant amount of R&D has been performed to characterize the microstructures and mechanical properties of directionally solidified eutectic composites. In parallel with this, development work has been performed on commercial pseudo-binary eutectic composites for industrial applications such as turbine blades. The work in government

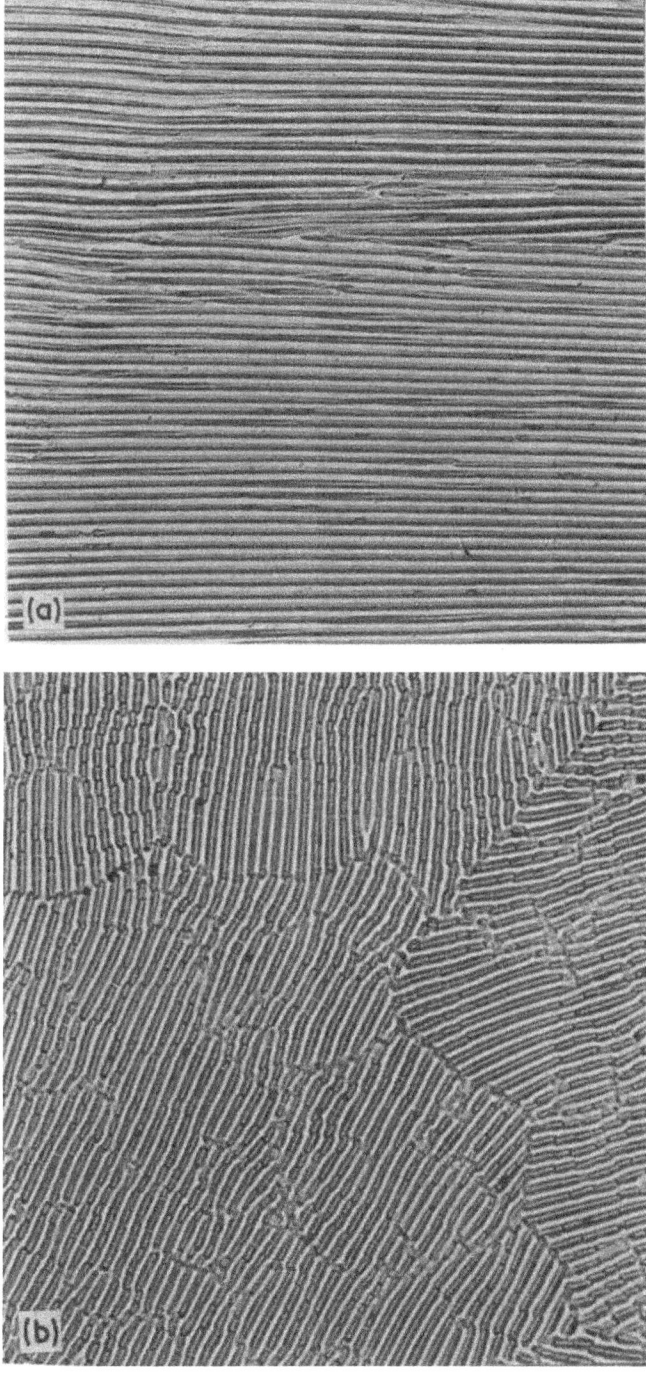

FIGURE 4.19   Optical micrographs of the Cd-Zn eutectic system showing the lamellar morphology. (a) Growth rate of 2.9 mm/h, longitudinal; (b) growth rate of 2.9 mm/h, transverse. (From Sahoo et al. [13].)

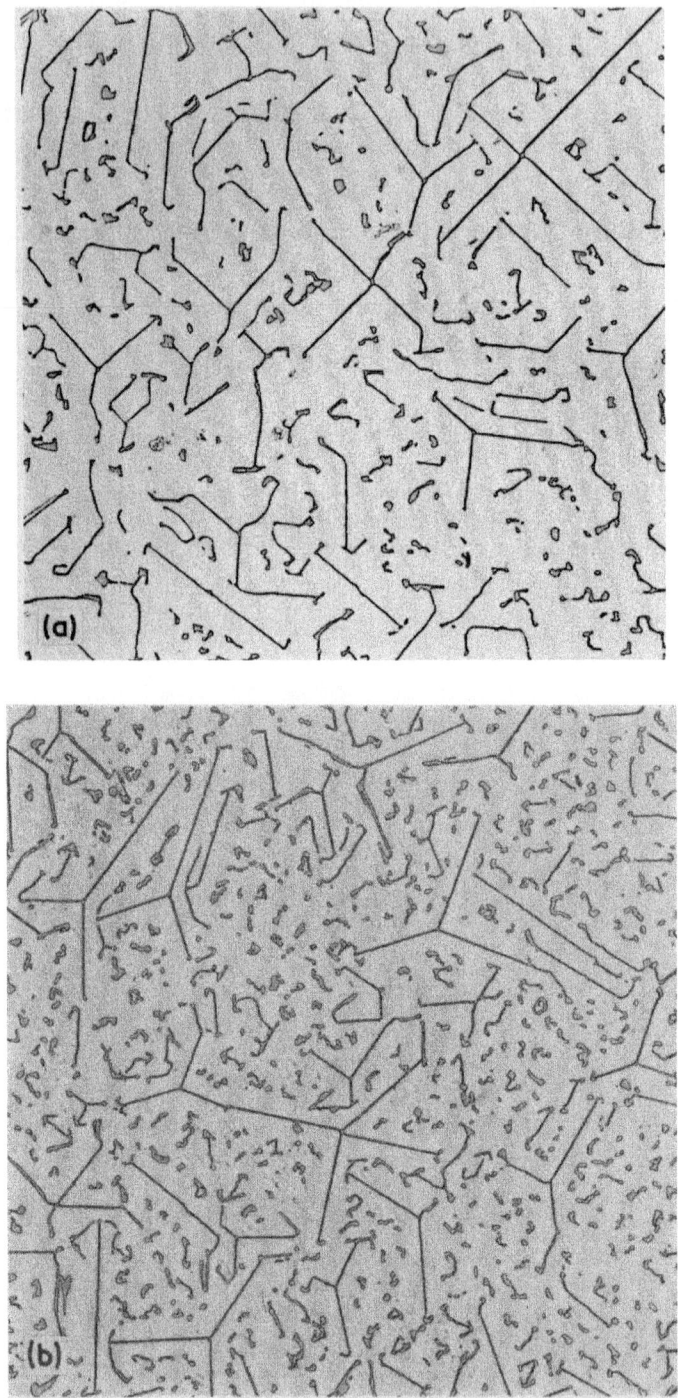

Figure 4.20 Optical micrographs of Zn-Ge eutectic alloys directionally solidified at different growth rates. (a) Growth rate of 5.4 mm/h, ×200; (b) growth rate of 16 mm/h, ×200; (*continued*)

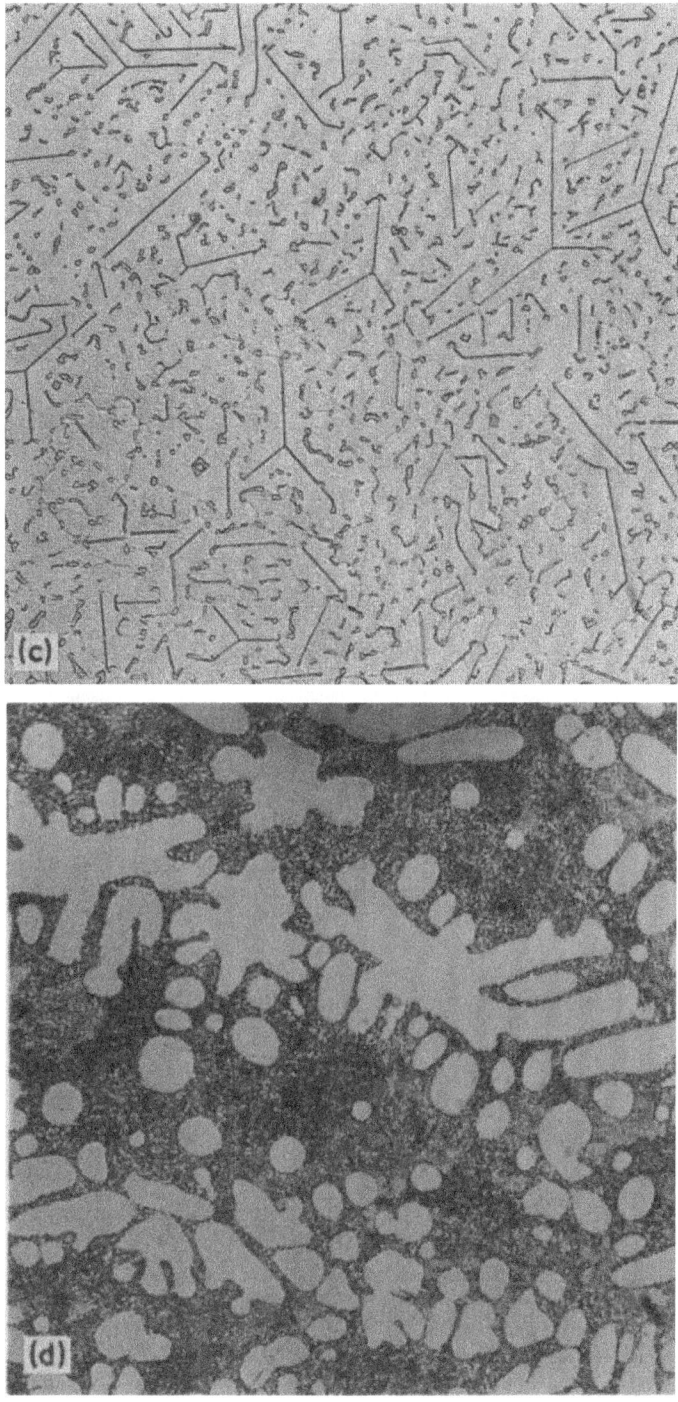

Figure 4.20 (*Continued*) (*c*) growth rate of 160 mm/h, ×500; and (*d*) growth rate of 4800 mm/h, ×500. (From Sahoo and Smith [12].)

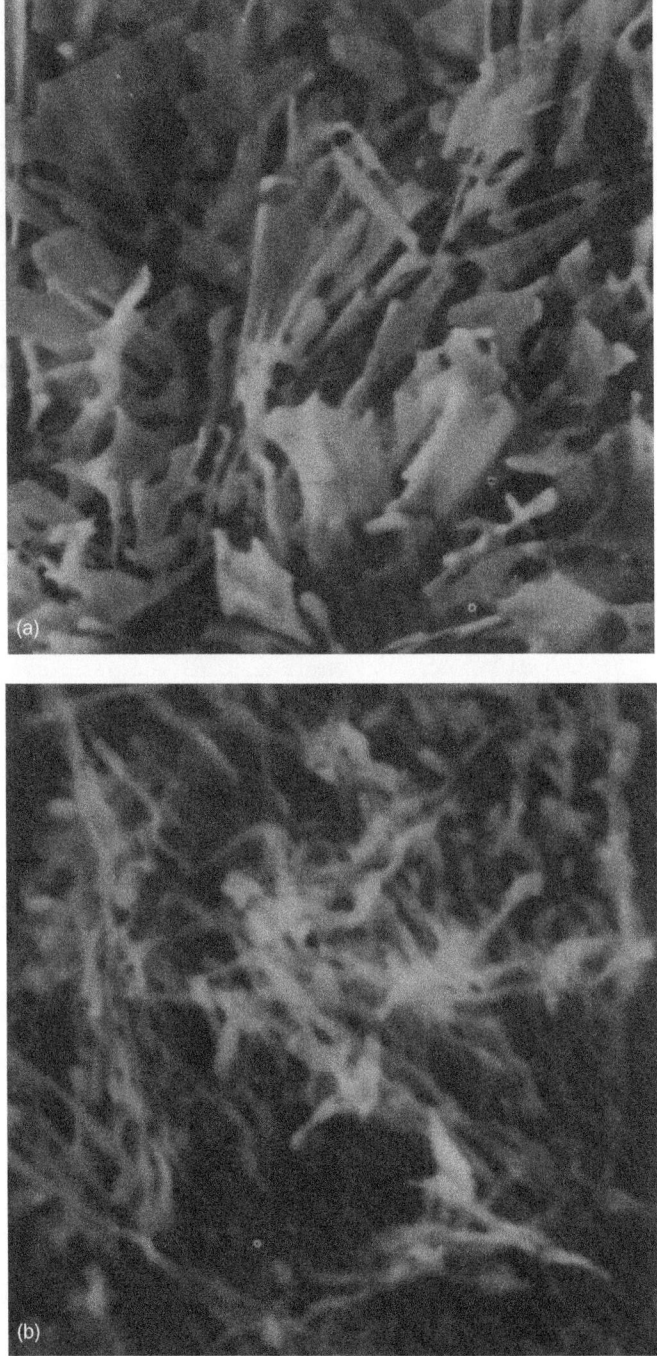

Figure 4.21 Scanning electron micrographs of the Zn-Ge eutectic system showing morphology of the minor phase (Ge) in the unmodified and sodium modified conditions. (*a*) Unmodified, ×2000; (*b*) sodium modified, ×5000. (From Sahoo and Smith [12].)

and industrial laboratories has been aimed at discovering the industrially useful eutectic composites to replace the nickel-base superalloys as high temperature turbine blade materials. One example is the NiTaC-based eutectic composites. Work at the General Electric Company, Cincinnati, Ohio Aircraft Engine Group has shown that the longitudinal strength properties such as ultimate tensile strength, creep-rupture, thermal fatigue, low-cycle fatigue and high-cycle fatigue are superior to the best conventionally cast superalloys [14, 15].

Other Casting Characteristics Related to the Freezing Mechanism

Fluidity

Fluidity is defined as the ability of a liquid metal to flow readily in a standard mold with constant cross section before the flow is terminated by solidification [3, 4, 6]. This is different from the physicist's definition of fluidity, which is the inverse of viscosity.

There is no standard test for fluidity testing. Fluidity in sand casting is usually measured by the length of a standard-spiral casting. A similar spiral mold has been used for the permanent mold casting. Figure 4.22 shows the fluidity spiral castings in sand and permanent molds [16, 17]. In this case, fluidity is taken as the length of the casting from the point where the metal enters the mold cavity to the limit which it reached. A simple plate mold has been used by Sahoo and colleagues to measure casting fluidity during permanent mold casting. In this case, the metal height in the metal mold with a cavity of 51 mm wide × 152 mm high has been taken as an indication of fluidity. Casting thickness was either 1.6 or 3.2 mm [18].

Both metal and mold characteristics together with casting variables can affect fluidity. These can be categorized as follows [6]:

- Metal variables
  - Chemical composition
  - Freezing range
  - Viscosity
  - Heat of fusion
- Mold and mold/metal variables
  - Heat transfer
  - Mold and metal thermal conductivity
  - Mold and metal mass density
  - Specific heat
  - Surface tension
- Casting variables
  - Metal head
  - Channel diameter
  - Pouring temperature (superheat)
  - Melt quality (inclusion content)

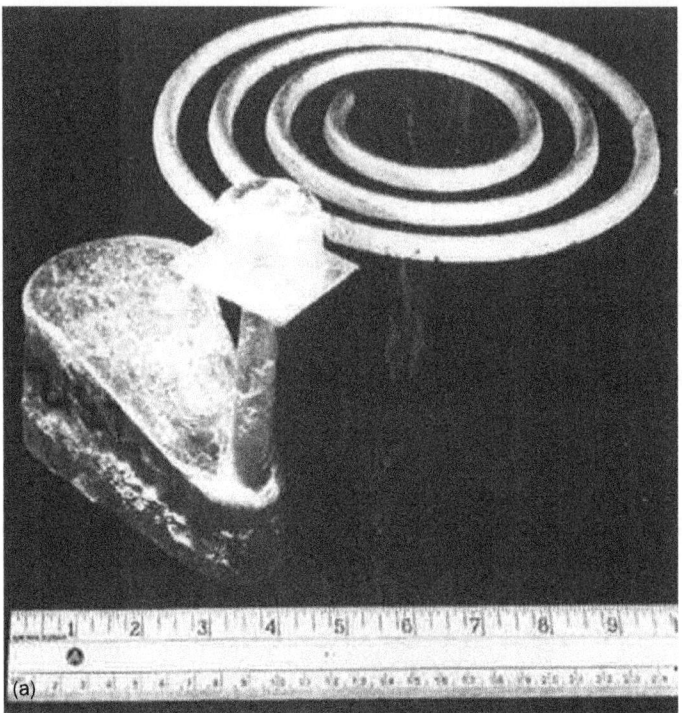

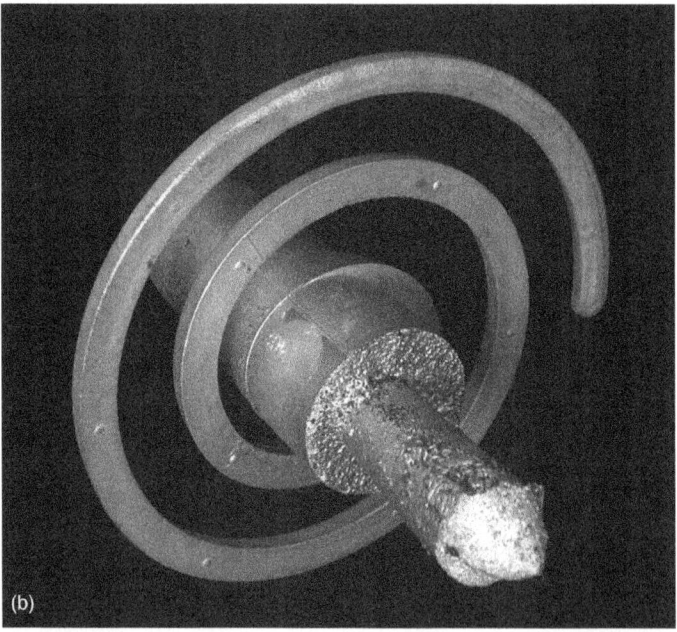

Figure 4.22   Fluidity spiral castings. (a) Sand casting and (b) permanent mold casting. (From Sahoo and Whiting [16], Thomson et al. [17].)

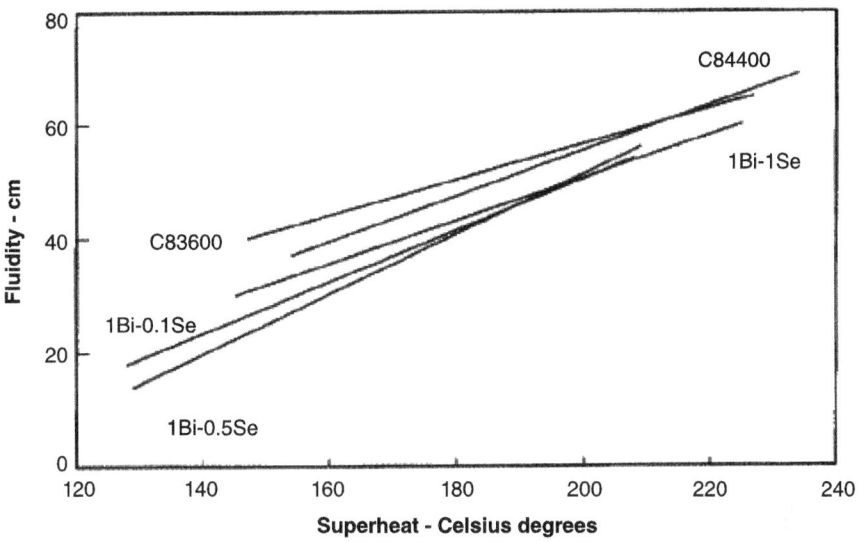

FIGURE 4.23 Fluidity of various copper-base alloys by the fluidity spiral as a function of superheat. (From Whiting et al. [19].)

Of these factors, important ones are melt superheat, metal composition, freezing range, and metal quality. In general, fluidity increases with melt superheat as the metal that is heated to a higher temperature will have a longer period in the mold in which it is liquid, and hence it will flow farther than the metal not so highly heated. An example showing the close correlation between fluidity and solidification temperature, as well as with superheat temperature, is given in Fig. 4.23 [19].

A short freezing range alloy has relatively higher fluidity compared with a long freezing range alloy. This has been verified for a group of aluminum alloys where a direct correlation was found between the fluidity of the alloys and their freezing range [20]. Since a long solidification range is indicative of a condition where the metal is in a mushy condition, consisting of interlacing dendrites surrounded by liquid at practically its freezing temperature, it would seem only natural that this condition would restrict fluidity.

Poor melt quality due to presence of oxide films have been shown to have an adverse effect on fluidity [3]. In case of aluminum alloys, a filtered (clean) alloy showed higher fluidity than the unfiltered alloy (Fig. 4.24). Use of filters in the gating system is known to remove inclusions from the molten metal [21–23].

Alloying can be used to improve casting fluidity. A good example is Al-Si alloys where addition of Si improves fluidity. Another example is addition of a very small amount of Al to yellow brass, silicon bronze, and silicon brass to improve fluidity due to modification of the microstructure [17, 24].

Dendritic Growth

Solidification morphology in alloys changes depending on the amount of constitutional undercooling and solute content as shown schematically in Fig. 4.25 [3]. As mentioned before, the planar interface is present when there is no constitutional undercooling,

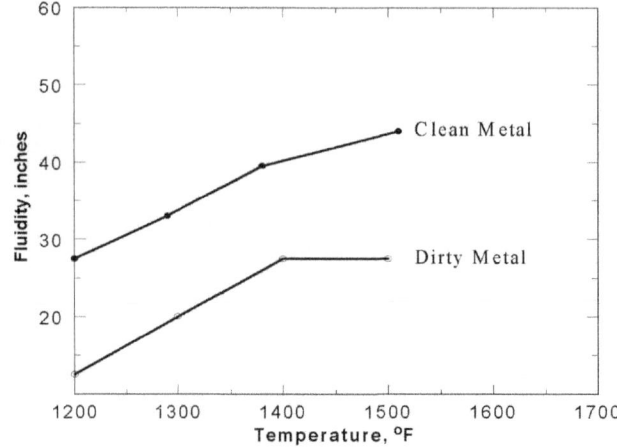

FIGURE 4.24 Comparison of the fluidity of a filtered (clean) and unfiltered (dirty) aluminum melt. (From Gruzleski [3].)

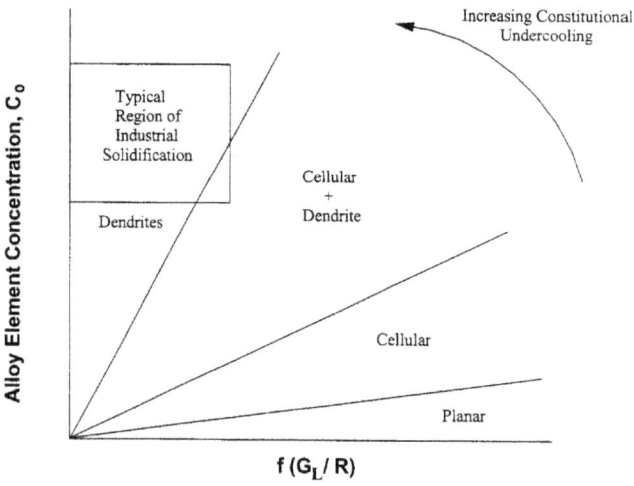

FIGURE 4.25 Schematic representation of solidification morphology as affected by constitutional undercooling which in turn is influenced by alloying element C_0, temperature gradient G_L, and growth rate R. (From Gruzleski [3].)

although this is very rare. With increasing alloy concentrations or decreasing G_L/R ratios, the morphology becomes progressively cellular, a mixture of cells and dendrites, and finally, fully dendritic. Invariably, a fully dendritic structure develops under industrial solidification conditions. Cellular dendritic structure develops when there is a small but positive temperature gradient in the liquid.

Sectioning a dendrite along its primary stalk reveals a branched structure with secondary, and sometimes tertiary arms as shown in Fig. 4.26 [25]. Branching is usually regular and follows definite directions. In cubic materials, branching is along the

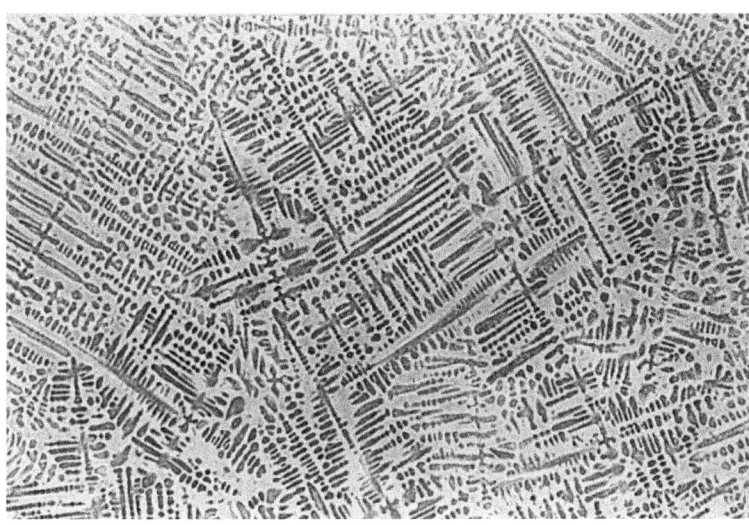

FIGURE 4.26 Dendritic structure in a permanent mold cast silicon bronze (alloy C87600) containing 6.68% Zn, 4.39% Si, and bal. Cu, 100X. (From Sahoo. [25].)

100 directions giving a fourfold symmetry. In the case of hexagonal materials, branching is along the $10\bar{1}0$ directions giving a sixfold symmetry.

The spacing of the dendrite arms called DAS (dendrite arm spacing) or SDAS (secondary dendrite arm spacing) is an important parameter which has been correlated with the mechanical properties, especially in aluminum alloys. Figure 4.27 illustrates the methodology to measure DAS [3]. It is usually the center-to-center distance between two cells. DAS decreases with increasing cooling rates as shown in Fig. 4.28. The relationship between SDAS and mechanical properties is given in Fig. 4.29 [26].

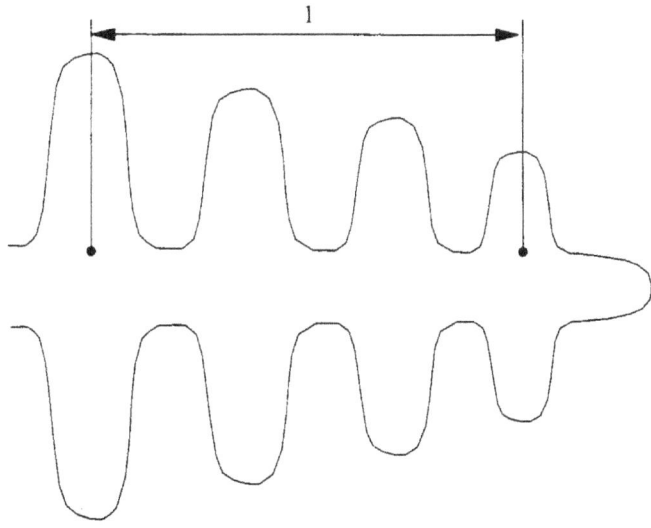

FIGURE 4.27 A methodology to measure DAS. (From Gruzleski [3].)

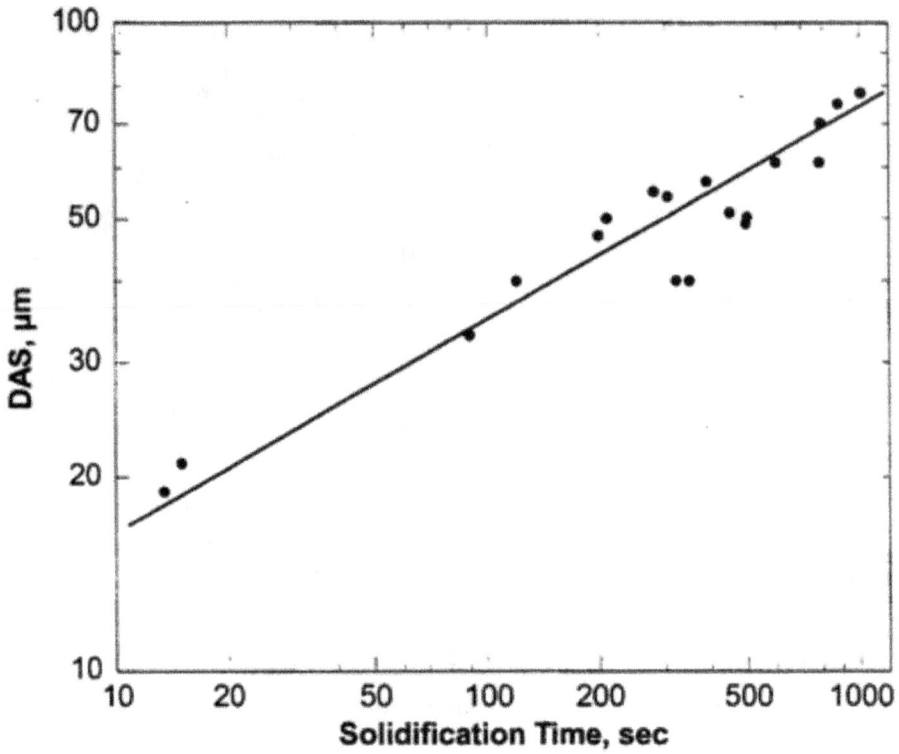

Figure 4.28 The relationship between DAS and local solidification time in a 356 aluminum alloy. (From Gruzleski [3].)

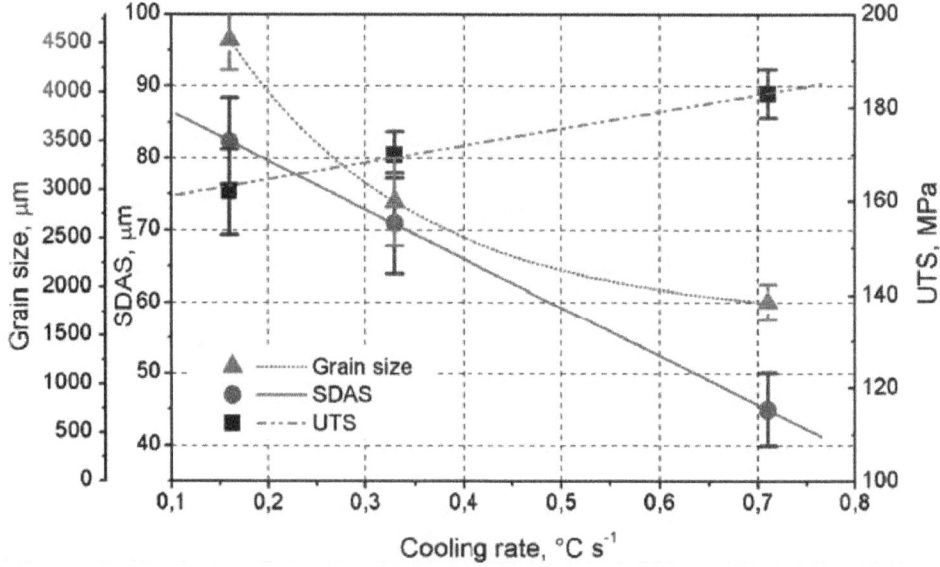

Figure 4.29 Effect of cooling rate on grain size, SDAS, and UTS of an Al-7Si-2Cu alloy. (From Dobrzanski et al. [26].)

Dendrite Coherency

The dendrite coherency point (DCP) is a useful concept in the study of solidification of alloys and refers to the instant when individual dendrites first impinge upon their neighbors during solidification. At this stage, an interconnected skeleton of solids forms through the mushy zone. The volume fraction of solids at this point is called the coherency fraction solid. Both alloy and solidification parameters such as solute concentration, grain refinement, stirring, modification, and cooling rate can affect the fraction solid at which dendrite coherency is reached [27–32].

The coherency fraction solid increases with increasing solute concentration and for a given solute concentration an alloy with a larger value of the slope of the liquidus line (m) has a larger value of coherency fraction solid. This is explained in terms of the growth restriction factor (GRF) which is given by

$$\text{GRF} = m(k - 1) \tag{4.6}$$

A higher GRF for an alloying element would contribute to a decrease in the dendrite growth rate which, in turn, would delay the DCP and hence, increase the coherency solid fraction. In the case of magnesium alloy AZ91E, addition of 0.005% Ti (added as Al-Ti-B master alloy) decreased the grain size from 1630 to 606 μm, the DCP was delayed, and the coherency solid fraction increased from 10.6 to 20.3% [33]. However, if the solute forms a new phase below the coherency temperature, a lower coherency solid fraction can be expected as less liquid metal is available at the DCP. This is the case with AZ91 alloy where addition of 1.03 and 2.05% Ca decreased the coherency solid fraction from 50 to 32 and 25%, respectively [34].

The coherency fraction solid increases with increasing degree of modification for hypoeutectic Al-Si alloys. The nucleation temperature and the internal volume fraction of solid decreases with increasing cooling rate [27]. This gives rise to a high volume fraction of grains. Thus, the temperature interval for dendrite coherency becomes shorter and the coherency fraction solid decreases.

Dendrite coherency influences the castability and casting defects of alloys. Early coherency and hence low solid fraction would tend to promote hot tearing. The coherency fraction solid is considered to mark the transition from mass to interdendritic feeding to compensate for shrinkage in casting processes. Casting defects such as macrosegregation, hot tearing, shrinkage, and gas porosity can result from contraction-induced stresses in the continuous network. The dendrite coherency point can be determined by rheological determination [27, 29, 33–35], where a viscometer is used to measure the torque required to rotate a spindle in the crucible during solidification and thermal analysis [36] methods, where one or two thermocouples can be used to measure the thermal conductivity.

Segregation

Segregation is a result of the inhomogeneous distribution of solute at the solid-liquid interface during solidification. This can occur in the microscale (microsegregation) or the macroscale (macrosegregation). Microsegregation results from partitioning and redistribution of solute between solid and liquid phases. The inhomogeneities are associated with dendritic and cellular dendritic structures.

The segregation ratio S, the local maximum solute content divided by the local minimum and the fraction of nonequilibrium constituents f_{neq} are two important parameters to measure the extent of microsegregation. The distribution coefficient k_0 described previously, can be used to know the extent of microsegregation. For $k_0 < 1$, the liquid is

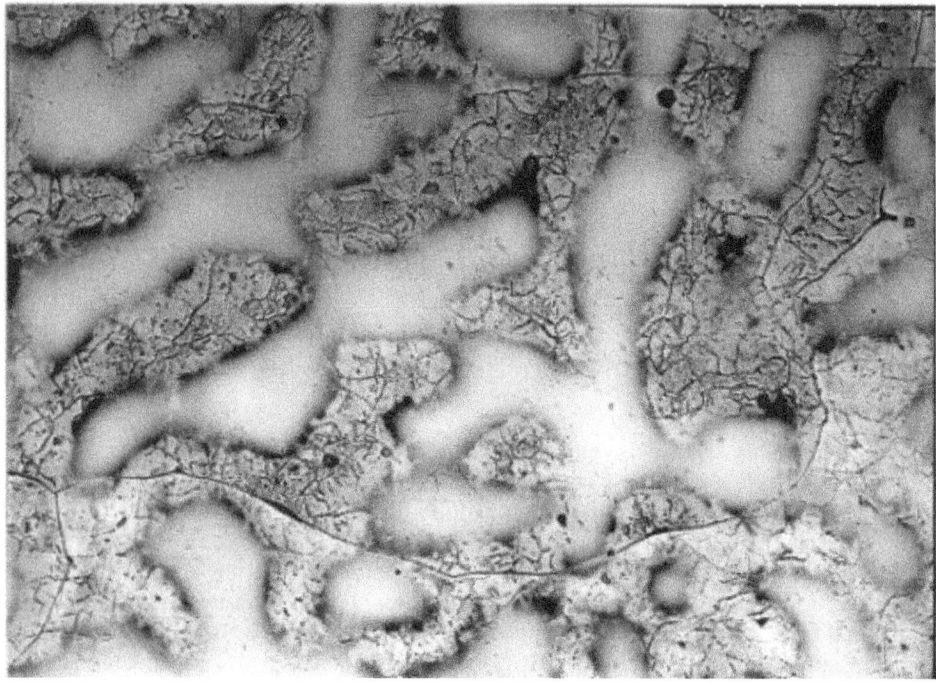

FIGURE 4.30   Optical micrograph showing microsegregation in a 70/30 Cu-Ni cast alloy containing 0.26% Si and 1.4% Cr. 500×. (From Sahoo et al. [37].)

enriched in solute and for $k_0 > 1$, the liquid is depleted in solute during solidification. The further the departure of k_0 from unity, the greater potential for microsegregation.

An example of microsegregation is shown in Fig. 4.30 [37], which is the cored micro-structure of a 70/30 Cu-Ni alloy containing 0.26% Si and 1.4% Cr. Microsegregation in this Cr/Si containing Cu-Ni alloy has been determined by measuring the solute concentration at dendrite and interdendritic centers using an electronprobe microanalyzer (EPMA) with energy dispersive unit and wavelength spectrometer. The composite scan is shown in Fig. 4.31 [37]. It should be noted that in the 70/30 Cu-Ni alloy, although Cu is the major constituent, the "spine" of the dendrite is rich in Ni as evident from the phase diagram (Fig. 8.15), as this is the first phase to solidify. The dendritic areas show vein-like silicides. Typical analyses from the EPMA are given in Table 4.1, which clearly show concentration of Ni, Fe, Si, and Cr at the dendrite centers. The distribution coefficient k_0 for these elements is also shown in Table 4.1.

Microsegregation cannot be eliminated completely. However, it can be minimized by enhancing the solidification process to produce a fine dispersion of segregates as in permanent mold casting. In the case of sand casting, chills should be used to promote rapid cooling.

Macrosegregation refers to compositional variations in castings or ingots and ranges in scale from several millimeters to centimeters or even meters. This long-range segregation is caused by mass movement of the solute-enriched liquid ($k_0 < 1$) and not by diffusion alone. Tin sweat phenomenon, which is observed in large tin bronze castings where drops of tin are found on the surface, is an example of macrosegregation.

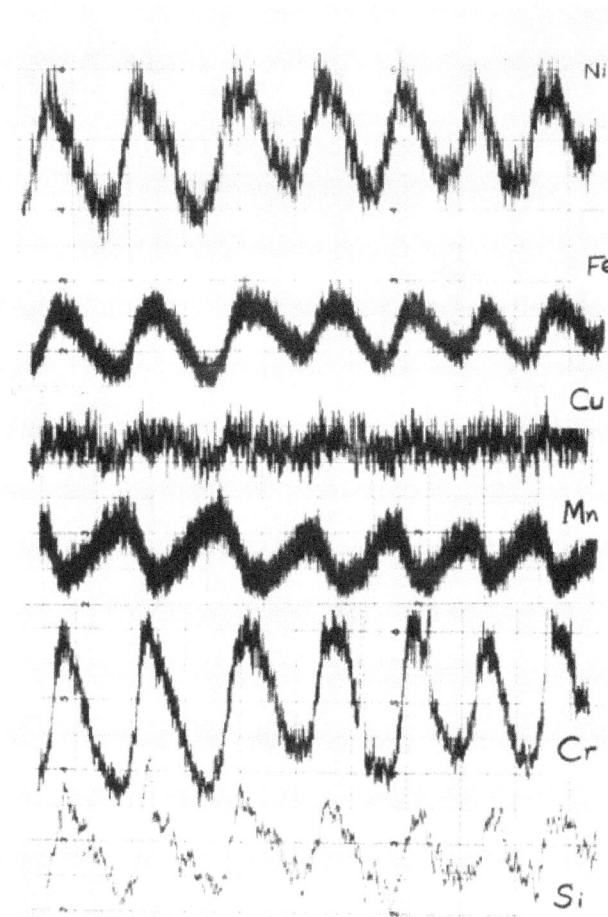

Figure 4.31 Solute distribution across dendritic and interdendritic regions. (From Sahoo et al. [37].)

Element	Composition %, C_0	Solute Concentration, %		Segregation Ratio, C_i/C_c	Effective Distribution Coefficient, $K_E = \dfrac{C_c}{C_0}$
		Dendrite Center, C_c	Interdendrite Center, C_i		
Cu	66.05	55.63	72.40	1.30	–
Ni	30.46	39.83	25.74	0.65	1.31
Fe	0.67	1.03	0.41	0.40	1.54
Mn	0.65	0.616	0.534	0.87	0.95
Si	0.34	0.418	0.230	0.55	1.23
Cr	1.74	2.835	1.110	0.39	1.63

Source: From Sahoo [37].

Table 4.1 Solute Concentrations at Dendrite and Interdendrite Centers

Inverse Segregation

Inverse segregation is opposite to what can be expected from the distribution coefficient. When $k_0 < 1$, inverse segregation results in movement of the interdendritic liquid toward the outer surface and hence, enrichment of solutes. If $k_0 > 1$, the outer surface of the casting becomes solute depleted compared with the inner regions.

Inverse segregation is observed in long freezing range alloys. Aluminum alloys containing high-density elements such as Cu, Fe, Cr, and Mn show severe inverse segregation.

"A" and "V" Segregation

"A" and "V" segregates are usually formed in the steel ingots due to the distribution of C, P, and S. "A" segregates are pipe shaped, solute-rich areas inclined to the axis of the ingot and extending from near the outer surface toward the center of the ingot. The C-, P-, and S-rich interdendritic fluid rises because of the lower density as compared to liquid iron and forms a channel which, when frozen, appears as a pencil of high solute materials in the ingot structure, giving the appearance of "A" segregates. "V" segregates, on the other hand, lie along the thermal center of the ingot and point toward the base. C, P, and S segregate into the fissures which are formed when slow sinking iron-rich liquid sinks to the bottom. These areas are the last to freeze and when sectioned, reveal the "V"-shaped segregates. Schematic representation of A and V segregates is given in Fig. 4.32. Figure 4.33 is a macrophotograph of both A and V segregates in a steel casting [3].

Hot Tearing

Hot tearing is one of the major defects in castings. It can be defined as a strain-induced fracture that occurs during solidification or subsequent cooling of a metal casting as a result of hindered contraction [38]. Pure metals and alloys of eutectic compositions are not prone to hot tearing while alloys of intermediate compositions exhibit susceptibility to hot

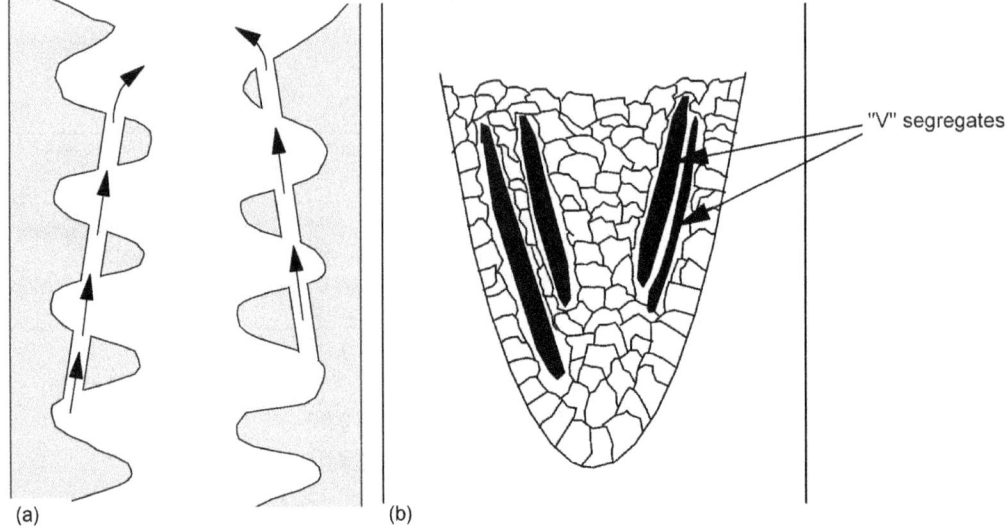

(a)   (b)

Figure 4.32 Schematic of A and V segregates in steel ingots. segregates form from channels of slowly flowing liquid rich in C, P, and S. These channels form in the mushy zone of large ingots and lead to a slow upward motion of the lighter liquid. Fissures formed between grains in the ingot center result in segregates. (a) segregates; (b) segregates. (From Gruzleski [3].)

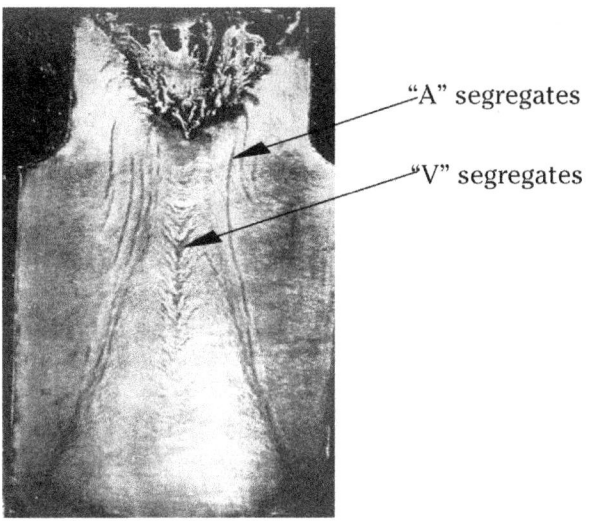

"A" segregates

"V" segregates

FIGURE 4.33   "A" and "V" segregates in a steel ingot. (From Gruzleski [3].)

tearing. Hot tears can be large and visible to the naked eye and sometimes very small and only visible after die penetrant inspection. This complex phenomenon depends largely on alloy composition, part design, feeding system, and casting parameters. Hot tearing defects arise partly due to the inability of the casting to shrink freely during cooling, and can be more severe with the constraint of a rigid metal mold. During early stages of solidification, mass feeding of solidification shrinkage can occur relatively easily to feed developed shrinkage cavity to relieve contraction stresses without hot tearing. Toward the final stages of solidification, the casting structure becomes a coherent mass due to the formation of a dendrite network, and liquid metal cannot flow easily to fill the voids caused by solidification and contractual shrinkage. Therefore, hot tear defects can be mitigated by implementing good casting design principles and solidification conditions to compensate for strains in the casting caused by both types of shrinkage. The need to fill the void space can produce significant stresses in parts of the casting, and if these stresses are greater than the ultimate strength of the metal at any time during solidification, hot tearing can occur.

The general observations on hot tearing of alloys are [39–45]:

- Alloys prone to hot tearing tend to be those with wide freezing ranges and/or little eutectic content that cannot completely surround the grains.

- Fine grain size promotes good interdendritic feeding, low interlocking stresses, and hot tear healing.

- High surface tension of the interdendritic liquid, which impedes interdendritic channel healing, would increase hot tear resistance.

- Oxide inclusions impede interdendritic feeding and reduce the wettability of the interdendritic fluid thereby adversely affecting hot tear tendency.

- Nature of interdendritic second phases and compounds has an effect of grain boundary strength and brittleness.

- In case of permanent mold casting, control of mold temperature in addition to grain refinement is critical to improving the hot tear resistance.

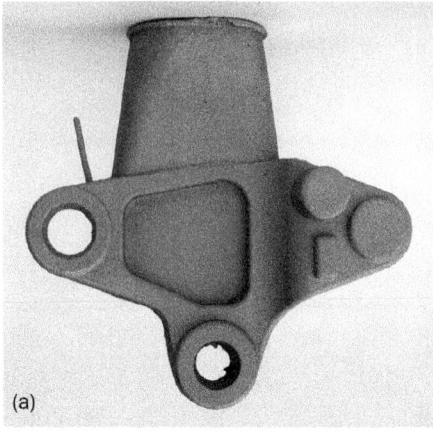
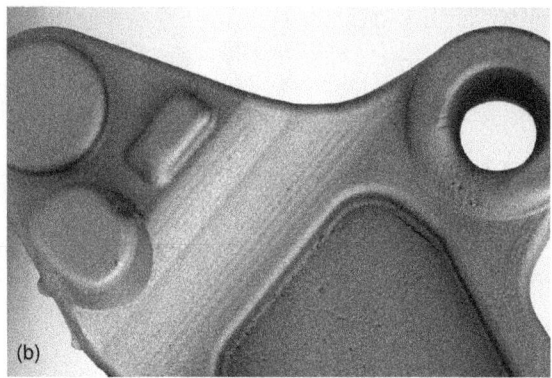

FIGURE 4.34 Hot tearing in a rocker arm casting produced in alloy 535 by gravity permanent mold casting. (a) Rocker arm casting with the riser; (b) alloy 535 showing hot tear crack. (From Fasoyinu et al. [41].)

Computer simulation using the MAGMASOFT software package has been performed to show the effect of mold temperature and grain refinement of a restrained metal mold on hot tearing of Al alloy A206 (Al-Cu type with a solidification range of 80°C). Typical hot tear cracks in two automotive components are shown in Fig. 4.34 [41]. The metal mold halves for the three-rod casting, typical castings, and hot tear cracks are shown in Fig. 4.35 [42, 43]. This mold design has been used extensively to evaluate the influence of casting parameters on the hot tear tendency of aluminum and magnesium alloys. The maximum principal strain developed during solidification was calculated by simulation and correlated with fraction solid as a function of mold temperature (290 and 450°C). The simulation results showed that the maximum principal strain developed during the last stage of solidification was lower at a higher mold temperature of 450°C and no hot tearing was observed under this condition. It has been demonstrated that a combination of grain refinement and preheating of metal mold to ≥400°C is necessary to prevent the formation of hot tearing during the solidification of alloy 206.0 in the restrained metal mold. The computer simulation results closely match the hot cracking trends observed experimentally during the casting trials. Similar conclusions have been obtained in another study on effect of mold temperature on hot tearing susceptibility where it has been shown that the mold temperature for aluminum alloys 319, 535, and 206 should exceed 350, 220, and 400°C, respectively, to eliminate hot tearing [46].

In addition to grain refinement and maintaining optimum mold temperature in permanent molds, hot tearing can also be eliminated/minimized by modifying alloy composition or adding other compounds such as misch metal. In case of alloys 535 and 206, it has been shown that during permanent mold casting of two automotive components such as the engine mount (low-pressure casting) and rocker arm (gravity casting), lowering of the magnesium level in 535.0 from 7.2 to 5.0% and copper level in 206.0 from 5.0 to 2.85% improved hot tear resistance [41]. Typical hot tear cracks in the rocker arm casting are shown in Fig. 4.34. Examining the Al-Mg and Al-Cu phase diagrams indicate that such reduction in Mg content in the Al-Mg alloy reduces the freezing range to 60°C. In case of the Al-Cu alloy, this reduction in the freezing range is 30°C.

Another method of eliminating/minimizing hot tearing is to add misch metal. In case of alloy 201, adding 1.5 to 2% misch metal, improved the hot tear resistance

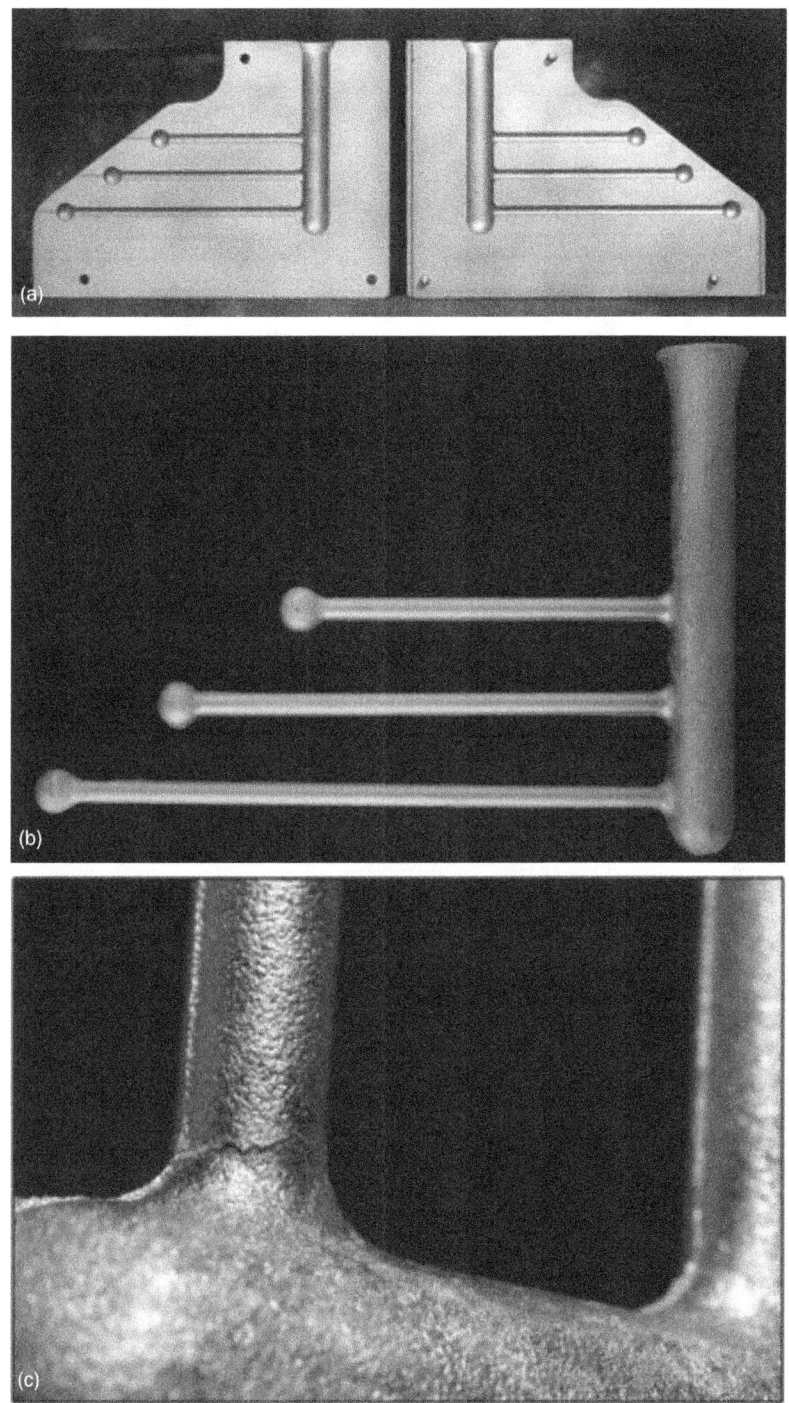

Figure 4.35 Photographs of (a) three-arm mold halves; (b) three-arm restrained casting; (c) & (d) showing the location of hot tear cracks. (From Fasoyinu et al. [43].) (*Continued*)

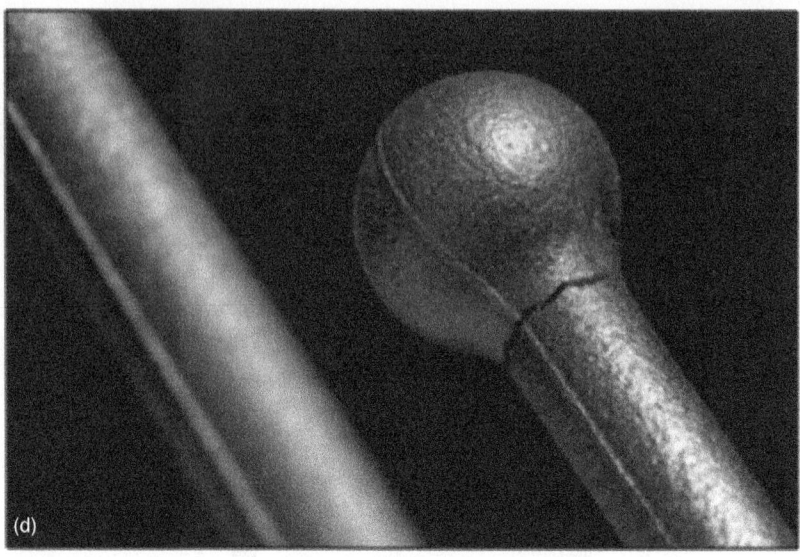

(d)

Figure 4.35 *(Continued)*

significantly. Metallographic examination indicated that addition of such misch metal refined the grain size (average secondary dendrite arm spacing reduced from 132 to 51 μm), even though the alloy had been grain refined using standard Al-Ti-B practice and the eutectic phase got modified [47].

Various test methods have been developed to evaluate the hot tearing tendency of alloys. These have been reviewed by Eskin and colleagues [48]. However, each test method has its limitations and there is no universal test method yet. Li and colleagues have recently reviewed hot tearing in aluminum alloys focusing on theories and methods, hot tearing variables, and test methods [49]. Following this, they produced a quantitative test method in collaboration with CANMET Materials Technology Laboratory as shown schematically in Fig. 4.36, where the test casting has two arms with a riser at the center. The right arm is constrained and the left arm is connected to the load cell. The data acquisition system measures the temperature of the casting and the stress generated. Parameters affecting hot tearing are alloy composition, casting temperature, mold temperature,

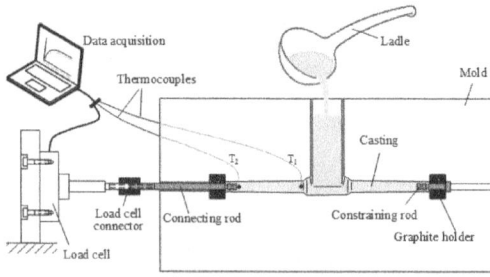

Figure 4.36 Schematic of the experimental setup to determine hot tear resisitance. (From Li et al. [49].)

and grain refinement. Details of the test method can be found in Refs. 49 and 50. A typical plot for alloy 206 is given in Fig. 4.37 showing load temperature and load derivative with time. Quantitative information on crack initiation and propagation can be obtained by analyzing the data in Fig. 4.37. Similar data on alloy 356 predicted no hot tearing.

Hot Isostatic Pressing

Hot isostatic pressing, also known as HIP or Hipping, is a process to eliminate porosity and improve mechanical properties in castings. An example is shown in Fig. 4.38 [6] for an aluminum alloy (A356-T6 condition), where shrinkage and gas porosities in Fig. 4.38a have been eliminated (Fig. 4.38b). The process involves surrounding the casting with a pressurized fluid, usually argon gas, while simultaneously heating the

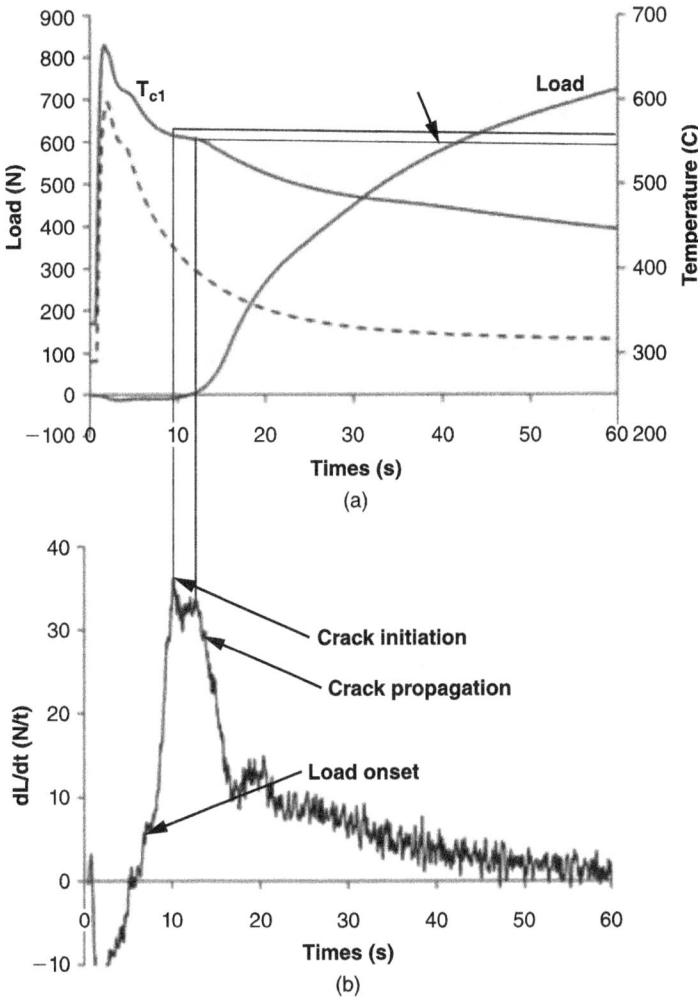

FIGURE 4.37 This illustrates: (a) temperatures and load development as a function of time for alloy M206; T_{C1} and T_2 are measured by thermocouples located at the centerline of the rod; (b) derivative of load versus time curves; (*Continued*)

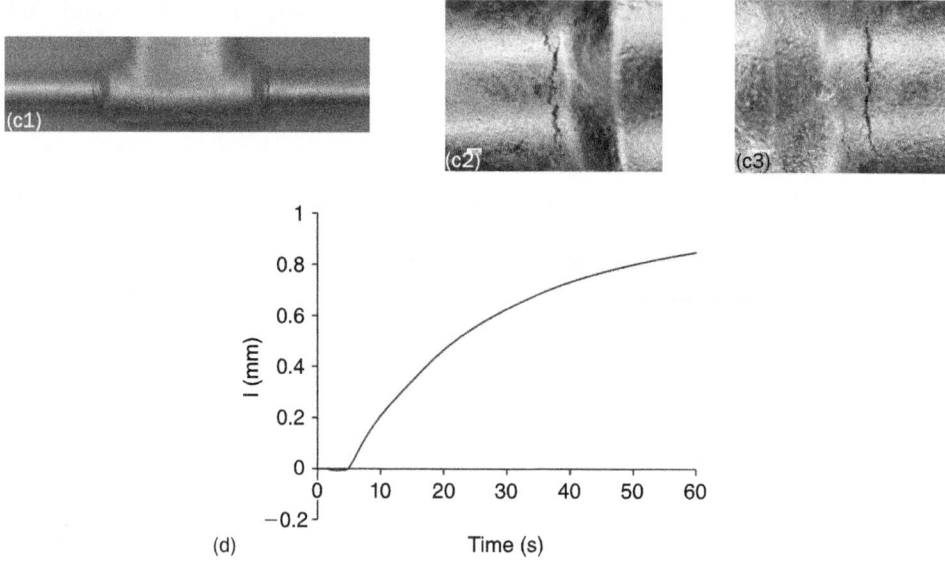

Figure 4.37 (*Continued*) (*c*) photographs of the constrained casting shows cracking locations; and (*d*) displacement measured as a function of time. (From Li et al. [49].)

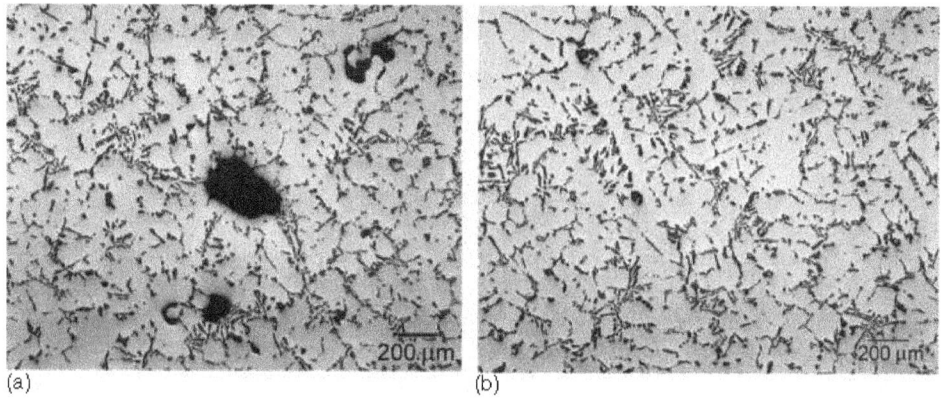

Figure 4.38 Effect of HIP on microstructure of A356 aluminum casting (*a*) in the as-cast and T6 condition showing porosity, and (*b*) in the as-cast, HIPed, and T6 condition showing elimination of porosity. (From ASM Handbook [6].)

casting to a temperature below the solidus but high enough to promote plastic flow and diffusion within the material. Internal void and microporosities are eliminated by plastic deformation, creep, and diffusion at elevated temperatures. The driving force to achieve densification is associated with the reduction in surface area and, hence, surface energy of the pores. This improves the mechanical properties of the castings. The applied pressure ranges from 51 MPa (7350 psi) to 310 MPa (45,000 psi) in a pressure vessel. The low end of the pressurization is good enough for most of the castings. The pressure is applied from all directions, hence, the term isostatic.

Material	Melting Point, T_m (°C)	Yield Stress at Room Temperature, MPa	Hipping Temperature, °C	Hipping Pressure, MPa
Al alloys	660 (Al)	100–627	500	100
Cu alloys	1083 (Cu)	60–960	800–950	100
Superalloys	1453 (Ni)	200–1,600	1,100–1,280	100–140
Ti alloys	1670 (Ti)	180–1,320	920	100
Steels	1536 (Fe)	500–1,980	950–1,160	100

Source: From Atkinson and Davies [51].

TABLE 4.2   Typical Hipping Temperatures and Pressure

There are no rams, dies, or external frictional forces. A typical pressure of 100 MPa (approximately 1000 atm) is roughly equivalent to the pressure at the bottom of the ocean's deepest trench. Commercial Hipping units are available with operating temperatures between 500 and 2000°C. Typical hipping temperatures and pressures for some casting alloys are given in Table 4.2 [51]. Work zone diameter can vary from 75 mm to 3 m.

The driving force for closure of an isolated spherical pore can be expressed as

$$P = 2\gamma/r \qquad (4.7)$$

where γ is the surface energy (in joules per square meter) of the internal surface of the pore and r is the radius of curvature of the pore surface [51]. For a pore of 0.1 mm in diameter, the driving force (with γ of 1 J/m^2) is 40 kPa. At a typical Hipping pressure of 100 MPa, the pore can be completely dissolved in the matrix.

Elimination of internal porosity leads to improvement in mechanical properties as illustrated in Figs. 4.39 and 4.40 [6]. These two figures show the improvement in high-cycle fatigue properties in alloy 356 and the improvement in UTS and percent

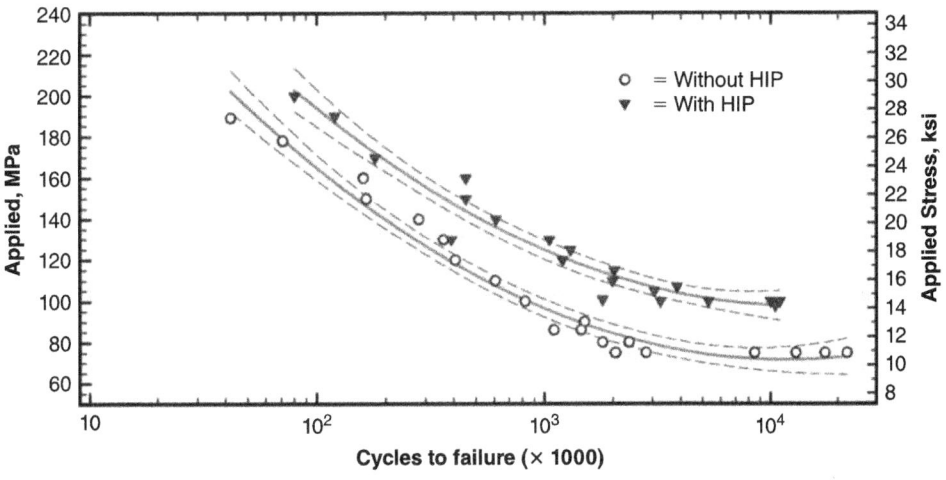

FIGURE 4.39   Improvement in fatigue life of aluminum alloy AlSi7MgCu0.5 (gravity die-cast automotive diesel-engine block) when HIP is used. (From *ASM Handbook* [6].)

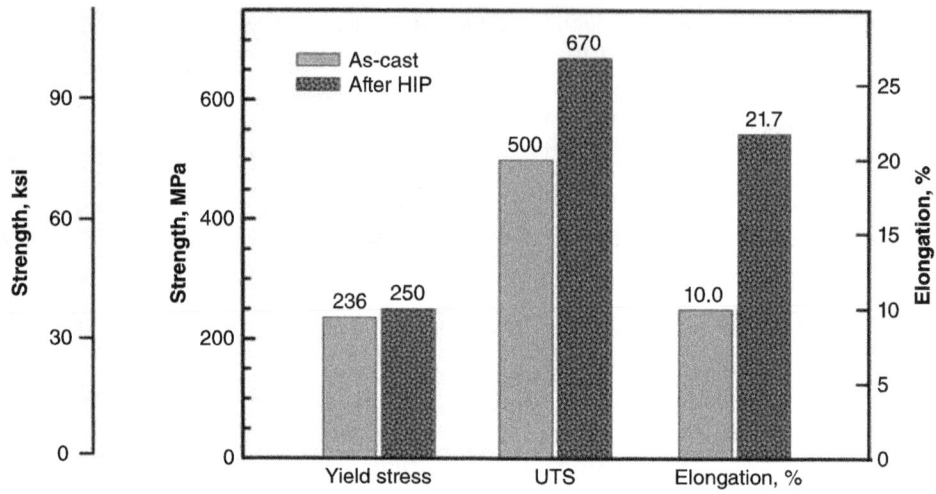

FIGURE 4.40 Improvement in mechanical properties of a cast nickel-aluminum bronze (C95800) following HIP. (From *ASM Handbook* [6].)

elongation of a nickel-aluminum bronze (C95800). In the bronze casting, the UTS increases by 34% and the percent elongation is almost doubled.

Solidification under Pressure

Another approach to eliminate/minimize porosity in castings is to solidify under pressure which may be less expensive than Hipping. This nonconventional technique has been applied to aluminum alloys A356, A319, A206, where solidification under 10 atmospheres pressure has led to reduction in porosity by 63 to 84% and consequently improvement in mechanical properties [52–54]. Amount of porosity present after solidification was less than 0.5% in A356 alloy cast in chemically bonded sand molds as a result of which UTS increased by about 6 to 11% and percent elongation increased by 50 to 130%. In the case of A319 and A206 alloys, increased soundness of wedge castings improved percent elongation by 13 to 45% and 112 to 120%, respectively. UTS of A206 alloy increased by about 18%.

The theory behind the reduction in porosity due to solidification under pressure is based on Sievert's law and Boyle's law, which are

$$C_s = K\sqrt{P} \qquad (4.8)$$

where C_s is the solubility of hydrogen which is the main dissolved gas in solid Al alloys, K is a constant, and P is pressure inside the pore which can be equal to the applied pressure.

$$V_p = C(1/P) \qquad (4.9)$$

where V_p is the volume of the hydrogen pore, and C is the volume available to form a pore [52]. If the volume of gas available for the formation of voids is constant, as pressure increases, volume of the voids must decrease. Thus, application of 1 to 10 atmospheres should lead to reduction in pore size by tenfold.

A pressure vessel is required for pressurizing during solidification. A combination of compressed air and nitrogen has been used successfully to do pressurization. This solidification technique is being used commercially. An example of this technique can be seen at Mercury Marine in Fond Du Lac, Wisconsin, where cylinder blocks for the marine industry are being cast using the lost foam process [55].

References

1. R. W. Heine, C. R. Loper, and P. C. Rosenthal, *Principles of Metal Casting*, McGraw-Hill Publishing Company, New York, 1967.

2. *Basic Principles of Gating and Risering*, 2d ed., American Foundry Society, Schaumburg, IL, 2008.

3. J. E. Gruzleski, *Microstructure Development during Metal Casting*, American Foundry Society, Des Plaines, IL, 2000.

4. B. Chalmers, *Principles of Solidification*, John Wiley & Sons, Inc., New York, 1964.

5. L. Backerud, L. M. Liljenvall, and H. Steen, "Solidification Characteristics of Copper Alloys," *INCRA Series on the Metallurgy of Copper*, International Copper Association, 1982.

6. *ASM Handbook*, Vol. 15, *Casting*, 2008, pp. 309, 408.

7. S. Ducharme, M. Sahoo, and K. Sadayappan (eds.), *Casting Copper-Base Alloys*, American Foundry Society, 2007.

8. F. Hudson and D. A. Hudson, *Gunmetal Castings*, Hart Publishing Company, Inc., 1967.

9. M. N. Croker, R. S. Fidler, and R. W. Smith, "Characterisation of Eutectic Structures," *Proceedings of Royal Society (London)*, A335: 15–37, 1973.

10. M. Sahoo and R. W. Smith, "Mechanical Properties of Unidirectionally Solidified Al-Si Eutectic Alloys," *Metal Science*, 9: 217–222, 1975.

11. M. Sahoo and R. W. Smith, "Structure and Mechanical Properties of some Modified Eutectic Composites," *Canadian Met. Quarterly*, 15: 1–7, 1976.

12. M. Sahoo and R. W. Smith, "Structure and Mechanical Properties of Unidirectionally Solidified Zn-Ge Eutectic Alloys," *Journal of Materials Science*, 11: 1125–1134, 1976.

13. M. Sahoo, R. A. Porter, and R. W. Smith, "Mechanical Behaviour of the Cd-Zn Eutectic Composite," *Journal of Materials Science*, 11: 1680–1690, 1976.

14. G. A. Chadwick, "Structure and Properties of Eutectic Alloys," *Metal Science*, 9(1): 300–304, 1975.

15. C. A. Brunch, R. C. Haubert, M. F. X. Gigliotti, and M. F. Henry, "Eutectic Composite Turbine Blade Development," *Final Report*, General Electric Co, Ohio Aircraft Engine Group, Cincinnati, Nov. 1976.

16. M. Sahoo and L. V. Whiting, "Foundry Characteristics of Sand Cast Zn-Al Alloys," *Transactions of the American Foundry Society*, 92: 861–870, 1984.

17. J. Thomson, F. A. Fasoyinu, M. Sadayappan, and M. Sahoo, "Casting Characteristics of Permanent Mold Cast Mg-Alloy AZ91E," *Transactions of the American Foundry Society*, 110: 1181–1189, 2002.

18. F. A. Fasoyinu, J. L. Dion, D. Cousineau, R. A. Matte, K. G. Davis, and M. Sahoo, "Fluidity of Permanent Mold Cast Copper-Base Alloys," *Transactions of the American Foundrymen's Society*, 100: 547–559, 1992.

19. L. V. Whiting, P. D. Newcombe, and M. Sahoo, "Casting Characteristics of Red Brass Containing Bismuth and Selenium," *Transactions of the American Foundry Society*, 103: 683–691, 1995.

20. E. E. Stonebrook and W. E. Sicha, "Correlation of Cooling Curve Data with Casting Characteristics of Aluminum Alloys," *Transactions of the AFS*, 57: 489, 1949.

21. M. Sahoo, J. R. Barry, and K. Kleinschmidt, "Use of Ceramic Foam Filters in the Brass and Bronze Foundry," *Transactions of the American Foundrymen's Society*, 89: 611–620, 1981.

22. M. Sahoo, J. L. Dion, C. Bibby, and A. L. Matthews, "Use of Ceramic Filters for Copper-base Alloys," *Transactions of the American Foundrymen's Society*, 102: 33–43, 1994.

23. A. L. Matthews, F. A. Fasoyinu, J. L. Dion, M. Popescu, and M. Sahoo, "A Comparison of Cellular and Foam Filter Effectiveness in Gravity Permanent Mold Castings of Aluminum Alloys," *Proceedings of the 4th International Conference on Molten Aluminum Processing*, Orlando, 1995, pp. 103–120.

24. M. Sadayappan, F. A. Fasoyinu, D. Cousineau, and M. Sahoo, "Effect of Minor Alloy Additions on the Fluidity of Permanent Mold Cast Copper-Base Alloys," *Transactions of the American Foundrymen's Society*, 106: 735–742, 1998.

25. M. Sahoo, CANMET Materials Technology Laboratory, Unpublished Work

26. L. A. Dobrzanski, R. Maniara, J. Sokolowski, and W. Kasprzak, "Effect of Cooling Rate on the Solidification Behavior of AlSi7Cu2 Alloy," *Journal of Materials Processing Technology*, 191: 317–320, 2007.

27. L. Arnberg, *Solidification Characteristics of Aluminum Alloys*, American Foundry Society, Des Plaines, IL, 1996.

28. Y. C. Lee, A. K. Dahle, and D. H. St. John, "The Role of Solute in Grain Refinement of Magnesium," *Metallurgical and Materials Transactions A*, A31: 2895–2906, 2000.

29. C. M. Gourlay, B. Meylan, and A. K. Dahle, "Rheological Transition at Low Solid Fraction in Solidifying Magnesium Alloy AZ91," *Materials Science Forum*, 561–565: 1067–1070, 2007.

30. N. Veldman, A. K. Dahle, D. St. John, and L. Arnberg, "Dendrite Coherency of Al-Si-Cu Alloys," *Metallurgical and Materials Transactions A*, A321: 147–155, 2001.

31. M. Malekan and S. Shabestari, "Effect of Grain Refinement on the Dendrite Coherency Point during Solidification of A319 Aluminum Alloy," *Metallurgical and Materials Transactions A*, A40: 3196–3203, 2009.

32. G. Chai, L. Backerud, T. Rolland, and L. Arnberg, "Dendrite Coherency during Equiaxed Solidification in Binary Aluminum Alloys," *Materials Transactions A*, 26(4): 965–970, 1995.

33. C. Ravindran, Ryerson University, Toronto, Ontario, Canada, unpublished work.

34. S. Liang, R. Chen, J. Blandin, M. Suery, and E. Jan, "Thermal Analysis and Solidification Pathways of Mg-Al-Ca System Alloys," *Materials Science and Engineering A*, A480: 603–608, 2008.

35. L. Arnberg, G. Chai, and L. Backerud, "Determination of Dendrite Coherency in Solidifying Melts by Rheological Measurements," *Materials Science and Engineering A*, A173: 101–103, 1993.

36. R. Chavez-Zamarripa, J. A. Ramos-Salas, J. Talamantes-Silva, S. Valtierra, and R. Colas, "Determination of the Dendrite Coherency Point during Solidification by Means of Thermal Diffusivity Analysis," *Metallurgical and Materials Transactions A*, A38: 1875–1879, 2007.

37. M. Sahoo, K. C. Wang, and J. O. Edwards, "Foundry Characteristics and Mechanical Properties of both Niobium-Modified and Chromium-Modified High-Strength 70/30 Cu-Ni Alloys," *Transactions of the American Foundrymen's Society*, 87: 529–536, 1979.

38. A. L. Keaney and J. Raffin, *Hot Tear Control Handbook for Aluminum Foundrymen and Casting Designers*, AFS Publication, Des Plaines, IL, 1971.

39. M. O. Pekguleryuz and P. Vermette, "A Study on Hot-Tear Resistance of Magnesium Diecasting Alloys," *Transactions of the American Foundry Society*, Vol. 114, Paper No. 06-092, 2006.

40. S. Lin, C. Aliravci, and M. O. Pekguleryuz, "Hot-Tear Susceptibility of Aluminum Wrought Alloys and the Effect of Grain Refining," *Metallurgical and Material Transactions A*, 38A: 1056–1068, May 2007.

41. Y. Fasoyinu, J. P. Thomson, M. Sahoo, P. Burke, and D. Weiss, "Permanent Mold Casting of Aluminum Alloys A206.0 and A535.0," *Transactions of the American Foundry Society*, Vol. 115, Paper No. 07-095, 2007.

42. Y. Fasoyinu and M. Sahoo, "Factors Influencing Hot Tearing of Aluminum Alloy 206.0 Poured in Metal Molds," *Proceedings of the International Symposium on Materials Development and Performance of Sulphur Capture Plants*, Edited by C. Barry and P. Wanjara, MetSoc, COM 2009, pp. 3–17.

43. Y. Fasoyinu, M. Sahoo, and S. Sikorski, "Hot Tearing of Aluminum Alloys 206 and 535 Poured in Metal Mold," *International Conference on Permanent Mold Casting Aluminum and Magnesium*, American Foundry Society, Schaumburg, IL, Feb. 2008, pp. 11–25.

44. M. Sadayappan, M. Sahoo, and R. W. Smith, "Influence of Alloying Elements and Melt Treatment on The Hot Tearing Resistance of Aluminum Alloy A201," *Light Metals 2001 Métaux Légers, Conference of Metallurgists—COM 2001*, Aug. 26–29, 2001, Toronto, Ontario, pp. 455–466.

45. Q. Liu, R. W. Smith, and M. Sahoo, "Control of Hot Tearing in Cast-to-Shape Products," *Proceedings of the 4th Decennial International Conference on Solidification Processing*, Sheffield, UK, 1997, pp. 213–216.

46. M. Sadayappan, M. Sahoo, and D. J. Weiss, "Evaluation of the Hot Tearing Susceptibility of Selected Magnesium Casting Alloys in Permanent Molds," *Transaction of the American Foundry Society*, Vol. 115, Paper No. 07-154, 2007.

47. M. Shkuka, B. J. Yang, R. W. Smith, M. Sadayappan, and M. Sahoo, "Microstructure and Mechanical Properties of Al-Cu Casting Alloys—Effect of Addition of Mischmetal," *Transactions of the American Foundry Society*, Vol. 112, Paper No. 04-130, 2004.

48. D. G. Eskin, Suyitno, and L. Katgerman, *Progress in Materials Science*, 49: 629–711, 2004.

49. S. Li, D. Apelian, and K. Sadayappan, "Hot Tearing in Cast Al Alloys: Mechanisms and Process Control," *International Journal of Metal Casting*, Paper No. 12-007, 2012.

50. S. Li, D. Apelian, and K. Sadayappan, "Quantitative Investigation of Hot Tearing of Al-Cu Alloy (206) Cast in a Constrained Bar Permanent Mold," *Materials Science Forum*, 618–619: 57–62, 2009.

51. H. V. Atkinson and S. Davies, "Fundamental Aspects of Hot Isostatic Pressing: An Overview," *Metallurgical and Materials Transactions A*, 31A: 2981–3000, Dec. 2000.

52. S. B. Ghanti, E. A. Druschitz, A. P. Druschitz, and J. A. Griffin, "The Effects of Solidification under Pressure on the Porosity and Mechanical Properties of Cast Aluminum Alloy," *Transactions of the American Foundry Society*, Paper No. 10-063, 2010, p. 7.

53. S. B. Ghanti, E. A. Druschitz, A. P. Druschitz, and J. A. Griffin, "The Effects of Solidification under Pressure on the Porosity and Mechanical Properties of A206-T6 Cast Aluminum Alloy," *Transactions of the American Foundry Society*, Paper No. 11-048, 2011, p. 8.

54. P. P. Chintalapati, J. A. Griffin, and R. D. Griffin, "Improved Mechanical Properties of Lost Foam Cast A356 and A319 Aluminum Solidified under Pressure," *Transactions of the American Foundry Society*, Paper No. 07-058, 2007 p. 17.

55. R. Donahue, Mercury Marine, Fond du Lac, private communication.

CHAPTER 5

Gating and Risering of Castings

Andrew Adams
Foseco

Definition of Runner System

Runner systems are the method of introducing liquid metal into part of the mold, the casting cavity, which forms the final shape required by the foundry customer. Since the runner system is the passageway to the casting cavity, it should be designed based on the requirements of the casting. The system should minimize the damage to the metal while it is moving inside the casting cavity. Where possible the runner system should assist the risering system in producing a sound casting, or at a minimum, not hinder it. Inside of these constraints, the runner system should be designed to meet the pouring time required for production on automatic molding machines and pouring units [1].

Under certain circumstances, a runner system may provide a casting that meets the customer's requirements without the use of additional risering systems. This is determined by the type of metal, type of casting, and the customer's specifications. However, in many cases the runner system works in concert with the risering system to produce a sound casting.

Runner systems may be designed differently for different types of molding systems. Two major types of mold systems are sand or ceramic, generally used for all metals, or metal or permanent molds, most commonly used for nonferrous metals, but may also be used for some ferrous castings. Generally, the terminology used for each molding is the same, but some of the calculations and designs utilized to obtain the final system may be different.

There are two other major categories in runner systems based on the molding equipment in the foundry; horizontally parted molding and vertically parted molding. Traditionally, molds were produced with their parting line parallel to the ground or horizontal molding. Equipment has been developed to produce molds with their parting line perpendicular to the ground or vertical molding. There are also molding equipment types that transition from one orientation to another during or directly after pouring. The same runner system terminology is generally used, but equipment-specific calculations may be needed when designing the runner system.

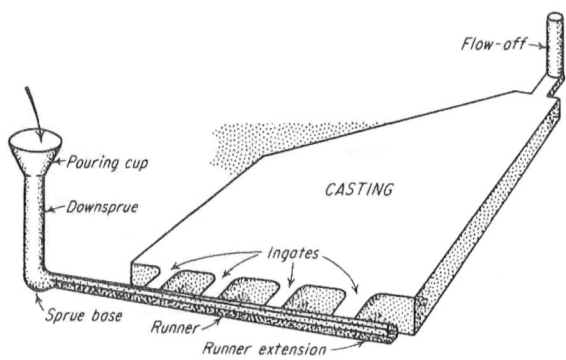

FIGURE 5.1   Finger-gated casting with flow-off. (From Heine et al. [2, p. 214].)

The elements of a basic and very common gating system include the downsprue, through which metal enters the runner and in turn passes through the ingates into the casting cavity (Fig. 5.1). That part of the gating system, which most restricts or regulates the rate of pouring, is the primary choke, more often called simply the choke. At the top of the downsprue, there may be a pouring cup or pouring basin to minimize splash and turbulence, and to promote the entry of only clean metal into the downsprue. To further prevent the entry of dirt or slag into the downsprue, the pouring basin may contain a skim core, a delay disk or screen, or a sprue plug. To prevent erosion of the gating system when a large amount of metal is to be poured, an impact core may be placed in the bottom of the pouring basin, at the bottom of the downsprue, or wherever the flowing metal impinges with more than normal force.

Runner System Design

A runner system design can be characterized by the relationship of the cross-sectional sizes of various components of the system. Foundry engineers determine these characteristics using a variety of methods: rigorous calculations, comparisons to previous systems, established tables, or in some cases approximations. In each instance, the designer is intending to optimize the quality and costs associated with producing the casting.

Regardless of the design process, runner systems can generally be separated into three categories:

1. Pressurized
2. Nonpressurized
3. Hybrid

Each approach has its own positive and negative points, which contribute to the performance obtained under specific casting configurations and metal types.

Pressurized

Pressurized runner systems (Fig. 5.2) are designed with the restriction or choke at the ingates. For horizontal molding, the runner is generally in the cope and the ingates are in the drag or in the cope at the parting line. The cope runner cross section is generally large, in an effort to slow down the velocity of the metal during filling and assist in slag

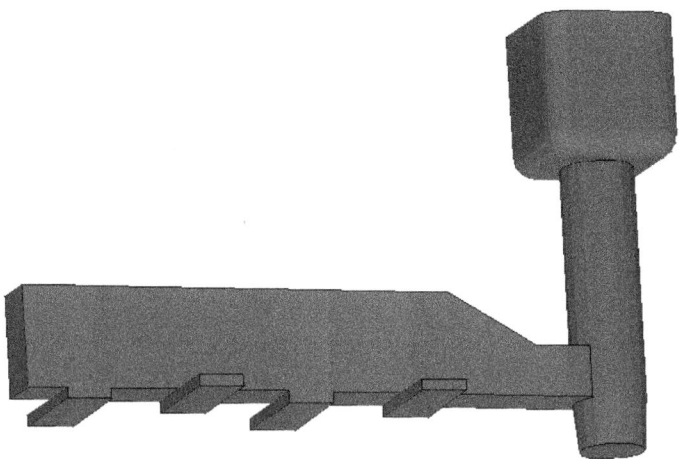

FIGURE 5.2   Model representing a pressurized runner system. (From Adams [1].)

flotation. The ingates are the smallest total area in the system and work to force the molten metal back up into the system. They are located near the parting line to allow the cleaner metal, at the bottom of the runner bar, to enter the casting cavity.

The greatest advantage of the pressurized system is the large runner that allows slag floatation. The most important disadvantage of a pressurized system is that it creates a high-velocity metal flow region in the ingate as well as immediately inside the casting. These high molten metal velocities are generally associated with slag formation and sand erosion. A secondary disadvantage is that the initial metal, which may be damaged during that initial flow and filling, can directly enter the casting cavity before the runner system is filled.

When pattern plate area was not at a premium, and the costs of remelting were not considered important, runner bars could be large and long without much concern to the foundry. However, with every square inch of pattern plate now needed for sellable castings, runners have to be shortened and their cross-sectional areas must be decreased in order to add castings to the mold. Additionally, because the runners are now much shorter, the castings are not as clean as they once were and the balance between quality and cost is no longer favorable.

Nonpressurized

Nonpressurized runner systems (Fig. 5.3) have the choke either in, or directly next to the sprue. This concept was to provide space for slag floatation outside the casting cavity by using a larger runner bar to reduce metal velocities. The ingates are also large in order to reduce velocity as the liquid metal enters the casting cavity. For horizontal molding, this means that the runners are in the drag with the ingates in the cope.

Pattern plate area can also be a downfall of the nonpressurized system. Increasing the yield by placing more casting cavities on the plate has become the rule of the foundry. Something is sacrificed when a tradeoff is made, and in this case, it was pattern plate area for the runner system. As the system became shorter, less time was available for slag flotation during pouring and the casting quality suffered.

Pressurized and nonpressurized runner systems are used in vertical molding applications based on the same principles that apply to horizontal molds. Although the

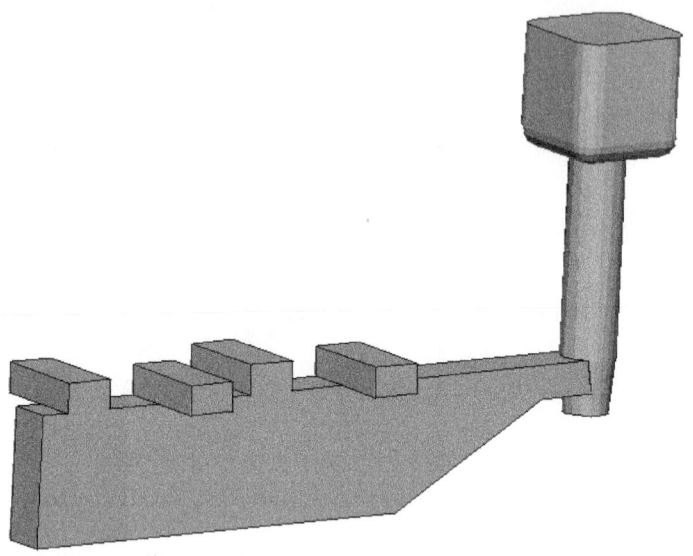

FIGURE 5.3   Model representing a nonpressurized runner system. (From Adams [1].)

cross-section calculations are basically the same, they have to be more rigorous due to the effect of gravity at the different levels of the casting cavities. An added problem in vertical molding, compared to horizontal molding, is the location of the ingate on the casting (Figs. 5.4 and 5.5). In some cases, top gating will make feeding a casting easier, but this causes the molten metal to waterfall through the casting cavity. Bottom gating makes feeder design somewhat more complicated, but usually provides lower turbulence during filling. The best designs are always a matter of judgment.

Hybrid

Hybrid systems try to combine the advantages of both pressurized and unpressurized systems, and sometimes include additional features specifically designed for slag

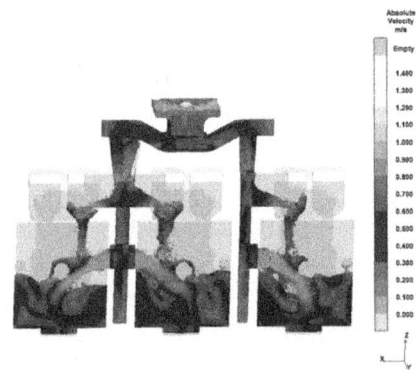

FIGURE 5.4   Top filling in vertical molding system. (From Adams [1].)

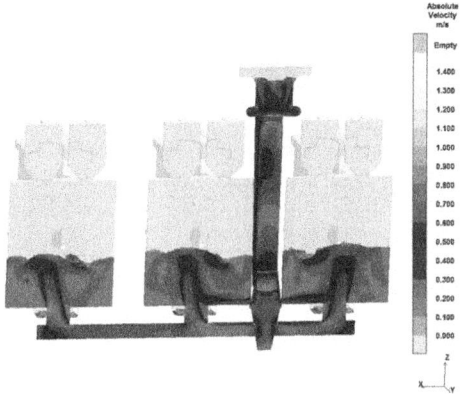

FIGURE 5.5   Bottom filling in vertical molding system. (From Adams [1].)

removal. Some of these hybrid systems include runners that change from drag to cope to drag to try to separate slag by stopping it in the first drag section, while some other systems feature swirl chambers which centrifugally separate slag.

Hybrid systems generally require more pattern land area than conventional systems. Finding space on the pattern plate can be a major concern for these systems because a sellable casting may have to be replaced by the system itself. The systems also may appear to give variable results that do not seem to be controllable.

Types of Gates

Metal can be directed into a mold cavity in various ways. The design of each gating system depends upon its primary objectives. Thus, a gate may be designed for ease of molding, to avoid turbulent flow, or to prevent washing of sand from the mold walls. Again, a principal objective might be to avoid inclusion of dross or slag with the metal entering the mold. Naturally, other factors are not disregarded when a gate is designed with a particular purpose in mind. Various designs of gates, which have been used in the past both successfully and unsuccessfully, are shown in Fig. 5.6, and the following discussions pertain to the gates illustrated.

Parting Gate

These gates enter the mold cavity along the parting line, separating the cope and drag portions of the mold. They may contain devices such as skim bobs or relief sprues to collect dross or slag (*a*, *b*, and *c*, Fig. 5.6) or to relieve pouring pressure. Design *d* illustrates the use of a pouring basin to serve this function; design *e* contains a shrink bob serving the dual function of slag or dross collector and metal reservoir, to feed the casting as it shrinks. Designs *f* and *g* illustrate the use of core inserts to filter the metal or prevent erosion of the mold.

Bottom Gate

The bottom gate enters the casting cavity at the bottom of the drag half of the mold. It is illustrated by design *h* in Fig. 5.6, although other variations are also used. For example,

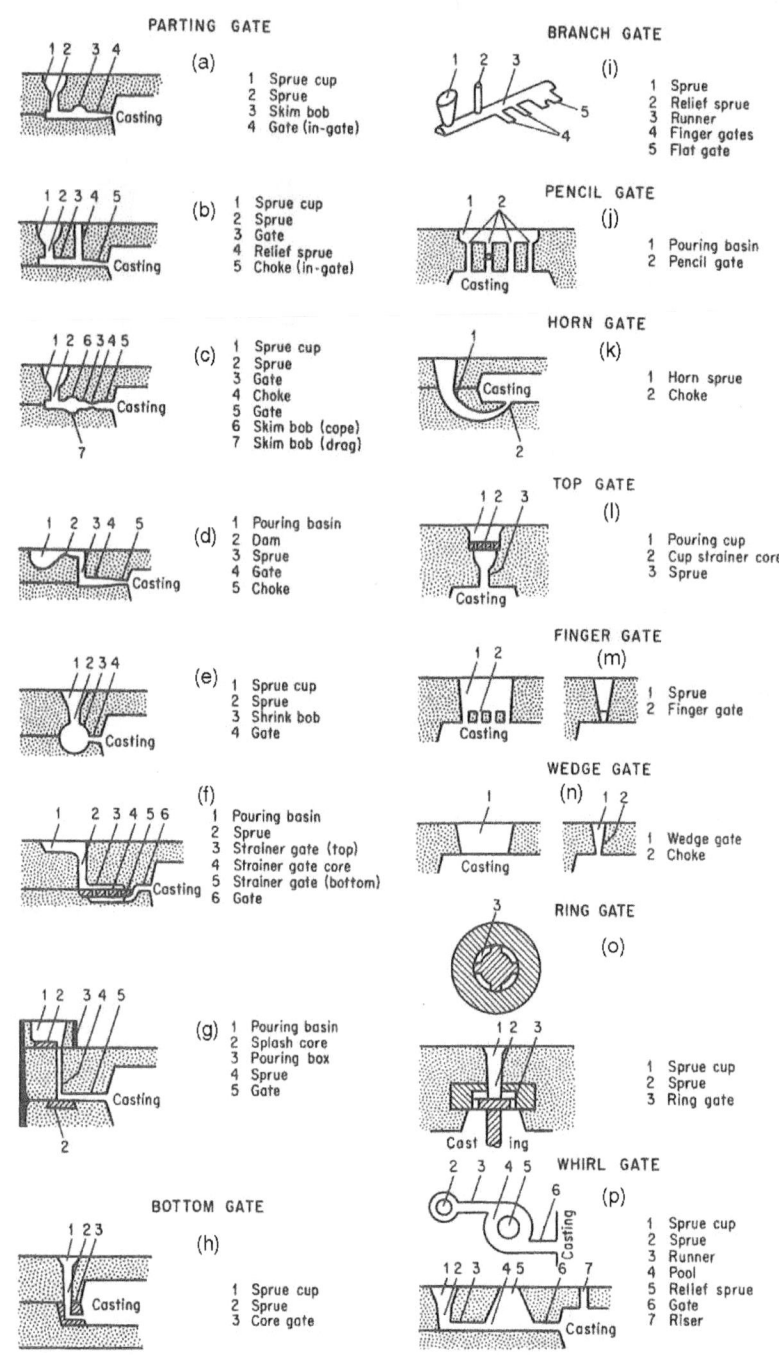

FIGURE 5.6   Examples of gating systems. (From Heine et al. [2, p. 216].)

a well at the base of the sprue or a change in the direction of metal flow may be incorporated, to reduce flow rates in the systems. A bottom gate is advocated for steel castings in particular, to reduce sand erosion and gas entrapment and to prevent splashing, which can result in cold shots.

Branch Gate

A branch gate is designed either to feed a single casting at several points or a number of individual castings (*i*, Fig. 5.6).

Horn Gate

This is a variety of bottom gate (*k*, Fig. 5.6). One objection to its use is a tendency to produce a fountain effect in the casting. However, it is a means of bottom gating without the necessity of a core for the gate.

Others

With the exception of the whirl gate, the remaining gate types illustrated in Fig. 5.6 are essentially variations of a top-gating system, since the metal enters the mold from above. The designs are intended to break up the metal stream and minimize the potential sand erosion from the high impact of the full metal stream on the mold surface. The increased surface area of each separated stream in contact with air, which could generate slags, must be considered. The whirl gate accomplishes somewhat the same purpose as the parting gates illustrated in *a*, *b*, and *c* of Fig. 5.6. The step gate is intended to have hot metal enter the various gates successively from the bottom to the top of the casting. If this objective were accomplished, the situation would promote directional solidification. Unless the step gates are properly designed, however, this gating system does not function as desired. Each gate must be slanted upward and properly proportioned relative to the other gates to attain the desired goal.

Top gating of castings (Fig. 5.7) and mold-reversal manipulation, as demonstrated in Fig. 5.8, are additional methods used for filling to favor directional solidification. The shrink bob illustrated in Fig. 5.6*e*, which is really a form of riser, is another method that is used frequently in malleable iron work, to provide proper feeding of a casting.

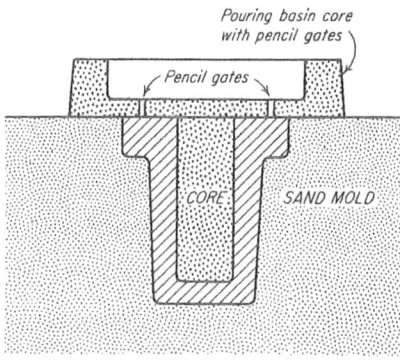

FIGURE 5.7 Top gating through pencil gates. (From Heine et al. [2, p. 229].)

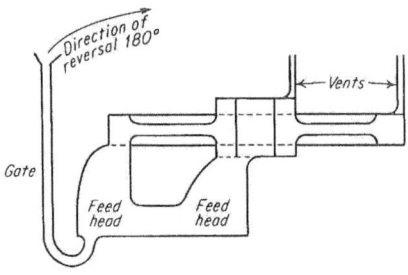

Figure 5.8   Mold-reversal method for securing proper feeding. (From Heine et al. [2, p. 229], Batty [3].)

Design of Gating System
Mold Materials

Wherever possible, the gating system is made of the same molding material as the casting cavity. Often in the production of large castings and alloys poured at high temperatures, ceramic and some other materials are used for a part of the gating system or for the complete system, either to minimize or eliminate the erosion of sand from the runner system or for convenience. The pouring basin or pouring cups are frequently made from core binders or ceramic material. Downsprues and gate cores may be obtained in ceramic materials called tile. Skim gates and impact cores can also be made of core binders or ceramic material, since they must withstand a considerable amount of erosive action and pressure from the metal, which could not be sustained by similar constructions made of green sand. Occasionally, a certain portion of the gating system may be constructed of high density sand, such as zircon sand, to prevent washing or metal penetration. The design of systems constructed with ceramic tile, or other materials, normally follows the rules used to design a sand runner system.

Factors Involved in Runner System Design

The physical aspects of runner systems have already been considered. How these are to be used to produce a sound casting is a question of gating design. Improper design of a runner system can cause one or more of the following defects in the casting:

- Sand, slag, dross, or other impurities
- Rough surface
- Entrapped gases
- Excessively oxidized metal
- Localized shrinkage (pipe shrinkage, or macroshrinkage)
- Dispersed porosity, or microporosity
- Incomplete fusion of liquid metal where two streams meet (cold shuts)
- Entrapped globules of presolidified metal (cold shots)
- Unfilled molds (misruns)
- Metal penetration into sand mold and/or core

The runner system must therefore be designed to accomplish the following objectives as quoted from Wallace and Evans [4]:

- Fill the mold rapidly, without laps or requiring excessively high pouring temperatures.
- Reduce or prevent agitation or turbulence and the formation of dross in the mold.
- Prevent slag, scum, dross, and eroded sand from entering the casting cavity by way of the gating system.
- Prevent aspiration of air or mold gases into the metal stream.
- Avoid erosion of molds and cores.
- Aid in obtaining suitable thermal gradients to attain directional solidification and minimize distortion in the casting.
- Obtain a maximum casting yield and minimum grinding costs.
- Provide for ease of pouring, utilizing available ladle and crane equipment.

It is evident that not all of these requirements are compatible, and compromises may have to be made to get as close as possible to the desired goal.

Bernoulli's Theorem

The flow of a liquid in a mold is governed by a number of variables, which are best summed up in terms of Bernoulli's theorem, which is based on the first law of thermodynamics. It states that the sum of the potential energy, the velocity energy, the pressure energy, and the frictional energy of a flowing liquid is equal to a constant [5]. This theorem can be expressed in the following equation:

$$wZ + wPv + \frac{wV^2}{2g} + wF = K \tag{5.1}$$

where

w = total weight of fluid flowing, lb
Z = height of liquid, ft
P = static pressure in the liquid, lb/ft^2
v = specific volume of the liquid, ft^3/lb
g = acceleration due to gravity, 32 ft/s^2
V = velocity, ft/s
F = frictional losses, ft
K = a constant

If this equation is divided by w, all the terms have the dimensions of length and may be considered to represent:

$$Z + Pv + \frac{V^2}{2g} + F = K \tag{5.2}$$

1. Potential head (Z)
2. Pressure head (Pv)
3. Velocity head ($V^2/2g$)
4. Frictional loss of head (F)

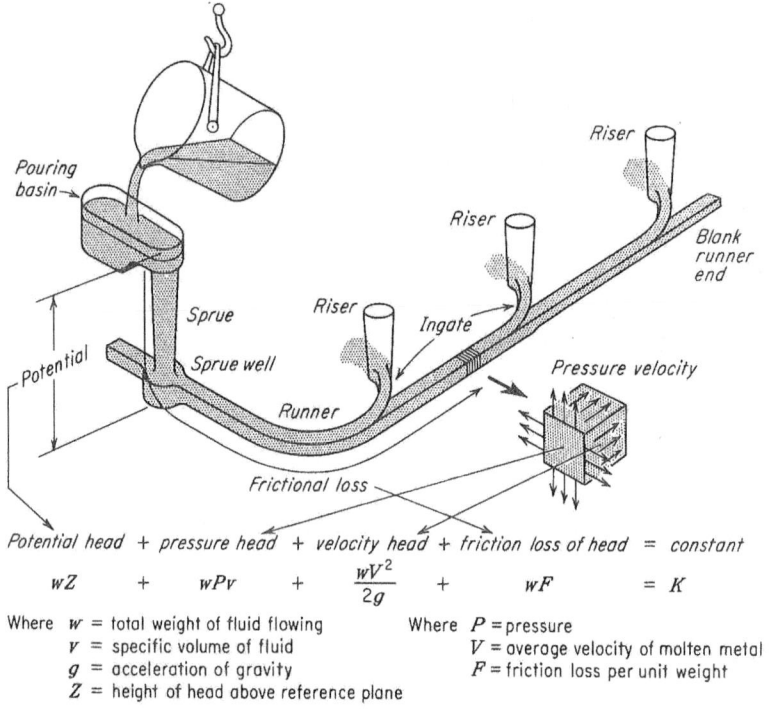

Potential head + pressure head + velocity head + friction loss of head = constant

$$wZ \quad + \quad wPv \quad + \quad \frac{wV^2}{2g} \quad + \quad wF \quad = K$$

Where w = total weight of fluid flowing   Where P = pressure
v = specific volume of fluid V = average velocity of molten metal
g = acceleration of gravity F = friction loss per unit weight
Z = height of head above reference plane

FIGURE 5.9   Application of Bernoulli's theorem to a gating system. (From Heine et al. [2, p. 220], Wallace and Evans [6].)

Bernoulli's theorem can be applied to a proper understanding of the flow of metal in a mold. The potential energy of the metal can be considered a maximum as the metal enters the pouring cup or basin. This potential energy is then rapidly changed to kinetic or velocity energy and pressure energy as the metal passes through the runner system. Once flow is established, and the potential and frictional heads are virtually constant, the velocity is high when the pressure is low, and the opposite is true. While metal is flowing, there is a constant loss of energy in the form of fluid friction between the metal and mold wall. Heat is also lost, which eventually leads to solidification of the metal, but it is not represented in Bernoulli's theorem.

A schematic illustration is given in Fig. 5.9 of the application of this principle to a typical gating system. In the past, it had been impossible to consider all of the implications of this theorem and its application to the design of gating systems, but a number of examples have been given in the literature [5–7]. Recently, this principle is applied through the use of computer simulation.

Metal Velocity in the Mold

Gravity is the driving factor in the filling of the mold, and therefore runner system designs. Metal is required to fall or flow in the vertical direction down the sprue of the runner system. Due to the effect of gravity, as the metal falls or flows down the sprue, the velocity of the stream will increase from essentially zero to some rate. The velocity obtained by the metal is related to the distance the metal falls along with a few other factors. Therefore, the greater the fall distance the greater the velocity. The velocity of a free falling body or stream is given as [8]:

$$V = \sqrt{2\,g\,h} \qquad\qquad (5.3)$$

where g is acceleration due to gravity, 386.4 in/s^2 and h is the distance of fall in inches.

This equation is theoretical and does not take into consideration losses due to friction. Some velocity of the metal is lost due to friction in a mold. The initial friction loss is related to the friction of the air and metal in the unfilled sprue. Once the sprue is filled with metal, the friction loss is related to the contact of the metal with the mold surface.

Law of Continuity

In most runner system designs, the shape and size of specific parts change due to requirements of the casting, pattern making, molding equipment, or other process steps. A round downsprue could be connected to square runners that are connected to rectangular ingates. The cross-sectional area of each of these must be calculated to relate to each other and provide a seamless and smooth flow of metal from one section to the next section. The Law of Continuity equation is used to reach that goal in the runner system design.

The Law of Continuity is a mass balance equation, which states, in a full channel, metal cannot be made or metal cannot be lost. Therefore, in a full runner system the volume per second flow in one section must be equal to the volume per second flow in another section [9].

$$A \text{ constant} = \frac{\text{volume in section 1}}{\text{time}} = \frac{\text{volume in section 2}}{\text{time}} = \cdots \qquad (5.4)$$

or

$$Q = A_1 V_1 = A_2 V_2 = A_3 V_3 = \cdots \qquad\qquad (5.5)$$

where

> Q = volumetric flow rate, in^3/s
> A_1 = cross-section area at point 1, in^2
> V_1 = metal velocity at point 1, in/s
> A_2 = cross-section area at point 2, in^2
> V_2 = metal velocity at point 2, in/s
> A_3 = cross-section area at point 3, in^2
> V_3 = metal velocity at point 3, in/s
> … and so on

In a full channel, if a certain volume of fluid flows past one point in the channel in a certain period of time, then that same volume will flow past another point downstream in the channel in the same time period. The velocity of the metal and cross-sectional area may be different at both points, but the volume (amount) is the same. If the cross-sectional area of the channel is constant, then the velocity of the liquid metal in the channel will also remain constant. However, if the cross-section area of the channel changes, the velocity of the liquid metal will also change. If the cross-sectional area of the channel is decreased, the velocity of the liquid metal will increase and the opposite statement is true.

Fluid Flow

Problems such as gas contamination, inclusion of dross or slag, and aspiration of gas are factors that must be considered when designing a gating system. A little examination will show that these problems are connected with the major problem of having the metal enter the mold in a quiet and uniform manner. In other words, these problems are related to fluid flow. Applying the laws governing fluid flow can improve any runner system design.

It should be noted that liquids flow either in a streamlined laminar fashion or in a turbulent manner. Smooth or turbulent flow will develop based upon the velocity of the liquid, the cross section of the flow channel, and the viscosity of the liquid. The relationship is expressed as the Reynolds number:

$$R_n = \frac{\text{mean velocity of flow} \times \text{diameter of tube} \times \text{density of liquid}}{\text{kinematic viscosity of liquid}} \qquad (5.6)$$

When the Reynolds number reaches a certain critical value, turbulent flow prevails. Apparently, most metals reach turbulent flow conditions quite readily. Investigations of runner systems, applied in most foundries, show that steel always flows under turbulent conditions ($R_n > 3500$) [10]. Turbulent flow creates such problems as inclusion of dross or slag, aspiration of air into the metal, erosion of the mold wall, and roughening of the casting surface.

Pouring Time

The manner in which the metal enters the casting cavity is important to the final properties of the casting. A slight trickle of metal, or metal poured too cold, is undesirable because the metal will freeze too fast to fill out the casting cavity or will develop cold shuts. Very rapid filling of the mold can also cause erosion of the mold wall, rough surface, excessive shrinkage, and other possible defects. There is, therefore, an optimum pouring rate, or pouring rate range, for most castings that must be established by experience. In die-castings or special casting techniques where metal is forced into a mold under pressure, this upper limit is probably set by the fluidity of the metal itself.

In conventional sand casting, establishing the optimum pouring rate is the first step in the design of the runner system. Once this is done, the next step is the proper proportioning and distribution of the various parts of the gating system, in order to achieve this rate. The characteristics of the various foundry alloys have a strong influence on the importance of this first step. Geometry of the casting is, of course, also a factor.

Traditionally it was thought some metals, like cast iron, are not as sensitive to pouring rate as others. Yet even for cast iron, an optimum pouring rate, which is a function of the casting size and shape [11], is advocated. A metal like steel is traditionally poured fast to avoid premature freezing, because it has a high freezing range compared with most other casting alloys. Metals like aluminum or magnesium alloys can be poured more slowly. The problem here is one of avoiding turbulence, drossing, and gas pickup.

Effective pouring rates for all commercial casting alloys have not been published, but do exist for many alloys. Pouring rates would be expected to reflect, to some extent, the practices in a given foundry or the limitations of the available equipment, as well as the casting geometry. Some data is available for cast iron, steel, brass, and bronze [4, 12, 13].

    a. Gray-iron castings <1000 lb:
      Pouring time *t* in seconds is

$$t = K\left(0.95 + \frac{T}{0.853}\right)\sqrt{W} \qquad (5.7)$$

where

T = average thickness in inches, in

W = weight in pounds, lb

K = fluidity factor determined by dividing the fluidity of the specific iron obtained from Fig. 5.10 by 40 (the fluidity value for iron of CE = 4.3 at a temperature of 2600°F)

As seen from Fig. 5.10, this factor is affected by iron composition and pouring temperature.

b. Gray-iron castings >1000 lb:
Pouring time t in seconds is

$$t = K\left(0.95 + \frac{T}{0.853}\right)\sqrt[3]{W} \tag{5.8}$$

c. Shell-molded ductile iron (vertical pouring):
Pouring time t in seconds is

$$t = K_1 \sqrt{W} \tag{5.9}$$

where K_1 = 1.8 for sections from $\frac{3}{8}$ to 1 in, 1.4 for thinner sections, and 2.0 for heavier sections.

d. Steel castings:

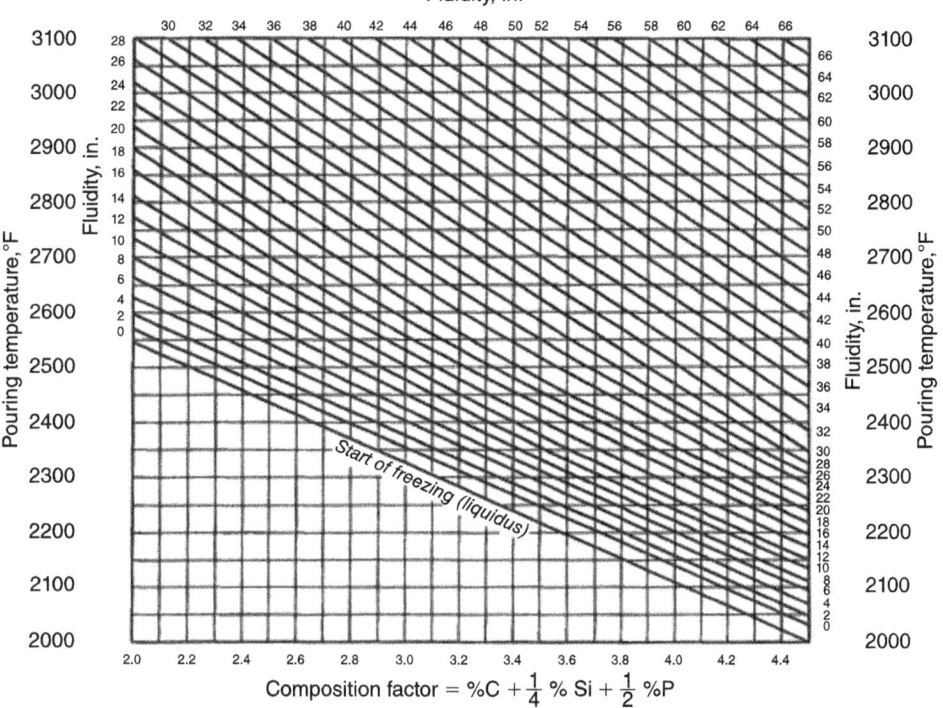

FIGURE 5.10 Fluidity related to pouring temperature and composition of gray and malleable cast iron. (From Heine et al. [2, p. 204], Porter and Rosenthal [14].)

Pouring time t in seconds is

$$t = K\sqrt{W} \tag{5.10}$$

where K varies from about 1.2 for 100-lb castings to about 0.4 for 100,000-lb castings when casting weight is plotted on a log scale.

Comparing these statistics for a 400-lb casting of 1 in average wall thickness, the following pouring-time values would be obtained:

a. Gray cast iron of 4.0 carbon equivalent poured at 2700°F:

$$t = 1\left(0.95 + \frac{1}{0.853}\right)\sqrt{400} = 42 \text{ seconds} \tag{5.11}$$

b. Shell-molded ductile iron:

$$t = 1.8\sqrt{400} = 36 \text{ seconds} \tag{5.12}$$

c. Steel:

$$t = 1.0\sqrt{400} = 20 \text{ seconds} \tag{5.13}$$

For comparison, a pouring time of 15 to 45 seconds is recommended for brass or bronze castings of less than 300 lb [15].

In many cases, the choice of pouring times is based on factors other than the best fluid flow conditions. Pouring times may be based on the constraints of the automatic pouring unit or the pouring ladle design. The pouring time may be set for a portion of the cycle time of an automatic molding machine line, to reach the maximum production rate for the equipment. In these and other cases, the impact of the required pouring time on the casting filling characteristics, and ultimately the casting quality, should be evaluated.

With the optimum pouring time established by whatever means, the next step is to proportion the gating system properly to achieve the desired rate while complying as closely as possible with the other desired characteristics of the gating system previously enumerated.

Choke Calculation

Choke Area

The smallest relative area in the runner system controls not only the flow rate into the mold cavity, but also the pouring time. Usually, this choke area occurs at the bottom of the sprue to establish the metal velocity as soon as possible, but this is not always the case [16]. If the choke area occurs at the base of the sprue, the proper area can be calculated by applying a formula based on the application of Bernoulli's theorem. For example, the choke area can be determined by using the following formula [8]:

$$A = \frac{W}{t\,d\,C\,\sqrt{2\,g\,H}} \tag{5.14}$$

where
      A = choke area, in^2
      W = casting weight, lb
      d = density of molten metal, lb/in^3 [3]
      H = effective height of metal head, in
      C = efficiency factor or nozzle coefficient (this is a function of the gating system used)

g = acceleration of gravity, 386.4 in/s [2]

t = pouring time, s

Sprue Design

As the metal gains velocity in passing through the sprue, it loses its pressure energy, or head. This is demonstrated by the constriction in cross section that occurs in a metal stream at distances down the sprue from the pouring spout. The loss of pressure head in a sprue may result in a tendency to form a vortex on the metal in the sprue, or a negative pressure effect in the metal column, so that gas from the mold is sucked into the metal stream. The remedy is to taper the sprue shape. This also reduces mold erosion and metal turbulence.

As mentioned in the preceding section, sprue size is often selected so that it controls the pouring rate; i.e., the occurrence of a major restriction to flow in the gating system in the sprue. This has the advantage of early establishment of the proper flow characteristics and reduction of the rate of flow of metal entering the mold cavity from the gates.

In alloys such as aluminum that are subject to drossing, the incorrect use of a restricted area to reduce the velocity of the metal may lead to drossing of the metal as it moves through the runners. A suggested sprue and gating design to overcome these effects is shown in Fig. 5.11.

Traditionally it was thought some metals, such as cast iron and steel, are not as prone to dross formation, and therefore, these precautions may not be necessary. However, it has been shown that casting cavities for all alloys should be filled with as little turbulence as possible to avoid defects. Furthermore, since steel is poured at a high temperature and chills very rapidly, the factor that may determine the sprue design is

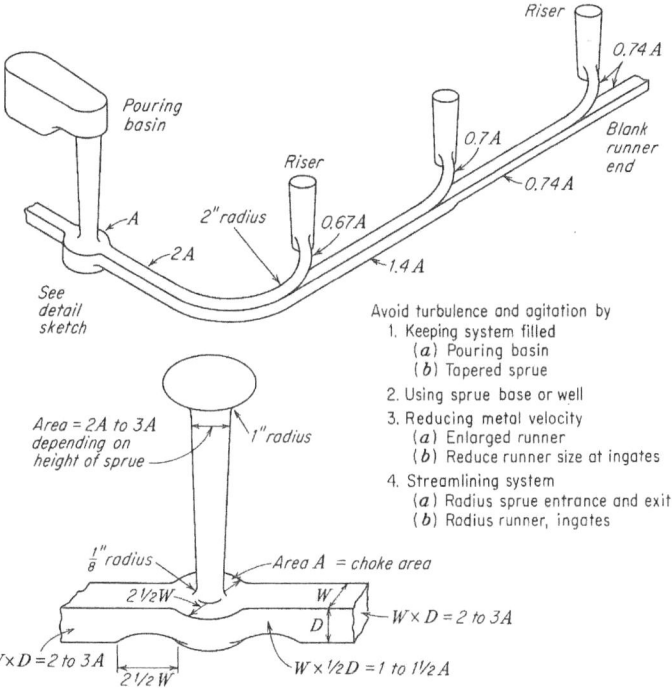

FIGURE 5.11 Gating system and sprue design developed for light-metal systems using horizontal gates. (From Heine et al. [2, p. 225], Wallace and Evans [6], Eastwood [17], Kura [18].)

not drossing and quiet entry of metal into the mold, but the need to get the metal into the mold fast enough to avoid excessive chilling. If this is the predominant consideration, sprue-runner ratios may differ from those advocated for the light-metal alloys.

Sprue-Runner-Gate Ratios

The runner system is commonly expressed as ratios of the choke area to the cross-section areas in the other parts of the system (runner and ingate). Thus, in some of the succeeding references to aluminum alloys, a ratio of 1:2:4 is used. The deviations that can be encountered in pouring practices for other metals are indicated by the ratios given in Table 5.1.

Thus, the gating system can vary widely from one leading to a nonpressurized system, such as 1:2:4 or 1:3:3, to one where the choke and pressure are at a maximum at the ingate, such as 4:8:3 or 1:2:1. If more than one ingate is used, the ratios pertain to the total area of all the ingates. In other words, in changing from one ingate to two while maintaining the same ingate ratio, the areas of the two ingates should equal that of the single ingate system.

Runners and Gates

Runners and gates should be designed to obtain the following characteristics:

- Absence of sharp corners or changes to sections that may lead to turbulence or gas entrapment
- Proper relation between cross-sectional areas of the several gates, between gates and runners, and between the runners and the sprue
- Proper location of the gates to ensure adequate feeding of low-velocity metal into the mold cavity

Metal	Ratio	Ref.
Steel	1:2:1.5	19
	1:3:3	19
	1:1:0.7	19
	1:2:2	20
Fin-gated	1:1:1	21
Gray cast iron	1:4:4	4
Pressurized system	1:1.3:1.1	22
Ductile iron, dry-sand molds	10:9:8	16
Shell-molded, vertical pouring	1:2:2	12
Pressure system	4:8:3	23
Reverse choke	1.2:1:2*	20
Aluminum	1:2:4	10
Pressurized system	1:2:1	24
Unpressurized system	1:3:3	24, 20
Brass	1:1:1–1:1:3	25

*With enlargements in runner varying from 3 to 6.
Source: From Heine et al. [2, p. 224].

TABLE 5.1   Selected Sprue-Runner-Gate Area Ratios

Studies of the gating systems now employed show that in some instances the systems do not function as anticipated. For example, whirl, riser, and horn gates were found to be ineffective in preventing turbulence in the molds when pouring steel [26]. In multiple-finger gating systems, it has been found that often most of the filling is accomplished by the fingers farthest from the sprue. This is the result of improper proportioning of the cross section of the gate and runners. Because of frictional losses and the abrupt change in cross section at these points, the liquid metal has a relatively low velocity and a fairly high pressure. Hence, it will readily flow into the farthest gate. The gates nearer the sprue will have less metal flowing through them because of higher velocities and lower pressures. This effect is demonstrated by Fig. 5.12*a*, which shows the proportion of liquid, which flows through gates supplying a block casting [27]. In this instance, the total sprue area to total runner area to total gate area was 1:2:4. Some correction was obtained by cutting down the ingate area to a 1:2:2 ratio because of the tendency to maintain a more uniform distribution of the metal in the filling system and hence more constant velocity and pressure conditions (Fig. 5.12*b*).

To be completely satisfactory, however, the runner beyond each gate should also be reduced in cross section in order to balance the flow in all parts of the system and thereby to further equalize the velocity and pressure. Such designs are illustrated in Figs. 5.11, 5.13, and 5.14. In Fig. 5.11, the runner is proportioned so that reasonably constant velocity and pressure are maintained. Furthermore, the design is streamlined to avoid sudden changes in direction that might create turbulence. In Figs. 5.13 and 5.14, the concepts of stepped runners and tapered runners, to balance metal flow through all ingates, are illustrated.

Thus, to satisfy the demands imposed by Bernoulli's theorem, it is necessary that the runner be reduced in cross section as it passes each ingate, or those gates farthest from the sprue be of the smallest cross section. Thereby, the volume of metal through distant gates is the same as that through those closer to the sprue. If such proportioning is not done, there is a tendency for the pressure to be at a maximum at the farthest gate and, therefore, for the flow to be greatest at that point. In some instances, there may even be a negative pressure existing at some of the first gates, leading toward an actual flow of metal out of the mold and back into the runner.

Other measures in addition to proper proportioning of ingate are

- To develop enlargements in the gating system to dissipate momentum effects
- To bend the runner away from the casting
- To use a tapered runner [29]

In summary, streamlining of a gating system reduces turbulence to a minimum, and proper proportioning of the various parts of the gating system adjust the pressure and velocity head of the metal so that it flows as desired.

Avoid turbulence and agitation by

- Keeping system filled
  - Pouring basin
  - Tapered sprue
  - Using sprue base or well

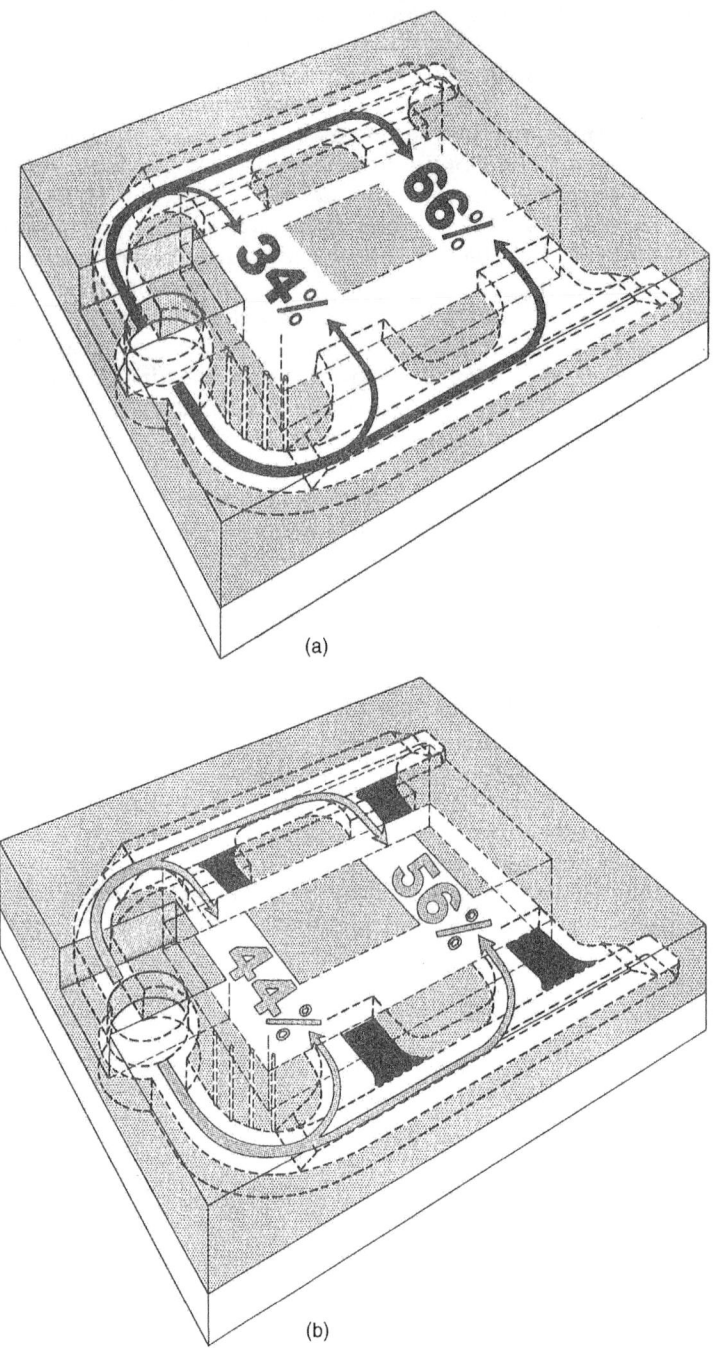

Figure 5.12 (a) Uneven distribution of flow in a gating system having uniform gate sizes and a 1:2:4 ratio for total sprue area to total runner area to total gate area, (b) improved flow conditions obtained by changing sprue-runner-gate ratio to 1:2:2. (From Heine et al. [2, p. 226], Grube and Eastwood [27].)

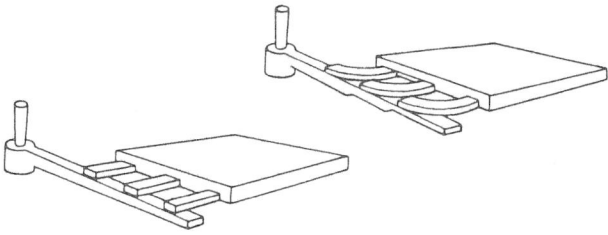

FIGURE 5.13   Stepped versus tapered runner for a single cavity gating system. (*Casting Copper-Base Alloys* [28].)

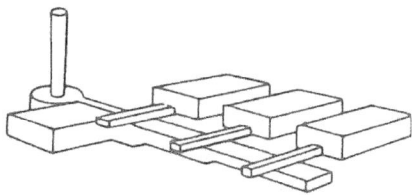

FIGURE 5.14   Multicavity gating system with stepped runner. (*Casting Copper-Base Alloys* [28].)

- Reducing metal velocity
  - Enlarged runner
  - Reduce runner size at ingates
- Streamlining system
  - Radius sprue entrance and exit
  - Radius runner, ingates

Vertical Gating Systems

Vertically parted molds provide another complexity to runner system designs. Generally, the runners and ingates are not on one horizontal plane, but instead on different levels on a vertically parted mold. Therefore, the size of the various components must take into account their vertical location in the mold to produce consistent, even filling of castings or a casting in the mold.

The two equations most often used in the calculation of a vertical runner system are velocity of a free falling stream and the law of continuity. These along with the choke area equation are the basis for vertical runner system designs.

The first major equation is velocity of a free falling stream and is given as

$$V = \sqrt{2\,g\,h} \tag{5.15}$$

where g is the acceleration due to gravity, 386.4 in/s [2] and h is the distance of fall in inches.

As noted in the horizontal system discussion, this equation is theoretical and does not take into consideration losses due to friction. Some velocity of the metal is lost due to friction in a mold.

The second important equation is the Law of Continuity, which is a mass balance equation. It states that in a full channel, metal cannot be made or metal cannot be lost. Therefore, in a full runner system, the volume per second flow in one section must be equal to the volume per second flow in another section.

$$A \text{ constant} = \frac{\text{volume in section 1}}{\text{time}} = \frac{\text{volume in section 2}}{\text{time}} = \cdots \qquad (5.16)$$

or

$$Q = A_1 V_1 = A_2 V_2 = A_3 V_3 = \cdots \qquad (5.17)$$

where
Q = volumetric flow rate, in³/s [3]
A_1 = cross-section area at point 1, in²
V_1 = metal velocity at point 1, in/s
A_2 = cross-section area at point 2, in²
V_2 = metal velocity at point 2, in/s
A_3 = cross-section area at point 3, in²
V_3 = metal velocity at point 3, in/s
... and so on

As with horizontal runner systems, the sprue and runner system design is important to maximize flow while minimizing turbulence in a vertical runner system. The use of tapered sprues and stepped runners are illustrated [22] in Figs. 5.15 and 5.16. The tapered sprue and tapered runner bar, calculated with the Law of Continuity, can be seen in Fig. 5.15, where the castings are bottom filled. In Fig. 5.16, a simultaneous filling of each casting is controlled by the tapered sprue and the appropriately sized ingates based on the Law of Continuity.

Vertical Runner System Design Example

The inputs for the ductile iron disc-casting example shown in Fig. 5.17 are

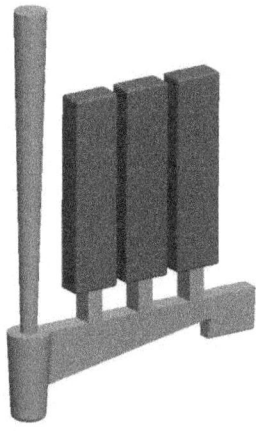

FIGURE 5.15 Vertical system with multiple cavities. (*Casting Copper-Base Alloys* [30].)

Figure 5.16 Vertical system with multiple cavities, bottom runner. (*Casting Copper-Base Alloys* [30].)

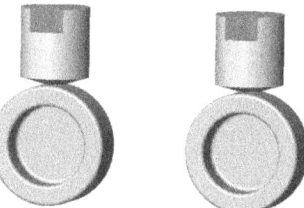

Figure 5.17 Two disc castings with risers. (From Adams [1].)

Casting			Ductile Iron Disc Casting	
Pour weight per side			49.62 lb	W
	Casting weight	36.55 lb		
	Riser	12.71 lb		
	Neck	0.36 lb		
Pour time			10 seconds	T
Sprue height			16 in	h
Casting height above runner			14.75 in	c
Friction factor (from previous results)			0.3	C
Density of liquid iron			0.22 lb/in^3	D
Acceleration of gravity			386.4 in/s [2]	G

Step 1. Head Height Calculation for Bottom Filling

The head height must take into consideration the location of the metal entry into the casting cavity. Since the casting is filled from the bottom, represented in Fig. 5.18, the head height must be reduced by a factor related to the height of the casting.

$$H = h - \frac{C}{2} \qquad (5.18)$$

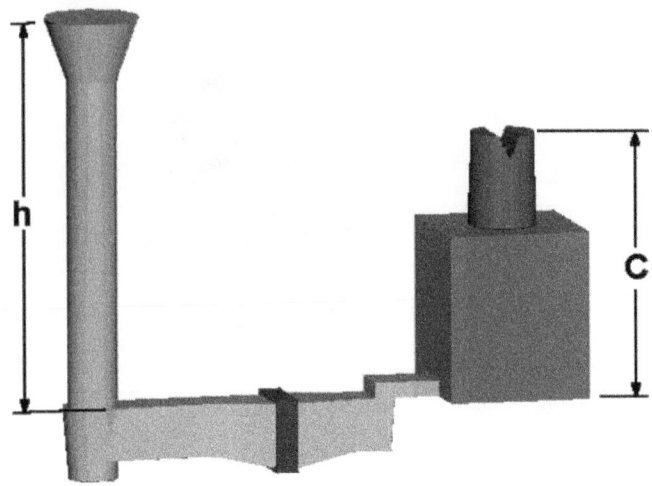

Figure 5.18 Example of variables used in bottom filling calculation. (From Adams [1].)

$$H = 16 - \frac{14.75}{2} \qquad (5.19)$$

$$H = 8.625 \text{ in} \qquad (5.20)$$

Step 2. Choke Calculation
The choke is calculated for one half side of the mold.

$$\text{Choke area} = A = \frac{W}{T \times D \times C \times \sqrt{2 \times G \times H}} \qquad (5.21)$$

$$\text{Choke area} = A = \frac{49.62}{10 \times 0.22 \times 0.3 \times \sqrt{2 \times 386.4 \times 8.625}} \qquad (5.22)$$

$$\text{Choke area} = A = 0.921 \text{ in}^2 \qquad (5.23)$$

Step 3. Runner System Calculation
The runner system is based on the choke using a ratio of 1.0:1.1:1.2 for the sprue, runner, and ingates, as shown in Table 5.2. The runner is calculated by increasing the choke area

Location	Ratio	Areas, in
Choke	1.0	0.921
Runner	1.1	1.013
Ingate	1.2	1.105
Sprue base (for two runners)	2 × 1.1	2.026

Table 5.2 Runner System Cross-Section Areas

by 10% and the ingates are calculated by increasing the choke area by 20%. The sprue bottom is based on adding the two sides together and increasing the area by 10%.

System Areas
Step 4. Sprue Calculation
The minimum sprue diameter at different levels is calculated by the law of continuity

$$Q = V_1 A_1 = V_2 A_2 = V_N A_N = \cdots \tag{5.24}$$

$$V = \sqrt{2\,G\,H} \tag{5.25}$$

Top of filter print

H_1	17 in
V_1	$\sqrt{2 \times 386.4 \times 17} = 114.62\,\text{in/s}$
A_1	2.026 in^2
H_2	14 in
V_2	$\sqrt{2 \times 386.4 \times 14} = 104.02\,\text{in/s}$
A_2	?

$$V_1 A_1 = V_2 A_2 \tag{5.26}$$

$$A_2 = \frac{V_1 A_1}{V_2} = \frac{114.62 \times 2.026}{104.02} = 2.232\ \text{in}^2 \tag{5.27}$$

Bottom of cup

H_1	14 in
V_1	$\sqrt{2 \times 386.4 \times 14} = 104.02\,\text{in/s}$
A_1	2.232 in^2
H_2	3 in
V_2	$\sqrt{2 \times 386.4 \times 3} = 48.15\,\text{in/s}$
A_2	?

$$V_1 A_1 = V_2 A_2 \tag{5.28}$$

$$A_2 = \frac{V_1 A_1}{V_2} = \frac{104.02 \times 2.232}{48.15} = 4.822\ \text{in}^2 \tag{5.29}$$

Step 5. Filter Choice
For ductile iron, the foam $50 \times 75 \times 22/10$ ppi filter has a flow rate of 8 to 15 lb/s from supplier literature.

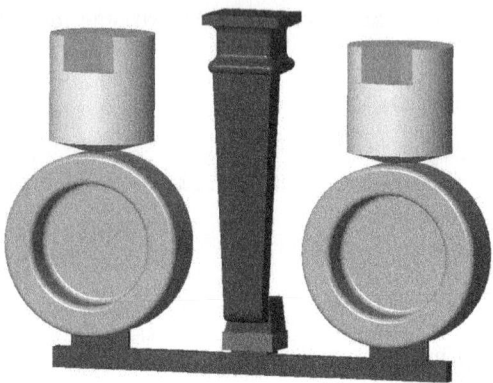

FIGURE 5.19   Runner system design. (From Adams [1].)

The total flow before potential blockage is 85 to 175 lb from supplier literature. Therefore, a single $50 \times 75 \times 22/10$ ppi filter will support this mold.

Step 6. Filter Print Selection
The horizontal, square in the ram side, filter print was chosen as an appropriate filter print design for the foam $50 \times 75 \times 22/10$ ppi filter.

The complete system is shown in Fig. 5.19.

Elimination of Slag and Dross

Traditionally, pouring basins, strainer cores, and suitable dams have been used in gating systems to help remove slag and dross from the metal stream before it enters the mold cavity. Some examples of such methods have been given previously. In the case of light-metal alloys, the difficulties are somewhat greater than for copper or ferrous alloys, since there is so little difference in specific gravity between the impurities and the metal.

Enlargements in the runner system to reduce the velocity of flow, or special devices such as the whirl gate, which whirls the dross and slag into the center of the riser, are other means of cleansing the metal. For aluminum castings, it has been recommended that the runners be placed in the drag with the ingate in the cope of the mold to reduce the inclusion of dross in the mold cavity. Traditionally for ferrous metals, on the other hand, the reverse situation has been suggested, with the runner in the cope and the ingate in the drag. In most cases and alloys, these techniques have been replaced with filters, which have improved the results.

Economy and Ease of Removal of Gates and Sprues

In addition to the factors already discussed, it is quite obvious that changes in design of the gating system, to reduce costs without affecting the quality of the casting, are something for which one should strive. Quite frequently, it is found that some modifications in practice, such as the inclusion of a chill in a strategic location, may greatly reduce the amount of metal required to feed a casting. In other instances, it may be found that gating in a particular location will result in much greater ease of removal of the gate than if located in another part of the casting. The immediate connection between the ingate and the mold is sometimes reduced in order to permit the gate to be removed readily from the casting; or a neck-down, or Washburn, core may be used for this purpose. There are so many factors in connection with each metal, and each particular casting, that it is a necessity for each particular job to be studied individually, with the viewpoint of improving the gating system to achieve a quality casting at a minimal cost.

Filtration

To meet the ever-tightening specifications of the foundry customers, foundries are continually modifying and improving their processes and techniques. One of the process tools employed is the application of filters in the molds or dies. Filters have moved from a method to assist in the production of a problem casting to being required to meet the mechanical and fatigue properties of a casting.

Types of Filters

There are three basic filter structures (Fig. 5.20) applied in metal casting, with variations in exact shape and material formulation. Foam structures have irregular passageways through which the metal flows. Three-dimensional cellular and two-dimensional cellular are two other basic configurations. The three-dimensional cellular structure has straight holes of some substantial length that act as passages for metal flow. The two-dimensional cellular part has holes for metal flow, but the part is relatively thin.

The material used in the manufacturing of the filter can vary and thus produce different results in the application. Some two-dimensional parts are produced from refractory sheets, metal sheets, or woven refractory fiber. The three-dimensional parts are based on various formulations of alumina-silica refractories. Foam filters are produced from alumina-silica refractories, zirconium-based refractories, carbon-bonded refractories, and nonceramic refractories. The choice of filter formulation (Fig. 5.21) is related to the metal that is cast and the foundry process techniques being utilized. Refractories that have a higher melting point are applied in steel casting, while lower melting point material can be used with aluminum and other nonferrous alloys. Figure 5.21 shows examples of the various filters.

Filtration Properties

Different filter structures and formulations provide different nonmetallic material removal. However, the basics of filtration can be separated into three categories; sieving,

Figure 5.20   Example of materials used for filtration in iron. (From Adams [1].)

FIGURE 5.21   Example of various filtration materials for different alloys. (From Adams [1].)

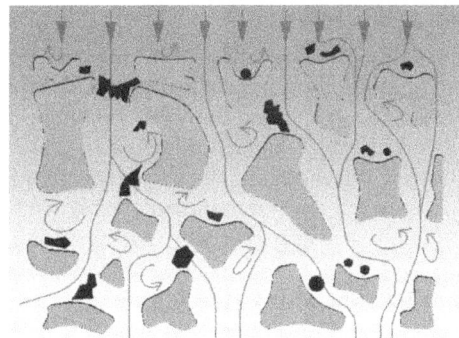

FIGURE 5.22   Flow of metal through a foam filter. (From Adams [1].)

filter cake formation, and deep bed (Fig. 5.22). Each category provides some level of nonmetallic material removal, depending on the metal poured and the general size of the nonmetallic in the metal.

Sieving

Sieving is simply the removal of particles that are too large to fit through the holes in the filter. A particle that is too large for the hole in the part will not travel along with the metal stream into the casting cavity, where it could produce a defect. Smaller holes capture smaller particles of slag on their front face.

Filter Cake

Filter cake (the second form of filtration) occurs for two reasons. Slag that is held on the front face of the filter holds more and finer slag because like materials stick together. The basis of the process though is that some slag must initially stick to the filter material. Once that first slag adheres to the filter material, more slag can adhere to it. Subsequent and finer slag continues to build on the earlier slag throughout the filling process.

Deep Bed

Deep bed filtration can occur only under two conditions: mechanical or chemical. Mechanical filtration is driven by the configuration of the filter. Chemical filtration is driven by the chemistry of the filter material.

Deep Bed Mechanical Mechanism

Mechanical filtration occurs by having a structure that splits and breaks up the metal stream as it is passing through the filter. The walls and strands of the filter develop turbulence in the liquid metal, thus forming areas where slag can separate. After separation, the slag is retained in these areas by the general metal flow. Since the turbulence and separation of the stream is taking place in a full filter, oxygen is not available to cause slag by reoxidation. On the outlet side of the filter, the metal recombines to form a coherent steam exiting the filter.

Deep Bed Chemical Mechanism

Chemical filtration arises from the design of the material used to make the filter. By choosing the correct combination of ceramics, the filter material can enhance slag movement to the walls of the filter and can increase the adhesion of the slag once it is near the wall. The filter material utilized creates a condition inside of the filter where the metal does not "wet" the filter material, but the slag does. When the filter materials are similar in composition to slag, the slag in the molten metal will easily wet the filter material producing a condition where small, fine slag particles are captured and held within the filter.

Horizontal Runner System with Filter Example

Step 1. Head Height Calculation for Middle Filling

The head height must take into consideration the location of the metal entry into the casting cavity. Since the casting is filled from the middle, represented in Fig. 5.23, the head height must be reduced by a factor related to a portion of the height of the casting.

$$H = h - \frac{p^2}{2c} \tag{5.30}$$

$$H = 12 - \frac{1^2}{2 \times 3.5} \tag{5.31}$$

$$H = 11.86 \text{ in} \tag{5.32}$$

Step 2. Choke Calculation

$$\text{Area} = A = \frac{W}{T \times D \times C \times \sqrt{2 \times G \times H}} \tag{5.33}$$

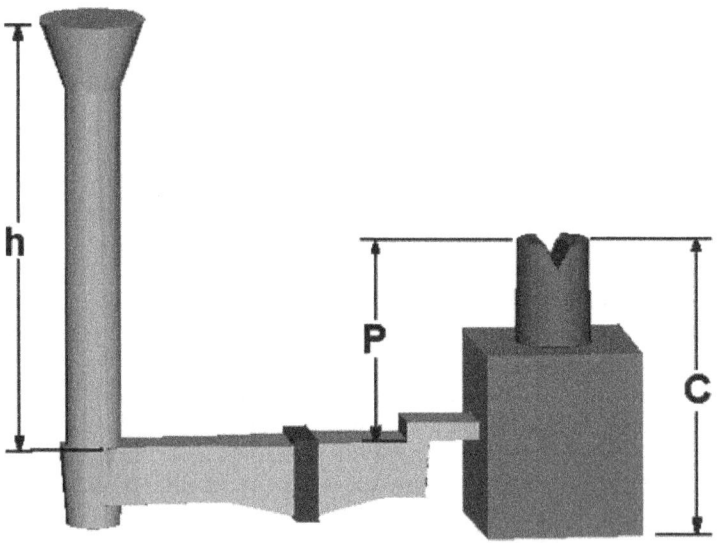

FIGURE 5.23   Example of variables used in middle filling calculation. (From Adams [1].)

Casting	Gray Iron Cylinder Head	
Casting weight	51.3 lb	
Pour weight	189 lb	W
Pour time	12 seconds	T
Cope height	12 in	h
Casting height in cope	1 in	p
Total casting height	3.5 in	c
Friction factor (from previous results)	0.6	C
Density of liquid iron	0.22 lb/in³	D
Acceleration of gravity	386.4 in/s²	G

TABLE 5.3   Gray Iron Cylinder Head Casting Information

The data from Table 5.3 is put into the equation along with the results from the previous calculation.

$$A = \frac{189}{12 \times 0.22 \times 0.6 \times \sqrt{2 \times 386.4 \times 11.86}} \tag{5.34}$$

$$A = 1.246 \text{ in}^2 \tag{5.35}$$

Step 3. Runner System Calculation

The runner system is based on the choke using a ratio of 1.0:1.1:1.2 for the sprue, runner and ingates. The system must be divided properly to provide equal iron to each casting at the same time. Since the layout, Fig. 5.24, is set to have two casting filled from one

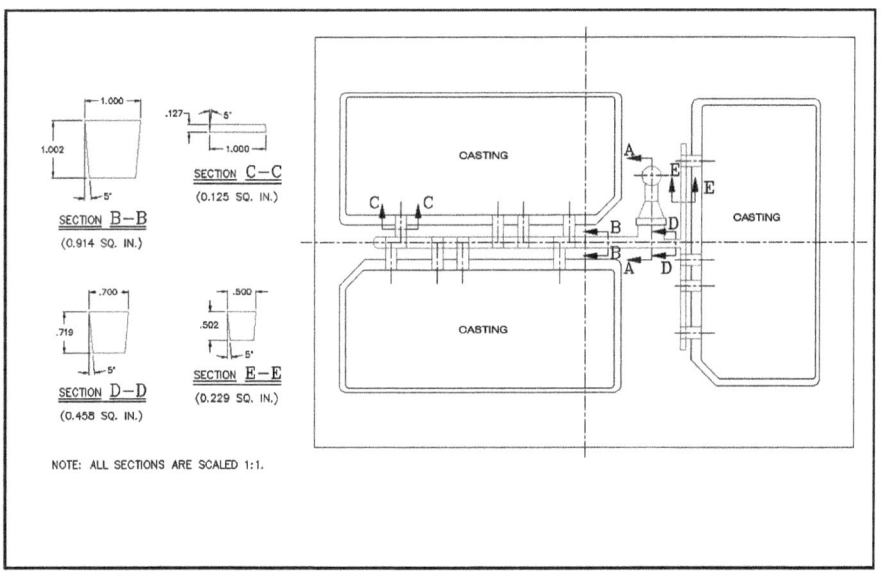

FIGURE 5.24   Layout of three castings and their runner system with a filter. (From Adams [1].)

runner bar and one casting filled from another runner bar the runner system area is split 2/3 and 1/3 after the filter system. The calculations are

Subsystem 1

$$\frac{1}{3} \times 1.246 \times 1.1 = 0.458 \text{ in}^2 \tag{5.36}$$

Subsystem 2

$$\frac{2}{3} \times 1.246 \times 1.1 = 0.914 \text{ in}^2 \tag{5.37}$$

Step 4. Ingate Calculation

Each casting required four ingates. The choke area is increased by 20% for the total ingates and the total area is divided by 12.

$$1.246 \times 1.2 \times \frac{1}{12} = 0.125 \text{ in}^2 \tag{5.38}$$

The runner is stepped when each ingate is passed.

Step 5. Sprue Calculation

The minimum sprue diameter, at the top, is calculated by the law of continuity

$$Q = V_1 A_1 = V_2 A_2 = V_N A_N = \cdots \tag{5.39}$$

$$V = \sqrt{2GH} \tag{5.40}$$

H_1	12 in
V_1	$\sqrt{2 \times 386.4 \times 12} = 96.30 \text{ in/s}$
A_1	1.246 in^2
H_2	3.5 in
V_2	$\sqrt{2 \times 386.4 \times 3.5} = 52.01 \text{ in/s}$
A_2	?

$$V_1 A_1 = V_2 A_2 \tag{5.41}$$

$$A_2 = \frac{V_1 A_1}{V_2} = \frac{96.30 \times 1.246}{52.01} = 2.307 \text{ in}^2 \tag{5.42}$$

A sprue with a diameter of 1.714 in would meet this requirement.

Step 6. Filter Choice

The foam $50 \times 75 \times 22/20$ ppi filter has a flow rate of 10 to 18 lb/s from supplier literature.

The total flow before potential blockage is 175 to 350 lb from supplier literature. Therefore, a single $50 \times 75 \times 22/20$ ppi filter will support this mold.

Step 7. Filter Print Selection

The vertical filter print was chosen as an appropriate filter print design for the foam $50 \times 75 \times 22/20$ ppi filter.

Risers/Risering Systems

Risering Systems

Earlier chapters described the solidification characteristics of metals and alloys, and how each was influenced by composition and external variables. All of these factors must be accounted for in designing a risering system for a casting. More specifically, the shrinkage behavior (Fig. 5.25) and the crystal-growth morphology must be recognized if the risering design is to be effective. In considering the freezing characteristics in the earlier chapter, the usual growth from the outside to the interior of the casting was revealed. This condition of having a partially solid, partially liquid zone growing from the outside inward is what is referred to as progressive solidification. Risering design must interact with this progressive solidification in such a way that no part of the casting is isolated from active feed channels during the entire freezing cycle. This is referred to as directional solidification.

Progressive solidification is a product of the freezing mechanism and cannot be avoided. However, the degree of progressive solidification can be controlled. Thus, a rapidly cooled casting, which results in a short distance between the start and end of freezing, is said to have a high degree of progressive solidification. Therefore, one which is slowly cooled would possess a low degree of progressive solidification.

Directional solidification is a product of alloy type, casting design, location of gates and risers, and the use of chills and other means for controlling the freezing process. Therefore, it is subject to the controls available to the foundry engineer. In principle, it means that if a casting is so proportioned and disposed, with respect to the feeding system that the sections most distant from the available liquid metal will solidify first, the result is a successive feeding of the contracting metal by still liquid metal until the heaviest and last-to-freeze section is reached. This, in turn, can be fed by extra

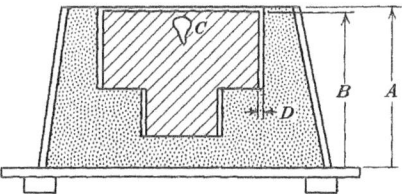

Figure 5.25   If superheated metal is filled to the top of the mold (level A), it will shrink somewhat on cooling to the freezing temperature (level B). During liquid-solid contraction, further reductions in volume take place, usually localized near the top of the casting, in the region, which freezes last (area C). Finally, the solid metal pulls away from the mold wall as it contracts (distance D). (From Heine et al. [2, p. 184].)

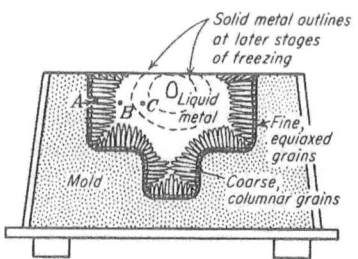

FIGURE 5.26 Cross section through a freezing casting of pure metal poured in an "open-face" mold. (From Heine et al. [2, p. 181].)

reservoirs of metal provided for that purpose and referred to as risers, or heads. These risers, or heads, are attached to the casting at the right locations so they can continually supply hot liquid metal to the shrinking casting until it is completely solidified. For example, imagine what would happen had the casting in Fig. 5.26 been poured in the reverse position, with the small section on top. Freezing would have occurred first in the small section, as before, but then there would be no liquid metal available to feed the heavier section by gravity, and it would have developed a general porosity that could not be eliminated. On the other hand, additional metal provided by an extra head, or riser, on top of the heavy section in Fig. 5.25 would completely eliminate the localized shrinkage shown. The system devised to feed the casting cavity serves the dual function of delivering the metal to this cavity, as well as of serving as a reservoir for the additional metal required as shrinkage takes place. In a very general way, delivery of the metal is accomplished by the gating system, whereas risers, or heads supplies reserve metal. When certain system designs are used, either one of these parts of the mold may serve the function of filling and feeding, therefore no clear-cut distinction can be made.

Risering System Defined

The Gating and Risering Committee of the American Foundry Society has done much toward standardizing the nomenclature in connection with the feeding of castings. Therefore, the definitions evolved by these groups serve as a useful reference for this purpose.

Castings of heavy section or of high shrinkage alloys commonly require a riser or reservoir where some riser metal stays liquid longer than the metal in the casting while both are solidifying. The riser thus provides the feed metal, which flows, from the riser to the casting to make up for the shrinkage, which takes place in the metal as it changes from liquid to solid. Depending on the location, the riser is described as a top riser or side riser and may be either an open riser (open to the top of the mold) or a blind riser (enclosed by the mold). Since risers are designed to contain liquid until the casting solidifies, riser height and riser neck are important dimensions, as are those of the body of the riser itself. Riser distance to the casting and the shape of the riser base are additional important details that pertain only to side risers.

Gates and risers are often designed to take advantage of the principle of controlled directional solidification, which requires that freezing start farthest from the riser and proceed toward the riser. To accomplish this, the castings and riser are gated with

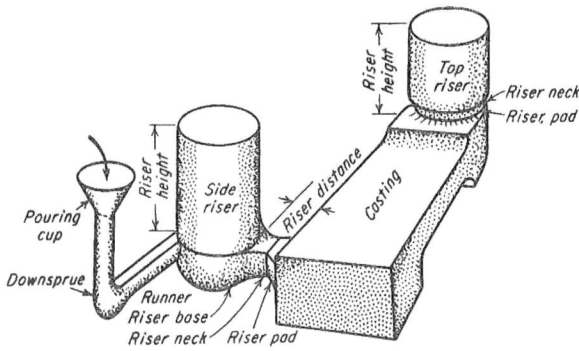

FIGURE 5.27   Riser-gated casting with side riser and top riser. (From Heine et al. [2, p. 215].)

metal entering the riser through a downsprue and runner, thus heating both the riser base and riser neck while flowing into the mold cavity (Fig. 5.27).

Additional definitions are found in publications of the AFS Gating and Risering Committee.

Primary Function of a Riser

The primary function of a riser is to feed metal to the casting as it solidifies. In some instances, it may also be considered as a part of the gating system. The riser requirements depend considerably on the type of metal being poured. Gray cast iron needs less feeding than some alloys because a period of graphitization occurs during the final stages of solidification, which causes an expansion that tends to counteract metal shrinkage. Steel or white cast iron, and many of the nonferrous alloys which have an extended solidification range, require large and sometimes elaborate feeding systems to obtain sound castings. The variation that can be expected in the volumetric shrinkage of some metals is shown in Table 5.4.

The values given in Table 5.4 represent the minimum requirements that must be satisfied by the riser. Any bulging or extension of a casting beyond its normal limits, because of a soft mold or excessive metal pressures in the mold, will require additional feed metal that must be provided by the riser.

Material	Volumetric Shrinkage, %
Medium carbon steel	2.5–3.0
1% carbon steel	4.0
Pure aluminum	6.60
Pure copper	4.92
Gray cast iron	2.5 to negative depending on graphitization, composition, etc.
Ductile cast iron	3.5 to negative depending on graphitization, composition, etc.
White cast iron	4.0–5.50

Source: From Heine et al. [2, p. 231].

TABLE 5.4   Approximate Solidification Shrinkage of Some Foundry Alloys

Although shrinkage and mold-wall movement are important factors in determining riser size, they are not the only factors. The susceptibility of various alloys to shrinkage defects, and hence the need for risering, is also related to the freezing mechanism.

Papers or books by R. Ruddle [31], J. Campbell [32], R. A. Flinn [33], and R. Wlodawer [34] should be examined to understand the various freezing mechanisms of different alloys. Another notable publication is "Feeding and Risering Guidelines for Steel Castings" by the Steel Founders' Society of America [35], which contains methods of feeding and updated data from various sources.

Their information also illustrates the importance of heat-transfer rates as a factor in the feeding of castings. Generally, the greater heat-transfer rate obtained from the chill mold improves the possibility of getting a sound casting.

Theoretical Considerations

The riser and the casting it feeds should be considered an integral system, because a sound casting cannot be made without adequate feed metal, no matter how much attention may be paid to other details. Since Table 5.4 indicates that only a relatively small amount of feed metal is necessary, one might conclude that risering is fairly simple and that only small reservoirs are necessary to compensate for shrinkage. However, the metal in risers is subject to the same laws of solidification as the metal in the castings. A little reflection will show that, to be effective, a riser must stay fluid at least as long as the casting and must also be able to feed the casting during this time. Consequently, the problem of providing this feed metal during the entire solidification period of the casting involves a few important variables, which are listed below and discussed in succeeding paragraphs:

- Riser shape
- Riser size as a function of casting shape
- Location of risers
- Grouping of castings
- Riser connections to the casting
- Use of chills
- Use of insulators and exothermic compounds
- Special conditions arising from joining sections

Riser Shape

A casting loses its thermal energy by transferring it to its surroundings by radiation, conduction, and convection. Without establishing the relative importance of these three modes of heat transfer, it is apparent that the surface area of the casting, relative to its volume, is important in determining the rate of this heat transfer. This concept was expressed mathematically by Chvorinov [36] as follows:

$$\text{Solidification time, } t \approx \frac{\text{square of volume}}{\text{square of area}} \approx \frac{V^2}{A^2} \tag{5.43}$$

Although this equation is somewhat oversimplified, it does indicate that, for a riser to have a solidification time equal to or greater than that of the casting, the minimum riser size would be obtained from a sphere.

Spheres, however, are usually difficult to mold, and would present feeding problems as well, since the last metal to freeze would be near the center of the sphere, where it could not be used to feed a casting. Practicalities dictate the use of cylinders for most risers, and the discussion hereafter will refer to such shapes unless noted otherwise. The base of a side riser may be hemispherical in shape (Fig. 5.27), and a blind riser, i.e., one which is enclosed by sand, may have a hemispherical top in order to provide the smallest possible surface-area-volume ratio.

Riser Size as a Function of Casting Shape

Two simple examples can be used to define the relationship between riser size and casting shape. If a cylindrical casting poured on end is to be fed by a riser, it is obvious that the riser must have a diameter at least as large as that of the cylinder. On the other hand, if the same volume of metal used in the cylindrical casting is distributed over a greater area in the form of a plate, having a thickness less than the diameter of the cylinder, the riser needed to feed this plate will not be as large as the one for the cylinder, since it will not have to remain molten as long as the riser on the cylinder. Obviously, then, the surface-area-volume ratio of the riser can be related to the surface-area-volume ratio of the casting.

Thus Caine [37] developed an equation for steel, which expresses the relative freezing time of riser and casting, in terms of the relative volume of the riser and casting:

$$X = \frac{0.10}{Y - 0.03} + 1.0* \tag{5.44}$$

*In the original equation, the constants were slightly different.
where

$$X = \text{freezing ratio or relative freezing time} = \frac{\text{casting area/casting volume}}{\text{riser area/riser volume}} \tag{5.45}$$

Y = riser volume/casting volume

An example of one of these curves is plotted in Fig. 5.28 [38]. This curve provides the theoretical locus of points, which separate sound (right) and unsound (left) castings.

This relation, together with the additional information, provided by the other curves in Fig. 5.28, has contributed much to the basic understanding of risering principles, but this technique may require some trial-and-error calculation to arrive at the desired riser size.

Another factor is the shape of the cavity, or pipe, formed in the riser. If the pipe should extend into the casting, it may be necessary to enlarge the riser sufficiently to avoid the situation, even though the casting may otherwise be sound. Consequently, the nature of shape of the shrinkage cavity generated in the riser must be observed.

Various alternative procedures are available to calculate riser size [39–43]. One of these takes into account the shape factor of the casting, which is expressed as the sum of the length and width of the casting divided by the thickness $(L + W)/T$. Research on this approach has shown that the riser-volume-casting-volume ratio is related to the shape factor as presented in Fig. 5.29a [42]. Ready conversion to riser diameter is made with the help of Fig. 5.29b. Others have developed riser-size data based on casting geometry and shrinkage factors [41].

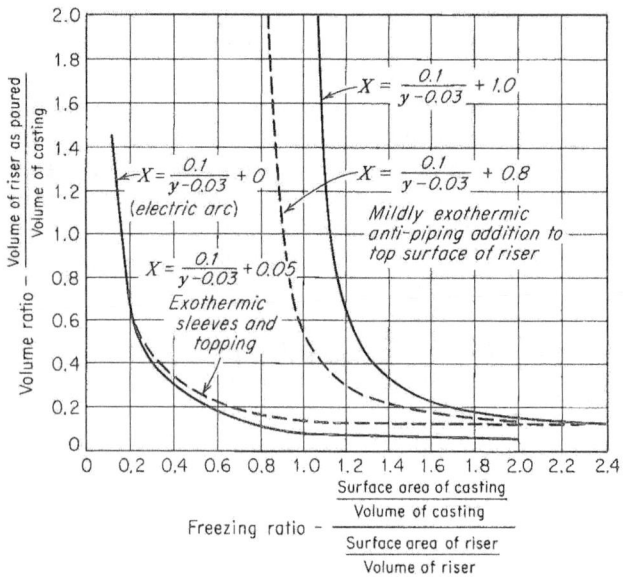

FIGURE 5.28 Plot of riser-volume to casting-volume ratio versus freezing ratio for various conditions existing in the riser. (From Heine et al. [2, p. 234], Wallace [38].)

Modulus Risering Method

The continual requirement for improved and consistent casting properties drove the need for a greater understanding of solidification and the need for techniques to predict accurately consistent results. Work by R. Wlodawer [34] began to address these needs by expanding the concept of Modulus for the solidification of steel castings based on the earlier ideas of Chvorinov. Additional work by R. Ruddle [31] in steel and other alloys rounded out the base information. The ultimate goal was to accurately understand how long it takes for a casting or a casting section to solidify. From that point, providing the right amount of feed metal at the right time became an easier and more consistent task.

$$\text{Modulus} = \frac{\text{volume}}{\text{cooling surface area}} \qquad (5.46)$$

Modulus Application in Example

The example shown in Fig. 5.30 illustrates the basics of modulus risering using sand risers [31]. It includes a check of the feeding distance. The modulus is corrected for riser contact area. Finally, the feed metal demand of the casting is examined.

Rectangular $16 \times 8 \times 4$ in casting:

1. *Determine number of risers needed*
   The feed distance equation for a $T \times 2T$ plate when $T = 4$ in is

$$\text{Feeding distance} = 11.6\sqrt{T} - 5.2 \qquad (5.47)$$

Feed distance = 17.9 in.
Therefore, a single centrally located riser will suffice.

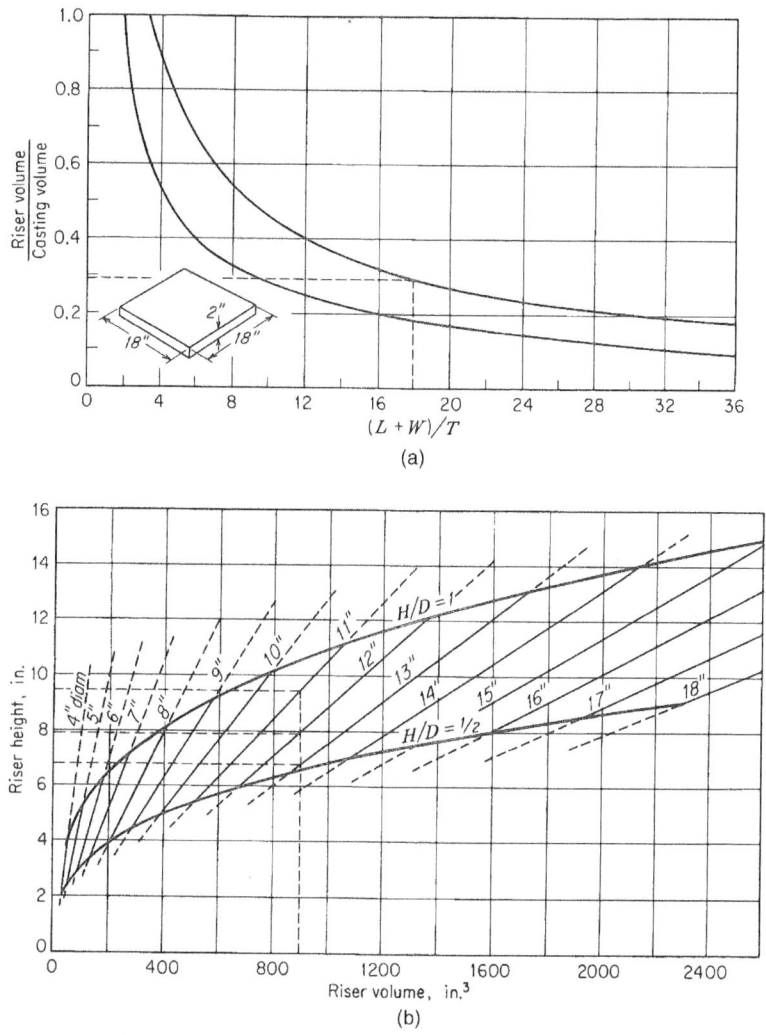

FIGURE 5.29   (a) Riser-volume to casting-volume ratio as a function of the shape factor, (b) chart for determining riser diameter. (From Heine et al. [2, p. 235], Myskowski et al. [39].)

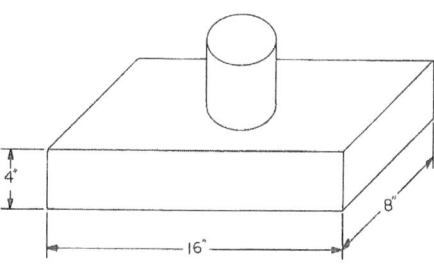

FIGURE 5.30   Plate casting with central top riser. (From Adams [1], Ruddle [31].)

2. *Calculate casting modulus*

$$\text{Modulus}_{\text{casting}} = \frac{\text{volume of casting}}{\text{cooling surface area of casting}} \tag{5.48}$$

$$\text{Modules}_{\text{casting}} = \frac{(\text{length})(\text{width})(\text{thickness})}{2[(\text{length})(\text{width}) + (\text{width})(\text{thickness}) + (\text{thickness})(\text{length})]} \tag{5.49}$$

$$M_C = \frac{16 \times 8 \times 4}{2[(16 \times 8) + (8 \times 4) + (4 \times 16)]} \tag{5.50}$$

$$M_C = \frac{512}{448} \tag{5.51}$$

where $M_C = 1.143$ in

$A_C = 448$ in^2

$V_C = 512$ in^3

3. *Determine size of a sand riser required*
   Generally, for ferrous alloys the riser modulus is 20% larger than the casting modulus and for some short freezing range copper based alloys, a 30% increase is preferred.

$$M_R = 1.2 \times M_C \tag{5.52}$$

$$M_R = \frac{V_R}{A_R} = \frac{\pi R^2 H}{2\pi RH + \pi R^2} \tag{5.53}$$

For a riser with an H:D ratio of 1:1
$D = 6.86$ in

4. *Calculate riser contact area*
   Increase riser diameter by a factor of 1.15 for approximation of contact effect:

$$6.86 \times 1.15 = 7.89 \text{ in} \tag{5.54}$$

Therefore, riser contact area is

$$\frac{\pi}{4} \times 7.89^2 = 49 \text{ in}^2 \tag{5.55}$$

5. *Recalculate modulus*

$$M'_C = \frac{512}{448 - 49} = 1.28 \text{ in} \tag{5.56}$$

Required riser diameter is now recalculated using step 3 above

$$M'_R = 1.2 \times M'_C \tag{5.57}$$

$D' = 7.68$ in

Round to the nearest inch in order to use an 8-in diamter × 8-in head.

6. *Check that the feed metal requirements of the casting are met*
The volume of steel contained in an 8-in diameter × 8-in riser, after allowing space for a 1-in top cover, is

$$V_R = \frac{\pi}{4} \times 8^2 \times 7 = 352 \text{ in}^3 \tag{5.58}$$

Hence the riser to casting volume ratio is

$$\frac{352}{512} = 0.69 \,(69\%) \tag{5.59}$$

The casting is "chunky"; so from published tables, based on foundry results, a volume ratio of 106% is needed [31]. Therefore, the volume of the riser must be increased to at least 543 in^3. A 9-in diameter × 10-in head (effective height 9 in) has a volume of 572 in^3 and would be satisfactory. Alternatively, an 8-in diameter × 12-in head could be used; thus having the advantage of a reduced contact area and riser removal costs.

It is interesting to note that if an insulated riser had been selected, it would have to be only 6 in in diameter × 6 in high (effective height 5 in) due to its efficiency of feed metal delivery [31]. Such a riser has a volume of 141 in^3, thus the use of insulation would reduce the riser weight by 60%, in turn saving melting costs.

One factor that must not be overlooked is the effect of pouring temperature on the feeding of castings. Changing the pouring temperature not only affects the thermal gradients in the casting; it also affects the extent of nucleation as well. For every surface-area-volume ratio formed in casting, there is an optimum pouring temperature. Pouring a casting of a given surface-area-volume ratio at too low a temperature will cause skulling over of the riser as well as ineffective feeding. Pouring at too high a temperature tends to cause the riser to dish rather than to pipe, which will reduce feeding action unless the riser is specifically designed to account for this. The dishing results from extensive dendritic growth during the freezing process.

Location of Risers

When a long bar or plate is cast without a riser, it is found that a certain length from each end of the bar or plate is sound. This result is due to the directional solidification that develops at the ends because of the greater heat extraction from those points compared with others. An example of the application of this directional solidification with a riser is given in Fig. 5.31. However, this end effect will occur even in the absence of a riser.

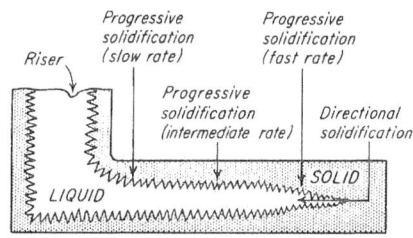

FIGURE 5.31 An example of progressive solidification, directional solidification, and feeding of a casting with a riser. (From Heine et al. [2, p. 213], Myskowski et al. [44].)

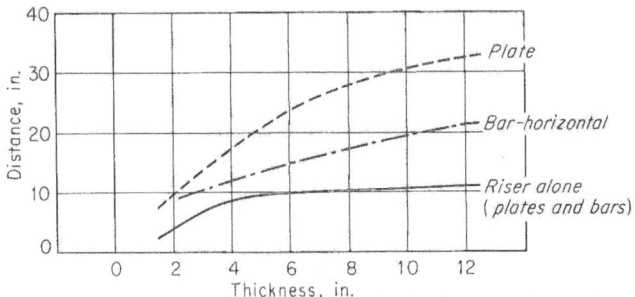

Figure 5.32 Maximum feeding distance of risers only and of total riser plus end effect on steel plates and bars. (From Heine et al. [2, p. 239], Wallace [38].)

Similarly, if a long bar or plate is cast horizontally, with one adequate riser at the center, it will be found that, for a certain distance from the riser, the casting will be sound because of the feeding action of the riser. Beyond this point, some form of shrinkage will be evident. These two effects can be referred to as the end effect and riser effect, respectively. Figure 5.32 illustrates the consequence of these effects in feeding steel plates and bars [38]. Feeding distances have been investigated by multiple researchers utilizing various alloys. The exact results differ due to processes and alloys employed in the experimentation. Therefore, the feed distance equations may not exactly agree but they do agree in principle. To obtain the most accurate results, utilize feed distance information from the most closely matched alloy.

If castings having a variety of section thicknesses are produced, the adjacent sections will have an effect on the soundness of the casting in a specific section. Thus, if a light section is attached to a heavy section, the extent of the sound region in the heavy section will not be as great as in the absence of the light section. Conversely, the presence of the heavy section next to the light section would increase the length of sound metal in the light section as compared to the case where the heavy section is absent.

Another similar situation is the case where both a light and a heavy section are attached to a section of intermediate thickness. Table 5.5 gives formulas to calculate feeding distances for a number of these possibilities for steel castings, but similar equations exist or could be developed for other metals. One of the results of these studies is to demonstrate that it is easier to produce a sound tapered section than a section of uniform thickness.

With the effect from risers and ends established, it is possible to use that data to determine the location of risers and effect the complete soundness in a bar or plate. Only a simple case will be used as an example. More specific examples can be found in Ref. 25 and *Feeding and Risering Guidelines for Steel Castings* by the Steel Founders' Society of America.

Problem
What is the theoretical length of a 4-in-thick bar of steel that can be cast soundly with the two top risers? In this case, the riser size is assumed to be adequate to feed the casting if spacing is correct.

Condition	Without Chill	With Chill
Plate, no end effect	$D = 3.6\sqrt{T}$	$D = 11.6\sqrt{T} - 3.2^*$
Bar, no end effect	$D = 3.6\sqrt{T}$	$D = 6\sqrt{T} + T$
Plate with end effect	$D = 11.6\sqrt{T} - 5.2$	$D = 11.6\sqrt{T} - 3.2$
Bar with end effect	$D = 6\sqrt{T}$	$D = 6\sqrt{T} + T$
Bar, vertical	$D = 7.15\sqrt{T}$	
Plate with parasite section	$D_H = 3\left(T_H - T_L\right) + 4.5$	
Plate with both light and heavy section attached	$D_M = 3.5\left(T_H - T_L\right)$	
Plate with parent section	$D_L = 3.5T_H$	

*Chill spaced uniformly between two risers.
D = distance from one riser to chill. Therefore total feeding distance between risers is 2D.
Source: Wallace [38].

TABLE 5.5   Feeding Distance of Risers in Terms of Thickness T

Solution

It will be noted that each riser will feed to one end on one side of the riser and toward the center of the bar on the other. Therefore, from Table 5.5, the end effect would be

$$6\sqrt{T} = 6\sqrt{4} = 12 \text{ in} \tag{5.60}$$

On the other side of the riser the feeding distance would be

$$D = 3.6\sqrt{T} = 3.6 \times 2 = 7.2 \text{ in} \tag{5.61}$$

Therefore, each riser would feed a total distance of 19.2 in plus their diameter (the distance under the riser), and the total length of bar that could be cast soundly would be 38.4 in plus 2 times the riser diameter (the distance under each riser).

Problems of this type may be more complex, depending on casting geometry. In some instances, graphical solutions using a compass to inscribe the circle of influence of a given riser are easier to handle. In the case of side risers, the radius of the riser must be added to the feeding distance to locate the circle of influence from the center of the riser. A schematic example of the development of an adequate riser and padding system is provided in Fig. 5.33.

Blind Risers

The conventional riser is open to the atmosphere. The so-called blind riser is enclosed by the sand mold and is usually designed for a minimum surface area per unit volume (Fig. 5.34). In the case of steel, which forms a solid outer skin of metal during solidification, the sprue solidifies early, and the casting and blind riser constitute a closed shell of metal, which develops a partial vacuum by virtue of the shrinkage that occurs during solidification. As shrinkage takes place in the casting, metal is drawn in from the riser to compensate for it. This can occur even though the riser is no higher than the casting, but the temperature gradient must be such that the casting freezes first. Of course, if the riser is of greater height than the casting, additional benefits are gained. If the skin of the

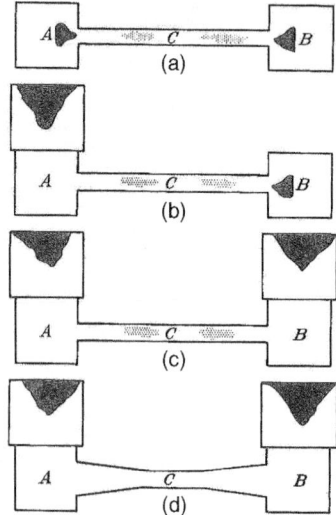

FIGURE 5.33 The development (*a* to *d*) of a riser and padding system to ensure casting soundness. Shaded area represents microporosity. (From Heine et al. [2, p. 241], Caine [37].)

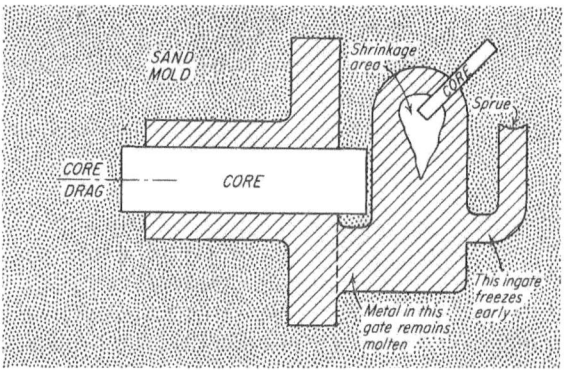

FIGURE 5.34 Cross-sectional diagram of a casting fed by a blind riser with atmospheric vent produced by a pencil core. (From Heine et al. [2, p. 241].)

casting is strong enough, and the skin of the blind riser is pierced to allow atmospheric pressure to exert an influence, the riser can then feed a casting of greater height than itself and the atmospheric pressure forcing liquid metal into the shrinkage areas as they develop. The skin of the riser can be kept open by using sand or graphite core inserts, as illustrated in Fig. 5.34. When a blind riser is used in this manner, it is referred to as an atmospheric or pressure riser.

The blind riser has a number of advantages, among which are

- When the riser is attached to the runner system, the hottest metal is in the riser, and the coldest is in the casting. This promotes directional solidification.

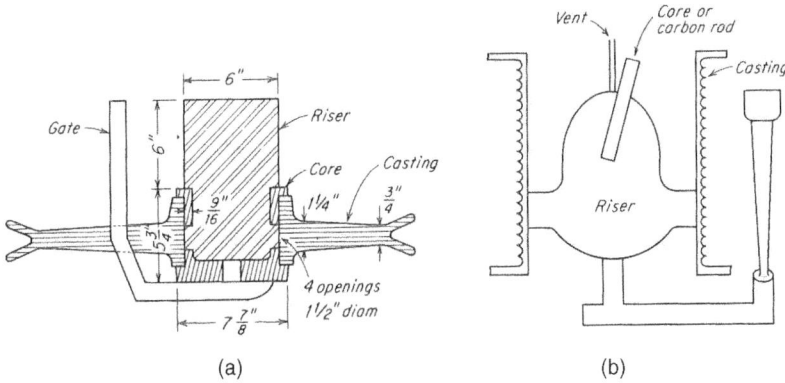

Figure 5.35 Examples of (a) an open-top internal riser and (b) a blind internal riser. (From Heine et al. [2, p. 242], Wallace [38].)

- Considerable latitude is allowed in positioning the blind riser.
- A blind riser can be smaller than a comparable open riser.
- Blind risers can be removed more easily from a casting.

One disadvantage, particularly if it is to feed a section of the casting at a greater height, is that a break in the skin of the casting itself will tend to equalize the pressure and reduce the effectiveness of the riser.

Internal Risers

Risers surrounded in whole, or in part, by the casting to be fed are referred to as internal risers. Risers of this type can be employed inside circular or cylindrically shaped castings. Since the risers are partially surrounded by the casting, their cooling rate is lower than that of risers, which are located above or to one side of a casting. This means that they can be made smaller than in a conventional case, thereby contributing to casting yield. Two examples of the internal riser are given in Fig. 5.35.

Grouping of Castings

Closely related to the use of internal risers is the improved efficiency obtained when several castings can be grouped about a single riser. Not only does one riser do the work of several, but also the grouping of castings near the riser lowers its cooling rate so that a smaller riser can be used. This principle is illustrated in Fig. 5.36.

Riser Connections to the Casting

How the riser is attached to the casting is important because it determines, first, how well the riser can feed the casting, and second, how readily the riser can be removed from the casting. It may also control, to some extent, the depth of the shrinkage cavity by solidifying just before the riser freezes, thereby preventing the cavity from extending into the casting. Riser-neck dimensions for three types of risers are given in

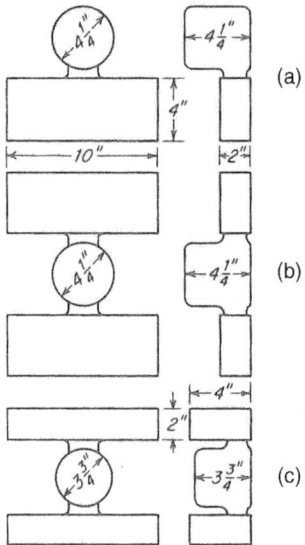

Figure 5.36 Improving casting yield by grouping castings about a single riser. Note the smaller riser diameter required for case c, where the proximity of the castings has lowered the riser cooling rate. (From Heine et al. [2, p. 243], Wallace [38].)

Type Riser	Length L_N	Cross Section L_N
General side	Short as feasible, not over $D/2$	Round, $D = 1.2\,L_N + 0.1D$
Plate side	Short as feasible, not over $D/3$	Rectangular, $H_N + 0.6$ to $0.8D$; as neck length increases, $W_N = L_N + 0.18D$
Top	Short as feasible, not over $D/2$	Round, $D = L_N + 0.2D$

Source: Wallace [38].

Table 5.6 Riser-Neck Dimensions

Table 5.6, which is to be used with reference to Fig. 5.37. These dimensions are for cases where the material surrounding the neck has the same thermal properties as the molding material used elsewhere. If insulating necks or necks made from core sand are used, the dimensions may be smaller.

Chills

Use of Chills

The foregoing discussion of risers deals largely with methods for securing directional solidification by extending the solidification process in some part of the mold system. It is entirely possible, however, that the same objective of directional solidification can be accomplished by the reverse procedure of chilling the metal in those portions of the

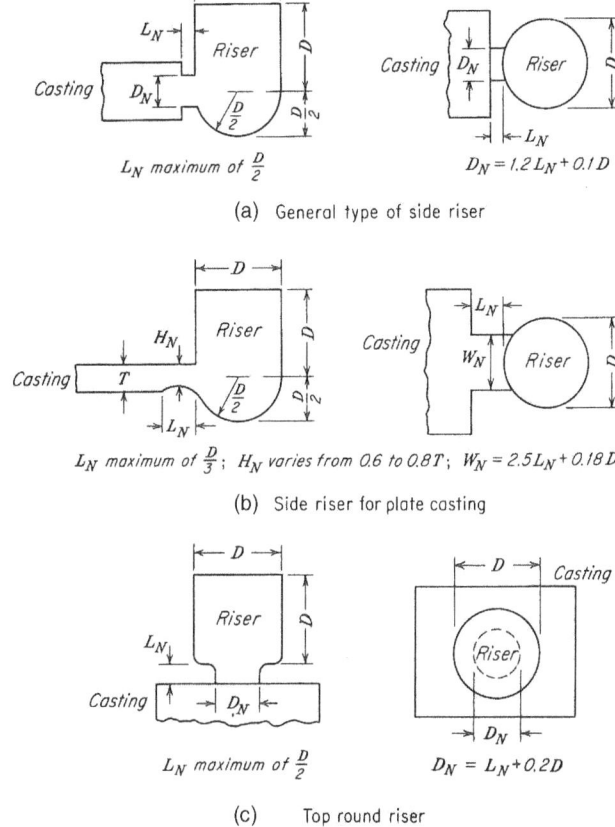

L_N maximum of $\frac{D}{2}$

$D_N = 1.2\,L_N + 0.1\,D$

(a)   General type of side riser

L_N maximum of $\frac{D}{3}$; H_N varies from 0.6 to 0.8 T; $W_N = 2.5\,L_N + 0.18\,D$

(b)   Side riser for plate casting

L_N maximum of $\frac{D}{2}$

$D_N = L_N + 0.2\,D$

(c)   Top round riser

Figure 5.37   Location and dimensions used in Table 5.6 for three types of risers. (From Heine et al. [2, p. 244], Wallace [38].)

casting that are more remote from the liquid-metal source. Both external and internal chills can be used for this purpose. However, feed metal requirements of the casting must still be considered.

External chills are placed in the mold walls at the mold-metal interfaces, whereas internal chills are placed in the casting cavity.

External Chills

External chills are metal inserts of steel, cast iron, or copper that are placed at appropriate locations in the mold to increase the freezing rate of the metal at those points. They may be standard shapes or, in special cases, may be shaped to conform to the required casting cavity dimensions; however, their size is determined by the cooling requirements. They can be used effectively at junctions or other portions of a casting that are difficult to feed by risers. A number of examples of the use of external chills are given in Fig. 5.38. The effect of chills in altering the freezing time of steel castings is demonstrated in Fig. 5.39. This figure showed that the gap between the start and end of freezing was

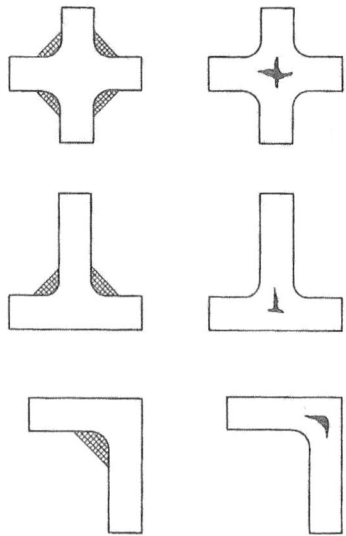

Figure 5.38 External chills of appropriate size can be used to eliminate porosity at casting junctions. (From Heine et al. [2, p. 245].)

drastically reduced by the use of external chills. This effect causes a notable improvement in the chance for getting a sound casting [25].

Advantages of external chills are

- It sets up steep temperature gradients.
- It promotes good directional and progressive solidification.
- It reduces the incidence of macro- and microporosity.

Chills must be dry to avoid forming blowholes in the metal. They are frequently given a protective wash of silica flour or other refractory material. This wash should be thoroughly dried before the chill is inserted into the mold. When chills are placed in green sand, moisture from the mold may condense on them if they are allowed to stand too long in the mold. This can be prevented by preheating the chill before its insertion, or pouring the mold shortly after it is made. Condensation of moisture on the chill should be avoided because it leads to gassing of the metal.

Although not strictly in the category of an external chill, the same effects can be accomplished by a variety of molding materials that will change the cooling characteristics of the mold. An example of this is shown in Fig. 5.40. Here, the use of a chill plus crushed magnesite and bonded silicon carbide, in the manner shown, results in the proper directional solidification. In some instances, similar effects can be obtained by varying the thickness of the molding material.

Internal Chills

Internal chills are placed in the mold at locations that cannot otherwise be reached effectively with external chills. They are also used in spots that are subsequently machined

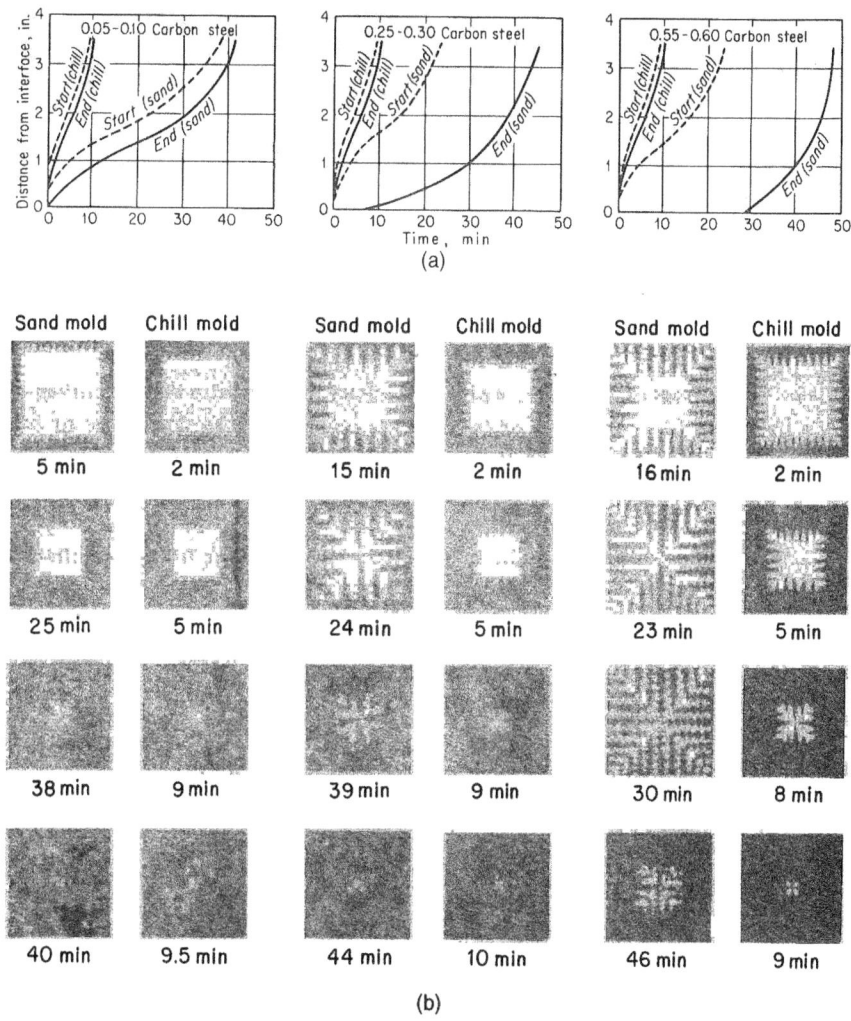

Figure 5.39 Effect of chill and sand molds on the solidification of various carbon content steels, shown (a) as TTT solidification curves and (b) as graphic representations. (From Heine et al. [2, p. 191], Bishop and Pellini [45].)

out, such as in bosses and lugs that are to be drilled or bored. Examples of internal chills are shown in Fig. 5.41.

The use of internal chills is somewhat more critical than external ones for the following reasons:

• The chill may not fuse with the casting, thereby establishing points of weakness.

• Cleanliness of the chill is more important since it will be completely surrounded by metal, and any gas that is created cannot readily escape.

• The chill may alter the mechanical properties of the casting where it is used.

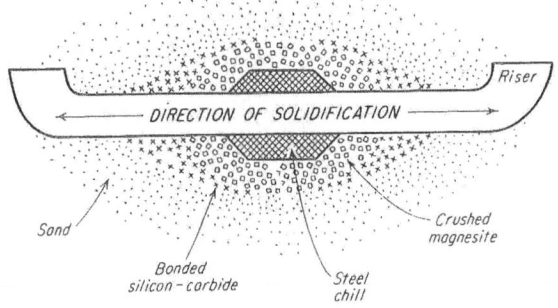

Figure 5.40   Directional solidification can be secured by the use of a variety of mold materials that change the cooling characteristics of the mold. (From Heine et al. [2, p. 246], Briggs [10].)

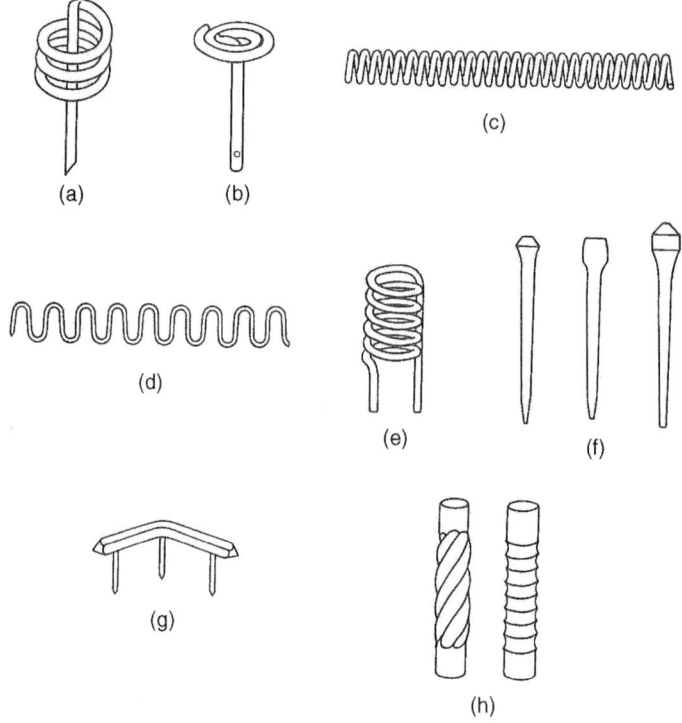

Figure 5.41   Examples of internal chills. (a) Chill coil nail, (b) flathead chill coil nails, (c) chill coil, (d) grid chill, (e) hub chill, (f) chill nails, (g) spider chill, (h) chill rods. (From Heine et al. [2, p. 247].)

The composition of the chill must be compatible with the metal being poured. Thus, a cast-iron internal chill would not be used for steel or a nonferrous casting. Usually, the chill should have approximately the composition of the metal in which it is to be used.

Special Conditions Arising from Joining Sections

Figure 5.38 gives examples of the shrinkage that can develop at section junctions and the use of chills to prevent it. Section junctions present special feeding problems that can be handled by using chills or by redesign of the casting. Figure 5.42 gives an example of a redesign to accomplish the same objective of eliminating shrinkage at a junction.

Use of Insulators and Exothermic Compounds

A riser can be made more efficient by employing some additional means to keep the top of the riser from freezing over so that the molten metal beneath can be exposed to atmospheric pressure. This can be done by the use of certain additions made to the surface of the molten metal in the riser, preferably as soon as possible after the metal enters the riser.

These additions serve as antipiping compounds through an insulating effect or from heat given off by an exothermic reaction in the compound. Insulating effects are obtained by such additions as powdered graphite or charcoal, rice or oat hulls, and refractory powders. These carbonaceous compounds, as well as other mixes specifically designed for that purpose, also provide additional heat from exothermic reactions. In the case of carbonaceous materials, this effect may merely result from oxidation of carbon. With other mixes, the oxidation may be more rigorous, resulting from reactions of the type $2Al + Fe_2O_3 = 2Fe + Al_2O_3$, which is strongly exothermic. Other oxidants and oxidizers, together with other additions, may also be employed [46]. Figure 5.43 indicates the improvement in riser efficiency obtained from the use of insulating and exothermic type compounds. Figure 5.28 illustrates the effect of these compounds and the use of an electric arc on the riser-to-casting freezing ratio.

Besides supplying an insulating or exothermic condition on the top of the riser, it is also possible to use sleeves to form the sides of the riser. By the proper selection of shaped insulating and exothermic materials, it is possible to secure a lower solidification rate in the riser and hence better feeding of the casting. A general example of this is a chunky casting (cube-like shape) would require a sand riser volume that is 140% of the casting volume while the requirements utilizing an insulating riser would be a volume that is only 32% of the casting volume [31]. Similarly, with a rangy casting (plate-like shape), the sand riser volume would be at least 12% of the casting volume while the insulating

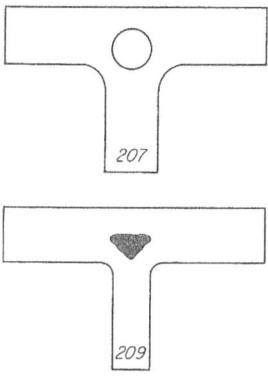

Figure 5.42 Effect of casting design on soundness at a junction. (From Heine et al. [2, p. 250], Wallace [38].)

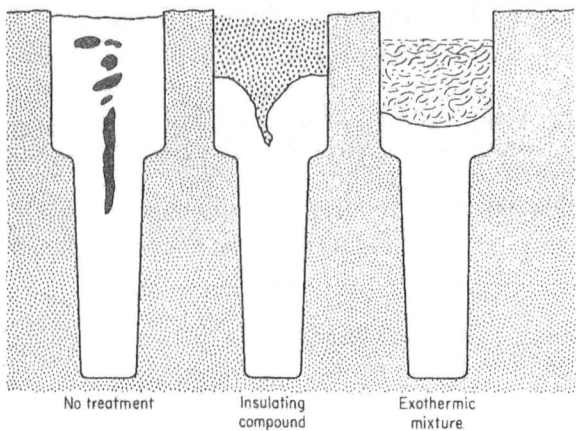

No treatment   Insulating   Exothermic
         compound   mixture

FIGURE 5.43   Use of an insulating compound or exothermic mixture in the riser reduces the piping tendency and decreases the amount of metal required in the riser. (From Heine et al. [2, p. 248].)

sleeve volume would be 8% of the casting volume. Therefore, the improved feeding characteristics obtained by utilizing insulating and exothermic materials are related to the composition of those materials and the configuration of the casting.

Insulating pads can also be employed in various parts of the mold, to decrease the cooling rate in thin sections where such an effect is desired to promote directional solidification. An illustration of the use of an insulating sleeve and pad is given in Fig. 5.44. Moldable exothermic compounds can also be used in place of the insulating pads [47, 48].

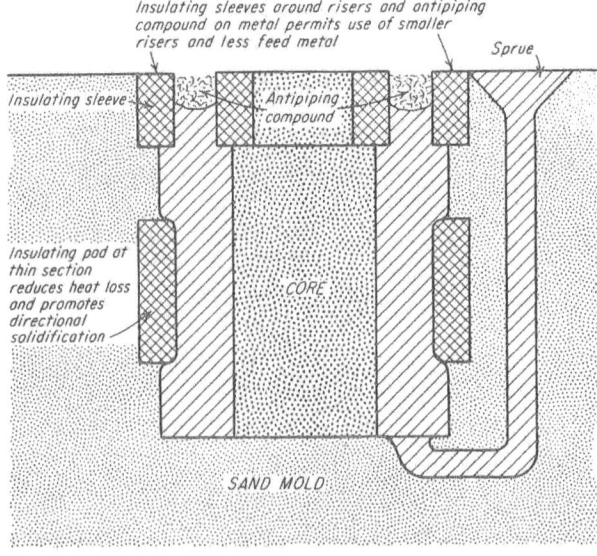

FIGURE 5.44   The combination of insulating sleeves around risers and insulating pads at thin sections of casting increases metal yield and promotes directional solidification. (From Heine et al. [2, p. 249].)

The net effect on the risers in using chills, insulators, or exothermic compounds on the casting is to reduce the size of the riser relative to the size of the casting. It is obvious that this means a higher casting yield. (Yield is the weight of casting relative to the total weight of casting plus the risers and the gates.) The use of chills, or the addition of insulating materials or exothermic compounds to the molds, means added costs, not only because of the use of these materials, but also with respect to any additional conditioning the system sand may require to remove these materials. Therefore, there are economic considerations, as well as technical ones, that will dictate whether or not these means of increasing casting yield are sound. There are, of course, certain instances where these practices are the only feasible ones to use, and in such instances, the question of cost is not involved.

Special Conditions Arising from Joining Sections

In designing the feeding system for a casting, the first step is to determine the location of risers and chills to assure that directional solidification is achieved. This can be done with the use of published information on feeding distances such as given in Table 5.5 and elsewhere, or on the basis of previous experience. The use of exothermic compounds, insulators, or other special techniques must be studied from both the technical and economic aspects. Once the riser locations are established, the riser sizes can be calculated, using the procedure outlined in this chapter, or one of the other methods given in chapters dealing with specific casting alloys.

With the risering system established, the gating system can be planned, using the principles presented in the first part of this chapter. These steps are not necessarily performed in the order given here, but the point to be made is that design of castings can be established on the basis of sound engineering principles without the necessity to be hit-or-miss.

In a very general way, a riser or gating system designed for steel ought also to be applicable to other metals. However, there are certain restrictions to this generalization.

Certain alloys, such as gray cast iron and some aluminum- and copper-base alloys, have an extended freezing range which delays the possibility of developing a gastight skin, such as is found in cast steel and some copper-base alloys. In these alloys, atmospheric or pressure risers must be applied with greater engineering care to avoid problems. In addition, gray cast iron undergoes some expansion during solidification as a result of graphitization occurring during the freezing period. As a consequence, gray cast iron may require less risering than other metals.

Customer requirements determine the extent of the gating and feeding systems used on a particular casting. As casting requirements increase and engineering techniques improve, a more judicious proportioning and positioning of the rigging systems will lead to reaching quality and economic goals.

References

1. A. Adams, "Review Iron Gating Design, Filter Use to Optimize Benefits," *Modern Casting*, March 2001.
2. R. W. Heine, C. R. Loper, Jr., and P. C. Rosenthal, *Principles of Metal Casting*, McGraw-Hill Book Company, New York, 1967.
3. G. Batty, "Controlled Directional Solidification," *Transactions of AFS*, 42: 237, 1934.
4. J. F. Wallace and E. B. Evans, "Gating of Gray Iron Castings," *Transactions of AFS*, 65: 267, 1957.

5. M. J. Berger and C. Locke, "A Theoretical Basis for the Design of Gates", *Foundry*, Vol. 79, February 1951, p. 112.

6. J. F. Wallace and E. B. Evans, "Principles of Gating," *Foundry*, 87: 74, Oct. 1959.

7. M. J. Berger and C. Locke, "Fluid Flow Mechanics of Molten Steel," Armour Research Foundation, Chicago, Apr. 12, 1951.

8. Report of IBF Subcommittee TS24, Ingots, *Foundry Trade Journal*, 99: 691, Dec. 15, 1955.

9. A. Adams, D. Westphal, et al. (eds.), *Basic Principles of Gating and Risering*, 2d ed., AFS, Schaumburg, Illinois, 2008, p. 4.

10. C. W. Briggs, *The Metallurgy of Steel Castings*, McGraw-Hill Book Company, New York, 1946.

11. H. W. Dietert, "How Fast Should a Mold be Poured?," *Foundry*, Vol. 81, Aug. 1955, p. 205.

12. H. O. Meriwether, "Shell Molded Ductile Iron Castings Gating and Risering I for Vertical Pouring," *Transactions of AFS*, 68: 516, 1960.

13. E. A. Lange and A. T. Bukowski, "Pouring Times for Steel Castings," U.S. Naval Research Laboratory Report, Washington, D.C., 1958.

14. L. F. Porter and P. C. Rosenthal, "Factors Affecting Fluidity of Gray Cast Iron," *Transactions of AFS*, 60: 725, 1952.

15. C. V. Knobeloch, "Choke that Gate, Modern Castings," Vol. 84, Dec. 1956, p. 48.

16. D. M. March, "Gating and Risering Ductile Iron Castings Poured in Dry Sand Molds," *Transactions of AFS*, 68: 512, 1960.

17. L. W. Eastwood, "Tentative Design of Horizontal Gating System for Light Alloys, in Symposium on the Principles of Gating," *American Foundrymen's Society*, 1951, p. 25.

18. J.G. Kura, "Calculation of Horizontal Gating Systems", *Transactions of AFS*, Vol. 27, May 1955, p. 123.

19. C. W. Briggs, "Gating Steel Castings," *Foundry*, Vol. 88, Jun. 1960, p. 124.

20. R. F. Polich, A. Saunders, Jr., and M. C. Flemings, "Gating Premium Quality in Castings," *Transactions of AFS*, 71: 418, 1963.

21. T. Finlay, "Fin Gating: New Cost-cutting Steel Techniques", *Modern Castings*, Vol. 40, September 1961, p. 53.

22. E. Bjorklund, "Calculating Ingate Dimensions for Gray Iron Castings," *Transactions of AFS*, 70: 193, 1962.

23. R. W. White, "Gating of Ductile Iron Castings," *Foundry*, Vol. 88, Feb. 1960, p. 101.

24. M. C. Flemings and H. F. Taylor, "Gating Aluminum Castings," *Foundry*, Vol. 88, Apr. 1960, p. 72.

25. C. W. Ward, Jr., and T. C. Jacobs, "Kiss Gating Brass Castings," *Transactions of AFS*, 70: 865, 1962.

26. J. G. Mezoff and H. E. Elliott, "A Study of Factors Affecting the Pouring m Rates of Castings," *Transactions of AFS*, 56: 279, 1948.

27. K. Grube and L. W. Eastwood, "A Study of the Principles of Gating," *Transactions of AFS*, 58: 76, 1950.

28. S. Ducharme, M. Sahoo, and K. Sadayappan, *Casting Copper-Base Alloys*, 2d ed., American Foundry Society, Schaumburg, Illinois, 2007, p. 172.

29. W. H. Johnson, W. O. Baker, and W. S. Pellini, "Principles of Gating Design: Factors Influencing Molten Steel Flow from Finger Gating System," *Transactions of AFS*, 58: 661, 1950.

30. S. Ducharme, M. Sahoo, and K. Sadayappan, *Casting Copper-Base Alloys*, 2d ed., American Foundry Society, Schaumburg, Illinois, 2007, p. 177.

31. R. Ruddle, *Risering of Steel Castings*, Foseco, Cleveland, Ohio, 1979, p. 3–1 and 1–19.

32. J. Campbell, *Castings*, 2d ed., Elsevier Butterworth-Heinemann, Oxford, UK, 2003.

33. R. A. Flinn, "Risering of Ductile Iron," *Transactions of AFS*, 63: 720–725, 1955.

34. R. Wlodawer, *Directional Solidification of Steel Castings*, Pergamon Press, Oxford, 1966, p. 5.

35. "Feeding and Risering Guidelines for Steel Castings," Steel Founders' Society of America, 2001.

36. N. Chvorinov, "Theory of the Solidification of Castings," *Geisserei*, 27: 177–225, 1940.

37. J. B. Caine, "Risering Castings," *Transactions of AFS*, 57: 66, 1949.

38. J. F. Wallace (ed.), "Fundamentals of Risering Steel Castings," Steel Founders' Society, 1960.

39. E. T. Myskowski, H. F. Bishop, and W. S. Pellini, "A Simplified Method for Determining Riser Dimensions," *Transactions of AFS*, 63: 271, 1955.

40. J. T. Berry and T. Watmough, "Factors Affecting Soundness in Alloys with Long and Short Freezing Range," *Transactions of AFS*, 69: 11, 1959.

41. H.D. Merchant, "Dimensioning of Sand Casting Risers", *Modern Castings*, Vol. 35, February 1959, p. 73.

42. J. F. Wallace, "Risering of Castings," *Foundry*, Vol. 87, Nov. 1959, p. 74.

43. C. W. Briggs, "Risering of Commercial Steel Castings," *Transactions of AFS*, 63: 287, 1955.

44. E. T. Myskowski, H. F. Bishop, and W. S. Pellini, "Application of Chills to Increasing the Feeding Range of Risers," *Transactions of AFS*, 60: 389, 1952.

45. H. F. Bishop and W. S. Pellini, "Solidification of Metals," *Foundry*, Vol. 80, Feb. 1952, p. 87.

46. S. L. Gertsman, "A Study of Insulating and Mildly Exothermic Antipiping Compounds Used for Steel Castings," *Transactions of AFS*, 57: 332, 1949.

47. H. F. Bishop, H. F. Taylor, and R. G. Powell, "Risering of Steel Castings with Exothermic Sleeves," *Foundry*, Vol. 86, Jun. 1958, p. 54.

48. T. C. Bunch and G. E. Dalbey, "Feeding of Castings," *Transactions of AFS*, 63: 503, 1969.

CHAPTER 6

Casting Process Simulation

Christof Heisser
MAGMA Foundry Technologies, Inc.

Traditional Methods of Gating and Riser Design

The casting process is probably the most elegant way to making a part. Unfortunately, it is also one of the most complex ones. Therefore, it was necessary to introduce simplifying rules for the gating and riser design process to provide foundry engineers a starting point. Traditionally, these rules separated the gating design from the riser design by generating rules for the former based on well-known fluid dynamics equations, and the latter on formulas derived from basic heat transfer and material transport phenomena. These simplifications appear to be very crude on the first look; however, if properly used and used consciously of the fact that they in fact are simplifying and crude, they are a valuable tool to provide a starting point to develop a proper gating and riser system. Even the use of sophisticated casting process simulation tools does not eliminate the need for a knowledgeable operator who is aware of the physical laws, the casting process, and what properties the sophisticated simulation tools are based on. Even if simulation tools might be available in the near future that propose gating and riser layouts based on just the casting geometry and some process boundary conditions, the knowledge of basic fluid flow and heat transfer laws, as well as their application on the layout of gating and riser configurations, is essential.

Based on more than 5,000 years of casting process history, foundry engineers eventually agreed on several basic requirements that a proper gating design needs to fulfill.

Melt needs to be supplied to the casting and adjacent risers during the filling process in a manner that ensures

- The complete filling of the entire mold cavity without misruns or areas not filled
- A quiet and controlled flow to eliminate or reduce defects that are caused by excessive melt velocities and turbulences
- The minimal use of melt volume and easy removal of all of its components for cost efficiency, which includes the avoidance of additional cores or mold pieces to accommodate gating and riser layouts

Even these three simple rules create contradictory objectives, i.e., rule 2 asks for a slow filling, which contradicts rule 1, the need for a fast filling to avoid premature solidification, not to mention the productivity requirements of high-production molding lines. Rule 3 also limits the complexity of gating systems potentially required by (2) and (1). "Controlled" also needs to be further explained, as it means a gating system should control the melt flow independently from outside boundary conditions, like the human who pours the melt out of the ladle or the automatic pouring furnace.

Furthermore, melt needs to be supplied to the casting during the solidification process to compensate for the volume deficit created by the density increase experienced by the casting during the temperature reduction during

- Cooling of the liquid phase of the metal (liquid contraction of the melt)
- Transition of the metal from its liquid phase to its first solid phase (solidification contraction)

However, the risers need to be

- As small as possible (lowest cost)
- And have the smallest possible contact area to the casting (for easy removal)

The different approaches of obeying these fundamental rules lead to a variety of casting processes, which each add their specific requirements and variations. It would be futile to discuss the gating design approach for each individual casting process here, hence only the most basic horizontal gravity sand casting example shall be used to provide a basic understanding.

A traditional gating system has the following components (Fig. 6.1):

- Ladle lip or furnace nozzle (outside the system in the mold, but part of the melt supply)
- Pouring basin or cup
- Sprue and transition to runner
- Runner and runner extensions
- Gates
- Risers
- Vents and flow-offs

The lip of the *pouring ladle* or the *nozzle* of a pouring furnace is often forgotten when a gating design is developed. However, it might have an impact on the filling of a casting if the gating system is not designed to control or eliminate the kinetic energy introduced into the melt by the fall height from the lip or nozzle until it enters the pouring basin or cup. If the gating system doesn't have a pouring basin or cup and the melt stream immediately enters the sprue, the velocities the system has to deal with are, at least initially, much higher than the actual height of the sprue would suggest. This fact also needs to be considered when simulation tools are used, as they usually require the operator to define an entry point for the melt and a flow rate, velocity, or pressure value(s). In rare cases the capacity of the gating system can exceed the capacity of melt a ladle or pouring furnace can provide. In that case, the ladle or nozzle becomes the *choke*. The choke is usually defined as the cross-sectional area that limits the flow of metal in a gating system in a manner that it controls the entire flow behavior of the system.

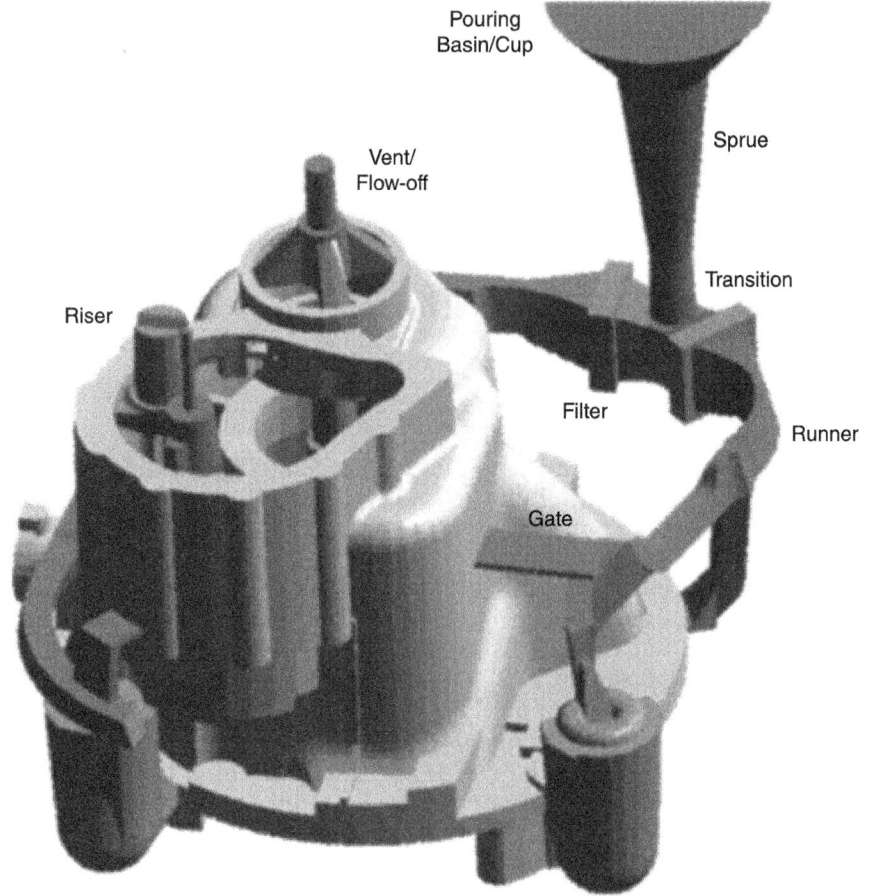

FIGURE 6.1 Basic components of a gating system. (MAGMA.)

The *pouring cup* or *basin* very often is used to take the initial inertia out of the melt stream. This is mostly accomplished by having the melt stream hit an area somewhat off-set from the sprue. Automatic molding lines often require the use of *reverse tapered sprues* (its cross section is smaller on top than at the bottom). This design can lead to having the choke at the top of the sprue, which is usually undesirable as it leads to air aspiration and the creation of oxides in the melt, which are usually not acceptable in the casting (Fig. 6.2).

As one of the main duties of a gating system is to *control the flow by taking inertia out of the melt stream*, traditional gating design methods simplify this process by defining cross sections in the sprue, the transition to the runner, the runner itself, or the gates. The most basic approach is to place the choke at the point in the gating system that is likely to experience the highest velocities and metallostatic pressure—the bottom of a tapered sprue. Sprues are usually tapered to follow the contraction of the melt stream as it accelerates due to the gravitational force. Based on Bernoulli's law, nomographs are used to depict the correct upper and lower cross-sectional area of the sprue based

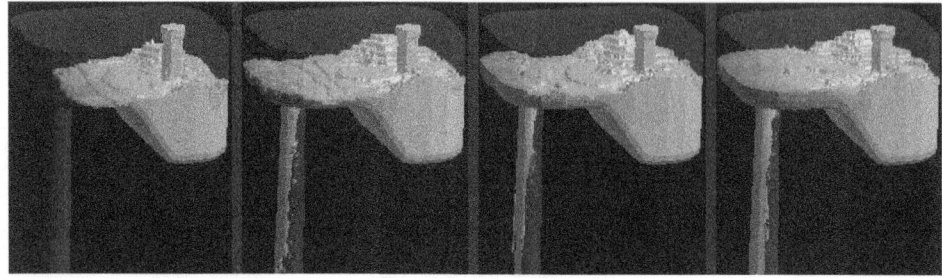

Figure 6.2 Choke created in upper cross section of a reverse tapered sprue. Inertia of flow leads to underutilization of upper cross section creating a choke different than expected from cross section. (MAGMA.)

on the desired flow rate and metal poured. For a reverse tapered sprue, the lower cross-sectional area is calculated the same way, but placed in the entry of the runner, as it cannot be located at the bottom of the sprue.

The *transition from the sprue to the runner* (sprue base or well) is supposed to ensure a smooth redirection of the vertical flow into the horizontal runner. With the ability of simulation tools to provide detailed displays of flow patterns many myths, like the creation of melt cushions in the sprue base, have been disproved and, therefore, contemporary designs for the sprue to runner transition focus on eliminating turbulences by channeling the melt flow in a laminar manner, often by the use of filters on the bottom of the sprue or in its vicinity.

The cross-sectional areas of the *runners* and *gates* behind the transition is again calculated based on Bernoulli's law to ensure the runners and gates are filled quickly and completely to avoid air entrapment and an equal flow rate in each gate. Bernoulli's principle states that for an ideal fluid, an increase in the speed of the fluid occurs simultaneously with a decrease in pressure or a decrease in the fluid's potential energy.

$$\frac{\vartheta^2}{2} + gz + \frac{p}{\rho} = \text{constant} \tag{6.1}$$

where ϑ = fluid flow speed at a point on a streamline

g = acceleration due to gravity

z = elevation of the point above a reference plane, with the positive z-direction pointing upward—so in the direction opposite to the gravitational acceleration

p = pressure at the chosen point

ρ = density of the fluid at all points in the fluid

Using this principle and describing liquid metal as incompressible fluid, velocity and pressure changes in the melt can be calculated based on changes in cross section and elevation (Fig. 6.3).

The idea of utilizing numerical models to predict the filling and solidification of castings came from physicists, mathematicians, and mechanical engineers.

It is also possible to place the choke in any area of the runner (usually at its entry) or in the gates. Terms like *pressurized* and *nonpressurized* gating systems are based on the location of the choke, but are abandoned, as this nomenclature is confusing, especially

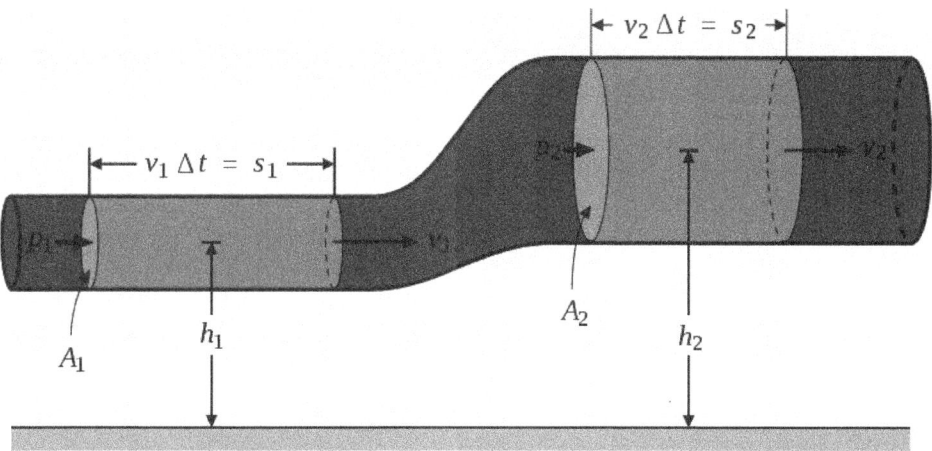

Figure 6.3 Graphic depiction of Bernoulli's law. (MAGMA.)(A color version of this figure is available at mhprofessional.com/pmc3.)

to novices. If it is necessary to characterize gating systems based on the location of the choke, they should be called that way (i.e., gate choked system, sprue choked system). There is an ongoing discussion about where to place the runner and the gates: runners in drag (bottom part of the mold), gates in the cope (upper part of a mold), or vice versa or both in the cope or both in the drag. As with all matters related to castings, the correct answer is probably, "It depends."

Unfortunately, Bernoulli's equations discount the effect of inertia, friction, and surface tension, so the runner or gate cross sections need to be adjusted not only for the melt volume taken out by each gate (easily done), but also for, at least, the inertia, which tends to provide the last gate with higher melt volumes than the first one (very complex calculations usually only feasible through the use of a fluid flow simulation tool (Fig. 6.4*a* and *b*). The variation of force with time at different ingate locations for the runner system is shown.

Runner extensions are used to capture the initial, often contaminated, melt that has entered the gating system first. Basic rules are used to design them; however, only fluid flow simulations can confirm if they actually work, as the basic rules, again, discount the effects of inertia, friction, and surface tension.

Risers

Risers are usually placed on top of or next to the casting. Sometimes risers need to be placed on the inside of the casting. In either case, they are supposed to provide liquid metal to compensate for the liquid and the solidification contraction (shrinkage) of the melt. Hence, they need to stay liquid longer than the area of the casting they are supposed to *feed* and they need to provide enough melt volume. As this is a fairly complex material transport phenomenon, foundries needed to develop simplifying feeding rules to closely estimate the necessary size and shape of risers and their connection to the casting (riser neck). The most basic one is to base calculations on the "modulus method," which is based on the theory that shapes with the same volume to surface ratio (modulus) solidify at the same time. The modulus can easily be calculated for simple shapes. It is sometimes possible to subdivide a casting geometry into several discrete simple shapes, which allow for the calculation of their individual moduli. In the past, charts linking moduli to riser sizes and dimensions (so-called nomographs) were used to determine the necessary modulus for a feeder (riser) based on the metal poured. These have been replaced by simulation programs and product databases of riser sleeve suppliers, which consider the insulating and exothermic (heating) values of their products. It also is necessary to ensure that the feeding path from the riser to the critical area of the castings stays open long enough. One method to determine the functionality of a feeding pass is the use of "Heuver's circles": a two-dimensional section of the casting and the riser is drawn. Now circles are fitted into the geometry of the casting. New circles are drawn on the way to the riser. Each subsequent circle needs to be bigger than the previous one. If

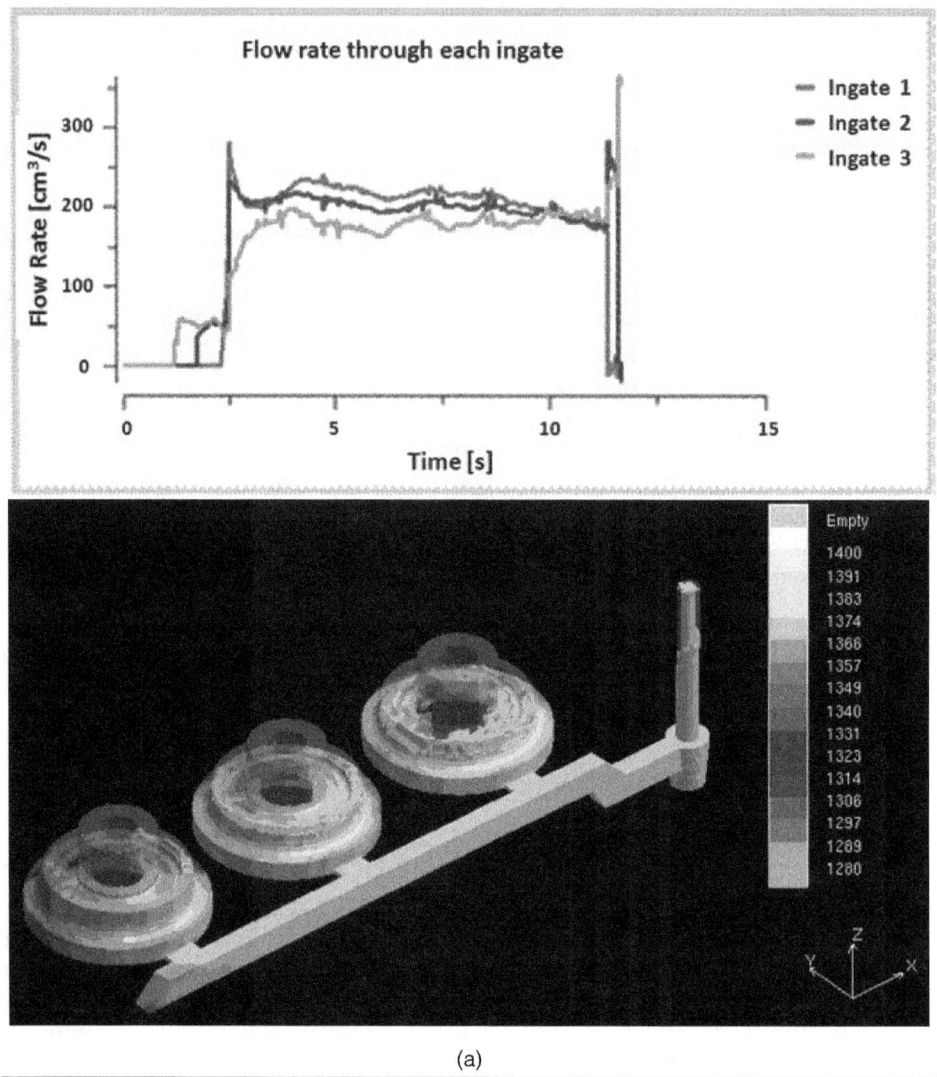

(a)

FIGURE 6.4 (a) Straight runner leads to different flow rates in each ingate, the uneven filling of parts, and an uneven temperature distribution. (MAGMA.) (A color version of this figure is available at mhprofessional.com/pmc3.) (*Continued*)

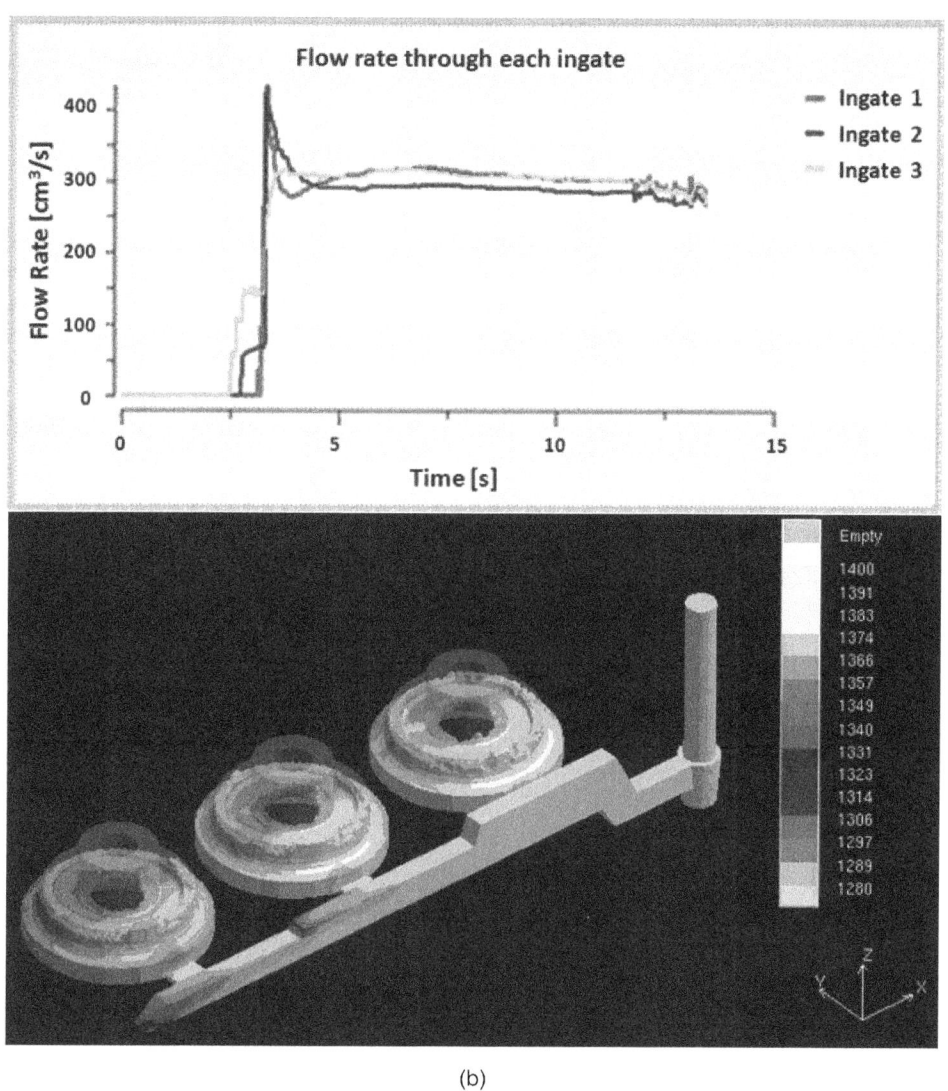

(b)

Figure 6.4 *(Continued)* (b) Properly stepped runner leads to equal flow rates in each ingate, the even filling of parts, and an even temperature distribution. (MAGMA.) (A color version of this figure is available at mhprofessional.com/pmc3.)

that is the case, the feeding path should work as desired. Nomographs were also used to determine the *feeding distance* of certain riser and geometry combinations. The case could be made that all the previously described methods are utterly useless as they discount:

- Temperature differences due to the filling process
- Temperature differences due to geometric features ("sand corner effect")
- Temperature differences due to cores
- Temperature differences due to convection and segregation effects
- Solidification morphology (i.e., planar or dendritic growth), etc.

However, their basic approach is still used in combination with simulation tools by determining more accurate initial calculations of moduli throughout a casting geometry. These values can be used as a starting point for a series of simulations to find the ideal riser volume (small as possible, smallest possible contact area).

Vents and Flow-Offs

Vents and flow-offs are used to allow the atmosphere entrapped in the mold cavity to escape. Besides open risers, gases can escape through the permeable mold sand and vents placed in molds and dies. The latter are usually placed on the highest elevations of a casting and dimensioned based on the experience of the foundry personnel. None of the previously described methods considers the effect of insufficient venting on the filling behavior of the casting process. Vents also potentially act as cooling fins on the casting surface and might impact the solidification behavior of a casting. Only simulation tools that consider two volumes, the incoming melt, and the entrapped atmosphere in the cavity can show the impact of vents on a particular casting process.

History of Casting Process Simulation

The description of the metal casting process in a physical-mathematical model and its calculation in a computer demanded the quantification of process parameters and process steps as they impact the casting quality. Instead of drawing conclusions from conventional casting results on potential process modifications, i.e., scrap castings showed that the last iteration didn't work, simulation results display casting quality as the combination of a multitude of parameters, which can be manipulated individually. Figure 6.5

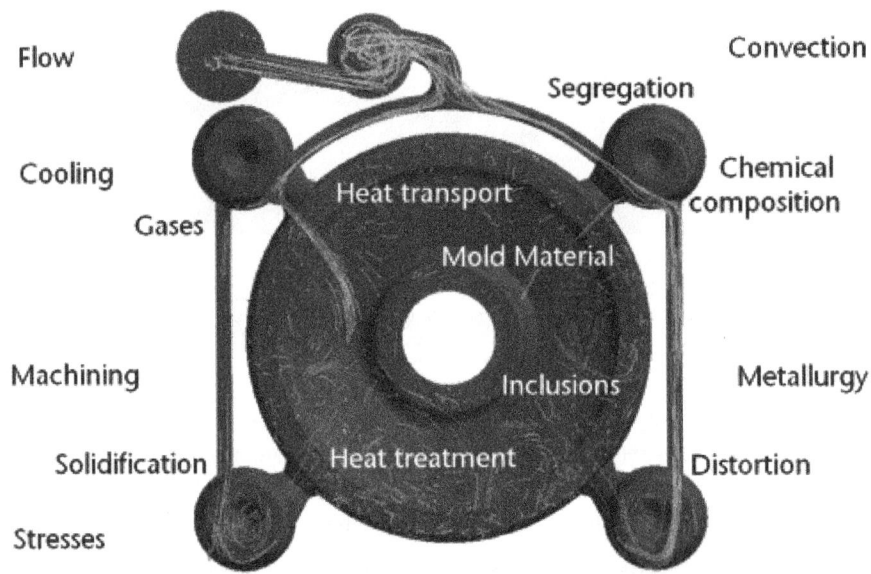

FIGURE 6.5 The biggest benefit of the casting process simulation is its ability to perform many tasks at the same time. However, it is also its biggest drawback, as many process parameters are linked to each other and have to be considered simultaneously. (MAGMA.)

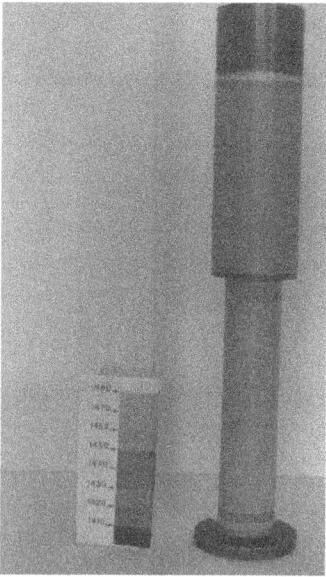

FIGURE 6.6   The first results of a temperature distribution in a hot tear test casting had to be painted on a wooden model, as there were no computers with color monitors in existence. (MAGMA.) (A color version of this figure is available at mhprofessional.com/pmc3.)

shows that many process parameters are linked to each other and have to be considered simultaneously.

In the 1950s, analog computers were used to predict the movement of a solidification front in one or two dimensions. The first digital computers were used in the 1960s to solve casting process–related problems (penetration of steel into mold sand), as well as to use two-dimensional simulations of steel casting solidification, which utilized a program developed by General Electric to simulate heat transfer.

In the 1970s, a program was used to simulate feeding distances in plate-like castings. P. N. Hansen described the prediction of hot tears in steel castings (Fig. 6.6) in 1975 using a three-dimensional model for the first time.

The research and development activities around the topic of casting process simulation increased substantially in multiple locations in the 1980s. Important milestones were the introduction of the term "criteria function" to depict centerline porosities, as well as another one to detect hot tears in steel castings. By the end of the 1980s, the first solutions to simulate the mold filling were provided.

Commercial programs were introduced to foundries at that time and called *solidification simulation* tools, even if some of them were already capable of simulating the entire filling and solidification process.

In the 1990s, development activities focused on the simulation of stresses and distortions in castings, as well as the first steps were taken to predict microstructures and mechanical properties.

The new millennium (2000 and later) saw the introduction of autonomous optimization tools that optimize casting designs and layouts by themselves in combining optimization tools, previously used by part developers and casting process simulation programs.

The Basics

The casting process can be divided into three main segments with the assignment of proper physical terms (Reference 11 is an excellent source of detailed information on which equations, methods, and algorithms are used in casting process simulation):

1. Mold filling is a fluid flow problem, combined with heat transfer problem.
2. Solidification is a heat transfer problem.
3. Distortion is a solid mechanical problem.

The goal of casting process simulation is to achieve the highest degree of accuracy within reasonable calculation times, which is accomplished by utilizing complex three-dimensional Navier-Stokes equations for the description of the fluid flow and three-dimensional heat conduction equations involving various metallurgically based models.

The following equations must, in general, be solved to properly describe a mold filling process:

- The three momentum equations (momentum conservation) together with a suitable constitutive law relating to stresses and velocities, usually incompressible, newtonian fluid for metals
- The continuity equation (mass conservation)
- The energy equation (energy conservation)

Modeling of the solidification process properly requires the solution of the energy equation (energy conservation).

The simulation of stress and distortion phenomena calls for the general solution of the following equations:

- The energy equation (energy conservation)
- The three equilibrium equations together with a suitable constitutive law relating stresses and displacements or displacement rates (typically, a rate independent thermo-elasto-plastic theory)

The actual equations and algorithms used in state-of-the-art casting process simulation programs are therefore highly complex, but nevertheless based on well-known fundamental equations, i.e., for heat transfer and fluid flow.

Fourier's law of heat conduction within a body is

$$q = -kA\frac{\partial T}{\partial x} \tag{6.2}$$

where

q = diffusive heat flow (heat flux) perpendicular through the surface of the area A
k = thermal conductivity
T = temperature
x = descriptive space parameter perpendicular to the surface

Note that the minus sign is inserted in order to satisfy the second law of thermodynamics, so the heat is calculated positive in the direction it actually flows (from higher to lower temperature area). Applying this to the three-dimensional world of most casting processes involving a liquid metal that is cooling in a process-dependent material (sand or metal mold), where conduction is the main contributor to energy transport, leads to the following Fourier equation:

$$\rho c_p \frac{\partial T}{\partial t} = \frac{\partial}{\partial x}\left(k\frac{\partial T}{\partial x}\right) + \frac{\partial}{\partial y}\left(k\frac{\partial T}{\partial y}\right) + \frac{\partial}{\partial z}\left(k\frac{\partial T}{\partial z}\right)$$ (6.3)

where ρ is the density and c_p is the specific heat of the material, assuming no internal heat generation within the material.

Depending on the casting process, Newton's law of cooling is applied to express the heat transfer through convection (sometimes used with large castings or when simulating cooling lines)

$$q = -hA\left(T_2 - T_1\right)$$ (6.4)

where T_1 and T_2 are temperatures of the surface of a body and the cooling fluid, and h is the convection heat transfer coefficient.

Literally the same equation is used to describe the heat transfer between two materials:

$$q = -hA\left(T_2 - T_1\right)$$ (6.5)

where h is the heat transfer coefficient between two materials at two different temperatures.

The Stefan-Boltzmann law is used to derive the net heat exchange of two surfaces via radiation (mostly interesting for investment casting simulation):

$$q = -A\varepsilon_{\text{mean}}\sigma\left(T_2^4 - T_1^4\right)$$ (6.6)

where $\varepsilon_{\text{mean}}$ is the mean emissivity of the two surfaces.

These basic equations are then combined with complex kinetic growth models to accurately simulate the solidification process of metal castings.

The flow of liquid metals can be described with the following Navier-Stokes equation for incompressible flow

$$\rho\left(\frac{\partial v}{\partial t} + v\,\nabla\,v\right) = -\nabla p + \mu\,\nabla^2 v + f$$ (6.7)

where

f = "other" body forces such as gravity or centrifugal force
$\mu\nabla^2 v$ = shear stress term of an incompressible, homogeneous newtonian fluid
μ = (constant) dynamic viscosity

Applying this to a three-dimensional cartesian coordinate system, as used in simulation programs, creates complex equation systems:

$$\rho\left(\frac{\partial u}{\partial t} + u\frac{\partial u}{\partial x} + v\frac{\partial u}{\partial y} + w\frac{\partial u}{\partial z}\right) = -\frac{\partial p}{\partial x} + \mu\left(\frac{\partial^2 u}{\partial x^2} + \frac{\partial^2 u}{\partial y^2} + \frac{\partial^2 u}{\partial z^2}\right) + \rho g_x$$

$$\rho\left(\frac{\partial v}{\partial t} + u\frac{\partial v}{\partial x} + v\frac{\partial v}{\partial y} + w\frac{\partial v}{\partial z}\right) = -\frac{\partial p}{\partial y} + \mu\left(\frac{\partial^2 v}{\partial x^2} + \frac{\partial^2 v}{\partial y^2} + \frac{\partial^2 v}{\partial z^2}\right) + \rho g_y$$ (6.8)

$$\rho\left(\frac{\partial w}{\partial t} + u\frac{\partial w}{\partial x} + v\frac{\partial w}{\partial y} + w\frac{\partial w}{\partial z}\right) = -\frac{\partial p}{\partial z} + \mu\left(\frac{\partial^2 w}{\partial x^2} + \frac{\partial^2 w}{\partial y^2} + \frac{\partial^2 w}{\partial z^2}\right) + \rho g_z$$

Again, these fluid flow equations need to be combined with heat transfer equations and growth models to consider heat transfer during the filling process and potential solidification within the filling process, not to mention the equations governing the free surface movement and cooling behavior.

Mathematical modeling of stress/strain phenomena in casting processes involves a coupled three-dimensional thermomechanical analysis including solidification and other phase transformations, shrinkage-dependent interfacial heat transfer due to relative motion between casting and mold, mold distortion, temperature, time-dependent plasticity, and hydrostatic pressure from liquid and crack formation—all these lead to quite complex equation systems. The basics of stress/distortion development during the cooling process of castings are described by the thermal strain:

$$\varepsilon = \varepsilon^{\text{Th}} = \alpha \Delta T \tag{6.9}$$

where

ε = strain
ε^{Th} = thermal strain
α = material-dependent coefficient
ΔT = temperature difference

Because here "strain" is simply defined as a relative elongation it can also be written as

$$\varepsilon = \frac{\partial u}{\partial x} \tag{6.10}$$

where u is the displacement of a point. Defining stress as force per area

$$\sigma = \frac{dF}{dA} \tag{6.11}$$

and assuming an elastic deformation, the one-dimensional Hooke's law becomes

$$\sigma = E\varepsilon^{El} = E\frac{\partial u}{\partial x} \tag{6.12}$$

or

$$F = EA\frac{\partial u}{\partial x} \tag{6.13}$$

Transferring this basic one-dimensional equation into three-dimensional applications is handled by describing stresses and strains as tensors.

The Simulation Methods

Numerical simulation is the process of solving a physical model through mathematical (differential) equations and the display of the calculated domain (the casting and the mold) through discrete single elements. In order to calculate the differential equations, several methods were developed, i.e., the finite element method (FEM), finite difference method (FDM), finite volume method (FVM), etc.

Each method has specific benefits and drawbacks and can yield good qualitative results depending on its area of application. The finite element methods have their roots in load simulations. The finite difference and finite volume methods come from the fluid flow simulation and show benefits in the description of heat and material transport phenomena.

The method to be selected has to be seen as independent from the subdivision of the calculation domain (i.e., the casting and the mold). FEM, as well as FDM or FVM, can utilize both unstructured (tetrahedrons, pentahedrons, or hexahedrons) and structured meshes (regular hexahedrons). As with the methods, all meshes have specific benefits and drawbacks. For example, while linear tetrahedrons experience numerical problems, they approximate the geometry very accurately. Hexahedrons, however, have superior calculation quality, but the transition between the casting and the mold has to be fitted using specific algorithms. Therefore mixed approaches, depending on application and physics, are currently used (i.e., tetrahedrons with additional gap nodes or regular hexahedrons with boundary adjustments) in commercial programs.

Eventually, the choice of which numerical method and mesh is used is driven by finding the best compromise between the quality (accuracy) of the calculation, optional automatic enmeshment, and calculation time.

Filling and Solidification Simulation

The first steps of describing the process in virtual terms were taken by focusing on heat transfer calculations and on the solidification process. As a result, the entire field of casting process simulation is often called solidification simulation. However, the whole process is called *casting*, not *solidification*. The mold filling is an integral part of the process and therefore must be considered. This is not only important for the gating layout but for the detection of filling related defects as well. Indeed, the inhomogeneous temperature distribution in the melt caused by the filling process has in many cases an impact on the solidification process (Fig. 6.7).

Even today, the dynamics of the mold filling process are often underestimated by practitioners. Key words like "quiet filling" and "laminar flow" are frequently used, but from a physical point of view, all filling processes from sand castings to high-pressure die castings are highly turbulent. This fact is based on the rheological properties of

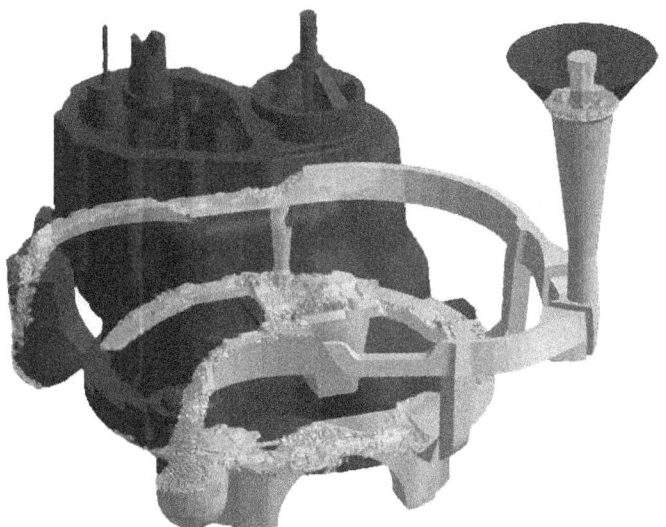

Figure 6.7   The prediction of the fluid flow in combination with the temperature loss of the melt is the key for the development of a robust gating system. Velocities and pressure distributions can be displayed quantitatively at any stage of the filling process. (MAGMA.)

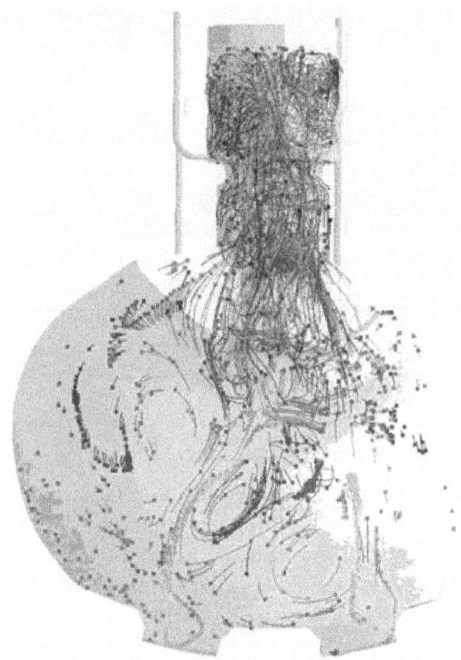

Figure 6.8   "Metal flows like water"—there is much truth to this. The kinematic viscosity of liquid melt is near the one of water. However, due to the multiple times higher density, liquid metal has a much higher energy content, which leads to big turbulence. The effects can be depicted in simulation tools by using virtual particles using colors to show the time duration they are inside the mold cavity. (MAGMA.) (A color version of this figure is available at mhprofessional.com/pmc3.)

molten metal. The energy that is created by the flowing melt is so high that it cannot be eliminated through foundry technological efforts. Therefore, strong turbulences and eddy currents are found inside the melt even when the melt surface appears to be rising quietly (Fig. 6.8). Many casting defects result from these undersurface movements, as well as reactions between melt and mold material. These defects include mold defects, air entrapments, oxidation defects, slag entrainments, or metallurgical challenges.

The fundamentals of the flow simulation provide quantitative information of velocities, pressures, and temperatures. These are the tools the casting expert uses to develop robust gating systems. The question of pressurized or nonpressurized gating systems can be answered quantitatively. Critical velocities (i.e., 0.5 m/s for aluminum) are immediately detectable. The filling simulation shows the cooling behavior of the melt all the way to a potential premature solidification (Fig. 6.9). Thereby, the deciding criteria to develop an effective gating system are available early in the gating design process (Fig. 6.10).

Current solidification simulation does not simply provide a description of the heat transfer process, however. Many of the criteria used to evaluate the solidification behavior are based on the information gained from the solidification simulation and thereby can provide the first clues of the casting quality. As soon as a three-dimensional geometry of a casting is available, a basic solidification and cooling simulation can be performed in minutes (Fig. 6.11). The prediction of hot spots and areas of final solidification do not only help the metalcaster in the engineering department, but also support the designer in evaluating the designs. The heat loss prediction is an important factor

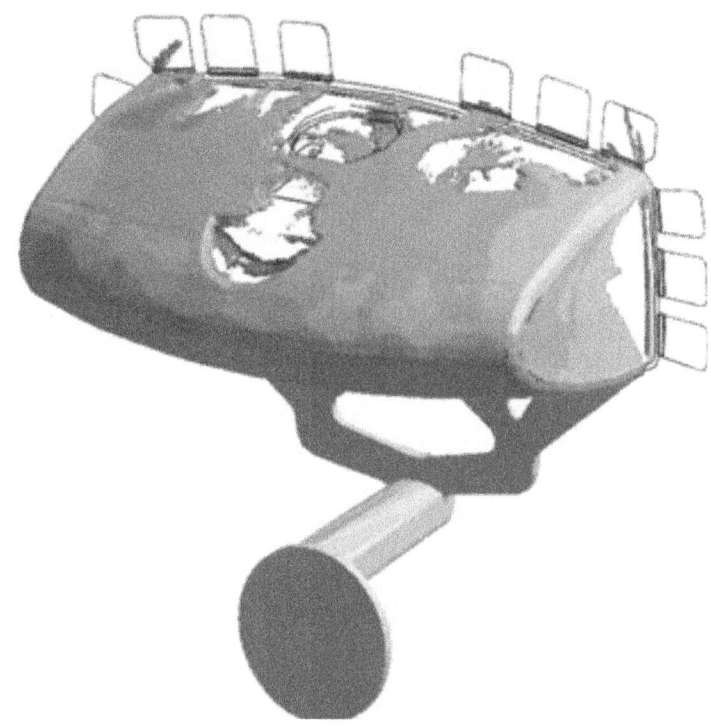

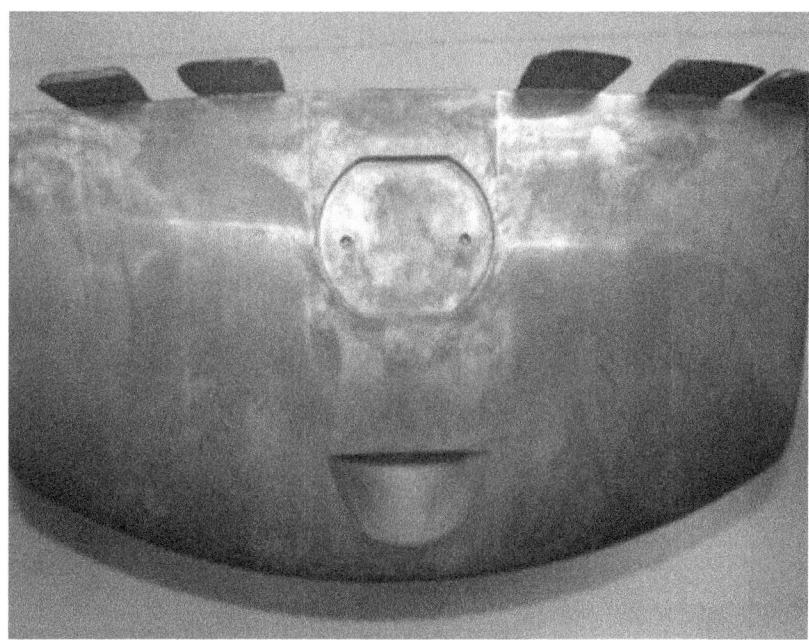

FIGURE 6.9 Typical ingate velocities of more than 10 m/s lead to a breakup of the melt surface, which lead to air entrapment. The filling simulation, therefore, not only considers the melt but also the air and gases in the die cavity. (MAGMA.)

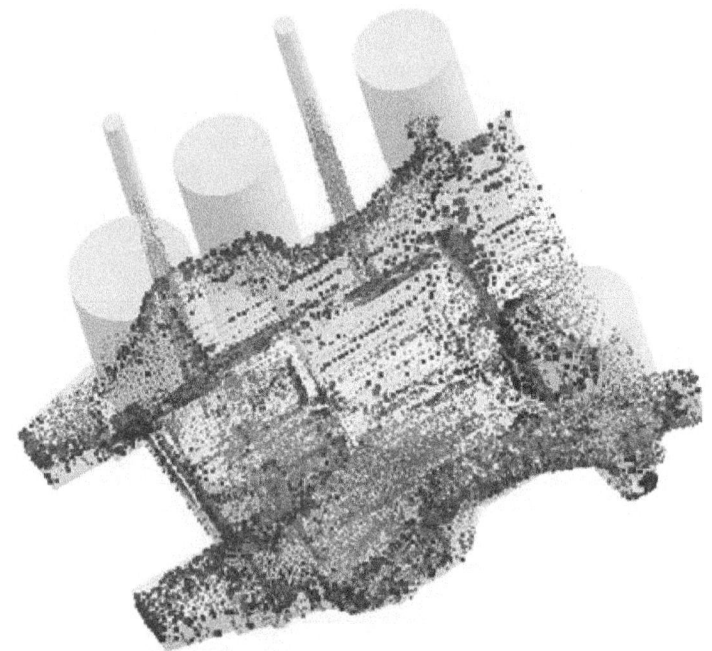

FIGURE 6.10 Current fluid flow simulation can predict the creation of oxides and inclusions. In this way, a mass and a volume are assigned to virtual particles that can grow, agglomerate, or adhere to the mold wall. Oxide distribution in a steel casting is shown. Colors highlight different size of oxides (blue: small, yellow: large). (MAGMA.) (A color version of this figure is available at mhprofessional.com/pmc3.)

FIGURE 6.11 The last areas to solidify in a ductile iron turbocharger housing. (MAGMA.) (A color version of this figure is available at mhprofessional.com/pmc3.)

for their layout, especially in permanent molds. The impact of boundary conditions such as preheating, coating, and heating or cooling, can be determined quantitatively in a short period of time. The knowledge of temperatures and solidification behavior leads to a quantitative prediction of the local thermal modulus in the casting, as well as solidification times, cooling rates, and temperature gradients (Fig. 6.12).

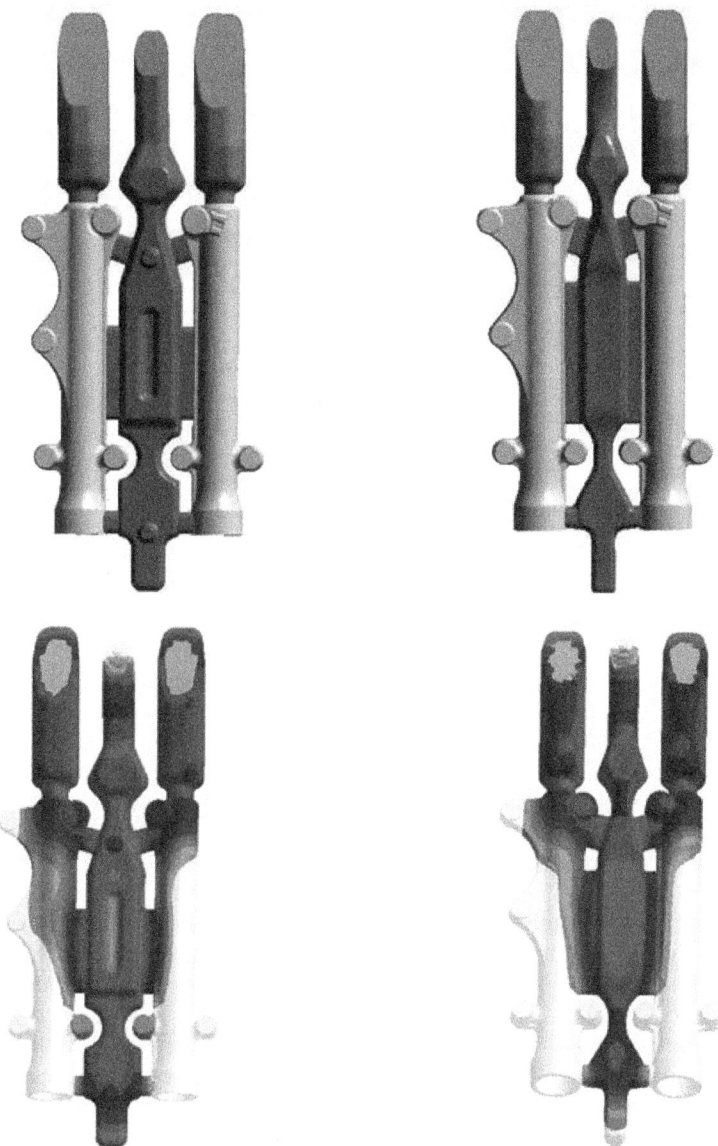

Figure 6.12 Casting process simulation is driven by comparing different solutions. This example shows two gating systems of a bicycle fork leading to differences in the solidification process. The right-hand geometry and resulting pictures show an optimized gating system leading to better directional solidification, eliminating a shrinkage defect. (MAGMA.) (A color version of this figure is available at mhprofessional.com/pmc3.)

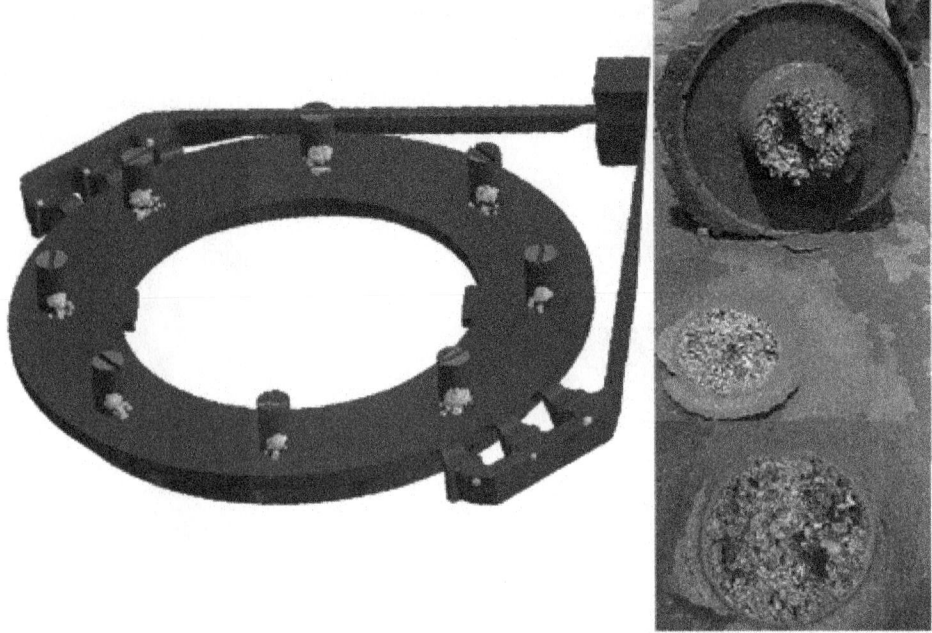

FIGURE 6.13 Secondary shrinkage below risers is shown for a ductile iron ring casting (dark areas in fracture surface are shrinkage defects). Only the combination of local shrinking and expansion behavior leads to a correct defect prediction (yellow areas at riser base in left picture). (MAGMA.)

The support of the feeding-related layout of the casting is still one of the most important duties for casting process simulation. Depending on the alloy poured, different feeding behaviors and self-feeding capabilities need to be considered to provide a defect-free casting. Therefore, it is not enough to base the prediction of shrinkage defects solely on hot spots derived from temperature fields, but also to be able to quantitatively predict them (Fig. 6.13). Solidification simulation had to be combined with density and mass transport calculations in order to evaluate the impact of the solidification morphology onto the feeding behavior, as well as to consider alloy-dependent feeding ranges. This is accomplished through the description of temperature-dependent thermophysical properties. Even if simple criteria functions like the ones provided by Niyama or Lee, both use combinations of local gradient and cooling rates to predict centerline or microporosity defects, offer important clues on shrinkage defects and microporosities, current developments go far beyond that approach. State-of-the-art simulation tools consider independent growth models for pores in combination with the interaction of pressures and gas concentrations (Fig. 6.14).

The *classic* thermophysical properties (conductivity, specific heat capacity, specific energy, latent heat, and density) are readily available for all typical alloys. All data have to be provided in a temperature-dependent format, as the information can change dramatically from pouring temperature through the solidification range to room temperature. With the introduction of mold filling simulation, additional rheological data needed to be provided, i.e., the viscosity of the melt throughout the solidification interval.

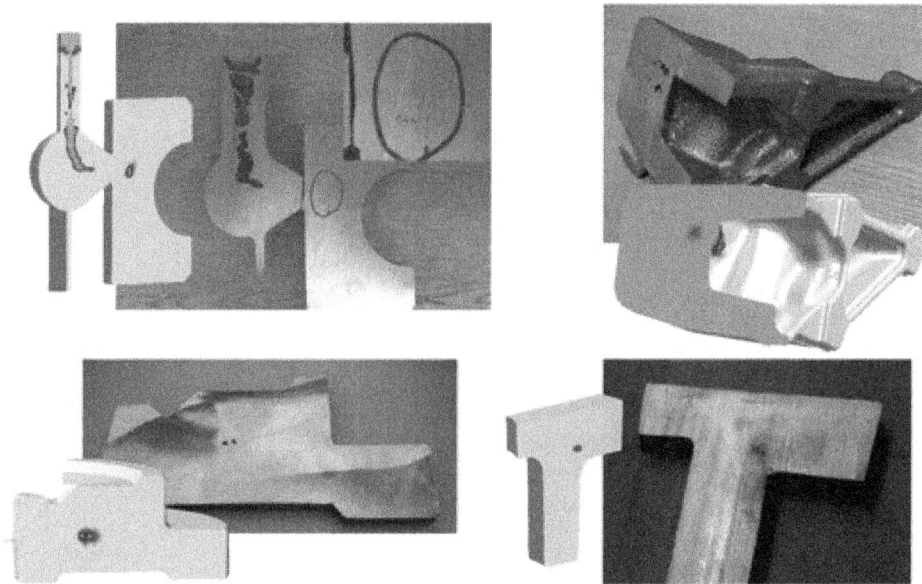

Figure 6.14 Examples displaying the accuracy of shrinkage prediction (the colored areas depict shrinkage defects, the white area are defect free) compared to cuts of real gray and ductile iron castings, as well as a steel casting. (MAGMA.)

The introduction of stress simulation has expanded the need for data regarding thermomechanical properties (temperature dependent, elastic modulus, area contraction, tensile strength, thermoplastic properties, time-dependent fatigue properties, etc.). The prediction of segregation requires information derived from phase diagrams, as well as distribution and diffusion coefficients. The micromodeling of microstructures and mechanical properties necessitate quantitative information about the growth of phases, the impact of alloying elements, as well as the quantitative consideration of metallurgical melt treatments like inoculation and grain refinement. Currently, the focus of data acquisition has moved toward the consideration of external process partners. These include mold materials and casting process aids like sleeves, filters, and coatings, where the suppliers of these products are required to provide these data.

After the development of the fundamental physical possibilities to simulate the pouring and solidification process of castings, the focus quickly switched to the specific requirements of different casting processes, including the evaluation of multiple cycles in permanent mold applications to consider the heat-up sequence of permanent molds and dies, as well as process interruptions. Even small process steps, such as the die opening sequence and spraying processes, are considered in detail by simulation tools. Heating and cooling channels cannot only be considered, but can be controlled by virtual thermocouples. The simulation of the high pressure die casting process allows for the input of the exact pressure curve to determine the filling process, as well as the air cooling of die parts (Fig. 6.15).

Vertically parted sand molds are simulated under differentiation between the "hot side" and the "cold side." It is also possible to consider the shakeout and cooling conditions in cooling drums.

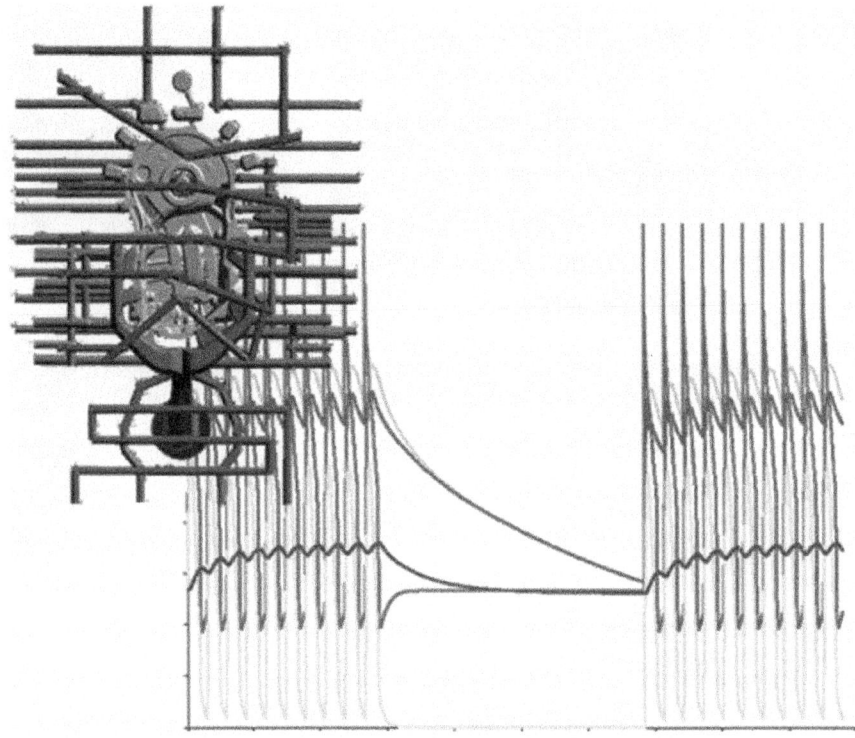

Figure 6.15   Permanent molds and dies provide a stable process only after they reach a quasi-steady-state condition. Simulation considers the entire mold and all predominant parameters with impact on the heat balance of a die, i.e., cooling and heating channels or coatings. The lines are temperature-time curves showing the influence of a process interruption on the time it takes to get back to the steady-state condition. (MAGMA.) (A color version of this figure is available at mhprofessional.com/pmc3.)

One focus of further development is the utilization of simulation results for the optimization of process steps and for the determination of control parameters for die-casting machines and molding lines. The consideration of a known cooling conveyor length and a desired shakeout temperature can be used for the prediction of the molding line's productivity. The final goal is to couple this data directly with the machines.

Microstructure Prediction

Even if the physical fundamentals for filling, solidification, stresses, and cooling process are the same for all alloys, the specific material behavior makes a difference, as displayed in Fig. 6.16 for aluminum alloys. Besides the process conditions, the nominal composition and metallurgical parameters (grain refinement) are defined. Based on this information, the program calculates the potential equilibrium phases, which are impacted by the accelerated cooling condition they experience (phase kinetics). The inhomogeneous solubility of alloying elements in the solid and the liquid phase leads to segregations and thereby to the potential creation of new, and sometimes undesired, phases. Only in the final

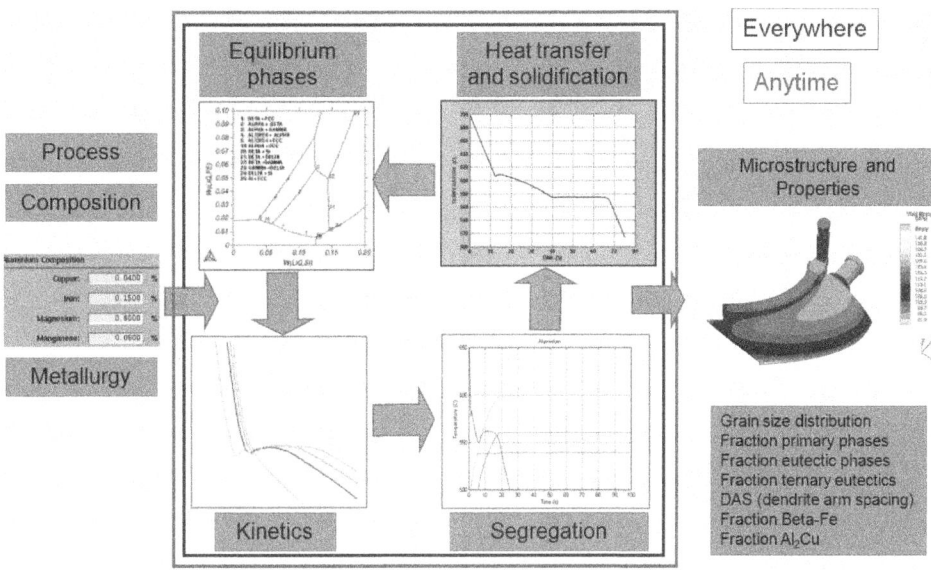

Figure 6.16 Overview of input parameters (*left*), calculation steps (*center*), and results for the prediction of microstructures in aluminum alloys (*right*). Beside process parameters and casting geometry, alloy composition and metallurgical parameters, i.e., grain refinement, are defined at the beginning (*left*). This information is used by the program to calculate stable phases, which can be moved by the accelerated cooling, the casting experiences (phase kinetics) (*center*). Different solubility of alloying elements in the solid and liquid phase leads to segregation and potential creation of undesired phases (*center*). Finally, this information is used to calculate the solidification progress by calculating the temperature distribution in each time step (*center*). These steps are repeated at any location and at any time to locally predict microstructure and mechanical properties (*right*). (MAGMA.) (A color version of this figure is available at mhprofessional.com/pmc3.)

step, based on this information, the solidification progress and the resulting temperature distribution are calculated in a time step. These steps are repeated for every location and every point in time before microstructures and mechanical properties are predicted.

The key to the development of material-specific simulation models was the specific feeding behavior of cast alloys and their strong dependence on the chosen metallurgy. A calculation of the feeding behavior based solely on temperature distributions was not sufficient. For example, large hotspots in iron castings can potentially completely feed themselves, but small hot spots can lead to shrinkage defects. The local shrinking and expansion behavior of a casting can only be calculated under the consideration of the locally developing phases (graphite, austenite, cementite) and their respective contribution to the local shrinking and expansion behavior. The creation kinetics of each phase is therefore considered throughout the entire progression of the solidification. This means that for cast iron not only is the dominant impact of the alloying elements considered, but also the inoculation and melt quality. Metalcasters use the impact of the inoculation or alloying elements for the creation or avoidance of white iron. These are overlaid by the local cooling conditions inside a casting. A simulation solely of the macroscopic solidification and cooling behavior cannot describe this interaction. Therefore, this so-called micromodeling is performed on many materials, considering the amount of any new phase created at any time based on the phenomena described above and shown in Figs. 6.17 and 6.18.

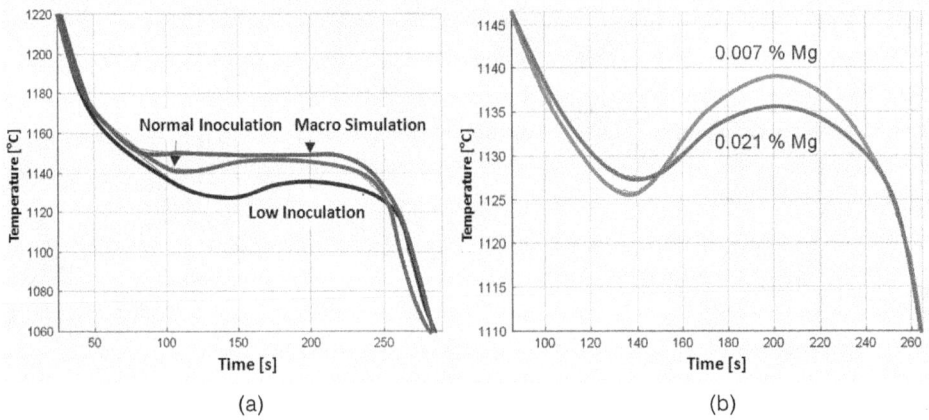

Figure 6.17 Differences between macroscopic and microscopic simulation (micromodeling) on simulated cooling curves in ductile iron. While the use of macroscopic heat transfer equations only modifies the shape of the cooling curve due to the released latent heat, micromodeling also considers the impact of different inoculation conditions (*a*). Even composition changes (i.e., change of effective magnesium content between 0.007 and 0.021%) modify the calculated undercooling, recalescence, and growth temperature (*b*). (MAGMA.) (A color version of this figure is available at mhprofessional.com/pmc3.)

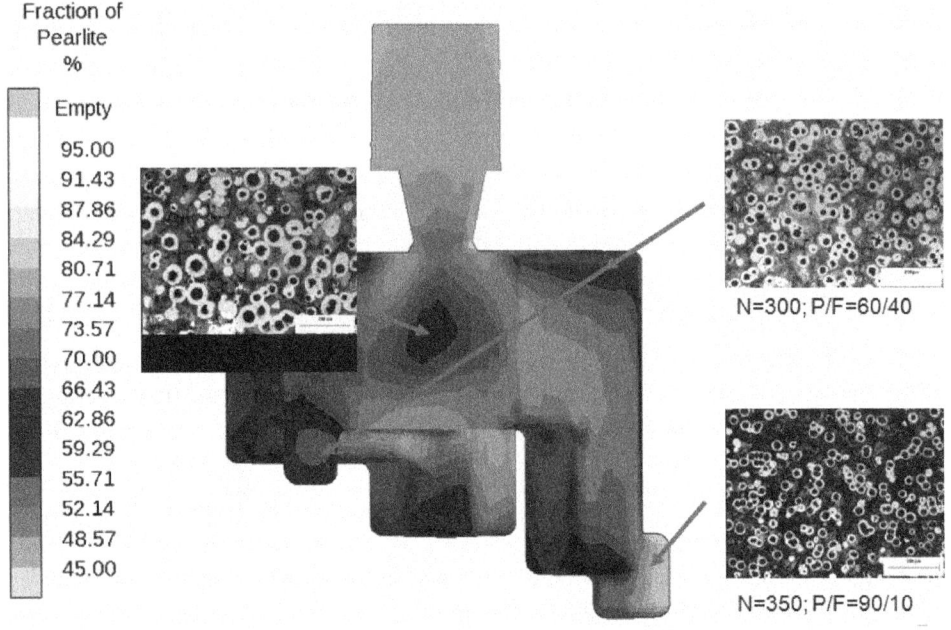

Figure 6.18 Microstructure simulation of cast iron alloys provides a multitude of quantitative information about local structures, i.e., ferrite/pearlite distribution. The software shows a reference picture of the microstructure found in the specific area at which the computer mouse is pointed. (MAGMA.) (A color version of this figure is available at mhprofessional.com/pmc3.)

Stresses and Distortion

Stress-related casting defects are as old as the casting process itself. In contrast to filling and solidification, stresses are initially invisible, contrary to the "predictable" behavior of filling and solidification. On top of that, they usually reverse during the cooling process.

The developments regarding the prediction of hot tears, as well as the creation of residual stresses and distortion behavior, created much more transparency (Fig. 6.19). The profound experiences made in the arena of load simulations in mechanical engineering provided a lot of help. However, the topic of stresses in castings is much more complex, as we need to deal with elastic-plastic or even visco-elastic material behavior, which need to be considered over a large temperature range. Additionally, the stress simulation has to consider not only the part itself, but also the impact of mold and cores, as they are often predominantly responsible for stress-related defects (Figs. 6.20 and 6.21).

The foundry engineer has learned this through experience and has attempted to avoid hot tears and cracks in castings by using various shrink factors, tear brackets (cooling fins), or even using straw in the mold material. Many of these options can now be evaluated quantitatively through stress simulation. However, the development of these tools is far from over. Today, research continues on improving the simulation models to better predict the complex influence of casting and mold material behavior on the stress development.

In addition to residual stresses, hot tears, crack formation, and the shrinkage and warpage of the castings, dies and permanent molds are moving into focus. Due to the high costs for permanent molds and dies, maintenance and repair efforts are very often

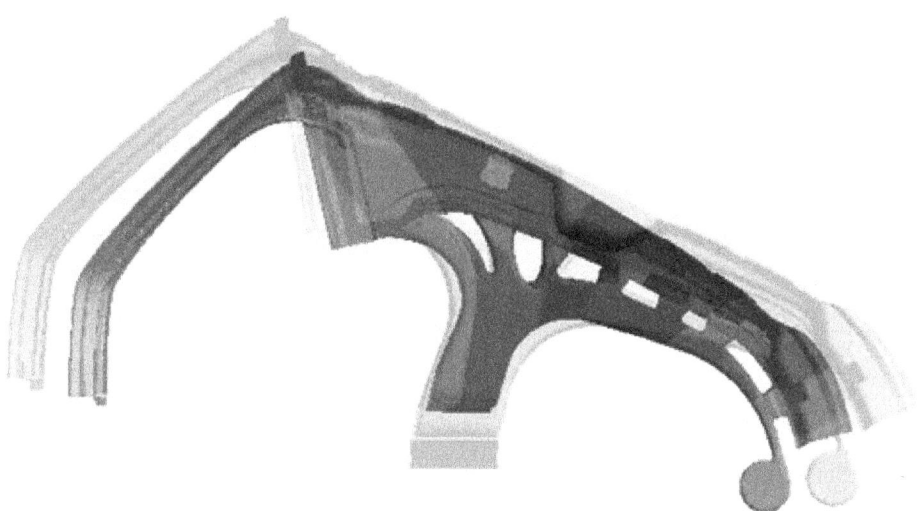

FIGURE 6.19 Distortion of a structural aluminum A380 part in a die-casting die. The alloy-specific contraction behavior in solidifying and cooling castings can lead to plastic (remaining) deformations, at room temperature (exaggerated display). (MAGMA.)

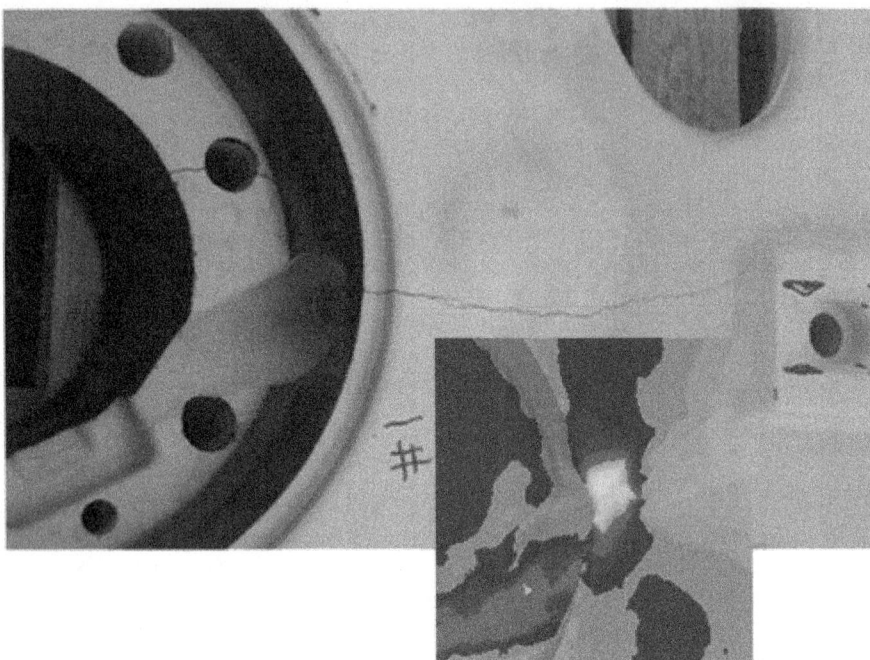

FIGURE 6.20 Prediction of cracks in a casting. The contraction of a casting and the constriction of the resulting movement of it at high temperatures can lead to very high strain rates. At the same time, the material is often brittle and cannot sustain any loads. This leads to cracks when high stresses (tension) are present at high temperatures (this often happens in places that show compression at room temperature, which is puzzling). (MAGMA.) (A color version of this figure is available at mhprofessional.com/pmc3.)

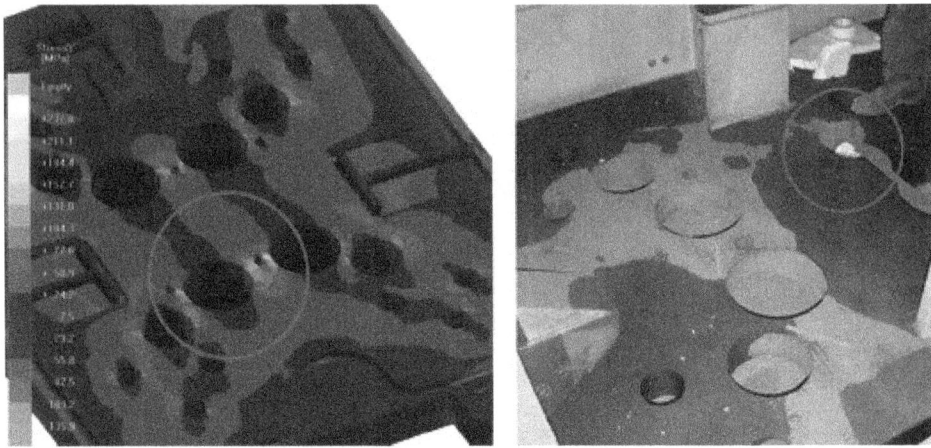

FIGURE 6.21 Prediction of potential cracks after machining. Simulation of stresses in castings can also consider the impact of machining. Stress redistribution can lead to stress concentration, which might lead to failures after machining. (MAGMA.) (A color version of this figure is available at mhprofessional.com/pmc3.)

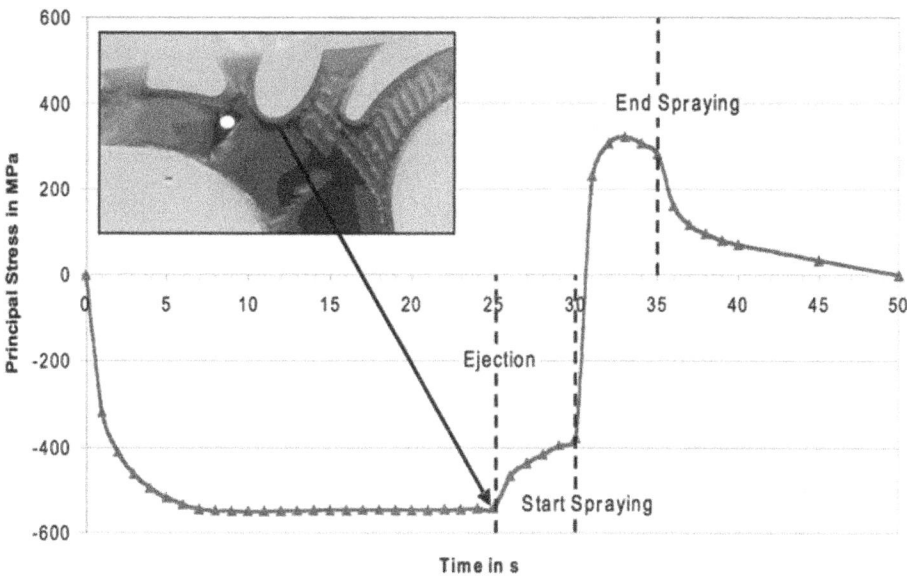

Figure 6.22 Die-life prediction. Premature heat-checking on the die surface or cracks around cooling channels are reasons for premature failures of dies or at least lead to expensive repairs. The main contributor to these effects is the stress cycling during the casting process. The die surface expands during the filling and solidification process. This leads to compressive stresses. The subsequent spraying and cooling leads to a contraction of the die surface, which leads to tension. This leads to a stress hysteresis over multiple casting cycles. Simulation of these phenomena leads to conclusions regarding crack initiation and lifetime of a die. (MAGMA.) (A color version of this figure is available at mhprofessional.com/pmc3.)

the deciding factor if a process is profitable. As the temperature behavior and the resulting stress development can easily be simulated, this application of simulation provides additional value and cost reduction potential (Fig. 6.22).

Mechanical Properties

Simulation results describing the filling, solidification, and stress behavior of castings are important sources for developing reliable processes. The foundry wants to provide their customers with defined properties of a casting and the customer wants to receive reproducible properties. The list of requirements for castings is therefore very long. Aside from questions in regard to dimensional tolerances, there are many mechanical properties at the center of interest. Therefore the final goal of casting process simulation is the prediction of casting properties. The base for such predictions is the quantitative information about local microstructures and potential casting defects. Due to its specific solidification behavior, cast iron was the pioneer in this area. The local microstructure and mechanical properties, including hardness, can today be simulated for all common graphite morphologies and compositions in a quantitative manner (Fig. 6.23). Similar developments are meanwhile available or will be available shortly for aluminum alloys.

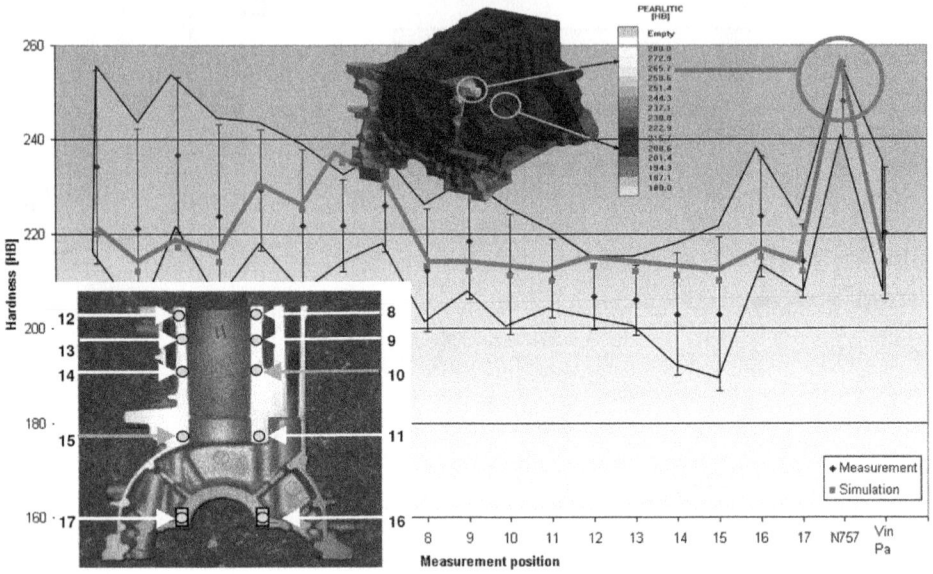

Figure 6.23 Prediction of local hardness (HB) for a gray iron engine block in comparison to measurements. The predicted values (red line) are within the tolerance band of measurements at several locations, highlighting a critical area (red circle) for the machining operation. (MAGMA.)

Heat Treatment and Machining Process

Most castings receive their final properties through processes after the casting process, i.e., heat treatment or machining. Therefore it is crucial for casting process simulation to consider these processes to reliably predict casting properties when the part is delivered (Fig. 6.24). One example, how this expanded process chain can be displayed, is the presentation of residual stresses in a casting as it is delivered to the customer (Fig. 6.25). Sometimes the residual stresses are completely eliminated by a stress-relieving heat treatment process. However, even the removal of the gating system can lead to a stress redistribution, which can lead to a measurable distortion of the casting. Rapid cooling or quenching, on the other hand, can again induce high stresses, which can be reduced or eliminated by a successive time or temperature-dependent tempering or aging process.

The machining process can add, redistribute, or relieve stresses. Indeed, the removal of material leads to a new stress equilibrium. Only the consideration of this aspect leads to useful results for the end user.

Process steps after the actual casting process also impact the properties of the microstructure and of the casting. It is possible today to couple a heat treat simulation with the consideration of diffusion and phase transfer phenomena to predict the microstructure and mechanical properties of the final product. This especially helps the steel foundries to recognize potential hardness immediately, as it helps aluminum foundries to find the parameters for an optimal aging process to achieve the desired strength and elongation.

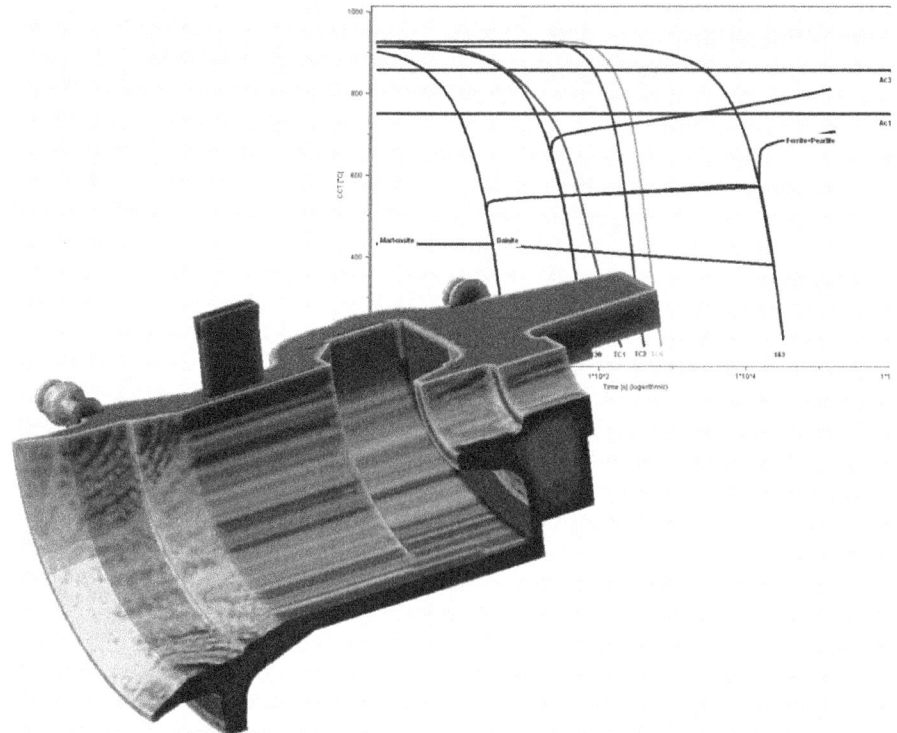

Figure 6.24 Microstructure prediction through heat treatment simulation of a steel casting. The coupling of diffusion calculation and TTT diagrams (background) provides the option to predict local microstructures after heat treatment, i.e., martensite distribution (foreground). (MAGMA.) (A color version of this figure is available at mhprofessional.com/pmc3.)

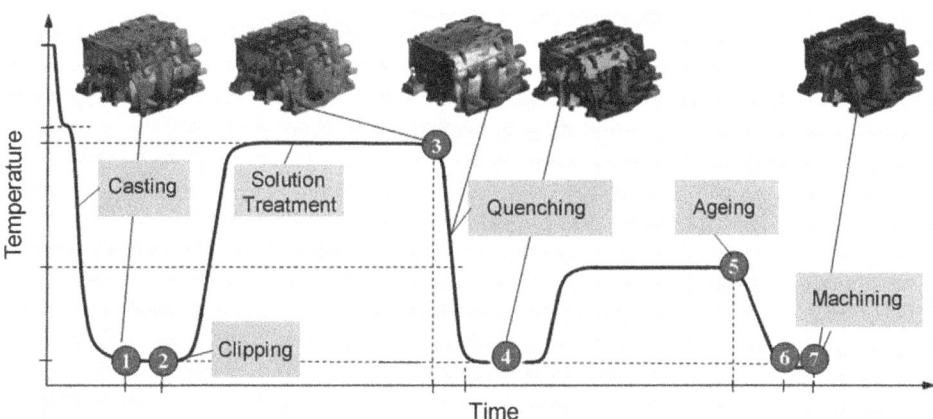

Figure 6.25 Temperature history in a cylinder head over the entire manufacturing process. The black line shows the temperature level a casting experiences throughout the casting and heat treatment process. Casting process simulation is nowadays capable of supporting all essential process steps and thereby capable of predicting residual stresses at the end of the machining process. This is the information a designer is looking for. (MAGMA.) (A color version of this figure is available at mhprofessional.com/pmc3.)

Sand Simulation

Clearly, the focus of casting process simulation development of the last 30 years has been the casting and its production. However, many production-related questions regarding molds and cores are still answered through a much bigger experimental effort (usually through trial and error) than used in the development of the rigging of a casting. The typical approach of optimizing a core box is to provide a maximum number of vents, many of which will then be closed based on the results of the trial and error process. This uncertainty in the mold and coremaking is caused by the complexity of the physics involved in the filling, shooting, and compressing of mold and core materials.

The coremaking process is, from a physical point of view, a multiphase fluid problem. After opening the valve, sand is engulfed by air and thereby accelerated. Air transports the sand into the core box. At the end of the filling process, the air needs to be separated (vented) from the sand. Recently, steps have been taken into this new complex world with its demanding physics. One challenge is the proper description of those physics. A bigger challenge is the multitude of internal and external boundary conditions (how many different valves and vents are there on the market?). Initial successes have motivated continuing development of simulations for the core room (Fig. 6.26). The final goal of the current development is the simulation of the entire process chain, with the consideration of shooting, gassing, venting, and degradation of binders during the casting process.

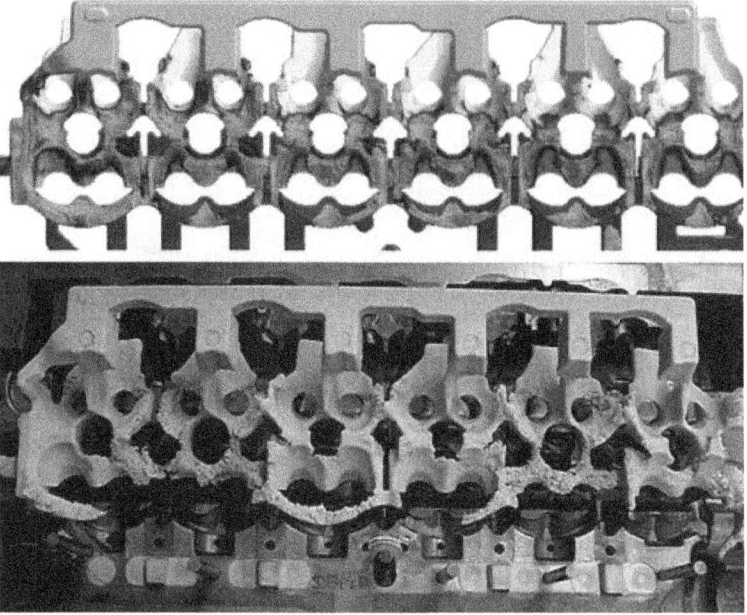

Figure 6.26 Simulation of coremaking and comparison to experiments. The filled areas (colored areas in upper picture) and the nonfilled areas (black) are matching the real-world trial of a water jacket core that could not be filled in the initial core box configuration. (MAGMA.) (A color version of this figure is available at mhprofessional.com/pmc3.)

Simulation as Development Tool

The initial area of utilization for casting process simulation tools was the engineering and/or the quality department of foundries. However, it became clear over the last 20 years that simulation can contribute more to cost reduction efforts the earlier it is used in the part development process. Only in the early stages of the part development process is it possible to consider potential design modifications to avoid problems at the end of the manufacturing process (Fig. 6.27).

This fact, unfortunately, generates a conflict: On the one hand, casting process simulation requires a detailed description of the part. On the other hand, the design still needs to be flexible enough to incorporate potential changes to find a better solution. The solution to this dilemma is the utilization of fast and flexible software tools, which use what data is available, but do not require full details (Fig. 6.28).

Casting designers and buyers have learned over the years that the prediction of local mechanical properties does not only support quality assurance (risk minimization), but it also provides them with the opportunity to reduce weight and optimize a part for a specific load scenario (design for manufacturing).

Hardware

"Simulation is always too slow." This statement, first voiced 20 years ago, is, in principle, still valid. If a foundry or their customer/designer has a question regarding the casting process or the properties of a casting, an answer is needed immediately. At the same

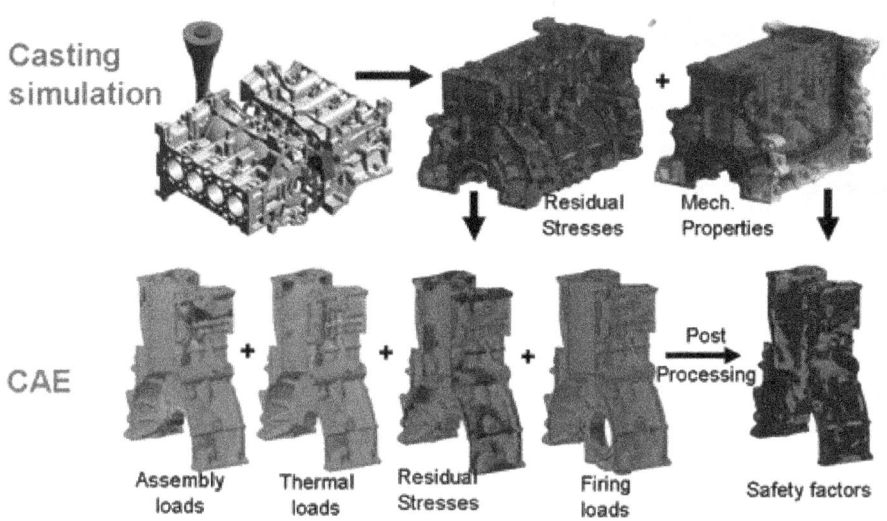

Figure 6.27 Integration of casting process simulation into CAE (computer-aided engineering) chain for the development of engines. Casting process simulation (*upper row*) is performed in parallel to application simulations (*lower row*). The information about local residual stresses and properties are then transferred from the casting process simulation into the application simulations to verify that the casting performs as designed and to assist in the determination of design safety factors to compensate for areas in the casting that are not as strong as initially expected. (MAGMA.) (A color version of this figure is available at mhprofessional.com/pmc3.)

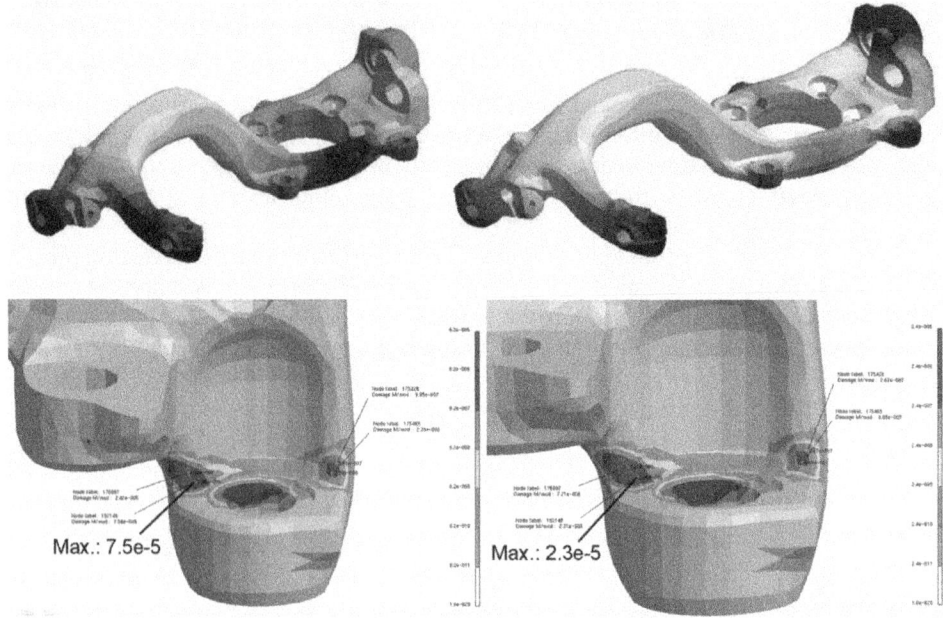

FIGURE 6.28   Different microstructures have an impact on fatigue life. In this case the prediction of the fatigue life was initiated for an aluminum A390 suspension part, benchmarking sand (prototype, *left column*) versus permanent mold (series production, *right column*) casting processes. The microstructure (dendrite arm space values, *upper row*) of the castings was hereby transferred into the fatigue life prediction tool. It was shown that the fatigue life of the permanent mold casting exceeded the one of the sand casting (*lower row*). (MAGMA.) (A color version of this figure is available at mhprofessional.com/pmc3.)

time, questions are getting more detailed and more comprehensive. Hence, the physics to answer these questions gets more and more complicated. This leads to an increase in calculation time, which literally eliminates the progress that has been made in computer speed over the last 20 years. The hardware development has made a significant contribution to the distribution and utilization of casting process simulation. Computers are able to calculate at least 1,000 times faster than 15 years ago, but at the same time, they cost less than 1/20th of what they did in 1990. In spite of this positive development, the challenges for future hardware are still equally daunting: 10 times higher resolution, using 10 times more physics, has to be calculated 10 times faster than today.

This only relates to requirements for simulation itself. The tool of the future is autonomous casting process optimization, meaning the sequential or parallel calculation of many designs to automatically find the optimum solution. The pathway out of this "poly-lemma" is the parallelization of the software. Future speed increases will no longer be defined by the frequency of a processor, but by the number of integrated CPUs (cores). Therefore, the challenge lies with the software developers. The hardware provides the platform, but the utilization of the architecture requires specific new ways of programming. A four-core CPU is already the standard configuration of mainstream PCs. In a few years, we will find cluster computers under the desk of every engineer, perhaps without him or her even knowing it. The vision of casting process simulation is to interactively modify geometries or process parameters on-screen and to receive an almost immediate answer from the optimization tool.

Autonomous Optimization

Casting process simulation always displays the status quo of its expert user. The user decides if the rigging system or process parameter set led to an acceptable result. Additionally, proposals for optimized solutions have to come from the operator.

One of the biggest benefits of the casting process is also its biggest downfall: Everything happens at the same time and is coupled. Changes in one process parameter impact many casting quality–defining features during the process, i.e., a change of the pouring temperature does not only change the solidification behavior, it also changes the fluidity of the melt, which can lead to a misrun. The metallurgy of the melt might be impacted, which could lead to changes in the temperature balance of the mold or die, which again can lead to problems with overheating or erosion. Multi-objective autonomous optimization offers a way out (Fig. 6.29). It uses the simulation tool as virtual experimentation field and changes pouring conditions, gating designs, or process parameters and tries this way to find the optimal route to fulfill the desired objective (Fig. 6.30). Several parameters can be changed and evaluated independently from each other (Fig. 6.31). Autonomous optimization tools take the classic approach of foundry engineers, to find the best compromise, and use validated physics. This does not only further reduce the need for trial runs to find the optimal process window, but allows for the detailed evaluation of many process parameters and their individual impact on providing a robust process.

Obviously, only that which can be simulated can be optimized. Optimization, therefore, is not a replacement for process knowledge and expertise. Despite beliefs to the contrary, the simulation user of the future needs to know the objectives and goals, and especially the quality criteria that are needed to reach these goals. The questions to ask a program are easy: What is a good gating system? To answer this question, quantitative solutions are required.

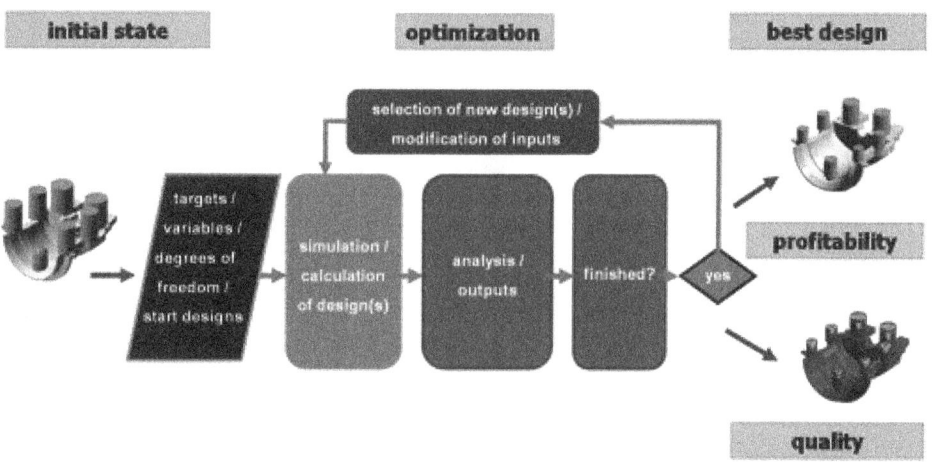

Figure 6.29 Optimization principle. Based on the initial designs, optimization variables are defined. Typical variables are geometric changes, i.e., for riser size and/or position, as well as gating systems. At the same time a multi-objective optimization goal is determined, i.e., to achieve the maximum yield in combination with the smallest shrinkage defect. The optimization project runs many simulations to find the best compromise. (MAGMA.) (A color version of this figure is available at mhprofessional.com/pmc3.)

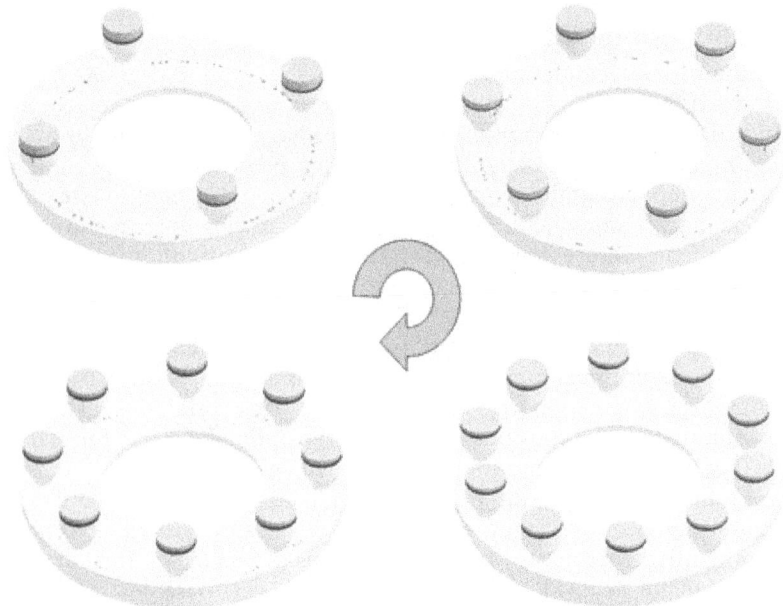

FIGURE 6.30 Optimization process of riser optimization for a steel casting. The optimization code evaluates the current shrinkage distribution and recognizes, initially, the increased feeding need. It might sometimes provide an overly conservative solution (*lower right*), but usually adjust for a final best compromise of riser volume and casting quality (*lower left*). (MAGMA.) (A color version of this figure is available at mhprofessional.com/pmc3.)

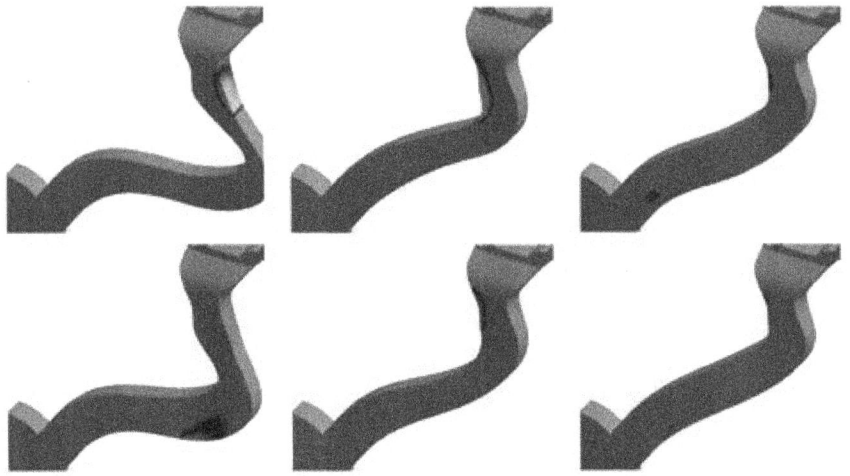

FIGURE 6.31 Multiple design variations of runners for a die casting, which were evaluated by an autonomous optimization run. These are 6 designs out of 257 actually calculated versions. The optimization goal was to avoid a loss of contact between the melt and the runner surface. The shot chamber location and the ingate angle at the casting were fixed. Otherwise, total freedom was given to the optimization tool to find a solution. The dark areas depict undesired areas of contact loss, which lead to turbulence and air entrapment. The final and best variation (*lower right*) shows no critical areas. (MAGMA.) (A color version of this figure is available at mhprofessional.com/pmc3.)

Further Reading

1. P. N. Hansen and P. R. Sahm, "A 3-D Geometric Modeler-Implicit FDM Solver Package for Simulation of Shaped Casting Solidification," "Modelling of Casting and Welding Process II," *Metallurgical Society of AIME* (1984), New England College Henniker, New Hampshire, 1983, pp. 243–247.

2. P. R. Sahm and P. N. Hansen, "Numerical Simulation and Modelling of Casting and Solidification Processes for Foundry and Cast-House," World Foundry Organization (WFO), Zurich, Switzerland, 1984.

3. E. Flender, P. N. Hansen, and P. R. Sahm, "Rechnerisches Simulieren und Modellieren des Warmrißverhaltens warmfester Stahlgußsorten bei de Erstarrung," *Giessereiforschung*, 39, Heft 4, S.137–149, 1987.

4. J. C. Sturm, W. Schäfer, and P. R. Sahm, "Modelling the Mold Filling and Solidification of a Steel Hammer Casting by Use of the Computer Aided Solidification Technologies (CASTS) Software System," *Modeling and Control of Casting and Welding Processes*, IV, S845, Herausgeber A. F. Giamei and G. J. Abbaschian, verlegt bei TMS, 1988.

5. A. Egner-Walter, "Berechnung der Entstehung von Spannungen beim Gießen," *Hoppenstedt, Gussprodukte `99*, Nov. 1999.

6. F. Bonollo and S. A. Odorizzi. "Foundry of Cast Irons: Processing and Simulation," In *Numerical Simulation of Foundry Processes*, F. Bonollo and S. Odorizzi (eds.), Padova: Servizi grafici editoriali, 2001, pp. 87–145.

7. G. Hartmann and R. Seefeldt, "Die zweite Generation von Simulations werkzeugen: Praktische Anwendung der rechnerischen Optimierung im Druckguss," *Giesserei*, Nr.2/2004, S.38–42, 2004.

8. J. C. Sturm, "Stand der Simulation für Gusseisen," *Giesserei*, 91, Nr. 6, *Special Simulation von Giessereiprozessen*, S.4, 2004.

9. G. Hartmann, A. Egner-Walter, and H. Dannbauer, "Simulation of Local Properties of Metal Cast Engine and Suspension Parts," Virtual Product Creation, Konferenz, 2004.

10. J. C. Sturm, "Vorhersage lokaler Eigenschaften von Gussteilen im Motorenbau," VDI Fachtagung Gießen im Motorenbau, Magdeburg, 2005.

11. J. Hattel (ed.), *Fundamentals of Numerical Modelling of Casting Processes*, Polyteknisk Forlag, Denmark, 2005.

12. K. D. Carlson and C. Beckermann, "Modeling of Reoxidation Inclusion Formation during Filling of Steel Castings," in Defect Formation, Detection, and Elimination During Casting, Welding, and Solidification (Proceedings of a Symposium Sponsored by Materials Science & Technology 2005), M. L. C. Clemens et al. (eds.), TMS, Warrendale, PA, 2005, pp. 35–46.

13. C. Midea, M. Burns, M. Schneider, and I. Wagner, "Advanced Thermo-Physical Data for Casting Process Simulation: The Importance of Accurate Sleeve Properties," *International Foundry Research*, 59(1): 34–43, 2007.

14. R. J. Menne, U. Weiss, A. Brohmer, A. Egner-Walter, M. Weber, and P. Oelling, "Implementation of Casting Simulation for Increased Engine Performance and

Reduced Development Time and Costs—Selected Examples from FORD R&D Engine Projects," 28th Internationales Wiener Motoren symposium, Vienna, Austria, Apr. 26–27 , 2007.

15. A. Pawlowski, R. Seefeldt, and J. C. Sturm, "Aus Eins mach Zwei, Sichere Übertragung einer bewährten Füllcharakteristik von einem Einfach- auf ein Zweifachdruckgusswerkzeug," *Giesserei*, 94, Nr. 4, S.34–42, 2007.

16. M. Schneider, C. Heisser, A. Serghini, and A. Kessler, "Experimentation, Physical Modeling and Simulation of Core Production Processes," *AFS Transactions*, 116: 419–432, 2008.

17. E. Flender and J. C. Sturm, "Giesstechnische Simulation," *Giesserei*, 96, Nr. 5, S.94–109, 2009.

18. J. Campbell, *Castings*, Elsevier, Butterworth Heinemann, Oxford, UK, 1995.

19. American Foundry Society, *Basic Principles of Gating and Risering*, 2d ed, American Foundry Society, Schaumburg, IL, 2008.

20. E. Niyama, T. Uchida, M. Morikawa & S. Saito; "A method of Shrinkage Prediction & its Application to Steel Casting Practice", *International Cast Metals Journal*, Vol.7, pp. 52–63, 1982.

21. Y. W. Lee, E. Chang & C. F. Chieu, "Modeling of feeding behavior of solidifying Al-7Si-0.3Mg alloy plate casting" *Met. Trans. B*, 1990, Vol. 21B, pp. 715–22.

CHAPTER 7

Aluminum and Aluminum Alloys

Mahi Sahoo
Suraja Consulting Inc.

Introduction

Aluminum castings are used for a wide range of functions, from decorative components, such as lighting fixtures, to highly engineered safety-critical components for automotive and aerospace applications. Following the Hall-Heroult electrolytic process of reduction of aluminum oxide in the nineteenth century, aluminum could be made available at a greatly reduced cost. Since about 1915, a combination of air transportation, development of specific casting alloys, improved properties, impetus provided by two world wars, and demand for lightweight components in automobiles to improve fuel efficiency have resulted in an ever-increasing use of aluminum castings. Alloy development and characterization of physical and mechanical characteristics provided the basis for new product development through the decades that followed. Casting processes were developed to extend the capabilities of foundries in new commercial and technical applications. The technology of molten metal processing, solidification processing, and property development has been advanced to assist the foundryman with the means of economical and reliable production of parts that consistently meet specified requirements. Today, aluminum alloy castings are produced in hundreds of compositions by all commercial casting processes including greensand, dry sand, lost foam, plaster mold, investment casting, permanent mold, counter-gravity low-pressure casting, and pressure die casting [1]. Typical cast components are shown in Fig. 7.1.

The market for aluminum castings for the past 10 years is shown in Table 7.1 [6].

Advantages of Aluminum Alloys

Certain engineering advantages are inherent in the use of aluminum alloys for castings. Light weight (per unit volume) is the one most commonly cited. With

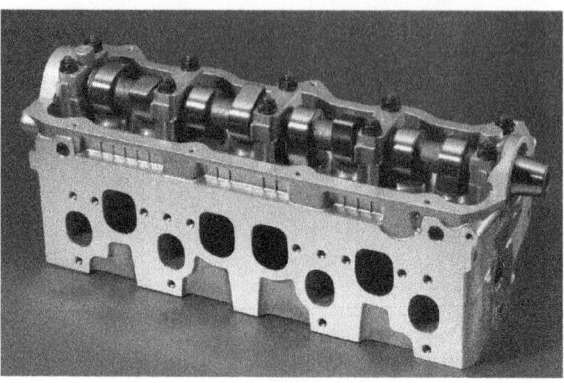

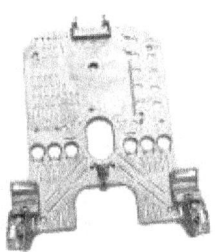
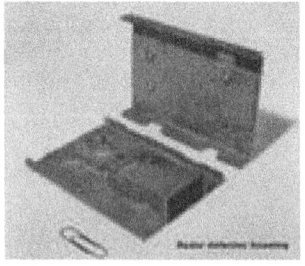

Figure 7.1 Typical aluminum alloy castings. (Courtesy of NADCA [2], American Foundry Society [3], Kasprzak [4], and of ECK Industries [5].) (*Continued*)

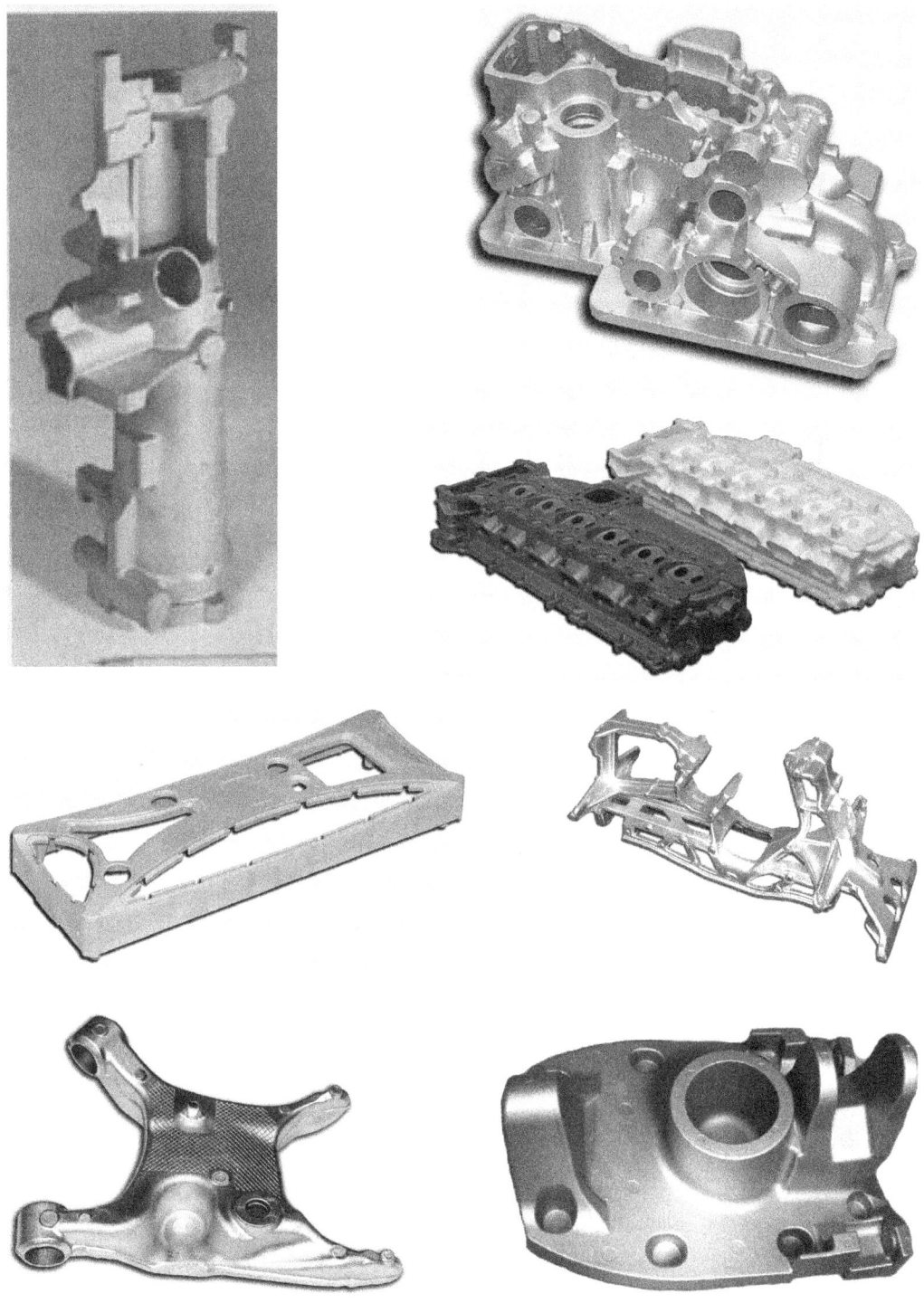

Figure 7.1 *(Continued)*

Figure 7.1   (Continued)

Year	Aluminum	Copper Base	Magnesium	Zinc	Other	Total Non-Ferrous
1999	2,042	329	68	400	48	2,887
2000	2,052	325	74	420	47	2,918
2001	1,910	300	78	322	45	2,655
2002	2,047	294	66	400	45	2,852
2003	2,152	305	81	380	50	2,968
2004	2,165	315	88	360	52	2,980
2005	2,293	322	111	345	54	3,125
2006	2,115	317	116	334	57	2,939
2007	2,036	312	121	328	59	2,856
2008	1,918	303	121	302	58	2,702
2009	1,191	179	66	166	120	1,722
2010	1,234	265	106	204	60	1,869
2011	1,523	263	99	181	60	2,126
2012	1,591	279	102	200	69	2,241

Source: From Spada [6].

Table 7.1   U.S. Nonferrous Casting Shipments (000 tons)

a density of 2.7 g/cc, aluminum castings can be handled easily in the foundry, compared to ferrous castings. Some of the numerous other desirable properties include:

- *A wide range of mechanical properties.* Strength, hardness, and other properties may be greatly altered by alloying and/or heat treatment. Properties of the strongest alloys can be favorably compared with those of the cast irons and lower-strength steels, especially if the weight factor is considered. Suitable strength for many engineering uses is thus available.

- *Architectural and decorative value.* Cast surfaces have good surface finish, are lustrous, and have no blemishes.

- *Corrosion resistance.* This property is, of course, relative, but resistance to atmospheric and water-corrosion conditions makes possible the use of aluminum for building construction, outboard-motor parts, food-handling equipment, etc.

- *Nontoxicity.* The use of aluminum castings for cooking utensils and other food-handling equipment requires that no chemical-reaction products, toxic to humans, be formed by action of the food on the aluminum alloy.

- *Electrical conductivity.* Rotor bars in induction motors are cast of aluminum because of its desirable electrical conductivity.

- *Ease of machining.* Most aluminum alloys are relatively easy to machine.

- *Casting properties.* Since aluminum has a relatively low melting point, about 660°C or 1220°F, and its alloys have a considerably lower operating temperature range (650 to 760°C or 1200 to 1400°F), the problems of melting and pouring are greatly simplified when compared with steels and cast irons. Both permanent metal molds and die casting are extensively practiced. Problems with furnace refractories and molding sands are reduced because of the lower pouring temperatures.

- *Lower casting shipping cost per piece.* This is due to the relatively lower density.

- *Good fluidity.* Aluminum alloys possess excellent casting fluidity and hence, thin sections can be filled easily.

- *Fast heat exchange.* Heat transfer from the molten aluminum alloy to the mold is relatively fast, providing faster casting cycles in metal molds.

- *Casting versatility.* In many cases, multicomponent welded or joined assemblies can be replaced with a single cast part.

Limitations

Cost is a major disadvantage since the cost of aluminum alloys in dollars per pound is greater than that of cast irons and many steels. However, the cost in dollars per pound is misleading unless it is recognized that the volume per pound of aluminum is about 2.90 times that of a pound of the ferrous alloys.

Engineering limitations include the following:

- Lack of resistance to abrasion and wear.

- Absence of aluminum alloys which can develop the combination of high tensile strength, toughness, and hardness obtainable in ferrous alloys.

- Lack of resistance to severe corrosion to the degree offered by numerous copper and nickel-base alloys and stainless steels. Obviously, in the selection of aluminum as a casting material demands that its advantages outweigh its limitations in any particular application.

Alloy Groups

Systems used to designate casting compositions are not internationally standardized. In the United States, comprehensive listings are maintained by general procurement specifications issued through government agencies (federal and military, for example) and by technical societies such as the American Society for Testing and Materials and the Society of Automotive Engineers. Alloy registrations by The Aluminum Association are in broadest use; its nomenclature is decimalized to define foundry alloy composition variations [1].

Designations in the form $xxx.1$ and $xxx.2$ include the composition of specific alloys in remelt ingot form suitable for foundry use. Designations in the form $xxx.0$ in all cases define composition limits applicable to castings. Prefix letters used primarily to define differences in impurity limits denote further variations in specified compositions. Accordingly, one of the most common gravity cast alloys, 356, is shown in variations A356, B356, and C356; each of these alloys has identical major alloy contents but has decreasing limits applicable to impurities, especially iron content. The Aluminum Association composition limits for registered aluminum foundry alloys used to cast shapes are given in Table 7.2, which does not include wrought alloys that are cast into ingots or billets intended for fabrication by mechanical deformation [1, 7].

Although the nomenclature and designations for various casting alloys are standardized in North America, many important alloys have been developed for engineered casting production worldwide. For the most part, each nation (and in many cases the individual firm) has developed its own alloy nomenclature. Excellent references are available that correlate, cross reference, or otherwise define significant compositions in international use [8–10].

The Aluminum Association designation system attempts alloy family recognition by the following scheme:

- $1xx.x$: Controlled unalloyed compositions
- $2xx.x$: Aluminum alloys containing copper as the major alloying element
- $3xx.x$: Aluminum-silicon alloys also containing magnesium and/or copper
- $4xx.x$: Binary aluminum-silicon alloys
- $5xx.x$: Aluminum alloys containing magnesium as the major alloying element
- $7xx.x$: Aluminum alloys containing zinc as the major alloying element. Also containing additions of either copper, magnesium, chromium, manganese, or combinations of these elements
- $8xx.x$: Aluminum alloys containing tin as the major alloying element

A separate four-digit Aluminum Aassociation designation system exists for wrought Al compositions.

AA No.	Products[a]	Si	Fe	Cu	Mn	Mg	Cr	Ni	Zn	Ti	Sn
						Composition[b], wt%					
201.0	S	0.10	0.15	4.0–5.2	0.20–0.50	0.15–0.55	–	–	–	0.15–0.35	–
A201.0	S	0.05	0.10	4.0–5.0	0.20–0.50	0.15–0.55	–	–	–	0.15–0.35	–
B201.0	S	0.05	0.05	4.0–5.0	0.20–0.50	0.25–0.35	–	–	–	0.15–0.35	–
204.0	S, P	0.20	0.35	4.2–5.0	0.10	0.15–0.35	–	0.05	0.10	0.15–0.30	0.05
206.0	S, P	0.10	0.15	4.2–5.0	0.20–0.50	0.15–0.35	–	0.05	0.10	0.15–0.30	0.05
A206.0	S, P	0.05	0.10	4.2–5.0	0.20–0.50	0.15–0.35	–	0.05	0.10	0.15–0.30	0.05
B206.0	S, P	0.05	0.10	4.2–5.0	0.20–0.50	0.15–0.35	–	0.05	0.10	0.15–0.30	0.05
319.0	S, P	5.5–6.5	1.0	3.0–4.0	0.50	0.10	–	0.35	1.0	0.25	–
A319.0	S, P	5.5–6.5	1.0	3.0–4.0	0.50	0.10	–	0.35	3.0	0.25	–
B319.0	S, P	5.5–6.5	1.2	3.0–4.0	0.8	0.10–0.50	–	0.50	1.0	0.25	–
354.0	P	8.6–9.4	0.20	1.6–2.0	0.10	0.40–0.6	–	–	0.10	0.20	–
355.0	S, P	4.5–5.5	0.6[c]	1.0–1.5	0.50[c]	0.40–0.6	0.25	–	0.35	0.25	–
A355.0	S, P	4.5–5.5	0.09	1.0–1.5	0.05	0.45–0.6	–	–	0.05	0.04–0.20	–
C355.0	S, P	4.5–5.5	0.20	1.0–1.5	0.10	0.45–0.6	–	–	0.10	0.20	–
356.0	S, P	6.5–7.5	0.6[c]	0.25	0.35[c]	0.20–0.45	–	–	0.35	0.25	–
A356.0	S, P	6.5–7.5	0.20	0.20	0.10	0.25–0.45	–	–	0.10	0.20	–
B356.0	S, P	6.5–7.5	0.09	0.05	0.05	0.25–0.45	–	–	0.05	0.04–0.20	–
C356.0	S, P	6.5–7.5	0.07	0.05	0.05	0.25–0.45	–	–	0.05	0.04–0.20	–
F356.0	S, P	6.5–7.5	0.20	0.20	0.10	0.17–0.25	–	–	0.10	0.04–0.20	–
357.0	S, P	6.5–7.5	0.15	0.05	0.03	0.45–0.6	–	0.05	0.20	0.04–0.20	–
A357.0	S, P	6.5–7.5	0.20	0.20	0.10	0.40–0.7	–	0.05	0.10	0.04–0.20	–
B357.0	S, P	6.5–7.5	0.09	0.05	0.05	0.40–0.6	–	0.05	0.05	0.04–0.20	–

Table 7.2 Chemical Composition Limits for Registered Aluminum Alloys in the Form of xxx.0 Casting, xxx.1 Ingot, and xxx.2 Ingot (*Continued*)

AA No.	Products[a]	Si	Fe	Cu	Mn	Mg	Cr	Ni	Zn	Ti	Sn
						Composition[b], wt%					
C357.0	S, P	6.5–7.5	0.09	0.05	0.05	0.45–0.7	–	–	0.05	0.04–0.20	–
D357.0	S	6.5–7.5	0.20	–	0.10	0.55–0.6	–	–	–	0.10–0.20	–
E357.0	S, P, I	6.5–7.5	0.10	–	0.10	0.55–0.6	–	–	–	0.10–0.20	–
F357.0	S, P, I	6.5–7.5	0.10	0.20	0.10	0.40–0.7	–	–	0.10	0.04–0.20	–
A380.0	D	7.5–9.5	1.3	3.0–4.0	0.50	0.10	–	0.50	3.0	–	0.35
B380.0	D	7.5–9.5	1.3	3.0–4.0	0.50	0.10	–	0.50	1.0	–	0.35
C380.0	D	7.5–9.5	1.3	3.0–4.0	0.50	0.10–0.30	–	0.50	3.0	–	0.35
D380	D	7.5–9.5	1.3	3.0–4.0	0.50	0.10–0.30	–	0.50	3.0	–	0.35
390.0	D	16.0–18.0	1.3	4.0–5.0	0.10	0.45–0.65	–	–	0.10	0.20	–
A390.0	S, P	16.0–18.0	0.50	4.0–5.0	0.10	0.45–0.85	–	–	0.10	0.20	–
B390.0	D	16.0–18.0	1.3	4.0–5.0	0.50	0.45–0.65	–	0.10	1.5	0.20	–
413	D	11.0–13.0	2.0	1.0	0.35	0.10	–	0.60	0.50	–	0.15
A413.0	D	11.0–13.0	1.3	1.0	0.35	0.10	–	0.50	0.50	–	0.15
B413.0	S, P	11.0–13.0	0.50	0.10	0.35	0.05	–	0.05	0.10	0.25	–
443.0	S, P	4.5–6.0	0.8	0.6	0.50	0.05	0.25	–	0.50	0.25	–
A443.0	S	4.5–6.0	0.8	0.30	0.50	0.05	0.25	–	0.50	0.25	–
B443.0	S, P	4.5–6.0	0.8	0.16	0.35	0.05	–	–	0.35	0.25	–
C443.0	D	4.5–6.0	2.0	0.6	0.35	0.10	–	0.50	0.50	–	0.15
444.0	S, P	6.5–7.5	0.6	0.25	0.35	0.10	–	–	0.35	0.25	–
A444.0	P	6.5–7.5	0.6	0.25	0.35	0.10	–	–	0.10	0.20	–
520.0	S	0.25	0.30	0.25	0.15	9.5–10.6	–	–	0.15	0.25	–
535.0	S	0.15	0.15	0.05	0.10–0.25	6.2–7.5	–	–	–	0.10–0.25	–
A535.0	S	0.20	0.20	0.10	0.10–0.25	6.2–7.5	–	–	–	0.25	–
B535.0	S	0.15	0.15	0.10	0.05	6.2–7.5	–	–	–	0.120–0.25	–

710.0	S	0.15	0.50	0.35–0.6	0.05	0.6–0.8	–	–	6.0–7.0	0.25	–
711.0	P	0.30	0.7–1.4	0.35–0.6	0.05	0.25–0.45	–	–	6.0–7.0	0.20	–
712.0	S	0.30	0.50	0.25	0.10	0.50–0.65	0.40–0.6	–	5.0–8.5	0.15–0.25	–
713.0	S, P	0.25	1.1	0.40–0.10	0.6	0.20–0.50	0.35	0.15	7.0–8.0	0.25	–
771.0	S	0.15	0.15	0.10	0.10	0.8–1.0	0.06–0.20	–	6.5–7.5	0.10–0.20	–
772.0	S	0.15	0.15	0.10	0.10	0.6–0.8	0.06–0.20	–	6.0–7.0	0.10–0.20	–
850.0	S, P	0.7	0.7	0.7–1.3	0.10	0.10	–	0.7–1.3	–	0.20	5.5–7.0
853.0	S, P	5.5–6.5	0.7	3.0–4.0	0.60	–	–	–	–	0.20	5.5–7.0

Note: Only composition limits that are identical to those listed herein or are registered with The Aluminum Association should be designated as "AA" alloys.

Source: From *ASM Handbook* [1, 7].

TABLE 7.2 Chemical Composition Limits for Registered Aluminum Alloys in the Form of *xxx*.0 Casting, *xxx*.1 Ingot, and *xxx*.2 Ingot (*Continued*)

Alloying Principles

The aluminum-base alloys may in general be characterized as eutectic systems, containing intermetallic compounds or elements as the excess phases [11]. Because of the relatively low solubilities of most of the alloying elements in aluminum and the complexity of the alloys that are produced, any one aluminum-base alloy may contain several metallic phases, which sometimes are quite complex in composition. These phases usually are appreciably more soluble near the eutectic temperatures than at room temperature, making it possible to heat-treat some of the alloys by solution and aging heat treatments.

All the properties of interest are, of course, influenced by the effects of the various elements with which aluminum is alloyed. The principal alloying elements in aluminum-base casting alloys are copper, silicon, magnesium, zinc, chromium, manganese, tin, and titanium. Iron is an element normally present and usually considered as an impurity. Some of the simpler effects of alloying can be considered as described below. The details can be found in Ref. 1.

Major Alloying Elements

Copper

The first and most widely used aluminum alloys were those containing 4 to 10% Cu. Copper substantially improves strength and hardness in the as-cast and heat-treated conditions. Alloys containing 4 to 6% Cu respond most strongly to thermal treatment. Copper generally reduces resistance to general corrosion and, in specific compositions and material conditions, stress corrosion susceptibility. Addition of copper also reduces hot tear resistance and decreases castability.

Silicon

The outstanding effect of silicon in aluminum alloys is the improvement of casting characteristics. Additions of silicon to pure aluminum dramatically improve fluidity, hot tear resistance, and feeding characteristics. The most prominently used compositions in all casting processes are those of the aluminum-silicon family. Commercial alloys span the hypoeutectic and hypereutectic ranges up to about 25% Si.

In general, an optimum range of silicon content can be assigned to casting processes. For slow cooling rate processes (such as plaster, investment, and sand), the range is 5 to 7%, for permanent mold 7 to 9%, and for die casting 8 to 20%. The bases for these recommendations are the relationship between cooling rate and fluidity and the effect of percentage of eutectic on feeding. Silicon additions are also accompanied by a reduction in specific gravity and coefficient of thermal expansion.

Magnesium

Mg is the basis for strength and hardness development in heat-treated Al-Si alloys and is commonly used in more complex Al-Si alloys containing copper, nickel, and other elements for the same purpose. The Mg_2Si-based precipitation hardening system displays a useful solubility limit corresponding to approximately 0.70% Mg, beyond which either no further strengthening occurs or matrix softening takes place. Common premium-strength compositions in the Al-Si family employ magnesium in the range of 0.40 to 0.70%.

Binary Al-Mg alloys are widely used in applications requiring a bright surface finish and corrosion resistance, as well as attractive combinations of strength and ductility.

Common compositions range from 4 to 10% Mg, and compositions containing more than 7% Mg are heat treatable.

Zinc

No significant benefits are obtained by the addition of zinc alone to aluminum. However, zinc is used as a principal alloying element in some alloys. Its chief beneficial effect seems to be that of making it possible to obtain a maximum of mechanical properties in the as-cast condition. Accompanied by the addition of copper and/or magnesium, zinc results in attractive heat treatable or naturally aging compositions. A number of such compositions are in common use. Zinc is also commonly found in secondary gravity and die-casting compositions.

Chromium

Chromium additions are commonly made in low concentrations to room temperature aging and thermally unstable compositions in which germination and grain growth are known to occur. Chromium typically forms the compound $CrAl_7$, which displays extremely limited solid-state solubility and is therefore useful in suppressing grain growth. Sludge that contains iron, manganese, and chromium is sometimes encountered in die-casting compositions, but it is rarely encountered in gravity casting alloys. Cr improves corrosion resistance in certain alloys and increases quench sensitivity at higher concentrations.

Manganese

Manganese is used for two main purposes in shape casting. The first is to correct the Fe phases, which form in secondary foundry alloy compositions, from β needles to α script thereby improving ductility. The second use is in advanced die-casting processes, where Mn can be substituted for Fe to reduce die soldering. Manganese's higher solubility in Al combined with its lower tendency to form brittle phases gives an alloy with better properties than those in which Fe is used as the exclusive remedy for die soldering. Some evidence also exists that a high volume fraction of $MnAl_6$ in alloys containing more than 0.5% Mn may beneficially influence internal casting soundness.

Tin

Tin is effective in improving antifriction characteristics, and is therefore useful in bearing applications. Casting alloys may contain up to 25% Sn. Additions can also be made to improve machinability. Sn may influence precipitation-hardening response in some alloy systems at levels in the hundreds of ppm.

Titanium

Titanium is extensively used to refine the grain structure of aluminum casting alloys, often in combination with smaller amounts of boron. Titanium in excess of the stoichiometry of TiB_2 is necessary for effective grain refinement of wrought alloys. Foundry alloys are effectively grain refined with ratios of Ti to B of 5:1. Titanium is often employed at concentrations greater than those required for grain refinement to reduce cracking tendencies in hot short compositions.

Iron

Iron improves hot tear resistance and decreases the tendency for die sticking or soldering in die casting. Increases in iron content are, however, accompanied by substantially decreased ductility. Iron reacts to form a myriad of insoluble phases in aluminum alloy melts, the most common of which are $FeAl_3$, $FeMnAl_6$, and $\alpha AlFeSi$. In Al-Si alloys the two most common are β-needles ($FeSiAl_5$) and the Mn corrected α-script [$(Fe,Mn)_3Si_2Al1_5$]. These essentially insoluble phases are responsible for improvements in strength, especially at elevated temperature. They do, however, reduce the ductility. As the fraction of insoluble phase increases with increased iron content, casting considerations such as flowability and feeding characteristics are adversely affected. Iron participates in the formation of sludging phases with manganese, chromium, and other elements; particularly at the concentrations and low holding temperatures normally associated with high-pressure die casting. In fact, the sludging factor (SF), a simple expression combining the concentrations of Fe, Mn, and Cr, is given as

$$SF = \%Fe + 2 \times \%Mn + 3 \times \%Cr \tag{7.1}$$

Plots of this factor versus temperature delineate the minimum temperature below which sludging can commonly be expected in high-pressure die casting.

Minor Alloying Elements and Impurities

The minor alloying elements or impurities present in the aluminum alloys greatly influence their properties. Two alloy characteristics which may be seriously impaired are ductility (and toughness) and corrosion resistance.

Antimony

At concentration levels equal to or greater than 0.05%, antimony refines eutectic aluminum-silicon phase to lamellar form in hypoeutectic compositions. The effectiveness of antimony in altering the eutectic structure depends on an absence of phosphorus and on an adequately rapid rate of solidification. Antimony also reacts with either sodium or strontium to form coarse intermetallics with adverse effects on castability and eutectic structure. Antimony will chemically combine with and precipitate common modifiers Na and Sr, and hence Sb has the potential to act as a modification poisoner if it enters the scrap recycling stream. This is compounded by the fact that Sb is very difficult to analyze for using even modern optical emission spectrometers due to the overlap which exists between its spectrum and that of Fe.

Antimony is classified as a heavy metal with potential toxicity and hygiene implications, especially as associated with the possibility of stibine gas formation and the effects of human exposure to other antimony compounds.

Beryllium

Beryllium additions of as low as a few parts per million may be effective in reducing oxidation losses and associated inclusions in magnesium-containing compositions. Studies have shown that proportionally increased beryllium concentrations are required for oxidation suppression as magnesium content increases.

At higher concentrations (>0.04%), beryllium affects the form and composition of iron-containing intermetallics, markedly improving strength and ductility. In addition to beneficially changing the morphology of the insoluble phase, beryllium changes its composition, rejecting magnesium from the Al-Fe-Si complex and thus permitting its full use for hardening purposes.

Beryllium-containing compounds are, however, numbered among the known carcinogens that require specific precautions in melting, molten metal handling, dross handling and disposition, and blasting, grinding, and welding. Standards define the maximum beryllium in welding rod and weld base metal as 0.008 and 0.010%, respectively. Both acute and chronic berylliolses are possible upon exposure to Be fumes. The chronic form can appear years after the exposure has ended. For these reasons there has been a movement to eliminate this element from aluminum alloys in general.

Bismuth

Bismuth improves the machinability of cast aluminum alloys at concentrations greater than 0.1%. It is also known to increase shrinkage tendencies in Al-Si alloys at concentrations greater than 30 ppm.

Boron

Boron combines with other metals to form borides present as solid particles in liquid Al, such as AlB_2 and TiB_2. The titanium boride particles then act as nucleation sites for the active grain-refining phase $TiAl_3$. Once nucleated from the molten aluminum this last phase serves as the nucleating point for new grains.

Intermetallic borides reduce tool life in machining operations, and as coarse particles, form objectionable inclusions with detrimental effects on mechanical properties and ductility. At high boron concentrations, borides contribute to furnace sludging, particle agglomeration, and increased risk of casting inclusions. However, boron treatment of aluminum-containing peritectic elements is practiced to improve purity and electrical conductivity in rotor casting. Higher rotor alloy grades may specify boron to exceed titanium and vanadium contents to ensure either the complexing or precipitation of these elements for improved electrical performance.

Cadmium

In concentrations exceeding 0.1%, Cd improves machinability. Precautions that acknowledge volatilization at 767°C (1413°F) are essential for toxicity reasons.

Calcium

Calcium is a weak aluminum-silicon eutectic modifier. It increases hydrogen solubility and is often responsible for casting porosity at trace concentration levels. Ca concentrations greater than approximately 0.005% also adversely affect ductility in aluminum-magnesium alloys.

Hydrogen

Hydrogen is the only gaseous species appreciably soluble in liquid aluminum. The fact that the solubility of hydrogen is considerably lower in solid aluminum means that there is the propensity for hydrogen to cause porosity through precipitation from the liquid in the form of bubbles as the metal solidifies. Commonly called *gas porosity*, this is avoided by degassing the metal with low hydrogen inert or active gasses using either a lance, porous media, or rotary impeller degasser in increasing order of cost and effectiveness.

The most common source of hydrogen is moisture from the atmosphere over the metal as aluminum will reduce water vapor to form alumina and dissolved hydrogen in the melt. For this reason air used to pump low-pressure furnaces is normally dried to a low dew point. Inert gas covers may be used over the metal in critical applications.

Indium

Indium is added, together with Zn, in the casting of sacrificial anodes. Indium helps to disrupt the normally tough coherent alumina surface coating typical of aluminum alloys and hence enhances its ability to sacrificially corrode.

Lead

Lead is commonly used in aluminum casting alloys at greater than 0.1% for improved machinability.

Mercury

Compositions containing mercury were developed as sacrificial anode materials for cathodic protection systems, especially in marine environments. The use of these optimally electronegative alloys, which did not passivate in seawater, was severely restricted for environmental reasons.

Nickel

Nickel is usually employed with copper to enhance elevated-temperature properties. It also reduces the coefficient of thermal expansion. Ni will accelerate filiform corrosion in applications such as coated automotive wheels if present in high enough trace quantities. Values up to roughly 100 ppm seem to be acceptable, however.

Phosphorus

Phosphorus reacts with Al to form AlP which nucleates and refines primary silicon phase in hypereutectic Al-Si alloys. At parts per million concentrations, phosphorus coarsens the eutectic structure in hypoeutectic Al-Si alloys. Phosphorus diminishes the effectiveness of the common eutectic modifiers sodium and strontium as well as the eutectic refiner antimony.

Silver

Silver is used in only a limited range of aluminum-copper premium-strength alloys at concentrations of 0.5 to 1.0%. Ag contributes to precipitation hardening and stress corrosion resistance.

Sodium

Sodium modifies the aluminum-silicon eutectic. Its presence is embrittling in aluminum-magnesium alloys. Sodium reacts with phosphorus, which reduces its own effectiveness in modifying the eutectic in hypoeutectic alloys as well as reducing the refining effect of phosphorus in hypereutectic alloys. The effects of sodium and phosphorous are mutually antagonistic.

Strontium

Strontium is used to modify the aluminum-silicon eutectic. Effective modification can be achieved at very low addition levels, but a range of recovered strontium of 0.004 to 0.02% is commonly used. Higher addition levels are frequently associated with casting porosity, especially in processes or in thick-section parts in which solidification occurs more slowly. This increase in porosity can be used to compensate for shrinkage in castings where increased distributed porosity is tolerable. Degassing efficiency as determined by vacuum solidification tests may also be adversely affected at higher strontium levels. The effects of strontium and phosphorous are mutually antagonistic in the same manner as discussed for sodium.

Physical Metallurgy

As mentioned before, the elements that are most commonly present in commercial aluminum alloys are copper, magnesium, silicon, zinc, and manganese. These elements have significant solid solubility in aluminum which increases with temperature. The liquid and solid solubility of various alloying elements in aluminum are listed in Table 7.3.

"Second-phase" constituents form when the solid solubility limit is exceeded. Elements such as Si, Sn, and Be remain as pure alloying ingredients in the microstructure. Intermetallic compound phases form in most of the ternary, quaternary, and multicomponent cast alloys. For example, alloys 356 (Al-Si-Mg, Fe) and 520 (Al-Mg-Si-Fe) can contain phases such as Si, Mg_2Si, and $Fe_2Si_2Al_9$, and $FeAl_3$, Fe_3SiAl_2, Mg_2Si, and Mg_2Al_3, respectively. The phases that appear in a cast structure depend on the rate of solidification.

Aluminum-Copper System

The structural effects of copper in Al-Cu base alloys are presented in the equilibrium diagram in Fig. 7. 2. The diagram shows solubility of copper in aluminum increasing in the solid state from less than 0.50% at room temperature to 5.65% at 565°C (1018°F). Copper above the solubility limit at any temperature appears microstructurally as the θ phase. The θ phase has a composition approximating the formula $CuAl_2$ (46.5% Al-53.5% Cu) and is a hard brittle constituent. By comparison, the solid-solution phase is relatively soft and ductile. Structurally, then, increasing copper content in Al-Cu base alloys results in an increasing percentage of the θ phase. The mechanical properties of hardness and strength can then be expected to increase as copper content increases while the ductility decreases. A limited percentage of copper thus has a beneficial effect of strengthening and hardening in Al-Cu base alloys. An excessive copper percentage will cause tensile properties to fall below the maximum values obtained. Furthermore, ductility is reduced to a very low level and brittleness results in alloys of high copper content. Therefore, copper percentages do not exceed 12% in most aluminum casting alloys. Actually, the copper percentages in aluminum casting alloys are adjusted so that the lower contents, 2 to 5% are used in alloys required to have optimum ductility (or toughness), whereas the higher percentages are used when greater hardness and strength are desired.

The sloping solvus line in Fig. 7.2*b* indicates that Al-Cu alloys can be heat treated. Solution treatment and age hardening can produce a high degree of strengthening in these alloys. Other alloying elements such as Mg, Si, and Zn confer heat-treating potentialities to Al-Cu base alloys and they greatly extend the range of properties available in aluminum castings. However, copper-containing aluminum alloys are less resistant to corrosion.

Alloys based in the Al-Cu system are susceptible to solidification cracking (hot tearing) and to interdendritic shrinkage because of their long solidification range.

Aluminum-Silicon System

Silicon is present in all commercial aluminum casting alloys. As an alloying element it is used in amounts up to about 14% Si. The binary Al-Si phase diagram is shown in Fig. 7.3. The solubility of Si in aluminum, the α phase, is limited to 1.65% at 577°C (1072°F)and less than 0.05% at room temperature [11, 13]. Undissolved silicon is

Element	Temperature(a)		Liquid Solubility		Solid Solubility	
	°C	°F	wt%	at%	wt%	at%
Ag	570	1060	72.0	60.9	55.6	23.8
Au	640	1180	5	0.7	0.36	0.049
B	660	1220	0.022	0.054	<0.001	<0.002
Be	645	1190	0.87	2.56	0.063	0.188
Bi	660(b)	1220(b)	3.4	0.45	<0.1	<0.01
Ca	620	1150	7.6	5.25	<0.1	<0.05
Cd	660(b)	1220(b)	6.7	1.69	0.47	0.11
Co	660	1220	1.0	0.46	<0.02	<0.01
Cr	660(c)	1220(c)	0.41	0.21	0.77	0.40
Cu	550	1020	33.15	17.39	5.67	2.48
Fe	655	1210	1.87	0.91	0.052	0.025
Ga	30	80	98.9	97.2	20.0	8.82
Ge	425	800	53.0	29.5	6.0	2.30
Hf	660(c)	1220(c)	0.49	0.074	1.22	0.186
In	640	1180	17.5	4.65	0.17	0.04
Li	600	1110	9.9	30.0	4.0	13.9
Mg	450	840	35.0	37.34	14.9	16.26
Mn	660	1220	1.95	0.97	1.82	0.90
Mo	660(c)	1220(c)	0.1	0.03	0.25	0.056
Na	660(b)	1220(b)	0.18	0.21	<0.003	<0.003
Nb	660(c)	1220(c)	0.01	0.003	0.22	0.064
Ni	640	1180	6.12	2.91	0.05	0.023
Pb	660	1220	1.52	0.20	0.15	0.02
Sb	660	1220	1.1	0.25	<0.1	<0.02
Sc	660	1220	0.52	0.31	0.38	0.23
Si	580	1080	12.6	12.16	1.65	1.59
Sn	230	450	99.5	97.83	<0.01	<0.002
Sr	655	1210	–	–	–	–
Ti	665(c)	1230(c)	0.0.15	0.084	1.00	0.57
V	665(c)	1230(c)	0.25	0.133	0.6	0.32
Y	645	1190	7.7	2.47	<0.1	<0.03
Zn	380	720	95.0	88.7	82.8	66.4
Zr	660(c)	1220(c)	0.11	0.033	0.28	0.085

Note: (a) Eutectic reaction unless designated otherwise; (b) monotectic reaction; (c) peritectic reaction.
Source: From Davis [12].

TABLE 7.3   Solubility Limits for Various Binary Aluminum Alloys

Al-Cu

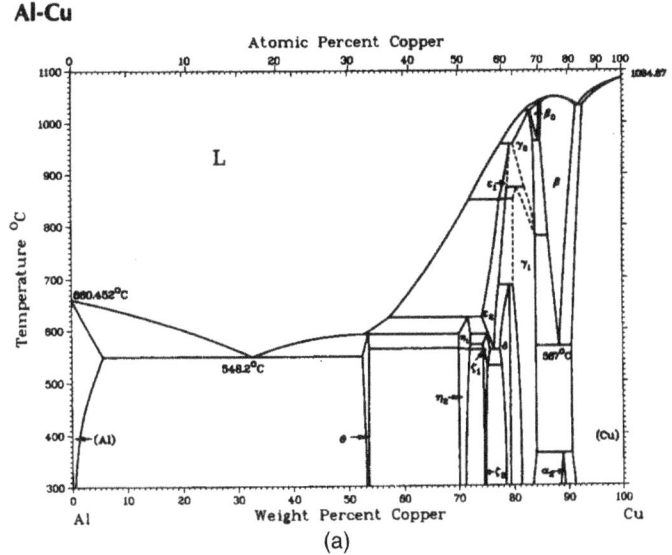

(a)

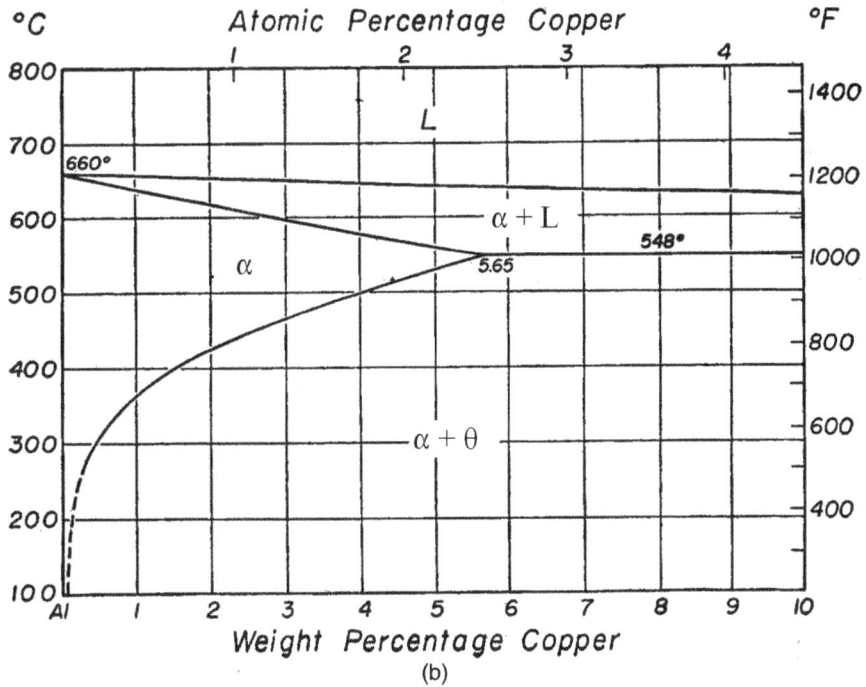

(b)

FIGURE 7.2 (a) The Al-Cu equilibrium system. (From *ASM Handbook* [14].) (b) Aluminum-rich corners of the phase diagram. (From *ASM Handbook* [17].)

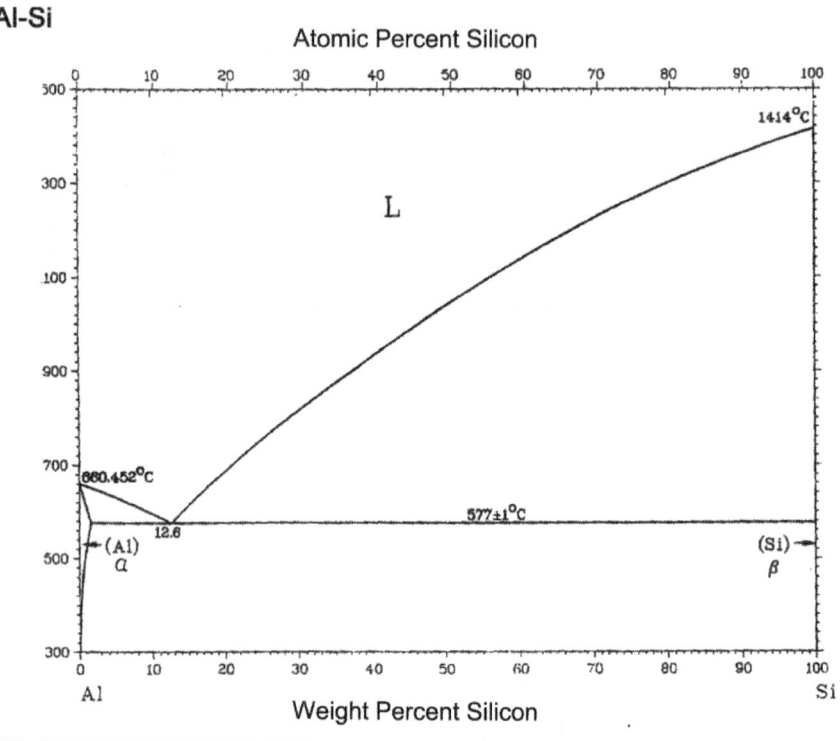

Al-Si

Fɪɢᴜʀᴇ **7.3**   The Al-Si equilibrium system. (From *ASM Handbook* [14].)

present as β, silicon particles containing an extremely small percentage of aluminum. Silicon phase in an aluminum alloy, containing 14.0% Si, is illustrated in Fig. 7.4 [15, 16]. The size of the silicon-rich β particles may be varied greatly. A rather coarse particle size occurs in normal melting and sand-casting practice. Structural refinement (as discussed later) by fast cooling rates (as in permanent mold and die-casting processes) or by addition of elements such as Na, Sr can modify the silicon morphology to produce from a flaky (as a hypoeutectic alloys) form a fine fibrous form.

The effect of silicon on the properties of Al-Si alloys is largely one of alloying since no significant benefits are obtained by attempts at solution heat-treating and aging. The percentage of silicon in the alloy is first in importance, closely followed by the micro-structural effects of modification by permanent-mold or die casting or special melting practices. The general effect of increasing silicon contents is shown in Fig. 7.5 to be that of increasing the strength until the eutectic silicon percentage is reached. Ductility, how-ever, is lowered. The beneficial effects of modification with elements such as sodium and by chill casting are also evident in Fig. 7.5. From these observations, it follows that aluminum-silicon alloys will be at their best when modified by suitable additions, or better, when cast in metal molds and best when modified by suitable addition together with permanent mold casting. Furthermore, since additional improvement cannot be obtained by heat treatment, these alloys will be used in the as-cast condition. Other elements used in aluminum alloys which, like Si, do not confer response to solution heat treatment are manganese and nickel.

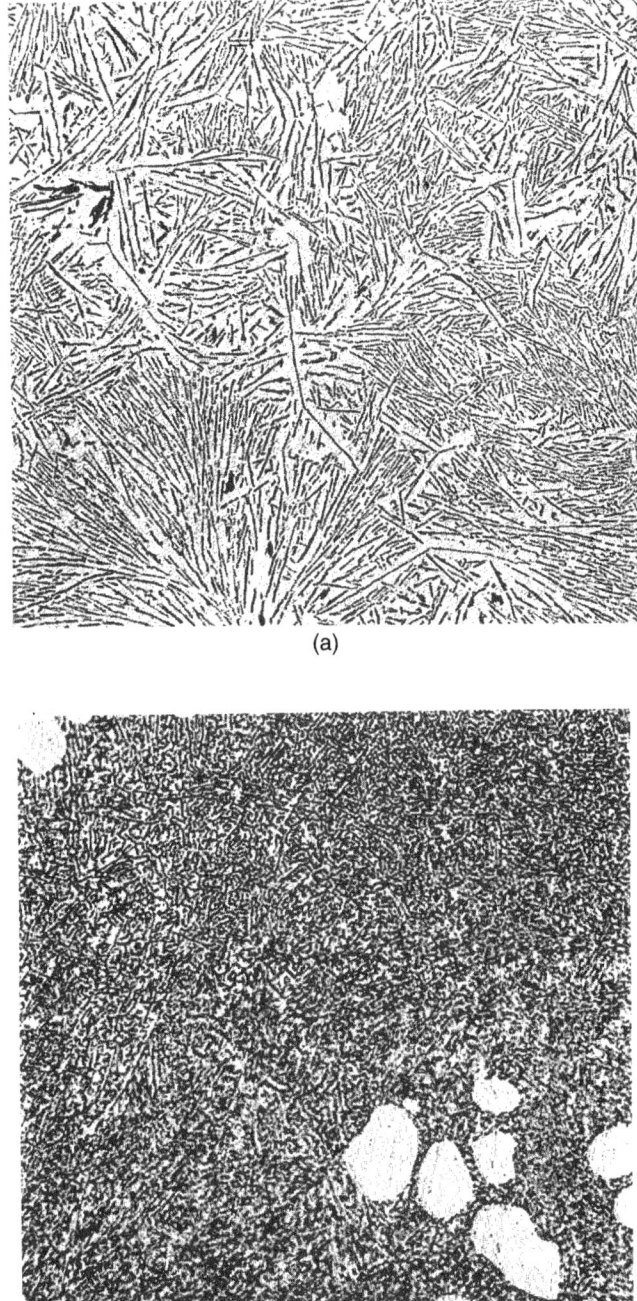

(a)

(b)

FIGURE 7.4 Optical micrographs of transverse sections through Al-Si eutectics. (a) Directionally solidified at 4.5×10^{-3} cm/s ($\times$150); (b) chill cast ($\times$650). (From Vinuk et al. [15], Sahoo and Smith [16].)

Aluminum-Silicon-Copper System

Aluminum-Silicon-Copper casting alloys are most widely used where copper contributes to strengthening and machinability and silicon contributes to castability and reduces hot shortness [13]. The amount of Cu and Si can vary in different alloys. Hypoeutectic Al-Si alloys with Cu are normally suited for complex castings and for permanent mold casting. Alloys with less than 5.6% Cu are heat treatable. Addition of Mg to Al-Si-Cu alloys enhances the strength properties, especially after heat treatment. Hypereutectic silicon alloys (12 to 30% Si) containing copper get the dual benefits of excellent wear resistance due to silicon and matrix hardening and elevated temperature strength due to copper.

Aluminum-Magnesium System

The alloying behavior of magnesium in aluminum is similar to that of copper. The equilibrium diagram for the binary system is shown in Fig. 7.6. The alloy system shows a solid-solubility change of the α phase with temperature, 14.9% Mg being soluble at

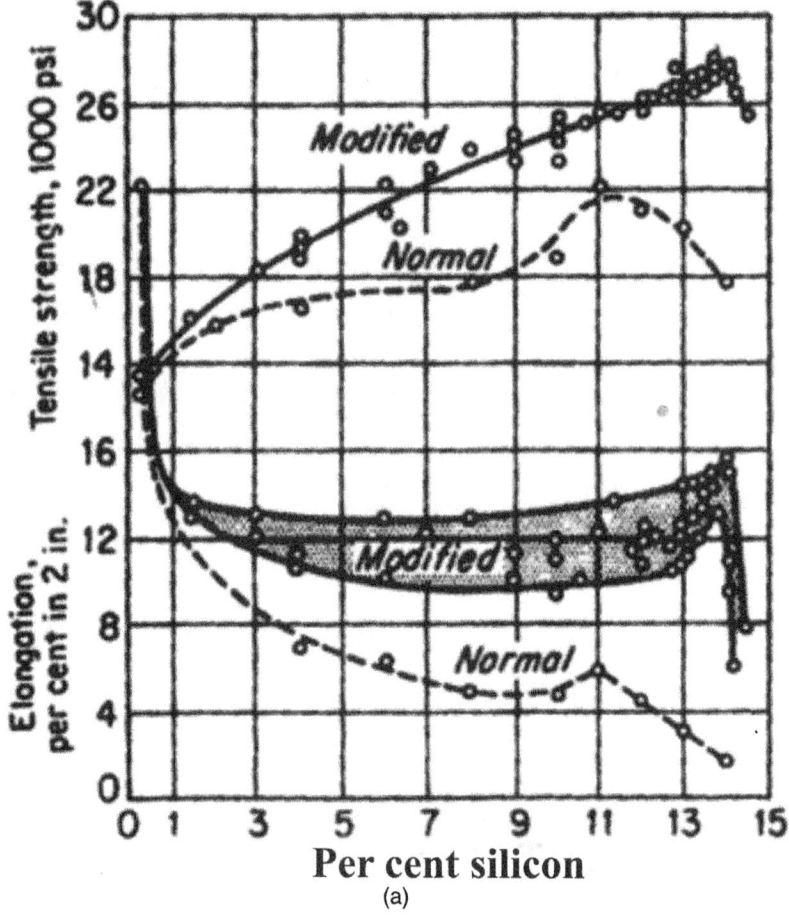

Figure 7.5   Properties of Al-Si casting alloys as a function of silicon in the alloy. (a) Applies to normal and modified alloys (sodium-treated) in sand castings;

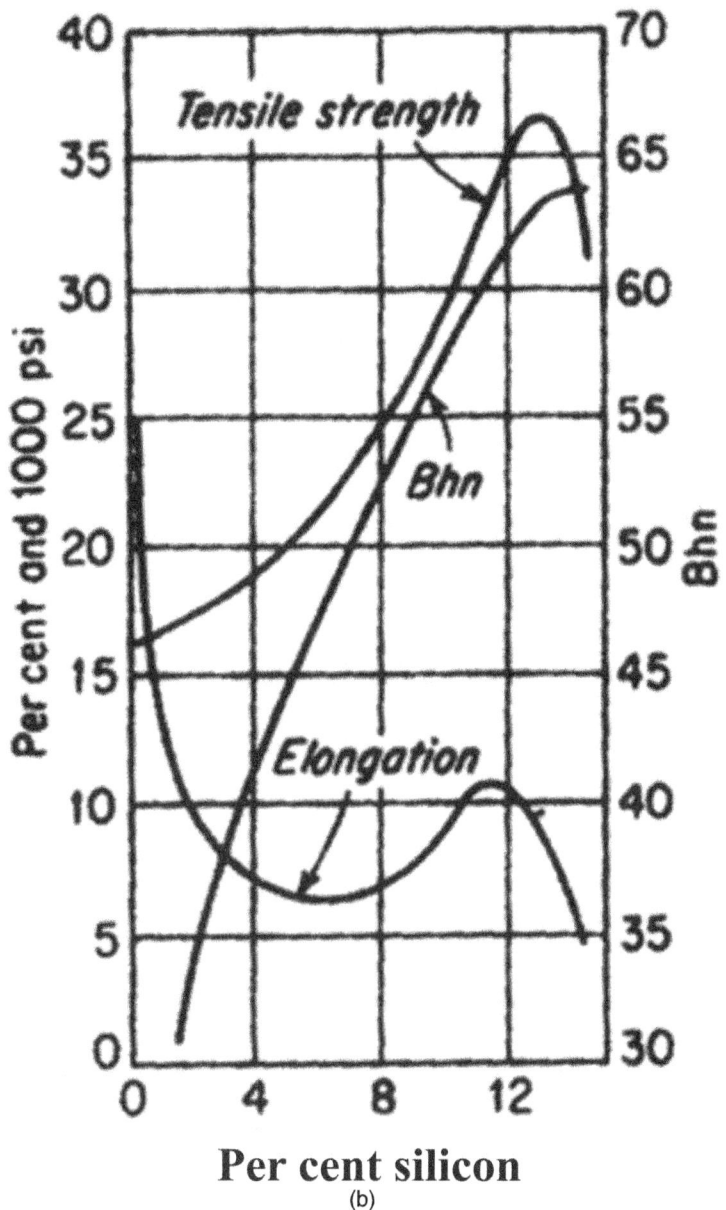

Per cent silicon
(b)

FIGURE 7.5 (*Continued*) (*b*) applies to chill castings. (From Heine et al. [11].)

451°C (844°F) and less than 2.90% at room temperature. A second, harder β (Al$_3$Mg$_2$) phase exists when the solid-solubility limit is exceeded. The opportunity for solution and aging heat treatments is present, and the mechanical property relationships with magnesium percentage are similar to those in the Al-Cu alloys. Several alloys are based on this binary system, and normally contain 4, 7, and 10% Mg. Complex alloys containing other elements, along with substantial percentage of magnesium are also available.

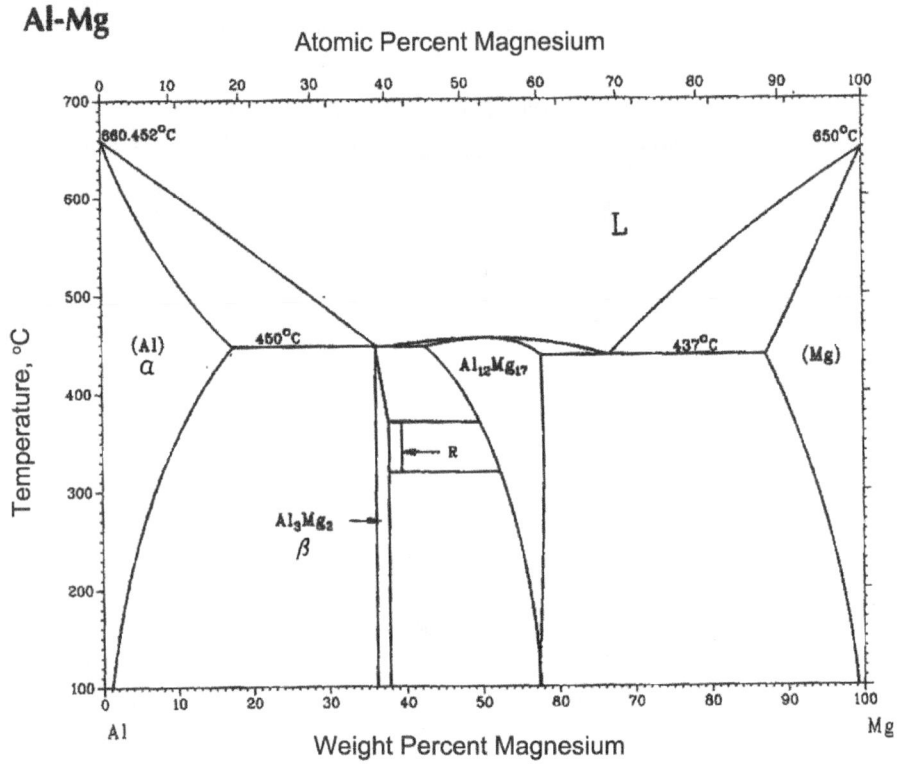

Figure 7.6   The Al-Mg equilibrium system. (From *ASM Handbook* [14].)

Their most important characteristic is corrosion resistance, including exposure to seawater and marine atmospheres. They have good machinability, weldability, and an attractive appearance in cast, machined, polished, and anodized conditions [11].

Aluminum-Magnesium Silicon System

Certain combinations of magnesium and silicon have been found to exhibit important alloying effects in aluminum. The two elements are able to combine and form the metallic compound Mg_2Si. They then behave as a quasi-binary alloy system, as illustrated in Fig. 7.7 [11, 17]. The Al-Mg_2Si system is also of the type permitting solution and aging treatments and their accompanying property changes. Ternary alloys taking advantage of this quasi-binary system and the beneficial effects of silicon contain small percentages of Mg, up to about 0.30%, and larger percentages of Si, 6 to 8.0%. The excess of silicon is present to improve casting properties of these alloys since it is not needed to form Mg_2Si.

In some alloys, the combined effects of Si and Mg are undesirable and they may then be limited as impurities [11]. Since all aluminum alloys contain silicon, the addition of magnesium is all that is necessary to obtain the hardening effect of Mg_2Si. The alloys may then become brittle. For this reason impurity limits for magnesium in alloys (the Al-Cu, Al-Si, and their complex alloys, for example) are set at 0.03 to 0.10% maximum.

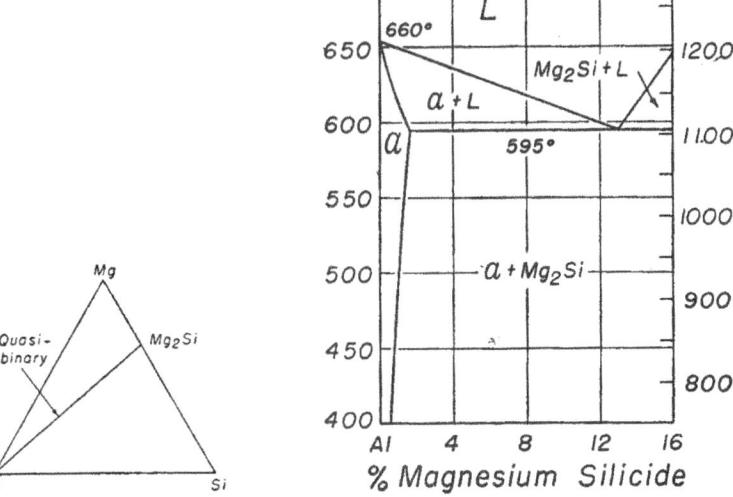

FIGURE 7.7 The Al-Mg-Si equilibrium system and the quasi-binary system Al-Mg-Si. (From Heine et al. [11], *ASM Handbook* [17].)

Aluminum-Zinc-Magnesium System

Many alloys of this type age naturally at room temperature, achieving full strength within 20 to 30 days after casting. Solution heat treatment is not typically necessary for property development. Rapid solidification in these alloys can result in microsegregation of magnesium-zinc phases that reduces hardening potential. Conventional solution heat treatments can be used when adequate property development does not occur through natural aging [13].

These alloys typically display moderate to good tensile properties in the as-cast condition. The melting temperatures of alloys of this group are high, an advantage in castings that are to be brazed.

Machinability and resistance to general corrosion is usually good. The chemistry of most alloys is controlled to minimize stress-corrosion susceptibility.

The castability of Al-Zn-Mg alloys is poor, and good foundry practices are required to minimize hot tearing and shrinkage defects [13].

Aluminum-Tin System

Tin is the major alloying element in compositions developed for bearing applications. It has also been employed with bismuth, lead, and cadmium at lower concentrations to provide free-machining properties. Aluminum and tin are essentially immiscible. Before and after solidification, tin is present in dispersed form. Mechanical agitation is required initially to achieve suspension of tin, and, because of density differences, gravity segregation may occur over time in the molten state [13].

Alloys containing 5.0 to 7.0% Sn are broadly used in bearings and bushings in which low friction, compressive strength, fatigue strength, and resistance to corrosion are important criteria. Additions of copper, nickel, and magnesium contribute to hardness and strength, and silicon is added to improve castability, reduce hot shortness, and increase compressive yield strength.

Element	Difference in Atomic Radii, $r_x - r_{Al}, \%^{(a)}$	Strength/Addition Values[b]							
		Yield Strength/% Addition[c]				Tensile Strength/% Addition[d]			
		MPa/ at%	Ksi/ at%	MPa/ wt%	Ksi/ wt%	MPa/ at%	Ksi/ at%	MPa/ wt%	Ksi/ wt%
Si	−3.8	9.3	1.35	9.2	1.33	40.0	5.8	39.6	5.75
Zn	−6.0	6.6	0.95	29	0.42	20.7	3.0	15.2	2.2
Cu	−10.7	16.2	2.35	13.8	2.0	88.3	12.8	43.1	6.25
Mn	−11.3	(e)	(e)	30.3	4.4	(e)	(e)	53.8	7.8
Mg	+11.8	17.2	2.5	18.6	2.7	51.0	7.4	50.3	7.3

Note: (a) Listed in order of increasing percent different in atomic radii; (b) Some property-percent addition relationships are nonlinear. Generally, the unit effect of smaller additions are greater; (c) Increase in yield strength (0.2% offset) for 1% (atomic or weight basis) alloy addition; (d) increase in ultimate tensile strength for 1% (atomic or weight basis) alloy addition; and (e) 1 at % manganese is not soluble.

Source: From Davis [12].

TABLE 7.4   Solid-Solution Effects on Strength of Principal Solute Elements in Superpurity Aluminum

Strengthening Mechanism

Aluminum alloys can be strengthened by solid solution formation, second-phase microstructural constituents, and dispersoid precipitates to increase strength, hardness, and resistance to wear, creep, or fatigue. These are affected by alloying elements, alloy phase diagrams, solidification microstructure (dendrite arm spacing, grain refinement, modification, porosity, etc.), and heat treatment. The solid solution hardening effect of the solute (alloying element) is governed by the difference in atomic radii of the solvent (Al) as shown in Table 7.4 for superpurity binary solid solution alloys [12]. For multiple solutes in solid solution, the effects are somewhat less than additive. This is mainly applicable to non-heat-treatable alloys. Strengthening by second-phase constituents are also applicable to non-heat-treatable alloys. Elements such as Ni, Fe, Ti, Mn, and Cr have low solid solubility and form second-phase constituents (intermetallic phases) in the as-cast condition.

Effect of solidification microstructure and heat treatments on strength properties are discussed later in detail.

Structure Control

Microstructural characteristics that strongly affect the mechanical properties of aluminum castings are

- Dendrite arm spacing
- Grain size and shape
- Eutectic modification and primary phase refinement
- Intermetallic phases (size, form, and distribution)

Dendrite Arm Spacing

In all commercial processes, with the exception of semisolid forming, solidification takes place through the formation of dendrites from liquid solution. The cells obtained

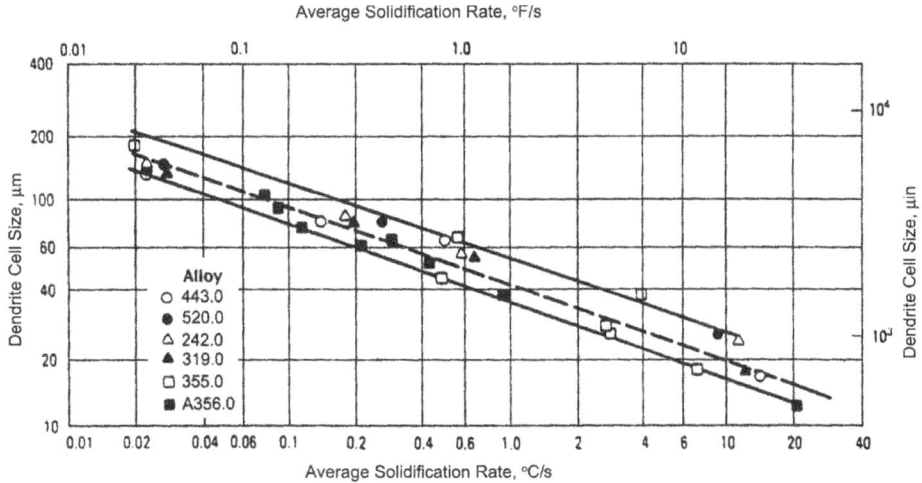

FIGURE 7.8   Dendrite arm spacing and dendrite cell size as a function of local solidification rate. (From Kaufman and Rooy [13].)

within the dendrite structure correspond to the dimensions separating the arms of primary and secondary dendrites and are exclusively controlled for a given composition by solidification rate (Fig. 7.8).

There are at least three measurements used to describe dendrite refinement:

- *Dendrite arm spacing.* The distance between developed secondary arms.

- *Dendrite cell interval.* The distance between centerlines of adjacent dendrite cells.

- *Dendrite cell size.* The width of individual dendrite cells.

The secondary dendrite arm spacing (SDAS) is a simple inverse cube root of the solidification time. The effect of the various conventional casting processes (and hence the cooling rate) on the dendrite cell size is shown in Table 7.5.

Figure 7.9 illustrates the improvement in mechanical properties due to change in dendrite cell size by solidification rate [1]. It is important to note that solidification rate

Casting Processes	Cooling Rate		Dendrite Arm Spacing	
	°C/s	°F/s	μm	Mils
Plaster, investment	1	1.80	100–1000	3.94–39.4
Greensand, shell	10	18.0	50–500	1.97–19.7
Permanent mold	100	180.0	30–70	1.18–2.76
Die	1000	1800	5–15	0.20–0.59

Source: From Kaufman and Rooy [13], p. 40.

TABLE 7.5   DAS for Various Conventional Casting Processes

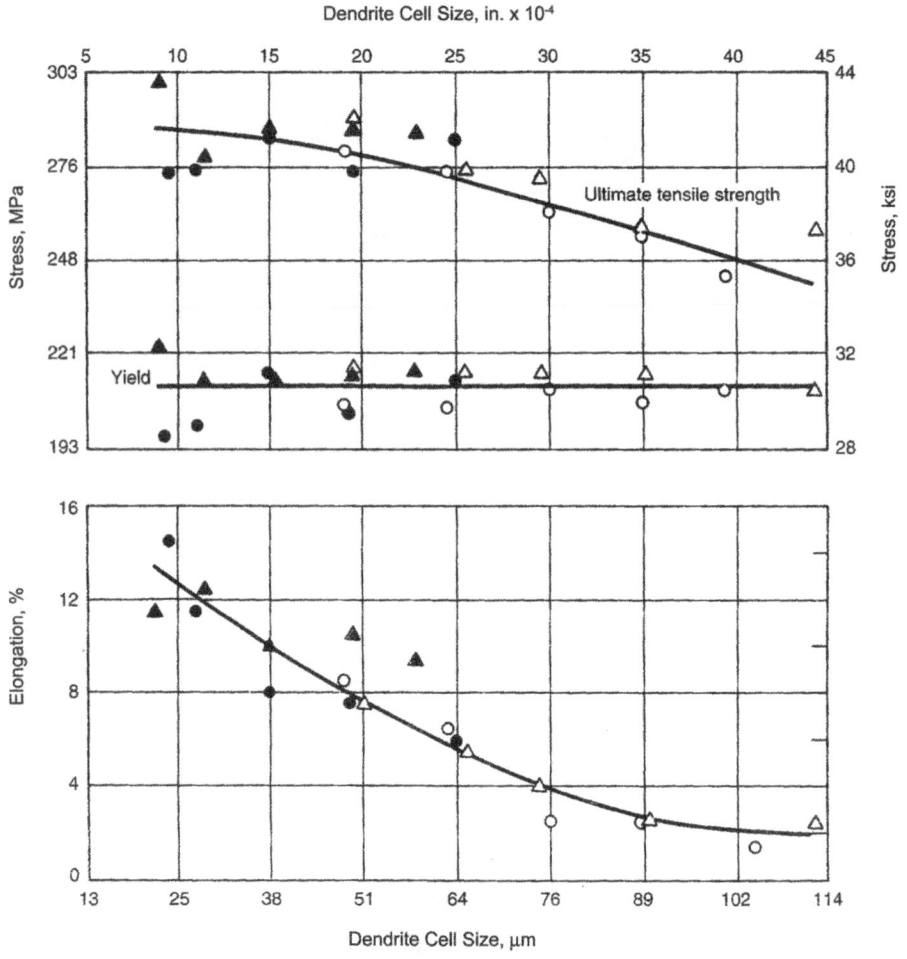

FIGURE 7.9   Tensile properties versus dendrite cell size for four heats of aluminum alloy A356-T62 plaster cast plates. (From *ASM Handbook* [1].)

also affects the morphology of the Al-Si eutectic including size and also the size and distribution of intermetallic phases. However, the DAS is taken as a measurable indication of the solidification rate.

Grain Structure and Grain Refinement

Under normal solidification conditions spanning the full range of commercial casting processes, aluminum alloys without grain refiners exhibit coarse columnar and/or coarse equiaxed structures. However, fine, equiaxed grains are desired for the best combination of strength and ductility by maximizing grain-boundary surface area and more finely distributing grain-boundary constituents [1]. Figure 7.10 illustrates the coarse and fine grain structures in an Al-Cu alloy [18]. A fine grain structure promotes the formation of

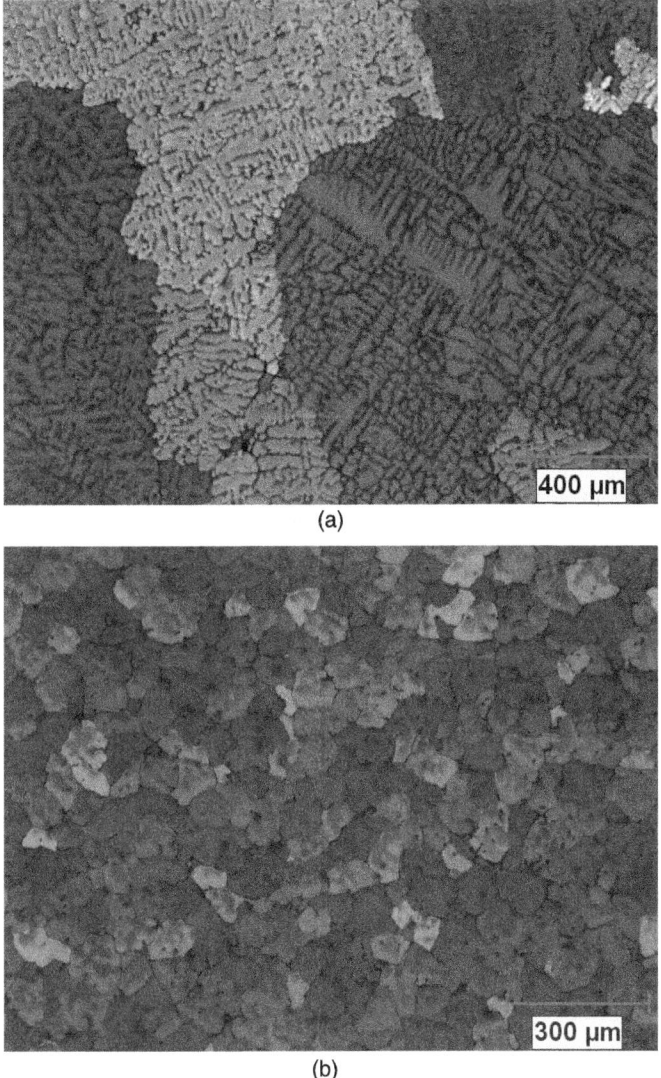

(a)

(b)

Figure 7.10 Anodized microstructures of 206.0 (*a*) before, and (*b*) after grain refinement. (From Fasoyinu et al. [18].)

finer, more evenly distributed intermetallic phases with corresponding improvement in feeding characteristics. Porosity, if present, is of smaller discrete void size in fine-grain parts. Other advantages of grain refinement are

- Improved feeding characteristics
- Increased hot tear resistance
- Improved mechanical properties
- Increased pressure tightness

- Improved response to thermal treatment
- Improved appearance following chemical, electrochemical, and mechanical finishing

The type and size of grains formed are functions of alloy composition, solidification rate, and the concentration of effective grain nucleation sites provided by the use of grain refiners. Grain refinement can be achieved by employing faster solidification rates as in permanent mold casting or high-pressure die casting and through the use of suitable grain refiners. The most widely used grain refiners are master alloys of titanium, or of titanium and boron, in aluminum. Aluminum-titanium refiners generally contain from 3 to 10% Ti. The same range of titanium concentrations is used in Al-Ti-B refiners with boron contents from 0.2 to 1% and titanium-to-boron ratios ranging from about 1 to 50 [13]. It should be noted that 1:1 ratio Ti:B master alloys, and even pure B, have been shown to be effective grain refiners of Al-Si alloys. They are less effective on non-Si- containing foundry or wrought compositions. Recent developments also include master alloys which include both grain refinement elements and compounds as well as Sr for modification. Titanium and aluminum carbides are also considered effective in the nucleation of grains. Ti-carbide-based grain refiners are used in operations such as direct chill casting in which there is a short residence time between addition in a trough and solidification as a billet as they will eventually convert to Al carbides if left in holding furnaces for long periods [1].

To be effective, grain refiners must introduce controlled, predictable, and operative quantities of aluminides and borides or carbides in the correct form, size, and distribution for grain nucleation. Refiners in rod form, developed for the continuous treatment of aluminum in primary operations and displaying clean, fine, unagglomerated microstructures are available in sheared lengths for foundry use. In addition to grain-refining master alloys in waffle or rolled rod form, salts, usually in compacted form that react with molten aluminum to form combinations of $TiAl_3$ and TiB_2, are also available.

Transduced ultrasonic energy has been shown to provide degrees of grain refinement under laboratory conditions [19, 20]. No commercial use of this technology has been demonstrated. The application of this method to engineered castings can be a problem.

Despite much progress in understanding the fundamentals of grain refinement, no universally accepted theory or mechanism exists to satisfy laboratory and industrial experience [1]. It is known that $TiAl_3$ is an active phase in the nucleation of aluminum crystals, ostensibly because of similarities in crystallographic lattice spacing. Nucleation may occur on $TiAl_3$ substrates that are undissolved or precipitate at sufficiently high titanium concentrations by peritectic reaction. Grain refinement can be achieved at much lower titanium concentrations than those predicted by the binary Al-Ti peritectic point of 0.15%. For this reason, other theories, such as conucleation of the aluminide by TiB_2 or carbides and constitutional effects on the peritectic reaction, are presumed to be influential. Other findings also suggest the active role of more complex borides of the Ti-Al-B type in grain nucleation [21, 22].

Dissolved Ti, and other elements, can influence the grain size via the growth restriction factor (GRF). Essentially, each dissolved element is treated as a constitutional undercooling agent and the effects are summed according to

$$\text{GRF} = \sum (k_i - 1)m_iC_0 \tag{7.2}$$

where k_i and m_i are the distribution coefficient between liquid and solid and the slope of the liquidus line for the ith element, respectively, and C_0 is the solute concentration [1].

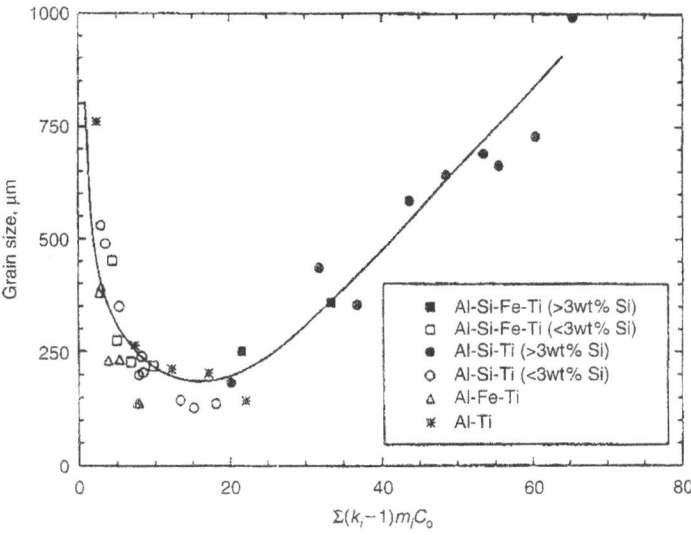

Figure 7.11 Growth restriction factor and grain size. Wrought aluminum compositions are generally to the left of the minimum in the curve and the high-silicon foundry alloys are to the right. (From *ASM Handbook* [1].)

Ti makes a strong contribution to the growth restriction factor. The impact of Ti is strongly influenced by the Si content, however, since at low Si levels characteristic of most common wrought alloys, the grains are growing in a plane front or cellular manner. Growth restriction will lead to greater undercooling and, hence, increased nucleation. High Si alloys, on the other hand, are solute rich and grow in a completely dendritic manner. As solute is efficiently directed into the interdendritic channels, there is less undercooling. The combination of factors leads to the result shown in Fig. 7.11 in which grain size decreases with increasing growth restriction factor for dilute compositions typical of most wrought Al alloys, and then switches over and increases as the growth restriction factor reaches the higher values typical of the more highly alloyed Al-Si and other common foundry alloys [1].

Grain refinement is the result of two separate processes: nucleation of new crystals from the melt, and the subsequent growth of these crystals to a limited size. Both these processes need a driving force that has to be supplied to the system via undercooling and/or supersaturation in relation to the equilibrium conditions of the actual system. During solidification, a sample undercools to a certain temperature, normally 3 to 4°C for aluminum alloys, in order for grains to nucleate. Once these grains have grown to a reasonable size, the sample temperature increases to the actual growth temperature; this is known as recalescence. At this point, new grains form. When a grain refining addition is made, there is no undercooling before the growth temperature is reached and therefore, there is no recalescence. Hence, observing the liquidus reaction of the time-temperature plot during solidification can indicate whether the resulting casting is grain refined or not. A schematic of the cooling curve is shown in Fig. 7.12 [23]. The cooling observed for grain refined and unrefined alloys are marked 1 and 2, respectively. The undercooling $\Delta\theta$ is an indication of the grain refinement. $\Delta\theta$ is large for an unrefined alloy as shown in Fig. 7.13 for alloy 356 [24].

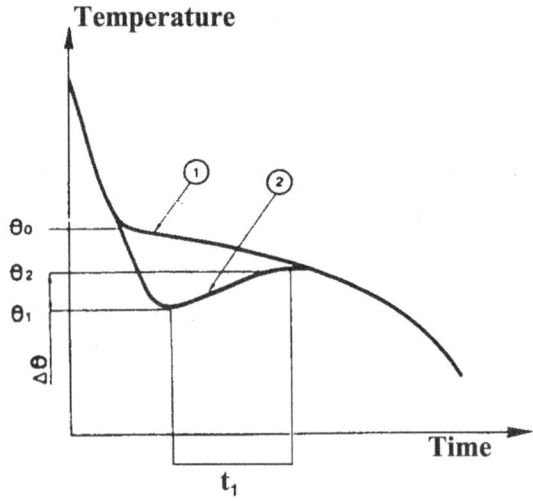

FIGURE 7.12   The cooling curve at the beginning of solidification: θ_0 = temperature at start of freezing of a well-refined alloy; θ_1 = temperature at start of freezing of an unrefined alloy; $\Delta\theta = \theta_2 - \theta_1$ the apparent supercooling; and t_1 = period of apparent supercooling. (From Gruzleski and Closset [23].)

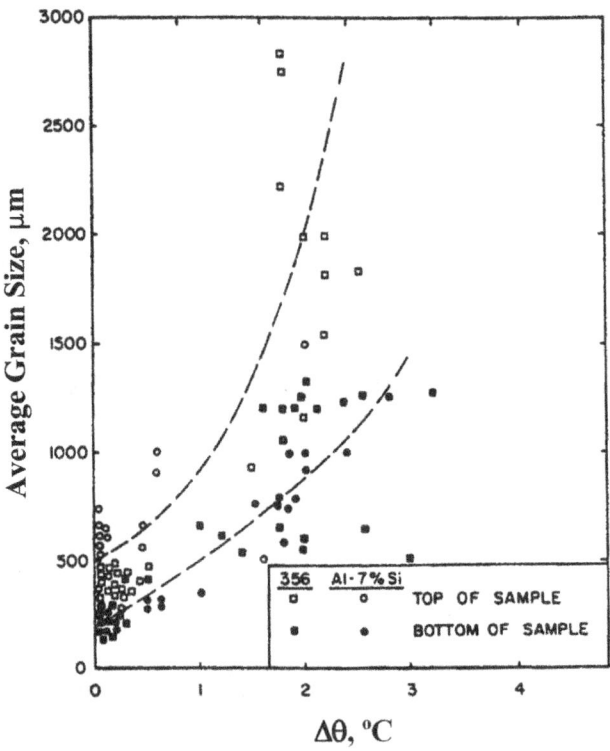

FIGURE 7.13   Average grain size as a function of $\Delta\theta$. (From Apelian et al [24])

Modification of Aluminum–Silicon Alloys

Cast components are produced from both hypo- and hypereutectic Al-Si alloys. The silicon morphology in these two groups of alloys should be modified to improve the mechanical properties. In case of hypoeutectic and eutectic alloys, the acicular and flaky form of silicon can be modified to a more fibrous eutectic structure by increasing solidification rate and by the addition of chemical modifiers (Figs. 7.4 and 7.14). Calcium, sodium, strontium, and antimony are known to influence the degree of eutectic modification that can be achieved during solidification. Figures 7.15 and 7.16 illustrate modification achieved by growth conditions and P-modifier additions for hypo- and hypereutectic Al-Si alloys [26].

Sodium is arguably the most potent modifier followed by strontium and calcium, but its effects are transient because of oxidation and vapor pressure losses. Strontium is less transient but may be less effective for modification under slow solidification rates (Fig. 7.17). Very low sodium concentrations (~0.001%) are required for effective

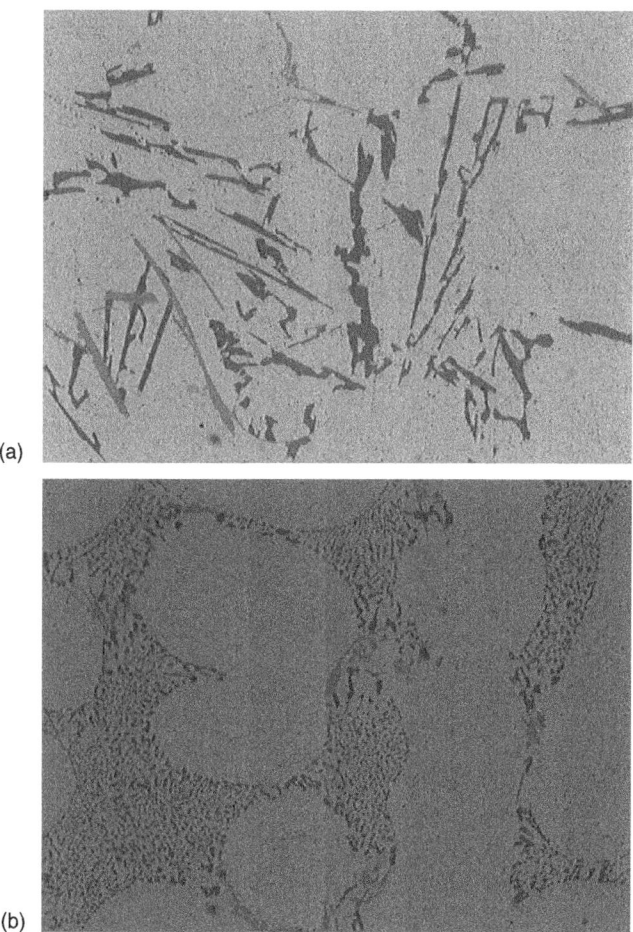

(a)

(b)

Figure 7.14 Optical micrographs of 356 alloy showing unmodified and Sr-modified conditions, ×500. (From Shankar [25].) (a) Unmodified, (b) 0.009% Sr and (*Continued*)

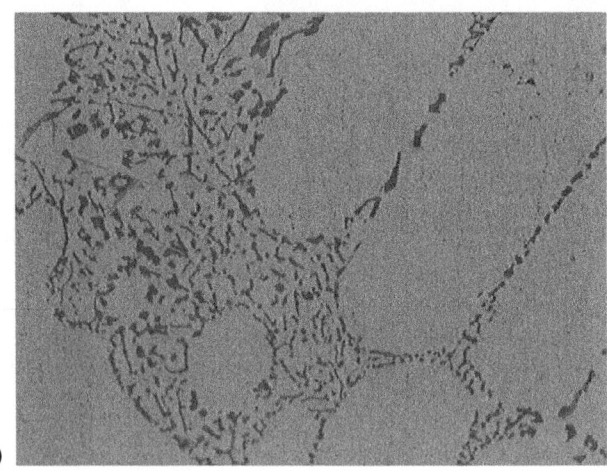

(c)

Figure 7.14   (c) 0.19% Sr. *(Continued)*

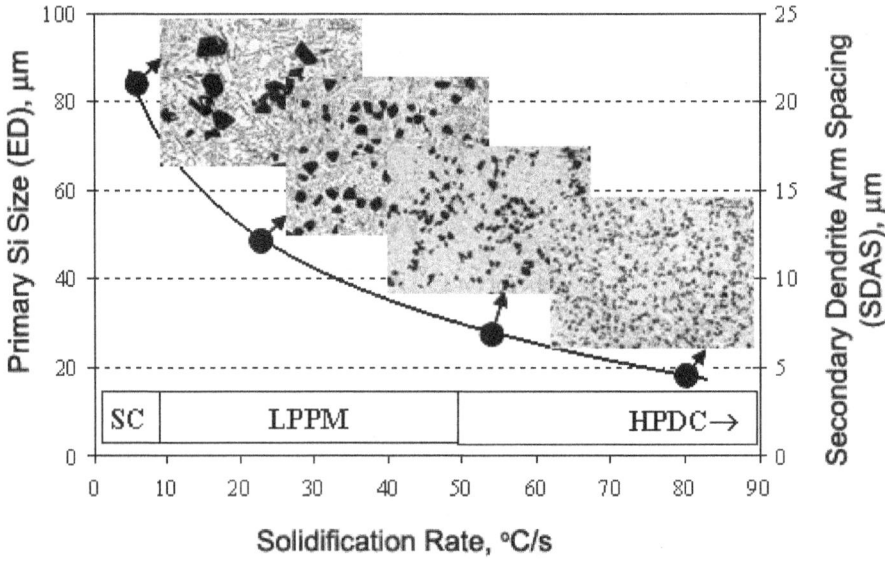

Figure 7.15   Microstructure refinement as a function of the cylinder block solidification rate established based on advanced laboratory solidification experiments. Note that the typical solidification rate for sand casting (SC), low-pressure permanent mold (LPPM), and high-pressure die casting (HPDC) are noted. (From Kasprzak et al. [26].)

modification. More typically, additions are made to obtain a sodium content in the melt of 0.005 to 0.015% in order to obtain practical casting times between remodification additions.

The combination of sodium and strontium offers advantages in initial effectiveness. Calcium is a weak modifier with little commercial value. Antimony provides a sustained effect, although the result is a finer lamellar rather than fibrous

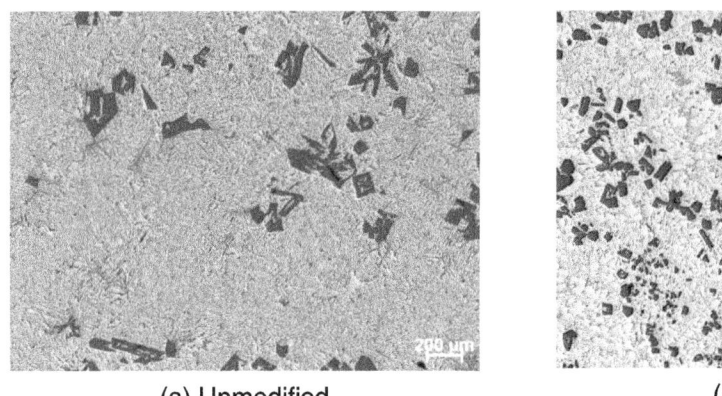

(a) Unmodified   (b) P-modifed

FIGURE 7.16 Optical micrographs of Al-20% Si alloy HPDC cylinder block showing microstructures for unmodified and P-modified conditions. (*a*) Unmodified and (*b*) P-modified (From Kasprzak et al. [4].)

eutectic. As a result, although permanent, the end, result obtained via antimony is not as good. In fact, particularly at low freezing rates, very large lamellar structures can arise. Antimony may give worse results than even unmodified alloys. Antimony is not compatible with other modifying elements.

A much wider range of strontium concentrations is in use. In general, addition rates far exceed those required for effective sodium modification. An operating range of 50 ppm wide between 0.004 and 0.020% (40 and 200 ppm) is the modern standard industry practice. Normally, good modification is achievable in the range of 0.004 to 0.015% Sr. Remodification through strontium additions may be required, although retreatment is less frequent than for sodium particularly at the lower Sr levels where fade rates can be quite low. Strontium additions are usually made through master alloys containing either 10 or 90% Sr in aluminum.

Modifier additions are usually accompanied by an increase in porosity. In the case of sodium and calcium, the reactions involved in element solution are invariably turbulent or are accompanied by compound reactions that by their nature increase dissolved hydrogen levels and entrained oxide levels. The use of hygroscopic salts including NaCl and NaF for modification also risks oxide formation and increased dissolved hydrogen content.

Typically, modified structures display higher tensile properties and appreciably improved ductility when compared to unmodified structures (Table 7.6). Property improvement is dependent on the degree to which porosity associated with the addition of modifiers is suppressed. Improved casting results include improved feeding and superior resistance to elevated-temperature cracking.

It is possible to achieve a state of overmodification, in which eutectic coarsening occurs, when sodium and/or strontium are used in excessive amounts (Fig. 7.14*c*). Reduced fluidity and susceptibility to porosity-related problems are usually encountered well before overmodification may be experienced.

It has been well established that phosphorus interferes with the modification mechanism. Phosphorus reacts with sodium and probably with strontium and calcium to

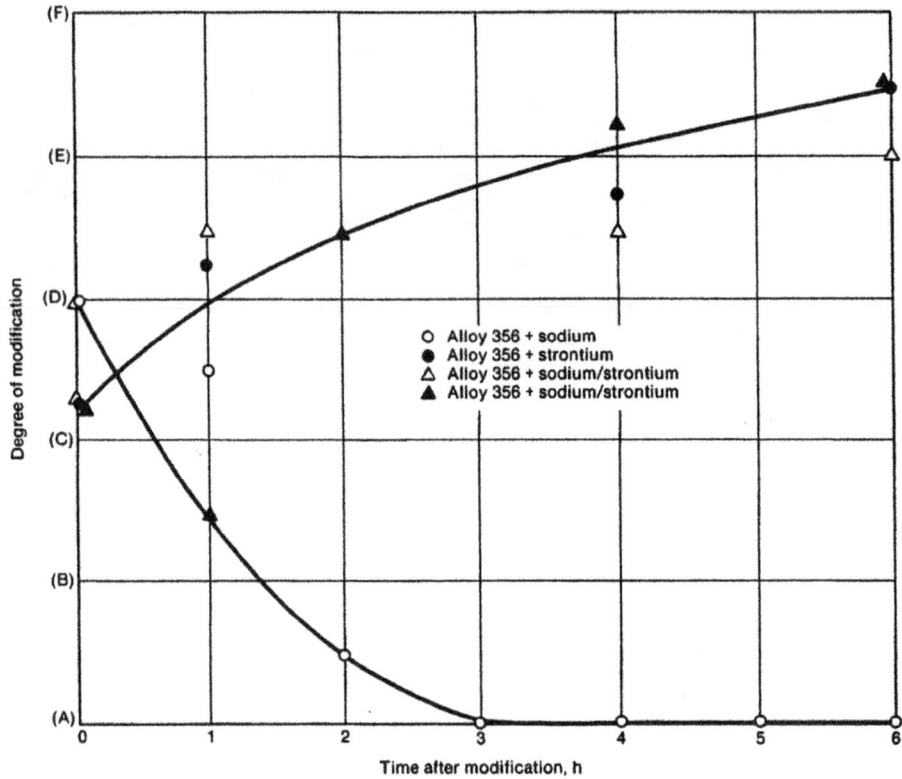

FIGURE 7.17 Effectiveness of sodium and strontium modifiers as a function of time. (From Kaufman and Rooy [13].)

form phosphides that nullify the modification additions. It is, therefore, desirable to use low-phosphorus metal when modification is a process objective and/or to make larger modifier additions to compensate for phosphorus-related losses. Primary producers control phosphorus contents in smelting and processing through the judicious choice of feed materials to provide less than 0.001% of phosphorus in alloyed ingot.

Thermal analysis is useful in assessing the degree of modification that can be displayed by the melt. This is shown schematically in Fig. 7.18 [23]. The effectiveness of modification treatment is defined by the degree and duration of undercooling at the solidus. With modification, the eutectic temperature is depressed, the undercooling for nucleation of the eutectic is increased, and the period of this undercooling is lengthened. Test results must be correlated with the degree of modification established metallographically for the castings since cooling rates for the sample will differ.

The feature most used in thermal analysis control of modification is the depression of the eutectic temperature by 6 to 8°C from the unmodified state (Fig. 7.19). Since the eutectic temperature is easy to measure, it is most often employed by foundries to assess whether a melt is properly modified or not.

Alloy and Temper	Product	Modification Treatment	Tensile Yield Strength		Ultimate Tensile Strength		Elongation %
			Ksi	MPa	Ksi	MPa	
13% Si	Sand cast test bars	None	–	–	18.0	124	2.0
		Na-modified	–	–	28.0	193	13.0
	Permanent mold test bars	None	–	–	28.0	193	3.6
		Na-modified	–	–	32.0	221	8.0
359.0	Permanent mold test bars	None	–	–	26.1	180	5.5
		0.07% Sr	–	–	30.5	210	12.0
356.0-T6	Sand cast test bars	None	30.1	208	41.9	289	2.0
		0.07% Sr	31.6	218	42.2	291	7.2
	Bars cut from chilled sand casting	None	30.9	213	41.2	293	3.0
		0.07% Sr	31.6	218	32.2	291	7.2
A356.0-T6	Sand cast test bars	None	26.0	179	40.0	226	4.8
		0.01% Sr	30.0	207	43.0	297	8.0
A444.0-T4	Permanent mold test bars	None	–	–	21.9	151	24.0
		0.07% Sr	–	–	21.6	149	30.0
A413.2	Sand cast test bars	None	16.3	112	19.8	137	1.8
		0.005-0.05% Sr	15.6	108	23.0	159	8.4
	Permanent mold test bars	None	18.1	125	24.4	168	6.0
		0.005–0.08% Sr	18.1	125	27.7	191	12.0
	Test bar cut from auto wheel	0.05% Sr	17.5	121	28.0	193	10.6
		0.06% SR	18.2	126	28.0	193	12.8

Source: From Kaufman and Rooy [13].

TABLE 7.6 Typical Mechanical Properties of Modified and Unmodified Cast Aluminum Alloys

Refinement of Hypereutectic Aluminum-Silicon Alloys

Hypereutectic Al-Si alloys commonly used in the foundry industry contain 14 to 20% Si. Microstructure of such alloys is characterized by coarse primary silicon crystals that are harmful in the casting and machining of components [13, 26–28]. Normal practice is to add phosphorus to molten alloys containing more than the eutectic concentration of silicon, made in the form of metallic phosphorus or phosphorus-containing compounds such as phosphor-copper and phosphorus pentachloride which lead to a marked effect on the distribution and form of the primary silicon phase (Fig. 7.16). Retained concentrations

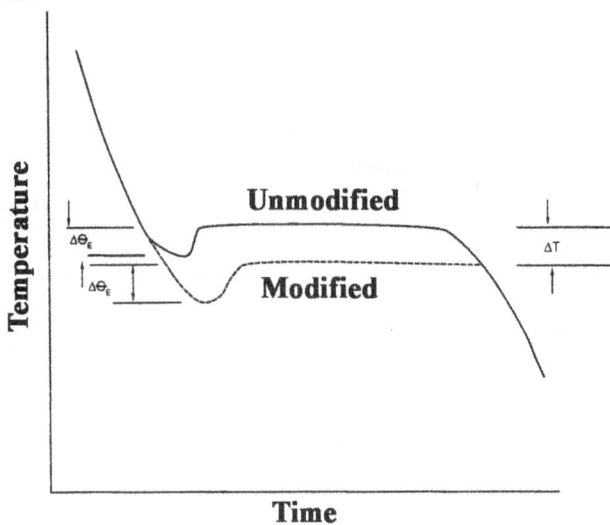

Figure 7.18 A comparison of the cooling curves of the eutectic regions of modified and unmodified alloys. (From Gruzleski and Closset [23].)

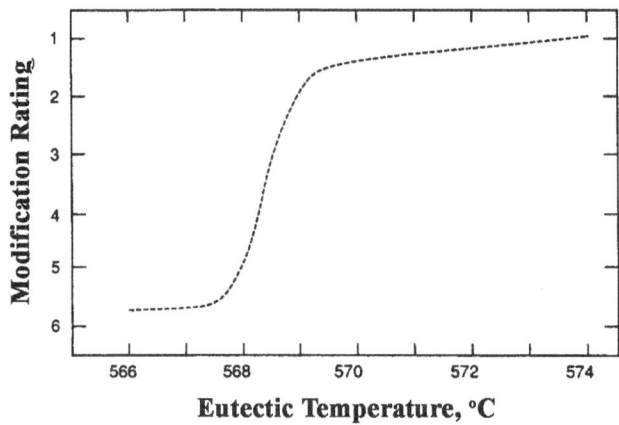

Figure 7.19 Relation between modification rating and the eutectic temperature for A356 alloy. (From Gruzleski and Closset [23].)

of phosphorus as low as 0.0015% are effective in achieving refinement of the primary phase. Primary silicon crystals can also be modified by increasing the cooling rate as shown in Fig. 7.15 and comparing Figs. 7.16a and b [28].

Phosphorus content can vary between 0.0015 and 0.03%. Solidification of phosphorus-treated melts, cooling to room temperature, reheating, remelting, and resampling in repetitive tests have shown that refinement is not lost; however, primary silicon particle size increases gradually, responding to a loss in phosphorus concentration. Common degassing methods accelerate phosphorus loss, especially when chlorine or freon is used. In fact, brief inert gas fluxing is frequently used to reactivate aluminum phosphide nuclei, (which is responsible for silicon refinement) presumably by resuspension.

Practices recommended for melt refinement are as follows:

- Melting and holding temperature should be minimum.

- Calcium and sodium contents should be controlled to low concentration levels.

- Brief nitrogen or argon fluxing after the addition of phosphorus is recommended to remove the hydrogen introduced during the addition and to distribute the aluminum phosphide nuclei uniformly in the melt.

Porosity

Internal porosity in complex aluminum castings has detrimental effects on the mechanical properties and pressure tightness. Porosity in aluminum is caused by the precipitation of hydrogen from liquid solution or by shrinkage during solidification, and more usually by a combination of these effects.

Hydrogen is the only gas that is appreciably soluble in aluminum and its alloys. The main source of hydrogen is the water vapor in the atmosphere, which reacts with aluminum by the reaction [23, 29]:

$$2Al(l) + 3H_2O(g) \rightarrow Al_2O_3(s) + 3H_2 \qquad (7.3)$$

Other sources of hydrogen include

- Fluxes, which are often hygroscopic and contain chemically attached water

- Solid additives, such as scrap, grain refiners, or even virgin ingots. All of these may contain adsorbed moisture or entrapped gases originating in a prior processing operation

- Tools, crucibles, furnace linings, and mold materials, which, if not preheated, may contain moisture

- Scraps, castings, machine turnings and borings, die-cast trim press scrap, gates and risers, and sand and other nonmetallic molding material debris.

- Fuel-fired furnaces (natural gas, oil) which have hydrogen available from fossil fuels

- Mold materials

Hydrogen solubility in both liquid and solid pure aluminum follows Sievert's law, which states that

$$\text{Solubility} = K_s\sqrt{P}_{GAS} \qquad (7.4)$$

where K_s is Sievert's constant and P_{GAS} is the partial pressure of the gas in the atmosphere above the melt.

Since solubility of hydrogen is directly proportional to the square root of the hydrogen partial pressure, a lowering of the partial pressure by 25 times results in only a fivehold solubility decrease [29].

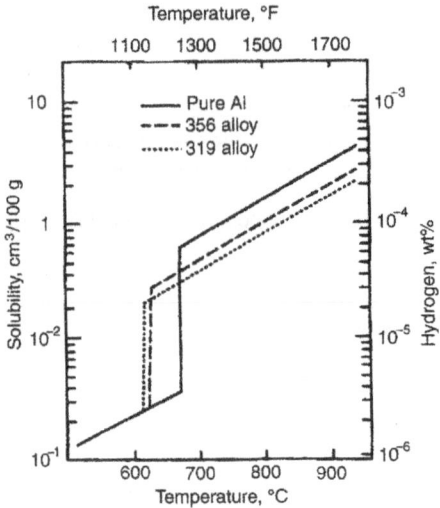

FIGURE 7.20 Hydrogen solubility in aluminum casting alloys. (From *ASM Handbook* [1, p. 185].)

Temperature is the most important factor affecting hydrogen solubility. As shown in Fig. 7.20 [1], hydrogen solubility is considerably greater in the liquid than in the solid state. Actual liquid and solid solubilities in pure aluminum just above and below the solidus are 0.69 and 0.04 ppm (cm^3/100 g) at the melting point (660°C or 1220°F) [29].

This confirms the facts that the solubility of hydrogen in liquid pure aluminum is more than 16 times the solubility of hydrogen in solid aluminum, and the equilibrium partition coefficient of hydrogen in aluminum is significantly less than 1.

Alloying elements have a strong and varying influence on the solubility of hydrogen in aluminum. For example, in liquid aluminum, silicon, copper, zinc, and iron decrease hydrogen solubility, whereas lithium, magnesium, and titanium increase it as shown in Fig. 7.21 [1].

Shrinkage Porosity

Shrinkage porosity is caused by volumetric contraction during solidification, especially from the liquid to the solid state which ranges from 3 to 8.5% for aluminum alloys. It is also affected by the solidification range of the alloy.

Shrinkage can appear as distributed voids or microshrinkage because of failure during the last stages of interdendritic feeding. The other form of shrinkage is centerline or piping voids resulting from gross directional effects. Short freezing range alloys such as 356.0 and 413.0 exhibit extensive piping as opposed to distributed shrinkage porosity due to directional solidification. By contrast, long freezing range alloys are prone to extensive microporosity resulting from interdendritic feeding. Effective gating and

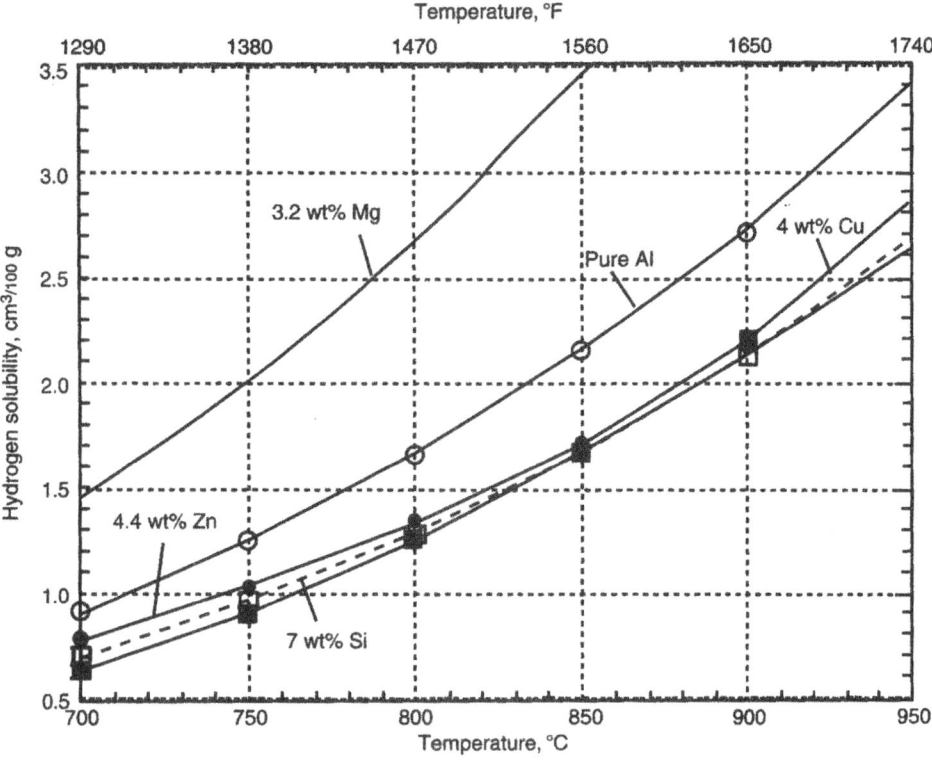

Figure 7.21 Hydrogen solubility in liquid pure aluminum and binary aluminum alloys at 1 atm hydrogen partial pressure. (From *ASM Handbook* [1, p. 68].)

risering design together with use of chills to promote directional solidification is likely to minimize the formation of microporosity.

Improved modification and refinement of aluminum-silicon alloys, improved grain refinement, and reduced oxide contents all improve feedability and therefore reduce shrinkage severity.

Typical porosity in cast aluminum alloys is shown in Fig. 7.22 [23].

Inclusions

Two types of inclusions are found in castings. These are exogenous and indigenous. The former originates from outside the bulk melt and consists of refractory bits broken off from furnace linings or crucible, flux particles or pieces of slags or dross, broken ceramic filters or loose sand left in the gating system. Indigenous inclusions are a result of the chemical reaction between the melt and some other chemical species.

Alumina (Al_2O_3) is the most common inclusion in aluminum alloy castings since aluminum oxidizes readily in liquid and solid states. Oxidation rate is greater at molten metal temperatures and increases with temperature and time of exposure. Magnesium

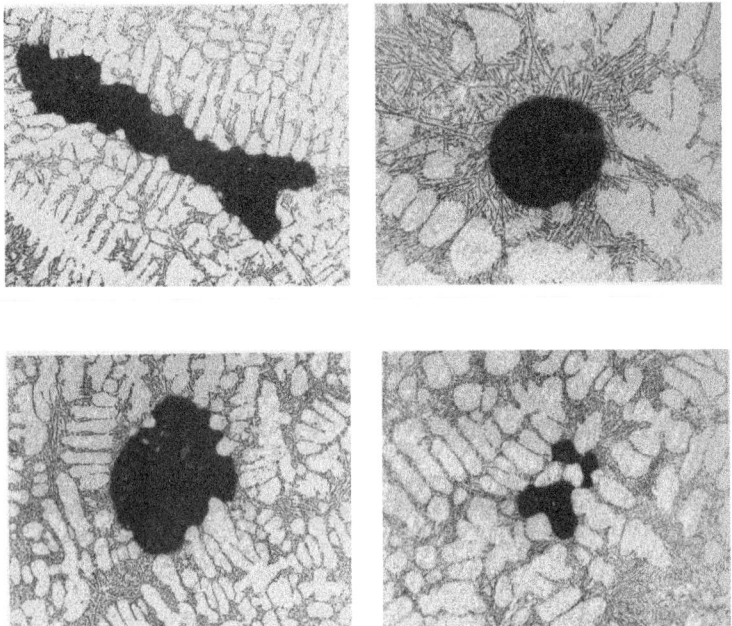

Figure 7.22 Typical porosity in cast aluminum alloys. (From Gruzleski and Closset [23].)

in aluminum alloys oxidizes and with time and temperature reacts with oxygen and aluminum oxide to form spinel as shown below [29]:

$$2Mg + O_2 \rightarrow 2MgO \tag{7.5}$$

$$MgO + Al_2O_3 \rightarrow MgAl_2O_4 \tag{7.6}$$

Liquid aluminum being a powerful oxidant can react with various refractory materials such as silica to form small alumina inclusions as shown below:

$$3SiO_2 + 4Al \rightarrow 2Al_2O_3 + 3Si \tag{7.7}$$

Other sources of inclusion due to chemical reaction are those between magnesium and aluminum chloride gas (due to halide flux treatment), aluminum—methane reaction due to combustion of natural gas and decomposition of hexachloroethane tablets [29].

$$2AlCl_3(g) + 3Mg \rightarrow 3MgCl_2(l) + 2Al \tag{7.8}$$

$$3CH_4 + 4Al \rightarrow Al_4C_{3(s)} + 12H \tag{7.9}$$

$$3C_2Cl_6 + 14Al \rightarrow 2Al_4C_{3(s)} + 6AlCl_{3(g)} \tag{7.10}$$

These inclusions are categorized in Table 7.7 [13].

While the oxide that initially forms on the surface of molten aluminum is highly protective and self-limiting, any agitation or turbulence in the treatment and handling of molten aluminum increases the risk of oxide entrainment and the immediate reformation of additional oxides. Oxide concentration can increase when alloying additions

Classification	Types Observed	Potential Source(s)
Nonmetallic exogenous	Various refractory particles, Al_4C_3, etc.	Refractory degradation, remelt ingot, refractory/metal reactions
Nonmetallic in situ	MgO, Al_2O_3 films, clusters, and dispersoids; $MgAl_2O_4$ films and clusters	Melting, alloying metal transfer turbulence
Homogenous halide salts	$MgCl_2$-NaCl-$CaCl_2$, etc.	Poor separation of fluxing reaction products
Particle/salt	$MgCl_2$-NaCl-$CaCl_2$/MgO, etc.	Salt generated during chlorine fluxing of magnesium-containing alloys, filter and metal-handling system releases

Source: From Kaufman and Rooy [13]., p. 51.

TABLE 7.7   Inclusion Sources and Types in Aluminum Alloy Castings

are stirred into the melt, when reactive elements and compounds are immersed, when metal is drawn for pouring, and when metal is poured and conducted by the gating system into the molt cavity. Induction melting is highly energy efficient and effective for melting fines and poor-quality scrap, but electromagnetically induced eddy currents result in a high level of entrained oxides.

Care should be taken in pouring, gating, and minimizing the reoxidation of the metal to control inclusions in castings. Molten metal filtrations can be effective in minimizing the inclusion level. This is needed to produce high-integrity, high-strength castings where fracture toughness characteristics are important.

Hydrogen pore formation can be suppressed by removing oxides from the melt.

Degassing

Gas purging is the most common method of degassing a melt. The principle is illustrated in Fig. 7.23. A bubble of some gas is introduced into the bottom of a melt. The gas may be inert or reactive with the metal concerned, but initially it must contain no atoms of the gas species to be removed from solution. Since partial pressure of the gas to be removed is initially zero in the bubble, there exists a driving force for diffusion of gas atoms from the liquid into the introduced bubble. As this bubble rises through the bath, it will pick up progressively more and more of the undesirable dissolved gas until it finally escapes from the free surface of the melt [29].

The first application of this principle was through decomposition of hexachloroethane (C_2Cl_6) tablet where the tablet is plunged into the aluminum melt using a perforated bell-shaped steel plunger. When submerged and plunged into the melt, the hexachloroethane decomposes into its respective components. The chlorine component immediately forms a metastable aluminum chloride gaseous compound, which is insoluble in the melt and serves as the purge gas. With any purge gas, the monatomic hydrogen diffuses into the purge gas bubble, combines with another hydrogen atom to form molecular hydrogen, and the bubble rises to the surface, where both gaseous species are released. Despite the environmental concerns because of the chlorine gas, this method of degassing is still used in some foundries.

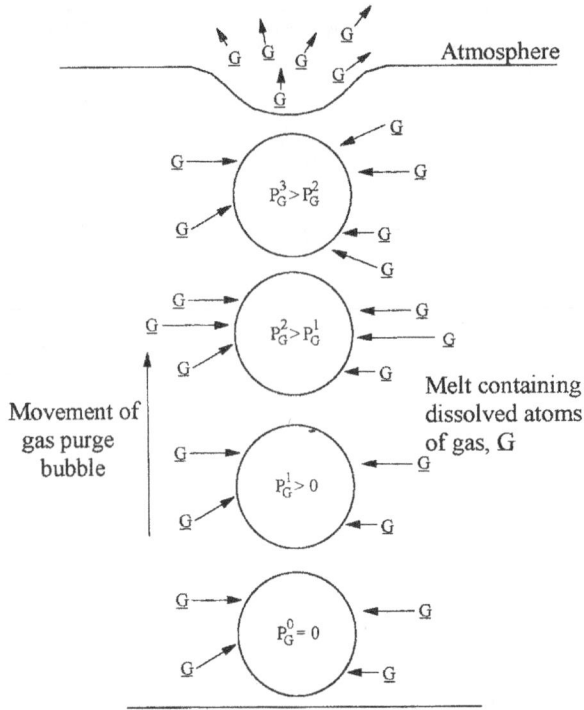

Figure 7.23 The principle of gas purging to remove dissolved gas. (From Gruzleski [29].)

The second method is purging an inert gas such as argon and nitrogen through a steel or graphite lance under low pressure (less than 207 kPa or 30 Psi) into the melt. The insoluble nitrogen or argon gas bubble then rises within the melt, and the hydrogen atoms diffuse to this process gas bubble and form a hydrogen molecule. When the bubble reaches the surface of the melt, the hydrogen is released. If the relative humidity in the atmosphere is particularly high, this will be a dynamic process because new hydrogen may become absorbed into the melt.

The lance degassing technology is relatively inefficient because of the large bubble size that emanates from the bottom of the lance, especially in the case of large vessels. It is important to produce a very fine dispersion of bubbles to be effective in a minimum of time. To this end, gas purging through a rotary impeller degasser consisting of a series of vanes which chops a gas stream into a mass of fine gas bubbles and then disperses them throughout the melt (Fig. 7.24). The effect of bubble size in degassing efficiency is shown in Fig. 7.25.

The rotary impeller degassing technique was introduced into the aluminum foundry industry in the mid-1980s and is used extensively in melting furnaces, transfer ladles, and in continuous-flow launder systems.

The basic components of a rotor degassing system include a mechanical drive unit to rotate the shaft/rotor assembly, a gas source, and controls to manipulate the gas injection into the system, and the shaft/rotor assembly itself. Equipment can be as simple as well-mounted drive unit or more complex, with a fixed-mast drive-up ladle treatment station or a mobile unit that can be moved from furnace to furnace.

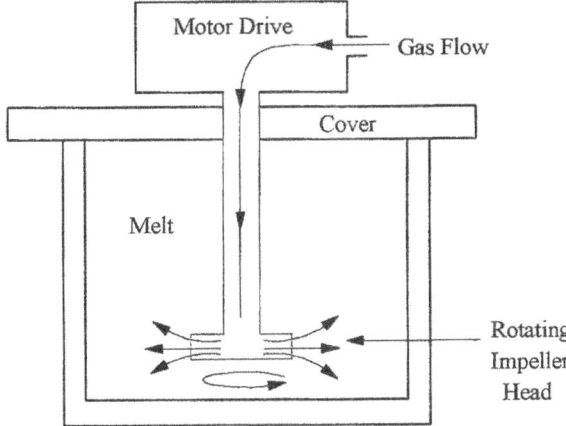

FIGURE 7.24 A schematic of a rotary impeller degassing unit is shown. The rotating impeller chops the gas stream into a mass of fine bubbles that provide very efficient degassing of the melt. (From Gruzleski [29].)

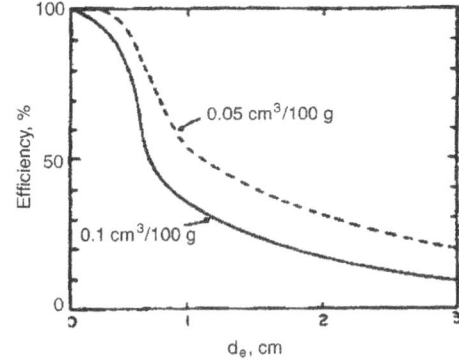

FIGURE 7.25 Degassing efficiency as a function of purge gas bubble size. (From *ASM Handbook* [1, p. 187].)

Drossing

Melting occurs most rapidly if a heel of molten aluminum is present when the charge is added. Melting down with minimum dross formation occurs when the charge is protected from combustion products and melting is rapid. Drossing is the formation of aluminum oxide and other oxides which accumulate on the melt surface. Complete separation of dross and metal would be favored by large differences in their specific gravities. Unfortunately, the specific gravities of the oxides and the molten metal are of similar magnitude, as revealed in Table 7.8. Some oxides float on the melt surface (i.e., dross) whereas others sink and form a sludge.

Compound	Specific Gravity (20°C)
Al_2O_3	3.99
$Al_2O_3 \cdot 3H_2O$	2.42
Al	2.70
MgO	3.65
Mg	1.74
Si	2.40
SiO_2	2.20–2.60
CuO	6.40
Cu_2O	6.0

Source: From Heine et al. [11].

TABLE 7.8   Specific Gravity of Some Materials in Drosses

Melting and Melt Handling

Melting Furnaces

The types of melting furnaces employed in aluminum foundries include the following [11]:

- Crucible furnaces, lift-out type
- Pot furnaces
  - Stationary, fuel fired
  - Titling, fuel fired
- Reverberatory furnaces, fuel fired, stationary, and titling types
- Barrel-type furnaces, fuel fired
- Induction furnaces, electrically operated
  - Low frequency
  - High frequency

Each of these furnaces has certain advantages. Fuel-fired furnaces are of two types: the indirect-flame type, in which the products of combustion do not come into direct contact with the metal, and the direct-flame type, in which there is direct contact between the combustion products and metal charge. Pot furnaces are usually indirect-flame or electrical-resistance furnaces whereas reverberatory furnaces are direct-flame furnaces. However, some constructions of pot and crucible furnaces approach the direct-flame conditions.

The charge material may range from prealloyed ingot of high quality to charge made up from low-grade scrap.

Crucible Furnaces

Crucible melting is a simple and flexible process. A cross section of a crucible furnace is shown in Fig. 7.26 [1] for a double pushout stationary type. Crucible capacity can vary

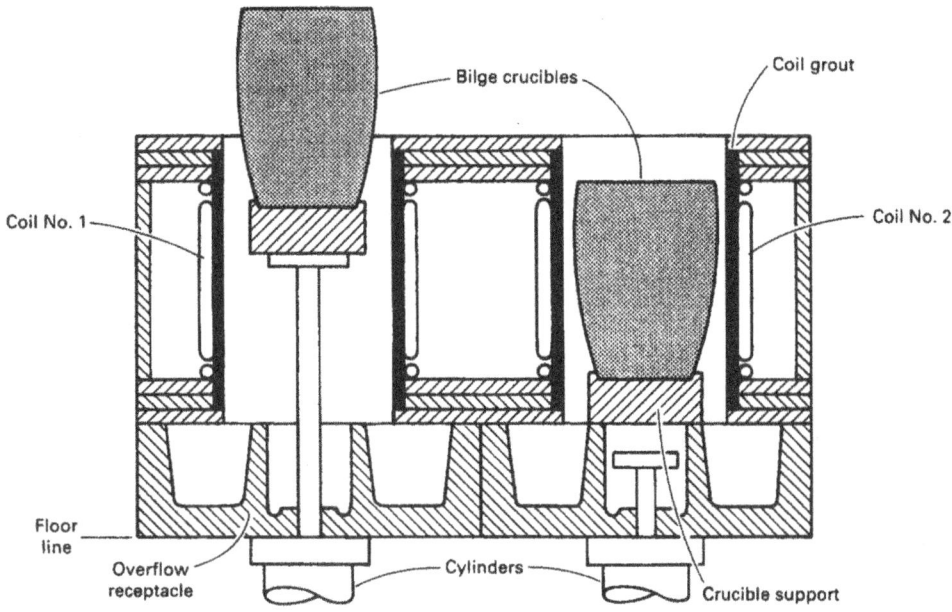

Figure 7.26 Cross section of a double push out stationary crucible furnace. (From *ASM Handbook* [1, p. 157].)

from a few grams for laboratory melts to 1360 kg (3000 lb) for large melts. Crucibles are made of a clay-graphite mixture or of silicon carbide. Fuel choices include electricity (resistance or induction), commercial gas (e.g., natural gas), and fuel oil.

Crucible furnaces can be classified as stationary, tilting, or movable. In case of stationary furnaces, molten metal is dipped out with hand ladles or the crucible containing molten metal is lifted out with tongs, placed on a ring shank, its temperature measured and casting is done directly from the crucible.

Tilting furnaces incorporate a crucible that is supported in the shell such that both are free to rotate on an axis. This allows the molten metal to be poured into ladles or molds. The axis of the pour can be either at the center of gravity or at the lip of the furnace (Fig. 7.27). The low electrical conductivity of clay-graphite crucible makes them ideal for use in such applications.

In the case of movable furnaces and/or pouring ladles, either melter or holder furnaces are cycled between the in-line pouring stations and melt and/or the postmelting temperature and quality adjustment operation. One can have two-, three-, or four-crucible melters that are arranged on a turntable. Usually, one furnace is in a holding mode and the remaining are used alternately for charging, stabilizing, and metal treatment [1, 7].

Pot Furnaces

Capacities of indirect-flame stationary-pot furnaces are limited by the cast iron or steel pot size to a relatively few hundred of pounds. Metal is ladled from the pot for pouring. Larger melts, up to 1360 kg (3000 lb), may be handled in tilting-pot furnaces. The melt is poured from the tilting furnace into ladles for distribution to the molds.

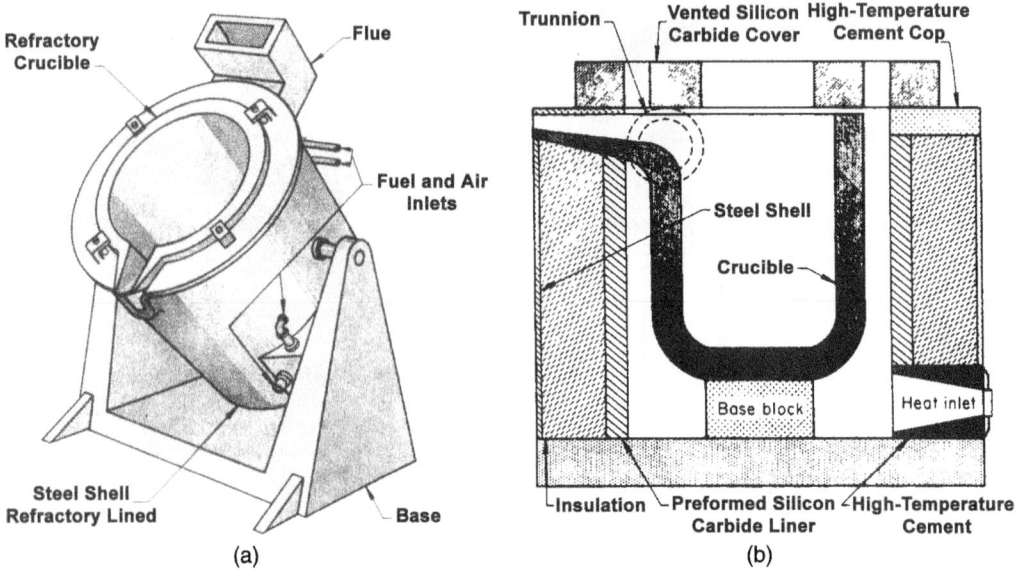

FIGURE 7.27 Two variations of a tilting crucible furnace: (a) center axis, (b) lip axis. (From *ASM Handbook* [1, p. 157].)

Reverberating Furnaces

Redesigned reverberating furnaces use natural gas and fuel oil for heating. Here the roof is slanted to bounce or reverberate the flame off the ceiling, down across the metal to be melted in the molten bath, and then on to the flux located at one end of the bath. Reverberatory furnaces are, of course, direct-flame furnaces, and the melt may therefore be subject to the extremes of drossing and gas absorption which can occur [7]. The capacity of the furnaces is typically 9070 to 45,350 kg (20,000 to 100,000 lb). Another widely used furnace for aluminum is the melter/holder furnace with a capacity of 2268 to 18,140 kg (5000 to 40,000 lb).

Induction Furnaces

Induction furnaces are more efficient and offer excellent metallurgical control coupled with its relatively pollution-free operation. These are of two types: coreless and channel.

In the coreless furnace, the refractory-lined crucible is completely surrounded by a water-cooled copper coil, while in the channel furnace the coil surrounds only a small appendage of the unit, called an inductor. The term "channel" refers to the channel that the molten metal forms as a loop within the inductor. It is this metal loop that forms the secondary of the electrical circuit, with the surrounding copper coil being the primary. In a coreless furnace, the entire metal content of the crucible is the secondary [1].

Coreless furnaces are used for melting and superheating, whereas a channel furnace is better suited for superheating, duplexing, and holding. Channel furnaces have been replaced by the coreless and resistance furnaces in aluminum-melting applications. A cross section of a coreless induction furnace is shown in Fig. 7.28 [1]. Coreless furnaces are of three types: high-frequency, low-frequency, and medium-frequency furnaces. Frequencies for induction furnaces vary depending upon the manufacturer of the equipment. In general, low frequency is 100 Hz or less, medium frequency is 200 to 500 Hz, and high frequency is more than 1000 Hz.

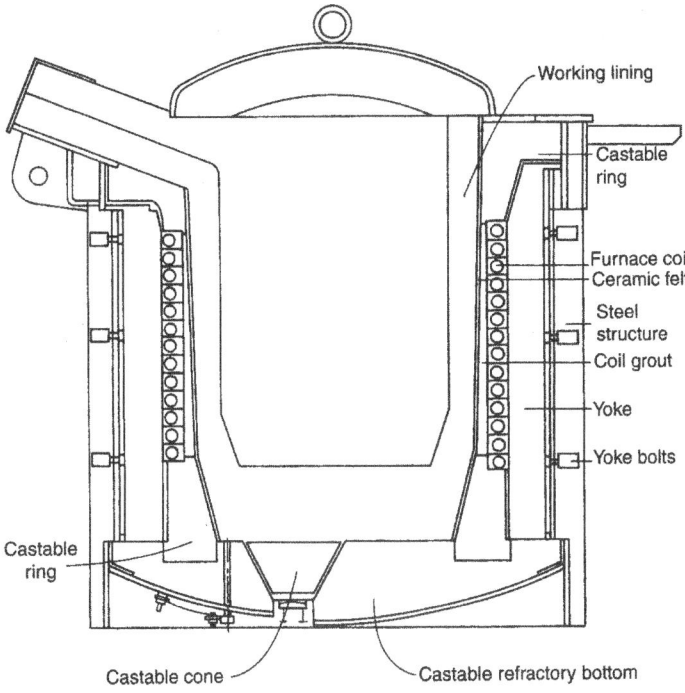

Figure 7.28 Cross section showing the coreless induction furnace. (From *ASM Handbook* [1, p. 109].)

High-frequency furnaces of the lift-coil type (Fig. 7.29) [1, p. 158] are limited by crucible size to about 90 lb heats of aluminum. A cross section of a tilt furnace for high-frequency induction melting is shown in Fig. 7.29 [1]. Motor-generator sets of 5 to 1000 kW providing frequencies up to 10,000 cycles may be used. Low-frequency furnaces in sizes ranging from 60 to 5000 lb of aluminum are available and are capable of melting 5 to 7 lb/h·kW rating of the furnace. Low-frequency furnaces have the characteristics that they must be started with a heel of molten metal, and so are emptied only when cleaning is necessary.

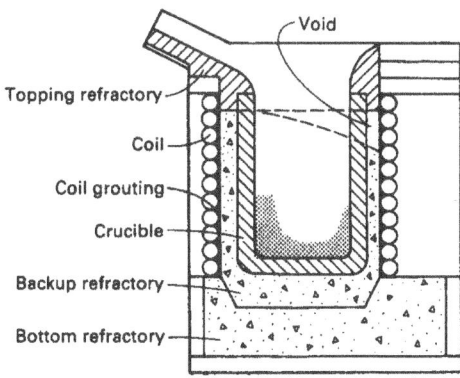

Figure 7.29 Cross section of a tilt furnace for the high-frequency induction melting of brass and bronze alloys. Crucible is of clay-graphite composition. Also shown are the locations of the molten metal buildup and the voids in the backup refractory, which shorten crucible life. (From *ASM Handbook* [1, p. 158].)

Casting Processes

Aluminum alloys can be cast by any of the commonly used processes: sand casting, plaster molding, lost foam casting, permanent mold and die casting, investment casting, and the recently developed ablation casting. Because of the low pouring temperature and specific gravity of aluminum alloys, molds are less affected by heat than in the case of iron and steel. Consequently, excellent surface finish and dimensional accuracy may be obtained even in large sand castings. The minimum section thickness for aluminum castings produced by the commonly used processes is illustrated in Table 7.9. In addition to minimum section thickness, one has to consider the casting (dimensional) tolerances for sand, permanent mold, die, investment casting, etc. These are dependent on section thickness. For detail information on tolerances, please refer to ASM Handbook, Volume 15, 2008, pages 629 and 1012–1018. Sand castings weighing several tons have been made. Permanent-mold castings weighing 354 kg (780 lb) and die castings up to 32 kg (70 lb) are in use. It is well known that the rapid chilling of the metal mold, and in the case of die casting, the effect of casting under pressure, produces the improved properties. However, it should be recognized that mere chemical specification of a certain alloy is no guarantee of mechanical properties. Casting process, casting design, melting practice, and the complete history of foundry processing must be considered. In spite of the improved mechanical properties obtained with metal molds, sand casting remains as a major process because of its inherent advantages. Different casting processes have been described in detail in Chapter 3.

Sand Casting

According to a recent AFS survey, aluminum sand castings represent about 20% of the total weight of aluminum castings produced in the USA using green-sand molding and chemically bonded sand molding. By comparison, die castings and permanent mold castings represent 60% and 20% respectively.

Greensand Molding

Greensand systems may be composed of components such as sand, clay, water, and other additives. Greensand molds are not baked or dried. The molding process is described in Fig. 7.30 [30]. The process starts with a pattern which is a replica of the finished casting, except that it is slightly larger to allow for aluminum shrinkage during solidification and cooling. The pattern (made of wood, plastic, or metal) can either be loose or attached to a plate.

A flask is placed around the pattern to contain the prepared, free-flowing molding sand that is placed on the pattern. Pressure is applied to the sand to compact it firmly against the face of the pattern. When it is compacted, the greensand mixture exhibits physical properties that allow it to hold exact shape of the pattern after the pattern is drawn from the mold.

After cores (if any) are placed in the bottom (drag) half of the mold, the top (cope) section is set in place. The closed mold is then ready for pouring the molten aluminum alloy.

Synthetic sands are silica, olivine [a solid solution of forsterite (Mg_2SiO_4) and fayalite (Fe_2SiO_4)], zircon, and chromite. Both zircon and chromite sands are used as special facing sands in an aluminum foundry.

After solidification, the casting is removed by mold shakeout and then sent for fettling to remove fins, gates, and risers before shipment or any postcasting operation such as heat treatment or machining.

Section Thickness, mm (in)	For Length of:
Sand Castings	
3.2 (1/8)	Under 76 mm (3 in)
4 (5/32)	76–152 mm (3–6 in)
4.8 (3/16)	Over 152 mm (6 in)
Permanent-Mold Castings	
2.5 (0.100)	Under 76 mm (3 in)
3.2 (0.125)	76–152 mm (3–6 in)
4.1 (0.1600	Over 152 mm (6 in)
Die Castings	
1.3 (0.050)	Small parts
2 (0.080)	Large parts
Plaster-Mold Castings	
1.6 (1/16)	51 mm (2 in) or less
2.4 (3/32)	76–152 mm (3–6 in)
3.2 (1/8)	Over 152 mm (6 in)
Shelle-Mold Castings	
1.6 (1/16)	Under 13 mm (½ in)
2.4 (3/32)	13–76 mm (½–3 in)
3.2 (1/8)	76–152 mm (3–6 in)
4 (5/32)	Over 152 mm (6 in)
Precision Sand Castings	
2 (0.080)	Under 76 mm (3 in)
2.5 (0.100)	76–152 mm (3–6 in)
2.5 (0.150)	Over 152 mm (6 in)
Investment Castings	
0.9 (0.035)	3.25 mm (¼ in)
1.5 (0.060)	51 mm (2 in) or less
2.3 (0.090)	51–102 mm (2–4 in)
3.2 (0.125)	102–203 mm (4–8 in)
3.8 (0.150)	Over 203 mm (8 in)
Centrifugal Permanent-Mold Castings	
1.8 mm (0.070 in) for up to 25.8 cm^2 (4 in^2) per casting; 2 mm (0.080 in) for 32.2–64.5 cm^2 (5–10 in^2); 2.3 mm (0.090 in) for 71–129 cm^2 (11–20 in^2); 2.5 mm (0.100 in) for 135.5–193.5 cm^2 (21–30 in^2); 2.8 mm (0.110 in) for 200–451.5 cm^2 (31–70 in^2); 3 mm (0.120 in) for 458–645 cm^2 (71–100 in^2); and 4 mm (0.156 in) for over 645 cm^2 (100 in^2) per casting	

Source: From Heine et al. [11].

TABLE 7.9 Minimum Section Thickness for Aluminum Castings Produced by Different Processes

Clays are the natural "glue" to hold sand molds together in green, dry, and hot conditions. Some sands, typically silica, occur in nature mixed with certain amounts of clay (kaolinite). These are called naturally bonded molding sands. The characteristics of synthetic sands suitable for aluminum castings are as follows:

AFS GFN	85
Southern bentonite (%)	6
Recommended moisture (%)	2.8–3.8
Permeability	45–75
Green compressive strength (kPa)	1.3–1.8 kg/cm^2, 18–25 psi

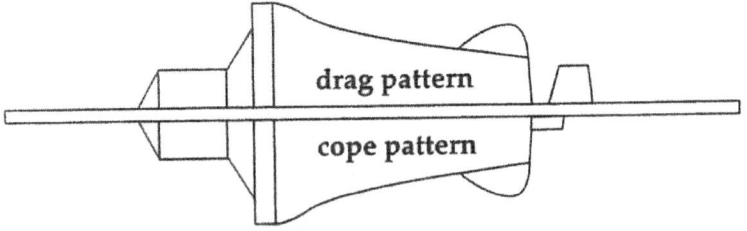

1. In this example, a cope pattern and a drag pattern are mounted onto a plate to form a match plate.

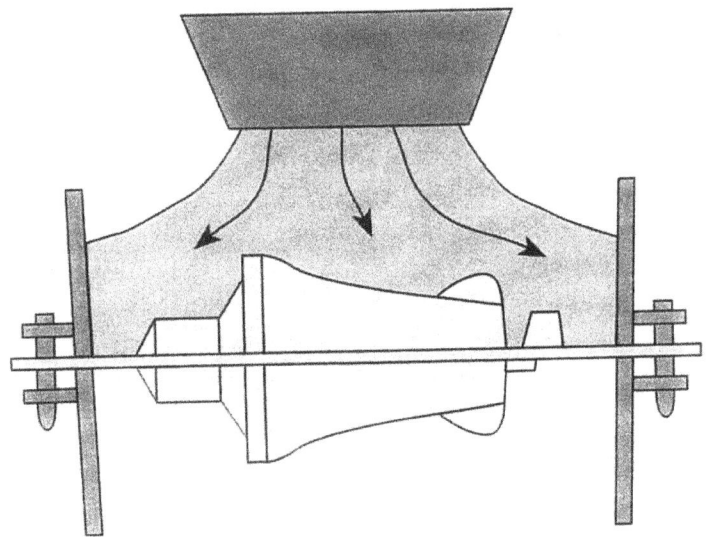

2. A flask is placed around the match plate, and prepared sand is dumped or blown in on top of it.

FIGURE 7.30 These steps illustrate how a greensand "match plate" mold typically is made. (From Zalensas [30, p. 188].) (*Continued*)

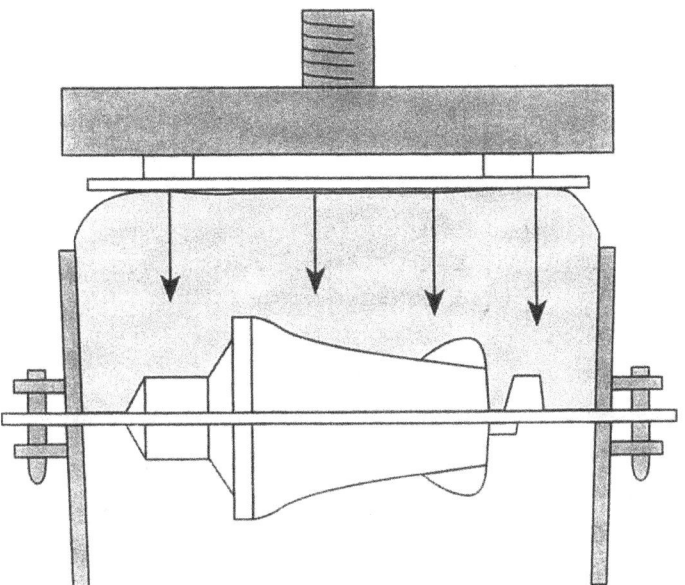

3. A squeeze board is placed over the sand, and the sand is compacted around the pattern.

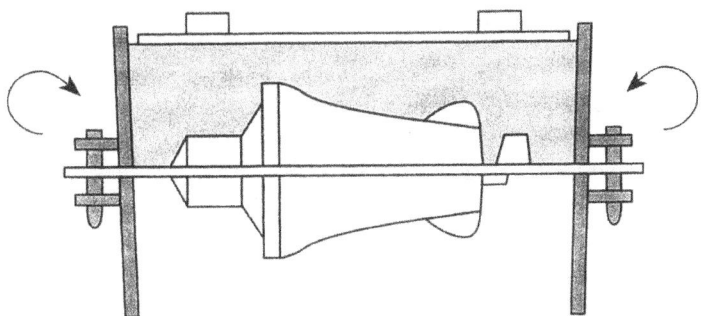

4. After compaction, the flask is turned over.

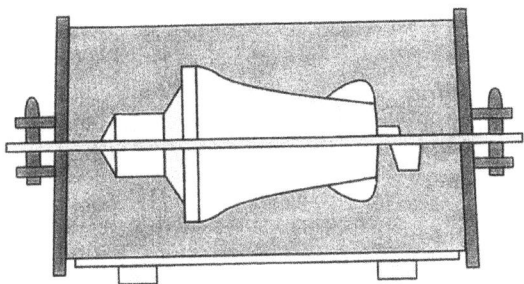

5. After rollover, steps 2 and 3 are repeated for the cope half of the mold.

FIGURE 7.30 (*Continued*)

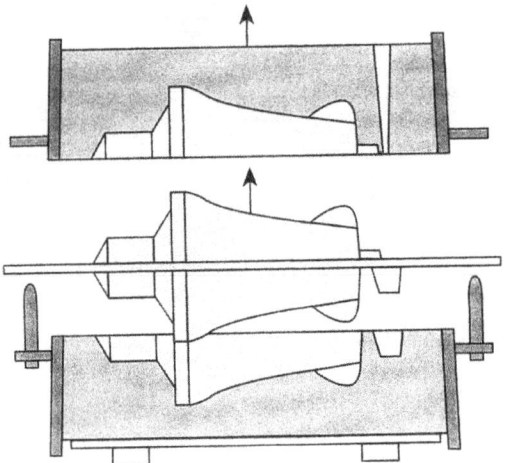

6. A sprue is cut into the cope. The cope then is lifted carefully off the drag, and the match plate is withdrawn from the mold.

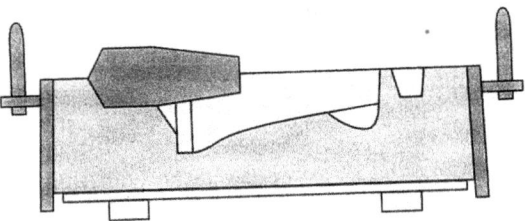

7. Cores (if any) are placed in core prints in the drag half of the mold.

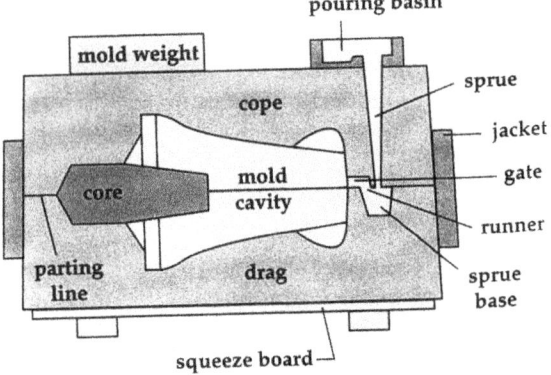

The cope is placed on top of the drag, and the flask is replaced with a mold jacket. Weights are placed on the assembled mold, and a pouring basin is added. The mold is ready for pouring.

FIGURE 7.30 *(Continued)*

Cold Box	Nobake	Heat-Activated
Phenolic/urethane/amine (PUA)	Furan/acid	Shell
Silicate/CO_2	Phenolic/acid	Core oil
Furan/SO_2	Phenolic/ester	Phenolic hot box
Acrylic/epoxy/SO_2	Oil urethane	Furan hot box
(Acrylic) FRC/SO_2	Silicate/ester	Urea formaldehyde hot box
Phenolic/ester (methyl formate)	Phenolic urethane (PUNB)	Warm box
Phenolic/CO_2	Phosphate/metal oxide	
	Polyol urethane	

Source: From Zalensas [30, p. 204].

TABLE 7.10 Categories of Resin Core/Mold Processes

Additives are added to molding sands to get improved sand characteristics. Cellulose acts as a cushioning material between sand grains to lower sand's expansion stresses. It also increases sand collapsibility, reduces hot tears and cracks, and improves sand flowability. Cereals are generally not added to aluminum sand mixtures. But small additions may be beneficial with certain heavy-section castings to increase deformation (ductility) making it easier to draw poorly maintained or difficult patterns.

Chemically Bonded Sand Molding

The core and mold making processes used for casting aluminum include heat-cured shell resin sand systems (which produce hollow shell-like cores) and carbon dioxide (CO_2) gas-hardened silicate systems, heat-cured hot-box binders, nobake systems (chemically cured without application of heat), vapor-catalyzed cold box processes, and vapor-reacted resins. Over 22 binder systems have been introduced in the last four decades. The various cold box, nobake, and heat-activated systems are summarized in Table 7.10.

Permanent Mold Casting

Permanent metal molds are made of steel or cast iron. Coring may be done with metal cores in permanent molds, but these must be removable after the metal has solidified. More intricate coring may be done by using sand, plaster, ceramics, or even salt. When nonmetal destructible cores are used, the process is called semipermanent mold casting. The mold surface is coated by spraying the hot mold with a suspension of a fine-particle-size refractory in water. The mold coating can be used to promote directional solidification. Sections required to freeze more slowly can be given a thicker refractory coating. Mold coatings serve to prevent sticking of the casting to the mold, provide a smooth surface, and assist in controlling solidification so that sound castings are obtained. Production of quality permanent-mold castings requires a careful control of pouring temperature, mold temperature (315 to 427°C or 600 to 800°F), mold coating, gating, pouring, mold manipulation, and continuity of operation. Only certain alloys are favorably cast in permanent molds because of hot tearing and other problems.

Molds are produced from gray iron, H13 tool steel, SAE 1020, and SAE 4140 steels. Gray iron has better heat transfer capacity (hence shorter cycle time, faster solidification rates, and more even die heating) and excellent machinability. H13 tool steel on the other hand is more resistant to wear and thermal stresses. Multiple-cavity molds are used to increase the production rate. Venting of the molds to permit air to escape during mold filling should be considered in mold design.

Mold coatings can be of two types: insulating and lubricating. Mold coatings are composed of (1) refractory fillers (titania, talc, mica, vermiculite, alumina, iron oxide, silicon carbide, and graphite); (2) binder (sodium and potassium silicate and clays); and(3) carrier which is usually water.

Metal molds have a production life of 10,000 to 120,000 or more castings [1, p. 689]. Advantages over sand castings include closer dimensional control, better surface finish, and improved mechanical properties (and hence, thinner wall castings and significantly lower-weight designs), less shrinkage, and gas porosity.

Typical applications are oil pans and engine cradles, pistons, engine blocks, cylinder heads, intake manifolds, and other functional parts of internal combustion engines for the automotive, trucking, diesel, and marine industries. Other major uses for permanent mold castings include internal and accessory parts for reciprocating and jet-type aviation engines, missiles, forms for concrete, textile machine parts, electric motor housings, portable and hand tool components, support members for outdoor light standards, electric griddles and kitchen pots and pans, and a host of other commercial applications.

Low-Pressure Permanent Mold Casting

Low-pressure casting is a process where molten metal is introduced to the mold by application of pressure to a hermetically sealed metal bath, forcing the molten metal up through a narrow-diameter fill (stalk) tube from a furnace usually residing below the casting machine (although there is a version using electromagnetic forces to lift metal into the mold, and then the furnace may be an open hearth located beside the casting machine). The process can be considered for low to high volumes of casting from 5 to 100 kg (11 to 220 lb) and usually incorporates the use of iron or steel permanent molds. Recent developments in sand molding technology have made precision sand molds a viable choice for high-volume low-pressure casting as well. A wide range of casting core options, such as expendable sand and shell cores and mechanical single or multipiece permanent cores, are successfully used in the low-pressure process.

The low-pressure process incorporates the use of a hydraulically operated machine to vertically open and close a casting mold that is situated over a ready supply of molten metal. A basic schematic of a low-pressure machine is given in Fig. 7.31. The machine consists of an upper (moving) and lower (stationary) platen, to which the upper (cope) and lower (drag) mold halves are attached, respectively. The moving platen is raised and lowered hydraulically to allow for casting and extraction of a component. Standard mold design allows the casting to remain in the upper mold as it is opened and subsequently ejected from the mold for manual or automatic removal. The stationary platen is typically positioned directly over a sealed, airtight furnace with a supply of molten metal.

To provide optimum performance and consistency to the process, almost all low-pressure machines are controlled by a programmable logistic controller (PLC). The PLC will control all moving functions of the casting machine, pressurization of the casting furnace, mold cooling/heating circuits (oil, water, or air), automatic casting extraction, and manipulation of casting furnace for refilling. The PLC can also track important historical data of the casting cycles, such as mold temperature, cycle time, metal temperature, pressure profiles, and cooling cycles, which may be used to gage productivity or troubleshoot casting problems.

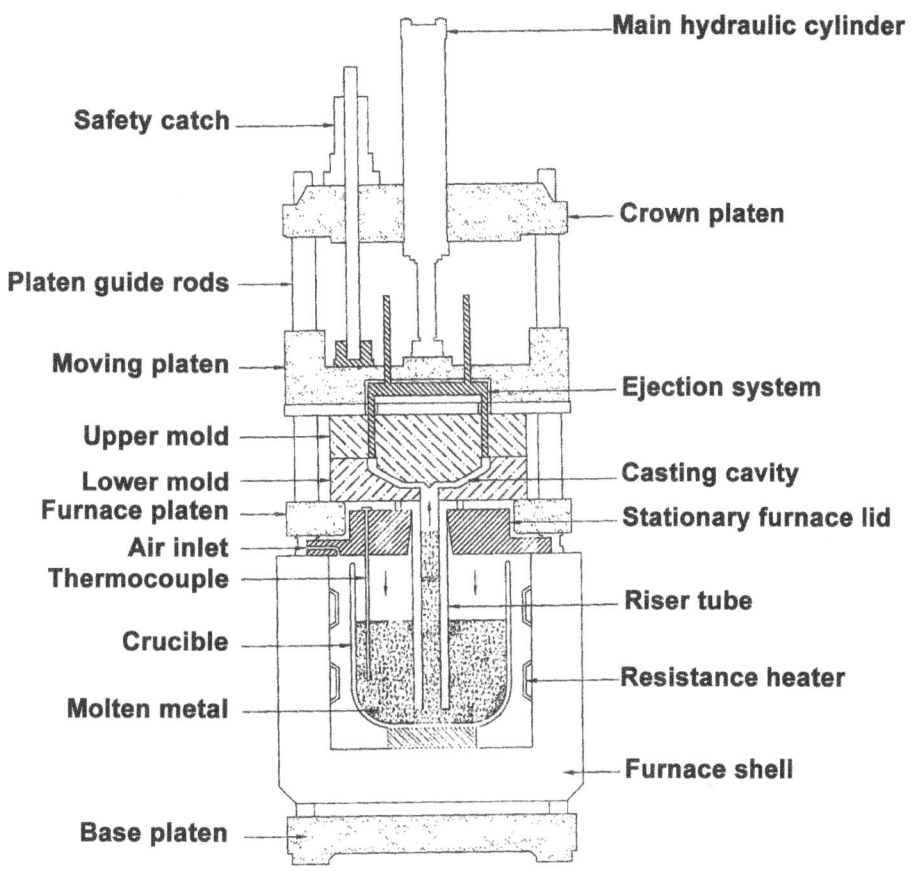

Safety catch

Platen guide rods

Moving platen

Upper mold

Lower mold
Furnace platen

Air inlet
Thermocouple

Crucible

Molten metal

Base platen

Main hydraulic cylinder

Crown platen

Ejection system

Casting cavity

Stationary furnace lid

Riser tube

Resistance heater

Furnace shell

FIGURE 7.31 Schematic of low-pressure casting machine with an electric resistance crucible furnace. (From *ASM Handbook* [1, p. 701].)

Most low-pressure machines include automatic casting extraction due to the size and number of castings produced by the process. A casting extractor can be a simple shuttle pan onto which the casting is ejected or a robotic arm that extracts the casting and takes it to a subsequent operation.

Advantages of the low-pressure casting process are very high casting yield, low inclusion level as the liquid metal is taken from the middle of the bath using a feed tube, and minimum turbulence.

High-Pressure Die Casting

High-pressure die casting (HPDC), also simply called *die casting*, consists of forcing molten metal under considerable pressure into metal dies. Die casting of aluminum alloys is the most rapidly developing metal casting process to produce larger, more complex, sounder, stronger, and more dimensionally accurate parts for industry groups such as transportation (especially automotive), agriculture, mining, construction, electronic, home appliance, sporting goods, etc. Because of the relatively lower melting point of aluminum alloys, thousands of castings can be produced in any die. The tolerances can be held unusually close for the cast parts (0.08 to 0.13 mm). The rapid chilling inherent in the die-casting process produces a very fine-grained structure that gives superior mechanical properties. The cast-to-shape parts require very little machining. The other advantages include very high production rate compared to conventional casting processes, better surface finish, and thinner wall components. The main disadvantages are the high tooling cost and costs for die-casting machines and their maintenance. In addition, choices of alloys are limited and internal porosity during conventional die casting restricts heat treatment or welding. However, vacuum HPDC has been successfully used to evacuate the air and gases from the die cavity and metal delivery system before and during molten metal injection.

Die casting of aluminum can be done either by the hot chamber or by the cold chamber high-pressure method. The hot chamber die-casting method is illustrated in Fig. 7.32 [31, 32].

In the hot-chamber die-casting method, the molten metal is held in an enclosed steel crucible, under a protective atmosphere. A valve allows a controlled volume of molten

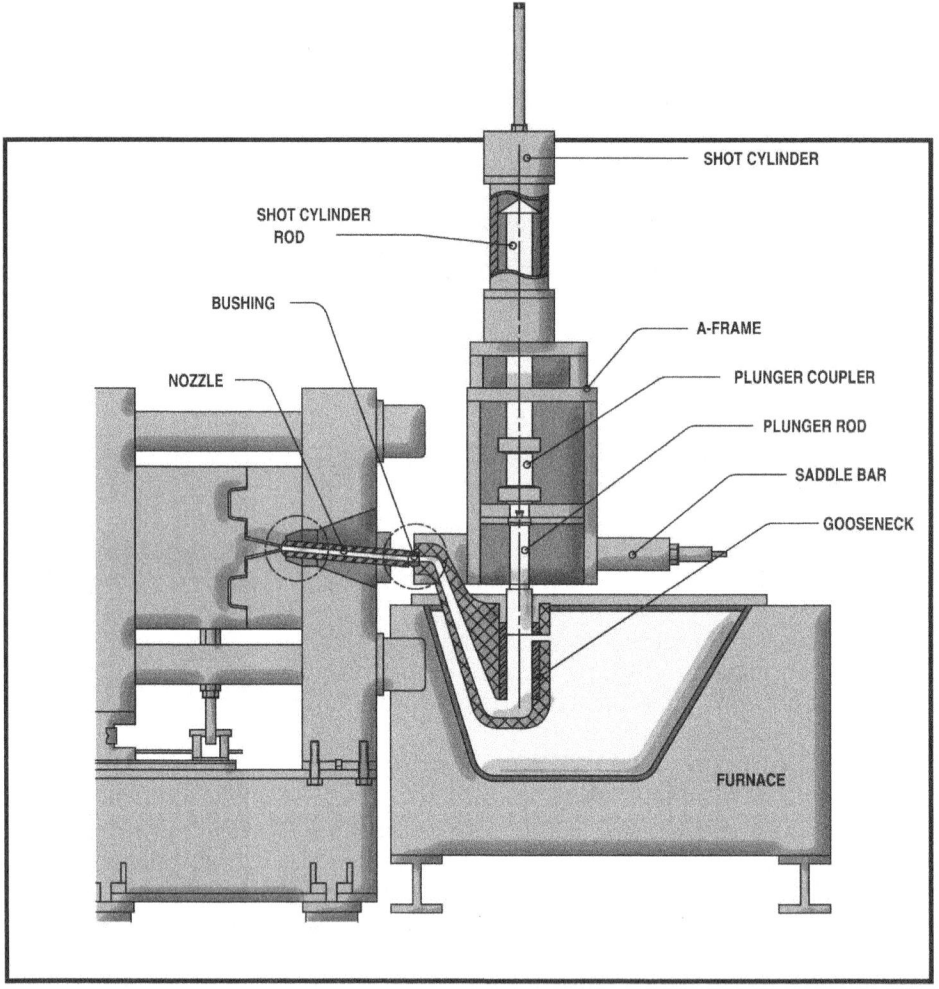

Figure 7.32   Schematic of the hot-chamber die-casting method. (From Udvardy [31], NADCA [32].)

metal into the gooseneck that is immersed in the molten metal. A plunger injects this metal into the cavity of the die through a nozzle. To prevent freezing of the metal, the nozzle is heated to 427 to 482°C (800 to 900°F) with gas or by induction heating.

The pressure applied on the molten metal during injection is lower than in cold-chamber die casting, and limits the size of parts made by the hot-chamber method. The main reason is the high operating temperature of various components, such as the nozzle, that limits the pressures that can be applied. A typical 400-ton magnesium hot-chamber machine makes parts that weigh up to 2.7 kg (6 lb). The machine has a clamping force of 400 tons, and applies about 0.7 MPa (5 ksi) maximum pressure on the metal. Due to the short cycle time (up to six parts per minute), the hot-chamber die-casting process is very competitive for small parts.

The cold-chamber die-casting method is depicted in Fig. 7.33 [31, 32]. The molten metal is fed into a shot cylinder either by hand ladling, auto-ladling, or by a pump. It is then injected fast (5 to 100 m/s) by a plunger into the cavity, where it solidifies into a net shape part under high pressure 0.3 to 2.9 MPa (5 to 20 ksi). If used to form undercuts, cores are retracted. Finally, the casting is ejected, and the part is trimmed by separating it from the gating system and the biscuit. The entire cycle takes usually less than 1 minute. It is a very cost effective near-net-shape casting process, especially for the automotive industry.

While the fast injection of molten metal into the die cavity allows filling of very thin sections, it also contributes to air entrapment in the castings. The resulting porosity can impair the mechanical properties. If high mechanical properties are targeted, special precautions need to be taken to prevent air entrapment. Eliminating the air from the shot sleeve and the cavity by application of vacuum can reduce the air entrapment in the casting [33]. This process not only removes the air from the cavity, but also eliminates excessive handling of the molten magnesium. The metal is drawn by the vacuum,

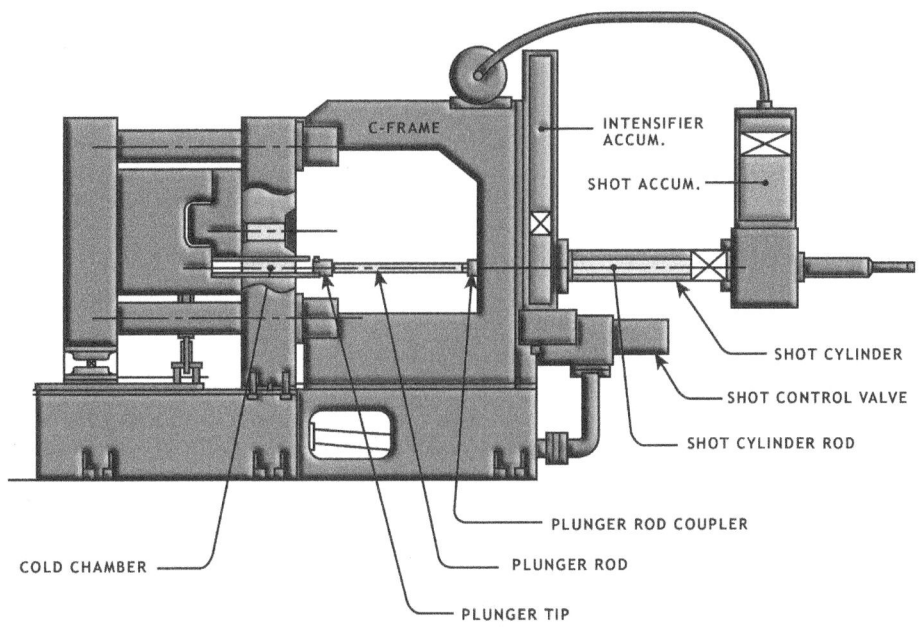

Figure 7.33 Cold-chamber machine. (From Udvardy [31], NADCA [32].)

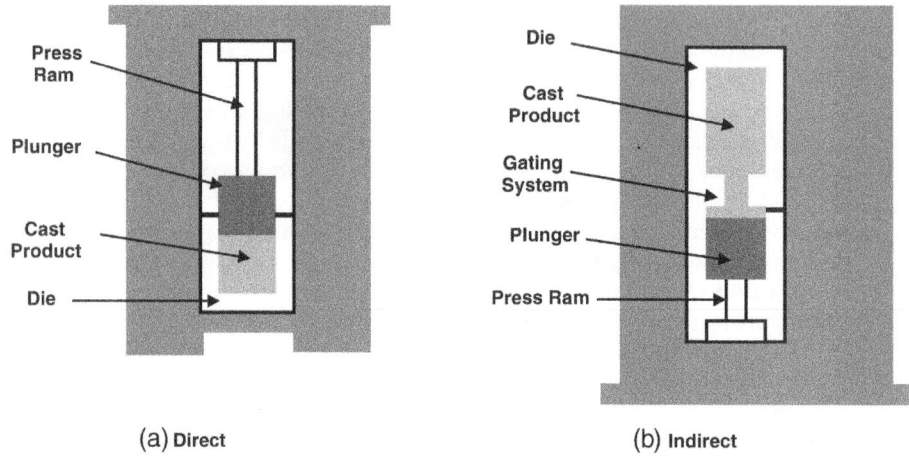

FIGURE 7.34 Schematic diagram of the squeeze casting processes. (*a*) Direct squeeze casting (DSC); (*b*) indirect squeeze casting (ISC). (From Ghomashchi and Vikhrov [34].)

directly from the furnace to the shot sleeve without exposure to air. The reduced level of oxides and porosity produces parts with superior mechanical properties.

A variation of the die-casting process is the squeeze casting to produce high integrity and heat-treatable components in aluminum alloys. The process involves slow metal filling into the die cavity with minimal turbulence followed by the application of pressure during solidification. There are two major squeeze casting processes: direct and indirect. In the direct squeeze casting process, the pressure is applied directly to the liquid or semisolid metal from a hydraulic source to produce a casting with minimum porosity. There is no gating system. In the indirect squeeze casting process, liquid metal is injected into the mold through gates located below the die cavity. Schematics of the two processes are shown in Fig. 7.34. Pressure levels of up to 70 to 100 MPa (482 to 689 ksi) are generally used.

Two types of squeeze casting machines are commercially available: vertical and horizontal-vertical. Typical cast components for the automotive industry are axle carrier, axle cover, front steering knuckle, and suspension links. Mechanical properties of squeeze-cast aluminum alloys are summarized in Table 7.11, together with those from other casting processes such as gravity permanent mold and die casting. In general, the strength properties are comparable in all the casting conditions. However, squeeze-cast alloys exhibit better ductility and impact toughness, probably due to lack of porosity in the matrix and development of a more refined microstructure.

Lost-Foam Casting

The use of foam patterns to produce castings was patented almost 50 years ago by Shroyer [35], and variants of the original invention have been used successfully to produce engineering components from both ferrous and nonferrous alloys. The lost-foam casting (LFC) process is a cost-effective method for producing complex castings using an expendable polystyrene pattern and unbonded sand. The fundamentals of the lost-foam casting process consist of placing a molded and refractory-coated polystyrene foam pattern and its gating system in a flask, surrounding it by unbonded sand, and vibrating it to achieve maximum compaction of the sand around the foam pattern assembly. Molten metal is poured directly on the top of the coated pattern through a

| Alloy | Process | Yield Strength[a] | | Tensile Strength[a] | | Elongation,[a] % | Hardness, (HRB) | Impact Strength[b] |
		MPa	Ksi	MPa	Ksi			J (ft·lb)
A356, 2-T6	Squeeze	221–234	32–34	296–310	43–45	10–14	48–63	14–18 (10–13)
A356.2-T6	GPM[c]	207–228	30–33	283–303	41–44	3–5	45–58	8–10 (6–7)
357-T6	Squeeze	241–262	35–38	324–338	47–49	8–10	52–58	N/A
357-T6	GPM[c]	241–262	36–38	331–338	48–50	5–7	50–65	N/A
383-F[d]	HPD[e]	152–172	22–25	241–262	35–38	1–2	50–60	N/A
383-F[d]	Squeeze	145–159	22–25	241–262	39–42	2.75–3.5	50–60	N/A
383-T4[d]	Squeeze	234–255	34–37	359–386	52–56	5–7	55–70	7–11 (5–8)
383-T4[d]	Squeeze	234–255	34–37	359–386	52–56	5–7	55–70	7–11 (5–8)
383-T4[d]	Squeeze	296–317	43–46	379–421	55–61	3–5	73–84	5–8 (4–6)
390-F	HPDC[e]	172–193	25–28	193–228	28–33	0.5–1.0	50–60	<1 (<1)
290-T6	Squeeze	N/A	N/A	352–396	51–57	<1	80–90	<1 (<1)
206-T4	Squeeze	248–269	36–39	400–421	58–61	21–26	63–75	47–56 (35–41)
206-T4	GPM[c]	248–283	36–41	407–441	59–64	14–18	60–75	35–48 (26–35)

Note: (a) All tensile data based on samples machined from actual castings; (b) cross section of impact specimens is 10 × 3.3 mm (0.4 × 0.1 in), all specimens machined from actual casting; (c) GPM, gravity permanent mold; (d) modified chemistry; and (e) HPDC, high-pressure die casting.

Source: From *ASM Handbook* [1, p. 730].

TABLE 7.11 Various Properties of Squeeze-Cast Aluminum Alloys

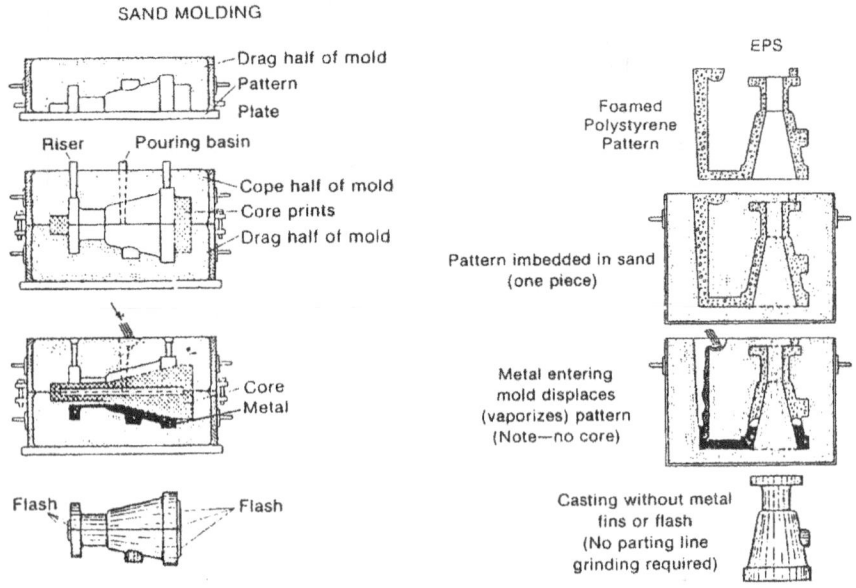

SAND MOLDING

Drag half of mold
Pattern
Plate

Riser   Pouring basin

Cope half of mold
Core prints
Drag half of mold

Core
Metal

Flash   Flash

EPS

Foamed
Polystyrene
Pattern

Pattern imbedded in sand
(one piece)

Metal entering
mold displaces
(vaporizes) pattern
(Note—no core)

Casting without metal
fins or flash
(No parting line
grinding required)

FIGURE 7.35   A schematic comparing the lost foam and sand casting processes. (From Monroe [37].)

gating system. This serves to vaporize the polystyrene pattern leaving a casting that is a replica of the foam pattern. The gases that form during the vaporization of the pattern permeate through the coating and sand and exit through the flask vents [36–38]. A schematic of the lost-foam process is shown in Fig. 7.35.

Some unique advantages of the LFC process are close dimensional tolerance, part consolidation, high casting yield, capability to produce complex castings, and elimination of sand cores and binders. The flexibility of gluing foam sections together to form one component with complex internal passages—as in engine blocks and heads—without the use of cores common in the sand-casting process is one of the major advantages of the LFC process. Some automotive engine blocks and heads are cast from aluminum alloy A356 using the LFC process [38].

Foam permeability, density, surface quality, fusion, and dimensional stability are the key control areas to produce premium quality castings. The foam pattern is prepared for a casting by attaching a gating system (sometimes molded as part of the pattern) of material of the same type and density. Patterns for aluminum castings are usually coated to eliminate metal penetration and burned-on sand and to improve as-cast surfaces. Sand used can be either washed and dried silica sands, lake sands, or other good sands.

An effective way of improving the fillability of aluminum alloys is the use of vacuum to assist removal of pyrolysis products and to improve the apparent fluidity [39–41]. Figure 7.36 shows a vacuum box that can be used for LFC. The degree of vacuum to be applied should be optimized depending on the thickness and complexity of the castings.

Molten aluminum shrinks by approximately 6% on solidification. This leads to about 1% porosity in conventional sand castings. However, application of pressure during lost-foam casting can reduce porosity to less than 0.1%. The pressure can be applied

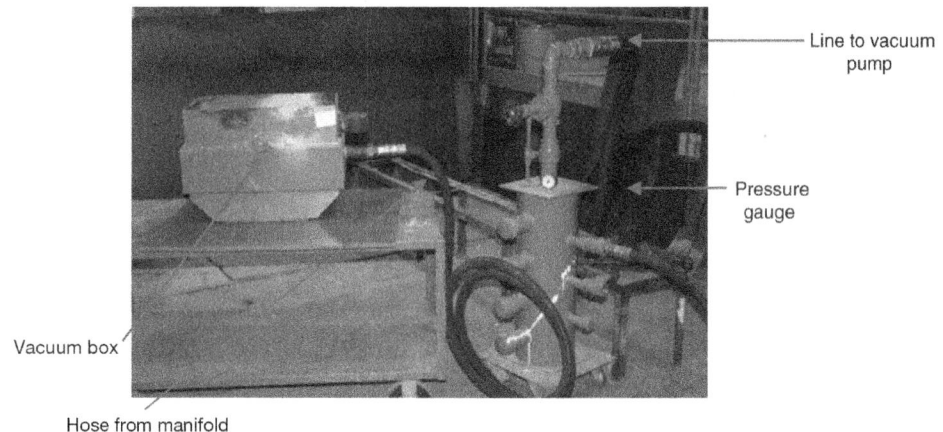

Figure 7.36   Vacuum box and pressure manifold system. (From Fasoyinu et al. [39].)

to 1 to 1.5 MPa (0.15 to 0.22 ksi or 10 to 15 atm) in approximately 60 to 90 seconds to eliminate hydrogen porosity and decrease unfed shrinkage porosity by over an order of magnitude compared to conventional casting processes.

The most common defect in lost-foam casting is the formation of folds [42, 43] due to oxide/liquid system films in the molten metal which prevent "knitting" of two metal fronts coming in contact with an intervening dry-sided oxide surface. Figure 7.37 shows fold defects in aluminum alloy 356 during lost-foam casting [41].

Ablation Casting

The ablation casting process is the latest emerging technology in the casting industry which takes advantages of the high cooling rate due to water spray on a sand aggregate with a water-soluble binder [44, 45]. In this process, the liquid metal is poured into the mold, and the mold is progressively ablated away with the molten metal in the mold cavity (Fig. 7.38). The mold may be tilted to control metal fill. Since the mold is progressively removed, the water can contact the metal casting directly, eliminating the air gap. As a result, the heat transfer is more rapid compared with other conventional casting processes and hence, high temperature gradients are established that help to eliminate shrinkage porosity, especially in thick section. The final outcome is very high solidification rates in a sand mold leading to fine microstructures and improved mechanical properties. A typical casting produced by this process in alloy B206 is shown in Fig. 7.39.

The rapid cooling effect of the ablation fluid on the solidifying alloy produces a unique microstructure and very high mechanical properties (see Table 7.12, where the mechanical properties for different alloys are compared). The SDAS after ablation is rather coarse (same as sand-cast products). However, the center sections have a fine DAS and the porosity level is significantly low.

Heat Treatment

Most of the aluminum alloys can be heat treated (annealed, solution treated, quenched, precipitation hardened, overaged, etc.) to obtain the desirable combination of physical

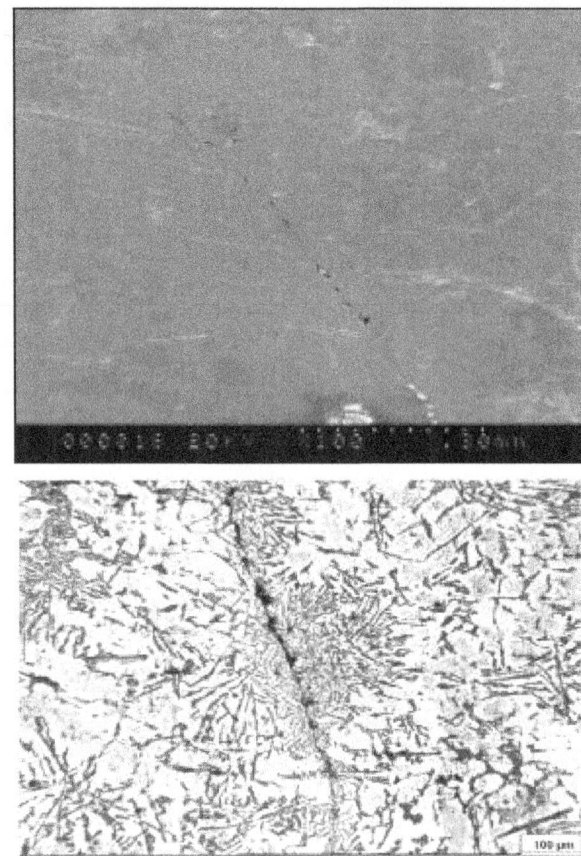

Figure 7.37   Fold defects in aluminum 356 alloy during lost-foam casting. (From Jagoo et al. [43].)

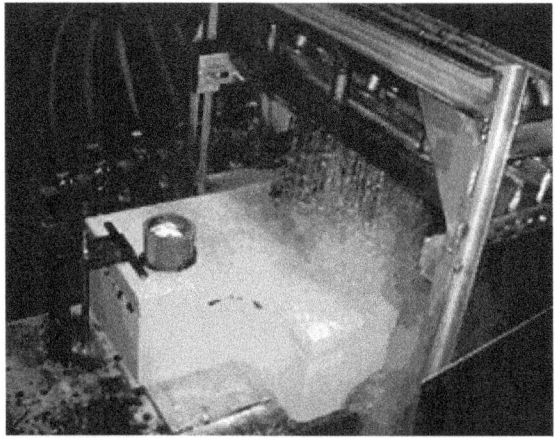

Figure 7.38   The ablation of the mold in progress. (From Grassi et al. [44], copyright © 2008 by The Minerals, Metals and Materials Society, reprinted with permission.)

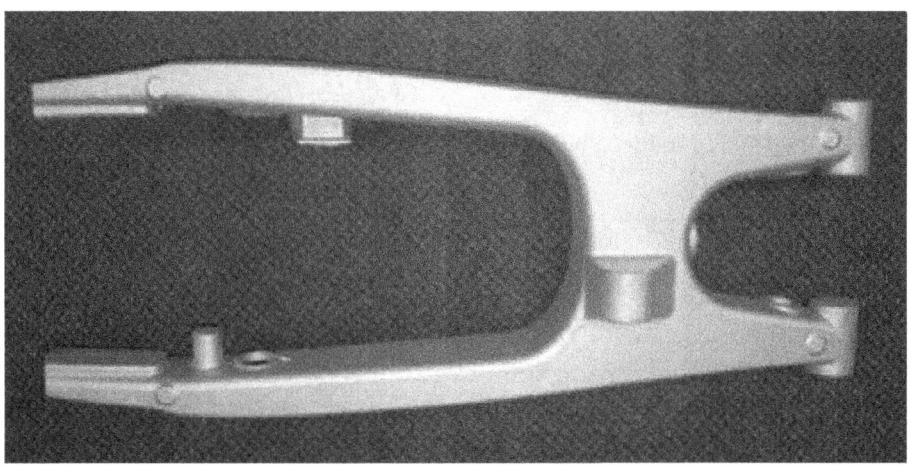

Figure 7.39 An aluminum alloy B206 (Al-4.5Cu) alloy swing arm casting, 2.5 mm thick wall, with isolated unfed bosses, notably free from hot tearing, characteristic of this alloy. (From Grassi et al. [44], copyright © 2008 by The Minerals, Metals and Materials Society, reprinted with permission.)

	Sand Cast	Perm. Mold	Squeeze Cast	Pressure/ Counter Pressure	Ablation Cast
UTS MPa	228	262	312	334	325
Yield MPa	179	207	243	239	261
Elongation %	3.5	4	11.0	14.3	12.5

Source: From Grassi et al. [44].

Table 7.12 Comparative Mechanical Properties for Alloy A356

and mechanical properties for different applications. One or more of the following objectives form the basis for thermal treatment [13, p. 61]:

- Increase strength and/or produce the mechanical properties associated with a particular material condition
- Stabilize mechanical and physical properties
- Increase hardness for improved machinability
- Relieve residual stresses induced by casting, quenching, machining, welding, or other operations
- Minimize or eliminate microsegregation
- Reduce susceptibility to corrosion by altering insoluble phases

The Aluminum Association has standardized the definitions and nomenclature applicable to thermal practice and maintains a registry of standard heat treatment practices and designations for industry use. Standardized temper designations applicable to castings are as follows:

- O (formerly T2, T2x): annealed (thermally stress relieved)
- T4: solution heat treated and quenched

- T5: artificially aged
- T6: solution heat treated, quenched, and aged
- T7: solution heat treated, quenched, and overaged
- T8: cold reduced before aging to improve compressive yield strength (bearings only)

There are variations to those thermal treatments with second and third digits in the standard designation, for examplem, T61, T62, T572, etc., to get different levels of strength and hardness. Recommended heat treatment practices for aluminum casting alloys are listed in Table 7.13.

The most common heat treatment practice is solution treatment, quenching and precipitation hardening or aging. This is because of the nature of the equilibrium diagram (shown in Fig. 7.40) for the Al-Cu system. Here, the solubility of the eutectic

Alloy	Temper	Type of Casting[a]	Solution Heat Treatment[b] Temperature °C	°F	Time, h	Aging Treatment Temperature °C	°F	Time, h
201.0	T6	S	510–515	950–960	2	Room temperature		12–24
	T7	S	525–530	980–990	14–20	Then 155	310	20
204.0	T4	S or P	520	970	10			
A/B206	T4	S or P	510	950	2			
			Then 530	990	14–20	Room temperature		5 days
			Then 530	990	14–20	Then 160	320	0.5–1.0
	T7	S or P	510	950	2	Room temperature		12–24
			Then 530	990	14–20	Then 190	370	4–5
319.0	T5	S	–	–	–	205	400	8
	T6	S	505	940	6–12	155	310	2–5
		P	505	940	4–12	155	310	2–5
355.0	T51	S or P	–	–	–	225	440	709
	T6	S	525	980	12	155	310	3–5
		P	525	980	4–12	155	310	2–5
	T62	P	525	980	4–12	170	340	14–18
	T7	S	525	980	12	225	440	3–5
		P	525	980	4–12	225	440	3–9
	T71	S	525	980	12	245	475	4–6
		P	525	980	4–12	245	475	3–6
C355.0	T6	S	525	980	12	155	310	3–5
	T61	P	525	980	6–12	Room temperature		8 min.
			–	–	–	155	310	10–12

TABLE 7.13 Typical Heat Treatments for Aluminum Alloy Sand and Permanent Mold Castings

| Alloy | Temper | Type of Casting[a] | Solution Heat Treatment[b] | | | Aging Treatment | | |
| | | | Temperature | | | Temperature | | |
			°C	°F	Time, h	°C	°F	Time, h
356.0	T51	S or P	–	–	–	225	440	7–9
	T6	S	540	1000	6–12	155	310	3–5
	T6	P	540	1000	4–12	155	310	2–5
	T7	S	540	1000	6–12	225	440	8
		P	540	1000	4–12	225	440	8
	T71	S	540	1000	6–12	245	475	3–6
		P	540	1000	4–12	245	475	3–6
		–	–	–	–	155	310	10–12
357.0	T6	P	540	1000	8	165	330	6–12
	T61	S	540	1000	10–12	155	310	6–12
520.0	T4	S	430	810	18			
535.0	T5	S	400	750	5			
712.0	T5	S	–	–	–	Room temperature		21 days
			–	–	–	155	315	6–8
771.0	T53	S	415	5		180	360	4
	T5	S	–	–	–	180	360	3–5
	T6	S	590	1090	6	130	265	3
	T71	S	590[j]	1090	6	140	285	15
850.0	T5	S or P	–	–	–	220	430	7–9

Note: (a) S = sand, P = permanent mold; (b) Unless otherwise indicated, solution treating is followed by quenching in water at 65–100°C (150–212°F).
Source: From *ASM Handbook* [1, p. 1072].

TABLE 7.13 Typical Heat Treatments for Aluminum Alloy Sand and Permanent Mold Castings (*Continued*)

increases with temperature and this is the case for other systems such as Al-Mg, Al-Zn, Al-Sn, etc. The solution treatment temperature is usually above the solvus temperature but less than the eutectic temperature and is governed for a specific composition. For multicomponent alloy systems as in the Al-Si-Cu-Mg and Al-Zn-Cu-Mg systems, stepped solution treatment is sometimes recommended to avoid the melting of lower melting phases. In the case of alloy 319, a typical stepped heat treatment is a first step at 495°C (923°F) for two hours followed by a second step at 515°C (959°F) for four hours [46]. This modified heat treatment could produce improved mechanical properties in thick sections (in excess of 215 MPa UTS, 170 YS, and 1.8% elongation). A two-step heat treatment has also been developed for the AC2A alloy (5.3% Si, 3.7% Cu, 0.14% Mn, and 0.04% Ti), which is used in light duty cylinder heads. It consists of one hour at 505°C (941°F) and then two hours at 525°C (977°F) to produce mechanical properties of 278 MPa UTS, 250 MPa YS, and 1% El [47, 48]. The stepped solution treatment temperature can be obtained from advanced thermal

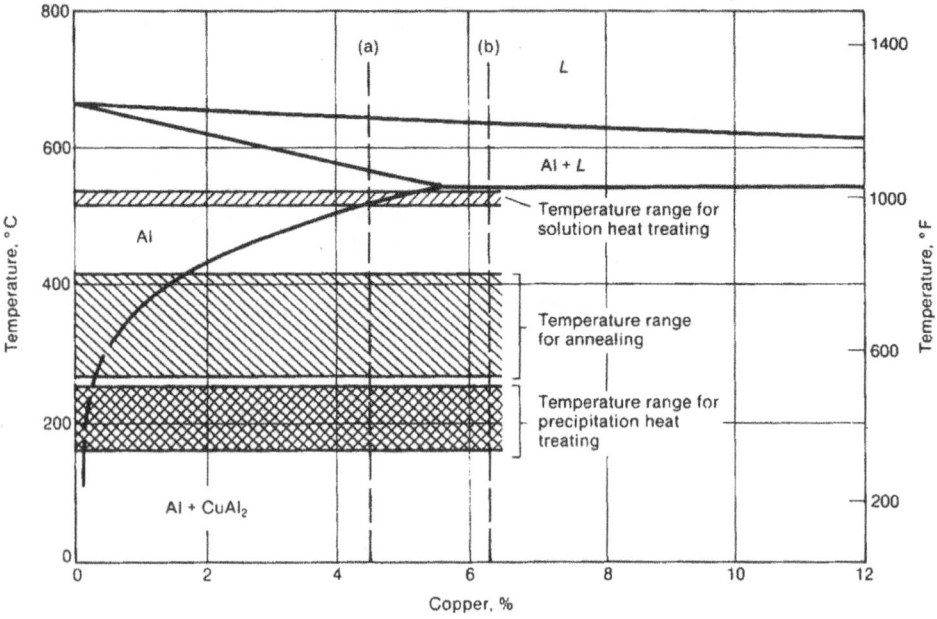

FIGURE 7.40 Portion of aluminum-copper binary phase diagram. Temperature ranges for annealing, precipitation heat treating, and solution heat treating are indicated. (From Kaufman and Rooy [13, p. 61].)

analysis as shown in Fig. 7.41 for an Al-Si-Cu alloy. Thermal analysis can also be done to modify the solution treatment temperature together with shorter solution treatment time [48].

The solution treatment time to achieve complete solution of the phases depends on the fineness of the microstructures. The coarse microstructure associated with slow solidification rate as in sand castings requires relatively longer times.

Rapid cooling by quenching from the solution treatment temperature is essential to maintain the metastable solution. Water is the quench medium of choice for aluminum alloys and the recommended water temperature for most commercial quenching is room temperature, 65°C (150°F), 80°C (180°F), or water near the boiling point. Quenching has also been accomplished in oil, salt baths, and organic solutions.

Precipitation hardening or aging (T6 condition) is performed after the solution treatment and quenching (T4 condition) to get the increased strength and hardness (although at a loss of ductility). Natural aging at room temperature following quenching can produce some hardening [49]. In case of alloy 356.2, natural aging can be varied between one and eight hours to get desirable mechanical properties in castings such as engine blocks (one hour natural aging) and steering knuckle and cylinder heads (four to eight hours natural aging). In the case of B206 alloy, natural aging of one day is recommended to get the benefits of artificial aging [50]. However, artificial aging at temperatures ranging from 95 to 260°C (200 to 500°F) depending on alloy and properties desired accelerates the aging process. Aging curves have been developed to facilitate process selection [13]. In general, longer times at lower aging temperatures result in higher peak strength. Overaging (T7 condition) consists of carrying out the aging cycle to a point beyond peak hardness.

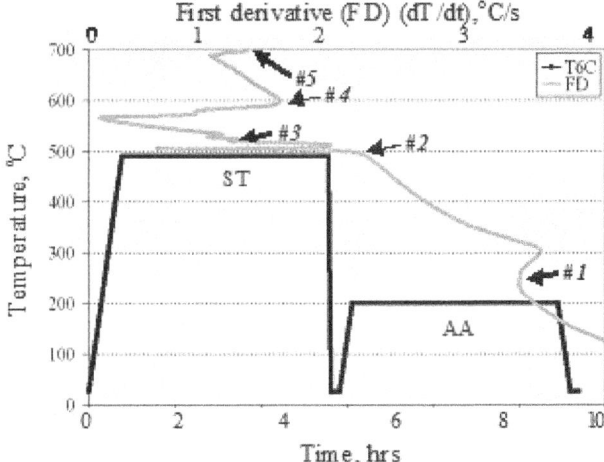

Superimposed First derivative (FD) of the Al-20%Si alloy showing transformation temperatures during the melting cycle on the conventional heat treatment schedule (T6C#1: solution at 490°C/4hrs and aging at 200°C/4hrs). The numbered arrows correspond to the following metallurgical reactions:

#1 - solid state precipitation (beginning of the over-aging reaction).

#2 - beginning of the Cu, Mg based phase melting (503.6°C).

#3 - end of the Cu, Mg based phases melting (534.1°C).

#4 - end of the Al-Si eutectic melting (580.2°C).

#5 - end of the primary Si crystals melting (701.9°C).

FIGURE 7.41   Thermal analysis of Al-20% Si alloy. (From Kasprzak et al. [48].)

Annealing of castings is done to relieve residual stress. A typical annealing practice consists of two to 4 hours, exposure at a minimum temperature of 345°C (650°F).

Mechanical and Physical Properties

Typical mechanical and physical properties of some common aluminum casting alloys are summarized in Tables 7.14 and 7.15, respectively. The mechanical properties for sand and permanent mold casting conditions are for both cast and heat treated conditions and are aimed at comparing with other alloys. In the case of die-casting alloys, as-cast mechanical properties are included except alloy 390, where T5 condition mechanical properties have been added. In general, the mechanical properties of casting alloys are dependent on alloying, heat treatment, and casting conditions.

Other mechanical properties such as fatigue strength, creep, endurance limit, compressive strength, and shear strength are also important from a design point of view. These data can be obtained from pertinent specifications, design standards, and producers.

The physical properties include solidification range, electrical conductivity, thermal conductivity, and coefficient of thermal expansion.

Alloy	Temper	Ultimate Tensile Strength		0.2% Offset Yield Strength		Elongation in 50 mm (2 in)	HB*
		MPa	Ksi	MPa	Ksi		
Sand Casting Alloys							
201.0	T43	414	60	255	37	17.0	–
	T6	448	65	379	55	8.0	130
	T7	467	68	414	60	5.5	–
A206.0	T4	354	51	250	36	7.0	–
319.0	F	186	27	124	18	2.0	70
	T5	207	30	179	26	1.5	80
	T6	250	36	164	24	2.0	80
355.0	F	159	23	83	12	3.0	–
	T51	193	28	159	23	1.5	65
	T6	241	35	172	25	3.0	80
	T61	269	39	241	35	1.0	90
	T7	264	38	250	26	0.5	85
	T71	241	35	200	29	1.5	75
C355.0	T6	269	39	200	29	5.0	85
356.0	F	164	24	124	18	6.0	–
	T51	172	25	138	20	2.0	60
	T6	228	33	164	24	3.5	70
	T7	234	34	207	30	2.0	75
	T71	193	28	145	21	3.5	60
A356.0	F	159	23	83	12	6.0	–
	T51	179	26	124	18	3.0	–
	T6	278	40	207	30	6.0	75
	T71	207	30	138	20	3.0	–
357.0	F	172	25	90	13	5.0	–
	T51	179	26	117	17	3.0	–
	T6	345	50	296	43	2.0	90
	T7	278	40	234	34	3.0	60
A357.0	T6	317	46	248	36	3.0	85
A390.0	F	179	26	179	26	<1.0	100
	T5	179	26	179	26	<1.0	100
	T6	278	40	278	40	<1.0	140
	T7	250	36	250	36	<1.0	115
443.0	F	131	19	55	8	8.0	40
A444.0	F	145	21	62	9	9.0	–
	T4	159	23	62	9	12.0	–
520.0	T4	331	48	179	26	16.0	75
A535.0	F	250	36	124	18	9.0	65

Table 7.14 Typical Mechanical Properties of Aluminum Casting Alloys (*Continued*)

Alloy	Temper	Ultimate Tensile Strength		0.2% Offset Yield Strength		Elongation in 50 mm (2 in)	HB*
		MPa	Ksi	MPa	Ksi		
712.0 (h)	F	241	35	172	25	5.0	75
850.0	T5	138	20	76	11	8.0	45
Permanent Mold Casting Alloys							
A206.0	T4	431	62	264	38	17.0	–
	T7	436	63	347	50	11.7	–
319.0	F	185	27	125	18	2.0	85
	T6	248	36	165	24	2.0	95
A356.0	T61	283	41	207	30	10.0	90
357.0	F	193	28	103	15	6.0	–
	T51	200	29	145	21	4.0	–
	T6	359	52	296	43	5.0	100
A357.0	T6	359	52	290	42	5.0	100
A390.0	F	200	29	200	29	<1.0	110
	T5	200	29	200	29	<1.0	110
	T6	310	45	310	45	<1.0	145
	T7	262	38	262	38	<1.0	120
A444.0	F	165	24	76	11	13.0	44
	T4	159	23	69	10	21.0	45
513.0	F	186	27	110	16	7.0	60
711.0	F	248	36	130	36	8.0	70
850.0	T5	159	23	76	11	12.0	45
Die-Casting Alloys							
360.0	F	324	47	172	25	3.0	75
A360.0	F	317	46	165	24	5.0	75
364.0	F	296	43	159	23	7.5	–
380.0	F	331	48	165	24	3.0	80
A380.0	F	324	47	159	23	4.0	75
384.0	F	324	47	172	25	1.0	–
390.0	F	279	40.5	241	35	1.0	120
	T5	296	43	265	38.5	1.0	–
392.0	F	290	42	262	38	<0.5	–
413.0	F	296	43	145	21	2.5	80
A413.0	F	241	35	110	16	3.5	80
513.0	F	276	40	152	22	10.0	–
515.0	F	283	41	–	–	10.0	–
518.0	F	310	45	186	27	8.0	80

*500 kg (1100 lb) load on 10 mm (0.4 in) ball.

Source: From *ASM Handbook* [1, p. 1080] (sand casting), Davis [12, p. 113] (permanent mold and die casting).

TABLE 7.14 Typical Mechanical Properties of Aluminum Casting Alloys (*Continued*)

Alloy	Temper and Product Form	Approximate Melting Range		Electrical Conductivity, % IACS	Thermal Conductivity at 25°C (77°F), W/m·K	Coefficient of Thermal Expansion per °C × 10⁻⁶ (per °F × 10⁻⁶)	
		°C	°F			20–100°C (68–212°F)	20–300°C (68–570°F)
201.0	T6 (S)	570–650	1060–1200	27–32	0.29	34.7 (19.3)	44.5 (24.7)
206.0		570–650	1060–1200	–	0.29	–	–
A206.0		570–650	1060–1200	–	0.29	–	–
319.0	F (S)	520–605	970–1120	27	0.27	21.6 (12.0)	24.1 (13.4)
	F (P)	520–605	970–1120	28	0.28	21.6 (12.0)	24.1 (13.4)
354.0	F (P)	540–600	1020–1150	32	0.30	20.9 (11.6)	22.9 (12.7)
355.0	T51 (S)	550–620	1020–1150	43	0.40	22.3 (12.4)	24.7 (13.7)
	T6 (S)	550–620	1020–1150	36	0.34	22.3 (12.4)	24.7 (13.7)
356.0	T6 (S)	560–615	1040–1140	39	0.36	21.4 (11.9)	23.4 (13.0)
A356.0	T6 (S)	560–610	1040–1130	40	0.36	21.4 (11.9)	23.4 (13.0)
357.0	T6 (S)	560–615	1040–1140	39	0.36	21.4 (11.9)	23.4 (13.0)
A357.0	T6 (S)	555–610	1030–1130	40	0.38	21.4 (11.9)	23.6 (13.1)
A380.0	F (D)	520–590	970–1090	27	0.26	21.1 (11.7)	22.7 (12.6)
390.0	F (D)	510–650	950–1200	25	0.32	18.5 (10.3)	–
413.0	F (D)	575–585	1070–1090	39	0.37	20.5 (11.4)	22.5 (12.5)
A413.0	F (D)	575–585	1070–1090	39	0.37	–	–
443.0	F (D)	575–630	1070–1170	37	0.35	22.1 (12.3)	24.1 (13.4)
A444.0	F (P)	575–630	1070–1170	41	0.38	21.8 (12.1)	23.8 (13.2)
520	T4 (S)	450–600	840–1110	21	0.21	25.2 (14.0)	27.0 (15.0)
535.0	F (S)	550–630	1020–1170	23	0.24	23.6 (13.1)	26.5 (14.7)
A535.0	F (D)	550–620	1020–1150	23	0.24	24.1 (13.4)	26.1 (14.5)
B535.0	F (S)	550–630	1020–1170	24	0.23	24.5 (13.6)	26.5 (14.7)
710.0	F (S)	600–650	1110–1200	35	0.33	24.1 (13.4)	26.3 (14.6)
711.0	F (P)	600–645	1110–1190	40	0.38	23.6 (13.1)	25.6 (14.2)
712.0	F (S)	600–640	1110–1180	40	0.38	23.6 (13.1)	25.6 (14.2)
713.0	F (S)	595–630	1110–1170	37	0.37	23.9 (13.3)	25.9 (14.4)
850.0	T5 (S)	225–650	440–1200	47	0.44	–	–
851.0	T5 (S)	230–630	450–1170	43	0.40	22.7 (12.6)	–
852.0	T5 (S)	210–635	410–1180	45	0.42	23.2 (12.9)	–

Note: S = sand cast, P = permanent mold, D = die cast. The specific gravity and weight data in this table assume solid (void-free) metal because some porosity cannot be avoided in commercial castings; their specific gravity or weight is slightly less than theoretical value.

Source: From *ASM Handbook,* [1, p. 1078].

TABLE 7.15 Typical Physical Properties of Aluminum Casting Alloys

Corrosion Resistance

Alloy composition and service environment affect the corrosion behavior of aluminum casting alloys. In general, resistance to many types of corrosion is an important virtue of aluminum alloys. Cooking utensils, food containers, food processing equipment, and outboard motors which operate in a variety of corrosive media are good examples. Alloys containing substantial amounts of copper have the poorest corrosion resistance.

Special corrosion problems may arise when aluminum castings are painted or coated. Filiform corrosion is encountered in alloys that have been clear coated.

Postcasting Operations

Commonly performed postcasting (secondary) operations include trimming and cleanup, inspection and testing, finishing, machining, heat treating, impregnating and straightening. Not all these operations are performed on every casting. Many castings, especially those produced by permanent mold or high-pressure die casting are finished with only a trimming operation to remove gates and flashes. Heat treating is one of the most crucial postcasting operations to get the desired mechanical properties.

Table 7.16 lists the various cleaning and trimming operations for aluminum castings [13].

Method	Purpose of Cleaning			
	Removing Refractories	Removing Gates and Risers	Finish Cleaning	Preparation for Welding or Repair
Wire brushing	X			
Air blastcleaning (sand, shot, glass)	X		X	
Mechanical blastcleaning	X		X	
Water blastcleaning	X		X	
Shearing		X		
Band sawing		X		X
Hack sawing		X		X
Grinding		X		X
Hand filing			X	X
Rotary filing			X	X
Polishing			X	
Brushing			X	
Buffing			X	
Tumbling (with or without media)			X	
Thermal deburring, etc.			X	

Source: From Zalensas [30, p. 326].

TABLE 7.16   Cleaning Methods Used for Aluminum Castings

Inspection and testing include visual inspection, dimensional inspection, and non-destructive testing such as fluorescent powder and dye penetrant inspection, radiographic inspection, and ultrasonic testing, pressure tightness testing, and mechanical and physical property tests.

Repairing of castings are sometimes needed to salvage defective castings. This involves straightening (to take care of warping and distortion), peening (to remove surface irregularities, small holes, and leaks), welding, and impregnation. Welding or weld repairing is usually done by the gas tungsten arc (GTA) and gas metal arc (GMA) processes because of the tenacious oxide layer which forms on aluminum when exposed to air. Impregnation is used to seal leaks caused by microporosity. Although the pressure impregnation technique is used extensively, the vacuum impregnation technique may be preferred to eliminate incomplete filling of the voids.

Hot isostatic pressing (HIP) can be used to improve the internal soundness or integrity, increase the density, and improve mechanical and fatigue properties of castings. However, it adds to the cost of the component. The casting in the pressure vessel is heated and pressure up to 105 MPa (15,000 Psi) is applied to perform "HIPing."

Protective or decorative finishes may be needed for some of the aluminum castings. Chemical conversion coatings or anodizing is done to produce a protective coating. Aluminum castings may be finished with a variety of lacquers or enamels for both decorative and protective purposes.

References

1. J. Fred Major *ASM Handbook*, Vol. 15, *Casting*, "Aluminum and Aluminum Alloys," 2008, pp. 1059–1084.

2. S. Udvardy, North American Die Casting Association (NADCA), private communication.

3. "Aluminum Alloys, Metal Casting Design and Purchasing," *2009 Casting Source Directory*, American Foundry Society, pp. 25–30.

4. W. Kasprzak, CANMET Materials Technology Laboratory, Ottawa, Ontario, Canada, private communication.

5. D. Weiss, ECK Industries, private communication.

6. A. Spada, American Foundry Society, private communication.

7. *ASM Handbook*, Vol. 15, *Casting*, "Aluminum and Aluminum Alloys," 1988, pp. 741–770.

8. R. C. Gibbons (ed.), *Woldman's Engineering Alloys*, 6th ed., American Society for Metals, 1979.

9. "Handbook of International Alloy Compositions and Designations," Metals and Ceramics Information Center, Battelle Memorial Institute, 1976.

10. J. Datta, *Aluminum-Schlüssel—Key to Aluminum Alloys*, 5th ed., Aluminium Verlag, Dusseldorf, Germany, 1997.

11. R. W. Heine, C. R. Loper, Jr., and P. C. Rosenthal, *Principles of Metal Casting*, McGraw-Hill Book Company, New York, 1967.

12. J. R. Davis (ed.), "Aluminum and Aluminum Alloys," *ASM Specialty Handbook*, ASM International, 1993.

13. J. G. Kaufman and E. L. Rooy (eds.), "Aluminum Alloy Castings, Properties, Processes, and Applications," *American Foundry Society and ASM International*, 2004.

14. *ASM Handbook*, Vol. 3, "Alloy Phase Diagrams," 1997.

15. F. Vinuk, M. Sahoo, and R. W. Smith, "The Hardness of Al-Si Alloys," *Journal of Materials Science*, 14: 975–982, 1979.

16. M. Sahoo and R. W. Smith, "Structure and Mechanical Properties of Modified Eutectics," *Canadian Metallurgical Quarterly*, 15: 1–6, 1976, Maney Publishing.

17. *ASM Handbook*, 1948 edition. Metals Park, OH.

18. F. A. Fasoyinu, M. Sahoo, and S. Sikorski, "Hot Tearing of Aluminum Alloys 206 and 535 Poured in Permanent Molds," *Proceedings of the 6th International Conference on Permanent Mold Casting of Aluminum and Magnesium*. Organized by AFS, Dallas, Texas, Feb. 11–12, 2008.

19. X. Jian, C. Xu, T. T. Meek, and Q. Han, "The Effect of Ultrasonic Vibration on the Solidification Structure of A356 Alloy," *AFS Transactions*, Paper No. 05-085, 2005.

20. X. Jian, Q. Han, H. Xu, and T. T. Meek, "Solidification of Aluminum Alloy A356 under Ultrasonic Vibration," *Solidification of Aluminum Alloys*, M. G. Chu, D. A. Granger, and Q. Han (eds.), The Minerals, Metals and Materials Society, Warrendale, PA, 2004, pp. 73–79.

21. L. Bakerud and Y. Shao, "Grain Refining Mechanisms in Aluminum as a Result of Additions of Titanium and Boron," *Part I, Aluminum*, 67(6–8): 780–785, Jul.-Aug. 1991.

22. L. Bakerud, M. Johnson, and P. Gustafson, "Grain Refining Mechanisms in Aluminium as a Result of Additions of Titanium and Boron," *Part II, Aluminium*, 67(9): 910–915, Sep. 1991.

23. J. H. Gruzleski and B. M. Closset, "The Treatment of Liquid Aluminum-Silicon Alloys," *American Foundry Society*, Des Plaines, IL, 1990.

24. D. Apelian, G. K. Sigworth, and K. R. Whaler, "Assessment of Grain Refinement and Modification of Al-Si Foundry Alloys by Thermal Analysis," *Transactions of AFS*, 92: 297–307, 1984.

25. S. Shankar, McMaster University, private communication.

26. W. Kasprzak, H. Kurita, J. H. Sokolowski, and H. Yamagata, "Energy Efficient Tempers for Aluminum Motorcycle Cylinder Blocks," *Advanced Materials and Processes*, Mar. 2010, pp. 24–27.

27. W. Kasprzak, M. Sahoo, J. Sokolowski, H. Yamagata, and H. Kurita, "The Effect of the Melt Temperature and the Cooling Rate on the Microstructure of the Al-20% Si Alloy Used in Monolithic Engine Blocks," *International Journal of Metal Casting*, 3: 55–71, 2009.

28. H. Yamagata, W. Kasrzak, M. Aniolek, H. Kurita, and J. H. Sokolowski, "The Effect of Average Cooling Rates on the Microstructure of the Al-20% Si High Pressure Die Casting Alloy Used for Monolithic Cylinder Blocks," *Journal of Materials Processing Technology*, 203: 333–341, 2008.

29. J. E. Gruzleski, "Microstructure Development during Metal Casting," American Foundrymen's Society, Inc., Des Plaines, IL, 2000.

30. D. L. Zalensas (ed.), *Aluminum Casting Technology*, 2d ed., American Foundry Society, 1993.

31. S. Udvardy, Des Plaines, IL NADCA, 2008.

32. NADCA, *Introduction to Die Castings*, 1999.

33. D. Schwam, "Vacuum Die Casting," NADCA Publication No. 528, 2007.

34. M. R. Ghomashchi and A. Vikhrov, "Squeeze Casting: An Overview," *Journal of Materials Processing Technology*, 101(1–3): 1–9, 2000.

35. H. F. Shroyer, "Cavityless Casting Mould and Method of Making Same," U.S. Patent No. 2830343, Apr. 15, 1958.

36. R. Bailey, "Understanding the Evaporative Foam Casting Process (EPC)," *Modern Casting*, Apr. 1982, pp. 58–61.

37. R. W. Monroe, "Expendable Pattern Casting," American Foundry Society, 1992.

38. Staff Report, "Engine Design Freedom with Lost Foam Casting," *Engineered Casting Solutions*, 6(4): 28–32, 2004.

39. Y. Fasoyinu, P. Newcombe, and M. Sahoo, "Lost Foam Casting of Magnesium Alloys AZ91D and AM50," *AFS Transactions*, 114: 707–718, 2006.

40. J. L. Dion, R. D. Warda, R. K. Buhr, J. R. Emmett, and M. Sahoo, "Production of Aluminum Alloy Castings Using Evaporative Pattern and Vacuum Techniques," *AFS Transactions*, 98: 131–140, 1990.

41. Y. Fasoyinu, H. Chioren, P. Newcombe, and M. Sahoo, "Vacuum Assisted Lost Foam Casting of Magnesium Alloy AZ91E," *AFS Transactions*, Vol. 113, Paper 05-53, 2005.

42. M. Sands and S. Shivkumar, "EPS Bead Fusion Effect on Fold Defect of Aluminum Alloys," *Journal of Materials Science*, 41: 2373–2379, 2006.

43. S. Jagoo, C. Ravindran, and D. Nolan, "Fold Defects in Aluminum A356 Lost Foam Casting," *"Transactions of the American Foundry Society"*, Vol. 115, Paper No. 07-068, 2007, pp. 1–6.

44. J. Grassi, J. Campbell, M. Hartlieb, and F. Major, "Ablation Casting," *Proceedings on Aluminum Alloys: Fabrication, Characterization and Applications*, W. Yin and S. K. Das (eds.), *The Minerals, Metals & Materials Society*, 2008, pp. 73–77.

45. J. Grassi, D. Weiss, and B. Cox, "Ablation Casting," *Technology for Magnesium Castings: Design, Products and Applications*, M. Sahoo as principal author, American Foundry Society, 2011, pp. 197–201.

46. J. H. Sokolowski, M. B. Djurdjevic, C. A. Kierkus, and D. O. Northwood, "Improvement of 319 Aluminum Alloy Casting Durability by High Temperature Solution Treatment," *Journal of Materials Processing Technology*, 109: 174–180, 2001.

47. W. Kasprzak, H. Onda, M. Aniolek, J. H. Sokolowski, and K. Akiyama, "Characterization of the AC2A Cylinder Head and Development of Its Heat Treatment Process," *Proceedings of the International Conference, on Aluminum Alloys*, Yokohama, Japan, Sep. 5–9, 2009.

48. W. Kasprzak, J. H. Sokolowski, H. Yamagata, and H. Kurita, "Development of Energy Efficient Heat Treatment Processes for Light Weight Automotive Castings," *Heat Treating Proceedings of the 25th ASM Heat Treating Society Conference*, F. Specht, S. MacKenzie, and D. Weires (eds.), Sep. 14–17, 2009, Indianapolis, IN, USA.

49. J. Manickaraj, G. Y. Liu, and S. Shankar, "Effect of Incubation Coupled with artificial Aging in T6 Heat Treatment of A356.2 Aluminum Casting Alloy", Vol. 15, Issue 4, p. 17–36, 2011.

50. D. Jean, J. F. Major, J. H. Sokolowski, B. Warnock, and W. Kasprzak, "Heat Treatment and Corrosion Resistance of B206 Aluminum Alloy," *AFS Transactions*, 117: 113–129, 2009.

CHAPTER 8

Copper and Copper Alloys

Mahi Sahoo
Suraja Consulting Inc.

Introduction

Historically, copper-alloy castings were among the earliest metallic objects made by man from molten metal [1]. Since copper could be found as native metal, it has been worked into artifacts, far back into antiquity. Its melting point and that of its alloys with aluminum, gold, tin, and zinc are low enough to be within the range of temperatures which can be reached by wood and charcoal fires. Copper melting and casting by artisans is known to have occurred as early as 3000 BC. The full value of copper alloys as casting materials, however, had to await the metallurgical developments of the past several hundred years, discoveries that made the metal more abundant, with a greater variety of useful properties. Foundry-production data for copper-alloy castings during the past few years are presented Table 8.1 [2]. It may be noted that sand-casting processes account for the greatest percentage of castings produced in these alloys.

Advantages of Copper Alloys

Certain engineering advantages are inherent in the use of copper alloys for castings. Some of these include

- High electrical and thermal conductivity
- Good corrosion resistance
- Appearance (pleasing color)
- Nontoxicity
- Good bearing qualities
- Antimicrobial properties
- Good biofouling resistance
- Alloying to produce industrial alloys
- Better wear and abrasion resistance

Year	Aluminum	Copper Base	Magnesium	Zinc	Other	Total Non-Ferrous
1999	2,042	329	68	400	48	2,887
2000	2,052	325	74	420	47	2,918
2001	1,910	300	78	322	45	2,655
2002	2,047	294	66	400	45	2,852
2003	2,152	305	81	380	50	2,968
2004	2,165	315	88	360	52	2,980
2005	2,293	322	111	345	54	3,125
2006	2,115	317	116	334	57	2,939
2007	2,036	312	121	328	59	2,856
2008	1,918	303	121	302	58	2,702
2009	1,191	179	66	166	120	1,722
2010	1,234	265	106	204	60	1,869
2011	1,523	263	99	181	60	2,126
2012	1,591	279	102	200	69	2,241

Source: From Spada [2].

TABLE 8.1 U.S. Nonferrous Casting Shipments (000 tons)

Although the above items are all favorable to the use of copper-base casting alloys, it must be recognized that certain limitations prevail. Cost is a factor of great import, being sufficiently high so that copper-base castings are not used unless their special advantages, as listed above, present a real engineering or economic advantage over other metals. Figure 8.1 shows some characteristic uses of copper-base castings [3].

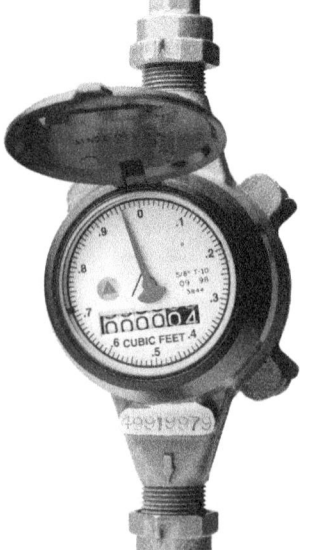
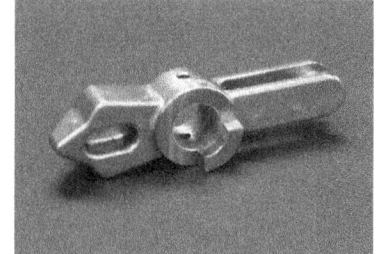

FIGURE 8.1 Typical copper castings. (From Michel [3], courtesy of PIAD.)

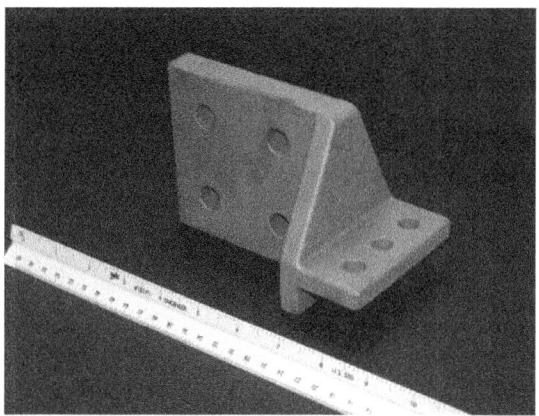
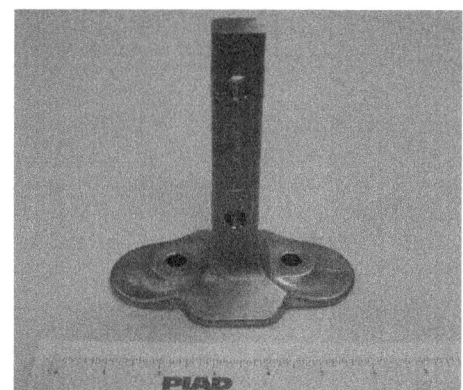

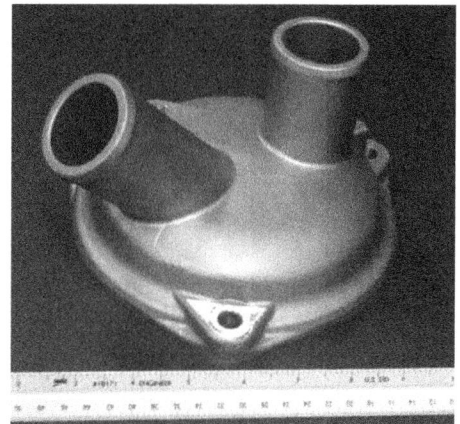

FIGURE 8.1 *(Continued)*

Alloying Principles

Pure copper is very soft. Higher strength properties are required for structural and various industrial applications and to compete with other metals and alloys such as steel, ductile iron, aluminum, etc. Pure copper is also difficult to cast as it is prone to cracking, gas porosities, etc. The casting characteristics, mechanical properties, and other physical properties of copper are improved by the addition of alloying elements such as Zn, Sn, Al, Ni, Si, Pb, Cr, Ag, Be, etc. Some of the major and minor alloying elements that are usually added to copper to get desirable properties are described below [4].

Zinc

Zinc is a relatively inexpensive, yet potent, alloying element in copper, imparting strength and hardness by the mechanism of solid-solution strengthening. Zinc is the major alloying element in brasses. It is either added alone or in combination with other elements. Copper alloys may contain between 1 and 45% zinc. The solid solubility of zinc in copper is about 36% which results in a single phase microstructure. Higher levels of zinc promote mixed microstructure in alloys known as alpha-beta brass.

Tin

Tin is the oldest known alloying element in copper and is a key alloying element in many bronzes. Tin is also a potent solid-solution strengthener in copper, even more so than zinc, as less tin is required for an equivalent increase in strength. Unlike zinc, tin also improves corrosion resistance. Tin reduces the melting temperature of copper, increases the fluidity, thereby making these alloys easier to melt and cast. However, tin increases the solidification range of copper alloys. The level of tin and zinc in thin section greensand casting alloys should not exceed 11.5% and even less in thicker section castings, otherwise porosity becomes an issue. Strength can be regained by addition of nickel.

Lead

Although lead is insoluble in copper at room temperature, many commercial brasses and bronzes contain significant amounts of lead. Lead solidifies last, so it is found at the grain boundaries or in interdendritic areas. The addition of lead helps to seal the normal shrinkage porosity formed toward the end of solidification. Leaded copper alloys have improved pressure tightness as a result of this sealing of these interdendritic spaces. However, lead is detrimental to pressure tightness in Cu-Sn-Ni alloys. High-lead alloys are used for bearings, where the lead on machined surfaces eliminates the tendency of copper parts to gall or bond to each other. Lead also improves machinability by providing a natural break, so that the removed metal forms chips rather than metal ribbons that foul the tool bits.

Aluminum

The solubility of aluminum in copper is limited to 9%. Aluminum improves the strength of copper, so high-strength copper alloys such as aluminum bronzes contain significant quantities of aluminum. The high-strength yellow brasses are strengthened by aluminum and contain 1 to 6% aluminum (nominal). Small amounts of aluminum (0.3%)

improve the fluidity of copper alloys, such as yellow brasses and silicon brasses for permanent mold casting [5].

Silicon

Copper-silicon alloys contain less than 5% silicon. Like aluminum, silicon forms a solid solution with copper. The excellent fluidity of copper-silicon alloys enables them to be used for art castings and permanent molding.

Nickel

Copper and nickel are completely soluble in both the liquid and solid phases. Copper alloys with higher nickel contents are known as nickel silvers due to their white color. Copper-nickel alloys are highly resistant to corrosion, so the alloys are used in the chemical, petroleum, food, and dairy industries. Nickel improves the quality, strength, and creep resistance of tin bronze and semi-red brass castings, and is more effective than lead in improving pressure tightness.

Beryllium

Beryllium is added to copper in small quantities to increase strength by precipitation hardening.

Chromium

Chromium is also added to high-copper alloys to increase strength by precipitation hardening with only a minor loss in the electrical conductivity.

Iron

Iron is not soluble in copper. However, iron is used as a grain refiner in yellow brasses, high-strength yellow brasses, and aluminum bronzes. Although the mechanism of the refinement is not well understood, iron increases the melting temperature of the copper alloys and iron additions to the liquid alloys at higher temperatures result in precipitates in the liquid which persist in the casting. Excessive iron levels can cause segregation and hard spots in these alloys. In tin bronzes and semi-red brass, iron increases the strength and hardness, but at great cost to ductility. For these alloys, iron and tin should be less than 0.01 and 4%, respectively, if minimum magnetic susceptibility is required.

Minor Additions

Antimony

Antimony is used in small quantities to reduce the specific corrosion problem, known as dezincification, when high-zinc brasses are to be used for valves, etc. Antimony (up to 0.25%) reduces the strength and ductility of tin bronzes. Recently, new low-lead semi-red brasses and tin bronzes have been developed where 0.1 to 1.5% Sb has been added along with 0.1 to 0.65% S to get better mechanical properties and good machinability in comparison with the leaded counter parts.

Bismuth

Bismuth, when present in pure copper and some high-strength alloys such as aluminum bronzes, is considered an impurity. Bismuth embrittles these alloys even when present as low as 0.01%. However, bismuth can be used as an alternate for lead in low-lead red brasses and yellow brasses to enhance machinability.

Selenium

Selenium is added to red brasses, dairy metal, and copper-silicon alloys to improve the subsequent machining of the castings for use. Selenium can also be used along with bismuth.

Manganese

Manganese is used to neutralize deleterious effects of iron in red brasses. Also, it can increase the strength of yellow brasses and aluminum bronzes to a certain extent, but is deleterious to 88-10-2 Cu-Sn-Zn and 85-5-5-5 leaded red brass.

Phosphorus

Phosphorus is used to deoxidize copper, bronzes, red brasses, and bismuth-containing brasses, and bronzes. Although phosphorus reduces conductivity, it is used to lower the oxygen level in high-copper alloys then followed by the addition of other costlier elements, such as boron, to eliminate the last traces of oxygen. Phosphorus is added to red brasses to reduce oxygen levels, but residual levels are limited to 0.02% when the metal is to be cast in greensand because of potential reactions with water in the sand. The structure, as well as fluidity and castability of copper alloys, are dramatically improved when oxides are eliminated. Although 0.05% phosphorus improves the machinability of red brasses and copper-silicon alloys, it is only used in copper-silicon alloys when cast in permanent molds.

Sulfur and Arsenic

Sulfur and arsenic levels are no longer issues for alloys made from primary copper, since the advent of electrolytic purification of copper. Sulfur, if present, can react with oxygen at higher pouring temperatures and cause gas porosity. In recent years 0.2–0.6 % sulfur has been added as an alloying element to red brasses to replace lead in view of the Environmental Protection Agency (EPA) restriction on lead content in potable water. These alloys exhibit good mechanical properties and machinability.

Impurities

Impurities in copper alloys occur as a result of processing issues of the original ore bodies and from processing and recycling. Arsenic, antimony, bismuth, and sulfur are examples of impurities from the original ores, while iron, silicon, and aluminum chiefly occur from recycling and processing. Nevertheless, different impurity elements affect each alloy family. Lead and bismuth are considered impurities in aluminum bronzes, while aluminum and silicon are harmful in red brasses. The tolerance to impurities depends on the alloy and the intended use. Most critical levels (if known) are included in the specifications.

Alloying Systems and Specifications

Copper alloys are identified by the unified numbering system (UNS). This system was developed during the 1970s to give a precise description of the composition range for each alloy, and is now an internationally recognized designation system. The UNS is managed jointly by the American Society for Testing and Materials, and the Society of Automotive Engineers. The Copper Development Association administers the registration, numbering, compilation, and compositions of specific alloys. Cast alloys are numbered from C80000 through C99999 [4]. The numbers for families of copper alloys are grouped by composition (Table 8.2). As the designations are based on precise compositional limits for the final castings, the analyses of commercial ingots may differ slightly from the designated composition. Such variances anticipate processing conditions in

Family	UNS Numbers
Coppers	C80001–C81399
High copper	C81400–C83299
Copper-tin-zinc (red brasses)	C83300–C83999
Copper-tin-zinc (semi-red brasses)	C84xxx
Copper-zinc (yellow brasses)	C85xxx
Manganese bronzes	C86xxx
Copper-silicon	C87xxx
Copper-bismuth	C88xxx, C89xxx
Copper-tin (tin bronzes)	C90xxx, C91xxx
Copper-tin-lead (leaded tin bronzes)	C92xxx
Copper-tin-lead (high leaded tin bronzes)	C93xxx–C94500
Copper-tin-nickel (nickel-tin bronzes)	C94600–C94900
Copper-aluminum-iron Copper-aluminum-nickel-iron (aluminum bronzes)	C95xxx
Copper-nickel-iron (nickel coppers)	C96xxx
Copper-nickel-zinc (nickel silvers)	C97xxx
Copper-lead	C98xxx
Special alloys	C99xxx

Source: From *Metals Handbook* [4].

TABLE 8.2   UNS Classification of Copper Casting Alloys

which some elements are lost by oxidation or volatilization, and impurities gained when the ingots and rejects are remelted to produce satisfactory castings.

Coppers

Physical properties of pure copper are given in Table 8.3 [6]. According to the International Annealed Copper Standard (IACS), the specific electrical resistance of pure annealed copper at 20°C is 1.7241 $\mu\Omega\cdot$cm, and its conductivity is, therefore, 0.5801 reciprocal microhm-cm. The conductivity of any particular batch or specimen of copper is defined as a percentage of this value. In the fully annealed condition, modern high conductivity copper frequently has a conductivity of 101% or even 102% IACS, but after heavy reduction by cold drawing it may be no more than about 97%. The temperature coefficient of resistance for constant mass is 0.0393 per °C at 20°C. Oxygen, as cuprous oxide, has little or no effect on the conductivity, and other elements are only present in very small amounts in high conductivity copper.

Conductivity copper castings are used for a wide variety of electrical and thermal conductivity applications. Fittings, cable connectors, cable dead ends, spacers, inductor heads, switch parts, etc., require differing degrees of conductivity of copper. These castings are ordinarily of high-copper content because other elements adversely affect the conductivity of copper. This idea is illustrated graphically in Fig. 8.2 [7]. Note the drastic effect of phosphorus on conductivity in Fig. 8.2. Deoxidation with P must therefore be carefully controlled to keep the residual percentage low, below 0.01%, preferably. Although the best conductivity is obtained with pure copper, the metal is soft, low in strength, difficult to machine, and has less desirable casting properties than many of its alloys.

	English Units	Metric Units
Melting point (liquidus)	1981°F	1083°C
Melting point (solidus)	1948°F	1065°C
Density	0.323 lb/in^3 at 68°F	8.94 g/cm^3 at 20°C
Specific gravity	8.94	8.94
Coefficient of thermal expansion	0.0000094 per °F from 68 to 572°F	0.0000169 per °C 68 to 572°F
Thermal conductivity	226 Btu/ft^2/ft/hr/°F at 68°F	0.934 cal/q cm/cm/s/°C at 20°C
Electrical conductivity*	100% IACS at 68°F	0.580 MΩ·cm at 20°C
Specific heat	0.092 Btu/lb/°F at 68°F	0.092 cal/g/°C at 20°C
Modulus of elasticity (tension)	17,000 Ksi	11,950 kg/sq mm

*Volume basis in as-cast condition except for precipitation hardening alloys which are in the full heat-treated condition.

Source: From *Standards Handbook* [6].

TABLE 8.3   Physical Properties of Pure Copper

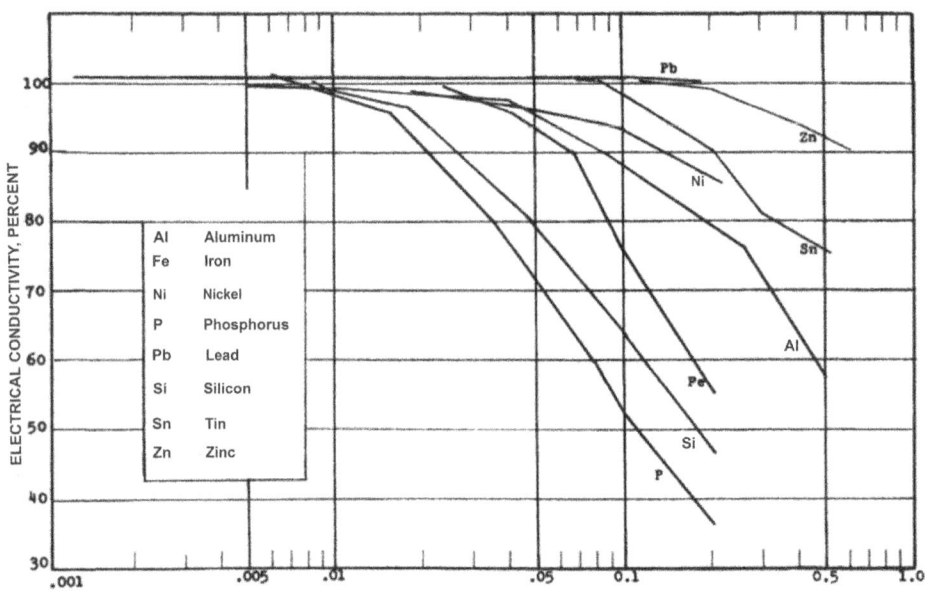

FIGURE 8.2   Effect of small percentages of elements on electrical conductivity of copper. (From American Foundry Society [7].)

Copper Alloys

Copper alloys can be classified into groups such as high-copper alloys, brasses, bronzes, copper-nickels, leaded coppers, and special alloys as described below [4].

High-Copper Alloys

The cast high-copper alloys have designated copper contents in excess of 94%. Copper is alloyed with other elements along with silver to obtain properties not achievable using pure copper. High-copper alloys are unique in that they combine high strength with high thermal and electrical conductivity, two properties which are seldom found together in the same material. These high coppers are alloyed with small amounts of elements including beryllium, silicon, nickel, tin, zinc, and chromium, constituted to have improved strength properties compared to high-purity copper, while maintaining a minimum of 85% conductivity, and are widely used for cast electrical conducting members.

Chromium Coppers

Chromium coppers are alloys containing up to 1.5% chromium. The strength of chromium copper is approximately twice that of pure copper, while the electrical conductivity is only 20% lower than that of pure copper. Other potential alloying agents are less effective and cause greater loss of conductivity. Typical applications for chromium copper are welding clamps and high-strength electrical connectors. Chromium-copper castings can be heat treated for precipitation strengthening.

Beryllium Copper

The beryllium coppers contain up to 2.7% Be along with some nickel and cobalt. These alloys can be precipitation hardened to obtain high strength and high conductivity. These alloys are to be handled carefully due to the toxic nature of beryllium with particular attention to any fumes and machining dusts.

Brasses

These alloys contain zinc as the principal alloying element, with or without other designated alloying elements such as tin, lead, iron, aluminum, nickel, and silicon. The cast alloys comprise three major families of brasses: copper-tin-zinc alloys (red, semi-red, and yellow brasses); manganese-bronze alloys (high-strength yellow brasses); and copper-zinc-silicon alloys (silicon brasses and bronzes). Alloys with more than 20% zinc are prone to selective corrosion known as *dezincification*.

Red Brasses

The red brasses are alloys with up to 8% zinc and may contain varying amounts of tin. The color of red brass is attributable to its relatively low zinc content. The highest volume cast red brass alloy, commonly known as 85-5-5-5, as it contains 85% copper, 5% tin, 5% lead, and 5% zinc, has been used commercially for several hundred years and accounts for more tonnage than any other cast copper alloy.

In foundry alloys, lead may be present in various amounts to promote pressure tightness in castings in service and to facilitate free machining during the manufacturing process. Most commercial red and semi-red brasses invariably contain more than 1% lead, although newer lead-free versions of these alloys are now available. Early versions of lead-free alloys contained both bismuth and selenium. These elements act synergistically to improve machinability. However, selenium has become expensive, and because of the low ductility of Cu-Bi-Se alloys, other alloys containing Bi are being promoted for potable water (plumbing) applications.

Semi-Red Brasses

Semi-red brasses contain 8 to 17% zinc in addition to tin and lead as other alloying elements.

Yellow Brasses

The yellow brasses are even lower in cost than the red brasses because their zinc content is higher, at 20 to 39% Zn. Other than zinc, these alloys contain low levels of tin, lead, and aluminum. Yellow brass has a pleasant yellow color which can be polished to a high luster. This accounts in part for its selection for decorative hardware. It is important to know the zinc equivalent of these brasses as given below:

$$\% \text{ zinc equivalent} = 100 - \frac{100 \times \% \text{ copper}}{100 + A} \tag{8.1}$$

where A is the algebraic sum of the values obtained from multiplying the percentage of each metal present (exclusive of Cu and Zn) by its "zinc replacement coefficient" diminished by one.

The zinc replacement coefficients are as follows:

	Zinc Replacement Coefficient (X)	Zinc Replacement Coefficient Diminished by 1
Tin	+2.0	+1.0
Aluminum	+6.0	+5.0
Manganese	+0.5	−0.5
Iron	+0.9	−0.1
Lead	+1.0	0.0
Nickel	−1.3	−2.3
Silicon	+10.0	+9.0
Other impurities	+1.0	0.0

Another formula for a zinc equivalent is [8]

$$\% \text{ zinc equivalent} = (100 \times X)/(X + Cu\%) \tag{8.2}$$

where X is the total of zinc equivalents contributed by the added alloying elements (which are the same as the zinc replacement coefficient shown above) plus the percentage of actual zinc present in the alloy.

The zinc equivalent calculation is important for high-strength yellow brasses used for casting application. For such application, the zinc equivalent should be below 45%.

High-Strength Yellow Brasses

High-strength yellow brasses, also known imprecisely as manganese bronzes, contain high zinc contents along with significant quantities of manganese, aluminum, and iron. Iron in these alloys causes fine grains which enhances the properties. Tin and lead may be present in these alloys as well. Ship propellers are the most important application for these alloys.

Silicon Bronze and Silicon Brass

The silicon bronzes and silicon brasses are essentially alloys with up to 16% zinc and up to 5.5% silicon. They have low melting points and high fluidity which are desirable for art castings, permanent mold, and pressure die-casting applications. Eco Brass® belongs to this family and it is a lead-free brass with good mechanical properties and mahcinability , suitable for drinking water applications.

Bismuth Brasses and Bronzes

In recent times, in response to the restrictions on the lead content in drinking water, brasses and bronzes with very low levels of lead (less than 0.09%) were developed. These alloys often contain bismuth to achieve the same machinability and casting characteristics as the leaded alloys. Sometimes selenium can be added to these alloys to enhance machinability further. Other alloying additions such as tin, zinc, iron, and nickel remain the same as in red or yellow brasses.

Antimony and Sulfur Containing Brasses and Bronzes

Sulfur-containing low-lead red brass and antimony and sulfur containing low-lead semi-red brass and tin bronze have been developed with good mechanical properties and machinability. Commercialization of these alloys are in progress. Their compositions and mechanical properties can be found by visiting the Website for the Copper Development Association, New York.

Bronzes

Bronze is now an imprecise classification, but originally bronze referred to alloys in which tin was the major alloying element. Today, the term bronze applies to a broader class of alloys in which the principal alloying element is neither zinc nor nickel. Nevertheless, the name bronze is in common use for a number of alloys which contain little, if any, tin. The term is generally used with a modifier rather than bronze alone. The foundry alloys include four main families of bronzes: copper-tin alloys (tin bronzes); copper-tin-lead alloys (leaded and high-leaded tin bronzes); copper-tin-nickel alloys (nickel-tin bronzes); and copper-aluminum alloys (aluminum bronzes). The alloys known as manganese bronzes, in which zinc is the major alloying element, are in fact high-strength yellow brasses, and are included in the brass category above.

Tin Bronzes

Modern cast tin bronzes are very similar to the alloys found in the many relics from the "Bronze Age," over 3500 years ago. They are basically alloys of copper and tin, where the tin content can be as high as 20%. The good corrosion resistance of tin bronzes to aqueous solutions accounts, in part, for the survival of these "Bronze Age" relics to this day. Additional attributes of tin bronzes include reasonably high strength, good wear resistance, and a lower coefficient of friction when compared to steel, making these alloys very useful for bearings, piston rings, and gears.

Leaded Tin Bronzes

High-leaded tin bronzes containing more than 7% tin and 7% lead are used for bearings, bushings, as well as pump and impeller components with high corrosion resistance and strength.

Aluminum Bronzes

Aluminum bronzes have complex metallurgical structures. Alloying elements always include aluminum and varying amounts of manganese, iron, and in some versions, nickel. Aluminum imparts both strength and oxidization resistance by virtue of the formation of alumina (Al_2O_3)-rich protective films. These alloys are very wear resistant, and exhibit good casting and welding characteristics. They are resistant to corrosion in many environments including seawater, chlorides, and dilute acids. Applications are varied and include propellers, valves, pickling hooks, pickling baskets, and wear rings. Aluminum bronzes and particularly nickel-aluminum bronzes are useful in fluid-moving applications, such as pump impellers, as these alloys have superior resistance to erosion, corrosion, and cavitation. They are also used to cast propellers for ships and submarines. Aluminum bronzes are used in permanent mold casting in addition to sand casting, as aluminum improves the fluidity of the liquid and enables complex shapes to be produced.

Copper-Nickel Alloys

The copper-nickel alloys are simple solid solutions of nickel in copper with or without other designated alloying elements. The nickel content varies from 9.0 to 33.0%. Small amounts of manganese (0.05 to 1.5%) and iron (0.4 to 1.8%) are also present. Silicon with either niobium or chromium improves the strength of the 30% nickel alloy, but silicon should be limited to 0.5% to facilitate welding. Their excellent corrosion resistance in seawater, combined with their high strengths, account for their wide use in pipes, heat exchangers, valves, ship tail-shaft sleeves, and other marine applications. They are also used in coins.

Copper-Nickel-Zinc Alloys

Known commonly as "nickel silvers," these alloys contain 11 to 27% nickel, 1 to 25% zinc, and 1 to 11% lead. The presence of nickel primarily accounts for their pleasant silver luster but in spite of their name, nickel silvers do not contain silver. Although nickel silvers are highly alloyed, they are simple solid solutions. They offer good corrosion resistance, and are easily cast and machined. Major uses include hardware for food processing, seals, architectural trim, musical instruments, valves, and door keys.

Leaded Coppers

These comprise a series of cast alloys of copper with 20% or more lead, usually with a small amount of silver present but without tin or zinc.

Special Alloys

Alloys whose chemical compositions do not fall into any of the above categories are categorized as "special alloys." For example, Inco has announced some new alloys containing chromium (1.4 to 2.0%) and silicon (0.3 to 0.6%) in 70/30 cupronickels (IN768) [8].

Another alloy is the Cu-Ni-Sn spinodal alloy which had been introduced by Bell Laboratories as a substitute for Cu-Be alloys in both as-cast and heat-treated conditions. The most common cast alloy is Cu-10-Ni-8Sn alloy which is strengthened by spinodal decomposition to develop attractive mechanical properties (930 MPa UTS and 830 MPa YS) comparable to those of cast Cu-Be alloys [9, 10].

Physical Metallurgy

Copper-base casting alloys can be subdivided into three groups based on their freezing ranges. Group I alloys are those with a narrow freezing range (<50°C). Group II alloys have an intermediate (medium) freezing range of 50 to 110°C. On the other hand, a freezing range of well over 110°C, even up to 170°C, can be expected in the Group III alloys. These alloys have been listed in Tables 8.4 to 8.6 [4, 6, 7, 10,].

There is a family of very important industrial alloys based on the Cu-Zn-Sn-Pb system called the leaded red brasses, leaded semi-red brasses, tin bronzes, leaded tin bronzes, high-leaded tin bronzes, and high-strength yellow brasses. All these except the high-strength yellow brasses exhibit wide freezing range. Lead is nearly or entirely insoluble in solid alloys, and thus has little relationship within the Cu-Zn-Sn system. In order to understand the physical metallurgy of this class of alloys, one must understand physical metallurgy of the binary Cu-Zn and Cu-Sn systems first.

Copper-Zinc Alloys

The alloying behavior of zinc in copper-zinc alloys is presented in the binary equilibrium diagram in Fig. 8.3 [7]. The diagram shows solubility of zinc in copper up to 32.5% at the solidus temperature and about 35% at room temperature. The solid-solution phase α is the major microstructure constituent of most brasses except for the high-zinc-content, high-strength type of brass (manganese bronze). The latter alloys contain a substantial amount of the β' constituent. The α phase is a relatively soft, ductile, and low-strength phase, and this is reflected in the hardness of cast alloys of varying zinc content. A really substantial increase in hardness (and strength) does not occur until the percentage of zinc in the alloy is high enough to cause an appreciable amount of the β' constituent to be present. The latter phase is a hard, brittle constituent, which, although increases the strength of $\alpha + \beta'$ mixtures, unfortunately also reduces ductility to the point of destroying the usefulness of the alloys if the zinc content is too high. Hence, it is evident that the zinc content of casting alloys is limited to a maximum of that which produces a desirable combination of hardness and strength without a harmful loss of ductility. The maximum zinc content is about 36%.

As can be seen in Fig. 8.3, the liquidus is lowered and temperature range of freezing is increased as the zinc content of the alloys increases. Some liquidus and solidus temperatures for various alloys are listed in Tables 8.4 to 8.6. The alloys in Tables 8.4 to 8.6 are not simple Cu-Zn brasses and have freezing ranges different from the diagram in Fig. 8.3.

Copper-Tin Alloys

The alloying behavior of tin in copper is similar to that of zinc. The equilibrium diagram for the binary system is shown in Fig. 8.4 [7]. The system shows solubility, the α phase, up to about 13.5% during solidification. At lower temperatures, the $\alpha + \delta$ eutectoid occurs. The α phase is a softer solid-solution phase, but the δ phase is exceedingly hard and brittle. It is evident from Fig. 8.4 that the alloys containing 5 to 15% tin have an unusually long freezing-temperature range over 204°C (400°F). The long solidification range makes castings of these alloys very hard to riser adequately and also promotes severe coring and microsegregation. The latter conditions are illustrated for a leaded tin bronze in Figs. 8.5 and 8.6.

Because so much segregation occurs during freezing, alloys as low as 7% tin contain the δ constituent and show the $\alpha + \delta$ eutectoid even though the "equilibrium" diagram does not indicate this fact. In sand castings, the nonequilibrium microstructures containing α or $\alpha + \delta$ eutectoid remain after cooling to room temperature. Only prolonged annealing at elevated temperatures would produce the equilibrium structures. Copper

Alloy Type	UNS No.	Nominal Composition, %										Yield Strength, 0.5%		Tensile Strength		Elongation	Liquidus Temperature	Solidus Temperature	Freezing Range
		Cu	Sn	Pb	Zn	Ni	Fe	Al	Mn	Si	Other	MPa	Ksi	MPa	Ksi	%	°C	°C	°C
Copper	C81100	100										62	9	172	25	40	1,083	1,064	19
Chrome copper	C81500 as cast	99									1 Cr	276	40	351	51	17	1,085	1,075	10
	C81400 as cast heat treated	99.1									0.8 Cr 0.06 Be	104 248 (HT)	15 36 (HT)	311 365	45 53 (HT)	15 11 (HT)	1,093	1,066	27
Leaded yellow brass	C85200	72	1	3	24							90	13	262	38	35	941	927	14
	C85400	67	1	3	29							83	12	234	34	35	941	927	14
	C85700	61	1	1	37							124	18	345	50	40	941	913	28
	C85800	62	1	1	36							207	30	379	55	15	899	871	28
Manganese bronze	C86200	63			27	3	3	4	3			331	48	655	95	20	941	899	42
	C86300	61			27	3	3	6	3			462	67	821	119	18	923	885	38
	C86400	58	1	1	38		1	0.5	0.5			172	25	448	65	20	880	862	18
	C86500	58			39		1	1	1			200	29	490	71	30	880	862	18
	C86700	58	1	1	34		2	2	2			290	42	586	85	20	880	862	18
	C86800	55.2			39.0	3.0	2.0		3.2			262	38	565	82	22	900	880	20
Aluminum bronze	C95200	88				3	3	9				186	27	552	80	35	1,045	1,042	3
	C95300 as cast heat treated	89					1	10				186 290	27 42	517 586	75 85	25 15	1,045	1,040	5
	C95400 as cast heat treated	85					4	11				241 372	35 54	586 724	85 105	18 8	1,038	1,027	11
	C95410 as cast heat treated	84				2	4	10				241 372	35 54	586 724	85 105	18 8	1,038	1,027	11
	C95500 as cast heat treated	81				4	4	11				303 469	44 68	689 827	100 120	12 10	1,054	1,038	16
	C95600	91						7		2		234	34	517	75	18	1,004	982	22
	C95700	75				2	3	8	12			310	45	655	95	26	990	950	40
	C95800	81				4.5	4	9	1.2			262	38	655	95	25	1,060	1,043	17

402

Nickel Silver	C97300	57	2	9	20	30			117	17	241	35	20	1,040	1,010	30
	C97600	64	4	4	8	35			165	24	310	45	20	1,143	1,108	35
	C97800	66	5	2	2	40			207	30	379	55	15	1,180	1,140	40
White Brass	C99700	56.5	1.5	22	25		1.8	13	172	25	379	55	25	902	879	23
	C99750	58		20	20		1.6	20	221	32	448	65	30	843	818	25

Source: From *Metals Handbook* [4], *Standards Handbook* [6], and *Metals Handbook* [10].

TABLE 8.4 Nominal Chemical Composition and Typical Mechanical Properties of Group I Sand-Cast Alloys

Alloy Type	UNS No.	Nominal Composition, %										Yield Strength, 0.5%		Tensile Strength		Elongation	Liquidus Temperature	Solidus Temperature	Freezing Range
		Cu	Sn	Pb	Zn	Ni	Fe	Al	Mn	Si	Other	MPa	Ksi	MPa	Ksi	%	°C	°C	°C
Beryllium copper	C82000 as cast heat treated	96.8									0.6 Be 2.6 Co	138 517	20 75 (HT)	345 662	50 96 (HT)	20 6 (HT)	1088	971	117
	C82200 as cast heat treated	97.9				1.5					0.6 Be 1.5 Ni	206 517	25 75 (HT)	380 655	55 95 (HT)	20 7 (HT)	1116	1038	88
	C82400 as cast heat treated	97.8									1.7 Be 0.43 Co	242 551	35 80 (HT)	483 690	70 100 (HT)	15 3 (HT)	996	899	97
	C82600 as cast heat treated	96.8								0.3	2.4 Be 0.5 Co	345 724	50 105 (HT)	552 827	80 120 (HT)	10 1 (HT)	954	827	97
	C82800 as cast heat treated	96.6								0.3	2.6 Be 0.5 Co	345 1070	50 155 (HT)	552 1139	80 165 (HT)	10 1 (HT)	932	885	47
Silicon brass	C87400	82			14					3.5	0.5 Pb	165	24	379	55	30	916	821	95
	C87500	82			14					4		207	30	462	67	21	916	821	95
Silicon bronze	C87300	95							1	4									
	C87600	91			5					4		221	32	455	66	20	971	860	111
	C87610	92			4					4		170	25	400	58	35	971	860	111
	C87800	82			14					4		345	50	586	85	25	916	821	95
Copper nickel	C96200	87.5				10	1.5		0.9	0.1	0.8 Nb	170	25	310	45	20	1149	1099	50
	C96400	67				30	0.7		0.8	0.5	1.0 Nb	255	37	469	68	28	1238	1171	67

Source: From *Metals Handbook* [4], *Standards Handbook* [6], and *Metals Handbook* [10].

TABLE 8.5 Nominal Chemical Composition and Typical Mechanical Properties of Group II Sand-Cast Alloys

Alloy Type	UNS No.	Nominal Composition, %										Yield Strength, 0.5%		Tensile Strength		Elongation	Liquidus Temperature	Solidus Temperature	Freezing Range
		Cu	Sn	Pb	Zn	Ni	Fe	Al	Mn	Si	Other	MPa	Ksi	MPa	Ksi	%	°C	°C	°C
Leaded red brass	C83450	88	2.5	2	6.5	1						103	15	255	37	31	1,015	860	155
	C83600	85	5	5	5							117	17	255	37	30	1,010	854	156
	C83800	83	4	6	7							110	16	241	35	25	1,004	843	161
Leaded semi-red brass	C84400	81	3	7	9							103	15	234	34	26	1,004	843	161
	C84800	76	2.5	6.5	15							97	14	255	37	35	954	832	122
Bismuth selenium brass	C89510	87	5		5						1 Bi 0.5Se	124	18	241	35	20	1,021	815	206
	C89520	86	5.5		5						1.9Bi 0.9Se	124	18	214	31	15	1,005	809	196
Bismuth brass	C89844	84.5	4		8						3Bi	103	15	234	34	30	1,010	843	167
Bismuth bronze	C89836	89.5	5.5	3	2 Ni							97	14	229	33	20	1,028	859	169
Tin bronze	C90300	88	8		4							145	21	310	45	30	1,000	854	146
	C90500	88	10		2							152	22	310	45	25	999	854	145
	C90700	89	11									152	22	303	44	20	999	831	168
	C91100	84	16									172	25	241	35	2	950	818	132
Leaded tin bronze	C92200	88	6	1.5	4.5							138	20	276	40	30	988	826	162
	C92300	87	8	1	4							138	20	276	40	25	999	854	145
	C92600	87	10	1	2							138	20	303	44	30	982	843	139
	C92700	88	10	2								142	21	300	42	20	982	843	139
	C92900	84	10	2.5	3.5 Ni							179	26	324	47	20	1,031	857	174

TABLE 8.6 Nominal Chemical Composition and Typical Mechanical Properties of Group III Sand-Cast Alloys (*Continued*)

Alloy Type	UNS No.	Nominal Composition, %										Yield Strength, 0.5%		Tensile Strength		Elongation	Liquidus Temperature	Solidus Temperature	Freezing Range
		Cu	Sn	Pb	Zn	Ni	Fe	Al	Mn	Si	Other	MPa	Ksi	MPa	Ksi	%	°C	°C	°C
High-leaded tin bronze	C93200	83	7	7	3							124	18	241	35	20	977	854	123
	C93500	85	5	9	1							110	16	221	32	20	999	854	145
	C93700	80	10	10								124	18	241	35	20	929	762	167
	C93800	78	7	15								110	16	207	30	18	943	854	89
	C94300	71	5	24								90	13	186	27	15	–	–	–
Nickel tin bronze	C94700	87.5	5	–	1.5	5.3						159	23	345	50	35	1,027	904	123
	C94800	86.5	–	0.7	1.8	5.3						159	23	310	45	35	1,027	904	123
	C94900	80	5	5	5	5						103	15	262	38	15	–	–	–

Source: From *Metals Handbook* [4], *Standards Handbook* [6], and *Metals Handbook* [10].

TABLE 8.6 Nominal Chemical Composition and Typical Mechanical Properties of Group III Sand-Cast Alloys (*Continued*)

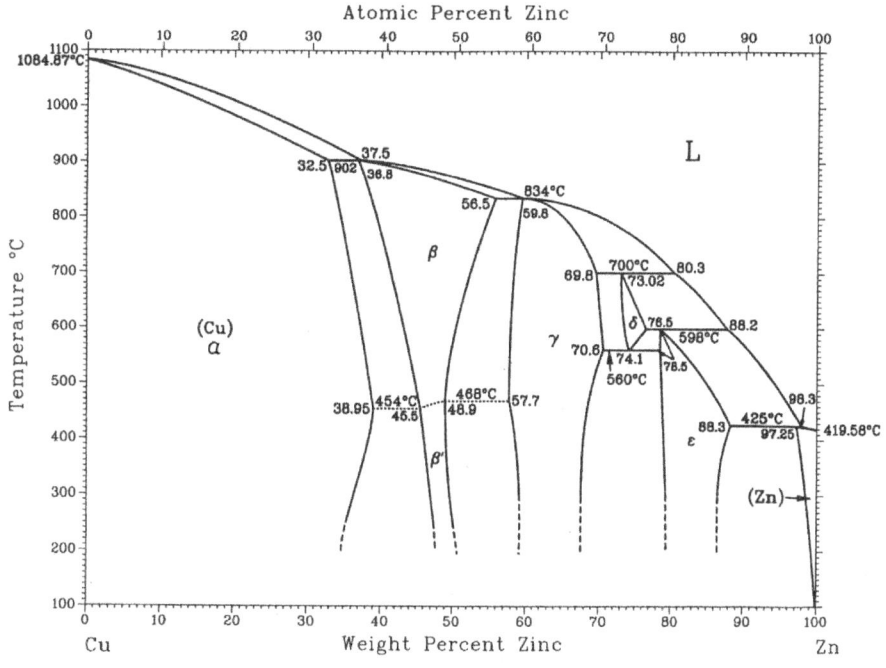

FIGURE 8.3 Copper-zinc phase diagram. (From *Casting Copper-Base Alloys* [7].)

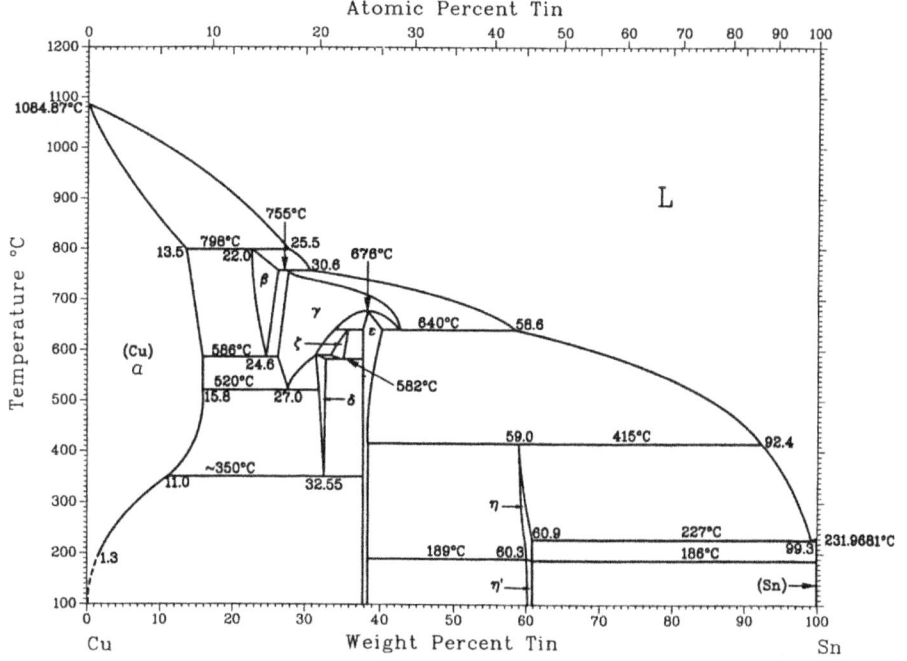

FIGURE 8.4 Copper-tin phase diagram. (From *Casting Copper-Base Alloys* [7].)

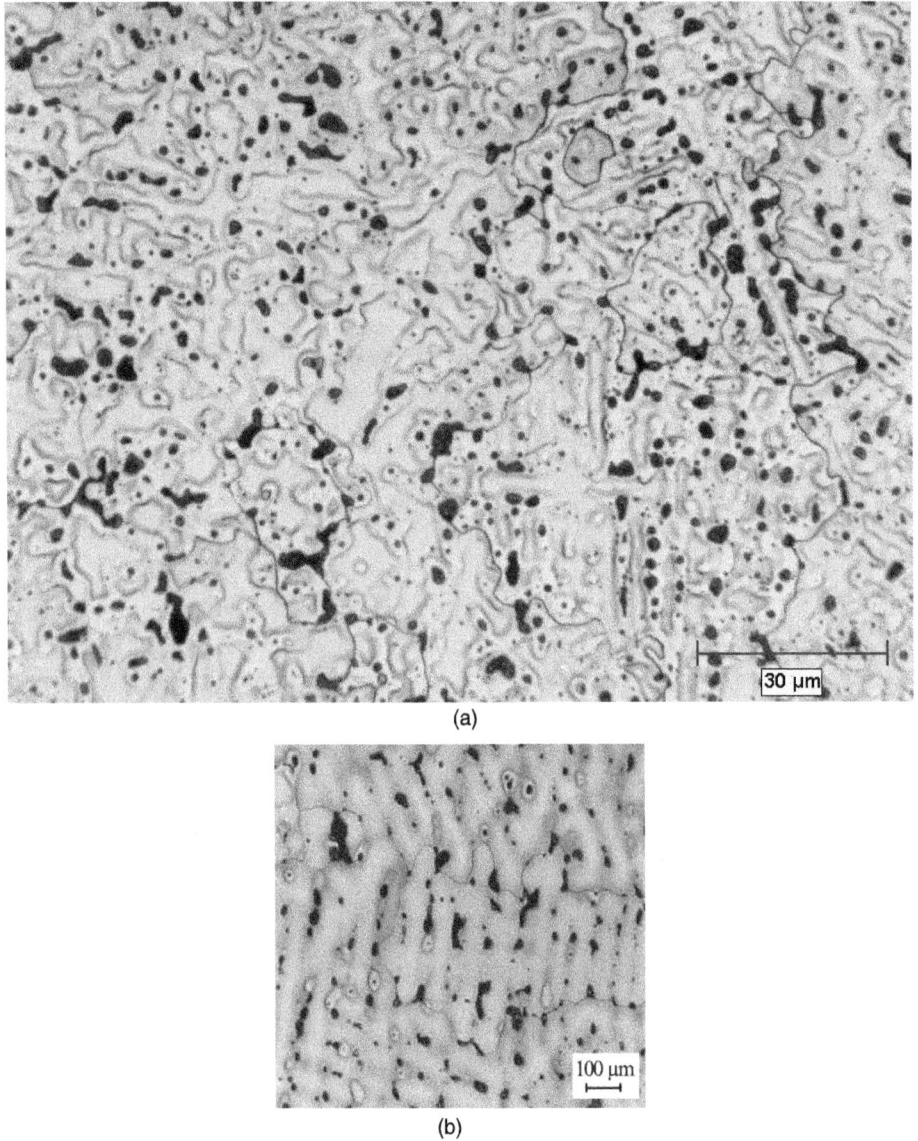

(a)

(b)

Figure 8.5   (a) Coring in a sand-cast red brass. (From *Casting Copper-Base Alloys* [7]). (b) Coring in a 77-mm-thick plate casting in 88-10-2 tin bronze. Rate of cooling through freezing range is 7°C/min. (Hudson [11].) (A color version of this figure is available at mhprofessional.com/pmc3.)

alloys are limited to lower maximum tin content than is the case for zinc, since embrittlement occurs with the presence of increasing amounts of the δ constituent at relatively low tin content. Tin is more effective percentage wise in strengthening copper than zinc is. The tin bronzes based on the Cu-Sn system have strength, hardness, and bearing qualities which make them suitable for gears, worms, bearing plates, turntables, sleeves, and liners. Because of good corrosion resistance, these applications find frequent use in marine construction, naval

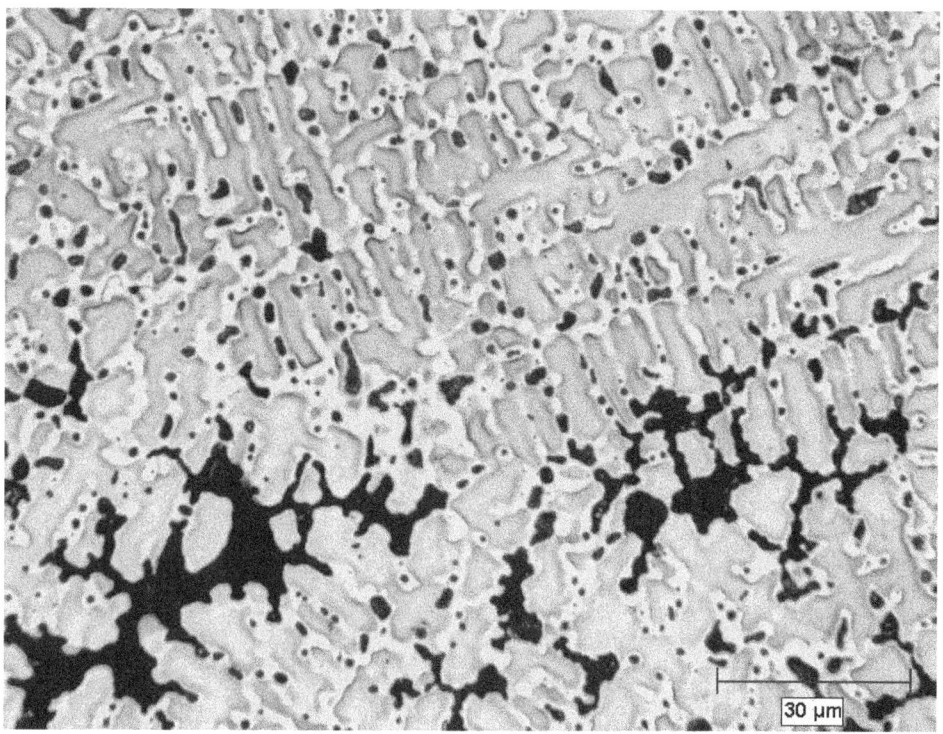

FIGURE 8.6 Localized interconnected microshrinkage in sand-cast red brass. (From *Casting Copper-Base Alloys* [7].) (A color version of this figure is available at mhprofessional.com/pmc3.)

vessels, bridges, dams, hydroelectric plants, chemical-processing industries, and the like. The tin bronzes are, of course, complex alloys modified by the presence of elements other than copper and tin. Table 8.6 presents the chemical and tensile requirements of tin-bronze casting alloys. The alloy 88% Cu, 10% Sn, 2% Zn is known as gunmetal. The alloys possess better all-around resistance to seawater than the brasses, so that they are used extensively in that field. Bell bronzes contain 20 to 23% tin, are relatively hard and brittle, and have the ability to produce musical tones when struck, which makes for their use in bell castings. Lead is added to tin bronzes for machinability or to obtain antifrictional properties. High-leaded tin bronzes are useful for sleeves, bushings, and bearings for railroad, rolling mill, and papermaking applications, where good bearing qualities against steel or iron surfaces are needed.

Copper-Tin-Zinc

The 500°C (932°F) isothermal section of the ternary copper-tin-zinc phase equilibrium diagram is shown in Fig. 8.7 [7], and represents the effective phase relationships existing at room temperature. The phases present, within the range of compositions of commercial interest, can be understood in terms of what has already been discussed in connection with the various binary copper alloy phase diagrams.

The principal phase occurring in all the commercial alloys is that designated as "alpha" (α) in Fig. 8.7. This field represents a continuous series of solid solutions analogous

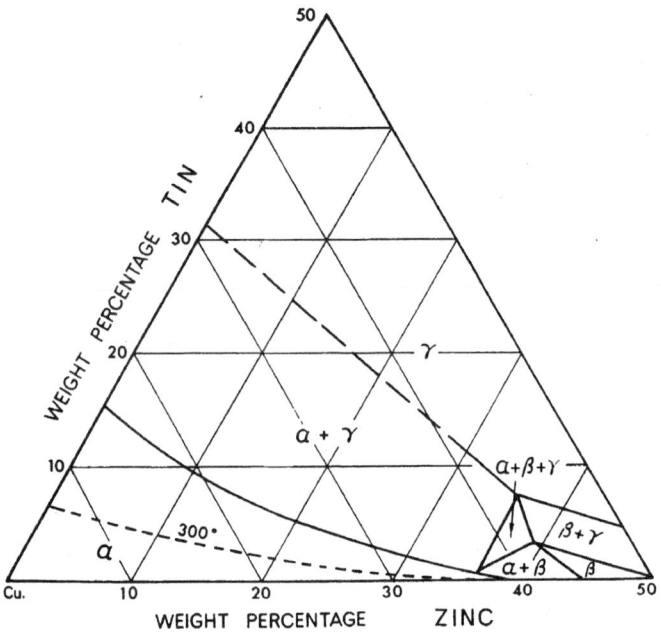

FIGURE 8.7   The 500°C (932°F) isothermal section of the Cu-Sn-Zn equilibrium diagram. (From *Casting Copper-Base Alloys* [7].)

to the alpha phases in the binary Cu-Zn and Cu-Sn systems. When both alloying elements are present simultaneously, each reduces the other element's solubility to some extent, and the solid curved line which bounds the alpha region shows the effective solubility limits. The wide freezing ranges of the ternary alloys result in as-cast structures wherein the alpha matrix shows pronounced coring.

The second phase which occurs consistently in all the leaded alloys is, of course, the lead itself. The quantity and distribution of the lead depends on the amount present in the alloy, the freezing range of the alloy, and its actual rate of solidification. Chill castings, for a given lead content, tend to exhibit finer grain size and, consequently, finer distribution of the lead particles than do sand castings. The lead inclusions are of a dark gray color and their shapes tend to conform to the dendritic grain boundaries of the matrix. Since the lead phase is virtually insoluble in the solid condition, there is no need to consider the constitution of these alloys in terms of a quaternary Cu-Sn-Pb-Zn equilibrium diagram. Instead, one can actually ignore the lead content and recalculate the percentage composition of the alloy in terms of the copper/tin/zinc ratio. For example, alloy C83600 (85-5-5-5) would be expected to solidify as would its unleaded equivalent −89½ Cu, 5¼ Sn, and 5¼ Zn.

Again, in the ternary diagram, the alpha field is bounded by a two-phased region of compositions consisting of alpha plus gamma (γ). The gamma phase in the ternary diagram, like the alpha phase, also consists of a continuous series of solid solutions of tin and zinc in copper. Metallurgically, this phase is said to be isomorphous with the gamma phase of the Cu-Zn system and the delta phase of the Cu-Sn system. When, due to nonequilibrium freezing conditions, the gamma phase appears in the solidified structures of the ternary alloys it has the identical appearance of the delta phase which constitutes the alpha-delta eutectoid in Cu-Sn alloys. For this reason, it is commonly referred to as the "delta" phase. In appearance, this phase forms a part of an alpha-delta eutectoid which occurs as bluish inclusions, often elongated, found at the grain boundaries of the matrix (Fig. 8.8).

The alpha-delta eutectoid occurs most often in the tin bronzes with 8 to 10% Sn, and is more prevalent during rapid freezing. It is frequently detected in leaded red brass compositions, but less often in the semi-red brasses. Excessive amounts of alpha-delta eutectoid are considered undesirable as they reduce the elongation of the alloy. The occurrence of this phase in commercial alloys is regulated by restricting the sum of the tin and zinc contents.

Copper-Lead Alloys

The copper-lead equilibrium system is illustrated in Fig. 8.9 [7]. The solid solubility of lead in copper is about 0.002 to 0.005% at room temperature. The presence of other elements may increase this limit slightly. However, it is known that leaded copper alloys have most of their lead present as islands of the element distributed throughout the microstructure. The lead islands dispersed throughout the microstructure of 85% Cu, 5% Sn, 5% Pb, and 5% Zn are illustrated in Fig. 8.10 [7]. Since lead is precipitated late during freezing of the metal, it segregates in areas which freeze last. At the end of freezing, it may fill in areas which might otherwise become shrinkage porosity. Lead in copper alloys thus often makes it easier to produce leakproof castings for valves and fittings. Another beneficial effect of lead is its use for improving machinability of copper alloys. The weak lead islands make machining soft, tough copper to a fine finish easier by causing machining chips to form, break, and flow more easily from cutting tools.

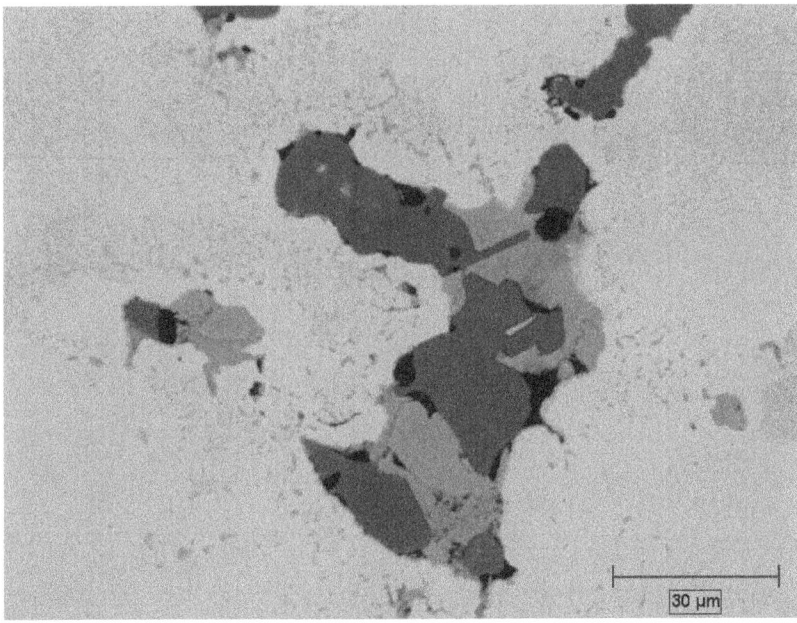

Figure 8.8 Distribution of lead and alpha-delta eutectoid phases in a sand-cast leaded red brass. (From *Casting Copper-Base Alloys* [7].) (A color version of this figure is available at mhprofessional.com/pmc3.)

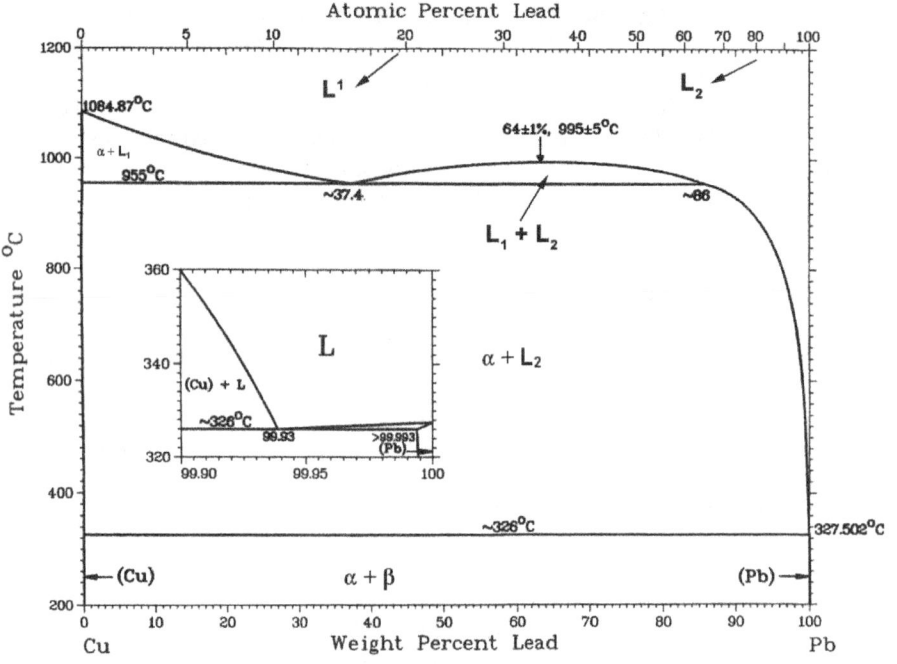

Figure 8.9   Copper-lead equilibrium phase diagram. (From *Casting Copper-Base Alloys* [7].)

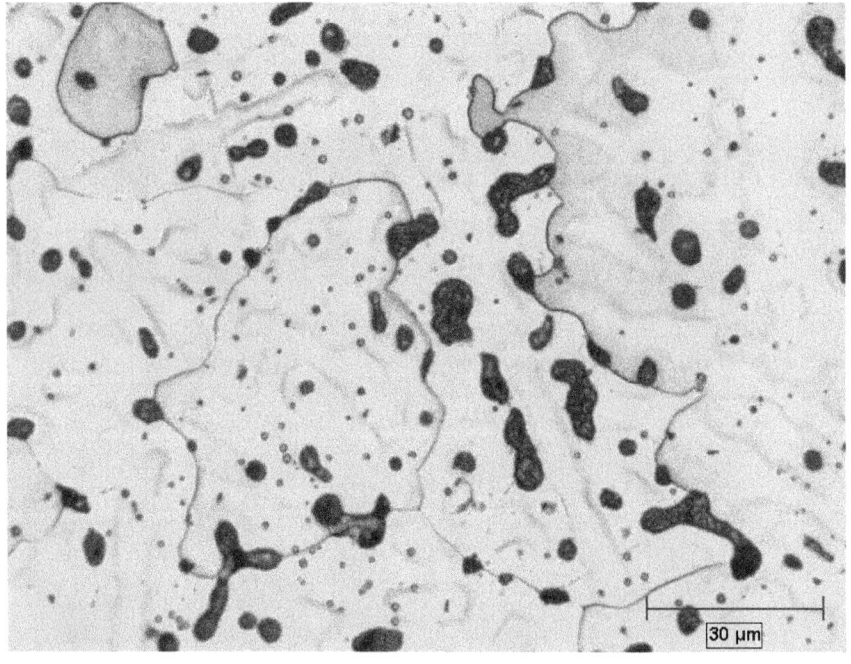

Figure 8.10   Distribution of lead and alpha-delta eutectoid phases in a sand-cast leaded red brass. (From *Casting Copper-Base Alloys* [7].) (A color version of this figure is available at mhprofessional.com/pmc3.)

Figure 8.9 shows that L_1, copper-rich liquid, begins to form as soon as the temperature drops below the liquidus temperature. Since L_1 has high density, it may separate and settle in melting pots or castings. As soon as the temperature drops low enough, below 953°C, L_2, which is over 90% Pb, begins to form, and because of its high density it sinks rapidly if free to do so. Because of this behavior of lead in copper, the alloys must always be heated adequately to ensure solution in the melt and must not be allowed to segregate during the cooling in ladles. Stirring is frequently used to disperse the lead. In castings, the distribution of the lead islands is greatly determined by the lead content, rate of solidification of the casting, and the presence of other elements. Segregated lead, in grain boundaries as large blobs in the lower portion of castings, causes poor mechanical properties. The strength of brasses and bronzes is lowered below minimum by the presence of excess lead, and it therefore may be considered as an impurity in some of the high-strength alloys, even though it is beneficial in other alloys.

Copper-Aluminum Alloys

The copper-aluminum phase diagram is shown in Fig. 8.11 [13]. Solid solubility of Al in Cu exists up to 7.4% at the solidus, with solubility increasing to 9.4% at 565°C. A unique feature of this system is the existence of a eutectoid at about 11.8% Al and 565°C. This enables the high-aluminum content alloys to be heat treated in a manner similar to steels involving solution treatment, quenching, and tempering.

The narrow freezing range of the copper-aluminum alloys is evident from Fig. 8.11. A range of 10 to 30°C in these alloys is seen to be very narrow compared to over 220°C in tin bronzes. This condition results in a large apparent solidification shrinkage and requires heavy risering to produce sound castings. The narrow freezing range, however, makes it possible to produce castings of maximum soundness with less tendency for microshrinkage than in long-freezing-range alloys.

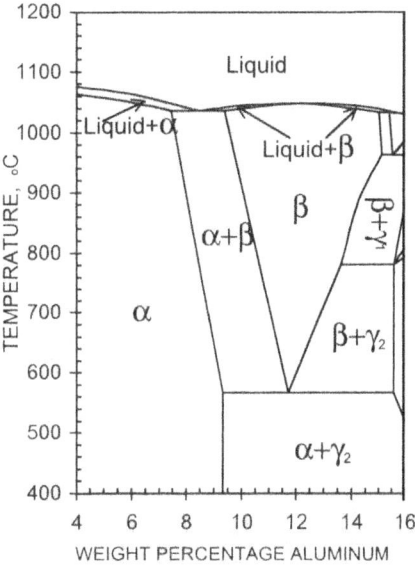

FIGURE 8.11   Copper-aluminum equilibrium phase diagram. (From Macken and Smith [13].)

Alloys based on the copper-aluminum system are known as aluminum bronzes which are usually of two types. Low-aluminum content bronzes are under 10% Al. They consist structurally of α solid solution and are softer, more ductile, and not heat treatable to high strength like the high-aluminum content ones. However, complex aluminum bronzes (C95800) containing 8.5 to 9.5% Al can be heat treated as discussed later.

Aluminum bronzes containing up to about 9.4% aluminum should exhibit a single-phase alpha (α) structure at room temperature. However, alloys containing more than 8% aluminum at the beginning of solidification always solidify as a two-phase material consisting of alpha + beta (α + β). On slow cooling, the high-temperature β phase should transform to α but as the transformation rate is too slow, alloys containing 8 to 9.4% aluminum eventually remain as a two-phase material. In alloys containing more than 9.4% Al, the β phase undergoes a eutectoid transformation, if cooled slowly in the region of 900 to 565°C, and forms a mixture of alpha + gamma2 ($\alpha + \gamma_2$).

The most common alloying element added to Cu-Al alloys is iron, which is added as a grain refiner to improve the toughness. Iron also reduces the formation of the γ_2 phase during solid transformation. The room temperature microstructure of an aluminum bronze with 5% iron consists of α and β phases with precipitates of Fe(δ) based on Fe_3Al. The other common alloying element, though not added alone in aluminum bronzes, is nickel. Nickel forms NiAl when the alloy is cooled slowly.

Nickel-aluminum bronzes contain nickel, iron, and manganese as alloying elements. Their microstructure always consists of various nickel- or iron-rich kappa (κ) phases in an alpha matrix. In addition, because of nonequilibrium solidification and cooling, which is characteristic of the casting process, the alloys with more than about 8.5% aluminum contain substantial β phase at higher temperatures. This phase further transforms to $\alpha + \gamma_2$ as the casting cools to room temperature as explained earlier and additional k phases also formed in this reaction. The phase diagrams of the above alloy systems are shown in Fig. 8.12 [12].

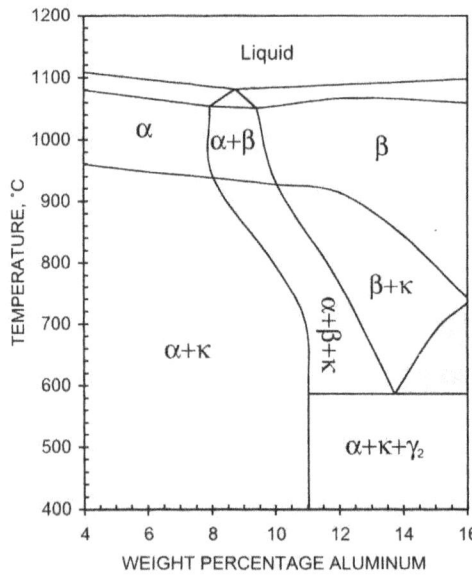

FIGURE 8.12 Cu-Al-5%Fe-5%Ni equilibrium phase diagram. (Harry Meigh [12].)

The κ phases, depending on the morphology and compositions, are identified as κ_I, κ_{II}, κ_{III}, and κ_{IV}. The schematic diagram indicating these various phases in nickel aluminum bronzes and a typical microstructure are shown in Fig. 8.13a and b [14].

The γ_2 phase is detrimental to both ductility and corrosion resistance. The amount of this phase in the microstructure increases with the aluminum content; therefore, the higher-strength, higher-aluminum-content aluminum bronzes C95300, C95400, and C95500 generally rely on heat treatment and/or alloying with nickel and manganese to prevent formation of excessive γ_2 phase in the microstructure. Both nickel and iron combine with the aluminum to form complex κ phases and thus increase the amount of aluminum necessary to produce the γ_2 phase.

The addition of manganese to the aluminum-bronze is also reported to slow down the transformation of β phase to the $\alpha + \gamma_2$ eutectoid. Hence, alloys such as Mn-Ni-Al bronze C95700 solidify in an all-β structure, although during slow cooling some alpha is formed. Although aluminum has the highest effect on properties (increasing strength and reducing ductility), the manganese effect is similar but less effective [15]. C95700 is often selected as heat treatment is not required for excellent properties. Slow cooling following stress relief increases magnetic permeability and reduces ductility and impact resistance due to low temperature precipitation of manganese-rich compounds [15, 16]. Silicon, like manganese, has a positive aluminum equivalent (1% Si is equivalent to 1.6% Al) and has the advantage of reducing the difficulties associated with the breakdown of the β phase to the $\alpha + \gamma_2$ eutectoid at slow cooling rates. This alloy has been used extensively in the past to cast propellers for ships. Optical micrograph of a cast propeller is shown in Fig. 8.14 [15].

Copper-Nickel

The Cu-Ni phase diagram is shown in Fig. 8.15 [17]. Cu and Ni have complete solubility in the entire composition range in both the liquid and solid phases. Casting alloys based on the Cu-Ni system are of three general classes: (1) cupronickel alloys, (2) leaded nickel brasses and bronzes, commonly known as nickel silvers, and (3) Ni-Sn type age-hardenable bronzes. Monel is a nickel-copper alloy.

The nickel content of cupronickels varies from 9 to 33%. Two types of cupronickels are of commercial importance for seawater applications. These are known as 90/10 (C96200) and 70/30 (C96400) cupronickels and are strengthened by Si and Nb which cause precipitation hardening. Small amounts of manganese (0.05 to 1.5%) and iron (0.4 to 1.8%) are also present. They are classified as medium freezing range alloys. Another class of 70/30 cupronickels (IN-768) contain Cr and Si to cause strengthening. Their typical mechanical properties are 496 MPa UTS, 325 MPa YS, and 18% elongation [7, 8].

The monels are basically 70/30 Ni-Cu alloys containing other alloying elements such as Si, Fe, Mn, C, and Nb. The main advantages of monels are high strength and toughness coupled with excellent resistance to mineral acids, salt solutions, food acids, strong alkalies, some marine environments, etc. Applicable specifications for monels are ASTM designations A494/A494M and U.S. federal specifications QQ-N-288. Composition E does not contain Nb. ASTM-specified minimum mechanical properties are 450 MPa UTS, 170 MPa YS, and 25% elongation for Grade M35-1. These mechanical properties can be easily met employing proper melting and casting procedures and good composition control [18]. A detailed investigation has been carried out to determine the effects of alloying elements such as C, Si, and Nb on the structure and mechanical properties and their critical contents to meet the ASTM and U.S. federal specifications for mechanical properties have been defined Sahoo et al. [19].

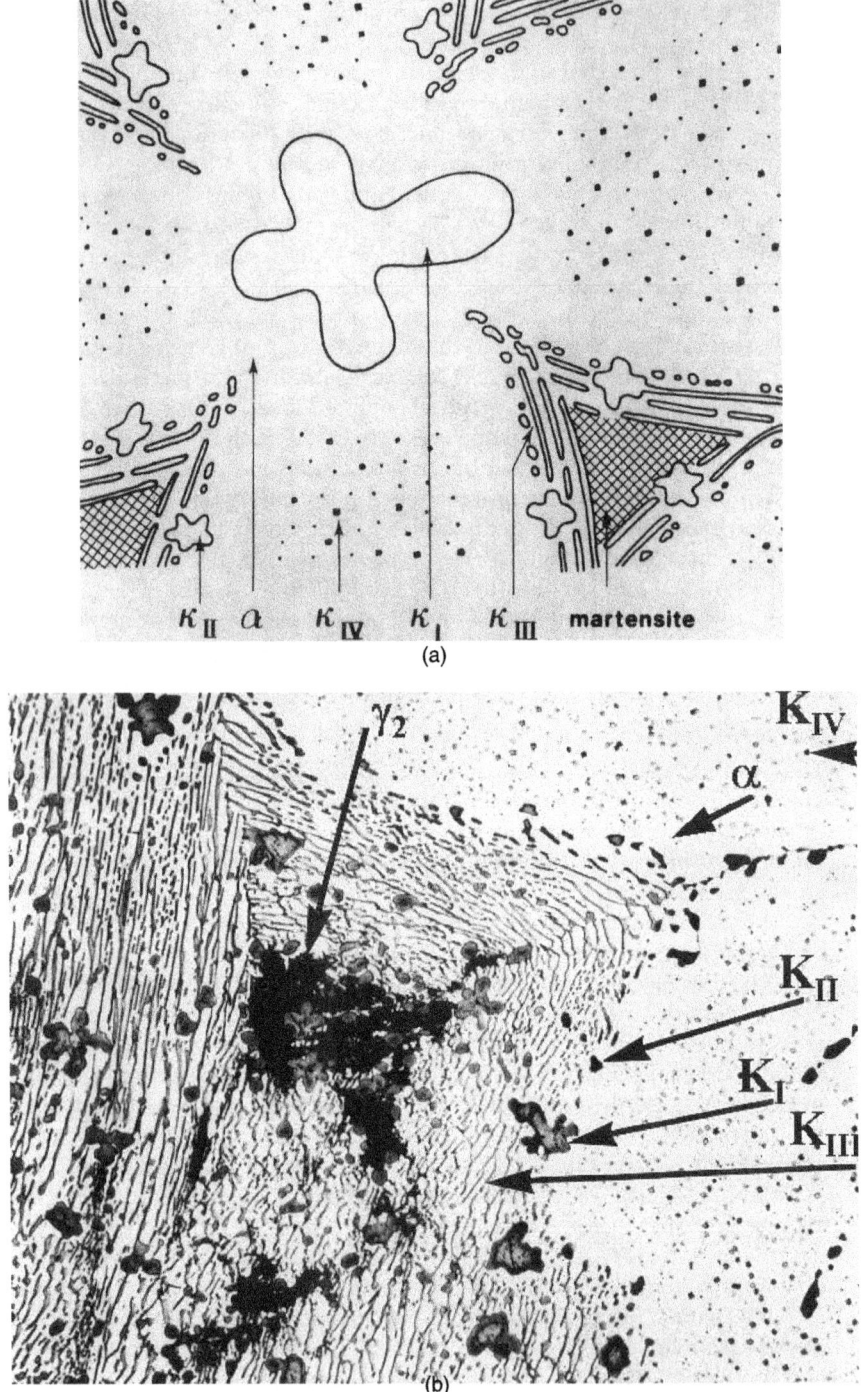

Figure 8.13   Effect of corrosion inhibiting treatment (700°C, six hours, air cool) on microstructure modification of a sand-cast nickel-aluminum bronze. (From Sahoo et al. [14].)

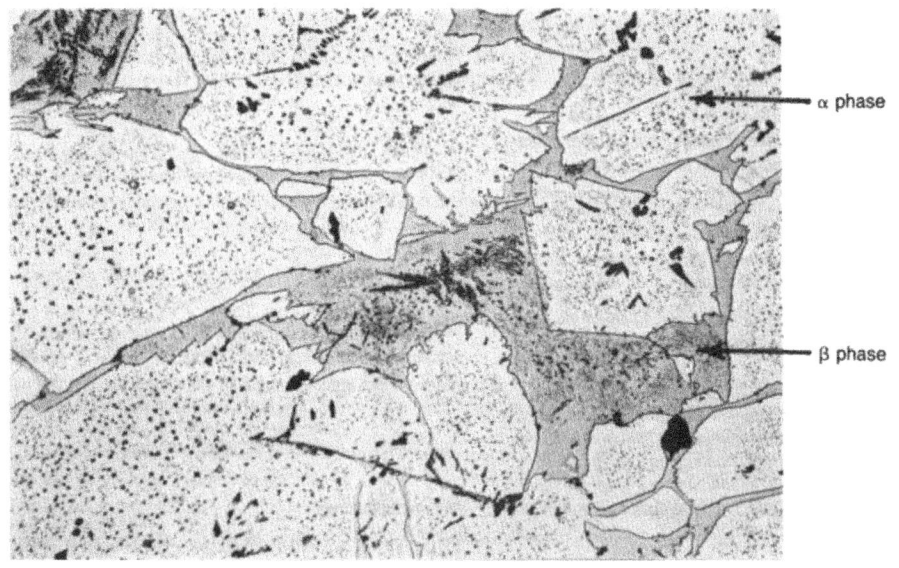

FIGURE 8.14 Optical micrograph of a propeller casting in a Mn-Ni-Al bronze alloy. (From Couture et al. [15].)

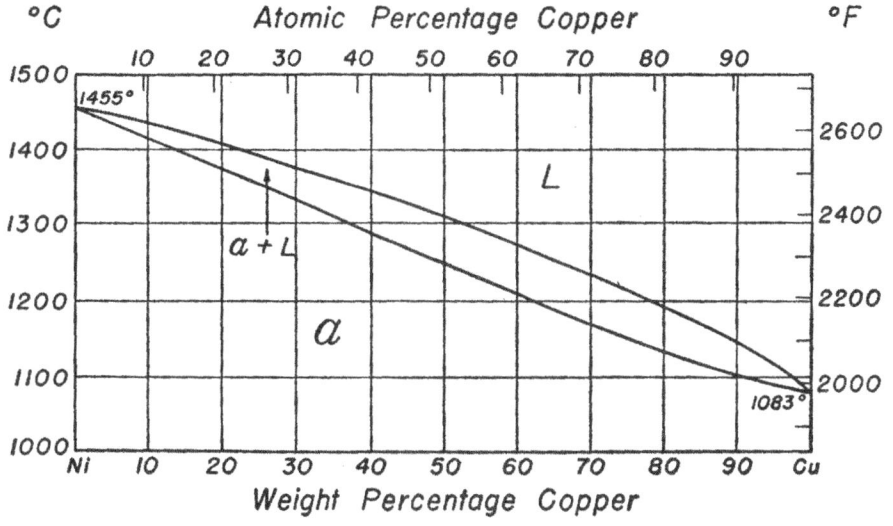

FIGURE 8.15 Copper-nickel equilibrium phase diagram. (From *Metals Handbook* [17].)

The nickel silvers (leaded nickel brasses and bronzes) are popularly known as "Dairy Metals" because of their initial use in milk processing around 1930. They are also known as "Dairy Brasses," "Dairy Bronzes," and "German Silvers." They have a pleasing silver or white appearance similar to stainless steel and are used in corrosion-resistant applications such as dairy and food machinery parts, restaurant kitchen equipment, soda fountains, steam fittings, hardware for marine, and building applications. Recently lead-free nickel silvers have been introduced. Mechanical properties and nominal chemical composition of these alloys (C97300, C97600, C97800, and C99970) are listed in Table 8.4.

Nickel brasses have only enough of their zinc content substituted by nickel to give a white color. Nickel bronzes contain more nickel and zinc. Lead is added to improve casting machinability and pressure tightness.

The nickel-tin bronzes (C94700, 94800, and C94900) provide a good combination of high strength, corrosion resistance, and wear resistance. Their nominal chemical compositions and mechanical properties are given in Table 8.6.

Heat Treatment

The only copper-base alloys susceptible to heat treatment are beryllium copper alloys, chromium copper alloys, and aluminum bronze alloys containing more than 10% aluminum.

Beryllium copper alloys can be heat treated by solution treating and aging. Solution-treating temperature limits must be observed if optimum properties are to be obtained from the precipitation hardening treatment.

After castings are solution treated, they are quenched in water. All castings, except those of Alloy C82000, may be solution treated in air and water quenched immediately after removal from the furnace. Alloy C82000 must be solution treated in protective atmospheres such as cracked ammonia or natural gas. The duration of solution treatment depends on section thickness.

Following solution heating, the castings are precipitation hardened. Table 8.7 shows the heat-treating cycles for beryllium copper alloys [4].

Chromium copper alloy containing 1% Cr can be heat treated in the same manner as beryllium copper. Here the solution treatment is 0.5 hour at 995 to 1010°C followed by a water quench. Next, the castings are precipitation hardened at 480°C for three hours. Because chromium is sensitive to oxidation, a protective atmosphere should be used to avoid an oxidized zone of approximately 3.05 mm (0.12 in) on the casting surface. If heat treating is done in an air furnace, the castings must be machined after treatment to remove this oxide in order to obtain accurate conductivity and hardness measurements. A salt bath can also be used for heat treatment to minimize oxidation.

Aluminum bronze casting alloys containing more than 10% aluminum are heat treatable. These are alloys whose normal microstructures contain more than one phase to the extent that beneficial quench and temper treatments are possible. The copper aluminum alloys normally containing iron are heat treated by procedures

Alloy	Solution Heat Treatment	Aging Treatment
C81400	0.5 h at 980–1010°C	2 h at 480°C
C82000	0.5 h at 910–930°C	3 h at 480°C
C82200	0.5 h at 900–930°C	3 h at 455°C
C82400	0.5 h at 765–800°C	3 h at 345°C
C82500	0.5 h at 765–800°C	3 h at 345°C
C82600	0.5 h at 765–800°C	3 h at 345°C
C82800	0.5 h at 765–800°C	3 h at 345°C

Source: From *Metals Handbook* [4, p. 1093.]

TABLE 8.7   Heat Treatment of Beryllium Copper Alloys

Alloy	Solution Treatment	Tempering Treatment
C95300	2 hours at 900°C	1 hour at 540–595°C
C95400	2 hours at 900°C	1 hour at 565–620°C
C95500	2 hours at 900°C	1 hour at 565–620°C

Source: From *Metals Handbook* [4, p. 1093].

TABLE 8.8 Heat Treatment of Aluminum-Bronze Alloys

somewhat similar to those used for heat treatment of steel, and have isothermal transformation diagrams that resemble those of carbon steels. For these alloys, the quench-hardening treatment is essentially a high-temperature soak intended to dissolve all of the α phase into the β phase. Quenching results in a hard room-temperature β martensite, and subsequent tempering reprecipitates fine α needles in the structure, forming a tempered β martensite. Table 8.8 shows typical heat treatments for three major aluminum bronze alloys [4].

The heat treatment procedure for the complex nickel-aluminum bronzes (C95800) is different from those shown in Table 8.8. These alloys contain the γ_2 phase which is more anodic and corrodes faster during seawater exposure. The accepted heat-treatment procedure is six to eight hours at 650 to 750°C followed by air cooling, which not only eliminates the γ_2 phase (or makes it discontinuous in high aluminum alloys) to improve their seawater corrosion resistance but also produces significant increases in the UTS and YS properties. Optical micrographs showing phase changes in as-cast and also heat-treated conditions are shown in Fig. 8.16 [20].

Sometimes the castings are subjected to a stress relief treatment which consists of 1 h/in at 260°C. However, the aluminum bronzes are stress relieved at 315°C.

Melting and Melt Handling

Gases in Copper

Gas porosity is a major factor in the quality and reliability of castings due to gas absorption during melting and gas evolution during casting and solidification processes. Although hydrogen is the most important dissolved gas in copper and its alloys, to understand the gas problem one must understand the hydrogen and oxygen solubilities and reactions.

Hydrogen in Copper Melting

The solubility of hydrogen in copper and copper alloys increases markedly with temperature, as shown in Fig. 8.17 [7]. A pronounced solubility increase is noted at the melting point of copper. Figure 8.17 also shows that the solubility of hydrogen is lower in copper-tin alloys (bronzes) than in pure copper. Hydrogen pickup by the molten metal can come from the furnace atmosphere, moisture, or oils on the furnace charge, ladles, and molding and core sands. Its effects in the metal are harmful since it can cause gas holes and microporosity. Dissolved hydrogen can add to the difficulties of dispersed shrinkage since the gas will readily diffuse to cavities, precipitate as molecular hydrogen gas bubbles, and prevent the cavities from being fed from riser or adjacent areas of the casting. Alloying elements have varying effects on hydrogen solubility (Fig. 8.18).

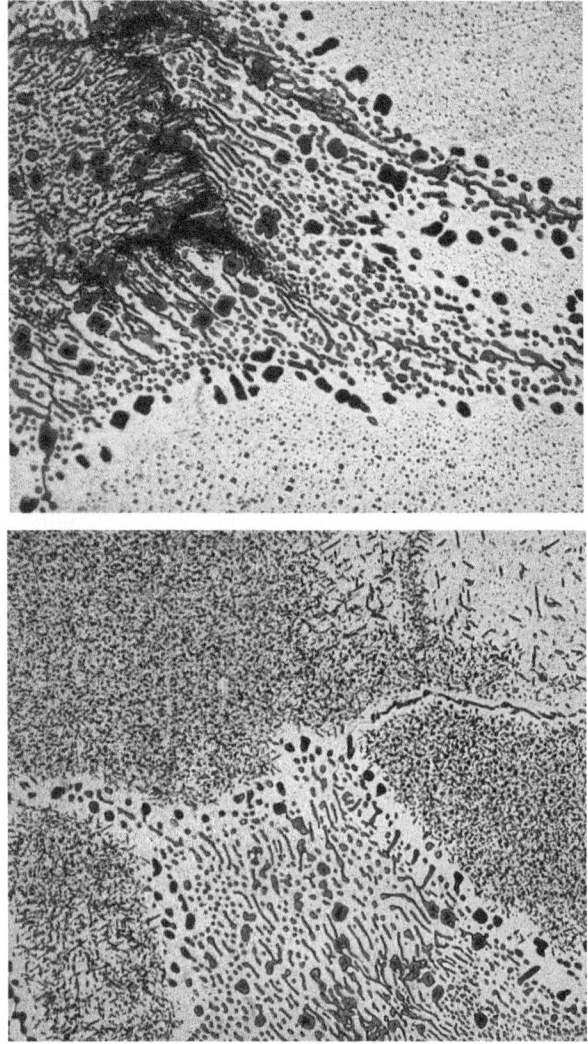

Figure 8.16   Effect of corrosion inhibiting treatment (700°C, six hours, air cool) on microstructured modification of a sand-cast nickel-aluminum bronze. (From Sahoo et al. [20].)

The solubility of hydrogen in pure copper has been experimentally determined by various investigators [21–28]. Combined linear regression analyses of the sets of reported data for hydrogen solubility in liquid and solid pure copper are given by the following equations:

Liquid copper-hydrogen [4, p. 70]:

$$\text{Log}_{10}S_l, \text{cm}^3/100 \text{ g} = \frac{-2327.59}{T} + 2.4649 \qquad (8.3)$$

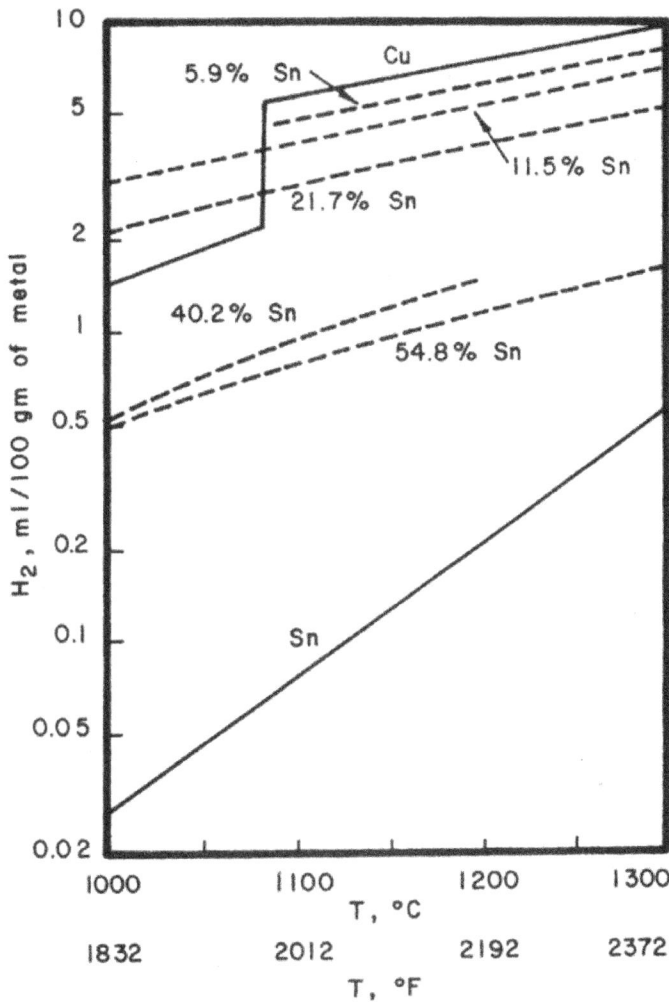

FIGURE 8.17 Solubility of hydrogen in copper, and copper-tin alloys related to temperature at 1 atm pressure. (From *Casting Copper-Base Alloys* [7].)

Solid copper-hydrogen:

$$\text{Log}_{10}S_s, \text{cm}^3/100\,\text{g} = \frac{-2057.76}{T} + 1.6820 \qquad (8.4)$$

where S_1 and S_s are the solubility of hydrogen in the liquid and solid states, respectively, and T is temperature in K. At the melting point, 1083°C (1981°F), the concentrations of hydrogen in equilibrium with the gas at atmospheric pressure are 5.6 and 1.46 cm^3/100 g for the liquid and solid, respectively.

Hydrogen solubility in both liquid and solid pure copper is directly proportional to the square root of the hydrogen partial pressure, in accordance with Sievert's law [22].

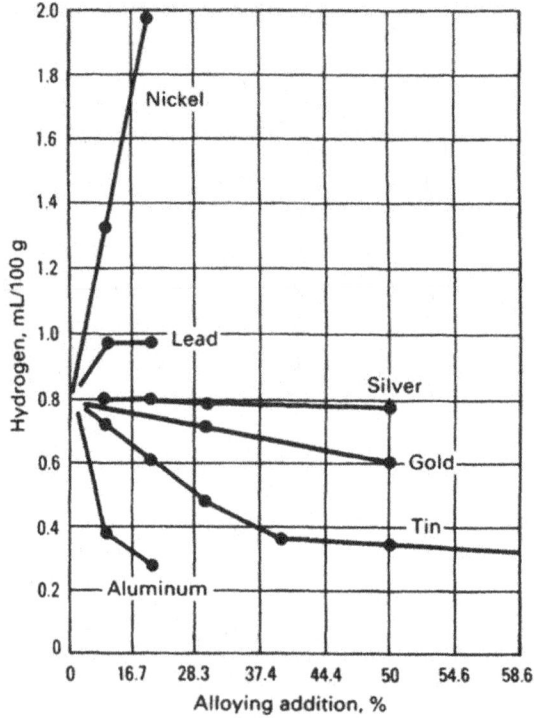

FIGURE 8.18 Effect of alloying elements on the solubility of hydrogen in copper. (From *Metals Handbook* [4].)

Published results on the effect of alloying elements on the solubility of hydrogen in copper are sparse. Nickel increases the solubility of hydrogen in liquid copper as shown in Fig. 8.18 [4]. Hydrogen can enter the copper directly from the atmosphere, but most likely, it enters according to

$$2Cu + H_2O = Cu_2O + 2\underline{H} \qquad (8.5)$$

$$H_2O = 2\underline{H} + \underline{O} \qquad (8.6)$$

Reactions given in Eqs. (8.5) and (8.6) are favored by copper with low oxygen contents. If there are high concentrations of alloying elements with more stable oxides than cuprous oxide, such as tin, the alloying element may react to put hydrogen into solution, for example:

$$\underline{Sn} + H_2O = SnO + 2\underline{H} \qquad (8.7)$$

Oxygen in Copper

Copper as an element is a metal which is readily oxidized in the molten condition. This possibility is illustrated in Fig. 8.19 [4, 7], which shows that the solubility of oxygen in

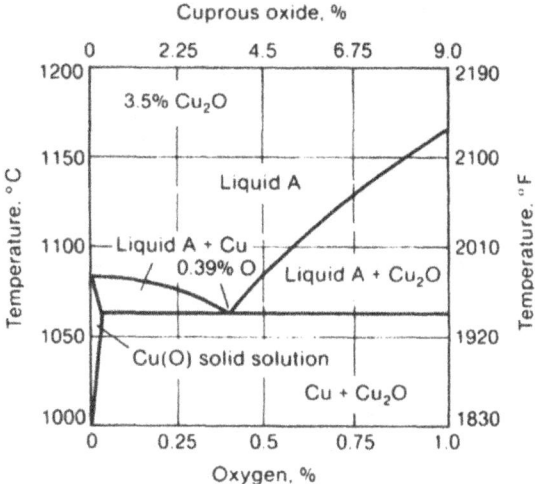

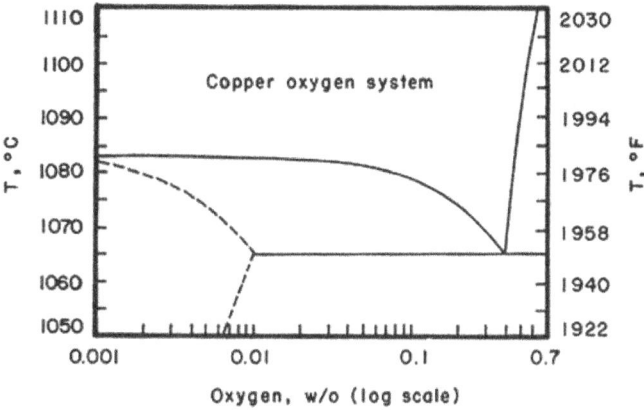

Figure 8.19 Copper-oxygen equilibrium phase diagram. The eutectic occurs at 0.39% oxygen. (From *Metals Handbook* [4], *Casting Copper-Base Alloys* [7].)

molten copper increases rapidly with temperature above its melting point. In the solid state copper can dissolve up to about 0.0035% of oxygen, with any excess occurring as Cu_2O, according to Fig. 8.19. Hence, it is evident that melting copper in the presence of free oxygen (or an oxidizing gas such as CO_2) favors oxidation or an increased percentage of oxygen in the molten copper. The actual percentage of dissolved oxygen reaches a value of 0.04 to 0.05% by weight when pure copper is melted under oxidizing atmospheres.

Oxygen forms a eutectic with copper at 0.39%. The Cu-O phase diagram shown in Fig. 8.19 shows that as little as 0.01% O_2 in the initial melt will produce 0.39% oxygen in the last liquid to solidity. This will form Cu-Cu_2O eutectic at grain boundaries.

Copper alloys contain substantially less oxygen under the same melting conditions than does pure copper. Many of the elements in the alloys, for example, zinc, tin, or aluminum, are deoxidizers to a degree, in that they may form oxides more strongly than copper and thus prevent the maximum dissolved oxygen content from being reached.

Other Gases in Copper

In addition to hydrogen, carbon and sulfur are gas-forming impurities which can be present in copper alloys. They may react as follows:

$$C \text{ (dissolved)} + O \text{ (dissolved)} \rightarrow CO \text{ (g)} \qquad (8.8)$$

$$S \text{ (dissolved)} + 2O \text{ (dissolved)} \rightarrow SO_2 \text{ (g)} \qquad (8.9)$$

The gaseous reaction products of Eqs. (8.8) and (8.9) may cause gas-hole defects, especially since these reactions occur more readily as temperature drops while a casting freezes. The solubility of carbon in copper is very low, under about 0.004%, so that CO does not appear to be a frequent source of gassing, especially since the oxygen needed for reaction in Eq (8.8) can be removed by the addition of deoxidizers. Considerable sulfur can be present in copper alloys, and it is therefore regarded as an impurity which must be held below certain limits, generally under 0.05 to 0.08% in most alloys. The most common means of preventing gassing from reactions (8.8) and (8.9) is to reduce the oxygen content by deoxidation below a percentage which will cause them to proceed to the right.

The gases that can be found in a number of copper alloys are listed in Table 8.9.

Degassing

As mentioned before, both hydrogen and oxygen can be present in copper alloys. There is a mutual relationship between hydrogen and oxygen solubility in molten copper as

Alloy Family	Gases Present	Remarks
Pure copper	Water vapor, hydrogen	Approximate hydrogen/water vapor ratio of 1. Higher purity increases the amount of water vapor and lowers hydrogen.
Copper-tin-lead-zinc alloys	Water vapor, hydrogen	Lead does not affect the gases present. Higher tin lowers total gas content. Increased zinc increases the amount of hydrogen, with a loss in water vapor.
Aluminum bronzes	Water vapor, hydrogen, CO	The presence of 5 wt% Ni in alloy C95800 causes CO to occur rather than water vapor. Lower aluminum leads to higher total gas contents.
Silicon brasses and bronzes	Water vapor, hydrogen	Approximate hydrogen/water vapor ratio of 0.5. Increased zinc decreases hydrogen and increases water vapor.
Copper-nickels	Water vapor, hydrogen, CO	All three gases are present up to 4 wt% Ni, after which only CO and hydrogen are present. Hydrogen increases with increasing nickel up to 10 wt% Ni but is decreased at 30 wt% Ni.

Source: From *Casting Copper-Base Alloys* [7, p. 132].

TABLE 8.9   Summary of Gases Found in Copper Alloys

illustrated in Fig. 8.20 [4], which shows that only one element can remain at a higher level in the melt. Thus, as the oxygen content is raised, the capacity for hydrogen absorption decreases. Hence, it is normal practice to melt copper alloys in a slightly oxidizing atmosphere to prevent hydrogen pickup and then deoxidize the melt to lower the oxygen content (see deoxidation). This is the oxidation-deoxidation practice to minimize hydrogen content in copper alloys.

Other degassing methods include (1) inert gas fluxing when argon or nitrogen is injected into the melt using a graphite tube, (2) zinc flaring, which applies to brasses containing more than 20% Zn. Zinc has a high vapor pressure (Table 8.10) and acts as a vapor purge to remove hydrogen present in the melt. In addition, the zinc oxide, being less dense, forms a tenacious, cohesive, and protective oxide skin to prevent hydrogen diffusion into the melt, and (3) solid degassing fluxes such as calcium carbonate, which liberates CO_2 upon heating and removes hydrogen.

Deoxidation

Elements which combine more effectively with oxygen than copper can be used to remove or decrease the oxygen content of a molten copper-base alloy melted under oxidizing conditions. Phosphorus, lithium, boron, calcium, magnesium, aluminum, silicon, and beryllium are such elements. Most commonly employed is a low-melting-point alloy of phosphorus and copper, 15% P-balance Cu. About 0.02% P or less is added to the

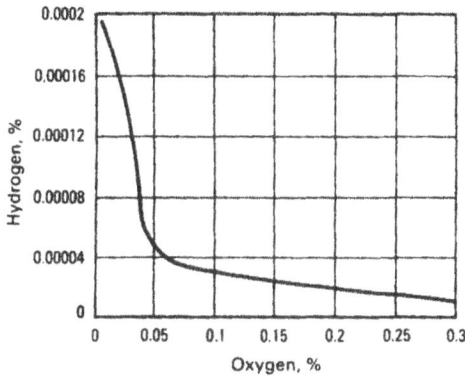

Figure 8.20 Equilibrium of water vapor with molten copper at 1083°C (1481°F). (From *Metals Handbook* [4].)

Temperature	Composition, Cu-Zn			
	60–40	**65–35**	**70–30**	**80–20**
900°C (1652°F)	160	125	90	30
1000°C (1832°F)	430	330	230	80
1100°C (2012°F)	980	760	540	180
1200°C (2192°F)	2000	1550	1100	370

Source: From Heine et al. [1].

Table 8.10 Vapor Pressure of Zinc in Molten Brass, mm Hg

melt for deoxidation purposes. This amounts to about 1 to 2 oz of 15% P-Cu per 100 lb of melt. After the addition is made, pouring should proceed at once. With the phosphorus residual in the alloy, the metal is no longer in an oxidized condition, and it can therefore pick up hydrogen again. The action of the phosphorus residual in reducing the oxygen content of a Cu-10% Sn bronze is illustrated in Fig. 8.21 [1]. It is evident from Fig. 8.21 that a residual of 0.02% P is as fully effective as a higher percentage of phosphorus in reducing the oxygen content to as low as 0.002%.

Drossing

Most, though not all, copper alloys contain readily oxidizable elements, such as zinc, tin, aluminum, magnesium, and manganese. The oxides separate more or less completely from the melt and form a dross. In many cases, the oxide has a low specific gravity and would be expected to float out of the melt. However, surface tension and other effects make the separation difficult in some alloys, such as high-zinc brasses and aluminum bronzes. The dross may then entrap considerable metal and cause high melting losses. Sometimes fluxes or charcoal covers are employed to minimize drossing. A cover of bottle glass thinned with borax is fluid and helps keep the metal surface clean. Charcoal as a protective layer is often used to minimize oxidation. Proprietary fluxes may be purchased which are claimed to cleanse the metal of oxides and prepare it for pouring. Undoubtedly, a minimum of agitation and melting under favorable combustion conditions decreases drossing.

Melting

The quality of copper-base alloy castings is greatly influenced by melting and pouring operations. The perfect mold produces a low-quality casting if correct metallurgical practices are not followed.

Copper has a low heat capacity and latent heat of fusion. Hence, less heat is required to melt copper (e.g., in comparison with aluminum). However, copper has a melting point of 1083°C and higher pouring temperatures are needed to pour copper and its alloys.

Other factors to consider in selecting the melting method are fuel (gas and oil prices for fuel-fired furnaces and electricity cost for electric melting), environmental regulation (foundry

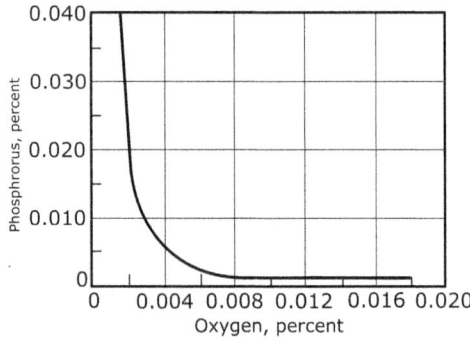

FIGURE 8.21 Effect of phosphorus residual on oxygen content of a 90-10 tin bronze. (From Heine et al. [1].)

emission controls and worker exposure to heat and noise), labor cost and availability of qualified and competent melt operators and increasing demand for premium quality castings.

Melting Equipment

Copper and copper alloys can be melted in either fuel-fired or electric furnaces.

Fuel-Fired Furnaces

Natural gas or fuel oil can be used for fuel-fired furnaces which include crucible or pit furnaces, rotary furnaces, or reverberatory furnaces.

Crucible furnaces can be of nontilting or tilting type. The former consists of a steel shell, refractory lining, refractory cover, and a burner (or burners). Either clay-graphite or carbon-bonded silicon carbide crucibles are used for melting for melt sizes of 5 to 700 lb that can be safely lifted out of the furnace. Tilting furnaces are usually used to melt larger amounts of copper alloys. Their design is similar to the pit furnaces. They are lined with a clay graphite or silicon carbide crucible cemented into the furnace shell with refractory cement.

Rotary melting furnaces are used to melt 1000 to 100,000 lb of metal. Most furnaces consist of a refractory-lined, cylindrical steel shell with the burner at one end and a flue at the other. Some furnaces can be tilted on the horizontal axis to facilitate charging and pouring.

Reverberatory furnaces are usually used by the ingot smelting industry to handle up to 200,000 lb of metal. These are stationary with a large, rectangular, shallow hearth. The flame source is at one end, and melting is accomplished by both direct flame impingement on the metal charge and by reflected heat from the refractory roof.

Electric Furnaces

Electric furnaces consist of induction furnaces, indirect-arc furnaces, and resistance furnaces. Medium or high frequency coreless induction furnaces are most common for melting copper alloys. Three types of coreless induction furnaces are used: box furnace with rammed refractory or crucible, push-up type with crucibles or lift coil and channel induction furnaces. The box furnace can be tilted for pouring. Most copper alloys are melted in 1000-cycle or 3000-cycle units (high frequency). Large melts are handled with 1000-cycle (or lower) frequencies.

Channel induction furnaces are used for brass mills or large, high-production brass foundries that use a single alloy and require a constant source of liquid metal.

Advantages of induction melting are reduced melting time and induction stirring but require large capital investment.

The electric indirect-arc furnace, known as a "rocking arc" furnace, is now rarely used with the advent of solid-state induction melting equipment.

Resistance furnaces are heated by resistance elements and look like crucible furnaces. Melting rate is slower than the induction furnaces, but clean metal can be obtained.

Casting Processes

Ninety per cent of the world's 2012 production of the 24 million tonnes of copper was processed by continuous casting, while only 1.8 million tonnes of copper three-dimensional shaped castings were processed by foundry type operations (Chapter 3 ref 14, 15).

Castings can be produced by a variety of processes such as sand casting, permanent mold casting, high-pressure die casting, centrifugal casting, investment casting, evaporative pattern casting, etc. The choice of the casting process is governed by the relative economics of the various processes, casting size, shape, complexity, quantity, surface finish, and dimensional accuracy together with alloy requirements and design specifications. The process capabilities of the casting processes are summarized in Table 8.11. Details of the casting processes are described in Chapter 3.

Sand Casting

In North America almost 85% of the annual tonnage of the copper alloy castings is produced by the green sand casting process. The process is relatively inexpensive, acceptably precise, and above all, highly versatile. Casting sizes can vary from a few grams to many kilograms. Further, it can be applied to simple shapes as well as castings of considerable complexity. The principal types of molds used for the production of copper alloy sand castings are as follows:

- Greensand
- Cement-bonded sand
- Waterless (oil/clay-bonded) sand
- Chemically bonded sand (shell, CO_2, cold-box)
- Metal molds (gravity, low-pressure).
- Ceramic/refractory molds (ceramic, investment).

Greensand Molding

Greensand molding is the most diversified and widespread process used for casting production, particularly for small- to medium-size castings. Molds are formed in (un-baked) greensand, which is most often silica (SiO_2), bonded with water and 6% southern bentonite (clay) to develop the required strength. Molding consists of compacting the tempered sand around the pattern (usually inside a flask of some type) by one or more mechanical means such as jolting, squeezing, ramming, vibrating, or blowing. Automatic molding machines have largely replaced jolt-squeeze where production

Casting Process	Weight Range, kg	Economical Quantity	Thin Section, mm	Cast Hole Dimensions, mm	Surface Finish, µm	Dimensional Accuracy, mm
Greensand	0.1–100	1–100,000	>5	>6	6–25	–
CO_2/silicate	0.1–100,000	1–1,000	>5	>6	6–25	–
Cold box	0.1–100,000	1–1,000	>5	>6	6–25	–
Shell mold	0.1–50	1,000+	>2	>3	3–12.5	±0.5
Ceramic	0.1–5,000	50+	>1.5	–	0.8–3.2	±0.5
Investment	0.1–50	1,000+	>1.5	>1	0.8–3.2	±0.5
Gravity PM	0.1–100	1,000+	>5	>6	3–6	±0.25
Low-pressure PM		1,000+				±0.5

Source: From *Casting Copper-Base Alloys* [7].

TABLE 8.11   Process Capabilities

numbers warrant. The plasticity of the bonding clay is sufficient to produce a mold which is rigid enough to hold its shape during pattern removal, core placement, pouring, and solidification. A simplified version of the process is illustrated in Fig. 8.22 [7]. A typical casting produced by this process, together with the mold from which it was made, is shown in Fig. 8.23 [7].

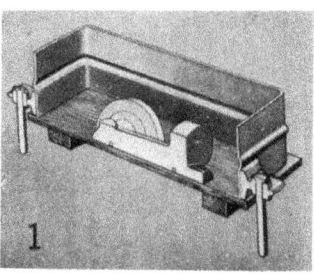

Cross section of the first step in making a green sand mold. Bottom half of the pattern is on the mold board and surrounded by the bottom or drag half of the flask.

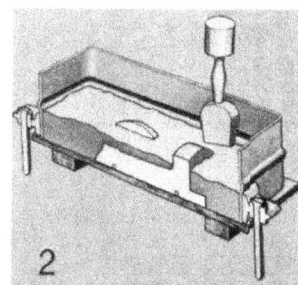

Molding sand is rammed around the pattern in multiple steps to provide uniform density.

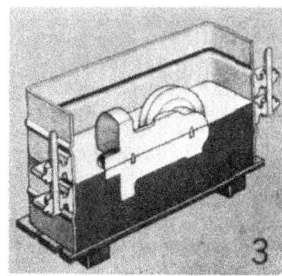

After the bottom half of the mold is filled, it is rolled upright and the top half of the pattern and flask are put in place to complete the mold.

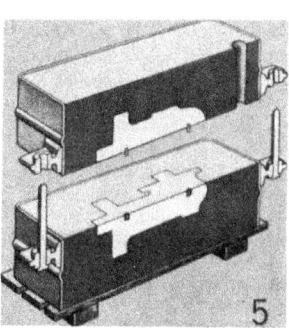

Section through the completed mold with pattern still in place and the sprue hole formed for entrance of molten metal.

Cope and drag halves of mold are separated in order that the pattern may be removed. The gate channel is then cut from the sprue to the mold cavity.

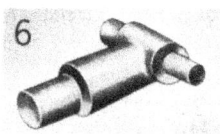

The core of bonded sand is made separately to form the internal passages of the casting.

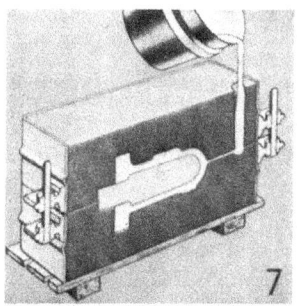

After placing core in the mold, it is closed and clamped to resist the pressure exerted by the molten metal when it is poured in the mold.

Figure 8.22   The process steps in making of greensand molds for a valve body. (From *Casting Copper-Base Alloys* [7].)

FIGURE 8.23 Typical greensand cope and drag halves of a valve body. (From *Casting Copper-Base Alloys* [7].)

Molding Sands

System sands in 40 Canadian greensand foundries were typically 95 AFS grain fineness. For foundries producing valves that are subsequently polished and plated, >95 gives a shiny finish, whereas <90 gives a noticeably darker, rough finish. Sand handling equipment can cause sand segregation, loss of fines, etc., so hoppers, silos and baghouses should be checked weekly for unnatural accumulations. Disposal of a significant fraction of the system sand, coupled with new sand additions of finer new sand (typically 105–120) can cause the size to drift.

The higher sand-to-metal ratio used in copper foundries results in less clay and carbonaceous additives being burnt out, so mainly southern bentonite is used, which readily absorbs water and spreads on the sand grains with high muller efficiency. Clay levels of 6% are adequate to produce 20 psi of green compression strength. If sand disposal is inadequate, then clay, dead clay and fines will exceed 10% (AFS Clay). This causes permeability to fall and moisture levels to exceed 3%, conditions which lead to gas defects and surface roughness.

Permeability of copper foundry sands should be 50–75. The older literature indicates much lower permeability, as well as issues with moisture and sand section thickness. However, sand recipes for copper castings contained high levels of natural sands (>20% fines and no bentonite). Synthetic sands and proper mulling gives marginal, but adequate permeability when coupled with appropriate additions of new sand and dust control (exhaust, and no floor sweepings), permeability.

It can be extremely revealing to plot sand properties and assorted defect rates over a period of up to a year. Very short periods will reveal correlations between sand properties and specific defects, although batch-to-batch, or property variation within batches produced with short mulling cycles can explain defect occurrence. The smaller equipment and large distances between mullers and molding equipment in copper foundries can cause significant variation in moisture and sand properties. Metal balls (plus 20 mesh), clay balls, extruded lead whiskers, zinc fume can build up in the sand and cause problems.

Average greensand properties for 40 synthetic sands in Canadian copper foundries over a 12 year period are shown in Table 8.12.

Core Sands

Cores for copper-base castings are made by conventional methods using core sands (shell, hot box, warm-box or nobake) using coarser core sands (AFS GFN 60–70) to decrease binder usage and improve porosity and core removal, which are required to allow for the properties of these alloys. Hot tearing can occur if the cores are too hard and resistant to collapse after the metal is poured. Typical mixtures are given in Chapter 3. Core coatings employing graphite, mica, or other washes may be employed for smoother surfaces.

Property	Value	Property	Value
Permeability	45–75	Green compressive strength	18–22 psi
Clay (MB clay)	5.5–7.5%	Fines-AFS clay	8–14%
AFS GFN	95	Moisture	2.8–3.8%

*Source: CANMET unpublished data from Mobile Foundry Laboratory visits

TABLE 8.12 Typical system sand 3-ram properties for brass and bronze castings*

Permanent Mold Casting

This process involves pouring molten metal into permanent molds as opposed to casting in sand or expendable molds. Permanent mold castings can be made by gravity pour, low pressure, and vacuum-cast methods. Gravity pouring, as the name implies, utilizes the force of gravity to run the metal down into the mold cavities. On the other hand, the low-pressure process uses air pressure to push the metal up into the mold cavity. In vacuum casting, the liquid metal is pulled into the cavity by the application of a partial vacuum. These two processes are also referred to as counter-gravity casting.

The advantages of the permanent mold process over many other casting processes are higher quality, higher precision, nearer net shape, higher mechanical properties, suitable for low- and high-volume production, and less environmental impact. The high quality is from the rapid directional solidification that the alloys undergo as they go from liquid to solid in a metal mold. Gravity pouring can be entirely manual, or by semi-automatic machines built specifically for the process. Smaller pieces (with weights as little as a few grams) and smaller molds are usually poured by hand. After extracting the casting, the mold halves are dipped in the water and graphite solution, and then transferred to the pouring station. A schematic diagram showing the successive tilt positions is presented in Fig. 8.24 [7]. In low-pressure casting, the furnace is surrounded by a metal jacket and lid which permit the chamber to be pressurized. A cast iron stalk or tube with a flange on top projects through the cover and is immersed at least 15 to 20 cm into the molten metal bath. The mold is moved directly over the tube. When the furnace chamber is pressurized, metal is forced up the stalk into the mold. The pressure is increased at a controlled rate to counteract the weight of metal in the assembly. A schematic of the low-pressure process is shown in Fig. 7.31.

Advantages of the low-pressure casting process are reduced turbulence during mold filling, better quality castings, and high casting yield. The drawbacks are reduced mold life due to thermal fatigue and limitations on the weight of castings (~15 kg).

Vacuum casting is similar to low pressure except that the liquid metal is pulled into the mold instead of a controlled push. The molds are mounted to a manipulator that has

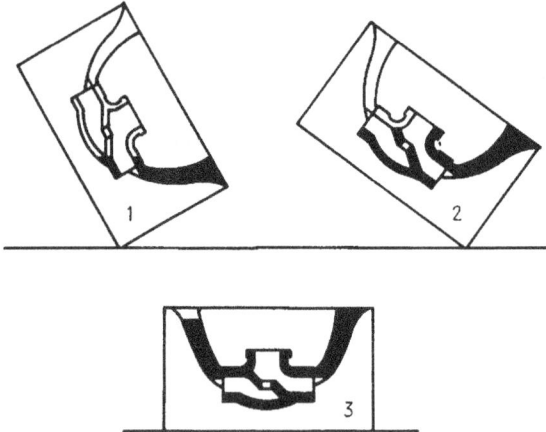

Figure 8.24   Successive tilt positions of a permanent mold during pouring. (From *Casting Copper-Base Alloys* [7].)

a vacuum pump. A tube extends down from the mold. When the tube is lowered into a molten metal bath, the vacuum pump reduces the pressure in the mold cavity causing the molten metal to be sucked into the mold cavity.

High-Pressure Die Casting

The high-pressure die-casting process is characterized by the use of high pressure to inject molten metal into reusable steel dies to form the cast shapes. The main advantage of the process is the close tolerance of cast components that require little or no machining other than simple trimming to remove gates and flash. Other advantages are high production rate, good surface finish, production of thinner wall thickness and complex shapes, and fine grain structure due to rapid solidification.

Die casting can be done either by the hot chamber or by the cold-chamber high-pressure method. In the hot-chamber process, the hydraulic actuator is in close contact with the molten metal to inject metal into the die cavity. In the cold-chamber process, the molten metal is fed into the shot cylinder by hand ladling, auto-ladling, or by a pump. It is then injected fast by a plunger into the cavity where it solidifies into a net shape.

Investment Casting

Investment casting utilizes expendable patterns which are completely surrounded or invested—that is, with no parting lines in the mold, using a molding aggregate in slurry form. The slurry hardens by chemical reaction. Patterns are most often made of fusible materials such as waxes, plastics, or fusible alloys (at one time, even frozen mercury was used) which permit their removal from the mold by melting out. Since wax was commonly used for centuries, the process has also come to be known as the lost-wax, or cire-perdue process.

As an artistic casting technique, the investment casting process has been used for centuries in the production of statuary and fine jewelry, utilizing fireclay washes to invest the wax patterns produced from the original art.

Post-casting Operations

Welding is an important post casting operation; especially from the point of defect elimination by weld repairing. Copper alloys such as cupronickels, aluminum bronzes, and manganese bronzes are readily weldable by a variety of methods. Other alloys are weldable only by the use of special techniques.

Cupronickels

Within the recommended chemistry ranges, the cupronickel alloys are weldable by all methods. These alloys may be welded by shielded metallic arc (SMA), gas metal arc (GMA), or gas tungsten arc (GTA) methods. The shielded-arc metal process is most often used, mainly due to the more convenient interpass temperature control and the ready availability of suitable filler metal. AWS Class 5.6 ECuNi filler metals are available for welding these alloys [29]. It is desirable to limit the silicon content in the chromium-modified 70/30 alloys between 0.3 and 0.5% to enhance weldability. The presence of certain impurities such as Pb, S, P, Bi, Se, Sb, and B can contribute to poor weldability and detrimental effect on weld properties. Interpass temperature should not exceed 120°C.

Aluminum Bronzes

Aluminum bronzes such as C95200, C95300, C95400, and C95800 can be joined by most welding processes using commercially available electrodes (Ampcotrode 10 for 952 and 953, Ampcotrode 150 for 954, and E10 for 958). Inert-gas shielded metal-arc (MIG) or tungsten-arc welding (TIG) produce the best results. Metal-arc processes using the correct grade of coated electrodes give satisfactory results. Carbon-arc welding has been reported to produce good results. However, oxy-acetylene welding is seldom successful. Weld repairing to rectify small defects, buildup, and hard facing are being performed on cast components. Interpass temperature can vary between 38 and 177°C for 958 and between 205 and 425°C for 952, 953, and 954 [30].

Oxide films which are always present in aluminum bronzes should be removed by using a flux or by grinding or scratch brushing before welding. Degreasing should also be done to remove oil films which may cause porosity.

Postweld stress relief or heat treatment is not performed on large components such as propellers unless a customer's specification requires it. However, castings for marine applications should be post-weld heat treated at about 700°C for 6 h followed by air cooling.

Cu-Mn-Al bronzes (alloy C95700) can be easily welded because of their low thermal conductivity and the absence of a brittle temperature range. The procedure is similar to aluminum bronzes. Flux-coated electrodes of matching composition are available for MIG and TIG welding. Preheating is not necessary but may be advantageous to minimize distortion. Postweld heat treatment at 650°C is recommended to get maximum corrosion resistance.

Manganese Bronzes

C86500 manganese bronze is welded using TIG or MIG. The parent metal must have an alpha-beta microstructure to be weldable. The surfaces are prepared as for aluminum bronzes but preheated to 150°C before welding. Interpass temperatures are kept between 150 and 427°C. A postweld stress relief is preformed, heating at 10°C (or less) per hour to 315 to 427°C, holding for six hours, and cooling at a rate of 10°C per hour or less.

Tin Bronzes and Leaded-Tin Bronzes

These bronzes have a long solidification range and are prone to hot-shortness which gives rise to cracking on cooling. Hence, they are difficult to weld or weld repair. Another welding problem is related to electrodes. Although flux-coated electrodes are available for metal-arc welding, the slag formed during welding is very strong and adherent. Unsound welds will result if such adherent slag is not removed. Hence, it may be necessary to perform interpass cleaning by grinding each pass. Use of inert-gas shielding process can minimize the formation of slag. Leaded-tin bronzes can be welded using ac welding current with thoriated tungsten electrode for the tungsten-arc process and dc electrode positive for the metal-arc and inert-gas-shielded metal-arc process. It is advisable to avoid sputtering during welding as it may cause porosity.

References

1. R. W. Heine, C. R. Loper, Jr., and P. C. Rosenthal, *Principles of Metal Casting*, McGraw-Hill Book Company, New York, 1967.

2. A. Spada, American Foundry Society, private communications.

3. J. Michel, Copper Development Association, private communications.

4. *Metals Handbook, Casting*, Vol. 15, ASM International, 2007: 1085–1094.

5. F. A. Fasoyinu, J. L. Dion, D. Cousineau, R. A. Matte, K. G. Davis, and M. Sahoo, "Fluidity of Permanent Mold Cast Copper-Base Alloys," *Transactions of American Foundrymen's Society*, 100: 547–559, 1992.

6. "Cast Products Alloy Data," in *Standards Handbook*, Part 7, Copper Development Association, Inc., 1996.

7. *Casting Copper-Base Alloys*, American Foundry Society, 2007. Edited by S. Ducharme, M. Sahoo and K. Sadayappan, Schaumburg, IL.

8. M. Sahoo, K. C. Wang, and J. O. Edwards, "Foundry Characteristics and Mechanical Properties of Both Niobium-Modified and Chromium-Modified High-Strength 70/30 Cu-Ni Alloys," *Transactions of American Foundrymen's Society*, 87: 529–536, 1979.

9. M. Sahoo and M. Wirth, "Grain Refinement of a Cast Cu-Ni-Sn Spinodal Alloy," *Transactions of American Foundrymen's Society*, 98: 25–33, 1990.

10. *Metals Handbook, Casting*, Vol. 15, ASM International, 1988.

11. F. Hudson, *Gunmetal Castings*, Hart Publishing Co., Inc., New York, 1968.

12. H. Meigh, *Cast and Wrought Aluminum Bronzes—Properties, Processes and Structure*, Copper Development Association, UK, 2000.

13. P. J. Macken and A. A. Smith, *The Aluminum Bronzes*, Copper Development Association, 1966.

14. M. Sahoo, J. O. Edwards, and R. Thomson, "Influence of Corrosion Inhibiting Heat Treatment on the Microstructure and Mechanical Properties of Nickel-Aluminum Bronze Alloy C95800," *Transactions of American Foundrymen's Society*, 87: 495–502, 1979.

15. A. Couture, M. Sahoo, B. Dogan, and J. D. Boyd, "Effect of Heat Treatment on the Properties of Mn-Ni-Al Bronze Propeller Alloys," *Transactions of American Foundrymen's Society*, 95: 537–552, 1987.

16. A. Couture and M. Sahoo, "Influence of Compositions and Heat Treatment on the Properties of Mn-Ni-Al Bronze Propeller Alloys," *Transactions of American Foundrymen's Society*, 96: 567–578, 1988.

17. *Metals Handbook*, 1948 edition, ASM International.

18. M. Sahoo, A. Taylor, and R. J. Dawson, "Foundry Characteristics and Mechanical Properties of Weldable Monels," *Transactions of American Foundrymen's Society*, 99: 507–517, 1991.

19. M. Sahoo, R. J. Lacroix, P. Newcombe, and B. Gracia, "Influence of Carbon, Silicon and Niobium Contents on the Structure and Mechanical Properties of Cast Monels," *Transactions of American Foundrymen's Society*, 100: 239–251, 1992.

20. M. Sahoo, J. O. Edwards, and R. Thomson, "Influence of Corrosion-Inhibiting Heat Treatment on the Microstructure and Impact Properties of Nickel-Aluminum Bronze Alloy C95800," *Transactions of American Foundrymen's Society*, 88: 769–776, 1980.

21. M. B. Bever and C. F. Floe, "Solubility of Hydrogen in Molten Copper-Tin Alloys," *Transactions of the Metallurgical Society of AIME*, 156: 149–159, 1944.

22. A. Sieverts, "The Solubility of Hydrogen in Copper, Iron, and Nickel," *Zeitschrift fur Physikalische Chemie*, 77: 591, 1911.

23. P. Rotgen and F. Moller, "On the Solubility of Gases in Copper and Aluminum," *Metallwirtschaft*, 13: 81, 97, 1934.

24. E. Kato, H. Ueno, and T. Orimo, "Solubility of Hydrogen in Liquid Copper Alloys," *Transactions of the Japan Institute of Metals*, 11: 351, 1970.

25. R. B. McLellan, "Solid Solutions of Hydrogen in Gold, Silver, and Copper," *Journal of Physics and Chemistry of Solids*, 34: 1137–1141, 1973.

26. C. Thomas, "Solubility of Hydrogen in Solid Copper, Silver, and Gold Obtained by Rapid Quench and Extraction Technique," *Transactions of the Metallurgical Society of AIME*, 239: 485–490, 1967.

27. F. G. Jones and R. D. Pehkle, "Solubility of Hydrogen in Solid Ni-Co and Ni-Cu Alloys," *Metallurgical and Materials Transactions*, 2: 2655–2663, 1971.

28. M. Mokaram, "The Solution and Diffusion of Hydrogen in Solid Copper and Copper Alloys," M65 Thesis, Brunel University, London, 1974.

29. M. Sahoo and W. P. Campbell, "Weldability of Both Nb-Modified and Cr-Modified High-Strength Cu-Ni Casting Alloys," *Transactions of American Foundrymen's Society*, 88: 727–736, 1980.

30. M. Sahoo, "Weldability of Nickel-Aluminum Bronze Alloy C95800," *Transactions of American Foundrymen's Society*, 90: 893–911, 1982.

CHAPTER 9

Magnesium and Magnesium Alloys

Mahi Sahoo
Suraja Consulting Inc.

Introduction

Magnesium with a density of 1.7 g/cc is the lightest of the structural metals. It has a weight about 35% lighter than aluminum and 60% lighter than steel. Considering strength to weight ratio, magnesium casting can compete with conventional high-pressure die-cast aluminum and plastic. Other advantages of magnesium include high thermal conductivity and thermal stability. These impressive properties make them suitable for producing cast components for the military and the, aerospace, aircraft, and helicopter industries. This is a market industry with good engineering capabilities and strong technology and vendor infrastructure. The other growth industry is the automotive sector, where the amount of magnesium cast components in passenger vehicles has increased to about 6 kg per vehicle and is expected to increase to about 25 kg per vehicle by 2020 with a projected growth rate of 10 to 15% per annum [1]. Magnesium cast components also find applications in markets for sporting goods, power tools, and the electronic industry. The market for magnesium castings for the past 10 years is shown in Table 7.1.

Most of the magnesium castings are produced by high-pressure die casting for the automotive sector and by sand casting for the aerospace sector. There are multiple casting processes available for the aluminum components. This is not the case for magnesium and there is the need for development of additional manufacturing processes. However, most of the casting methods used to produce aluminum castings can be applied to magnesium castings with development of optimum casting parameters. These include permanent mold technologies (gravity tilt pour and low pressure), squeeze casting, lost-foam casting, and new emerging technologies such as ablation casting [2]. Typical cast components are shown in Fig. 9.1. Other high-pressure die-cast components can be found in Ref. 4.

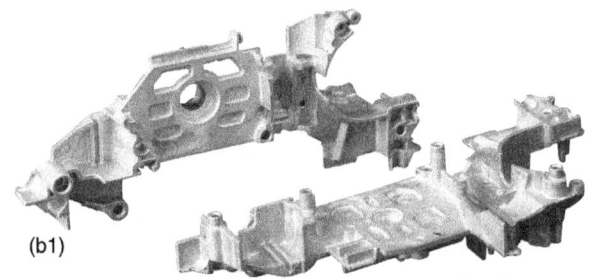

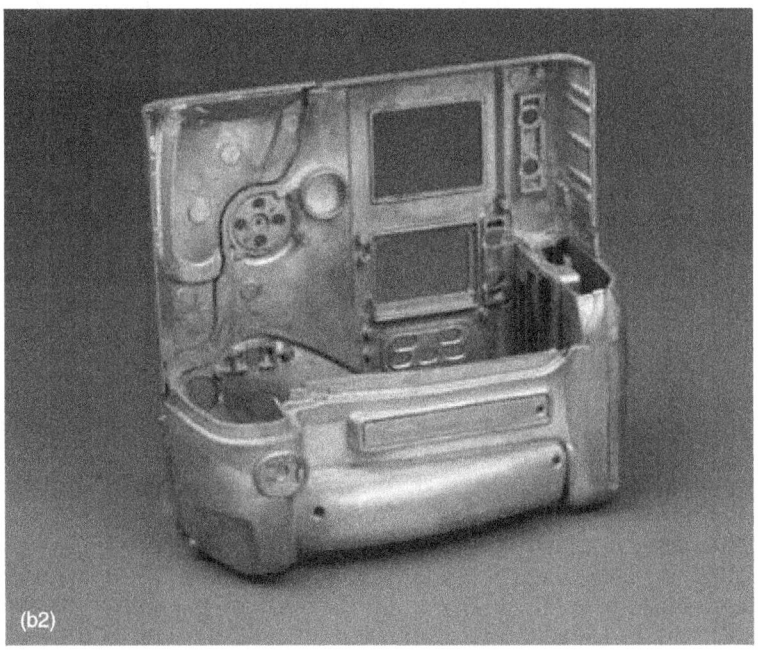

Figure 9.1   Typical magnesium cast products. (*a*) Sandcast, (*b*) hot-chamber die cast,

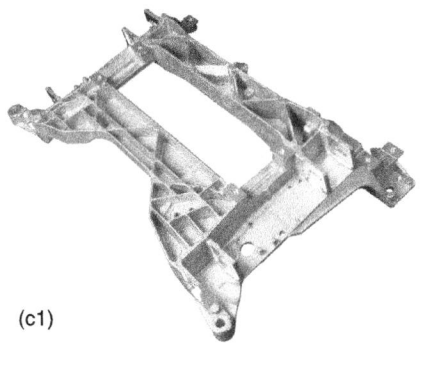

(c1)

(c2)

(c3)

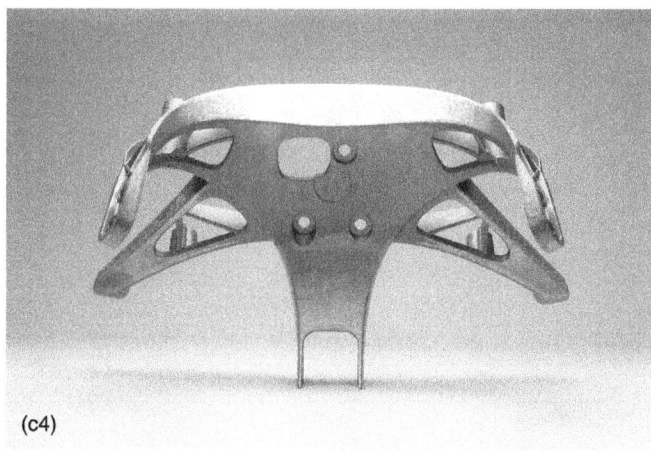

(c4)

FIGURE 9.1 (*Continued*) (*c*) Cold-chamber die cast. (Courtesy of ECK Industries [2], and NADCA [3].)

Magnesium Alloys

The physical and thermal properties of pure magnesium are summarized in Table 9.1. The crystal structure of pure magnesium is close-packed hexagonal. Its density is 2/3 than that of aluminum but the thermal expansion coefficient is 1/3 higher that of aluminum. Magnesium, like most pure metals, is soft with low mechanical properties such as 90 MPa UTS, 21 MPa YS, 2 to 6% elongation, and hardness of 30 HRB. Magnesium can be alloyed with many elements to enhance properties such as strength, formability, and corrosion resistance. Major alloying elements used in cast magnesium alloys are aluminum, zinc, zirconium, yttrium, and rare earth elements. Silver and thorium have also been used as major alloying additions for casting alloys. However, economics and health concerns make these elements unattractive. Lithium, which can be considered as a major alloying element, is restricted to wrought alloys and extensively used for aerospace applications.

Cast magnesium alloys contain other minor alloying additions such as beryllium, calcium, silicon, and manganese that are used to enhance properties such as castability, high-temperature strength, formability, and oxidation resistance.

Major Alloying Elements

Aluminum

Aluminum is the most widely used alloying addition in magnesium to increase the mechanical properties. Another important effect aluminum has on magnesium is the reduction in grain size. Mg-Al alloys can be heat treated (solution treatment and aging) to improve the mechanical properties because of the sloping solvus line (Fig. 9.2a).

Zinc

Zinc has very low solid solubility in magnesium and forms many intermetallic compounds (Fig. 9.2b). Alloys containing zinc can be strengthened through precipitation hardening. Zinc on its own improves the yield strength of magnesium linearly. However, both UTS and ductility exhibit a peak around 4% zinc and decrease at higher zinc levels. Zinc is used as a major element in combination with Zr and/or RE (rare earth

Property	Value
Density, gm/cc	1.74
Melting point, °C	650
Volumetric shrinkage on solidification, %	4.2
Linear shrinkage, %	1.5
Solid linear shrinkage (650–20°C), %	1.7
Coefficient of thermal expansion (20–100°C), um.	26.1
Thermal conductivity (25°C), W/m·K	156
Electrical conductivity, IACS	39
Latent heat of fusion, kJ/kg	360–377
Specific heat capacity at 20°C, kJ/kg·K	1.025

Source: From *ASM Specialty Handbook* [5], Sadayappan and Luo [5].

TABLE 9.1   Physical and Thermal Properties of Pure Magnesium

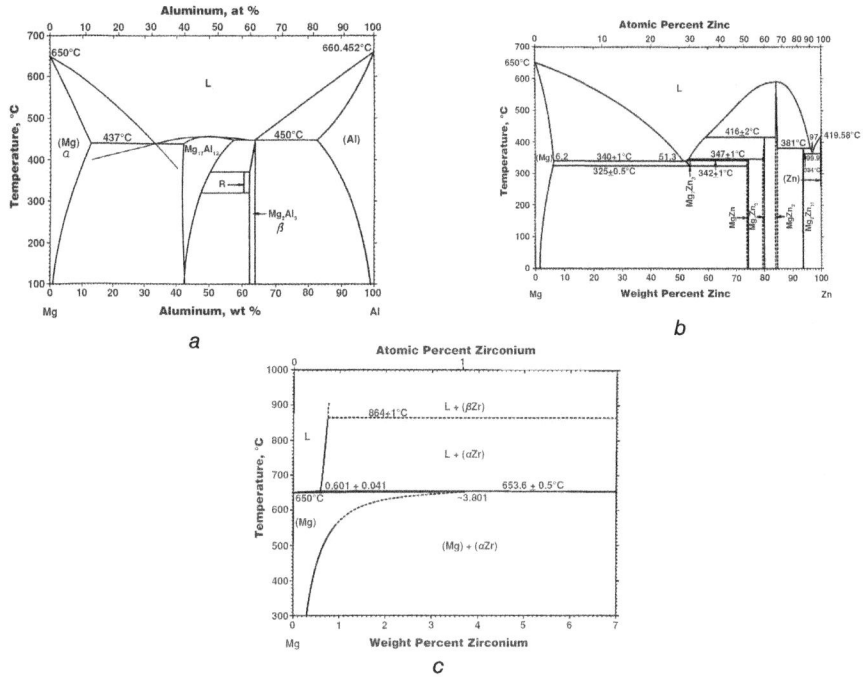

Figure 9.2 Phase diagrams for three magnesium alloys. (a) Mg-Al, (b) Mg-Zn, and (c) Mg-Zr. (From Sadayappan and Luo [6].)

elements). Zinc is also used as a minor addition in alloys with Al. In general, zinc addition increases strength but is limited to 1% due to hot shortness in Mg-Al alloys. Zinc can also be used to counteract the adverse effect of iron on corrosion.

Zirconium

Zirconium is less soluble in liquid as well as solid magnesium compared to other elements. At the melting temperature of magnesium, the liquid solubility of zirconium is around 0.7%, which falls rapidly as the temperature is lowered (Fig. 9.2c). Zirconium does not form intermetallic compounds at the useful range and is usually added for grain refinement in alloys with zinc, rare earth elements, or silver. At least 0.5 to 0.7% Zr is needed to obtain good refinement, which is beyond the liquid solubility levels in magnesium. However, this element is not an effective grain refiner in alloys containing Al and Mn as it forms stable compounds and is separated from the melt.

Rare Earth Metals

Rare earth elements are added to magnesium as misch metal or didymium. Misch metal is an alloy itself and contains cerium (Ce), lanthanum (La), and neodymium (Nd). On the other hand, didymium contains neodymium and praseodymium (Pr). Currently many new alloys are being developed with individual additions of rare earth elements such as gadolinium (Gd). Rare earth elements have very low solubility in magnesium and usually form intermetallic compounds with magnesium. Rare earth additions reduce the solidification range of the alloys, make feeding easier, and also improve the high-temperature strength of magnesium [6].

Yttrium

Yttrium can replace silver or thorium in alloys for high-temperature applications and better creep resistance. It has a good solubility of around 12% which is comparable to that of aluminum.

Silver

Silver is widely used for aerospace casting alloys due to its ability to provide high-temperature strength. Magnesium alloys with silver can be heat treated. Most of the common alloys with silver also contain rare earth elements.

Lithium

Lithium has very high solid solubility in magnesium. The hexagonal crystal structure of magnesium is modified to cubic with the addition of about 11% lithium [7]. However, lithium does not improve the strength or high-temperature stability of magnesium. Mg-Li alloys in combination with other alloying elements such as aluminum are used as wrought alloys for aerospace applications.

Minor Alloying Additions

Calcium enhances the oxidation resistance when magnesium is molten. It also reduces the oxidation rates during heat treatment. Calcium was originally added to wrought magnesium alloys to improve rolling and mechanical properties. A 0.05% Ca addition to casting alloys AZ92 and AZ63 just prior to pouring is used to reduce microporosity and shorten the solution heat-treatment time of alloys containing zinc. Calcium additions can also improve the creep resistance. Higher levels of calcium are known to reduce the oxidation rates but may cause heat-treatment difficulties.

Silicon improves the casting fluidity of magnesium alloys and is invariably used in combination with aluminum. However, the corrosion performance of magnesium is impaired in the presence of silicon.

Strontium is another element that can be added in combination with aluminum for improving the creep resistance. It can refine the grain structure of magnesium alloys and also improve the high-temperature strength.

Beryllium is used in very low levels, less than 5 ppm, especially in die-casting alloys to reduce the oxidation and fuming of molten magnesium alloys when melts are being held for prolonged periods of time. However, beryllium is known to reduce the efficacy of carbon in grain refinement and cause grain coarsening in sand and permanent mold casting alloys.

Manganese is used to control the effect of iron present in the system. It forms intermetallic compounds with iron that are less harmful when corrosion is considered. Manganese has very limited solubility in magnesium and is restricted to 0.3% in alloys containing aluminum.

Tin is sometimes used in small quantities to improve the ductility of Mg-Al alloys.

Copper can be used to increase the high-temperature strength of certain magnesium alloys. However, the detrimental effect of copper on corrosion overrides its benefits.

Impurities

Most of the common impurities present in magnesium are transition elements and reduce its corrosion resistance. These elements include iron, copper, nickel, and cobalt. The limits beyond which these elements cause severe damage are quite small, of the order of 0.005%.

Al - Zn - Mn	Zr	Zn - RE - Zr	Ag - RE - Zr	Y - RE - Zr
AZ63	K1A	EZ33	QE22	WE43
AZ81		ZE41	EQ21	WE54
AZ91				
AZ92			Nd - Gd - Zn - Zr	
AM100			EV31	

FIGURE 9.3 Magnesium casting alloy groups along with corresponding commercial names. (From Clark [8].)

The negative effect of iron can be neutralized to some extent by the addition of manganese. Fluxes are sometimes used in melting magnesium alloys, as protection against oxidation. Fluxes commonly contain chlorides of barium, sodium, and potassium and these elements can be picked up by the melt that can cause embrittlement in magnesium alloys.

Alloy Groups

The four basic groups of alloy systems which are currently being commercially produced for gravity (sand or investment) casting are based on the major alloying elements: aluminum, zinc, manganese, zirconium, and rare earths [6, 8]. These are subdivided as follows:

Magnesium-aluminum-zinc-manganese   Magnesium-zinc-rare earths-zirconium

Magnesium-silver-rare earths-zirconium   Magnesium-rare earths-zirconium

The alloy groups plus the individual alloys can be found in Fig. 9.3. Commercial magnesium alloys have been named using a practice established in 1952 by the

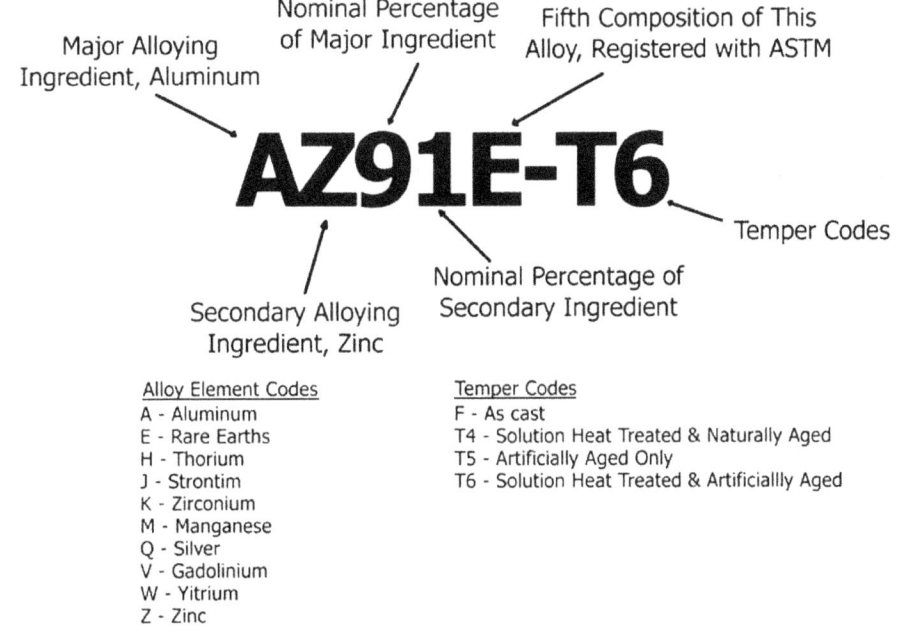

FIGURE 9.4 ASTM B951 magnesium alloy identification system.

A-Aluminum	Q-Silver
C-Copper	S-Silicon
E-Rare earths	V-Gadolinium
J-Strontium	W-Yttrium
K-Zirconium	Z-Zinc
M-Manganese	T-Tin*
H-Thorium*	L-Lithium*

*For historical reference purpose only; no commercial casting alloys are in use now.

Source: From Sadayappan and Luo [6].

TABLE 9.2   Alloy Designations

American Society for Testing and Materials, or ASTM (ASTM B951, "Standard Practice for Codification of Unalloyed Magnesium and Magnesium-Alloys, Cast and Wrought").

Code designations are based on chemical composition limits. For alloys, designations consist of two letters representing the alloying elements specified in the greatest amount (arranged in order of decreasing percentages), followed by the respective percentages (rounded off to whole numbers) and a serial or variant letter [8]. Figure 9.4 presents one typical example. Alloy designations for other alloying elements are shown in Table 9.2.

Physical Metallurgy

Mg-Al System

The phase diagram of Mg-Al system is presented in Fig. 9.2a [6]. The solid solubility of aluminum in magnesium is 12.7% at 435°C (815°F), which is the eutectic temperature and decreases to less than 2% at room temperature. Aluminum increases the strength and hardness of magnesium. Commercial casting alloys contain less than 10% aluminum. Aluminum-containing alloys have an intermetallic β phase ($Mg_{17}Al_{12}$) formed in the interdendritic areas and grain boundaries. This phase can reduce the ductility. However, by solution and aging treatments the morphology and distribution of this phase can be modified. For this reason, magnesium alloys with more than 6% Al can be heat-treated resulting in a range of properties. However, the slow response of the alloys to the treatment makes the process less attractive.

Aluminum can be alloyed with other elements such as zinc, manganese, and rare earths. Many of the casting alloys contain one or more of these elements. On the other hand, zirconium is not usually added in combination with aluminum as it reacts with aluminum and precipitates as an intermetallic compound.

The Mg-Al-Zn system (AZ91 type) is the workhorse of the Mg family of casting alloys. The phase constitution and equilibrium solidification characteristics of the Mg-Al-Zn-based alloys can be understood through the binary Mg-Al phase diagram shown in Fig. 9.2a. The small addition of zinc to Mg-Al alloys has little effect on the structure of the binary alloy, and a ternary Mg-Al-Zn phase appears only when the Zn to Al ratio exceeds 1:3 [10]. According to the phase diagram the equilibrium solidification sequence for the AZ91 alloy is as follows: (a) nucleation of primary magnesium at the liquidus, approximately 600°C (1112°F), (b) end of solidification at the solidus, approximately 500°C (932°F), and (c) eutectic solidification at 435°C (813°F). Thermal analysis for a cooling rate

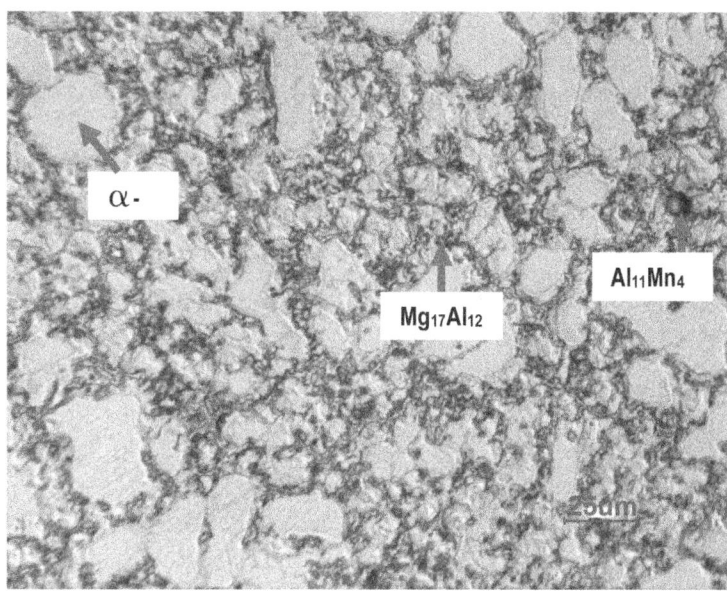

FIGURE 9.5 Optical micrograph showing the as-cast microstructure of AZ91 alloy. (From Sadayappan and Luo [6].)

of 0.06°C/s shows that the liquidus (T_L) and solidus (T_S) temperatures are 600°C (1112°F) and 435°C (815°F), respectively. With increasing cooling rates T_L and T_S decrease [9, 11].

Figure 9.5 shows the typical as-cast microstructures of AZ91. In the slow cooling rate condition (near-equilibrium), a partially divorced eutectic [Mg(α) + $Mg_{17}Al_{12}$(β)] structure can be expected [9, 11]. It is believed that the lamellar $Mg_{17}Al_{12}$ structure is formed after solidification via discontinuous precipitation from Mg(α) supersaturated solid solution. Such lamellar precipitates do not appear in specimens with higher solidification rates. With decreasing aluminum content in the alloys, the tendency for forming divorced eutectic structure increases. It is worth noting that the morphology of divorced $Mg_{17}Al_{12}$(β) precipitates leads to better mechanical properties than the lamellar structure.

Mg-Al-Mn-Based Alloys

The solidification sequence of Mg-Al-Mn-based AM series alloys is similar to that of AZ91 alloy. The nonequilibrium liquidus and solidus temperatures as well as the solidification sequence are given in Table 9.3.

Temperature	Transformation
T_L [601 – 593°C (1113.8 – 1099.8°F)]	Start of Mg(α) nucleation: L → Mg (α) + L
$T_L - T_E$	Growth of Mg(α) phase: L → Mg (α) + L
T_E [434°C (813.2°F)]	Divorced eutectic reaction: L → Mg (α) + $Mg_{17}Al_{12}$

Source: From Luo and Sadayappan [9].

TABLE 9.3 Nonequilibrium Liquidus and Solidus Temperatures and the Solidification Sequence for AM60 Alloy

Mg-Al-Si-Based Alloys

The solidification microstructure of Mg-4%Al-1%Si alloy AS41 is characterized by the presence of two intermetallic phases (i.e., $Mg_{17}Al_{12}$ and Mg_2Si). Under near-equilibrium conditions, the solidification of AS41 alloy starts with the nucleation of Mg(α) crystals at 621°C (1150°F), followed by the onset of Mg_2Si formation at 611°C (1132°F), and ends at the complete transformation of liquid magnesium to solid [Mg(α) + Mg_2Si] structure at 568°C (1054°F). $Mg_{17}Al_{12}$ forms as lamellar precipitates at grain boundaries via discontinuous precipitation from the Mg(α) phase. Under nonequilibrium solidification, the process starts with the nucleation of Mg(α) phase at 622°C (1152°F), followed by the onset of Mg_2Si formation at 613°C (1135°F), and ends at 424°C (795°F), forming the divorced eutectic structure of [Mg(α) + $Mg_{17}Al_{12}$] [9].

Mg-Al-RE-Based Alloys

The solidification of AE42 (Mg-4%Al-2%RE) alloy is more complicated than AZ, AM, or AS series alloys, as the addition of 2% RE significantly changes the solidification path of Mg-Al alloy due to the formation of Al_4RE intermetallic phase [9]. The solidification sequence is described in Table 9.4. A typical microstructure of alloy AE42 is shown in Fig. 9.6.

Temperature	Transformation
T_L [625°C (1157°F)]	Start of Mg(α) formation: L → Mg(α) + L
T_F [615°C (1139°F)]	Start of Al_4RE formation: L → Al_4RE + L
T_S [513°C (955°F)]	End of solidification: L → Mg(α) + Al_4RE

Source: From Luo and Sadayappan [9].

TABLE 9.4 The Nonequilibrium Solidification Sequence for AE42 Alloy

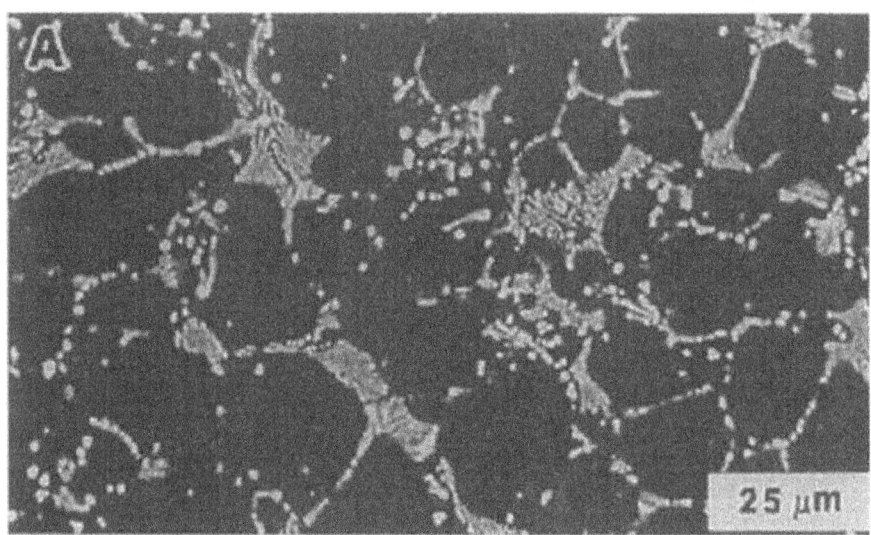

FIGURE 9.6 SEM micrograph showing the formation of $Al_{11}RE_3$ phase in die-cast AE42 alloy. (From Sadayappan and Luo [6].)

Mg-Zn Alloy Systems

The Mg-Zn phase diagram is presented in Fig. 9.2*b*. Zinc has a maximum solubility of 6% at the Mg-Zn eutectic temperature of 340°C (644°F). Some of the important magnesium sand and permanent mold casting alloys are also based on the Mg-Zn system. Due to the very low solubility of zinc in magnesium, Mg-Zn intermetallic phase appears even in alloys containing 2% Zn. Primary magnesium precipitation is soon followed by the precipitation of Mg_7Zn_3 phase during eutectic reaction. If slowly cooled after solidification, this eutectic phase will undergo eutectoid transformation as a magnesium solid solution and MgZn phase at 330°C [6].

Binary Mg-Zn alloys have a coarse grain structure and are prone to microporosity formation. Addition of Cu to Mg-Zn alloys promotes precipitation hardening. Progressive addition of Cu increases the eutectic temperature. The divorced eutectic in binary Mg-Zn alloys is replaced by a lamellar eutectic in the ternary Mg-Zn-Cu alloys. The commercial Mg-Zn-Cu alloys are known as ZC63.

Casting Processes

As mentioned before, magnesium castings are generally produced as high-pressure die castings or sand castings. Many of these techniques are already applied for special geometries or niche markets but broader applications are in development. Other options for the production of high quality structural magnesium castings include, but are not limited to permanent mold castings, low-pressure permanent mold casting, low-pressure sand casting, V-process casting, investment casting, lost-foam casting, ablation casting, and also thixomolding.

Sand Casting

Sand casting of magnesium follows the same basic practice as the sand casting of other metals with some important exceptions. These include (1) the need for chemical inhibitors, (2) changes in metal filtration materials, (3) compensation for the low fluid metal pressure of magnesium, relative to aluminum, and (4) issues related to the low-heat content of magnesium [12]. Magnesium oxidizes readily in a molten state, and oxidation continues in the mold during and after the metal is poured, both through direct interactions with the mold atmosphere and by reaction with the oxygen or moisture in the silica sand. Inhibitors are added into the sand to prevent the second reaction. Inhibitors work through a number of mechanisms, but basically, they promote the formation of a protective film on the surface of the molten magnesium as it enters the sand mold; they help to create a protective atmosphere in the mold cavity; and they form a protective coating around the sand grains forming the mold.

Inhibitor formulations vary widely, but sulfur is one of the oldest and universally used inhibitor. Sulfur works by volatilizing readily in the mold to form sulfur vapor. Usually, sulfur is used in combination with potassium fluoborate to reduce the amount of sulfur needed. Riser sleeves and pouring sprues are available with inhibitors already added, although they can also be produced by dipping standard inhibitorless sleeves and sprues in a mixture of alcohol and potassium fluoborate and drying thoroughly.

Most of these are produced in chemically bonded sand for the aerospace industry but some are made using a modified greensand technique. The molding sand and inhibitor type and amount are normally adjusted to suit the heaviest section being cast. A typical molding sand for small castings contains McConnesville 130 sand, Southern Bentonite 7%, sulfur 2%, sodium sulfur 2%, sodium silicofluoride 2% for greensand molding, and this has 4% moisture. For CO_2-bonded sand mold, only 1% sodium silico-fluoride is used and sulfur is not added [13].

The low-heat content of magnesium alloys creates a problem with respect to core collapsibility. The lower heat content means that breakdown of the sand binder is harder to achieve. To compensate for this, cores should be designed to be hollow when possible and be made from sands that demonstrate good collapsibility or good breakdown at low temperatures. For fragile or thin-sectioned castings, it may be necessary to bake the cores out of the cast part to prevent damage while removing the residual sand.

Gravity or Tilt Pour Permanent-Mold Casting

Higher volume requirements and lower cost make permanent-mold casting a viable alternative in many applications. Typically, castings will have a better surface finish than sand with better dimensional control. In order to make the mold cavity as free of oxygen as possible, purging with the cover gas before pouring is recommended. Recent developments in permanent mold coatings for magnesium make it easier to produce parts without constant recoating [14]. Mold coatings are usually applied between 175 and 230°C mold temperature. The concept of multiple coating applications (bottom coat and top coat) is recommended for magnesium castings. The base coat is insulating Dycote 6 from Foseco and the top coat is HGLS 07290402, from Hill and Griffith Company, which contains fluorspar. In one trial an additional 1% CaF_2 was added to the top-coat [15]. Some magnesium alloys are prone to hot tearing. This problem can be overcome in this process by careful control of mold temperature and by grain refining.

Low-Pressure Permanent Mold

Low-pressure permanent mold uses a horizontally parted mold with metal entering the mold from a furnace under the mold. As a result, clean metal is drawn from the middle of the melt and the entry of the metal from the lowest point in the mold reduces the turbulence normally encountered in gravity pouring [14]. It has many of the advantages of a standard permanent mold. Other advantages are high casting yield since less gating system is needed and the castings are relatively sound as opposed to high-pressure die casting and hence, mechanical properties can be improved by heat treatment. The geometry of the casting will determine if it is best produced in conventional permanent mold or low pressure. A schematic of the low-pressure process is shown in Fig. 9.7.

Low-Pressure Sand Casting

Sand molds can be mounted on standard low-pressure pouring devices to improve the flow of metal into the mold. Low-pressure feeding improves the apparent fluidity of the alloy and yields are much higher than in traditional sand casting.

Lost-Foam Casting

Lost-foam is a promising near-net-shape casting process for the production of magnesium casting. During lost-foam casting (LFC), a pattern formed of a polymeric foam material, such as polystyrene or polymethylmethacrylate, is supported in a flask and surrounded by an unbonded particulate material, such as silica sand or a synthetic

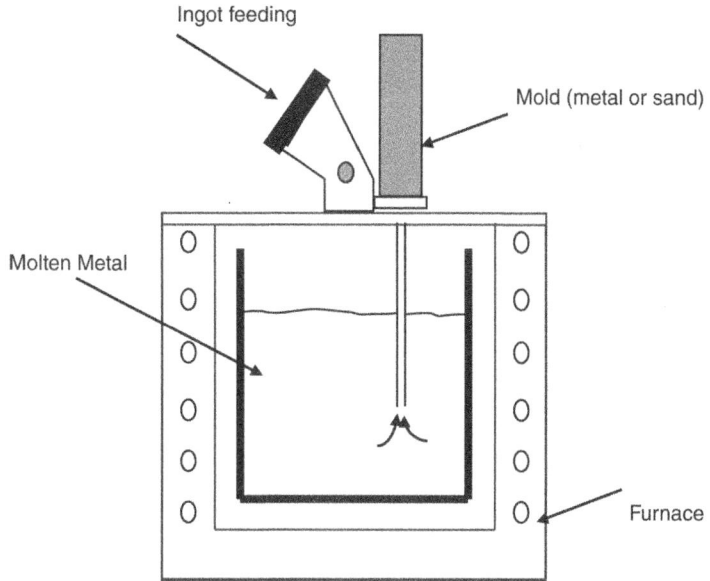

Figure 9.7 Schematic of the low-pressure casting process. (From Weiss [2].)

mullite sand [16]. When the molten metal contacts the pattern, the foam material decomposes while the metal replaces the foam material to produce a cast part that is identical in configuration to the pattern. LFC has significant cost and environmental advantages and enables metal casters to produce complex parts often not possible using other methods. The process allows designers to consolidate parts, reduce machining, and minimize assembly operations. Another distinct advantage of LFC is the use of unbonded sand around the coated foam, resulting in high collapsibility of the mold at the end of the solidification process of the alloys. This makes the lost-foam casting process uniquely suitable for casting of alloys with high hot-tearing susceptibility, such as creep-resistant Mg alloys. Significant development work has been done in the last ten years to tailor this process to the unique requirements of magnesium. Mechanical properties are equivalent to those of sand castings. This process may be the lowest-cost option for complicated castings. Figure 9.8 shows an engine block produced via LFC.

High-Pressure Die Casting

Die casting of magnesium can be done either by the hot chamber or by the cold chamber high-pressure method. The hot-chamber die-casting method is illustrated in Fig. 9.9 [3, 4].

In the hot-chamber die-casting method, the molten metal is held in an enclosed steel crucible, under a protective atmosphere. A valve allows a controlled volume of molten metal into the gooseneck that is immersed in the molten metal. A plunger injects this metal into the cavity of the die through a nozzle. To prevent freezing of the metal, the nozzle is heated to 427 to 482°C (800 to 900°F) with gas or by induction heating.

The pressure applied on the molten metal during injection is lower than in cold-chamber die casting, and limits the size of parts made by the hot-chamber method. The main reason is the high-operating temperature of various components, such as the nozzle, that limits the pressures that can be applied. A typical 400-ton magnesium

Figure 9.8   Engine block produced in magnesium alloy AZ91E by the lost-foam casting process. (From Weiss [2].)

hot-chamber machine makes parts that weigh up to 6 lb. The machine has a clamping force of 400 tons, and applies about 5 Ksi maximum pressure on the metal. Due to the short cycle time (up to six parts per minute), the hot-chamber die-casting process is very competitive for small parts. Typical magnesium parts made by the hot-chamber die casting include powertrain housings (gearbox, electric motors, hydraulic elements, cylinder head covers, brackets, and cross-beams), safety systems (steering wheel columns, steering wheel skeletons, and airbag housings), chassis and body (body elements, fixation elements, accessory elements, and decorative parts), car interior parts (cockpit cross-beams, dashboard elements, door handles, knobs, switches), and electronics components used in cell phones and computers.

The cold-chamber die-casting method is depicted in Fig. 9.10 [3, 4]. The molten magnesium is fed into a shot cylinder either by hand ladling, auto-ladling, or by a pump. It is then injected fast (5 to 100 m/s) by a plunger into the cavity, where it solidifies into a net shape part under high pressure (5 to 20 Ksi). If used to form undercuts, cores are retracted. Finally, the casting is ejected, and the part is trimmed by separating it from the gating system and the biscuit. The entire cycle takes usually less than one minute. It is a very cost-effective near-net-shape casting process, especially for the automotive industry. Typical cast parts include cradle, dash board, fluke shield main internal, etc.

While the fast injection of molten metal into the die cavity allows filling of very thin sections, it also contributes to air entrapment in the castings. The resulting porosity can impair the mechanical properties. If high mechanical properties are targeted, special precautions need to be taken to prevent air entrapment. Eliminating the air from the shot sleeve and the cavity by application of vacuum can reduce the air entrapment in

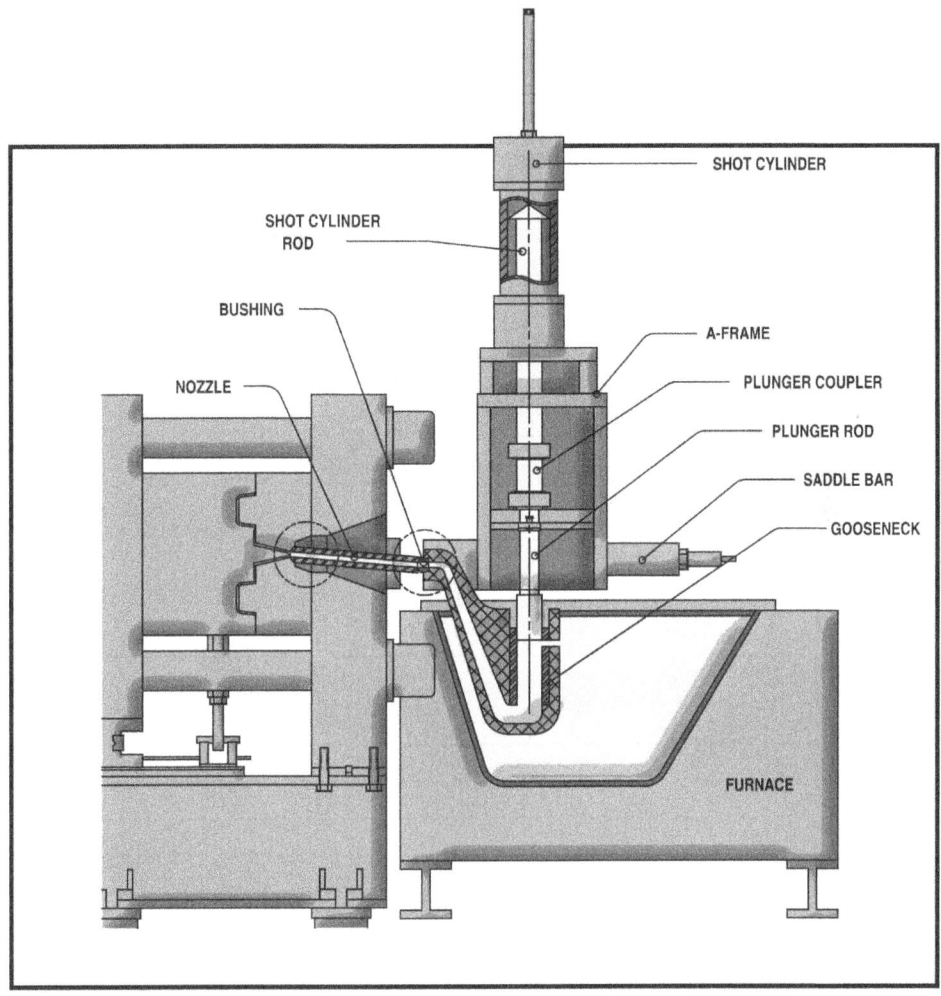

FIGURE **9.9** Schematic of the hot-chamber die-casting method. (From Udvardy [3], Schwam [4].)

the casting [4, 5]. This process not only removes the air from the cavity, but also eliminates excessive handling of the molten magnesium. The metal is drawn by the vacuum, directly from the furnace to the shot sleeve without exposure to air. The reduced level of oxides and porosity produces parts with superior mechanical properties. Typical diecast alloys and their compositions are listed in Table 9.5.

Thixomolding

Thixomolding is a semisolid metal (SSM) casting process, used exclusively with magnesium alloys using chips which are reheated and then injection molded, similar to plastic injection molding [18]. In this process, starting granules are subjected to mechanical shear as they are heated into the region between the liquidus and solidus and accumulated for injection molding in a self-contained and one-step unitized machine. Thus,

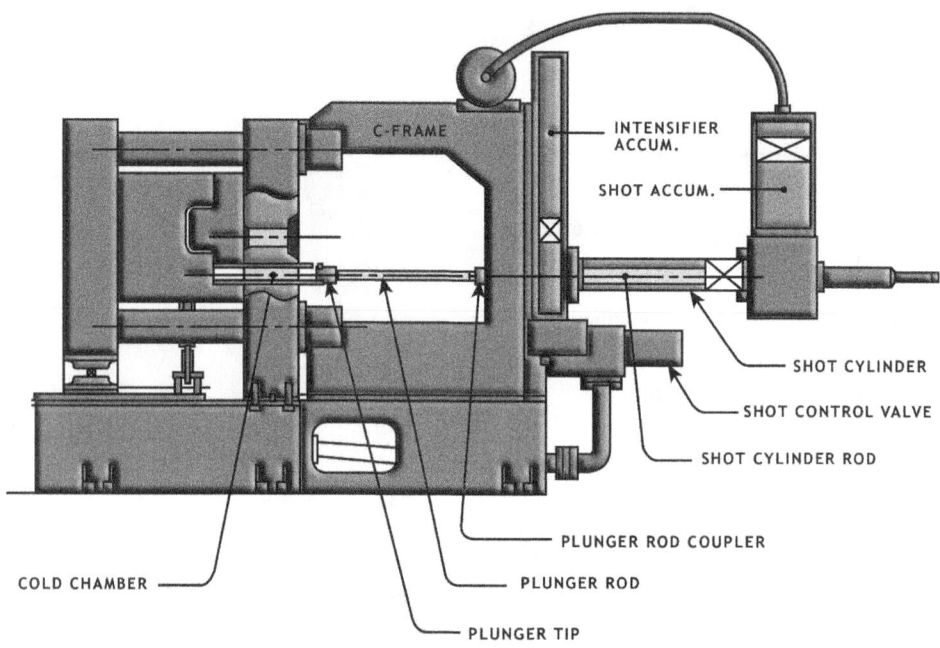

C-FRAME

INTENSIFIER ACCUM.

SHOT ACCUM.

SHOT CYLINDER

SHOT CONTROL VALVE

SHOT CYLINDER ROD

PLUNGER ROD COUPLER

COLD CHAMBER

PLUNGER ROD

PLUNGER TIP

FIGURE 9.10 Cold-chamber machine. (From Udvardy [3], Schwam [4].)

Alloy	Al	Mn	Zn	Si	Rare Earth
AE42-F	4.0	0.10			2.5
AM20-F	2.1	0.10			
AM50A-F	4.9	0.26			
AM60A and AM60B-F*	6.0	0.13			
AS21-F	2.2	0.10		1.0	
AS41A and AS41B-F†	4.2	0.20		1.0	
AZ91A, B and D-F9‡	9.0	0.13	0.7		

*AM60B castings have max. contaminant level of 0.005% Fe, 0.002% Ni, and 0.010% Cu.
†AS41B castings have max. contaminant level of 0.035% Fe, 0.002% Ni, and 0.002% Cu.
‡AZ91D castings have max. contaminant level of 0.0050% Fe, 0.002% Ni, and 0.030% Cu.
Source: From *ASM Specialty Handbook* [17].

TABLE 9.5 Nominal Composition of Die-Cast Magnesium Alloys

melting and melt handling problems are eliminated. Use of argon during injection molding enables 100% recycling of sprues, gates, runners, and scrap without secondary refining. The fraction solid can be varied between 0.05 and 0.6 during injection molding and thixotropic properties are developed in situ in the near-net-shape parts. Advantages of the process include better mechanical properties, good surface quality, and faster cycle time compared with conventional die castings. Microstructures of a thixomolded

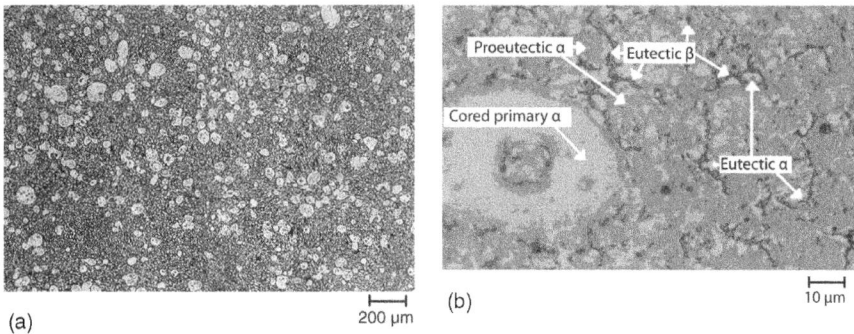

Figure 9.11 Micrographs of Thixomolded Magnesium Alloy AZ91D. (From Carnaham and Decker [18].)

AZ91D alloy are shown in Fig. 9.11. Applications include parts for the electronic and communication industry, auto industry, hand tools, and sporting goods.

Ablation Casting

The ablation casting process is the latest emerging technology in the casting industry which takes advantages of the high cooling rate due to water spray on a sand aggregate with a water-soluble binder [19]. In this process, the liquid metal is poured into the mold, and the mold is progressively ablated away with the molten metal in the mold cavity. The mold may be tilted to control metal fill. Since the mold is progressively removed, the water can contact the metal casting directly, eliminating the air gap. As a result, the heat transfer is more rapid compared with other conventional casting processes and hence, high-temperature gradients are established that help to eliminate shrinkage porosity, especially in thick sections. The final outcome is very high solidification rates in a sand mold leading to fine microstructures and improved mechanical properties. The ablation casting process and a typical cast component are shown in Figs. 7.38 and 7.39 respectively.

Both prototype and production type components have been produced in conventional aluminum casting alloys such as alloy 356 by the ablation casting process. A lower control arm has been produced in magnesium alloy AZ91 and AM60B for the High Integrity Magnesium Automotive Components (HIMAC) project supported by the U.S. Automotive Materials Partnership (USAMP) through the U.S. Council on Automotive Research (U.S. CAR).

Heat Treatment

Heat treatment of magnesium castings is performed in order to change the mechanical properties, alter the metallurgical structure, or relieve residual stresses. Sand and permanent mold castings are usually heat-treated. On the other hand, high-pressure die castings are normally used as-cast, because of economic considerations and the risk of blistering and surface damage associated with porosity. However, with the introduction of vacuum high-pressure die casting, the porosity levels in the castings can be minimized; as a result magnesium die castings can be heat treated to improve their mechanical properties.

The various types of heat treatments that can be performed on magnesium castings are designated by the following symbols [5, 20]:

F   As-cast

T2   Stress-relief annealed

T4   Solution-treated

T5   Artificially aged

T6   Solution treated and artificially aged

T7   Solution treated and stabilized

Of these, solution treating followed by quenching and aging (T6) is the most commonly used heat-treatment practice. Magnesium alloys can be heat treated in air. However, as a precaution, a protective atmosphere can be used to prevent surface oxidation as well as active burning should the furnace exceed the proper temperature. Government specification MIL-M-6857 requires a protective atmosphere for solution treating above 400°C (752°F). Table 9.6 summarizes the mechanical properties of some magnesium alloys in both as-cast and heat-treated conditions.

Alloy	Casting Type	Heat Treatment	YS, MPa	UTS, MPa	% Elongation	Ref.
AZ91, Grain Refined	Plaster	F	93	117	1	21
		T4 20 h/413°C (775.4°F)	88	168	4.7	21
		T6 16 h/169°C (336°F)	108	168	2.1	21
AZ91	Sand	F	110	183	4.5	22
AZ91		T4 24 h/410°C (770°F) + 16 h/200°C (392°F)	162	283	3.5	22
AZ91E	Low-pressure permanent mold*	F	92	180	3.4	23
		T4 18 h/420°C (788°F)	77	219	6.6	23
		T6 18 h/420°C (788°F) + 16 h/175°C (347°F)	138	224	1.7	23
ZK61	Sand	T6 2 h/500°C (932°F) + 48 h/130°C (266°F)	218	313	8.5	22
WE43		T6	180	250	7	24
WE54		T6	205	280	4	24
EZ33		T5	110	160	3	24

TABLE 9.6 Mechanical Properties of Cast and Heat-Treated Magnesium Alloys

Alloy	Casting Type	Heat Treatment	YS, MPa	UTS, MPa	% Elongation	Ref.
AZ91D	Squeeze cast	F	96	179	5	25
		T4	76	220	10.5	25
		T6	117	255	6.5	25
	Sand		83	154	2	25
	Gravity die cast	F	96	176	2	25
	RDC (rheo die-cast)	F	145	248	7.4	26
		T4 5 h/413°C (775°F)	91	230	11.2	26
		T5 5 h/216°C (421°F)	133	236	6.5	26
		T6 5 h/413°C (775°F) + 5 h/216°C (421°F)	134	255	6.7	26
MRI 201S	Sand/permanent mold	T6	170	260	6	27
MRI 202S	Sand/permanent mold	T6	150	250	7	27
MRI 203S	Sand/permanent mold	T6	125	210	4	27
AM50	Low-pressure permanent mold*	F	58	192	8.7	23
		T4 6 h/425°C (797°F)	64	211	9.5	23
		T6 6 h/425°C (797°F) + 16 h/175°C (347°F)	66	200	8.6	23
AM60B	Permanent mold	F	65	163	5.5	28
		T4, 16 h/410°C (770°F)	60.5	195.8	8.2	28
		T6 16 h/410°C (770°F) + 16 h/175°C (347°F)	67	189	7	28
		F Grain Refined with C_2Cl_6	83	212	8.1	28
		Grain Refined T4, 16 h/410°C (770°F)	79.2	200.5	7.5	28
		Grain Refined T6, 16 h/410°C (770°F) + 16 h/175°C (347°F)	80	209	8.7	28

*Subsize specimens (25 mm gauge length, 10 mm width, and 2 mm thickness) were removed from prototype cylindrical castings.

Source: From Sahoo and Keist [20].

TABLE 9.6 Mechanical Properties of Cast and Heat-Treated Magnesium Alloys (*Continued*)

Stress Relief Treatment (T2)

Residual stresses can be introduced into magnesium castings due to

- Mold restraint during solidification
- Nonuniform cooling after heat treatment
- Quenching from the solution treatment temperature

Postcasting operations such as machining and welding or weld repairing can also introduce stresses. Hence, stress relief heat treatments should be performed. Such treatments can also prevent stress corrosion cracking of magnesium castings. The temperature for stress relieving can be similar to that of aluminum alloys, i.e., in the range of 250 to 350°C (482 to 662°F).

Fluidized bed heat treatment of magnesium alloys has been performed to reduce heat treatment time. The process is not well established as in aluminum alloys [20, 29–31].

Gas in Mg Alloys

Magnesium dissolves more hydrogen than aluminum at comparable temperatures and hydrogen partial pressures and has the potential to form hydrides. Figure 9.12 shows the solubility of hydrogen in molten magnesium at 101 kPa [32]. However, the propensity for gas porosity formation in magnesium is less than that in aluminum. The main source of hydrogen in magnesium and its alloys is moisture in the surrounding atmosphere. Other sources are damp fluxes, tools, and corroded charge. At the melting point, 651°C (1204°F), the concentration of hydrogen in equilibrium with gas at atmospheric pressure are 44.2 and 29.9 cm^3/100 g for the liquid and solid, respectively.

Alloying elements affect the solubility of hydrogen in magnesium. With the exception of Ni, Zr, and Sr, the addition of alloying elements generally reduces the solubility of hydrogen in magnesium.

Hydrogen, when present in excess, causes microporosity in castings reducing the strength and ductility. However, gas porosity is significantly less than that in aluminum

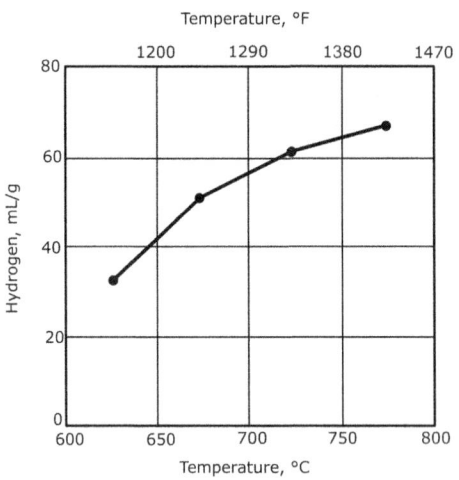

Figure 9.12   Solubility of hydrogen in molten magnesium at 101 kPa (760 torr). (From Neff [32].)

alloys. This is due to the higher partial pressure of hydrogen in magnesium and its alloys and nucleation of gas porosity during solidification becomes difficult.

Melting and Melt Handling

Magnesium casting alloys must be handled similar to other liquid metal handling techniques: with safety in mind, first and foremost. Magnesium multiplies these safety concerns due to its inherent nature to react and oxidize, at times, violently. Often, magnesium can be easier to work with than other liquid metals; for example, magnesium metalcasters use steel crucibles to hold liquid magnesium prior to pouring while aluminum metalcasters require refractory crucibles for similar processes.

Melting

Magnesium may be melted in gas-fired or electric furnaces, including induction furnaces. Steel crucibles equipped with an additional outer layer of heat-resistant high-alloy steel crucible for added lifetime are used. These crucibles have two important design features [33]:

1. *Firebrick material.* The refractory material on the walls and bottom of the furnace are made from clay firebrick material (high alumina, low silica) for less reactivity (i.e., if there is liquid magnesium leakage into the furnace chamber).

2. *Drag-out hole.* For access to allow cleaning of the furnace bottom for liquid magnesium drainage in an emergency (Fig. 9.13). Frequent periodic cleaning of the furnace bottom is performed to remove any oxidized metal scale that may have fallen off the sides of the crucible. This scale may combine with air and liquid magnesium (in case of a crucible leak) to form thermite, which is a very high temperature and exothermic reaction.

Before melting magnesium alloys, the following steps must be followed:

- Be sure the proper fire extinguishing materials are on hand and fully functional (i.e., such as class D fire extinguishers, M130 fused magnesium salt flux, or dry sand).

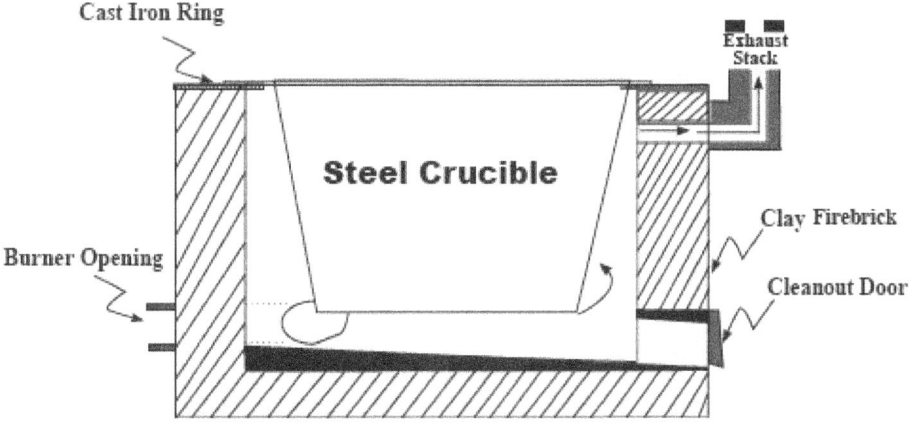

FIGURE 9.13 A schematic drawing of a magnesium melting setup for a gas-fired or electric furnace. (From Bailey [33].)

- Ensure that the protective cover gas is operating and in ample supply, if a complete fluxless remelting operation is desired.
- All materials (Mg metal, alloy additions, ladles, skimmers, pumps, etc.) that will be in contact with the liquid metal bath are preheated to at least 100°C (212°F).
- Check that no "bridges" are formed in the Mg feedstock in the initial charge that can form a dome during melting that may trap gas that can later escape, violently.

Cover Gas (Oxidation Prevention of Liquid Metal)

Commonly in magnesium sand-casting operations, specific casting cover flux is used to prevent oxygen (in the production facility air) to access the liquid magnesium surface. The choice of which salt flux (which are a blend of salts and insoluble materials, such as CaF_2) is very Mg alloy dependent. Many sand, permanent mold, low-pressure, and die-casting furnaces now employ a protective cover gas to inhibit magnesium oxidation (burning) in the casting crucible [33]. This is a requirement for automotive castings as salt inclusions from entrained flux particles are very detrimental to the corrosion resistance of castings. To this end, many protective cover gas solutions are in use globally that include

- *100% SO_2.* Older technology with significant occupational health and maintenance issues
- *1 to 2% SO_2 with nitrogen as carrier gas.* Gaining market acceptance for many small crucible applications. The SO_2 gas supply distribution system's temperature must be kept above 85°F (29°C) in order to prevent reliquification of SO_2.
- *SF_6/CO_2/dry air.* Common sand and permanent mold blended gas solution, although the excellent active inhibiting gas (SF_6) in this blended gas solution is likely to be prohibited and/or controlled in the future due to its grossly high global warming potential.
- *SF_6/dry air.* This solution is used predominantly in the die-casting industry. The lower metal casting temperatures (versus normal sand casting pouring temperatures) allows the elimination of CO_2 from the blended gas solution. This solution has the same environmental issue as the previous sand casting solution due to its active gas (SF_6).
- *SF_6 with nitrogen as carrier gas.* A more robust blended cover gas solution (less inhibition variance during normal operations) versus solutions using all or part of the carrier gas with dry air. This solution has the same environmental issue as the previous sand casting solution due to its active gas (SF_6).
- *AM-cover.* This relatively new blended cover gas solution uses as its active gas the common refrigerant HFC-134a, with either nitrogen or carbon dioxide as the carrier gas (never dry air, even as part of the blend). While a very good solution to the environmental issues associated with the SF_6-based cover gas solutions, this patented process technology has not been accepted uniformly in the magnesium industry at this time [34].
- *Novec 612.* This newer solution has also been satisfactorily tested as a suitable replacement for SF_6-based cover gas solutions in die-casting furnaces [35]. It is commercially available and will require training of foundry personnel since the active ingredient is a liquid that must be transformed prior to blending.

Degassing

Molten magnesium and alloys can be degassed with hexachloroethane, chlorine, or nitrogen. Addition of hexachloroethane by plunging a perforated steel plunger is more common and serves the dual function of degassing and grain refinement by supplying carbon through decomposition. Chlorine or chlorine/nitrogen gas purging is performed using clean, dry lances, preferably graphite, with thorough mixing for 5 to 10 minutes. Rare-earth-containing alloys should be degassed only with nitrogen or argon as chlorination leads to removal of expensive alloying elements. Treatment with a flux consisting of 55% KCl, 34% $MgCl_2$, 9% $BaCl_2$, and 2% $CaCl_2$ apparently decreases hydrogen content to 3 to 4 cm^3/100g.

Hot Tearing

Hot tearing characteristics of cast alloys have been described in Chapter 4 together with some general observations from Refs. 36 to 38. Because of the high interest in magnesium alloys for automotive applications, it may be of interest to know the experience gained in the casting of an engine cradle by the low-pressure permanent mold casting process in alloys MRI 202S and AM-SC1 by the project team within the Structural Cast Magnesium Development (SCMD) program under the leadership of the American Foundry Society (AFS) [15]. Photographs of the cradle casting are shown in Fig. 9.14 where the photograph on the bottom shows two hot tear cracks.

In order to understand the reasons for such hot tear cracks, the effect of mold temperature on hot tearing susceptibility of several magnesium alloys has been studied in detail and compared with aluminum alloys. Table 9.7 summarizes the minimum mold temperature required to eliminate hot tearing in the magnesium and aluminum alloys tested. Of all the creep-resistant magnesium alloy tested, MRI 206S can be cast at the lowest mold temperature (340°C) without hot tearing. All other creep-resistant magnesium alloys need a mold temperature of 390°C or higher to eliminate hot tearing. In case of the Al alloys, alloy 206, which is extremely sensitive to hot tearing, needs a mold temperature of 400°C to eliminate hot tearing. Alloy 206 has a solidification range of 80°C. By contrast, Almag 535, which also has a solidification range of 80°C, did not show hot tearing when cast at a mold temperature of 220°C. Alloy 319 has a solidification range of 90°C and hot tearing was not observed at a mold temperature of 350°C. The intergranular nature of the hot-tear cracks is shown in Fig. 9.15.

Grain Refinement

Several magnesium alloys (specifically sand-casting alloys) need grain refinement to achieve desired component mechanical properties and improve hot tear resistance. These alloys (such as ZE41, WE43, MRI 202S, ML-10, and AM-SC1) are usually grain refined with a magnesium-zirconium master alloy addition. The other group of alloys belongs to the Mg-Al series which are usually grain refined using the carbon hexachloride (C_2Cl_6) tablet.

Mg-Zr master alloys are added at a melt temperature of 790 to 815°C (1450 to 1500°F) with a ladle or skimmer and held below the metal surface followed by good stirring after the master alloy has dissolved.

When grain refining using the C_2Cl_6 tablet, it is advisable to plunge the tablet using a bell-shaped perforated steel plunger and hold for a few minutes until the bubbling (due to chlorine gas evolution) stops. Normal practice is to use one tablet (70 g) for every 20 kg melt. A good ventilation system is required in the melting area. Optical micrographs showing the grain refinement of alloy AM60B using the C_2Cl_6 tablet are shown in Fig. 9.16.

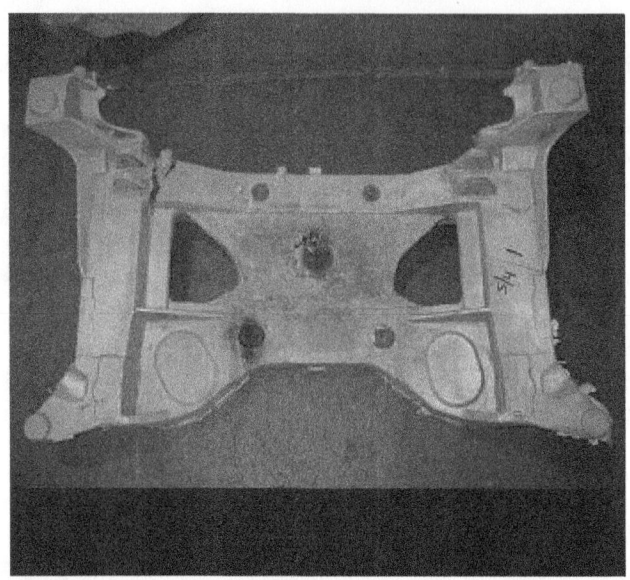

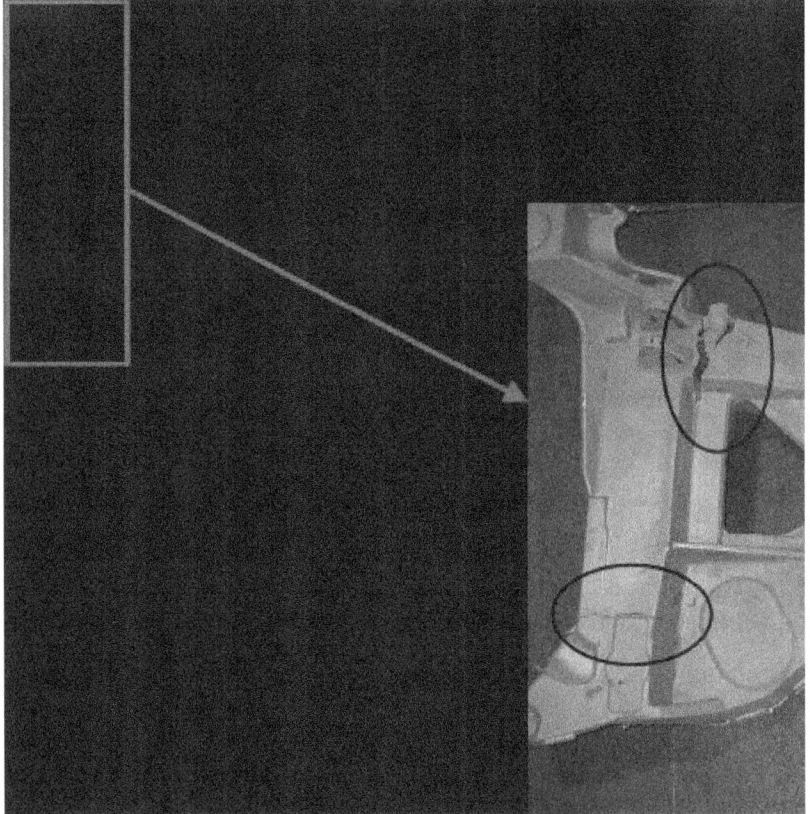

Figure 9.14 Engine cradle produced by low-pressure permanent mold casting. (From Sadayappan et al. [15].)

Alloy	Mold Temperature for No Hot Tearing (C)
AZ91	≥335
MRI 206S	≥340
AJ62	≥380
MRI 202S	≥390
AE63	≥395
AE44	≥395 (1 crack)
MRI 201S	≥420
AE35	≥415 (3 cracks)
319	~350
535	≥220
206	≥400

Source: From Sadayappan et al. [15].

TABLE 9.7 Hot Tearing Data for Aluminum and Magnesium Alloys

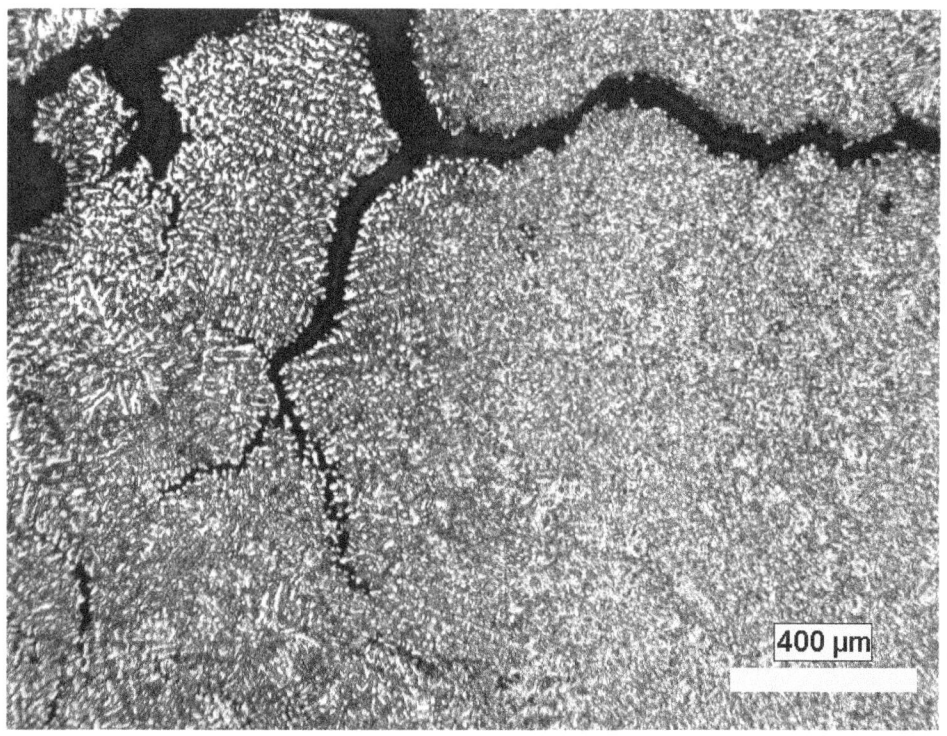

FIGURE 9.15 Micrographs showing likely intergranular cracking in alloy AZ91. (From Sahoo et al. [14], Thomson et al. [39].)

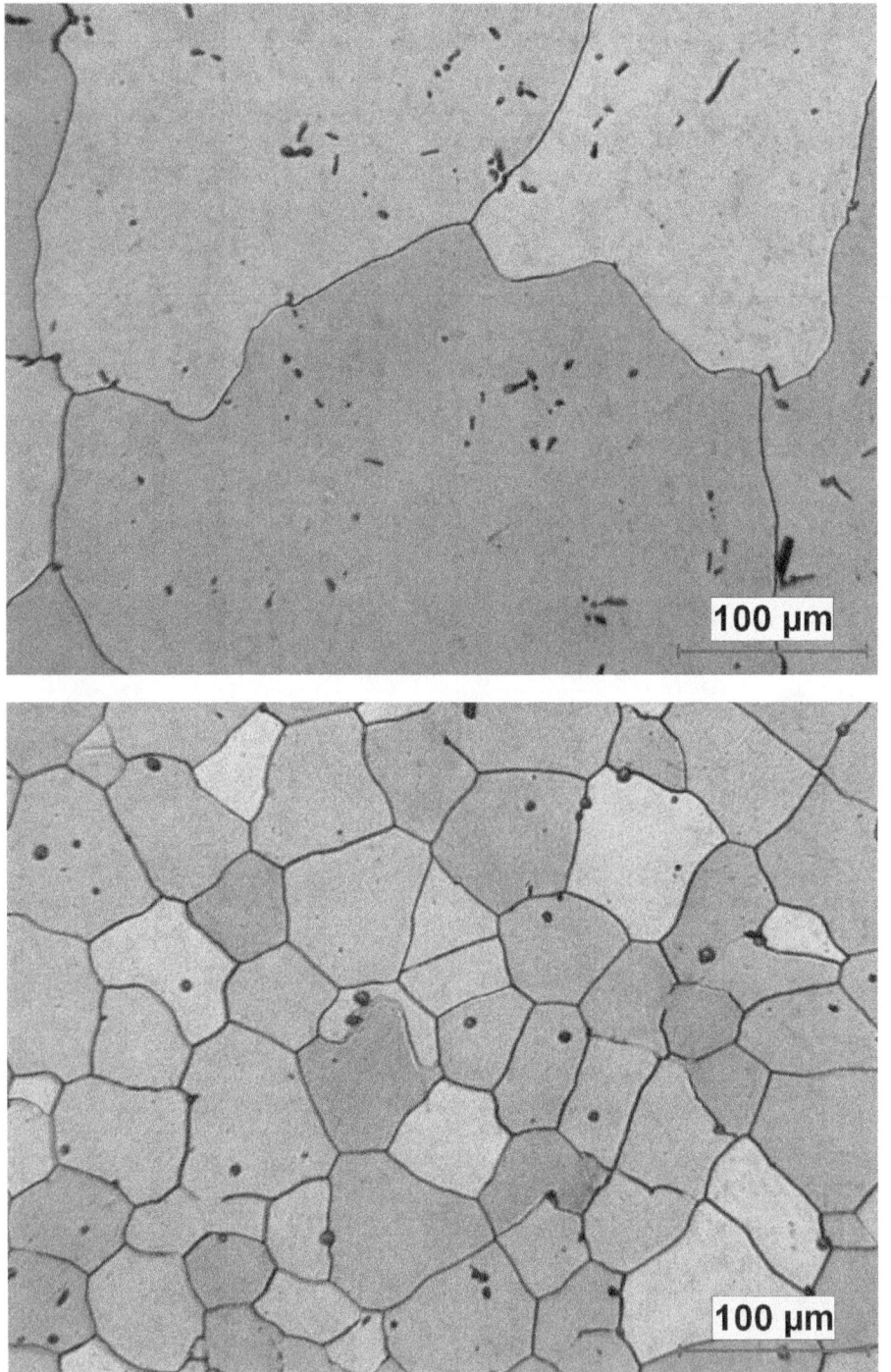

Figure 9.16   Optical micrographs of non-grain-refined and grain-refined AM60B alloy. (From Bailey [33].)

Mechanical and Physical Properties

Gravity Cast Alloys

Table 9.8 summarizes the composition and properties of some alloys used for sand and permanent mold castings [5, 6]. Since properties of sand and permanent mold castings are very similar they are included in a single table. The overall impression is of a rather narrow range of properties among those alloys still in production. The two outstanding alloys, ZE63A-T6 and ZK61A-T6, have the highest UTS and YS while retaining an elongation of 10%. However, in case of ZE63A, the 2.6% RE adds considerably to this alloy's cost, along with some increase in density. The long established alloy AZ91 has a highly satisfactory combination of room temperature strength and ductility, along with good corrosion resistance, and is usually considered a "benchmark" when comparing the properties of the other and newer alloys.

Magnesium alloys, for discussion purposes, can be grouped into two subdivisions, as alloys with and without zirconium additions. The general features of these two groups of alloys are presented below.

	Composition, wt%					Tensile Properties		
Alloy	**Al**	**Mn**	**Zn**	**Zr**	**Other**	**UTS, MPa**	**YS (0.2%), MPa**	**El%**
AM100A-T61	10.0	0.10				275	150	1
AZ63A-T6	6.0	0.15	3.0			275	130	5
AZ81A-T4	7.6	0.13	0.7			275	83	15
AZ91C, E-T6	8.7	0.13	0.7			275	145	6
AZ92A-T6	9.0	0.10	2.0			275	150	3
EQ21A-T6				0.7	1.5 Ag, 2.1 Di, 0.05–0.10 Cu	235	195	2
EZ33A-T5			2.7	0.6	3.3 RE	160	110	2
K1A-F*				0.7		180	55	19
QE22A-T6				0.7	2.5 Ag, 2.1 Di	260	195	3
WE43A-T6				0.7	4.0 Y, 3.4 RE	250		2
WE54A-T6				0.7	5.2 Y, 3.0 RE	250	172	2
ZC63A-T6		0.25	6.0		2.7 Cu	210		4
ZE41A-T5			4.2	0.7	1.2 RE	205	140	3.5
ZE63A-T6			5.8	0.7	2.6 RE	300	195	10
ZK51A-T5[†]			4.6	0.7		205	165	3.5
ZK61A-T5[†]			6.0	0.7		310	185	
ZK61A-T6[†]			6.0	0.7		310	195	10

*Good damping capacity but poor mechanical properties.
[†]Poor foundry characteristics; never used in permanent molds.

Source: From Sadayappan and Luo [6].

TABLE 9.8 Compositions and Room Temperature Tensile Properties of Some Sand-Cast and Permanent Mold Cast Alloys

Alloys without Zirconium

Zirconium (Zr) cannot be added to the magnesium alloys with aluminum. Most common alloys are Mg-Al-Zn (AZ) and Mg-Al-Mn (AM) which have many advantages as common casting alloys but suffer from microporosity in sand and gravity permanent mold castings. The AZ91 alloys are widely used in die casting.

The properties of aluminum-containing alloys, especially creep resistance, can be improved by the addition of calcium, tin, strontium, or rare earth elements. Most of these special alloys are intended for die casting and discussed separately later in this chapter.

Alloys with Zirconium

Zirconium has partial solubility in magnesium and refines the grain size of the metal. At least 0.5% zirconium should be available in the melt in soluble form for successful grain refinement. The reduced grain size enhances the ductility with small improvement on strength. In case of pure Mg-Zr alloys, zirconium is observed as particles in the grains. In other complex systems it forms compounds.

Zirconium is not used alone in magnesium alloys but combined with other elements such as zinc, rare earths (cerium, gadolinium), yttrium, and silver. The alloying element, other than zirconium, is selected to improve strength and foundry characteristics. Zirconium forms mostly compounds with many of the alloying elements. The alloy additions can improve strength, particularly yield strength and creep resistance, as pure magnesium can undergo creep deformation even at temperatures as low as 100°C (212°F).

Mg-Zn alloys are more susceptible to shrinkage and cracking when zinc content increases beyond 4%. Hence, alloys such as ZK61 are not used in permanent mold casting applications. On the other hand, alloys with zinc and rare earth elements (ZE or EZ systems) exhibit good high-temperature strength and can be cast using sand or permanent mold processes. Silver is used extensively in many magnesium alloys (QE alloys) for aerospace applications due to the high-temperature strength and castability. However, these alloys are being replaced with newer ones with yttrium (WE-type alloys). Both silver- and yttrium-containing alloys can be cast using sand or investment casting processes.

Die-Cast Alloys

Conventional Die-Casting Alloys

There are currently two major alloy systems, Mg-Al-Zn (AZ) and Mg-Al-Mn (AM), used for magnesium die-casting applications. AZ91 (Mg-9%Al-1%Zn) is used for nonstructural parts that are strength dominated and exposed to ambient temperatures like brackets, covers, cases, and housings; providing essentially the same functionality with significant mass savings compared to steel, aluminum, or zinc alloys. For structural applications such as instrument panels, steering systems, and radiator supports, where fracture behavior is important, AM50 (Mg-5%Al-0.3%Mn) or AM60 (Mg-6%Al-0.3%Mn), offer unique advantages due to their higher ductility (about 10% elongation) and higher impact strength compared to die-cast A380 aluminum alloy [40]. Typical die-cast alloys and their compositions are listed in Table 9.5 and some mechanical properties in Table 9.9.

Alloy	Tensile Strength, ksi (MPa)	Yield Strength, Ksi (MPa)	Elongation, %	Impact Strength, J (ft·lb)
AE42	34 (234)	21 (145)	11	9 (12)
AM20	31 (214)	13 (90)	20	13 (18)
AM50A	33 (237)	18 (125)	15	18 (18)
AM60A,B	35 (240)	19 (130)	13	13 (18)
AS21	32 (220)	17 (120)	13	9 (12)
AS41A,B	35 (240)	20 (140)	15	12 (16)
AZ91A,B,D	36 (250)	23 (160)	7	4.4 (6.1)

Source: From *ASM Specialty Handbook* [17].

TABLE 9.9   Typical Mechanical Properties of Die-Cast Magnesium Alloy (test bars, as-cast)

Creep-Resistant Alloys

A major disadvantage of current magnesium alloys is their poor creep resistance at temperatures above 125°C (257°F) [40], which is inadequate for elevated temperature applications. For example, automotive transmission cases operate at up to 175°C (347°F), engine blocks up to 200°C (392°F), while pistons are exposed to temperatures above 300°C (572°F). Creep mechanisms and recent magnesium alloy development for creep resistance have been reviewed extensively [40]. Both diffusion-controlled dislocation climb and grain boundary sliding are reported mechanisms for creep in magnesium alloys, depending on the alloy system, microstructure, and stress and temperature regimes. The poor thermal stability of the $Mg_{17}Al_{12}$ phase and its discontinuous precipitation can result in substantial grain boundary sliding at elevated temperatures. The accelerated diffusion of aluminum solute in the magnesium matrix and the self-diffusion of magnesium at elevated temperatures can also contribute to creep deformation in Mg-Al based alloys. Possible approaches to improving creep resistance in magnesium alloys include

- Suppressing the formation of $Mg_{17}Al_{12}$ phase
- Pinning grain boundary sliding
- Slowing diffusion in the magnesium matrix

Based on the above approaches, several new alloy systems have been developed in the last decade [6]. Table 9.10 summarizes the tensile properties, creep performance, castability, and microstructure of several creep-resistant magnesium alloys in comparison with the current workhorse die-cast aluminum alloy A380. The new magnesium alloys in Table 9.10 contain Ca, Sr, RE (rare earth), and Si, and offer varying degrees of improvement in creep resistance and other properties. RE additions are beneficial to creep due to the formation of $Al_{11}RE_3$ in the microstructure (Fig. 9.6), but should be controlled to the lowest possible level to minimize cost [41]. Calcium and strontium can significantly improve creep resistance of magnesium alloys due to the formation of $(MgAl)_2Ca$ [42] and Mg-Al-Sr intermetallic phase(s) [43], respectively. However, Ca and Sr additions need to be carefully controlled to avoid potential castability problems such as die-sticking and cracking. Silicon additions below the Mg-Mg_2Si eutectic point (1.34% Si) offer very limited improvement in creep resistance. Hypereutectic Mg-Si is a

Alloy System	Alloy	Tensile Properties			Total Creep Extension, %	Creep Strength at 175°C, MPa	Strengthening Phase(s)	Castability	Remarks
		YS, MPa	UTS, MPa	Elongation, %					
Mg-Al-RE	AE42 (4Al-2RE)	139 at RT 118 at 121°C 106 at 177°C	226 at RT 177 at 121°C 135 at 177°C	11 at RT 23 at 121°C 28 at 177°C	0.33(3 MPa/150°C/200 h) 0.11(83 MPa/150°C/200 h) 0.08(34 MPa/177°C/100 h)	50	Al_4RE	Good	Expensive Not for >175°C
Mg-Al-Ca	AX51 (5Al-0.8Ca)	128 at RT 102 at 150°C	192 at RT 161 at 150°C	7 at RT 7 at 150°C	0.26(35 MPa/150°C/200 h)	N/A	Al_2Ca	Poor	Die-sticking Cracking
	AX52 (5Al-2Ca)	161 at RT 133 at 175°C	228 at RT 171 at 175°C	13 at RT 23 at 175°C	0.05(83 MPa/150°C/100 h) 0.06(70 MPa/175°C/100 h) 0.26(56 MPa/200°C/100 h)	75	$(Mg,Al)_2Ca$	Good	Excellent combination of properties Very promising for powertrains
	AX53 (5Al-2Ca)	186 at RT 151 at 175°C	250 at RT 196 at 175°C	9 at RT 15 at 175°C	0.07(83 MPa/150°C100 h) 0.09(70 MPa/175°C/100 h) 0.28(56 MPa/200°C/100 h)	74	$(Mg,Al)_2Ca$	Good	
	AXJ (5Al-3Ca-0.07Sr)	190 at RT 146 at 175°C	238 at RT 196 at 175°C	8 at RT 15 at 175°C	0.05(83 MPa/150°C/100 h) 0.06(70 MPa/175°C/100 h) 0.20(56 MPa/200°C/100 h)	84	$(Mg,Al)_2Ca$	Good	
Mg-Zn-Al-Ca	ZAX8506 (8Zn-5Al-0.6Ca)	146 at RT 117 at 150°C	219 at RT 159 at 150°C	5 at RT 11 at 150°C	0.26(35 MPa/150°C/100 h)	N/A	Low eutectic Mg-Al-Zn-Ca phase	Fair	Not suitable at >175°C
Mg-Al-Ca-RE	ACM 522 (5Al-2Ca-2RE)	158 at RT 138 at 150°C 132 at 175°C	200 at RT 175 at 150°C 152 at 175°C	4 at RT 7 at 150°C 9 at 175°C	N/A	74	Al-Ce Mg-Al-Ca	Fair	Expensive Not suitable at >125°C
	MRI 153 (Al-Ca-RE)	165 at RT 118 at 150°C	250 at RT	5 at RT	0.15(50 MPa/150°C/100 h)	N/A	Al_2Ca $Mg_{17}Al_{12}$	Good	
Mg-Al-Sr	AJ52x (5Al-2Sr)	145 at RT 108 at 150°C 103 at 175°C	202 at RT 164 at 150°C 148 at 175°C	4 at RT 14 at 150°C 15 at 175°C	0.03(35 MPa/150°C/200 h) 0.09(35 MPa/175°C/200 h) 0.03(50 MPa/150°C/200 h)	N/A	M-Al-Sr	Fair	Higher casting temperature required
	AJ51x (5Al-1Sr)	145 at RT 108 at 150°C 103 at 175°C	202 at RT 164 at 150°C 148 at 175°C	4 at RT 14 at 150°C 15 at 175°C	0.07(35 MPa/150°C/200 h) 0.05(35 MPa/175°C/200 h) 0.07(50 MPa/150°C/200 h)				

Mg-Al-Si	AS21X (5Al-1Si-0.15RE)	120 at RT 90 at 150°C 80 at 175°C	235 at RT 125 at 150°C 115 at 175°C	12 at RT 35 at 150°C 32 at 175°C	0.1(40 MPa/150°C/100 h) 0.3(60 MPa/175°C/100 h) 0.7(80 MPa/200°C/100 h)	N/A	Mg$_2$Si Mg-RE-Mn	Good	Not suitable at >125°C
Mg-RE-Zn	MEZ (2.5RE-0.35Zn)	97 at RT 78 at 150°C 73 at 175°C	N/A	3 at RT 8 at 150°C 5 at 175°C	0.03(80 MPa/175°C/100 h)	N/A	Mg$_{12}$RE	Fair	Expensive
Al	A380	155 at RT 149 at 150°C 154 at 175°C	290 at RT 255 at 150°C 248 at 175°C	3 at RT 6 at 150°C 7 at 175°C	0.18(35 MPa/150°C/200 h) 0.15(35 MPa/175°C/200 h) 0.08(50 MPa/150°C/200 h)	93	Si Al$_2$Cu	Good	Workhorse alloy in powertrain

Source: Sadayappan and Luo [6].

TABLE 9.10 Summary of Creep-Resistant Magnesium Alloys

Alloy	Specific Gravity	Density (lb/in³)	Cofficient of Themal Expansion µ in/in·°F⁻¹ (68–212°F)	Themal Conductivity Btu/ft·h·°F (68°F)	Electrical Conductivity % IACS	Specific Heat Btu/ft·h·°F	Poisson's Ratio	Melting Range, ·°F	Casting Temperature Range, ·°F	Damping Index*	Pattern Makers Shrinkage %
AZ81A-T4	1.80	0.0650	13.8	41.8	12.0	0.25	0.35	915–1130	1300–1550	0.02	1.2
AZ91E-T6	1.81	0.0654	14.0	41.8	11.2	0.25	0.35	875–1105	1300–1550	0.2	1.2
AZ92A-T6	1.83	0.0661	14.0	41.8	12.3	0.25	0.35	830–1100	1300–1550	0.2	1.2
ZE41A-T5	1.84	0.0658	15.1	63	25.4	0.23	0.35	990–1180	1380–1510	1.0	1.3
EZ33A-T5	1.80	0.0659	14.5	58	23.6	0.25	0.33	1013–1185	1380–1510	4.5	1.5
QE22A-T6	1.82	0.0658	14.8	65	25.2	0.24	0.35	1022–1185	1380–1510	0.4	1.3
EV31A-T6	1.83	0.0659	14.2	67.0	18.2	0.24	0.27	1013–1184	1380–1510	–	1.5
WE43B-T6	1.84	0.0665	14.8	29.6	11.6	0.23	0.27	1013–1185	1380–1510	0.09	1.5
WE54A-T6	1.85	0.0668	14.8	29.6	10.0	0.23	0.27	1013–1185	1380–1510	0.17	1.5
K1A-F	1.75	0.0632	15.1	71	30.2	0.25	–	1200–1202	1200–1202	60	1.6

Note: Modulus of elasticity for all unreinforced alloys at 68°F: 6.5 × 10⁶.

*The damping index is the specific damping capacity of a material measured at a stress that is numerically equal to one-tenth the value of the 0.2% yield stress.

Source: From Clark [8].

TABLE 9.11 Magnesium Casting Alloy Physical Properties

potential alloy system for elevated temperature applications, but its high liquidus temperature makes it difficult for high-pressure die casting.

The physical properties of some magnesium alloys are listed in Table 9.11 [8].

References

1. "Challenges Facing the Magnesium Industry" chapter by G. Cole, *Technology for Magnesium Production: Design, Products and Applications*, M. Sahoo (as principal author), American Foundry Society, 2011, pp. 263–289.

2. "Process Selection" chapter by D. Weiss, *Technology for Magnesium Production Design: Products and Applications*, M. Sahoo (as principal author), American Foundry Society, 2011, pp. 131–138.

3. S. Udvardy, North American Die Casting Association (NADCA).

4. "High-Pressure Die Casting" chapter by D. Schwam, *Technology for Magnesium Production: Design, Products and Applications,* M. Sahoo (as principal author), American Foundry Society, 2011, pp. 183–196.

5. M. M. Avedesian and H. Baker, *ASM Specialty Handbook*—"Magnesium and Magnesium Alloys," 1999.

6. "Physical Metallurgy" chapter by K. Sadayappan and A. Luo, *Technology for Magnesium Production: Design, Products and Applications*, M. Sahoo (as principal author), American Foundry Society, 2011, pp. 9–27.

7. *ASM Handbook*, Vol. 3, Alloy Phase Diagrams, Materials Park, OH, 1997, p. 2.276

8. "Material Selection—Sand and Investment Casting Alloys" chapter by K. Clark, *Technology for Magnesium Production: Design, Products and Applications*, M. Sahoo (as principal author), American Foundry Society, 2011, pp. 49–65.

9. "Solidification" chapter by A. Luo and K. Sadayappan, *Technology for Magnesium Production: Design, Products and Application*, M. Sahoo (as principal author), American Foundry Society, 2011, pp. 29–47.

10. E. F. Emley, *Principles of Magnesium Technology*, Pergamon Press, New York, 1986.

11. A. A. Luo, "Understanding the Solidification of Magnesium Alloys," *Proceedings of the Third International Magnesium Conference*, G. W. Lorimer (ed.), The Institute of Metals, London, UK, 1996, pp. 449–464.

12. "Sand Casting" chapter by D. Weiss, *Technology for Magnesium Production: Design, Products and Applications*, M. Sahoo (as principal author), American Foundry Society, 2011, pp. 139–140.

13. M. Sahoo, CANMET Materials Technology Laboratory Work, unpublished.

14. M. Sahoo, D. Weiss, and M. Marlatt, "Factors Affecting Successful Permanent Mold Casting of Magnesium," *International Conference on Permanent Mold Casting of Aluminum and Magnesium*, Dallas, TX, Feb. 2008, pp. 11–12.

15. M. Sadayappan, M. Sahoo, and D. J. Weiss, "Evaluation of the Hot Tearing Susceptibility of Selected Magnesium Casting Alloys in Permanent Molds," *Transaction of the American Foundry Society*, Vol. 115, Paper No. 07-154, 2007.

16. "Lost Foam Casting" chapter by Q. Han C. Song and M. Marlatt, *Technology for Magnesium Production: Design, Products and Applications*, M. Sahoo (as principal author), American Foundry Society, 2011, pp. 161–182.

17. Magnesium and Magnesium Alloys—ASM Specialty Handbook, Edited by M. M. Avedesian and H. Baker, ASM International, Materials Park, OH, 1999, p.66-77.

18. R. D. Carnaham and R. F. Decker, "Thixomolding®, Semi-Solid Injection Molding of Mg Alloys," *ASM Handbook*, Vol. 15, Dec. 2008, pp. 777–782.

19. "Ablation Casting" chapter by J. Grassi, *Technology for Magnesium Production: Design, Products and Applications*, M. Sahoo (as principal author), American Foundry Society, 2011, pp. 197–201.

20. "Alloy Heat Treatment" chapter by M. Sahoo and J. Keist, *Technology for Magnesium Production: Design, Products and Applications,*. M. Sahoo (as principal author), American Foundry Society, 2011, pp. 87–106.

21. N. Fantetti, A. Thorvaldsen, and A. Couture, "Plaster Cast and Sand Cast Magnesium Prototypes—Advantages and Limitations," SAE Technical Paper No. 914043 (10 pages), 1991.

22. J. W. Meir, "The Effect of Various Factors on the Mechanical Properties of Magnesium Alloy Castings," *Congress International*, de Fonderie, Brussels, 1958, pp. 225–240.

23. A. Luo, P. H. Fu, Y. D. Yu, L. M. Peng, H. Y. Jiang, C. Q. Zhai, and A. K. Sachdev, "Low-Pressure Die Casting of AZ91 and AM50 Magnesium Alloys," *AFS Transactions*, 116: 805–815, 2008.

24. G. Pantelakis, N. D. Alexopoulos, and A. N. Chamos, "Mechanical Performance Evaluation of Cast Magnesium Alloys for Automotive and Aeronautical Applications," *Transactions of the ASME*, 129: 422–430, 2007.

25. H. Hu, "Squeeze Casting of Magnesium Alloys and Their Composites," *Journal Materials Science*, 33(6): 1579–1589, 1998.

26. Y. Wang, G. Liu, and Z. Fan, "A New Heat Treatment Procedure for Rheo-Die Cast AZ91D Magnesium Alloy," *Scripta Materialia*, 54: 903–908, 2006.

27. S. Bronfin, E. Aghion, N. Fantetti, F. Von Buch, S. Schumannand, and H. Friedrich, "High Temperature Magnesium Alloys for Sand and Permanent Mold Applications," SAE Technical Paper 2004-01-9656, 2004, pp. 266–270.

28. M. Sahoo, CANMET Materials Technology Laboratory, Unpublished work.

29. C. H. Bergman and I. S. Krause, "Fluidized-Bed Heat Treatment Process and Apparatus for Use in Manufacturing Line," U.S. Patent No. 6,042,369, Issued Mar. 28, 2000.

30. J. Van Wert, D. Apelian, and C. Bergman, "Fluidized Bed Heat Treating: The Answer to One-Piece Flow as a Commercial Application," *Proceedings of the AFS International Conference on Structural Aluminum Castings*, Orlando, FL, Nov. 2003, pp. 383–392.

31. J. Keist, "The Development of a Fluidized Bed Process for the Heat Treatment of Aluminum Alloys," *Journal of Metals*, Apr. 2005, pp. 34–39.

32. D. Neff, "Degassing," *ASM Handbook*, Vol. 15, Dec. 2008, pp. 185–193.

33. "Melt Control" chapter by R. Bailey, *Technology for Magnesium Production: Design, Products and Applications*, M. Sahoo (as principal author), American Foundry Society, 2011, pp. 107–120.

34. N. J. Ricketts, "Alternatives to SF$_6$ for Magnesium Melt Protection," *Proceedings of the Light Metals Technology Conference 2007*, Saint-Sauveur, Quebec, Canada, Sep. 24–26, 2007, pp. 249–254.

35. D. S. Milbrath, "3M™ Novec™ 612 Magnesium Protection Fluid, Its Development and Use in Full Scale Molten Magnesium Processes," *60th Annual World Magnesium Conference*, Stuttgart, Germany, May 11–12, 2003, pp. 26–30.

36. A. L. Keaney and J. Raffin, *Hot Tear Control Handbook for Aluminum Foundrymen and Casting Designers*, AFS Publication, Des Plaines, IL, 1971.

37. M. O. Pekguleryuz and P. Vermette, "A Study on Hot-Tear Resistance of Magnesium Diecasting Alloys," *Transaction of the American Foundry Society*, Vol. 114, 2006 Paper No. 06-092.

38. S. Lin, C. Aliravci, and M. O. Pekguleryuz, "Hot-Tear Susceptibility of Aluminum Wrought Alloys and the Effect of Grain Refining," *Metallurgical and Material Transactions*, 38A: 1056–1068, May 2007.

39. J. P. Thomson, S. Xu, M. Sadayappan, P. D. Newcombe, L. Milette, and M. Sahoo, "Low Pressure Casting of Magnesium Alloys AZ91 and AM50," *Transactions of the American Foundry Society*, Vol. 112, Paper No. 04-120, 2004.

40. A. A. Luo, "Magnesium: Current and Potential Automotive Applications," *Journal of Metals*, 54(2): 42–48, 2002.

41. B. R. Powell, A. Luo, V. Rezhets, J. J. Bommarito, and B. L. Tiwari, "Development of Creep-Resistant Magnesium Alloys for Powertrain Applications," Part 1 of 2, SAE Technical Paper No. 2001-01-0422, SAE, Warrendale, PA, 2001.

42. A. A. Luo, M. P. Balogh, and B. R. Powell, "Creep and Microstructure of Magnesium-Aluminum-Calcium Based Alloys," *Metallurgical and Materials Transactions*, 33A: 567–574, 2002.

43. D. Argo, M. O. Pekguleryuz, P. Labelle, M. Direks, T. Sparks, and T. Waltemate, "Die Castability and Properties of Mg-Al-Sr Based Alloys," *Magnesium Technology 2001*, J. N. Hryn (ed.), TMS, Warrendale, PA, 2001, pp. 131–136.

CHAPTER 10

Zinc and Zinc Alloys

Mahi Sahoo
Suraja Consulting Inc.

Introduction

Zinc and zinc alloys are used for a variety of functional and decorative applications such as castings, rolled sheets, drawn wire, forgings, extrusion, and coatings. The mechanical properties are comparable to some grades of cast iron, and other nonferrous alloys based on Al and Cu. Zinc-based alloys have good corrosion resistance in normal atmospheric conditions, in aqueous solutions, and when used in petroleum products. Their corrosion resistance is enhanced by painting, plating, or chromate or phosphate treatment and substantially improved by anodizing.

One of the major uses of zinc is in coatings for steel constituents to protect them from corrosion. This coating process is known as galvanizing and metallic zinc coatings can be applied as hot dip galvanizing, electrogalvanizing, metalizing (from a spray of molten metal), and mechanical galvanizing (in the form of zinc powder).

Other uses of zinc are as an alloying constituent in brass, bronze, aluminum, and magnesium alloys and a sacrificial anode for marine environments.

Although high-pressure die casting is the most common process to produce components for automotive, general hardware, agricultural equipment, electronic, and electrical fittings, domestic and garden appliances, computer hardware, business machines, recording machines, radios, and hand tools, sand casting, permanent mold casting, and continuous casting are also practiced. Some examples of cast components in zinc alloys are shown in Fig. 10.1 [1, 2]. Other casting methods include investment casting, squeeze casting, and semisolid casting. Metal-matrix composites have also been produced by various foundry methods [3–6]. All processes are based on the same family of alloys derived from the Zn-Al system. Foundry production data for zinc alloy castings during the past few years are given in Table 7.1 [7].

Alloying Additions

Aluminum and copper are the main alloying elements in zinc alloys to produce better mechanical properties. Magnesium is also an alloying element. Impurities such as Fe, Ni, Cr, Pb, Cd, Sn, etc., must be strictly controlled to prevent intergranular corrosion, dimensional changes, and loss of mechanical properties [3, 5]. Influence of the alloying elements and impurity elements on the alloy characteristics are described below.

473

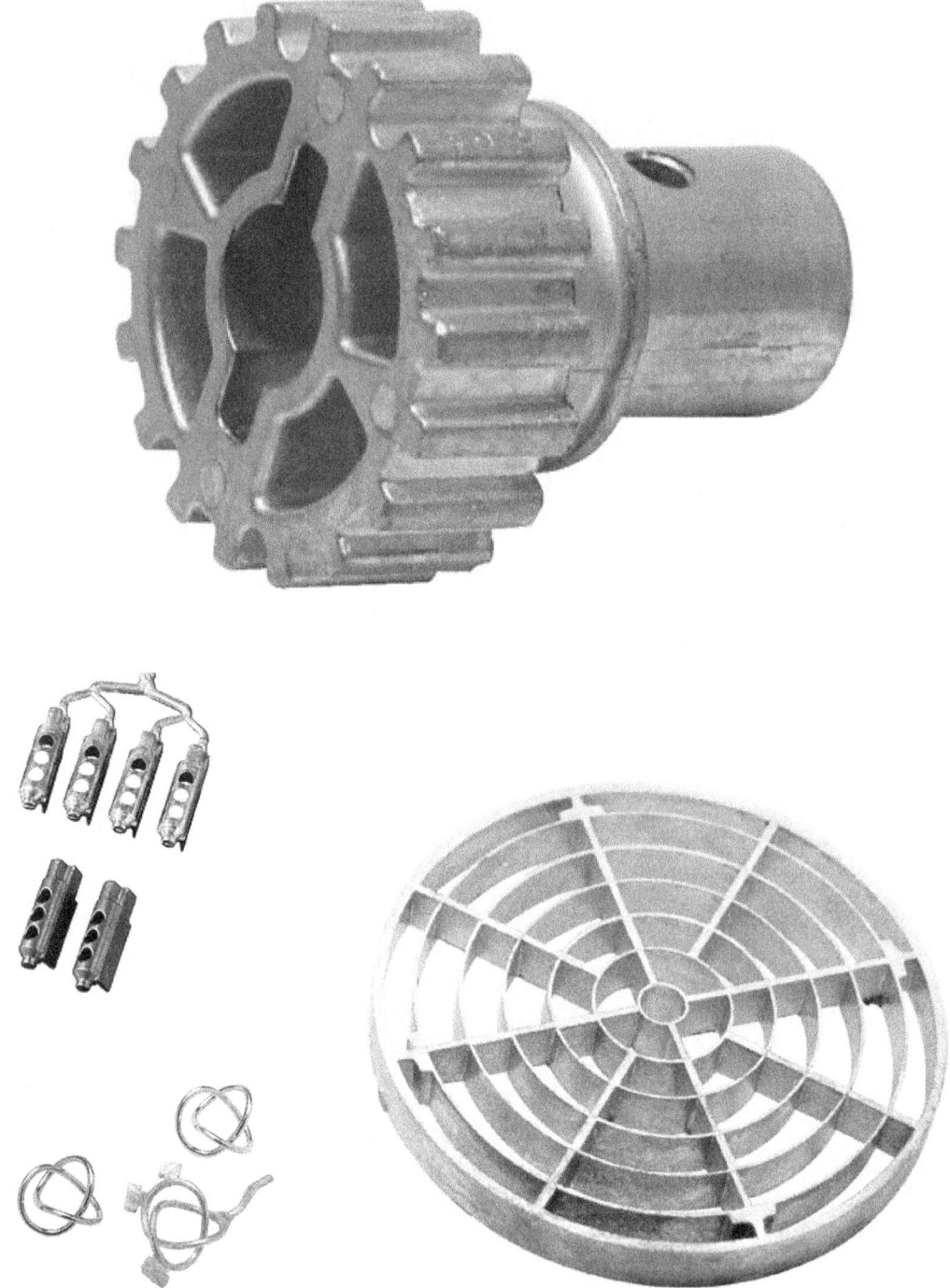

Figure 10.1 Typical cast components in zinc alloys. (Courtesy of NADCA [1], Custom Aluminum Foundry Limited [2].)

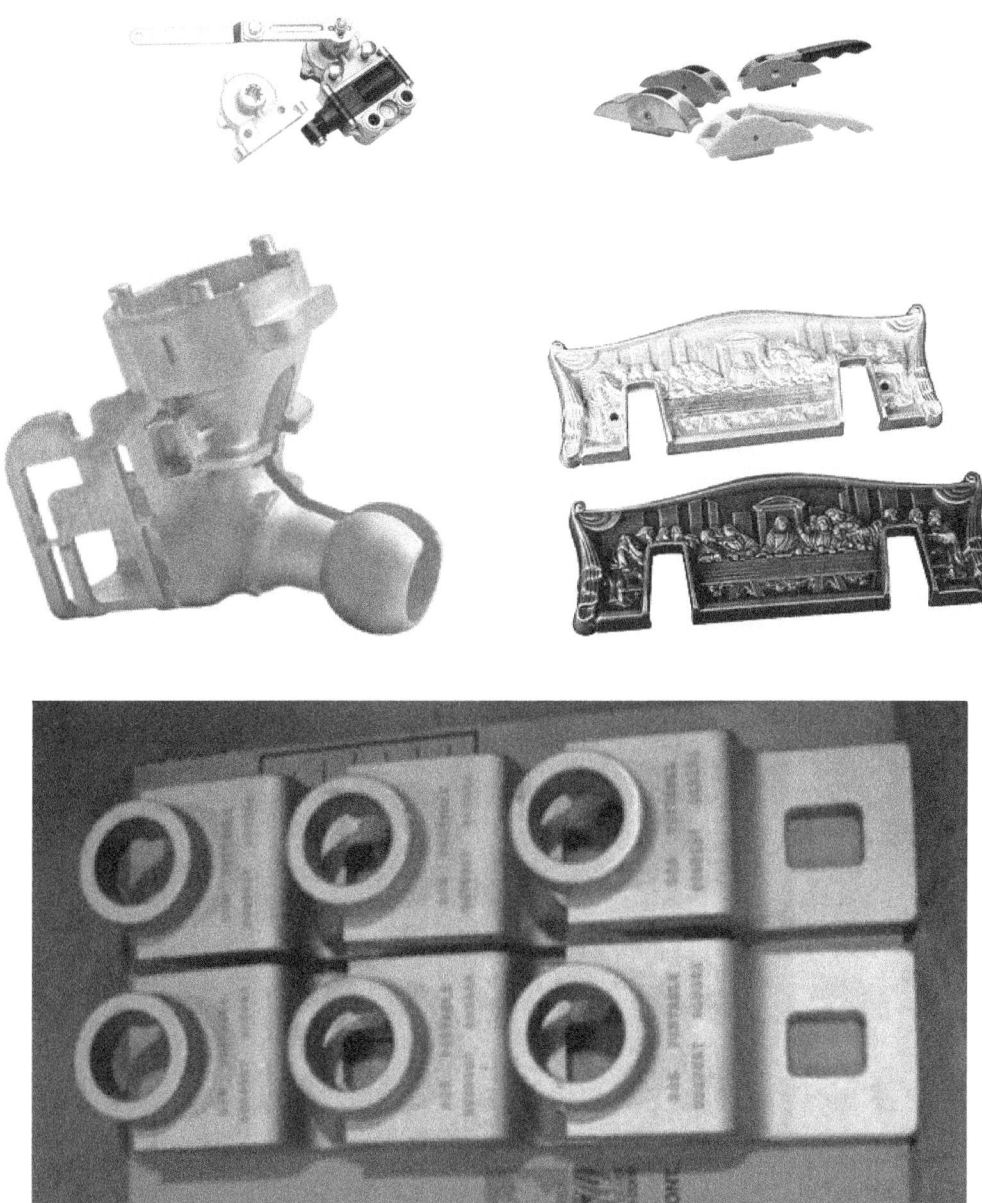

FIGURE 10.1 (Continued)

Aluminum

Addition of aluminum not only improves strength, but also reduces grain size, improves casting fluidity and castability, and minimizes the attack of the molten metal in the casting and handling equipment made of iron and steel.

Copper

Copper increases hardness and strength and minimizes the undesirable effects of impurities.

Magnesium

The primary function of magnesium is to minimize the susceptibility to intergranular corrosion caused by the presence of impurities. The magnesium content must be maintained within the range specified in Table 10.1. Excessive amounts of magnesium lead to lowering of casting fluidity and promoting of hot tearing. In addition, hardness increases with increasing magnesium contents but at the expense of ductility.

Impurity Elements

Iron

Iron remains in solid solution in zinc up to 0.02%. When this limit is exceeded, hard iron-aluminum compounds can form which can produce comet tails during buffing and can dull tools during machining.

Lead, Cadminum, and Tin

Die-cast parts can show swelling, cracking or distortion when the limits for Pb, Cd and Sn are more than those shown in Table 10.1. Of these three elements, the adverse effect of tin is more potent in promoting subsurface intergranular corrosion. The upper limits for lead, cadminium, and tin are 0.006, 0.005, and 0.003%, respectively.

Nickel, Chromium, Silicon, and Manganese

The solubility limit of each of these in zinc is as follows: 0.02% Ni, 0.02% Cr, 0.035% Si, and 0.5% Mn. Higher amounts lead to the formation of intermetallic compounds with aluminum that are less dense than the zinc melt and therefore can be skimmed off its surface before pouring or casting.

Types of Alloys

The common casting alloys are listed in Table 10.1 with their chemical composition [3]. Of these alloys, 2, 3, 5, and 7 belong to the hypoeutectic series. The characteristics of these alloys are described below [6].

Alloy No. 2 has the highest tensile strength, creep strength, and hardness of all alloys in the hypoeutectic series of die-casting alloys (see Table 10.2). The higher copper content (0.3% Cu) causes some dimensional instability and leads to a net expansion of approximately 0.0014% after 20 years. It also causes some loss of impact strength and ductility. Alloy No. 2 has good bearing properties.

Alloy No. 3 is the most widely used zinc die-casting alloy in the United States. It provides the best overall combination of strength, castability, dimensional stability, ease of finishing, and cost.

Alloy No. 5 produces castings that are both harder and stronger than those made from alloy No. 3. However, improvements in their properties come at the expense of ductility, and postforming operations such as riveting, swaging, or crimping must be done with care. The creep resistance of alloy No. 5 is second only to that of alloy No. 2 among the hypoeutectic zinc-aluminum alloys.

Alloy No. 7 is essentially a high-purity version of alloy No. 3. Because of its lower magnesium content, alloy No. 7 has even better castability than No. 3, enabling

| Alloy | | | Composition,* wt% | | | | | | | Liquidus Temp., °C/°F | Solidus Temp., °C/°F | Freezing Range, °C/°F |
Common Designation	UNS No.	Applicable Standard (ASTM)	Al	Cu	Mg	Fe	Pb	Cd	Sn			
No. 2	Z35541	B 86	3.5–4.3	2.5–3.0	0.020–0.050	0.100	0.005	0.004	0.003	390/734	379/715	11/19
No. 3	Z33530	B 86	3.5–4.3	0.25	0.02–0.03	0.100	0.005	0.004	0.003	387/728	381/718	6/10
No. 5	Z35531	B 86	3.5–4.3	0.75–1.25	0.03–0.08	0.075	0.005	0.004	0.003	386/727	380/717	6/10
No. 7	Z33523	B 86	3.5–4.3	0.25	0.005–0.02	0.10	0.003	0.002	0.001	387/728	381/718	6/10
ZA-8	Z35636	B 86	8.0–8.8	0.8–1.3	0.015–0.03	0.075	0.005	0.004	0.003	404/759	375/707	29/52
ZA-12	Z35631	B 86	10.5–11.5	0.5–1.25	0.015–0.03	0.10	0.004	0.003	0.002	432/810	377/710	55/100
ZA-27	Z35841	B 86	25.0–28.0	2.0–2.5	0.01–0.02	–	0.004	0.003	0.002	484/903	375/707	109/195
ACuZinc5	Z46541	B 894	2.9	5.5	0.04	0.075	0.005	0.004	0.003			

*Maximum unless range is given or otherwise indicated.

Source: From *ASM Handbook* [3, 6].

Table 10.1 Compositions and Physical Properties of Zinc Casting Alloys

excellent reproduction of surface detail in castings. Alloy No. 7 has the highest ductility among the hypoeutectic alloys.

Alloy ZA-8 is the only member of the hypereutectic alloys that can be hot-chamber die cast along with the hypoeutectic alloys. It is equivalent to alloy No. 2 in many respects, but AZ-8 has higher tensile, fatigue, and creep strengths, is more dimensionally stable, and has lower density. Alloy ZA-8 castings can be readily finished, thereby combining their high structural strength with excellent appearance.

Alloy ZA-12 has very good castability in cold-chamber die-casting machines. It is lower in density than all other zinc alloys except ZA-27, and it is frequently specified for castings that must combine casting quality with optimum performance. The plating quality of ZA-12 is lower than that of ZA-8, but it has excellent bearing and wear properties.

Alloy ZA-27 is the lightest, hardest, and strongest of all the zinc alloys, but it has relatively low ductility and impact strength when pressure die cast. Because of the wide freezing range of ZA-27, casting quality can suffer unless care is taken. The secondary creep strength of ZA-27 is better than that of all the other zinc alloys except for the now rarely used ILZRO (International Lead-Zinc Research Organization) 16; however, ZA-8 has better primary creep strength. Alloy ZA-27 demonstrates the highest sound and vibration damping properties of all the zinc casting alloys; as a group, zinc alloys have a damping resistance equal to that of cast irons at elevated temperatures.

The three ZA alloys are also suitable for sand casting and permanent mold casting.

Specialty Alloy

A higher-copper die-casting alloy is also used in commercial production and is listed in ASTM B 894. This alloy, commonly referred to as ACuZinc 5, has 5 to 6% Cu, 2.5 to 3.3% Al, 0.025 to 0.05% Mg, and limitations on impurity similar to the conventional low-aluminum casting alloys as shown in Table 10.1 [3]. The tensile strength of this alloy ranges between 310 and 355 MPa (45 and 51 Ksi) with a yield stress in the range of 240 to 284 MPa (35 to 41 Ksi). Elongations between 4.6 and 9.4% are measured. The Brinell hardness is 105 to 115 (500 kg). Alloys for slush casting are listed in ASTM B792.

Physical Metallurgy

The melting point of zinc is low, only 420°C (788°F). Aluminum is the principal alloying element as it forms a eutectic with zinc at 6% Al and 381°C (718°F), see Fig. 10.2 [8]. The casting alloys can be divided into two groups based on the aluminum content: (1) the hypoeutectic alloys containing about 4% Al and, (2) the hypereutectic alloy with ≥6% Al.

Three important phase reactions take place in this system: the eutectic as mentioned above, the eutectoid at 22.3% Al and 277°C and the monotectoid at 352°C. The first phase to solidify from the liquid in the case of hypoeutectic alloys (e.g., alloys 2, 3, 5, and 7) is very rich in Zn. These go through the eutectic reaction and hence, the microstructure consists of primary dendrites of zinc-rich solid solution with interdendritic eutectic structures rich in Zn. A typical microstructure for a permanent mold cast hypoeutectic alloy is shown in Fig. 10.3 [9]. Microstructures of a number of industrial alloys have been described in Ref. 10. The two hypereutectic alloys containing 8% Al (ZA-8) and 12% Al (ZA-12) alloys also go through the eutectic transformation; however, the first phase to solidify is rich in aluminum. The volume fraction of the eutectic phases will vary for the

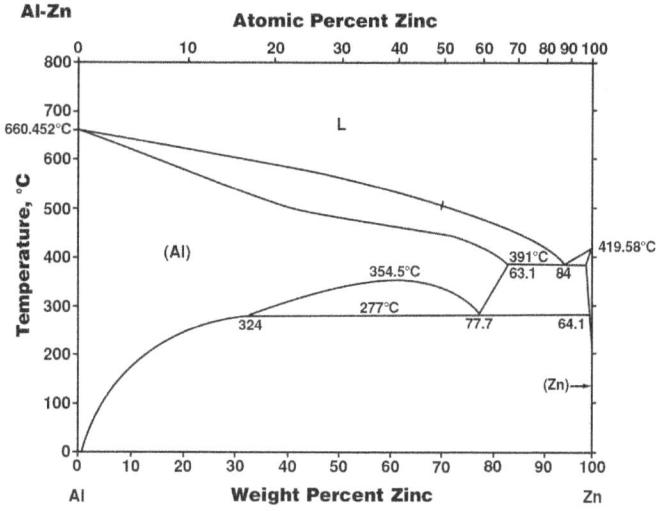

FIGURE 10.2   Zn-Al equilibrium diagram. (From *ASM Handbook* [8].)

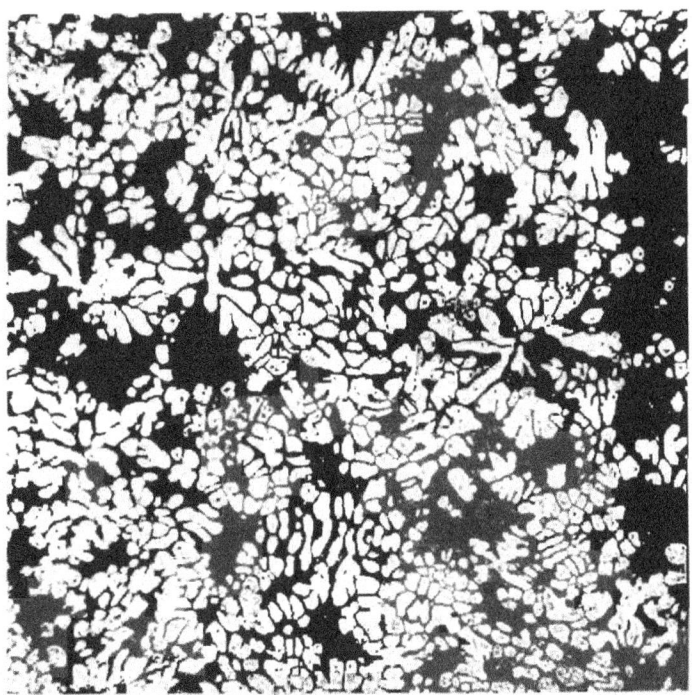

FIGURE 10.3   Optical micrograph of a hypoeutectic alloy (ZAMAK 3 containing 3.9 to 4.3 Al, 0 to 0.1 Cu, 0.03 to 0.06 Mg). Permanent mold cast and annealed at 250°C followed by slow cooling, showing dendritic structure (the white zone is Zn-rich phase surrounded by eutectic). (From Goodwin [9].)

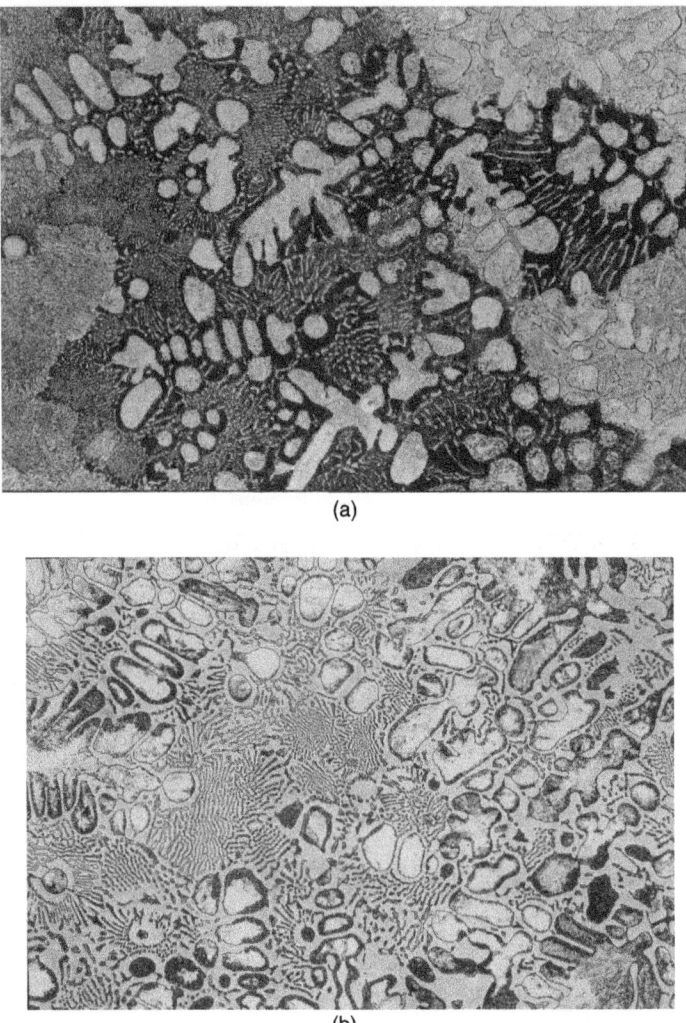

(a)

(b)

Figure 10.4   Optical micrographs of (a) ZA-8 and (b) ZA-12 alloys, ×200. (From Sahoo et al. [11].)

hypo- and hypereutectic alloys, as evident from their phase diagram in Fig. 10.2. Typical optical and scanning electron micrographs are shown in Figs. 10.4 and 10.5 for the ZA-8 and ZA-12 alloys.

The Zn-27% Al alloy (ZA-27) goes through the eutectoid and monotectoid transformation. The first phase to solidity is rich in Al and hence, the microstructure consists of aluminum-rich primary dendrites together with the eutectoid structure. Because of the presence of Cu, additional course phase such as $CuZn_4$ can appear in the interdendritic areas. When the iron content of the alloy is high (0.05 to 0.1%), isolated faceted crystals ($FeAl_3$) appear throughout the structure. These microstructural characteristics are shown in Fig. 10.6a to d.

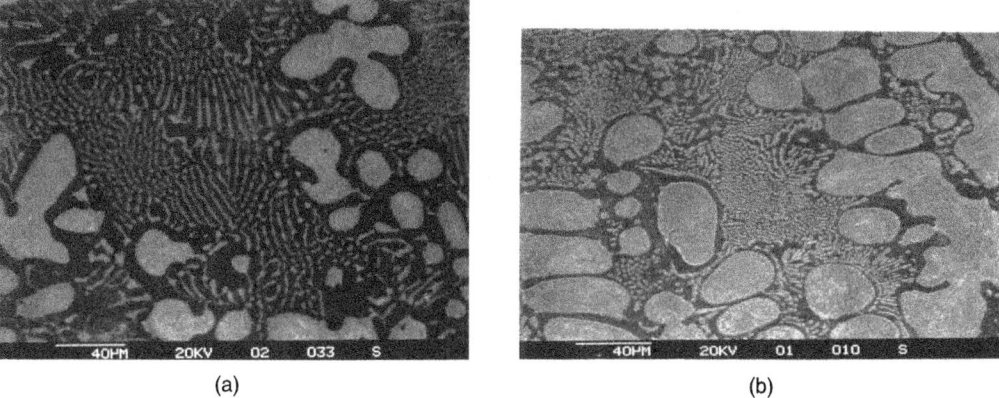

FIGURE 10.5   Scanning electron micrographs of (a) ZA-8 with 8% Al and (b) ZA-12 with 11% Al at higher magnifications. (From Sahoo et al. [11].)

The solidification characteristics of the alloys are given in Table 10.1. As shown, the hypoeutectic alloys have a very narrow freezing range of 6 to 11°C. ZA-8 is also a short-freezing-range alloy with a solidification range of 29°C. ZA-12, on the other hand, can be classified as a medium-freezing-range alloy with a freezing range of 55°C. By contrast, ZA-27 has a freezing range of 109°C and hence, is considered a long-freezing-range alloy.

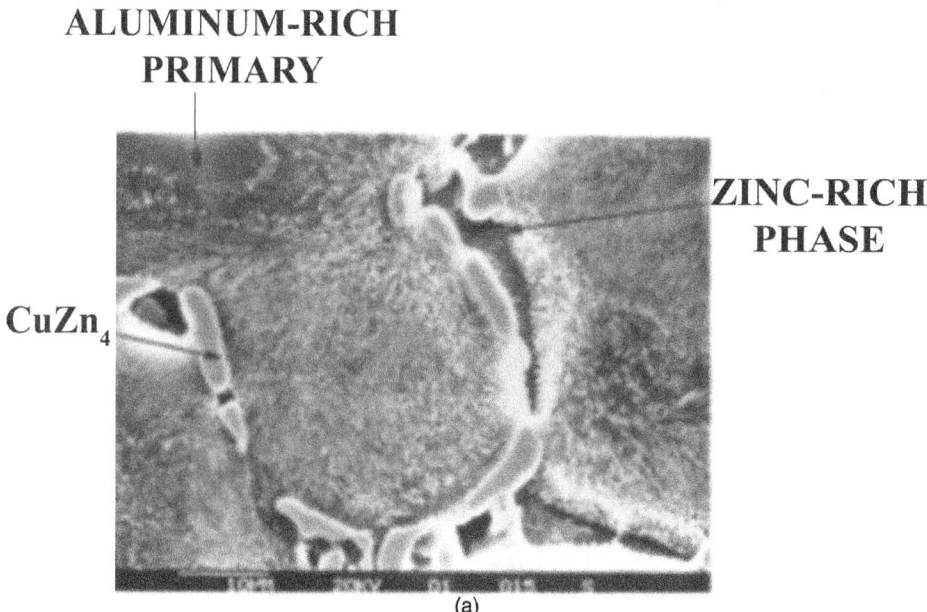

FIGURE 10.6   Scanning electron micrographs of a ZA-27 alloy showing the phases present. (From Sahoo et al. [11].) *(Continued)*

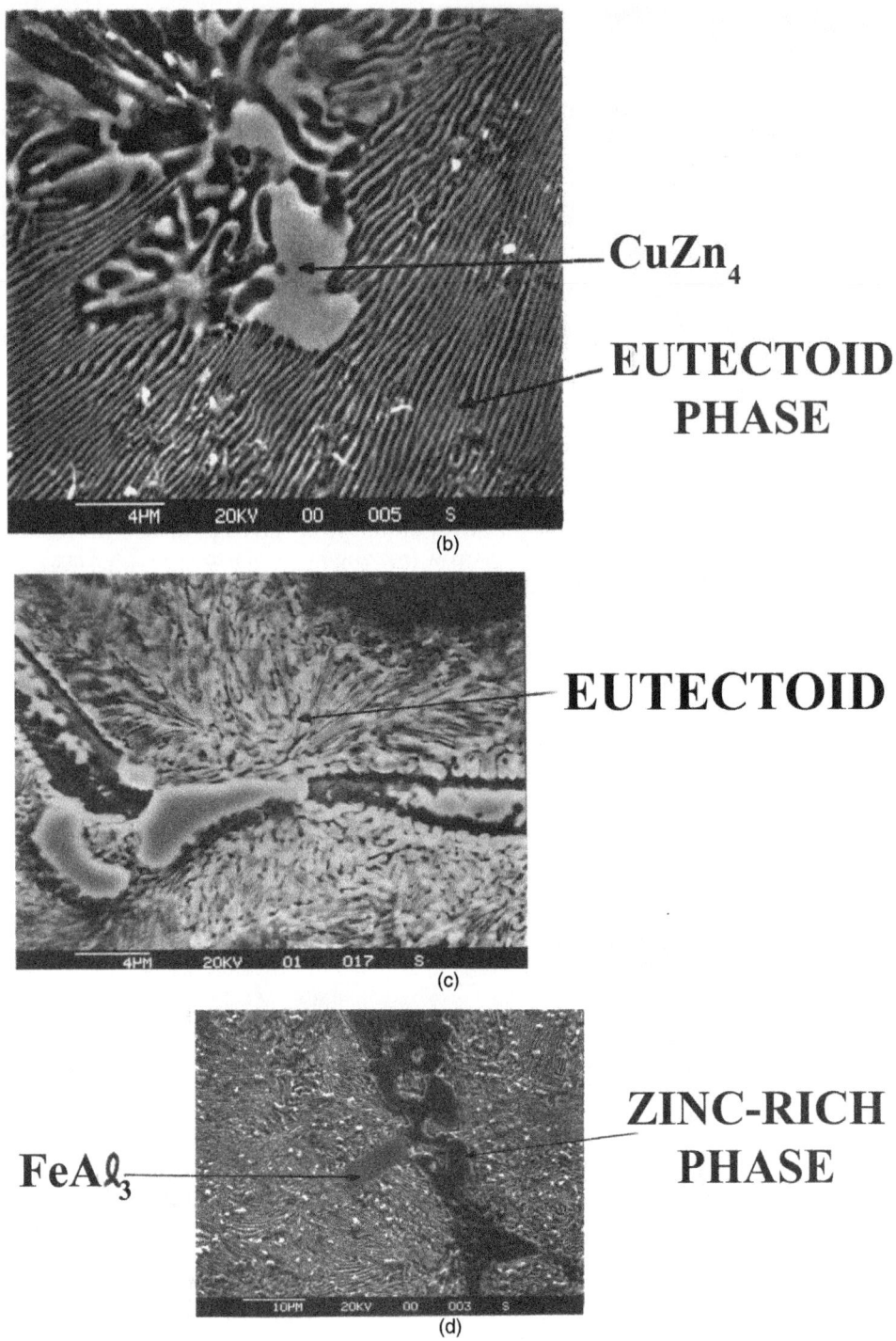

FIGURE 10.6 (Continued)

FIGURE 10.7   Underside shrinkage defect in a ZA-27 alloy. (From Sahoo and Whiting [12].)

One interesting phenomenon for the ZA-27 alloys is the observation of shrinkage defects on the bottom face (drag side) of plate and slab castings [12]. As shown in Fig. 10.7, these defects have a circular or spiral pattern below the risers of a plate casting. The gating and risering system of the plate casting is shown in Fig. 10.8. X-ray examination of the plate casting has shown considerable microporosity in the same circular pattern. In large castings, the underside shrinkage can be exaggerated to such an extent that the cavity in the bottom face of a $127 \times 203 \times 356$ mm block casting can extend well into the castings (Fig. 10.9). This phenomenon has been explained in the 27% Al alloy in terms of the dendritic form of solidification and the density difference between the first phase (aluminum-rich and hence relatively lighter) to solidify and the remaining liquid (relatively denser). Solidification proceeds from the sides in and from top to bottom: hence, the zinc-rich liquid is concentrated into the bottom of the casting. This is reflected in the composition gradient between the top and bottom of the large block casting. The zinc-rich phase is the last to solidify. Underside shrinkage results from surface tension forces which, to compensate for the solidification volume change, causes the remaining liquid to be drawn

FIGURE 10.8   Plate casting ($25 \times 152 \times 304$ mm) with the sprue, runner, and ingates. The plate from the other side of the runner has been removed. Gating ratio 1:5:3. (From Sahoo and Whiting [12].)

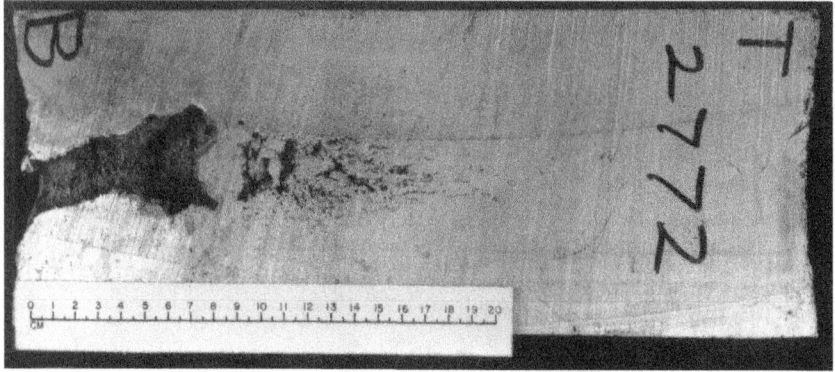

FIGURE **10.9** Half section of the 127 × 203 × 356 mm block casting in AZ-27 alloy showing the cavity on the bottom surface (B) rather than at the conventional top surface (T). (From Sahoo and Whiting [12].)

into the interdendritic spaces. The wide freezing range allows time for the gravity separation of aluminum-rich dendrites to occur.

Use of chills on the drag face did not eliminate this defect. However, addition of Li, K, and Cs from Group IA and Be, Ca, Sr, and Ba from Group IIA in the periodic table in the concentration range 0.02 to 0.1% eliminated the underside shrinkage defect in Zn-Al foundry alloys, especially the Zn-27% alloy [13,14]. Their mechanical properties and corrosion behavior remain unaffected following strontium addition up to 0.06% [11, 14]. Underside shrinkage defects have also been eliminated in Zn-Al alloys containing 20–40% Al by the addition of 0.25 to 2% mischmetal [15].

Heat Treatment

The most common heat treatment for zinc alloy castings is a stabilizing heat treatment to compensate for the volume change that the die-cast components experience due to the presence of copper. Rapid solidification during die casting causes copper to be enriched in the primary phase. Precipitation of the copper phase leads to volume change and hence, the stabilizing treatment, usually at about 100°C (212°F) for three to six hours followed by air cooling [3]. The heat treatment is longer for lower stabilizing temperature, e.g., 10 to 20 hours at 70°C (158°F). Mechanical properties are usually reduced by stabilizing heat treatment. For example, aging at 95°C (203°F) for 10 days reduced the UTS of low aluminum alloys by 40 MPa (6 Ksi) and that of higher aluminum ZA alloys by about 80 MPa (12 Ksi). In case of alloy 2, elongation and impact properties are adversely affected. The effects of aging on alloys 3, 5, and 7 is minimal due to their lower copper content.

Melting and Melt Handling

Zinc alloys have a relatively low melting point. Zinc does not oxidize appreciably and hydrogen absorption at typical zinc melting temperatures is not a problem. Hence, fluxes need not be used for melt protection. However, chloride-containing fluxes ($ZnCl_2$, KCl, and NaCl) can be used to minimize melt loss [4]. These fluxes also help to agglomerate nonmetallic residues and contaminants from dirty scrap. An exothermic flux, usually containing nitrate salts and silicofluoride double salts, can be used to recover much of the zinc entrapped in zinc alloy dross that forms during melting, holding, turbulent transfer, and remelt operations, especially for large melts for high-pressure die casting [4, 5].

Nonmetallic oxide inclusions in zinc-aluminum-based alloys are less of a problem than intermetallic inclusions such as iron-zinc and iron-aluminum. Intermetallic compound precipitation such as Mn_2Al_5, Ni_2Al_3, and $CrAl_4$ can form when the concentration for each element exceeds 0.02%. However, being of low density, they usually rise to the top of the melt and can be easily skimmed off.

Melting

Zinc alloys for sand, permanent mold, or investment casting can be melted in fuel-fired, resistance-heated, or induction furnaces using silicon carbide or clay-graphite crucibles. In the case of high-pressure die casting, the large melt size necessitates the use of a gas-fired furnace for melting and holding. These can also be induction heated. Pots for melting and holding can be cast from gray or ductile iron.

Casting Processes

High-pressure die casting (both hot and cold chamber) of zinc alloys is most common. Other casting processes include sand casting, permanent mold casting, plaster casting, squeeze casting, and semisolid casting.

As mentioned before, ZA-27 alloy has a long freezing range (109°C/230°F) and hence, special precautions should be taken to control porosity and underside shrinkage defects. It is interesting to note that ZA-27 is easy to cast in the low-pressure permanent mold process despite its long freezing range. By contrast, ZA-8 (short freezing range) and ZA-12 (medium freezing range) are not so easy to cast by low pressure. Typical casting defects for these two alloys are shrinkage porosity, hot tearing, and surface sinks [15].

Both metallic (cast iron or steel) and graphite molds can be used for permanent mold casting. Plaster molding is used for prototyping of die-casting alloys.

Postcasting Operations

Functional and/or decorative coatings can be applied to zinc alloys. Although zinc alloys have good corrosion resistance in normal atmospheric conditions, their corrosion resistance can be improved further by painting, plating, chromate, or phosphate treatment, and anodizing for use in petroleum products [3].

Repair welding of zinc die castings can be performed by using an alloy rod of the same compositions with a gentle heat of a slightly reducing oxyacetylene flame.

Mechanical and Physical Properties

Mechanical properties of typical Zn-Al alloys are summarized in Table 10.2. The die-cast hypoeutectic alloys have good strength, elongation, and impact toughness properties. The ZA series of alloys also have good strength properties, but elongation can vary from 1 to 10% depending on the alloy and casting process. As mentioned before, the ZA alloys deliver the highest strength among the most widely used nonferrous alloys and match or exceed the strength of some cast irons [5].

TABLE 10.2 Comparison of Typical Mechanical Properties of Zinc Casting Alloys

Alloy and Product Form(s)	Ultimate Tensile Strength		0.2% Offset Yield Strength		Elongation in 50 mm (2 in), %	Hardness HB	Impact Strength		Fatigue Strength		Young's Modulus	
	MPa	Ksi	MPa	Ksi			J	ft·lb/ft	MPa	Ksi	GPa	10⁶ psi
No. 2 (D)	358	52	–	–	7	100	48(b)	35	58	8.5	–	–
No. 3 (D)	283	41	–	–	10	82	58(b)	43	47.6	6.9	–	–
No. 5 (D)	331	48	–	–	7	91	65(b)	48	56.5	8.2	–	–
No. 7 (D)	283	41	–	–	13	80	58(b)	43	–	–	–	–
ZA-8 (S)	248–276	36–40	200	29	1–2	80–90	20	15	–	–	85.5	12.4
ZA-8 (P)	221–255	32–37	207	30	1–2	85–90	–	–	51.8	7.5	85.5	12.4
ZA-8 (D)	372	54	290	42	6–10	95–100	42(b)	31	–	–	…	…
ZA-12 (S)	276–317	40–46	207	30	1–3	90–105	25(b)	19	103.5	15	83.0	12.0
ZA-12 (P)	310–345	45–50	207	30	1–3	90–105	–	–	–	–	83.0	12.0
ZA-12 (D)	400	58	317	46	4–7	95–115	28(b)	21	–	–	–	–
ZA-27 (S) (a)	400–440	58–64	365	53	3–6	110–120	47(b)	35	172.5	25	75.2	10.9
ZA-27 (P)	421–427	61–62	365	53	1	110–120	–	–	–	–	75.2	10.9
ZA-27 (D)	421	61	365	53	1–3	105–125	–	–	–	–	–	–

D = die cast; S = sand cast; P = permanent cast; (a) As-cast; (b) Unnotched Charpy

Source: From *ASM Handbook* [3].

486

Alloys 3, 5, 7, and ZA-8 and ZA-12 are considered nonincendiary and spark proof; these will not ignite hazardous fuel-air mixtures, vapors, or particulate matter when struck by rusted ferrous materials. Their nonmagnetic properties make them suitable for electronic applications.

References

1. S. Udvardy, NADCA, private communication.

2. Custom Aluminum Foundry Limited, Cambridge, Ontario, Canada.

3. F. E. Goodwin "Zinc and Zinc Alloys Castings," *ASM Handbook, Castings*, Vol. 15, 2008 edition, 1095–1099.

4. "Casting of Zinc Alloys," *ASM Handbook, Castings*, Vol. 15, 2008 edition, pp. 1049–1055.

5. D. C. H. Nevison "Zinc and Zinc Alloys," *ASM Handbook, Castings*, Vol. 15, 1988 edition, pp. 786–797.

6. R. J. Barnhurst "Zinc and Zinc Alloys," *ASM Handbook*, Vol. 2, 1990 edition, pp. 527–542.

7. A. Spada, American Foundry Society, private communication.

8. "Alloy Phase Diagrams," *ASM Handbook*, Vol. 3, 1997 edition, pp. 2.56.

9. F. Goodwin, ILZRO, private communication.

10. "Atlas of Microstructures of Industrial Alloys," *ASM Handbook*, Vol. 7, 1972.

11. M. Sahoo, L. V. Whiting, V. Chartrand, and G. Weatherall, "Effect of Strontium on the Structure and Mechanical Properties of Zinc-Aluminum Foundry Alloys," *Transactions of American Foundrymen's Society*, 94: 225–242, 1986.

12. M. Sahoo and L. V. Whiting, "Foundry Characteristics of Sand Cast Zn-Al Alloys," *Transactions of American Foundrymen's Society*, 92: 861–870, 1984.

13. M. Sahoo, L. V. Whiting, and D. W. G. White, "Control of Underside Shrinkage in Zinc-Aluminum Foundry Alloys by the Addition of Trace Elements," *Transactions of American Foundrymen's Society*, 93: 133–144, 1985.

14. M. A. Savas and S. Altintas, "The Microstructure Control of Cast and Mechanical Properties of Zinc-Aluminum Alloys," Journal of Materials Science, vol. 28, 1993, p. 1775–1780.

15. M. Ghoreshy and R. W. Smith, "Castable Zinc-Aluminum Alloys," U. S. A. Patent 4789522, 1988

16. L. V. Whiting, M. Sahoo, V. Moore, and E. Valeriote, "Influence of Sr, Ca and Na on the Corrosion of Sand-Cast ZA-27 Alloy in Wet Steam," *Transactions of American Foundrymen's Society*, 97: 173–178, 1989.

17. J. L. Dion, J. R. Emmett, and M. Sahoo, "Low-Pressure Die Casting of Zinc-Aluminum Foundry Alloys," *Transactions of American Foundrymen's Society*, 95: 813–818, 1987.

CHAPTER 11

Cast Irons

Al Alagarsamy
Alagarsamy Consulting

Introduction

Cast irons of different types constitute the largest tonnage of all metal castings produced. Cast irons offer a wide range of properties, namely strength, hardness, machinability, corrosion resistance, abrasion resistance, damping, etc. Foundry properties such as high casting yield, good fluidity, low shrinkage, easily attainable casting soundness, and others make cast irons highly desirable material for engineering and other uses. Recent developments have added to the versatility of cast irons by offering enhanced engineering properties ensuring widespread use in the future. During 2010, a total of 94.1 million tons of castings were produced in the world, as per the *Modern Casting Magazine* (December 2011) World-casting census. All types of iron castings accounted for 73% of total casting production (gray iron 47%, ductile 25%, and malleable 1%). Malleable iron tonnage is decreasing while ductile iron and gray iron tonnage is increasing. For the history of cast iron starting in 2000 BC to the present, students are referred to the review by C. R. Loper Jr. [1, 2].

Examples of Iron Castings

Cast iron is a versatile engineering material, and economical to produce. In Fig. 11.1, a few of the multitude of castings made in different molding processes and meeting different specifications are shown. There is a wide range of sizes, shapes, and applications where iron castings are used.

Various Types of Cast Irons—Definitions

Cast irons can be defined as alloys of iron with carbon and silicon along with other elements depending on type and grade, where %C + 1/6% Si = 2.0% or higher. This represents maximum solubility of carbon in austenite. Above this level, carbon will precipitate as either free carbon (graphite) or as cementite (iron carbide). The term *cast iron* is a generic term that refers to a family of materials differing widely in their properties. Cast iron (unalloyed) is in general an alloy of iron, carbon (up to about 4.0%), and silicon (up to about 4.5%). Alloyed cast irons may contain elements like nickel, chromium, molybdenum, and manganese to varying percentages. Definitions of specific types are given below.

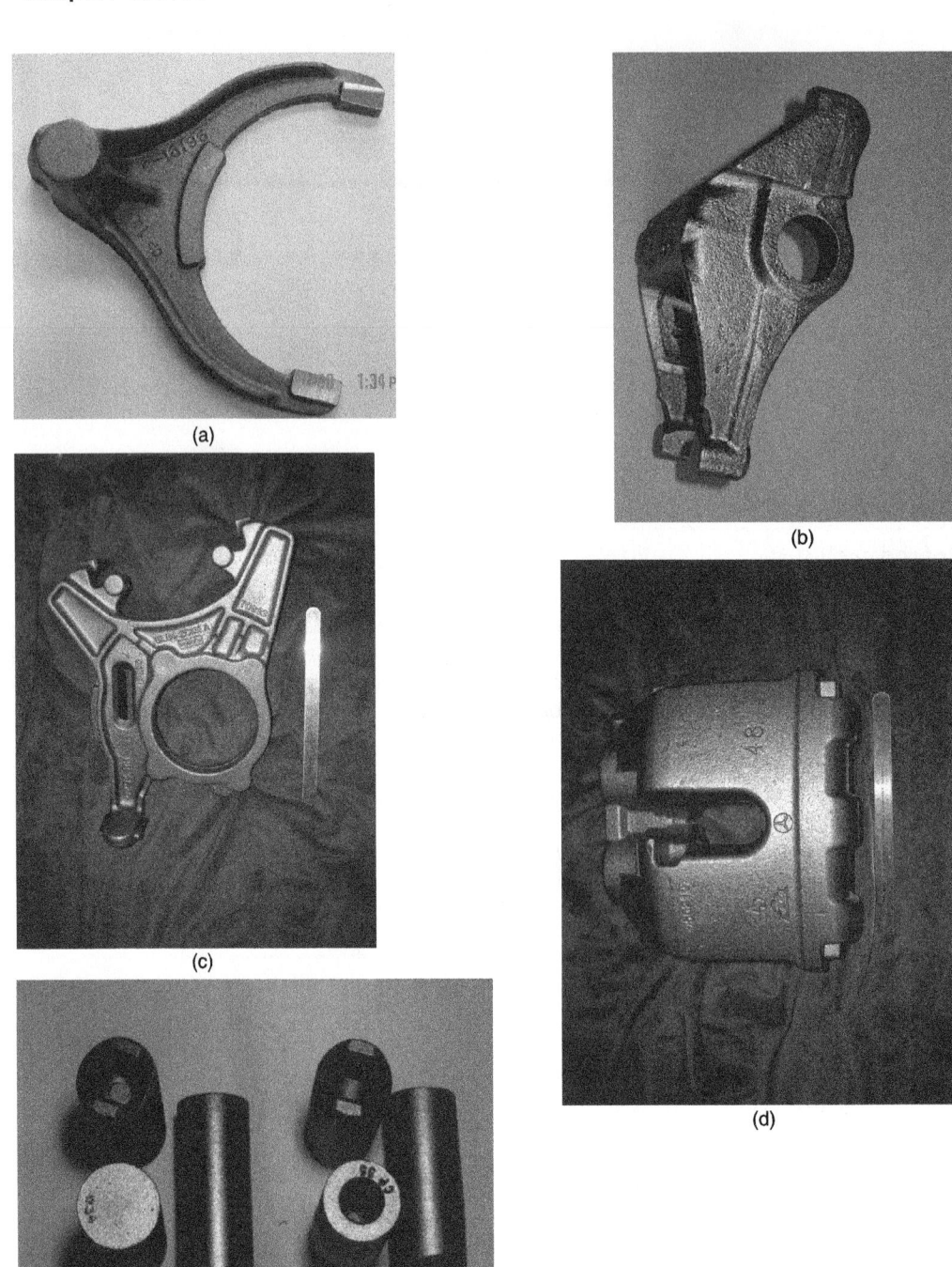

Figure 11.1 Typical castings produced in cast iron. (*a*) Gear shift fork, (*b*) rocker arm, (*c*) anchor plate, (*d*) caliper, and (*e*) valve lifters, shell moded. (*Continued*)

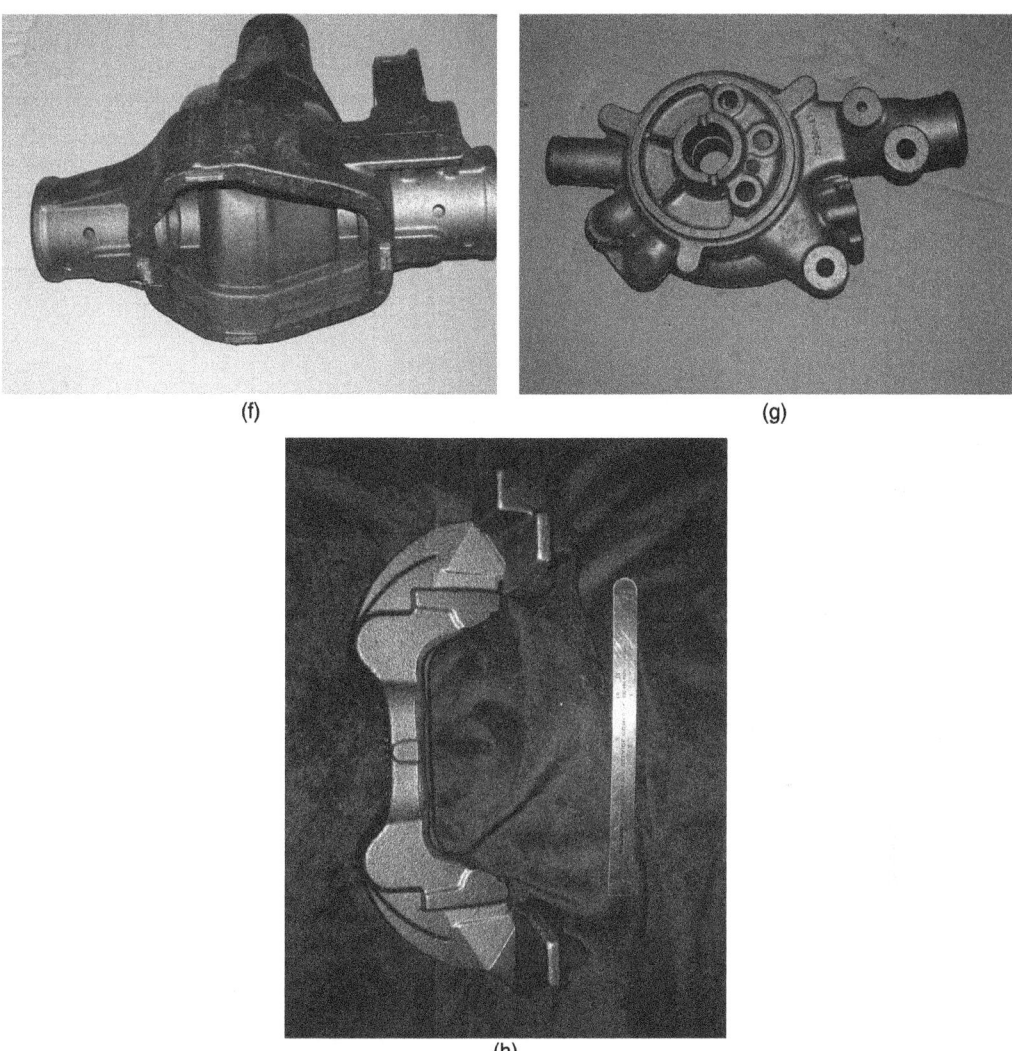

(f)   (g)

(h)

Figure 11.1 (*Continued*) (*f*) Differential carrier—lost-foam process, (*g*) fluid circulating pump—lost-foam process, (*h*) brake anchor. (Private communication *a*, *b*, and *e* from Castwell, Skokie, IL and *c*, *d*, *f*, *g*, and *h* from Alagarsamy Consulting.)

Gray Cast Iron

An iron in which most of the carbon appears as free or graphitic carbon in "flake" form (Fig. 11.2*a*). Matrix is generally pearlitic but could also contain ferrite. Gray cast iron when fractured appears as gray and sooty.

White Cast Iron

An iron having a composition such that after solidification, its carbon is present in combined form as cementite (iron carbide). White iron exhibits a white crystalline fracture (Fig. 11.2*b* and *c*).

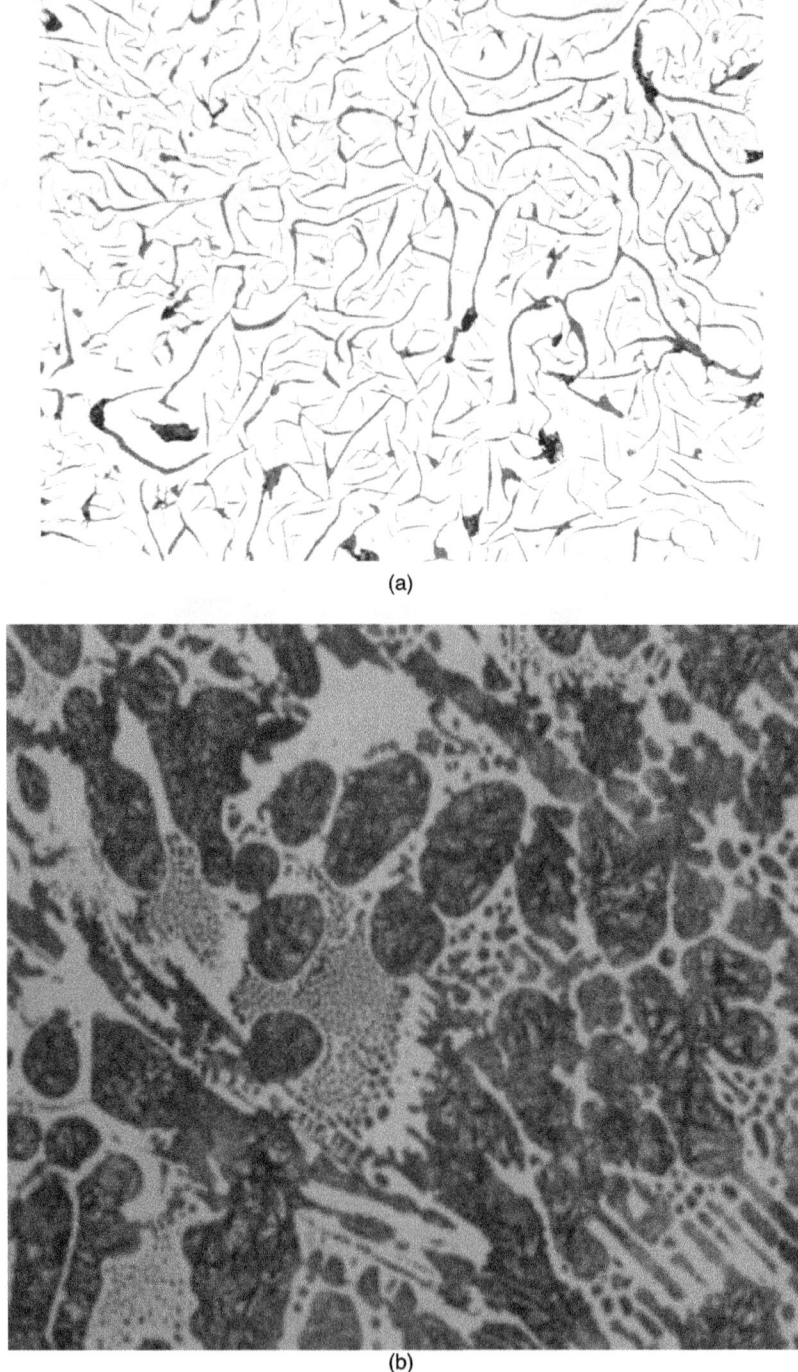

(a)

(b)

FIGURE 11.2   Micrographs of various forms of cast irons. (a) flake graphite in gray iron, (b) white iron (Ni-hard), etched. (*Continued*)

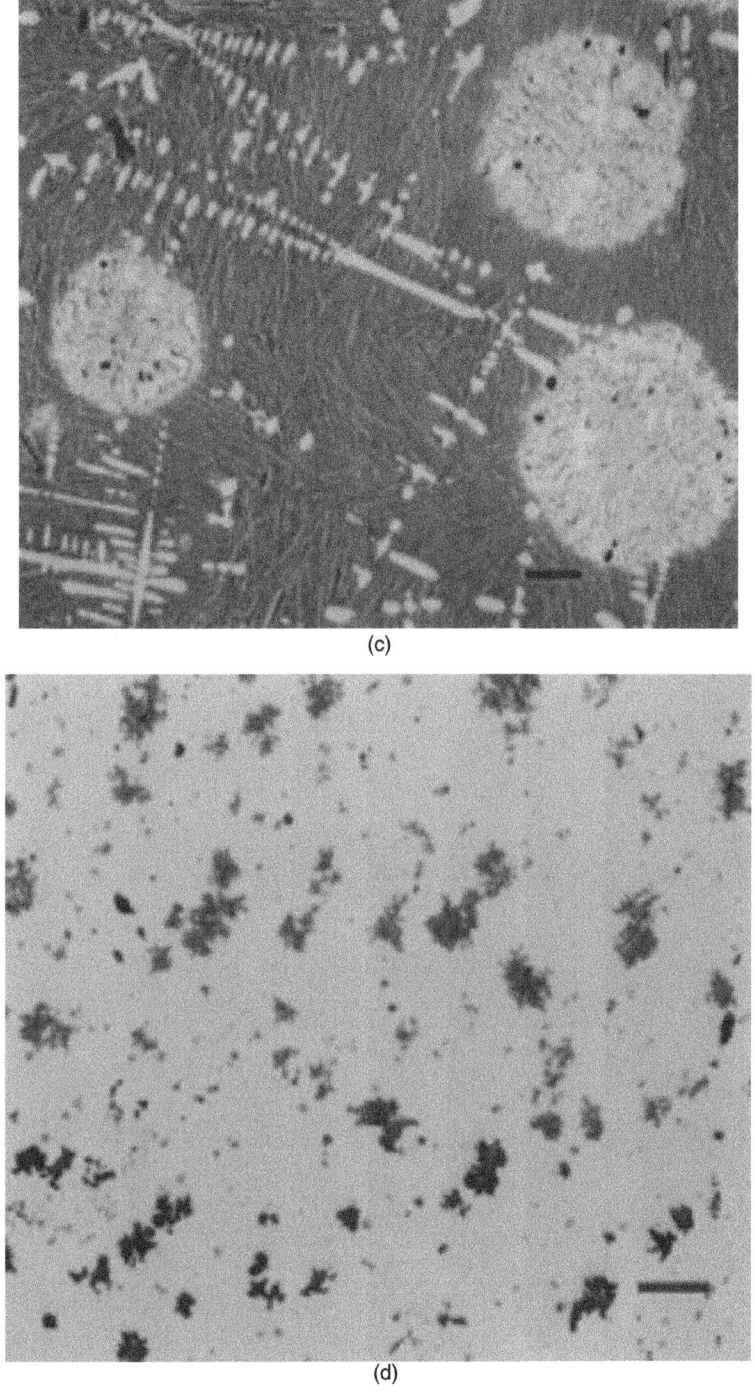

(c)

(d)

Figure 11.2   (*Continued*) (*c*) massive cementite in chill cast iron, etched, (*d*) temper carbon in malleable iron. (*Continued*)

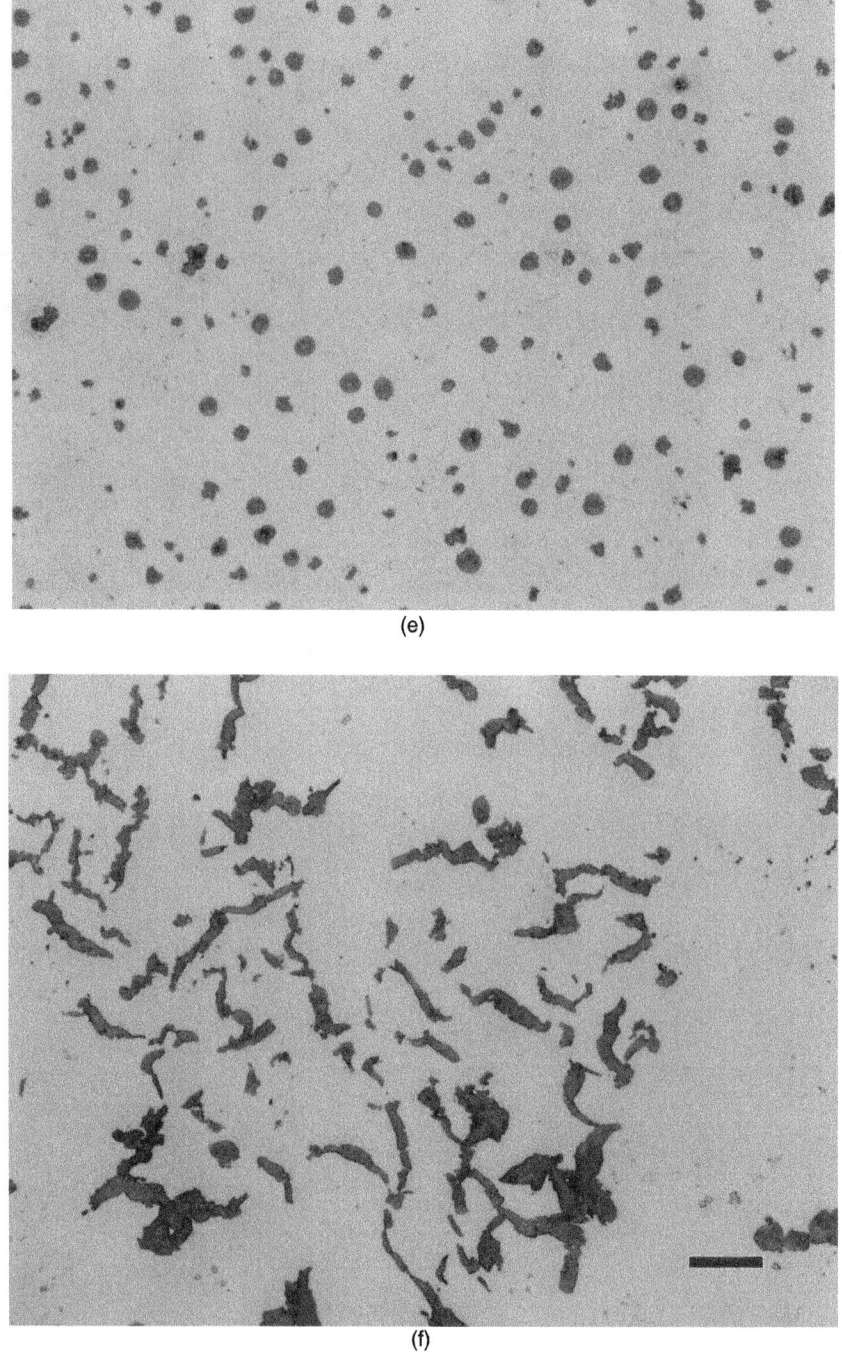

(e)

(f)

FIGURE 11.2 (*Continued*) (*e*) spheroidal graphite in ductile iron, and (*f*) vermicular graphite in CG iron. (From Vander Voort [3].)

Chilled Cast Iron

An iron that would freeze normally as gray iron with flake graphite is forced to solidify fast in certain locations by rapid cooling by use of metal or graphite chills. Fractured surfaces of chilled iron show areas of white iron where it cooled fast and areas of gray iron where it solidified slowly.

Malleable Iron

An iron that develops ductility or malleability after heat treating a white iron of suitable chemistry is classified as malleable iron. The carbon in malleable iron is present as nodular-shaped aggregates of graphite (Fig. 11.2d).

Ductile Iron (Spheroidal Graphite Cast Iron)

Molten cast iron after special treatment with alloys containing magnesium, cerium, lanthanum, etc., will solidify where most of the carbon will occur as spheroidal-shaped graphite particles (Fig. 11.2e) rather than flakes. Because of the shape of graphite, the matrix is interconnected imparting ductility to the iron.

Compacted Graphite Cast Irons

Iron that is melted and treated almost similar to ductile iron, but with much less alloy additions, solidifies with graphite of vermicular shape (shorter, stubbier flakes than flake graphite). This type of cast iron is considered to be in between flake graphite and nodular graphite cast irons with properties falling between the two (Fig. 11.2f).

Austenitic Cast Irons

Irons alloyed with nickel, chrome, copper, and molybdenum that could have either flake graphite or nodular graphite but with austenitic matrix at room temperature have special properties and are used in corrosion resistance and high-temperature applications.

The definitions given above illustrate important factors that affect the nature of cast irons. These are chemical composition, melting and conversion conditions, solidification process, cooling rate, and final microstructure. Besides these primary factors, there are other significant factors that affect the quality of cast irons. These are discussed later.

Various types of cast irons can be generally represented in the carbon-silicon diagram [4], modified to reflect the extended range of silicon, as shown in Fig. 11.3, as a broad range. Special irons may have different compositions than those depicted in Fig. 11.3. Table 11.1 shows ranges of key elements for different types of cast irons.

Microstructures of Cast Irons

Various members of the family of cast irons are very sensitive to processing variables, which affect the microstructure and thus the mechanical and other properties. The structural components of cast irons differentiate various types of irons—gray, malleable, white, nodular, and vermicular graphite. The more significant factors are defined below.

Graphite

Carbon in cast iron may precipitate as free or elemental carbon as graphite. In gray irons, this graphite precipitates in the form of flakes. Because of the low density of graphite and depending on the carbon content of the iron, the volume of free graphite

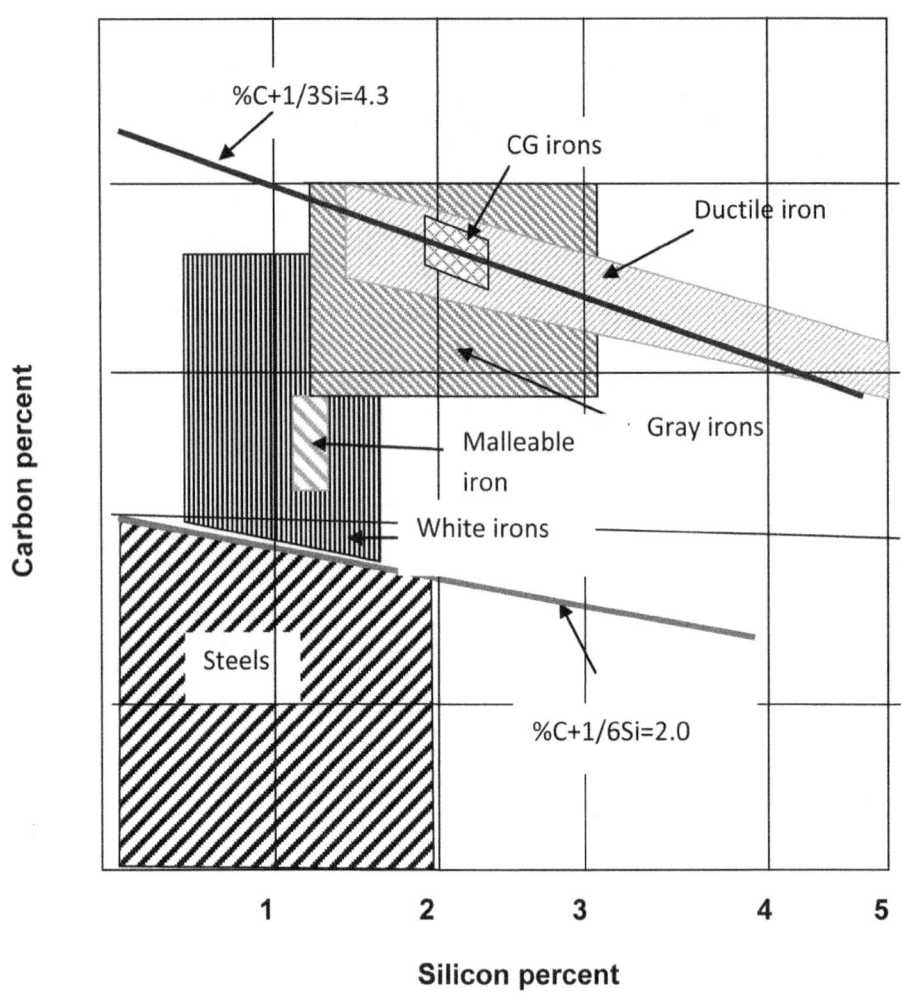

FIGURE 11.3 Carbon and silicon ranges for cast irons and steel. (From Walton [4].)

Element	Gray Iron, %	White Iron*, %	Ductile Iron	CG Iron
Carbon	2.5–4.0	1.8–3.6	3.2–4.0	3.4–3.7
Silicon	1.0–3.0	0.5–1.9	2.0–4.5	1.8–2.4
Manganese	0.4–1.0	0.25–0.8	0.1–0.5	0.1–0.5
Sulfur	0.05–0.25	0.06–0.20	0.004–0.015	0.004–0.015
Phosphorous	0.05–1.0	0.06–0.18	0.01–0.05	0.01–0.05

*Note: Such compositions may be converted from white to malleable iron by heat treatment.

Source: From Heine et al. [1].

TABLE 11.1 Broad Ranges of Composition of Key Elements in Various Cast Irons

may vary from 6 to 16% of the total volume. There are other forms of graphite in cast irons as shown in Fig. 11.2*d*, *e*, and *f*. Temper carbon is another form that is present in malleable iron as a result of heat treatment of white iron. Nodular or spherical graphite in ductile iron and vermicular graphite in compacted graphite irons develop when cast iron is treated with small quantities of magnesium and rare earths.

The amount, type, size, shape, and distribution of the graphite in cast irons greatly influence their properties. The type and size characteristics of flake graphite in gray irons have been described in standards adopted by ASTM and AFS (Fig. 11.4). In general, type A graphite has random orientation and is uniformly distributed, and is the desired

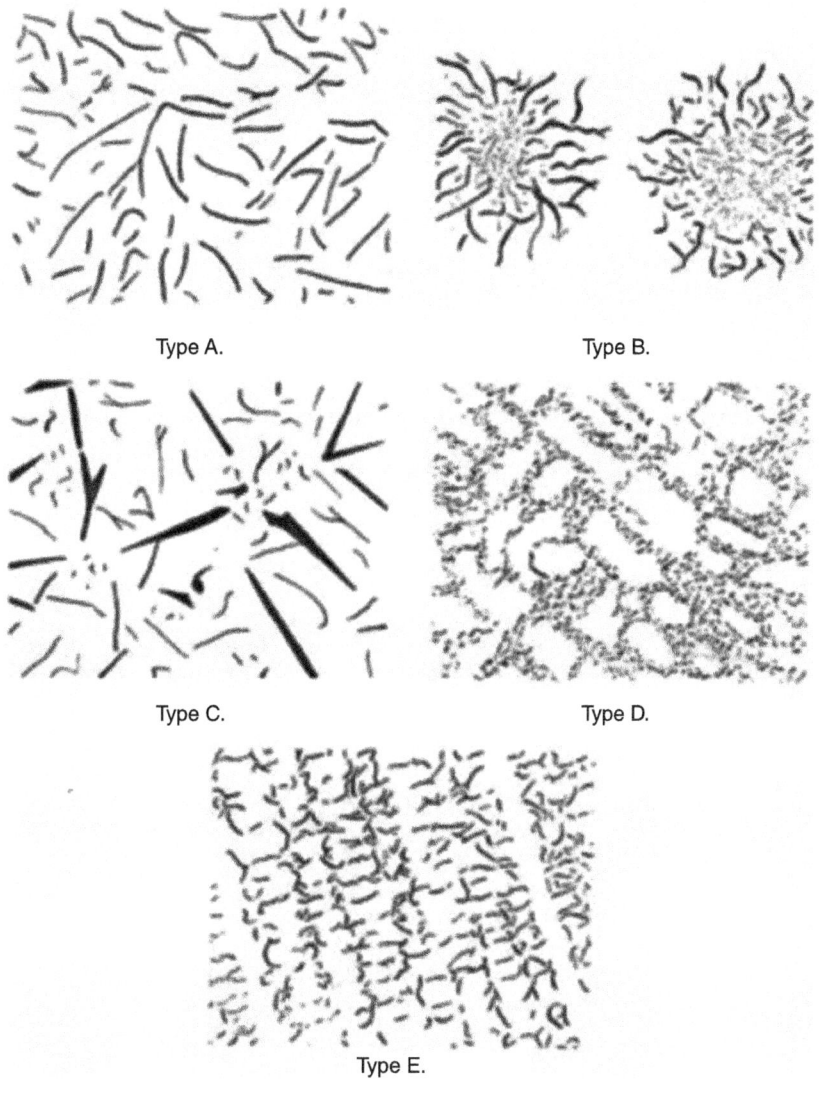

Type A.

Type B.

Type C.

Type D.

Type E.

FIGURE 11.4 Types of flake graphite—type A, B, C, D, and E. (From Walton [4].)

type in gray cast irons [5]. Flake graphite breaks up the continuity of the matrix, and is responsible for the lack of ductility observed in tensile testing of gray irons. Temper carbon (spherules) in malleable irons and spheroidal graphite in ductile irons do not reduce ductility, as does the flake graphite. Although vermicular graphite (in CG iron) reduces ductility to some extent, it still has significant ductility in tensile testing.

Cementite

Carbon in cast irons may occur entirely or in part in the chemically combined form as Fe_3C, cementite. Other elements like Mn and Cr will substitute for iron. Free or massive cementite develops during the solidification of white or chilled cast irons (Fig. 11.2b and c). Cementite is very hard and brittle and influences the properties greatly if present in significant quantities. The weight of Fe_3C is about 15 times the weight of the carbon. If a cast iron containing 2.5% carbon solidifies as white iron, it will contain 37.5% iron carbide and therefore will be very hard and brittle. Cementite is also one of the two constituents of pearlite.

Ferrite

Ferrite may be defined as the normal temperature body-centered cubic crystalline form of iron and as such is relatively soft, ductile, and of moderate strength. In cast irons, the silicon is in ferrite phase (free ferrite and ferrite in pearlite). Silicon and other elements harden and strengthen the ferrite, and the hardness can vary from 140 to over 200 BHN. Mechanical properties of ferrite depend on alloy composition.

Structurally, ferrite in cast irons may occur as free ferrite or as ferrite in pearlite. Free ferrite dominates in malleable irons and other irons of maximum ductility. In gray irons, ferrite occurs mainly as a part of pearlite unless chemistry or heat treatment favors ferrite formation (Fig. 11.5a and b). In ductile iron, ferrite forms around the graphite nodules first and extends all the way to the grain boundary when favorable graphitizing potential exists (Fig. 11.5c and d). In compacted graphite irons, ferrite is favored around the graphite particles.

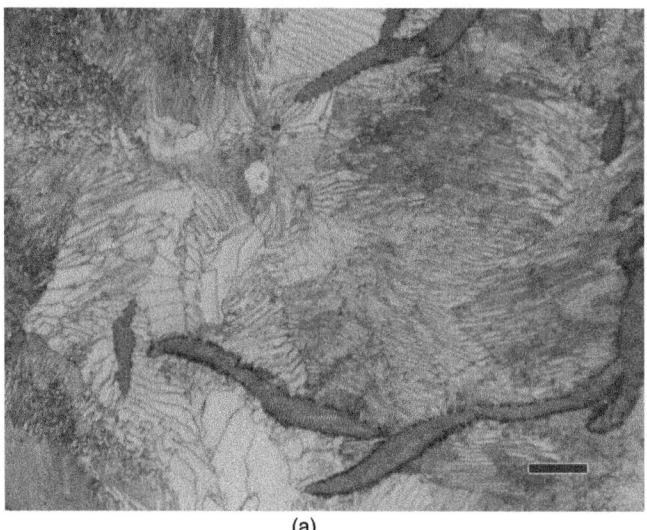

(a)

FIGURE 11.5 Gray and ductile iron morphologies; (a) pearlitic matrix (*Continued*)

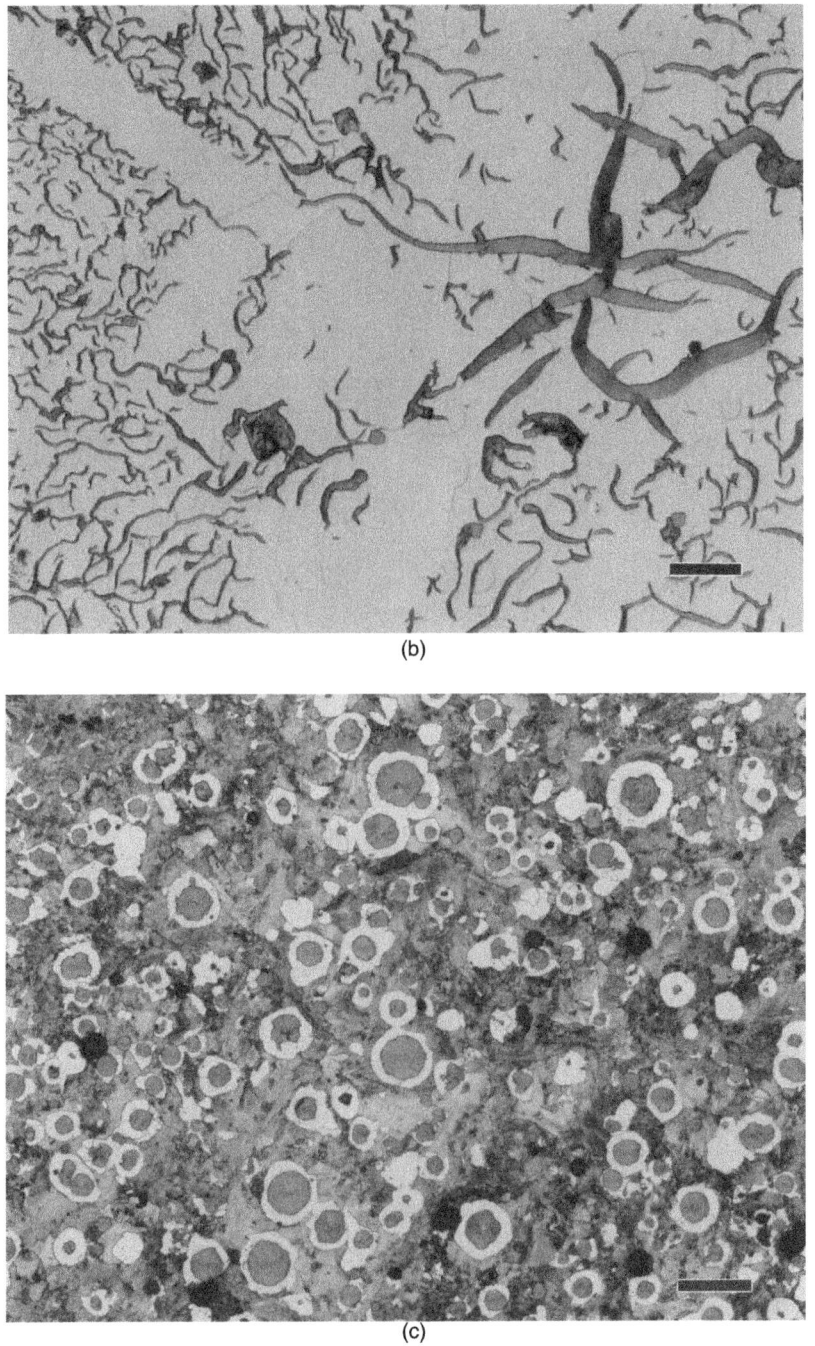

(b)

(c)

Figure 11.5   (*Continued*) (*b*) ferritic matrix, (*c*) pearlitic matrix and (*Continued*)

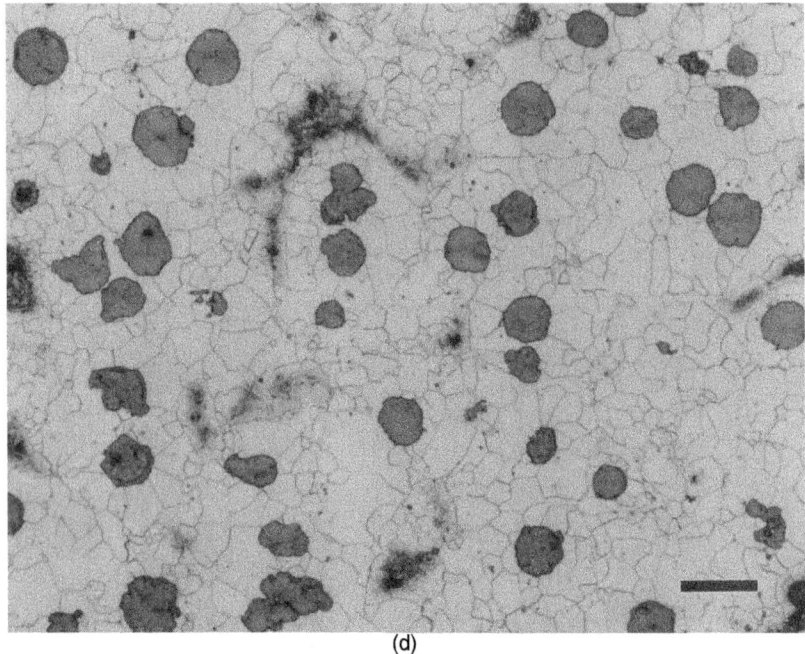

(d)

FIGURE 11.5   (*Continued*) (*d*) ferritic matrix. (From Vander Voort [3].)

Pearlite

Pearlite consists of a mixture of ferrite and cementite arranged in alternate lamellae. Pearlite in cast irons is strong and moderately hard (200 to 300 BHN) and has some ductility. The amount of pearlite present in cast irons depends on the degree of graphitization of the iron (Fig. 11.5*a* and *c*). In fully pearlitic iron, carbon is contained in the pearlite as iron carbide (also called combined carbon); it can be 0.5 to 0.9%. In white iron, pearlite is the predominant constituent besides cementite.

Steadite

The phosphorous present in cast irons, especially gray irons, occurs often as steadite, a eutectic of iron and iron-phosphide of low melting point, about 1750 to 1800°F. Phosphorous segregates into areas where the iron solidifies late in the freezing process. Steadite is distributed in grain boundary areas (Fig. 11.6) [3]. Iron phosphide, like iron carbide, is very hard and brittle. Excessive phosphorous content raises the hardness and brittleness of gray iron due to steadite formation.

Austenite

In cast irons, austenite may be defined as a high-temperature face-centered cubic crystalline form of iron occurring during solidification and which, during slow cooling, changes to pearlite, ferrite, or a mixture of the two. Austenite is present as a portion of the microstructure at room temperature in irons especially alloyed with nickel and chromium and in irons that are subjected to specialty heat treatment like austempering.

FIGURE 11.6 Optical micrograph of steadite (iron-iron phosphide eutectic in gray cast pig iron). (From Vander Voort [3].)

Ledeburite

Ledeburite is the eutectic of austenite and cementite. It forms during solidification without graphite and the austenite transforms into pearlite below the austenite transformation temperature while cementite retains its structure.

In addition to the above structural constituents, cast irons may contain nonmetallic inclusions. These are mainly sulfides of manganese due to the presence of sulfur and silicates of iron and manganese formed by the reaction with oxygen. For summary purposes, some properties of microconstituents of cast irons are presented in Table 11.2.

From Table 11.2, it can be seen that structural components of widely varying properties may occur in cast irons. Therefore, the properties of commercial iron castings are largely influenced by the microstructure developed during and after solidification. In this connection, it should be noted that chemical composition is not the only factor determining the microstructure. Cooling rates, freezing mechanism, presence of certain gases in the metal, and a host of other variables in the foundry practice may markedly affect the microstructures formed and consequently the properties. The actual manipulation in the foundry may cause drastic changes in the microstructure and properties of the iron. For example,

Constituent	Specific Gravity, g/cc	Tensile Strength, psi	Elongation, %	Reduction of Area, %	BHN	Remarks
Ferrite	7.86	39,500–42,000	61	30.9	75	
Silicoferrite	–	45,150	50	91.6	–	0.82% Si
Silicoferrite	–	63,500	50	85	–	2.28% Si
Silicoferrite	–	77,400	21	28.7	–	3.4% Si
Cementite	7.66	–	–	–	550	–
Pearlite	7.846	120,000	15	–	240	–
Pearlite	–	125,000	10	15	200	Pearlitic steel
Pearlite	–	105,000–125,000	–	–	–	0–0.8% manganese
Graphite	2.55	–	–	–	–	–
Ledeburite	–	–	–	–	680–840	–
Steadite	7.32	–	–	–	–	–
Manganese sulfide	4	–	–	–	–	–
Iron sulfide	5.02	–	–	–	–	–

Source: From Heine et al. [1].

TABLE 11.2   Selected Properties of Microconstituents in Cast Irons

in gray irons, the type of graphite, whether A, B, C, D, or E, may be greatly altered by many variables in the melting practice, and holding and handling of the molten metal. In malleable irons, the melting practice and heat treatment given to the white irons are major factors affecting the properties. Ductile and CG irons are also very sensitive to the foundry practice with respect to graphite particle number, shape, size, and distribution. In all irons, the influence of section size and cooling rate is very important. With all the factors involved, the family of cast irons can offer widely varying structures and properties for engineering use, each with its own advantages and limitations. It is the job of the trained foundry personnel to produce the kind of microstructure desired in the castings.

Melting and Melt Handling of Cast Irons

Furnace Equipment Used

The following melting practices are currently in use for cast irons having a wide variety of chemistry and casting requirements.

- Cupola
- Electric induction
- Electric arc
- Air

Some of these furnaces have been used for a long time and have been improved for efficiency and control of chemistry to achieve targeted metallurgical properties.

The highest tonnage of metal is being melted in cupola furnaces and then electric induction furnaces. Cast irons are also melted in electric arc furnaces and air furnaces. Highly alloyed irons are melted in electric induction furnaces.

Cupola Furnace

The largest tonnage of iron is melted in cupola furnaces. It is suited for melting metal at high rates continuously for modern high-production foundries. The basic melting principle in the cupola furnace has remained the same for more than 200 years. Cupola capacities vary from as little as 1 ton/h to over 100 tons/h. Advantages of cupola melting include

- *Continuous melting*. Foundry production is facilitated as iron can be accumulated in fore-hearths and/or holding furnaces for pouring large molds as well as supply of iron at regular intervals directly from the cupola or from accumulating devices. Flow of molten metal and molds may be synchronized for quantity production as required by high production automotive foundries and pipe casting machines, etc.

- *Low-cost melting*. Various types of low-cost raw materials could be used for melting with lower operating cost than any other melting furnace producing equal tonnage.

- Chemical-composition control is possible with proper and careful furnace operation and control with various devices available for continuous melting.

- Adequate temperature control for fluidity in pouring castings with varying thicknesses can be obtained.

The limitations with the cupola furnace are given below:

- Loss of alloying elements such as chromium, molybdenum, manganese, and silicon due to oxidation in the stack.

- Highest temperature obtained is limited to preferred melting conditions.

- Cost of installation for similar melt rates is higher for cupola furnaces than electric melting.

- Pollution-control equipment costs are higher for cupola melting.

- Lower carbon content (below 2.8%) is difficult to attain due to the direct contact of metal with carbonaceous fuel.

Structure

The structural features of a simple-concept cupola furnace and its cross-sectional view are shown in Figs. 11.7 and 11.8, respectively [8]. It consists of a vertical shaft or a steel shell. The shell could be lined with refractory brick. Essential features are tall vertical column for preheating the charge, opening bottom doors for dropping the bottom at the end of campaign, combustion air delivery system through tuyeres into the bottom section of the furnace, a tap hole for draining iron from the furnace either continuously or intermittently, and a slag port separate from iron tap hole or a slag notch at the iron trough for continuous flowing cupolas. The bottom of the furnace is rammed up with molding sand with fireclay and/or refractory topping for longer (up to 6 weeks) campaigns. Details of cupola design and dimensions are beyond the scope of this book and can be found elsewhere [6].

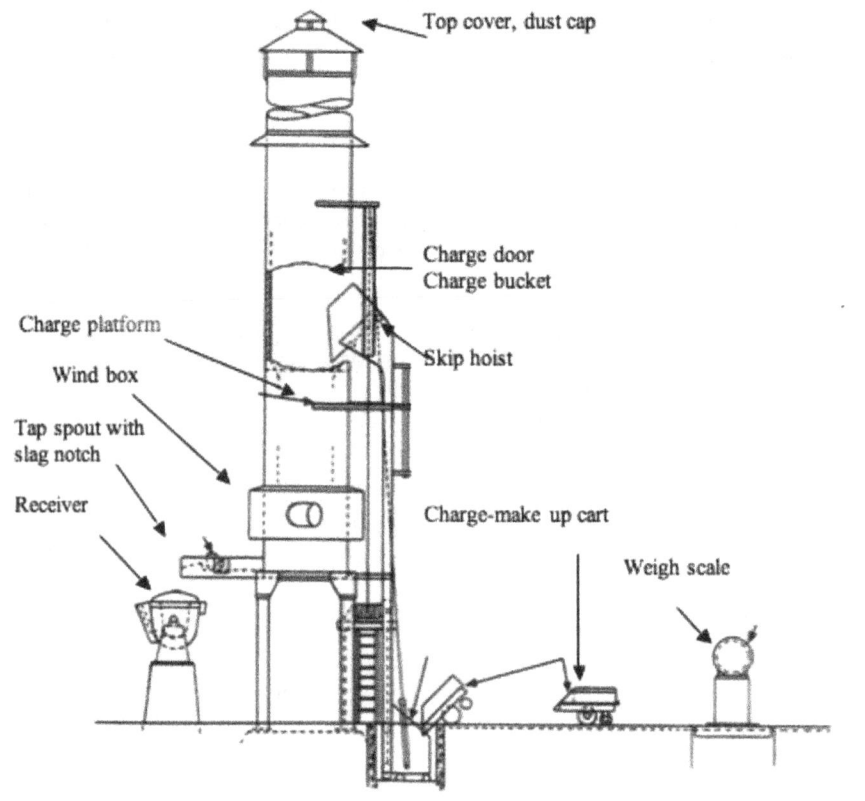

Top cover, dust cap

Charge door
Charge bucket

Charge platform

Wind box

Skip hoist

Tap spout with
slag notch

Receiver

Charge-make up cart

Weigh scale

FIGURE 11.7   Simple cupola installation. (From AFS [6].)

Steps in Cupola Operation

A cupola "heat," which may last one day to several weeks, includes operations that precede and follow the period during which iron is being melted. The cycle of events that occurs each time a "heat" is made are listed below:

- Preparation of the refractory lining, bottom, tap hole, and slag hole
- Lighting and burning in the coke bed
- Charging
- Melting
  - Starting the air blast
  - Charging
- Tapping and slagging (may be intermittent or continuous)
- Dropping the bottom at the end of the heat campaign

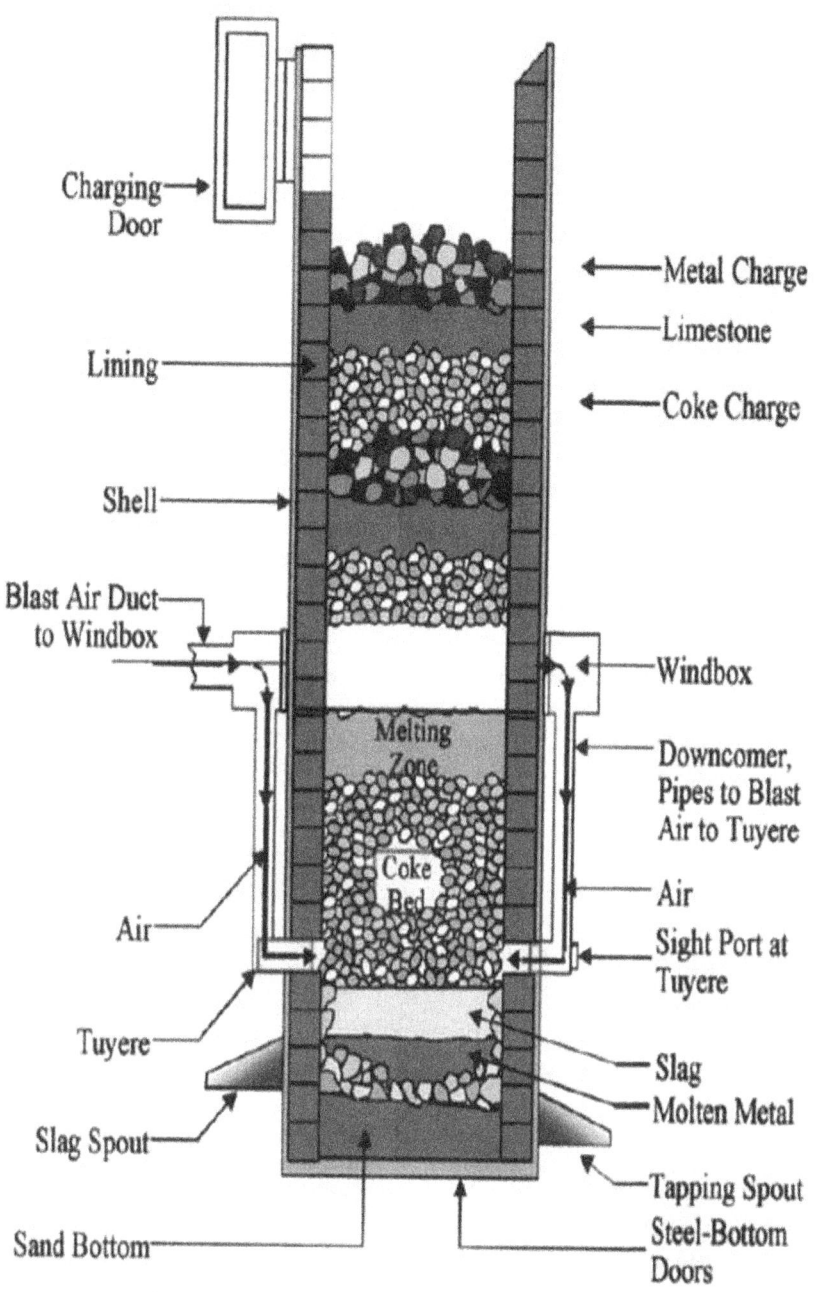

Charging Door

Lining

Shell

Blast Air Duct to Windbox

Air

Tuyere

Slag Spout

Sand Bottom

Metal Charge

Limestone

Coke Charge

Melting Zone

Coke Bed

Windbox

Downcomer, Pipes to Blast Air to Tuyere

Air

Sight Port at Tuyere

Slag

Molten Metal

Tapping Spout

Steel-Bottom Doors

FIGURE 11.8   Cross-sectional view of a simple cupola. (From AFS [6].)

Preparation for Cupola Melting

For successful cupola operation, there are several key steps that should be satisfied:

- After cleaning the slag inside the cupola and repairing any worn-out refractory lining the cupola is ready for preparing for another campaign.
- Bottom preparation—ramming the bottom after the doors are closed. The thickness of bottom lining and material depend on the length of the campaign. For single-day operation of the cupola molding-sand with extra fireclay can be rammed up to 25.4 cm (10 in). If operated for longer periods, the lining wears out and can result in metal leaking through the bottom. A complete refractory bottom or a mixture of sand bottom with refractory top is used for one week or several weeks of operation.
- The tap hole is repaired. The height of the tap hole to the slag hole (if the slag spout is separate from the iron spout) is an important measurement as it affects the amount of liquid metal it can hold before iron runs out of the slag hole.
- Bed preparation—after the bottom has been installed, coke with kindling is charged into the cupola. The coke is ignited and a fire is developed. When the temperature of the coke bed is high enough additional coke is placed on top. The height of the hot coke bed is important and is controlled by measuring the depth from the top by a chain and charging additional coke if necessary. If it is too high, the coke bed is burned off until the right height is reached.

  Once the bed is properly burned-in, the cupola is ready for charging and melting. Before continuous melting starts, the stack needs to be filled with charge materials, and this helps to preheat the charge for quick and efficient melting. Once the stack is filled with metal, coke, and limestone charges, blast can be turned on to start the melting operation.

Charging

Cupola charge essentially consists of metallics (steel scrap, returns, pig iron), fuel (coke), and flux. These are all weighed for each charge and placed in a charging bucket. The charge bucket is elevated by a skip hoist or other types of mechanical equipment to the charge door and dumped into the cupola shaft. To fill the shaft it takes around 10 to 15 charges. Once the cupola is filled with charge materials, cupola blast air is turned on. It takes about 10 to 15 minutes to start melting the metals in the charge. After a few more minutes of accumulating the melt in the cupola well, the tap hole can be opened and molten metal can be tapped out of the cupola into ladles or holding ladles or furnaces. This melting operation may continue for a few hours or a few weeks, which is called a campaign or heat.

Dropping the Bottom

At the end of a campaign, charging is stopped and most of the remaining charge in the cupola shaft is melted. At this time cupola melting is done, and any remaining materials inside the cupola may be dropped, by pulling the support to the bottom doors. Before dropping the bottom, the area must be prepared so that there is no water in the area and a dry sand bed is placed to prevent any metal splashing and fire from liquid metal falling down. The drop is cleaned and the cupola is allowed to cool before preparing for the next heat.

Charge Materials for Melting of Different Types of Cast Irons in a Cupola

Main metallic charge materials used in the cupola are virgin pig iron, steel scrap, cast iron scrap, in-house returns, ferrosilicon briquettes, ferromanganese briquettes, briquettes of iron, and steel machining chips. Silicon carbide briquettes of varying composition are being used to supplement silicon in the charge. Recovery of silicon is dependent on the silicon concentration [7]. For fuel and carbon pickup, coke, carbon briquettes, and coal are charged along with the metal through the charge door. Coke fines and natural gas may be used as additional fuel through the tuyeres in the melt zone area.

Along with the metal and coke charge, limestone, and/or dolomite stone is used as a fluxing agent. Calcined limestone and dolomite stones combine with oxides of iron, silicon, and other elements to form molten slag.

Gray Iron

Chemistry of the charge materials will depend on the grade of gray iron produced in the cupola. Generally, low-phosphorus, low-sulfur pig irons are preferred, but many different grades of pig iron are used depending on availability, price, and geographical location. If briquetted borings are used, it should have carbon mixed with the metal to minimize oxidation and facilitate carbon pickup. Cast scrap containing parts that may contain elements like antimony, tin, lead, and zinc may be harmful to the iron.

Steel scrap should be sized such that it will not bridge as the material slides down in the stack. Too thin steel scarp will oxidize in the upper stack and will increase melt loss. Material bulk density and size are important factors in selecting suitable charge material for the cupola, which depends on the size of the cupola. Large pieces of metals will hang up in the stack leading to bridging and operational problems.

White Iron

Pig iron for malleable iron melting is not as critical for phosphorous, but low sulfur is preferred. Chromium levels should be controlled to a low level in both steel scrap and other metallic scrap, due to the problem it poses in malleablizing heat treatment cycle.

Ductile Iron

Purchased materials (pig iron and steel scrap) need to be low in phosphorous, sulfur, manganese, and chrome as well as other deleterious elements like tin, antimony, lead, zinc, and titanium. These elements, even in small quantities, adversely affect the properties of ductile iron, which is discussed in detail later in this chapter.

Typical materials used in cupola melting include:

Inputs		Outputs	
Metallic materials	1 ton	Molten iron	0.98 ton
Coke and coke breeze	220 lb	Molten slag	100 lb
Flux (limestone)	60 lb	Effluent gases	1.25 tons
Combustion (blast) air	1.1 ton		

The most basic design consists of a weigh scale to weigh all the charge materials, a skip loader to charge the skip bucket, and a skip hoist to deliver the charge to the charge door.

A flame arrester or dust cap keeps the fly ash from scattering over the area. A wind box delivers combustion air through tuyeres (port holes) into the cupola just above the bottom.

Successful cupola operation hinges largely on combustion control. In order to have favorable melting conditions it is necessary to have a balanced combination of coke and air supplied at the proper rate. With proper combustion, the control of metal composition, temperature, and slagging can be accomplished.

Coke and Air Supply

Coke supply and air delivery is balanced such that the stack gases contain about 12 to 14% CO_2 and 11 to 15% CO. Even though it looks like the combustion is incomplete with this much quantity of CO, it is essential to maintain this ratio for controlling the quality of melt and slag. Excess CO means too much coke used, lower temperature, and lower melt rate. Whereas too little CO in the stack gases means a high level of oxidation in the melt. Amount of air supplied is compensated for the variation of humidity in the air.

With the proper balance between air and coke supply, the bed height inside the cupola is maintained within the proper range to result in closer chemistry and temperature control. With excess air supply, bed height will reduce, and melt temperature and CO content of the stack gases will decrease. Oxidation of melt will increase indicated by brown fumes emanating from the cupola stack. On the other hand, with insufficient air supply, bed height will increase, increasing the CO content in the stack and ultimately decreasing metal temperature and melt rate.

Metal Composition and Properties

Composition and property control depend on

- Composition of elements in the charge (weight and material selection)
- Known and consistent composition changes during melting operation
- Use of chill testing, carbon equivalent (CE) testing, measuring temperatures, and proper inoculation after melting

Composition of metal produced from the cupola can be estimated based on available knowledge as well as accrued knowledge in the plant. To estimate the output chemistry it is not only essential to know the chemical content of the charge and the weight in charge makeup but also expected losses and gains in certain elements during melting, which may vary depending on the melting conditions. Typical losses or gains are shown in Table 11.3.

Charge mix is calculated based on the information above and known chemistry of the charge adjusted for the particular operation based on current results and experience.

Slag Composition

Chemistry of iron out of the cupola depends not only on the metals charged but also on the nature of the slag. From limestone or dolomite used for fluxing, CaO and MgO, which are basic components of the slag, are generated. From oxides in the charge and from the result of oxidation in the cupola, SiO_2, FeO, and MnO_2, which are acid components, are formed. The relative quantities of these basic and acid materials in the slag affect the carbon and sulfur pickup in the iron. Basicity ratio expresses the relative quantities of basic and acid parts of the slag. It is expressed as the ratio of basic

Element	Losses or Gains	Comments
Silicon	Typical 10%, may be as high as 20%	Si loss ends up in slag
Carbon	Gain could be as much as 2% of metallic charge or higher	Depends on steel in charge mix, coke ratio, and air supply
Manganese	Typical 15%	Manganese loss enters the slag
Phosphorous	No loss or gain	
Sulfur	A pickup of 0.03–0.06%	Pickup comes from coke and depends on melting practice
Nickel, copper, tin	No loss or gain	
Chromium	Slight to 10% loss	

Source: From Heine et al. [1].

TABLE 11.3   Composition Changes during Cupola Melting (acid slag practice for gray iron)

components to acid components. As a first approximation, the basicity ratio is expressed as $\{(CaO + MgO)/SiO_2\}$.

When this ratio is greater than 1, the operation of the cupola is said to be basic. For gray iron melting and for most ductile iron melting, the basicity ratio is less than 1, which indicates that the slag is mostly acidic. With acid slags, sulfur is higher and carbon is lower in the melt.

Ductile Iron Charge Make up

Pig irons used for ductile iron melting must be of low residual iron—low in phosphorous, chromium, and manganese. The pig iron must also have low sulfur to minimize the sulfur in the base iron. The sulfur level in the ductile base iron needs to be controlled to a low level, preferably below 0.015%. This is not possible with acid slag melting in the cupola. However, sulfur can be reduced in the cupola well by the use of basic slag practice. This is achieved mainly by increasing the flux (limestone) in the charge. Operation of cupola with basic slag has many significant differences compared to acid slag operation. Typical characteristics of basic slag are as follows:

- Slag basicity—increased
- Silicon and manganese loss—increases with basic slag
- Coke consumption—increased
- Metal temperature—reduced
- Melt rate—reduced
- Sulfur level—reduced
- Carbon level—increased

Control of carbon and sulfur is rather difficult in a basic slag cupola. Carbon goes up as the sulfur comes down and vice versa. Mainly because of the inconsistency in C and S levels in the base iron, basic slag practice is not followed by ductile iron foundries using cupola for melting. Instead, most of them use external desulfurization to reduce the sulfur level or use special treatment methods that can tolerate higher sulfur levels.

Modern Cupolas

Recent installations of cupolas, especially the high capacity ones, have many of the following improvements which reduce cost of operation as well as provide better process control (Figs. 11.9 to 11.11).

- Water-cooled shell
- Water jacket
- Water spray

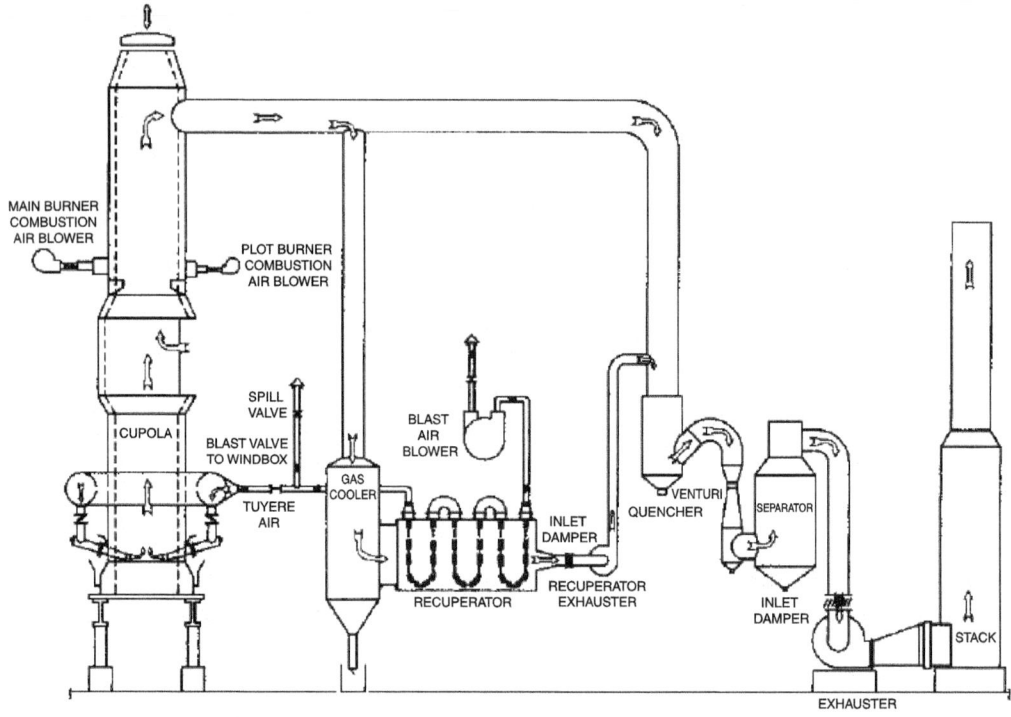

FIGURE 11.9   Recuperative hot blast installation. (From AFS [8].)

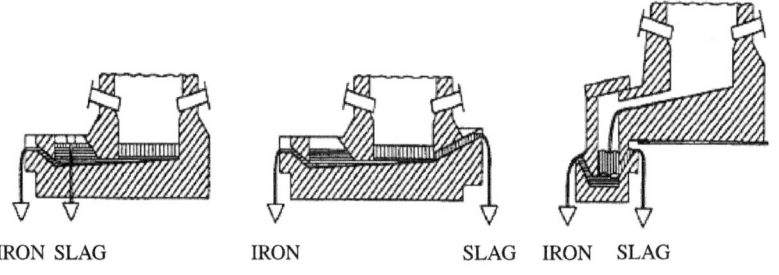

FIGURE 11.10   Different slagging methods for a continuously operating cupola. (From AFS [8].)

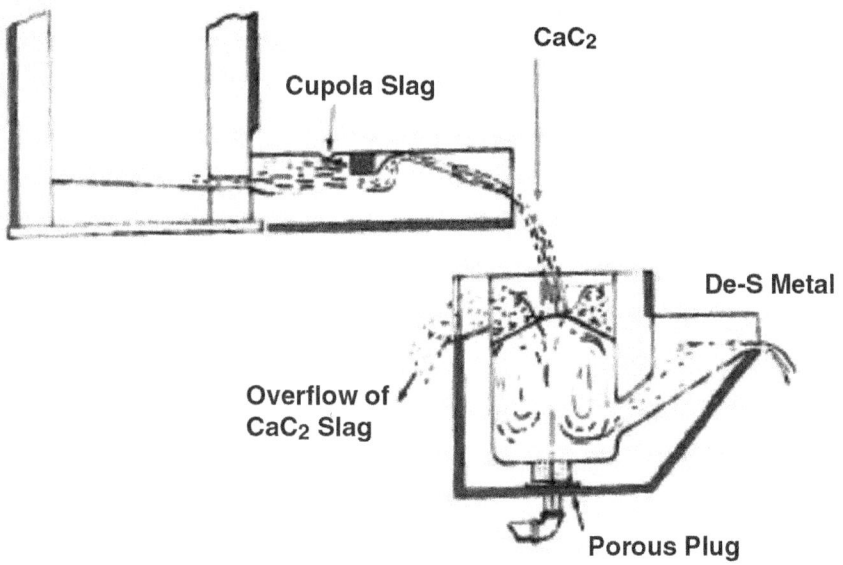

FIGURE **11.11** Continuous external desulfurization with porous plug ladle. Nitrogen gas is used for mixing the desulfurizing agent, either calcium carbide or lime. For large installations, multiple plugs are used. (From *The Sorrelmetal Book of Ductile Iron* [9].)

- Hot blast
- Dirty gas recuperative hot blast
- Cleaned gas, CO combusted hot blast
- Gas-fired hot blast
- Water-cooled tuyeres
- Oxygen enrichment
- Carbon injection through tuyeres
- Ferrosilicon and other fine material injection
- Oxy-fuel burners
- External desulfurization
- Cokeless cupolas

Apart from the above-mentioned improvements to the cupola equipment there have been many improvements in measuring different variables such as temperatures in different zones, CO/CO_2 ratios, metal chemistry, metal temperature, and using expert systems to control different inputs to get closely controlled liquid melt at the spout.

Air furnaces

Cast irons are also melted in air furnaces, mainly for white irons and malleable irons. The cold-melt air furnace is used for batch melting. The air furnace is a reverberatory furnace fired with pulverized coal or oil. Furnace sizes range from 15 to 40 tons. The furnace hearth is rectangular and provides a molten bath about 30 cm (12 in) deep. Tap holes are

Material	Batch Melting (air furnace charge)	Duplexing (cupola charge)
Pig iron (malleable), %	25–35	12
Sprue (in-house scrap), %	45–55	50
Malleable iron scrap, %	5–20	10
Steel scrap, %	0–10	38
Fuel:		160–220
Coke, lb/ton of melt	0	
Coal,lb/ton of melt	700–1000	180–220
Electricity, kWh/ton of melt	0	480 (duplex-arc furnace)
Flux, lb/ton of melt	0	60–80 (cupola)

Source: From Heine et al. [1].

TABLE 11.4 Typical Charge Materials for Different Melting Methods in Air Furnaces

provided on the side of the furnace. Sidewalls are made of firebrick supported by steel, and the bottom is either sand or firebrick. The furnace top consists of a series of removable firebrick arches known as "bungs." By removing some of the bungs, the furnace may be charged with cold metal through the top. A typical furnace charge is given in Table 11.4.

Smaller-size materials are usually placed at the bottom of the furnace. Since mold yield is about 50% for white iron castings, foundry scrap in the charge is around 50%. The balance of charge materials for air furnace is selected to get carbon around 2.65 to 2.85%, and to keep sulfur and phosphorous levels below the maximum, and silicon and manganese slightly below the final desired range. To minimize problems with annealing, chromium levels should be controlled below 0.07%, preferably below 0.03%. Fuel air mixture used for melting should be controlled to minimize oxidation. When the molten metal bath reaches 2600°F (1430°C), slag that is formed from oxidation of metal and attrition of refractory is skimmed. Bath temperature is then raised to tap temperatures of 2800 to 2900°F (1540 to 1600°C). At temperatures below 2700°F (1480°C), manganese and silicon will oxidize and above this temperature carbon losses can occur rapidly. At higher temperatures, carbon will deoxidize silica from the slag as well as from the refractory, resulting in a slight increase in silicon level (Table 11.5).

Carbon losses can be controlled by richer fuel to air mixture, or by the addition of graphite, pet coke, or other recarburizer.

Period of Heat	% °C	% Si	% Mn	% P	% S
Charge	2.8–3.2	1.10–1.25	0.45–0.55	0.14 max	0.09
After meltdown	2.7–2.9	0.90–1.10	0.3–0.4	0.14 max	0.09
Prelim analysis	2.5–2.6	0.96	0.37	0.14 max	0.1
Final analysis	2.3	1.05	0.35	0.14 max	0.1

Source: From Heine et al. [1].

TABLE 11.5 Typical Chemical Composition Changes of Air Furnace Heat

The analysis changes occurring during melting in an air furnace affect the structures seen in solidified iron. A sprue fracture test is often used to observe the condition of the iron (white, mottled, or gray). Mottling results from the formation of flake graphite during freezing. Before tapping the metal, a white iron fracture is ensured. The objective in melting quality malleable iron is to produce a completely white iron with no free flake graphite in the castings as free graphite lowers the mechanical properties. Oxidizing conditions during melting, high melt temperatures, and low carbon and silicon levels favor white iron formation. Tapping temperature range is 2800 to 2900°F (1538 to 1593°C) and molds are poured around 2600 to 2800°F (1427 to 1538°C).

Electric Induction Furnaces

More and more of cast iron melting is carried out in electrical induction furnaces due to lower capital cost and easier chemical composition and temperature control. The furnace sizes vary from 100 lb to more than 25 tons. New installations of smaller capacity are mostly electric induction furnaces. Depending on the frequency of power input to induction coils, they are classified into low- or line-frequency, medium-frequency, and high-frequency induction melting. Heat generation in induction furnaces is from the eddy current generated in the charge from the electromagnetic induction. There are two types of induction coils used in the furnaces: (1) coreless furnaces, where there is no separate magnetic core, the metallic charge acts as the secondary coil and generates currents from the magnetic flux from the primary coil, and (2) channel furnaces, where the primary coil is wound around a steel magnetic core. The secondary is a loop of iron surrounding one leg of the magnetic core. Concept drawings of these furnaces are shown in Figs. 11.12 and 11.13.

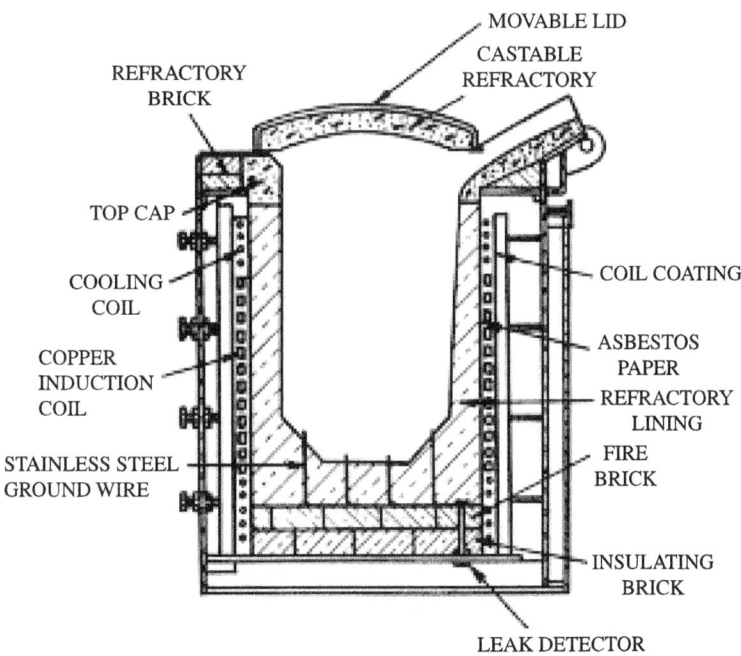

Figure 11.12 Cross-sectional view of a coreless induction furnace. (From AFS [8].)

Coreless Induction Furnace

This is a cylindrical furnace with a primary coil, which is made of many loops of hollow, water-cooled copper tube backed by magnetic yokes. These yokes concentrate the magnetic flux inside the furnace for maximum power generation for melting. Line frequency (60 cps in North America) in a low-frequency induction furnace generates heat in the charge at a much lower rate in solid metals of small pieces. This type of furnace does not empty all of the molten iron, but instead holds a portion of melt in the furnace to increase the melting rate. This is called heel melting. Metal charged into a liquid bath needs to be dry and enclosed metal tubes are not advisable due to explosion risk. Charge is generally preheated to dry out any moisture that may be trapped inside. Furnaces are emptied at the end of a campaign, which could be once a day or once a week.

In medium- and high-frequency furnaces heat generation is faster in the cold charge and hence they are usually cold charged and liquid metal is emptied every time. There are some medium-frequency furnaces that hold heel instead of emptying every time. Medium frequencies vary from 180 to 960 cps. High-frequency furnaces tend to be much smaller in size and the frequencies range from 3000 cps and up.

Coreless induction furnaces for iron melting are generally lined with silica refractory bonded with boric acid. For installation of the lining, a steel shell resembling the interior shape and size of the furnace is used to form the interior surface of the furnace. Loose refractory materials are placed in between the steel shell and the safety lining near the electrical coil. After consolidating the refractory with vibratory and other rammers, starter blocks (large pieces of cast metal, usually excess metal poured into pig molds during production) are placed inside the steel lining. Power is then turned on in steps of increasing power input, following a prescribed time/heat cycle. After the starter blocks are heated to red-hot temperature, more charge materials are placed inside the furnace. Heat from the melting iron inside the steel shell sinters the lining next to the steel shell. The lining away from the sintered layer stays granular, which helps in preventing cracks during heating and cooling of the lining.

Slag generated during the melting of iron in induction furnaces is generally acidic from the dirt and oxides charged along with sand adhering to the returns. Prior to removal of slag power is turned off, to let the suspended slag particles float to the top of the bath. Slag is removed by means of slag spoons in smaller furnaces and with mechanical aids like clamshell buckets suspended from overhead cranes in larger furnaces. After removal of the slag, iron is sampled for determining the chemistry, especially carbon equivalent. After the chemistry is checked and corrected if needed, iron can be tapped into pouring ladles or transport ladles. Generally, only a portion of the iron (10 to 25% of the furnace capacity) from the furnace is tapped and the furnace is recharged with preheated metallic materials.

Channel Furnaces

These furnaces are generally used for holding molten iron melted in cupolas or other induction furnaces for controlling temperature and chemistry prior to pouring of molds. The channel furnaces are also used for melting cast iron. A cross-sectional view of a channel induction furnace is shown in Fig. 11.13. Some are used to melt throughout the night to get a large volume of melt during daytime with back charging as metal is tapped out during the day. These types of melting and holding furnaces may have bath capacity upwards of 40 tons and the size is dictated by the need for liquid melt for a single pour for large castings. Power input in channel furnaces is not as high as coreless

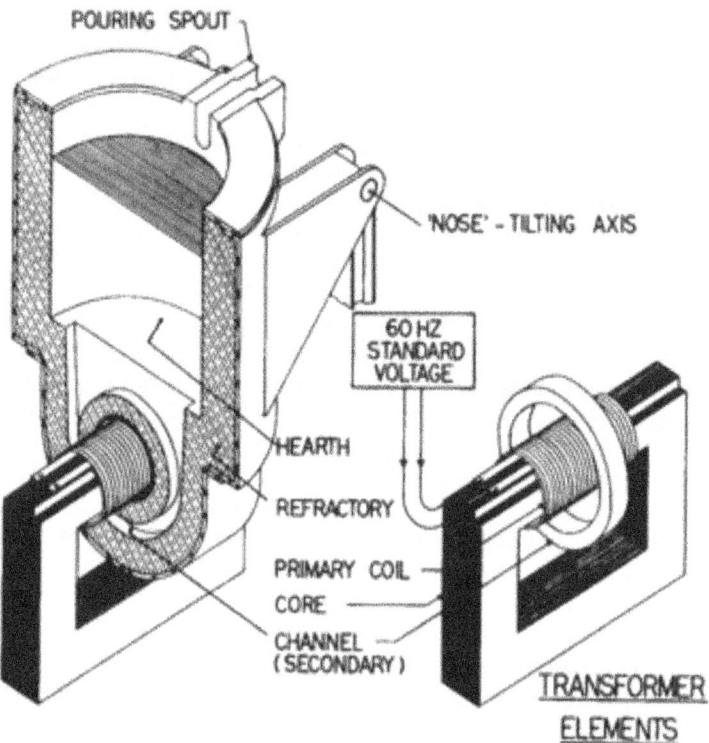

POURING SPOUT

'NOSE' - TILTING AXIS

60 HZ
STANDARD
VOLTAGE

HEARTH

REFRACTORY

PRIMARY COIL

CORE

CHANNEL
(SECONDARY)

TRANSFORMER
ELEMENTS

FIGURE 11.13 Cross-sectional view of a channel induction furnace (single loop coil). (From AFS [8].)

furnaces. Channel furnaces can be of single channel or twin channels. Heat is generated in the loops of metal, which circulates into the bath dispersing heat into the bath of liquid metal. Channel furnaces are efficient heating units as heat is generated entirely inside the metal bath with minimal losses.

Prior to the start of melting, liquid iron is poured into the furnace, the loop or channel is filled with liquid metal, and power is turned on to superheat the liquid. Due to the fact that liquid iron is necessary to start the furnace, channel furnaces are operated continuously without completely emptying the furnace. Operation times may extend from six months to more than one year. Channel furnaces are increasingly used as pressure-pour furnaces for gray iron and ductile iron production in high volume foundries.

The inductors where the electrical coils are placed are lined with alumina or basic lining, either cast or dry rammed. These linings may last from six months to more than two years depending on the operation. The body of the channel furnaces may be lined with refractory bricks or cast or rammed lining with high-alumina refractory for longer life. Continual operation of the furnaces may result in buildup of hard encrustations of slag in the entrance of inductor loops and metal intakes and outlet siphons, especially if the charge materials are dirty or transferred liquid iron contains certain types of inclusions.

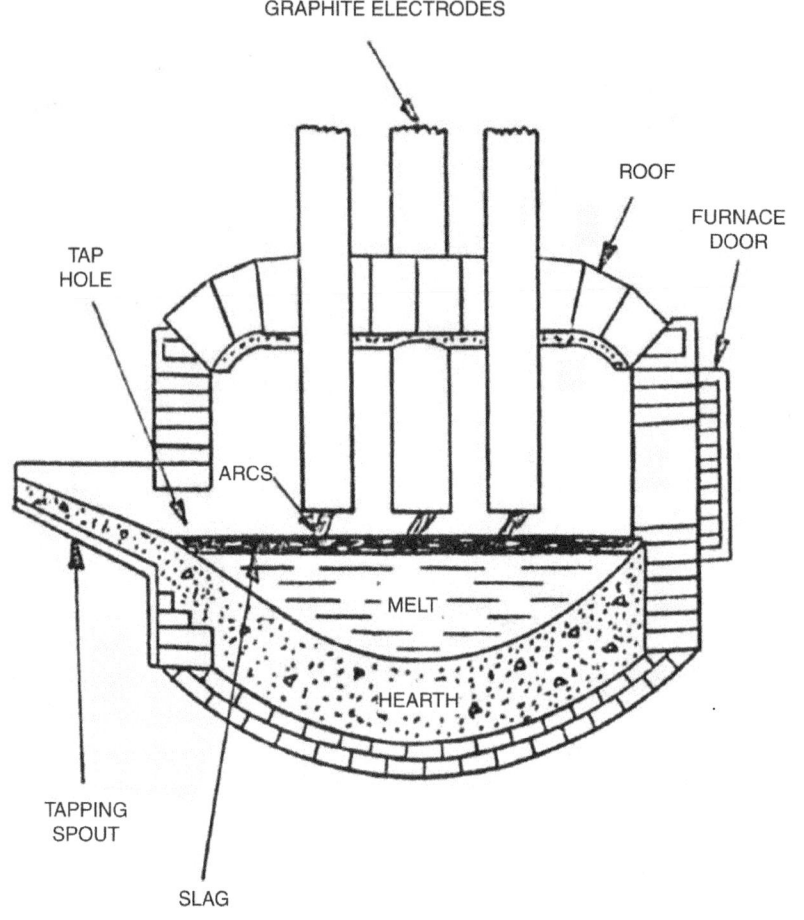

GRAPHITE ELECTRODES

ROOF

FURNACE
DOOR

TAP
HOLE

ARCS

MELT

HEARTH

TAPPING
SPOUT

SLAG

FIGURE 11.14   Cross-sectional view of a direct arc furnace. (From AFS [8].)

Electric Arc Furnace

It is not popular to melt iron in arc furnaces. Cross-sectional view of a direct arc furnace is shown in Fig. 11.14. But there are some advantages to melting in an arc furnace, such as low power consumption and oxidation of elements like Cr and Mn, if low residual iron is needed. Iron melted in arc furnaces needs to be conditioned in either duplexing furnaces or other induction furnaces with respect to chemistry, oxidation level, etc. It may also be necessary to have a holding furnace with arc melting due to batch melting (once per hour). A cross-sectional view of a direct arc furnace is shown in Fig. 11.14.

Principles of Cast Iron Melting

Many of the principles of chemical composition changes during the melting of cast irons can be delineated. Although much of the high-temperature chemistry of molten metal, slags, and temperature and atmosphere is known, the circumstances of cupola, air furnace, arc furnace, or induction melting require special interpretation of the principles.

Types of Chemical Reactions

Two principal types of chemical reactions, oxidizing and reducing, are encountered during melting of cast irons. Some typical ones are given below:

1 *Oxidizing reactions*

    a. $\underline{C} + O_2\,(g) = CO_2\,(g)$ (11.1)

    b. $2\underline{C} + O_2\,(g) = 2CO\,(g)$ (11.2)

    c. $\underline{Si} + 2\underline{O} = SiO_2\,(s)$ (11.3)

    d. $\underline{Si} + xFeO\,(\text{slag, solid}) = yFeO{\cdot}SiO_2\,(\text{slag}) + 2Fe$ (11.4)

    e. $\underline{Mn} + FeO\,(\text{slag, solid}) = MnO\,(l) + Fe$ (11.5)

2 *Reducing reactions*

    a. $SiO_2\,(\text{solid, refractory, slag}) + 2\underline{C} = \underline{Si} + 2CO\,(g)$ (11.6)

    b. $MnO\,(\text{liquid, slag}) + C = \underline{Mn} + CO\,(g)$ (11.7)

    c. $Al_2O_3\,(\text{solid}) + 3\underline{C} = 2\underline{Al} + 3CO\,(g)$ (11.8)

When the chemical symbol is underlined in the above equations, it refers to the element dissolved in the molten iron. All the reactions mentioned above are influenced by temperature and concentration.

Effect of Temperature

Temperatures encountered in cast iron melting extend from room temperature to about 3500°F (1930°C). Marked changes in chemical reactions occur over this temperature range. Oxidation reactions, especially involving carbon, are usually considered to progress more rapidly with increasing temperature. Some of the elements involved in cast iron melting and the free energy of formation of their oxides against temperature are shown in Fig. 11.15.

Higher negative free-energy change implies greater spontaneity of reaction. On the graph, a line with negative slope indicates decreasing reaction tendency with increasing temperature. Thus, the tendency of oxidation of silicon and manganese by oxygen decreases with increasing temperature. Carbon, on the other hand, oxidizes more readily with increasing temperature. Furthermore, reduction of oxides of silicon and manganese by carbon occurs more readily with increase in temperature.

As a result of these relationships, silicon and manganese are lost primarily at low temperatures during melting, less than 2600°F (1430°C), and carbon is lost at higher temperatures. A gain in silicon, in the iron occurs at high temperature because of the ability of carbon to reduce silica.

These composition trends may be most readily observed in induction-furnace melting of cast irons lined with silica refractory. Figure 11.16 illustrates composition changes in molten cast iron at various temperatures. The pronounced effect of carbon loss and silicon pickup is evident in Fig. 11.16. The influence of temperature on the formation of oxides may be readily observed by observing the melt surface during melting. At high temperatures where carbon is capable of reducing oxides, no slag scum forms and the melt surface is clear. As the iron cools, carbon loses its reducing power and a slag scum

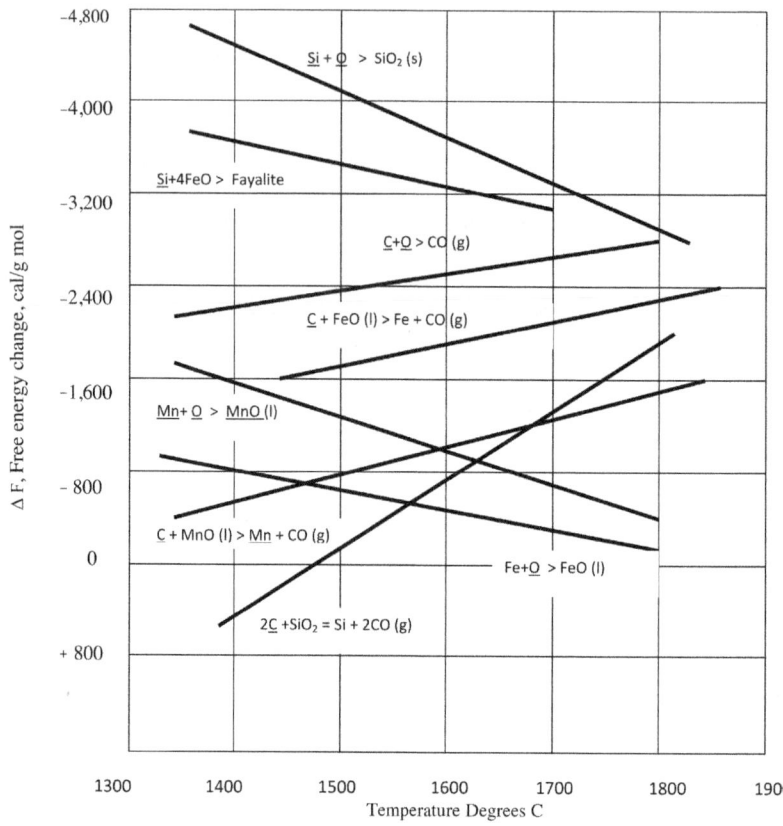

FIGURE 11.15 Approximate free energy changes of some iron-melting reactions as a function of temperature. Symbols underlined are elements dissolved in iron at standard concentrations of 1%. Negative values indicate that spontaneous reaction is likely to occur. (From Heine et al. [1].)

will form and cover the surface. Oxide formation proceeds readily, as evidenced by the appearance of the slag, which is an iron silicate. The aforementioned facts readily explain the nature of the curves in Fig. 11.16.

Effect of Concentration

The reactions during melting are not only influenced by temperature but they are also concentration dependent. When considering composition, other factors such as type of refractory—basic or acidic—slag composition, and gas atmosphere become especially important. In typical iron melting, acid refractories are used which are in contact with acid slags. At high temperatures, silica reduction becomes important. The concentration of silica and of silicon and carbon in the iron may be related by the chemical equilibrium constant (K) as follows:

$$SiO_2(s) + 2\underline{C} = \underline{Si} + 2CO \ (g) \tag{11.9}$$

$$K = \underline{Si}*(CO)^2/SiO_2 \ (s)*\underline{C}^2 \tag{11.10}$$

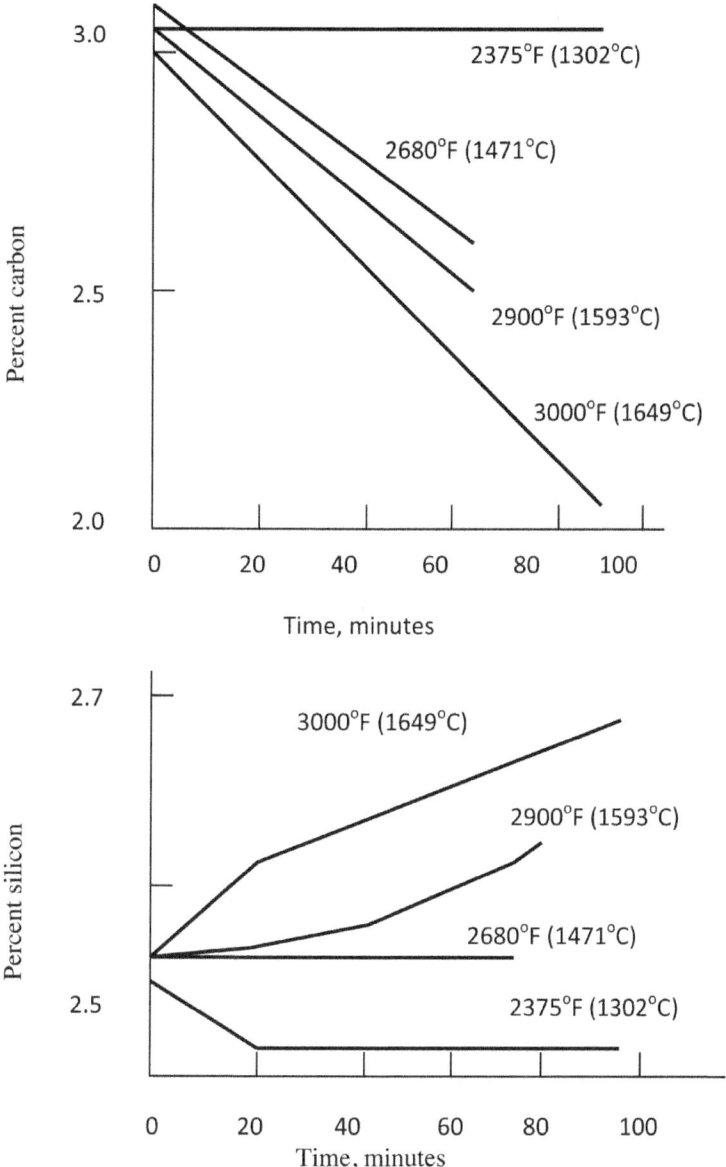

Figure 11.16 Curves showing changes of carbon and silicon percentages in molten cast iron held at that temperature indicated in a silica crucible in contact with air in an induction-melting furnace. (From Heine et al. [1].)

Using the equilibrium constant, equilibrium-concentration curves may be plotted for various temperatures, shown in Fig. 11.17. Certain limitations inherent in thermodynamic calculations apply to Fig. 11.17. However, the graph shows the directions of composition changes in molten irons at various temperatures. An iron of 3.5% C and 2.3% Si at 2372°F (1300°C) lies to the right of the concentration curve for that temperature. It therefore contains an excess of silicon, and would not pick up silicon from the refractory

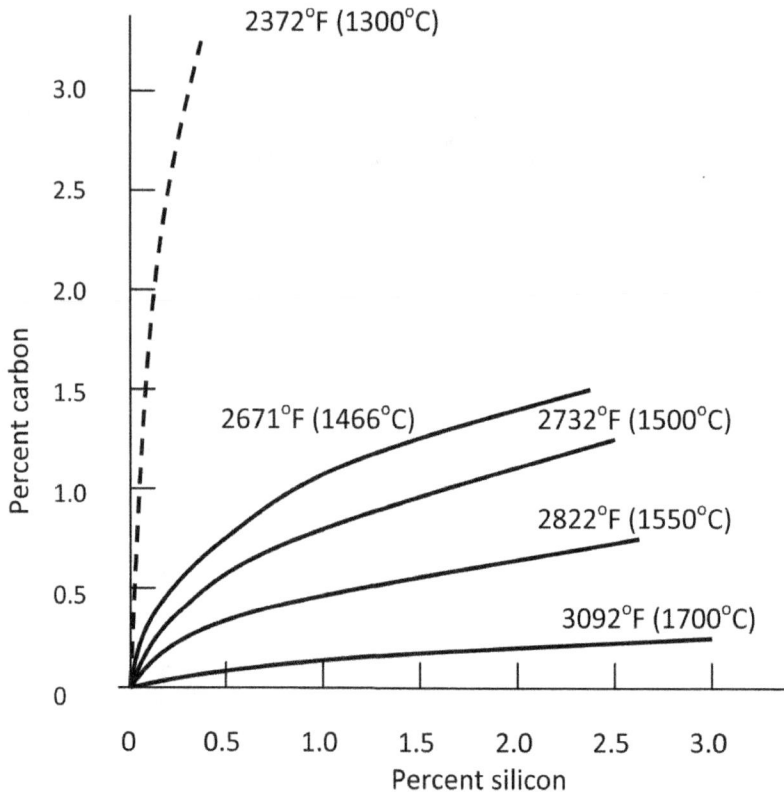

Figure 11.17   Calculated equilibrium concentration of percentage carbon and silicon for $SiO_2(s) +$ $2C = Si + 2CO(g)$ in molten iron-carbon-silicon alloys contained in a silica crucible under 1 atmospheric pressure of the CO. Solid curves indicate temperatures at which silica reduction will occur spontaneously if an excess of carbon is present. (From Heine et al. [1].)

lining or slag. For the same iron at 2822°F (1550°C), however, the composition lies to the left of the concentration curve and therefore contains an excess of carbon. Silica reduction will occur with the iron gaining silicon, if a sufficient supply of silica is available from the refractory or slag. These facts were illustrated in Fig. 11.16.

The temperature at which the change from oxidation of silicon to reduction of silica occurs may be calculated for standard concentrations. From "K" values given in Table 11.6, the temperature at which the reaction changes can be calculated as 2671°F (1466°C) for silica. Thus, melting temperatures span the range of the two types of reactions. At high temperatures, iron will pick up both silicon and manganese if their oxides are available in sufficient concentrations for reduction by carbon. With slag and refractory, there is adequate supply of silica but not of manganese oxide available for such reduction, and so only silicon pickup occurs. Concentration of carbon, silicon, manganese, and silica, and manganese oxides and temperature of the melting environment are primary factors in weight-composition changes.

Temperature		K value
°C	°F	
1400	2552	0.31
1466	2671	1.0
1500	2732	1.76
1600	2912	8.29

Source: From Heine et al. [1].

TABLE 11.6 The K Values Calculated Thermodynamically for Various Temperatures (11.1)

Effects of Iron Oxide

Oxidation may originate from many sources. Air or carbon dioxide may oxidize the molten iron. Iron oxide is also oxidizing, and it occurs in slags or as rust in the charge, or it may form by reaction with oxygen. The source of oxidation must be looked upon as a major factor in determining iron composition changes. Oxidation of silicon by iron differs from other sources of oxidation in that it is relatively insensitive to temperature [1]. Silicon losses caused by reactions with iron oxide occur as readily at high temperatures as they do at low temperatures. It has been proved that iron oxide in slag, as rust, or generated in any other way will cause silicon and manganese oxidation losses from cast iron melts even at high temperatures, where these losses would not normally occur because of protective action of carbon [1].

Melting Process

In any type of furnace, iron passes through a meltdown (liquification) stage at temperatures up to 2300°F (1260°C). During this period, furnace gases such as oxygen and carbon dioxide are important, whether in cupola, air furnace, or induction furnace, because they can form an iron oxide-rich skin on the iron. As the metal liquefies, silicon and manganese are oxidized by the aforementioned iron oxide. Rust initially present in the charge also contributes to this source of oxidation. Silicon losses from rusty and clean scrap during induction melting are shown in Table 11.7. The table also reveals that the oxidation loss during meltdown carries through the entire course of the heat.

Considering the temperature effects, one may state that the loss of silicon and manganese, about 0.1 to 0.30%, in induction and air furnace melting occurs during meltdown.

Material (sample during melting and heating)	Clean Scrap		Rusty Scrap	
	C, %	Si, %	C, %	Si, %
Melting stock	3.4	2.6	3.35	2.56
2375°F (1302°C)	3.28	2.42	3.20	2.38
2680°F (1471°C) 0 min	3.27	2.42	3.16	2.36
2900°F (1593°C) 0 min	3.21	2.47	2.94	2.30
2900°F (1593°C) 60 min	Pickup in 60 min = 0.13% Si		Pickup in 60 min = 0.13 % Si	

Note: Melted in induction furnace in silica crucible.

Source: From Heine et al. [1].

TABLE 11.7 Effect of Rust on Analysis of Melt

An added feature occurs in the cupola, since iron droplets must pass through an ascending gas stream and also a slag containing iron oxide. An oxidizing slag can cause silicon losses at high temperatures where silica reduction will occur simultaneously with the loss. The net change in analysis, a loss or gain, depends upon predominating reaction. Oxidizing conditions during meltdown thus favor increased silicon and manganese losses through an increased concentration of iron oxide in the melting environment.

During meltdown, the molten cast iron is covered with a saturated, solid, flaky iron silicate slag. At a certain temperature, the slag disappears due to the reducing power of carbon. For an iron containing 2.4% Si and 3.4% C, this temperature of slag-scum disappearance is about 2550°F (1399°C) in a CO_2 or air atmosphere. However, the temperature of slag disappearance varies with carbon and silicon concentrations, gas atmosphere, and refractory type. Appearance of slag is a significant fact: it means silicon is likely to be oxidized from the iron. The same slag also forms when molten irons are cooled from high to low temperatures.

High-Temperature Melting

Holding the molten iron at elevated temperatures, 2700 to 3200°F (1482 to 1760°C), is accompanied by well-defined changes. Molten cast irons decarburize rapidly at temperatures above 2550°F (1399°C) during induction melting when CO_2 or air is the furnace atmosphere. Decarburization is illustrated graphically in Fig. 11.16. No silicon or manganese losses will occur at temperatures 2700°F (1482°C) and above, unless iron oxide is introduced from some source [1]. Carbon dioxide, even at 100% concentration, will not cause silicon loss. Nor will raw air, unless it is directed at the melt surface at a high enough rate. Raw air will then react with molten iron to form iron oxide (FeO), and thus an iron silicate slag by reaction with the iron oxide. Silicon oxidation under these extreme conditions is accompanied by manganese loss in the ratio of about 10:4 and formation of slag cover on the molten-iron surface.

Rate of decarburization at high temperatures is mainly affected by the rate of change of atmosphere, if CO_2 or air exists over the melt surface. Oxidation by CO_2 is an endothermic reaction and is thus much milder than air oxidation, which is exothermic.

Silica reduction and silicon pickup by the iron are additional features of high-temperature melting when silica is available, as in acid slags and refractories. The gross pickup of silicon is of course counterbalanced by some oxidation when iron oxide is present. A high-iron-oxide slag may cause a net loss of silicon, whereas low-iron-oxide slags may cause no loss or even permit a gain. Highly reducing conditions, for example, the presence of coke, will cause the most rapid increase of silicon content of the iron at high temperature.

Melting Practice

Even when the concentration of elements and temperature favor reduction reactions reducing oxide content, kinetics and time factors are also important to determine the quality of melt during tapping of the heat. Apart from the oxides present in the charge materials, oxides can form during melting depending on the violence of movement of melt and reactions with moisture and volatiles present in other materials added to the melt. Molding sand, even if it seems dry, will contain moisture that is combined with bentonites to oxidize the iron. These oxides will be reduced by carbon and silicon if given an adequate amount of time. In high-production foundries, furnaces will be tapped as soon as it reaches the tap temperature, thus the iron may not be free of iron oxide slag particles suspended in the melt. These iron oxide particles may persist through the melt treatment and handling process. Reduction reactions will continue to take place throughout the casting process. As a result of these reactions, gas defects such as pinholes and dross defects may form in the castings. To achieve the highest quality level in the castings it is important to maintain low levels of oxides in the melt.

Principles of changes in chemical composition during melting of cast irons have been described with reference to C, Si, and Mn. Oxidation-reduction-type reactions also apply in the case of aluminum, titanium, and other elements. These elements are readily oxidized during most melting operations. It is known, however, that carbon in cast iron can reduce aluminum oxides at high temperatures and result in aluminum residual in the iron. Both aluminum and titanium may carry through the melting process from scrap materials because of the protective action of carbon. The principles of reactions, temperature, and concentration effect are applicable to many of the elements, although much remains to be learned in the case of less common elements.

Solidification of Cast Irons

Cast iron solidification follows in one of two modes, either as a metastable iron-iron carbide system or in a stable iron-graphite system. General principles of solidification have been presented in Chapter 4. In the stable system, iron can solidify with graphite in various forms such as flake, spheroidal, vermicular, and variations of these forms. To understand the formation of these microstructures, the iron-carbon diagrams for 0% and 2% silicon shown in Fig. 11.18 are helpful.

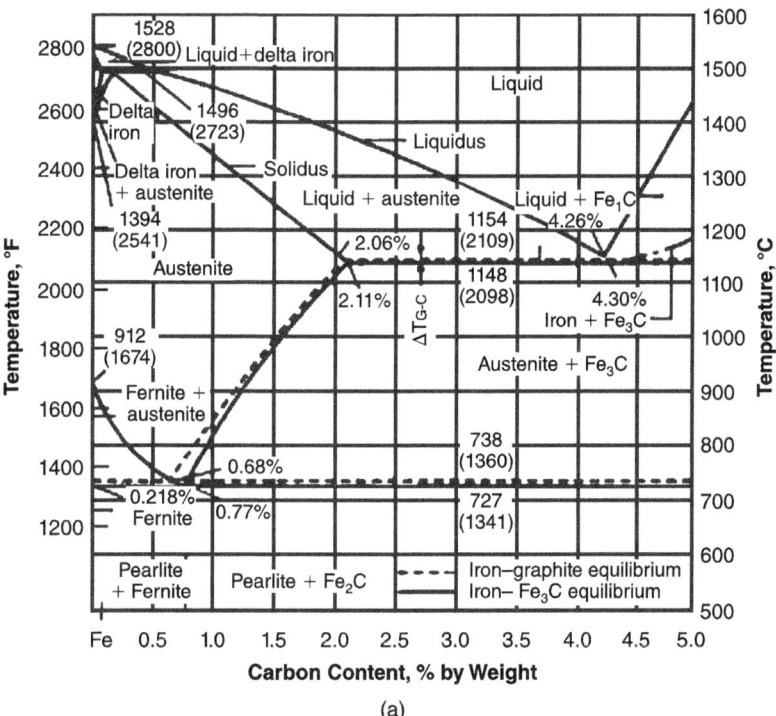

(a)

Figure 11.18   Fe-C phase diagrams (a) at 0% Si both stable and metastable phases (*Continued*)

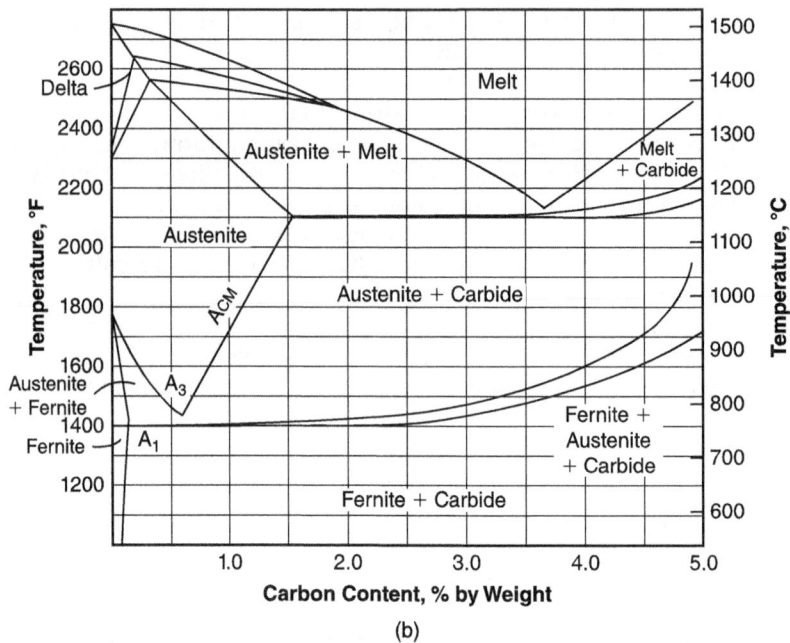

FIGURE **11.18** (*Continued*) (*b*) at 2.0% silicon only stable phase is shown. (From Stefanescu [10].)

White Cast Irons

In the iron-iron carbide phase system, carbon in the alloys occurs as the metastable compound iron carbide. Freezing of a hypoeutectic alloy, less than 4.3% carbon, with no silicon, begins with the formation of austenite dendrites and is completed by the solidification of the eutectic austenite-iron carbide (ledeburite). After solidification, cooling of the iron below the critical temperature results in the transformation of austenite to pearlite. The room temperature microstructure resulting from the aforementioned process is given in Fig. 11.2b, the primary proeutectic austenite has transformed into pearlite and austenite-carbide eutectic as a pearlite-carbide mixture. A eutectic Fe-C alloy containing 4.3% C should freeze without the formation of primary austenite dendrites and should consist only of the eutectic-type structure in the metastable system, and is shown in Fig. 11.19. Heating and cooling of alloys with the microstructure shown here are accompanied by changes predictable from the iron-iron carbide phase diagram. Under such circumstances, the iron carbide phase behaves as a stable phase although it is known as metastable. A *metastable phase* is one which behaves as though it was stable but which is actually unstable. Iron carbide becomes unstable when it is in contact with graphite at elevated temperatures. Prolonged exposure to high temperatures or the presence of certain elements in the alloy may cause the formation of graphite nuclei and thus promote the change from the metastable iron carbide to the stable graphite phase. Conversely, rapid cooling and certain elements such as tellurium in the alloy tend to prevent nucleation of graphite and thus cause the metastable carbide phase to persist. In a particular case, the tendency to approach the complete equilibrium represented by the iron-graphite phase system results from the balance reached by the effects of all the factors that promote graphitization and those that oppose graphitization.

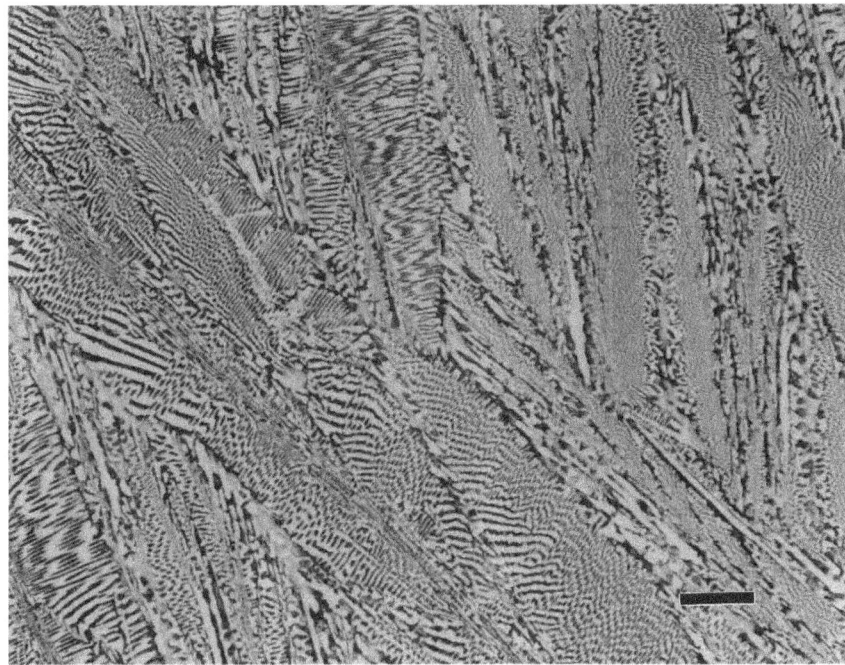

Figure 11.19   Ledeburite in white cast iron–eutectic of austenite and cementite, 4% C, 0.3% Si, 0.11% Mn, 0.96% Cr, magnification bar: 100 μm. (From Vander Voort [3].)

Malleabilization

Simple white irons are heat treated to break down the carbides to produce a temper carbon. This operation converts the brittle carbidic castings to malleable castings. This heat treatment involves two steps. Stage 1 heat treatment, also called annealing, dissolves the carbides into austenite and at the end of this step only austenite and graphite nodules are present. The temperature at which the annealing is carried out as well as the chemistry of the iron and nucleating agents present affect the time for full anneal. Higher temperature shortens the time. The range of temperature used for first stage heat treatment is 1650 to 1780°F (900 to 970°C). Even as the time is shortened at higher temperatures other factors such as casting distortion and oxidation at the surface should be considered.

Second stage heat treatment results in converting austenite to ferrite, pearlite, or even martensite. Once the castings are fully annealed, they are cooled down to just above the upper critical temperature 1360 to 1400°F (740 to 760°C). From this temperature cooling rate affects the final microstructure. Slow cooling at the rate of 5 to 20°F (3 to 11°C) per hour transforms the austenite into fully ferritic matrix. Carbon dissolved in the austenite is deposited onto the existing temper carbon particles.

If the castings are cooled fast from the upper critical temperature, the final matrix will be of pearlite and temper carbon. Castings may be cooled slowly from the annealing temperature to upper critical temperature to reduce the dissolved carbon in austenite prior to pearlite reaction. Pearlitic castings may be reheat-treated to get martensite or tempered martensite.

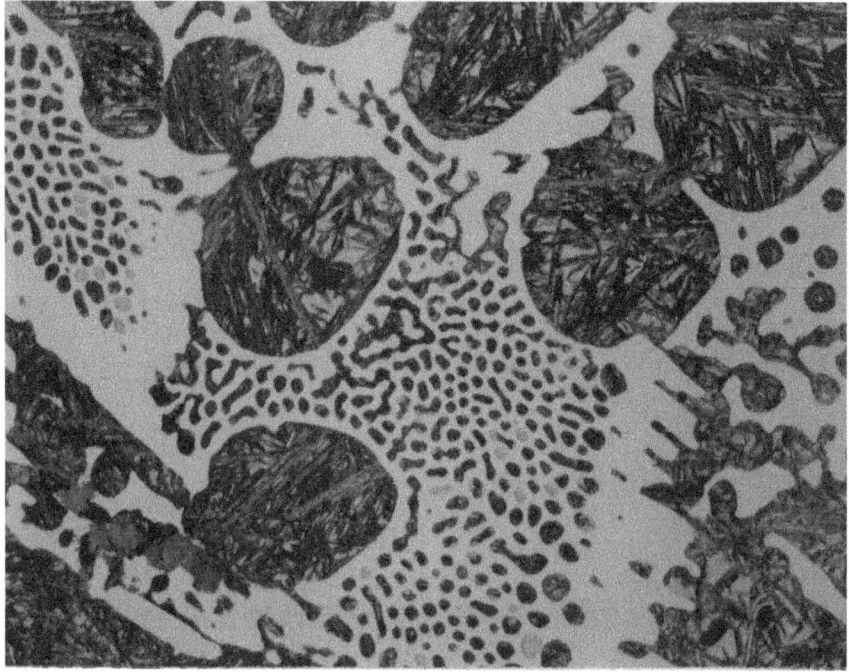

Figure 11.20 Microstructure of Ni-hard cast iron (Fe: 3.3%, C: 0.9%, Mn: 0.9%, Si: 1.8%, Cr: 4.4%, Ni: 0.4%, Mo: 0.4%) revealing massive cementite, ledeburite, and patches of plate martensite and retained austenite. Etched with aqueous 10% $Na_2S_2O_5$. Original at 500×. Ledeburite is a eutectic of cementite and austenite where, with cooling, the austenite transforms to ferrite and cementite in the form of pearlite. Some graphite is also present. (From Vander Voort [3].)

To stabilize carbides, alloying elements like chromium, manganese, and molybdenum are added to the melt at various concentrations listed later in this chapter. One of the popular commercial white irons produced for wear-resistant applications is "Ni-hard" and its microstructure is shown in Fig. 11.20. Ni-hard is white iron alloyed with nickel and chrome for wear applications such as mill liners.

Graphitic Cast Irons

Graphite Precipitation during Solidification

When cast irons solidify in stable eutectic mode, excess carbon is precipitated in the form of graphite, which can take different shapes. Main classifications for graphite shapes are (1) flake graphite, (2) spheroidal graphite, and (3) compacted or vermicular graphite. Some cases of coral graphite are seen as an anomaly. Various graphite forms and their orientation and growth axis are shown in Fig 11.21.

The shape of graphite affects the strength significantly and cast irons are classified according to the shape. Flake graphite irons are known as gray cast irons; spheroidal graphite irons are known as ductile irons; and vermicular graphite irons are called compacted graphite irons, or CG irons for short. Graphite shapes are determined during solidification and are not altered during subsequent cooling and/or heat treatments.

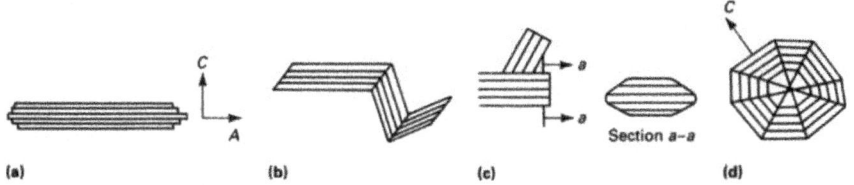

Figure 11.21 Schematic of various graphite shapes growing with austenite during eutectic solidification: (a) flake graphite, (b) vermicular graphite, (c) coral graphite, and (d) spheroidal graphite. (From Stefanescu [10].)

Flake Graphite

Flake graphite size and distribution can vary widely depending on the carbon equivalent, cooling rate, and processing variables and thus affect the mechanical properties greatly. Flake graphite irons should be considered as a family with tensile strengths ranging from 20 to 60 Ksi (138 to 413 MPa). Different classes of flake graphite are shown in Fig. 11.4. Type A through type E, classified by AFS and ASTM, show the distribution and shape of the graphite flakes [5].

From the evidence presented from the research work [11], using extracted graphite particles by deep etching and examining under SEM and TEM, flake graphite growth patterns can be seen clearly in Fig. 11.22.

Type A has randomly oriented flakes, which are desirable in gray cast irons. Even in type A, the size of graphite varies with carbon equivalent and cooling rate. Long flakes signify higher carbon equivalent and low cell count or larger cell size. When eutectic cells are large, the flakes are longer. It is not easy to distinguish the eutectic cells with type A graphite morphology without special etching or some other techniques such as direct austempering. If carbide stabilizing elements like Cr, Mo, and V are present they will form grain boundary carbides and macro etching can reveal the eutectic cells as shown in Fig. 11.23. Higher cell count, either from inoculation or from faster rate of solidification, decreases the length of the graphite flake. Classification of graphite sizes from 1 through 8 (longest to shortest) is provided in ASTM standards. Graphite flake length has direct correlation to tensile strength. When effective nucleation occurs, austenite-graphite eutectic grows in a cellular manner. Cellular growth continues until cells impinge on one another, or until the composition of the melt changes due to segregation of certain elements (Cr, Mn, Mo, V, P) so that other phases form in the intercellular areas.

Failure to nucleate austenite-graphite eutectic early in eutectic solidification results in eutectic undercooling, and eutectic nucleation at a lower temperature. At this lower temperature diffusion distances between graphite flakes are reduced such that flakes at the center of the eutectic cell are finer and the interflake distance is reduced. Then the heat of fusion raises the melt temperature nearby (recalescence) and continued eutectic growth occurs with coarser flakes and developing type B graphite (rosette graphite).

Type C graphite precipitates from hypereutectic composition surrounded entirely by liquid, so that it grows as a crystalline, blocky morphology prior to the eutectic solidification.

Increased volume fraction of austenite in hypoeutectic irons and faster cooling rates decreases the interdendritic volume of liquid. Development of this proeutectic

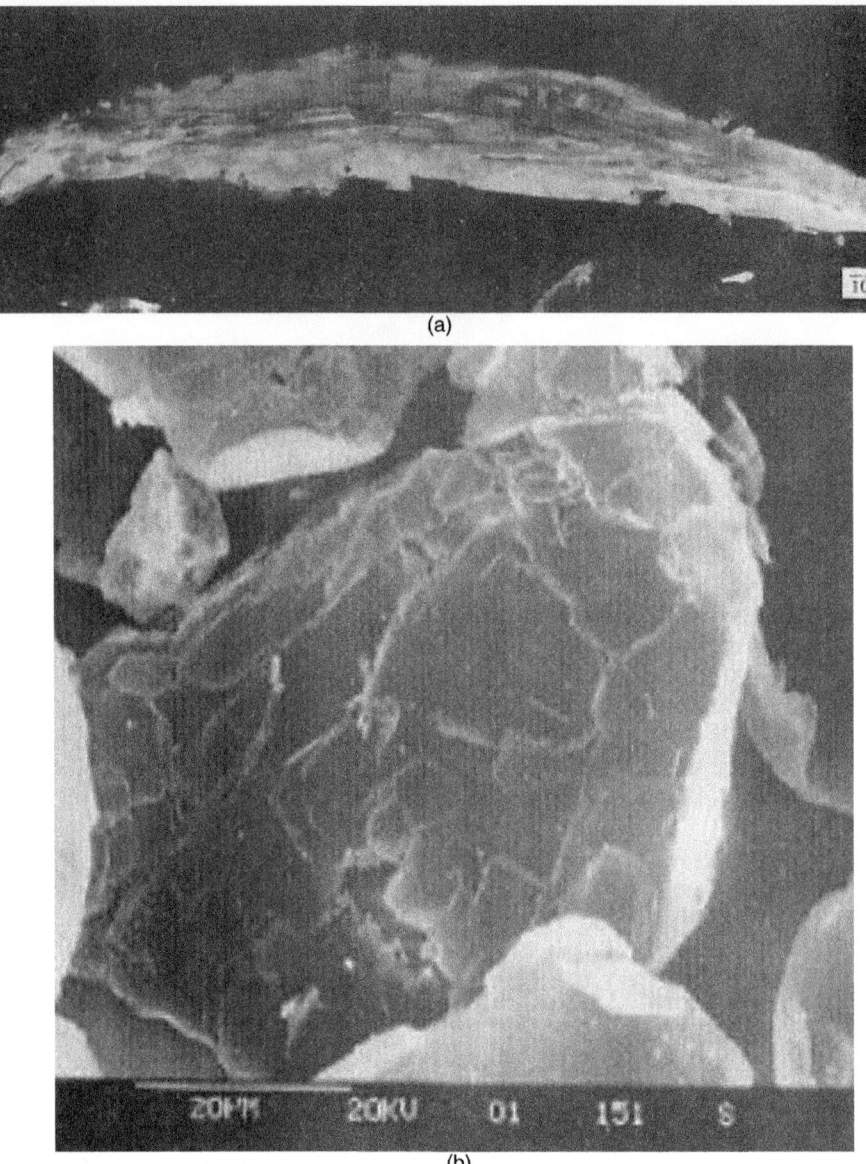

Figure 11.22 SEM micrograph of (a) flake graphite removed by deep etching and (b) details of the surface structure of the flake graphite. (From Loper [11].)

austenite is associated with delayed eutectic nucleation. Eutectic graphite flakes tend to be finer (similar to the core of the type B graphite) and grow in a cellular manner similar to type A graphite. Thus, the undercooled graphite is classified as type D graphite and has randomly oriented graphite flakes.

Type E graphite is an extreme form of type D graphite where the interdendritic areas are severely restricted by the development of proeutectic austenite dendrites. This

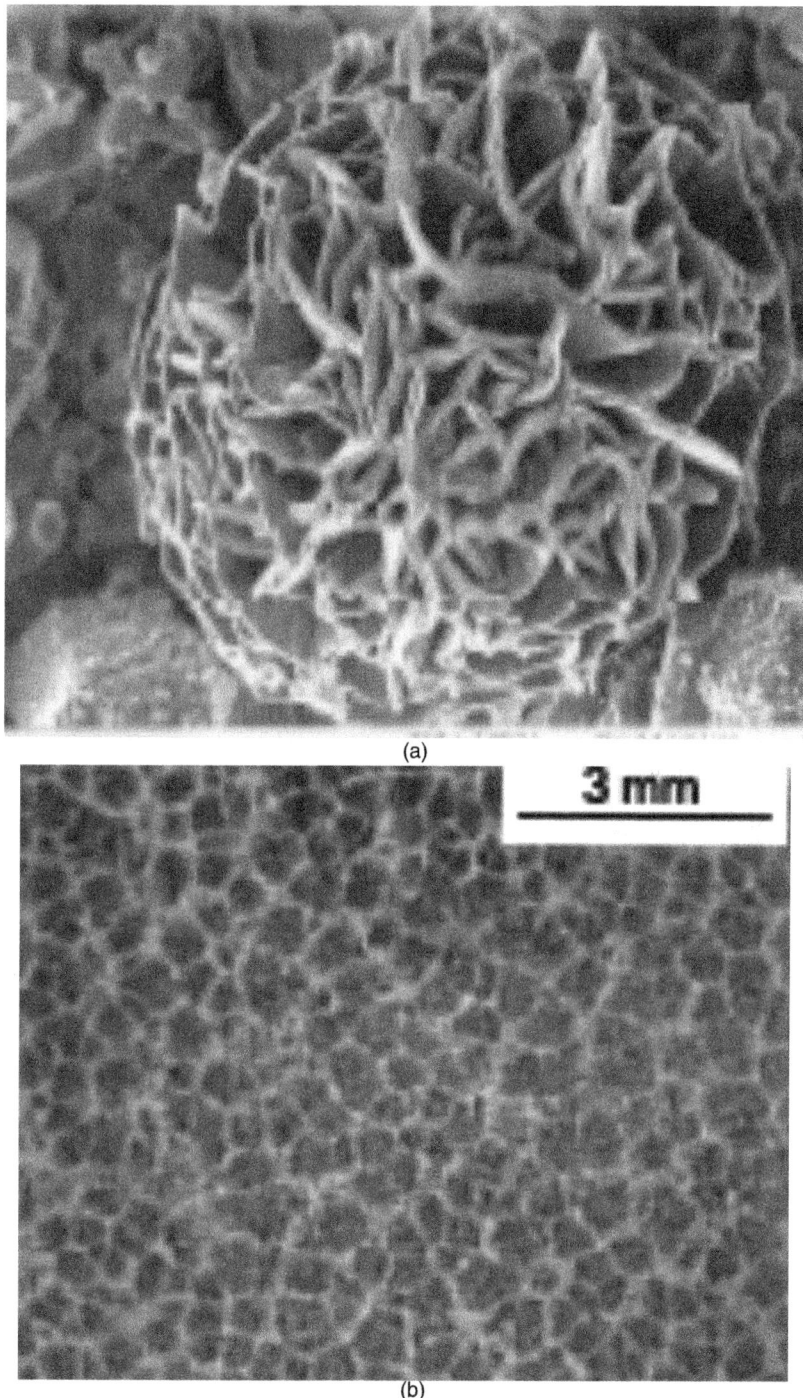

FIGURE **11.23** (*a*) One deep-etched single cell [12], (*b*) eutectic cells revealed by etching with Steads reagent [Alagarsamy Consulting], (*Continued*)

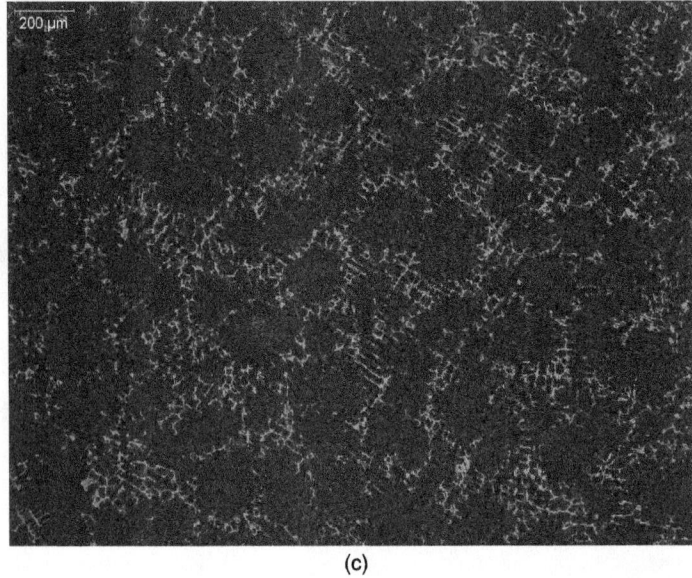

(c)

FIGURE 11.23 *(Continued)* *(c)* macro-etched to reveal grain boundary carbides. (From Gundlach, private communication [13].)

results in a more divorced eutectic formation so that the resulting graphite flakes are oriented within the austenite dendrites. Figure 11.24 shows type A through E graphites, discussed above. Anomalous graphite morphology known as "Widmanstatten" graphite is also shown in Fig. 11.24 [5].

Spheroidal or Nodular Graphite (SG) Cast Irons

Graphite shape is dramatically changed from flake graphite when sulfur and oxygen in the cast iron melt is tied up by shape-active elements like Mg, rare earths, and Ca. After the sulfur and oxygen are tied up as sulfides and oxides, a small amount of excess Mg will cause the graphite to precipitate in nodular (spheroidal) form. It may not be enough to have sufficient Mg to promote fully nodular graphite without some form of postinoculation to provide nuclei for graphite precipitation. Without the postinoculation, iron may solidify with carbides due to lack of graphite nucleation and growth resulting in the eutectic solidifying at lower temperature [below 1950°F (1066°C)]. In contrast to gray irons, most ductile iron castings are produced in the hypereutectic range. Within the narrow CE range, ductile iron strengths are not affected by small changes in carbon equivalent, as is the case with flake graphite irons.

Properties in ductile iron casting are mainly varied by controlling relative volumes of various components of the matrix such as ferrite, pearlite, and martensite by the use of alloying elements and to a lesser extent by the rate of cooling in the molds and outside the molds.

Spheroidal Graphite Development

Spheroidal graphite nucleates around inclusions introduced during melt conversion and inoculation steps in contact with liquid. As most ductile irons are of hypereutectic composition graphite grows in contact with liquid prior to encapsulation by austenite

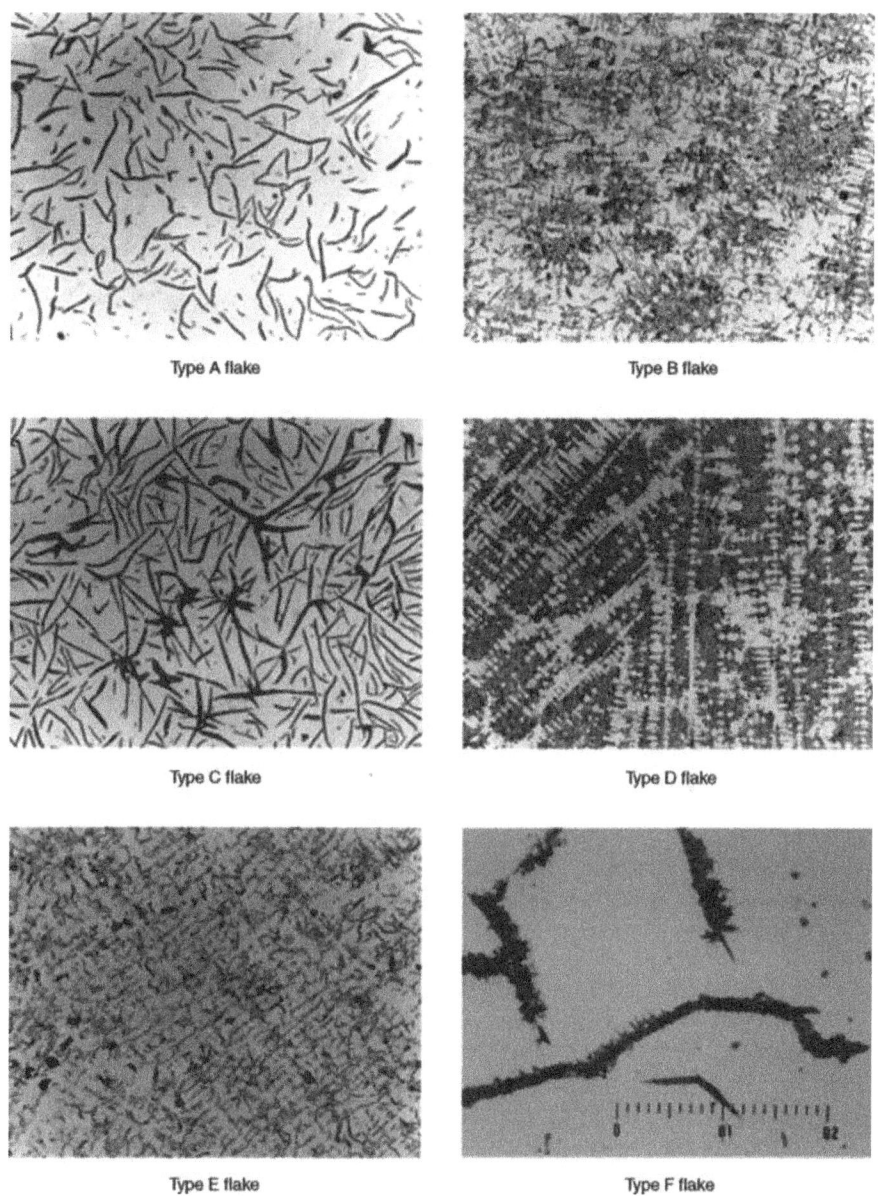

Type A flake

Type B flake

Type C flake

Type D flake

Type E flake

Type F flake

FIGURE 11.24   Flake graphite forms A through E and Widmanstatten graphite F. (From Goodrich [5].)

shell. Further growth occurs as divorced eutectic, carbon diffusing through austenite shell, and depositing on to the nodule surface. Spheroidal graphite development can be seen clearly as growth in three stages, the nuclei dark spot in the center, growth of spheroid in contact with liquid as dense graphite in the center, and then growth by diffusion of carbon through austenite in Figs. 11.25*a* and *b*. Radial and circumferential growth details can also be seen in Fig. 11.25.

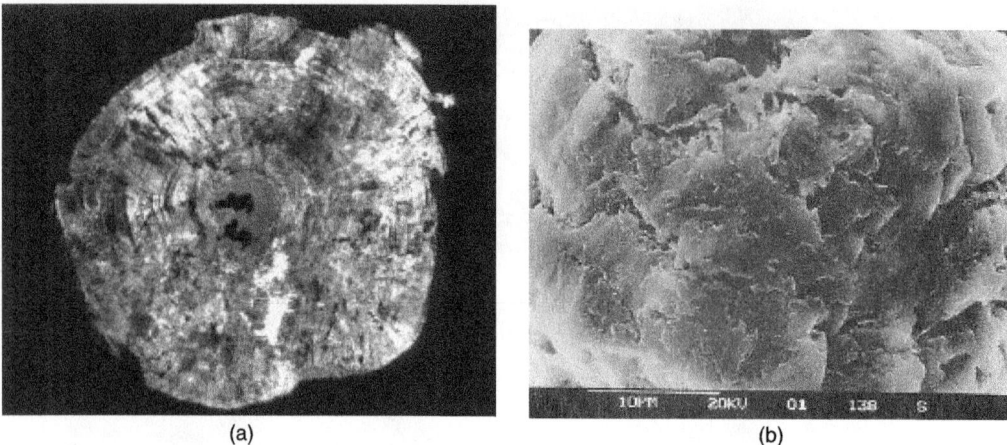

(a)   (b)

Figure 11.25   (a) Cross section of a spheroidal graphite removed by deep etching showing the radial and angular crystallographic planes, and (b) detail of the surface structure of the nodule. (From Loper and Fang [14].)

Vermicular Graphite

When the graphite shape is thicker than typical flake graphite and shorter in length as shown in Fig 11.2f, it is classified as vermicular graphite. Castings produced with this type of graphite are called *compacted graphite* (CG) cast irons. The aspect ratio (ratio of length of graphite particle to width) may vary between 3 and 8. This type of graphite results from treating the melt with magnesium and/or rare earths, and keeping the residual magnesium low (around 0.01 to 0.02%). This graphite morphology allows for better use of the matrix, yielding higher strength and ductility than flake graphite cast iron. Similarities between the solidification patterns of flake and compacted graphite iron explain the good castability of the later compared to ductile iron. In addition, the interconnected graphite provides better thermal conductivity and damping capacity than spheroidal graphite.

Vermicular graphite nucleates as spheroidal graphite but grows differently due to lower magnesium and/or higher sulfur, titanium, and oxygen levels. Figure 11.26 depicts the three-dimensional view of a deep-etched graphite particle.

Coral or Chunk Graphite

Even though this type of graphite is not desired, it can be seen in castings when processes are out of control. When excess rare earths are present, especially in heavier sections, chunk graphite can form. Growth of this type of graphite is shown in Fig. 11.27.

Factors Affecting Solidification of Cast Irons

Solidification of Gray Iron

In a hypereutectic gray iron, solidification begins with the precipitation of kish graphite in the melt. Kish grows as large, straight, undistorted flakes or as very thick, lumpy flakes that tend to rise to the surface of the melt because of their low relative density. When the temperature has been lowered sufficiently, the remaining liquid solidifies as a eutectic structure of austenite and graphite. Generally, eutectic graphite is finer than kish graphite.

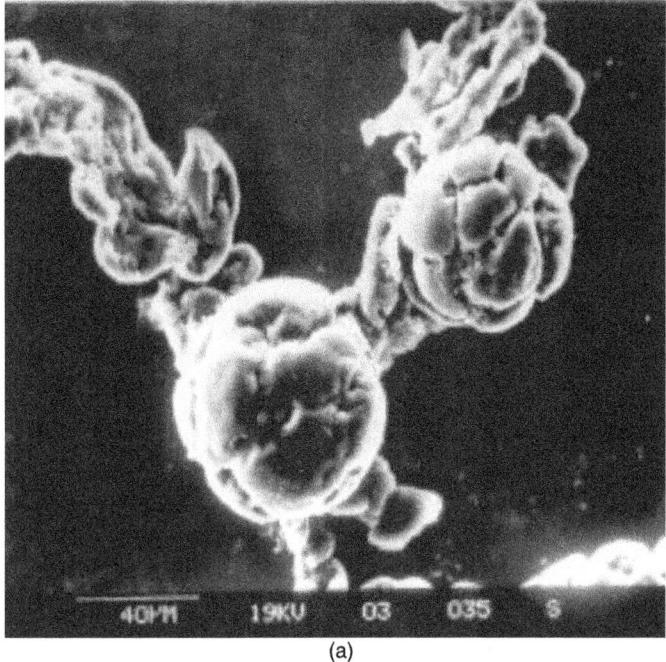

(a)

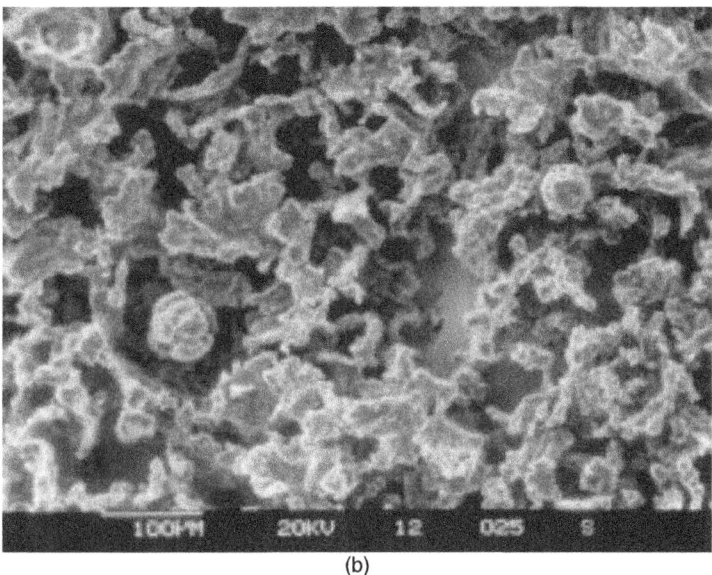

(b)

Figure 11.26 (a) Vermicular graphite extracted by deep etching showing thin stems and spheroid ends and (b) interconnected compacted graphite structure including the nodules. (From Loper and Fang [14].)

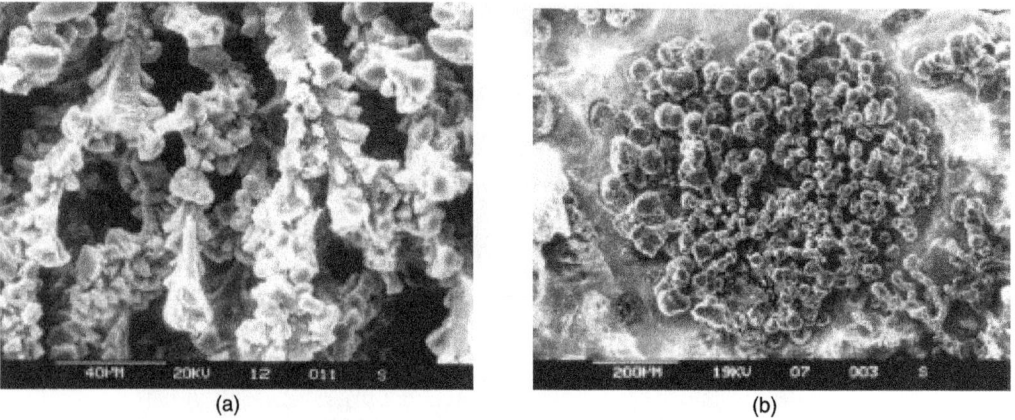

Figure 11.27   (a) Columns of coral graphite and (b) coral graphite cell interconnected with graphite columns, rounded ends. They are also referred to as chunk graphite. (From Loper and Fang [14].)

In hypoeutectic iron, solidification begins with the formation of proeutectic austenite dendrites. As the temperature falls, the dendrites grow, and the carbon content of the remaining liquid increases. When the increasing carbon content and decreasing temperature reach eutectic values, eutectic solidification begins. Eutectic growth from many different nuclei proceeds along crystallization fronts [10, 11]. During the eutectic solidification all three phases—austenite, graphite, and liquid, Fig. 11.28a—are in contact with each other until the end of solidification. This type of eutectic solidification is termed "coupled growth." Ultimately, the eutectic cells meet and consume the remaining liquid in the spaces between them. During eutectic solidification, the austenite in the eutectic becomes continuous with the dendritic proeutectic austenite, and the structure can be described as a dispersion of graphite flakes in austenite. Some irons with forward-segregating elements like Mn, Cr, and Mo, may solidify at the end with carbides, see Fig. 11.23c. After solidification, the eutectic cell structure and the proeutectic austenite dendrites cannot be distinguished easily, except by special etching or in strongly hypoeutectic iron.

With eutectic compositions, obviously, solidification takes place as the molten alloy is cooled through the normal eutectic temperature range, but without the prior formation of a proeutectic constituent. During the solidification process, the controlling factor remains the rate at which the solidification is proceeding (rate of heat extraction). The rapid solidification favored by thin section sizes or highly conductive molding media can result in undercooling (Fig. 11.29). Undercooling can cause the solidification to start at a temperature lower than the expected equilibrium eutectic temperature indicated by austenite graphite eutectic temperature for a given composition (Fig. 11.18a and b). This can result in a modification of the carbon form, from type A to B to D to type E, or can completely suppress its formation and form primary carbides instead.

Cooling curves associated with different types of flake graphite are shown in Fig. 11.30. Faster cooling, as is the case with thin sections, increases undercooling, which in turn may increase eutectic cell count (Fig. 11.31), decreasing flake length. Elements like silicon, aluminum, and titanium are strong graphitizers and they tend to reduce undercooling and their relative effects are shown in Fig. 11.32. They tend to decrease undercooling, whereas carbide-stabilizing elements may increase undercooling (Fig. 11.33) [15]. As the carbon equivalent (CE) which is approximately $C + 1/3$ Si increases, the tensile strength decreases (Fig. 11.34).

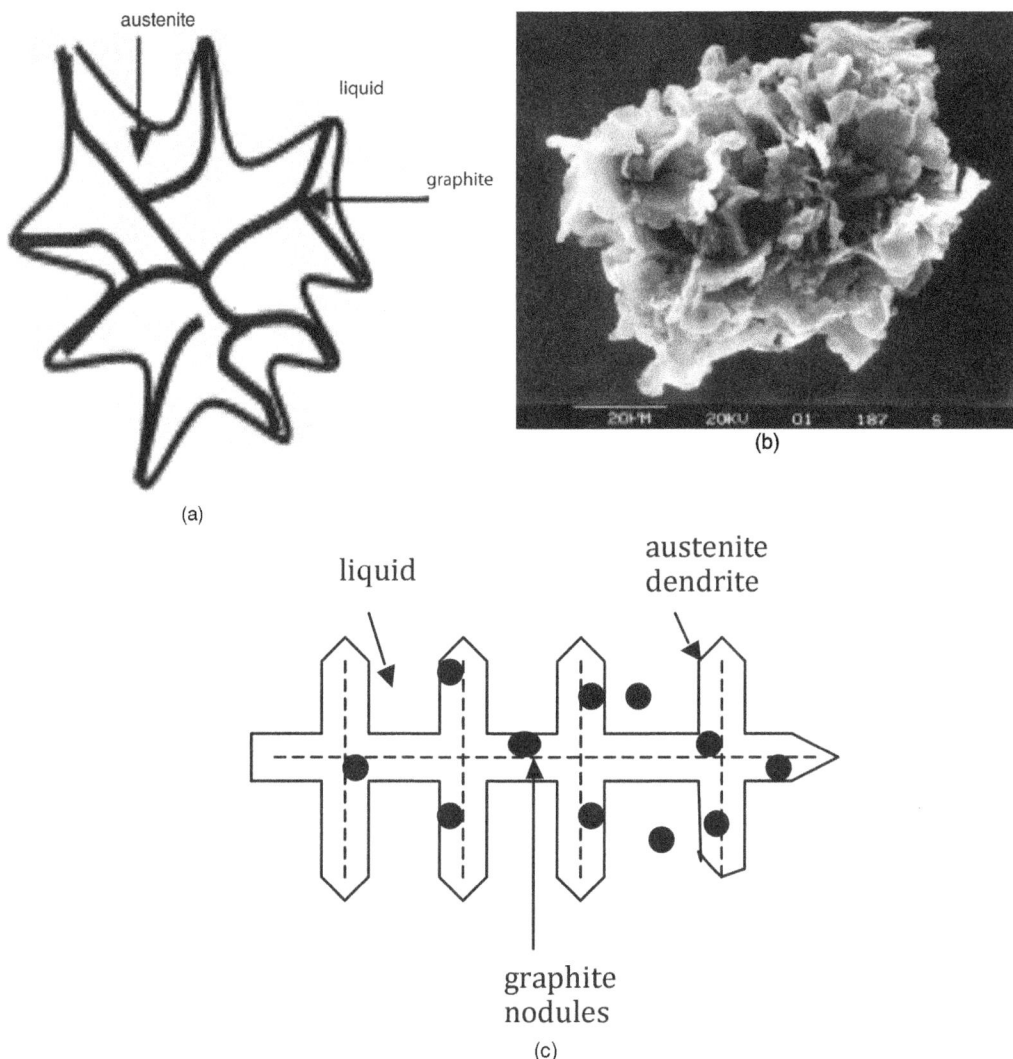

Figure 11.28 Differences in eutectic solidification of gray and ductile irons—coupled [11, 15] and divorced growth of graphite/austenite eutectic. (a) Schematic of flake graphite-austenite eutectic cell and (b) extracted cell by deep-etching technique. Gray iron type A eutectic cell growth—liquid is in contact with austenite and graphite—coupled growth. (c) Schematic of spheroidal graphite-austenite eutectic growth, multinodular model liquid is not in contact with graphite nodule for later part of the growth, divorced eutectic. (From Loper [11], Davis [15], Rivera et al. [16].)

Presence of silicon in the alloy is the single most important composition factor promoting graphitization in gray irons. The effect of silicon may be visualized with the aid of vertical sections of the ternary alloy system Fe-C-Si, as shown in Fig. 11.18*b* [9]. Consider the freezing process for an alloy Fe-C-Si with 2% Si and 3.5% C. Under equilibrium freezing conditions, primary austenite dendrites are formed in the temperature range from the liquidus curve (about 2150°F) (1177°C) to the curve indicating the beginning of eutectic freezing, at about 2100°F (1149°C). Simultaneous solidification of the eutectic

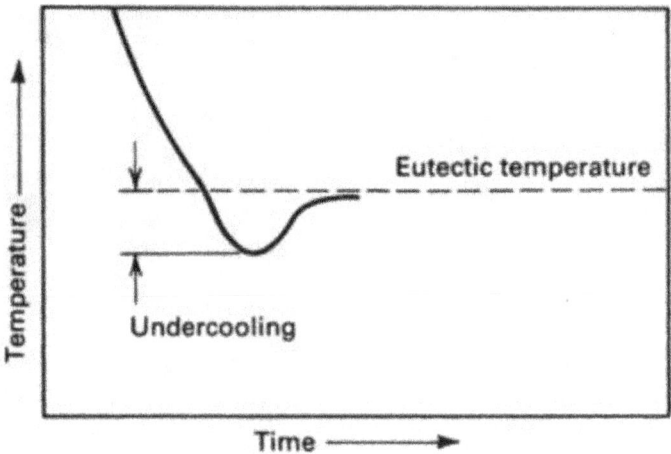

FIGURE 11.29 Cooling curve showing eutectic undercooling. Degree of undercooling depends on melt conditions and inoculation. (From Davis [15].)

austenite plus graphite completes the freezing process. The eutectic freezing occurs in a temperature range of about 2100°F (1149°C) to about 2060°F (1127°C). When the solidification is complete, the microstructure of the alloy under consideration consists of about 20% primary austenite and 80% austenite graphite eutectic. At the solidus temperature, the austenite is saturated with carbon of about 1.5%.

Ductile Iron Solidification

Many theories of spheroidal graphite nucleation and solidification models have been postulated since the discovery of ductile iron by many researchers around the world. Most of these are critically reviewed in the *Ductile Iron Handbook* [10]. No theory has universal support by the academic community and the industry. Extensive work done on the center graphite nuclei using transmission electron microscopy and scanning electron microscopy reveal the compounds found in the center to be complex oxides and sulfides of silicon and magnesium.

Ductile iron solidification is controlled by the precipitation of two phases, austenite and graphite, similar to that of flake graphite cast irons. Chemistry, particularly carbon and silicon, and inoculation affect the nucleation and growth of these phases and resulting mechanical properties.

The difference between gray iron (flake graphite) solidification and ductile iron is the contact between austenite, graphite, and liquid. In gray iron, during eutectic solidification all three phases are in contact (coupled growth). Austenite and graphite precipitate alongside each other as the solidification progresses, as shown in Fig. 11.28; until all the liquid is consumed, this mode is maintained. Even though the base chemistry is similar to the gray iron, ductile iron solidification of the eutectic is different. When the eutectic solidifies beginning with either hypoeutectic or hypereutectic irons, regardless of which phase is precipitated first, most researchers believe that the graphite spheroids are enveloped by austenite soon after it forms. Then the eutectic solidifies as divorced eutectic; graphite is not in contact with the liquid, only austenite is in contact with the liquid. Graphite grows inside the austenite as the solidification progresses.

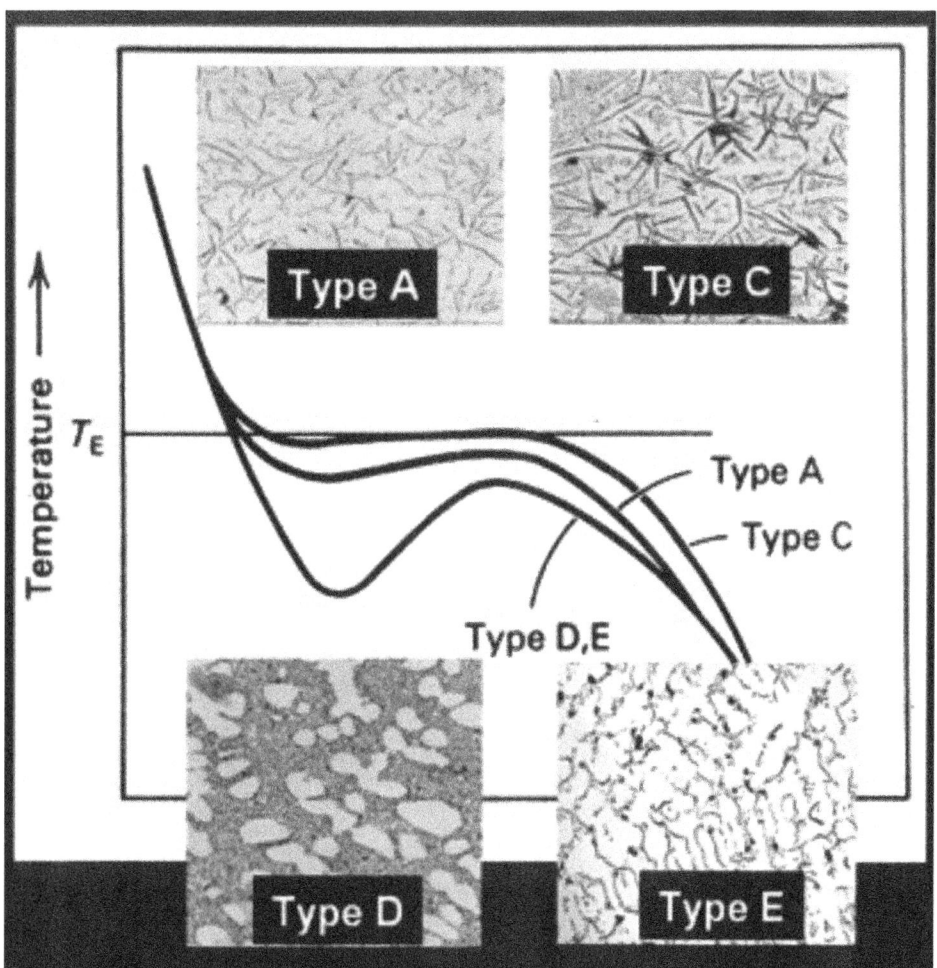

Figure 11.30   Schematic cooling curves of gray irons with different degree of undercooling resulting in different types of graphite flakes. (From Alagarsamy Consulting [21].)

When the chemistry is hypoeutectic, the first phase to precipitate is austenite. As the remaining liquid progresses toward eutectic chemistry, austenite may grow from the outer mold walls toward the center of the sections. These austenite grains will be columnar at the surface and equiaxed in the interior sections of the casting, revealed by direct austempering experiments [16]. Figure 11.35 shows the macrostructure of gray and ductile irons of hypoeutectic, eutectic, and hypereutectic compositions.

The austenite grain size is influenced by carbon equivalent [16], the size decreasing as the CE goes from hypo- to eutectic to hypereutectic. Studies on nucleation of graphite and austenite [17] show that the austenite is a poor nucleating agent of graphite, whereas graphite is a very good nucleating agent for austenite. These factors seem to agree, that in hypereutectic irons, the first phase to precipitate is graphite and then immediately austenite is nucleated and surrounds graphite and grows in a dendritic form.

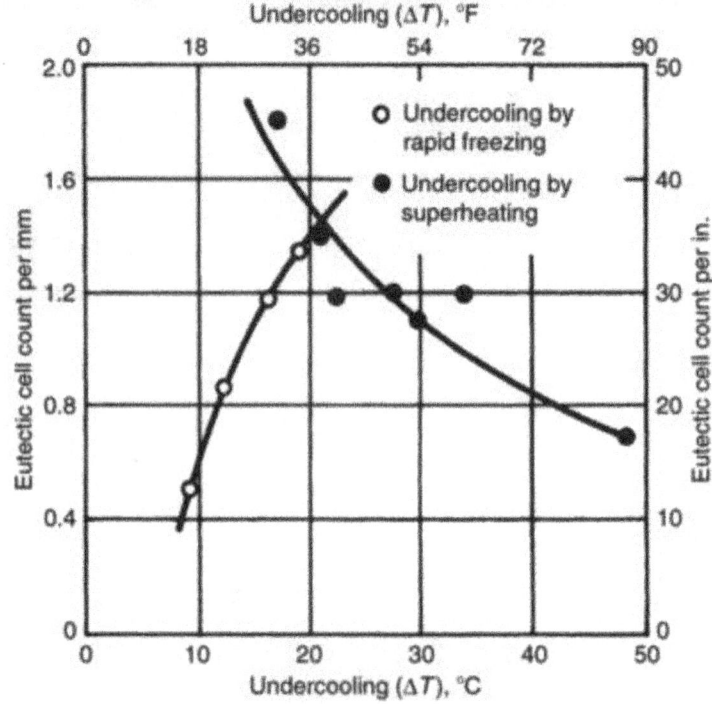

FIGURE 11.31 Degree of undercooling increases with faster cooling, increasing cell count (decreasing flake length). Superheating the melt delays nucleation of graphite, increasing undercooling, but the cell count decreases with increasing flake length. (From Davis [15].)

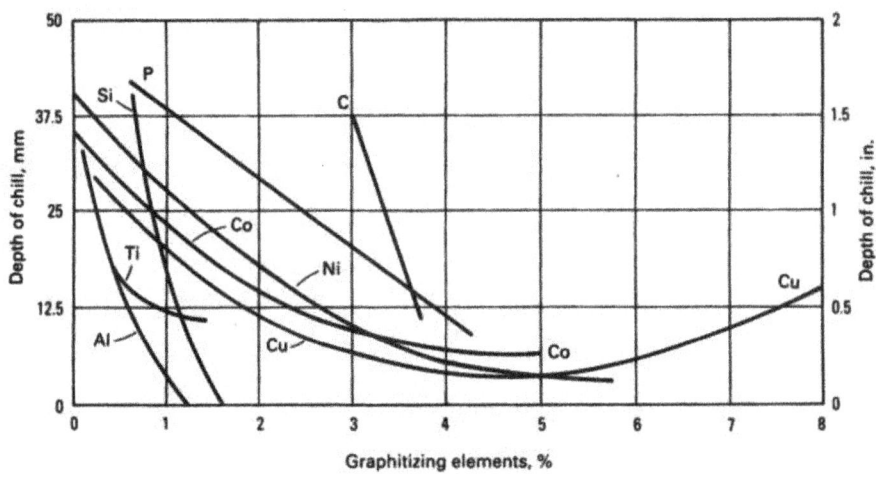

FIGURE 11.32 Effect on chill reduction by graphitizing elements. (From Walton [4].)

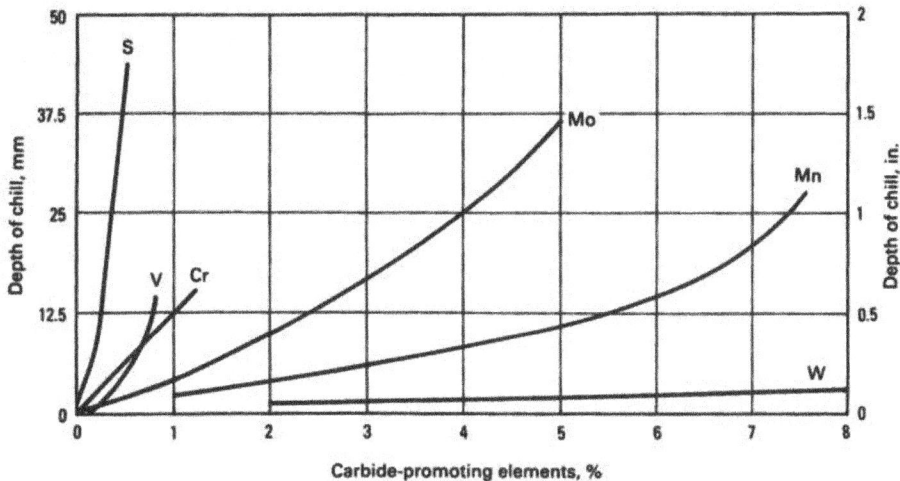

FIGURE 11.33 Effect on chill formation by some of carbide forming elements. (From Walton [4].)

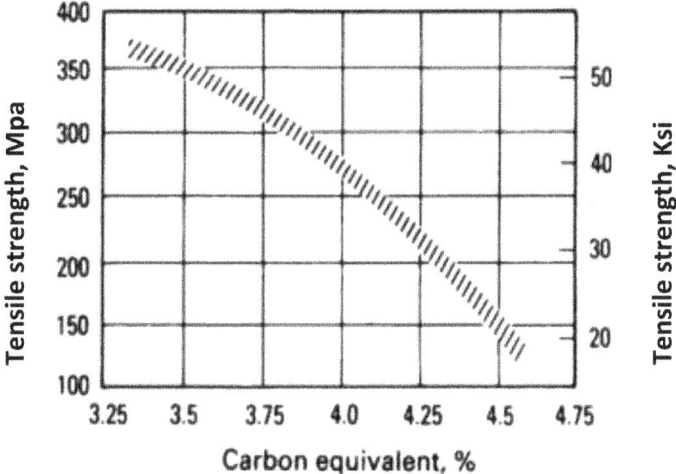

FIGURE 11.34 Tensile strength decreases as the carbon equivalent increases with increasing graphite flake length. (From Walton [4].)

A somewhat similar model has also been proposed [18]. As the eutectic solidification progresses, austenite grains grow and sometimes the dendrite arms grow linearly and graphite particles nucleate in the liquid pools taking the configuration of dendrite shapes; this is termed as nodule alignment. Nodule distribution seems to be greatly affected by austenite growth. As per this model, many nodules are encompassed by a single austenite grain (Fig. 11.36). This is different from the uninodular model. With increased number of graphite nuclei there will be increased number of austenite grains, however, they will be smaller. Figure 11.35 illustrates the grain size distribution in ductile irons. As most ductile irons are hypereutectic, the solidification is termed "mushy" as it proceeds on a wide front compared to gray iron solidification, which is termed "skin-forming" or

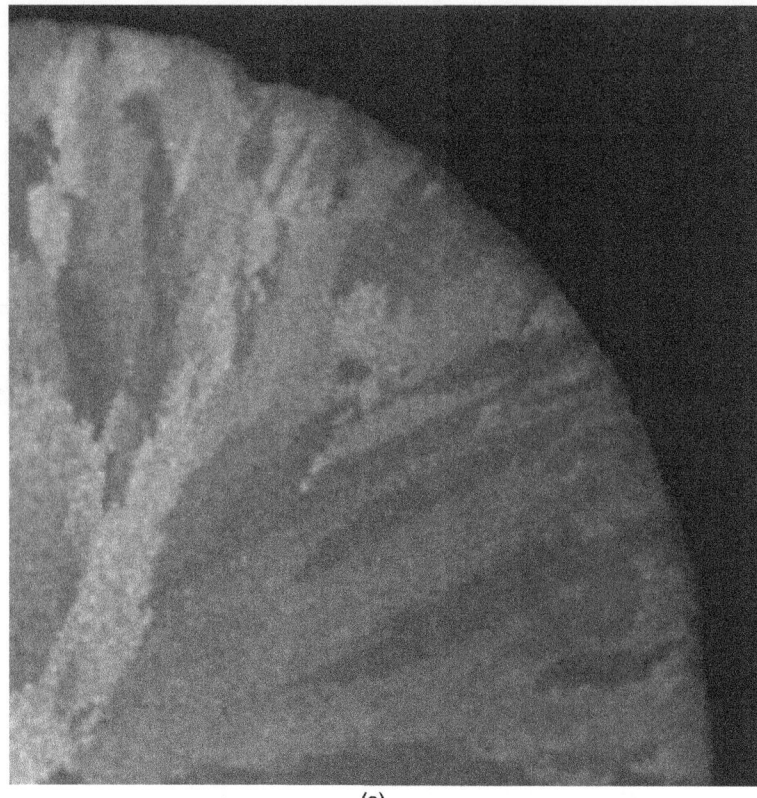

(a)

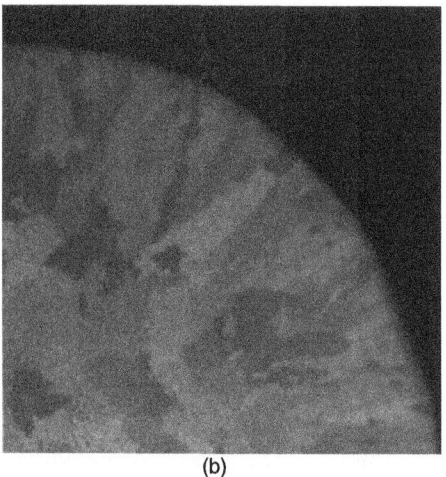

(b)

FIGURE 11.35   Prior austenite grains in both gray and ductile irons revealed by direct austempering after solidification (DAAS). Maximum number of grains occurs in eutectic composition in gray iron and in hypereutectic composition in ductile. (*a*) G-1: hypoeutectic gray; (*b*) G-2: eutectic gray

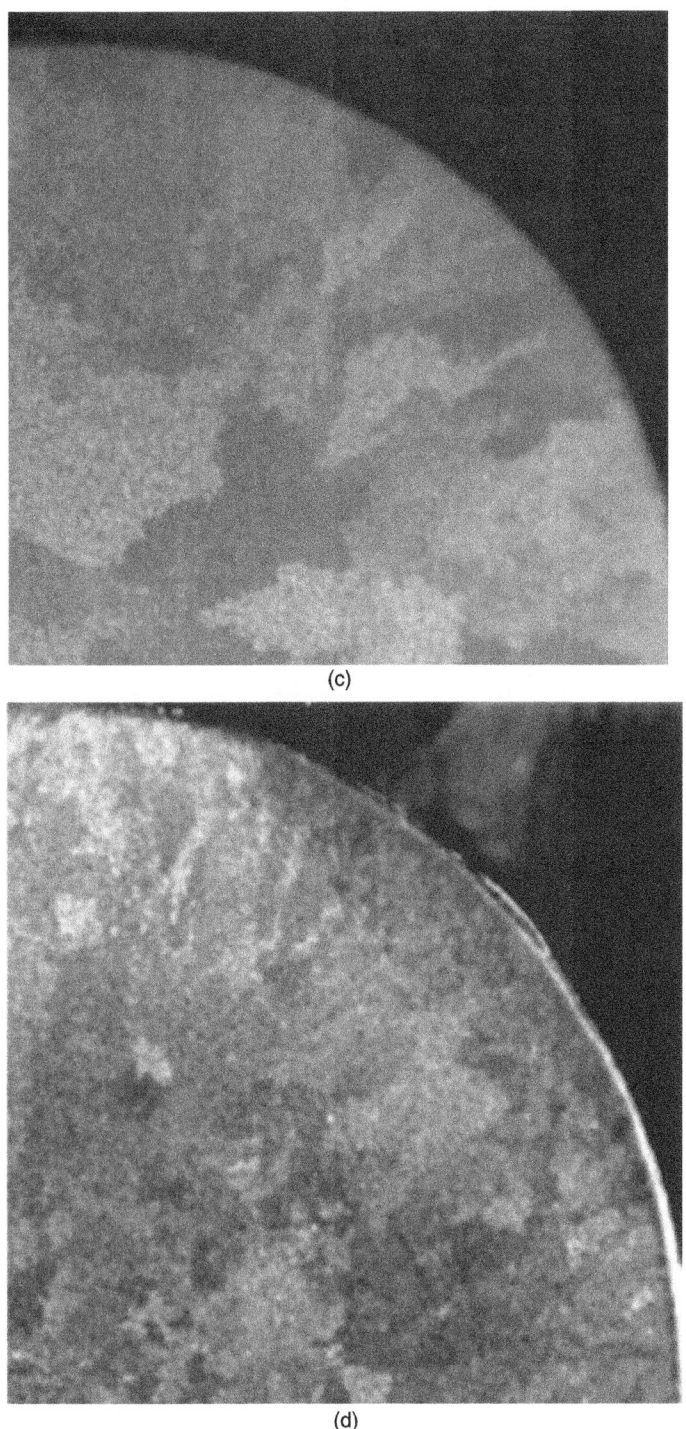

(c)

(d)

FIGURE 11.35 *(Continued)* (c) G-3: hypereutectic gray; (d) S-1: hypoeutectic ductile

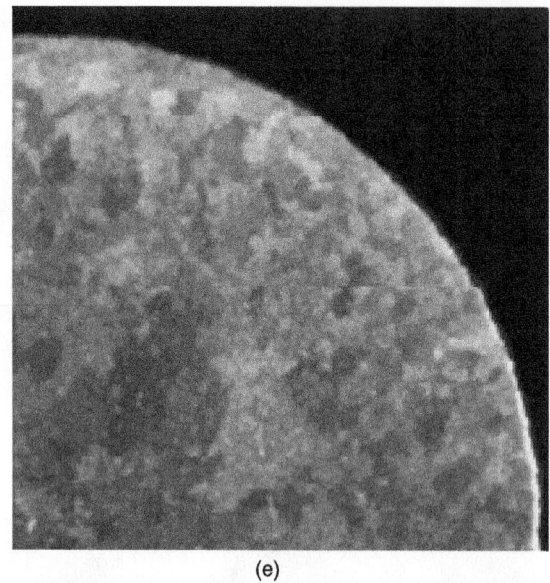

(e)

(f)

FIGURE 11.35   (Continued) (e) S-2: eutectic ductile; and (f) S-3: hypereutectic ductile. (Courtesy of R. E. Boeri, Director, INTEMA, Argentina.)

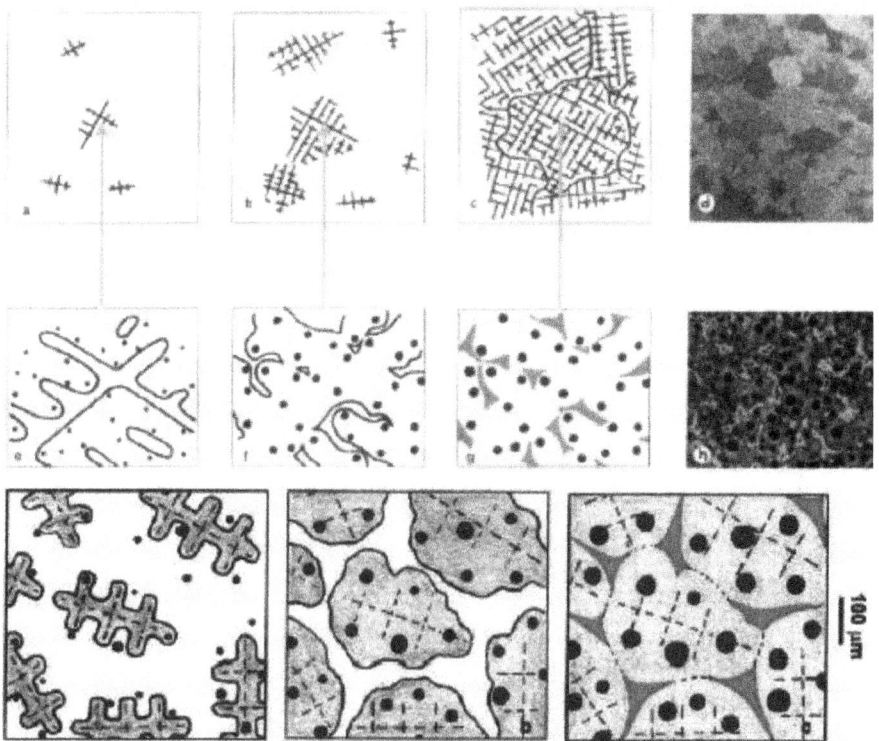

FIGURE 11.36 Ductile iron solidification model proposed by Rivera et al. (From Rivera et al. [16].)

narrow front solidification. The solidification model shown in Fig. 11.36, proposed by Rivera and others [16], seems to agree with the observations in the industry.

As stated earlier, not all graphite nodules end up as different austenite grains, as would be the case if every nodule is enveloped independently by the austenite shell. In each austenite grain, there are multiple graphite nodules, and this can be seen by segregation patterns of forward-segregating elements like Mn and Cr. Many researchers have shown that copper and silicon are found in higher concentrations around the graphite nodules and at lower concentrations along austenite grain boundaries. This fact of forward segregation of carbide-forming elements and reverse segregation of silicon and copper has a significant effect on several areas of ductile iron production such as shrinkage development, hardenability, austemperability, fatigue endurance, and impact properties.

The magnitude or severity of segregation is dependent on the austenite grain size, which in turn may depend on the nodule count or the number of graphite particles.

Volume of shrinkage decreases as the CE increases toward eutectic range and there is a range of CE where castings can be produced without shrinkage [19]. As the CE increases much above the eutectic, shrinkage starts to increase. This range of zero shrinkage is reduced with increasing amounts of forward-segregating elements, as shown in Fig. 11.37. Other studies [20] have also shown that increased amounts of forward-segregating elements increase shrinkage in ductile iron castings.

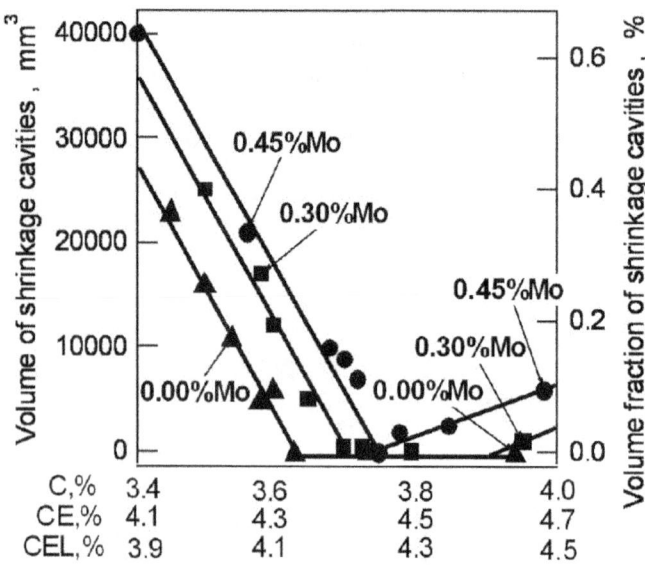

Figure 11.37   Effect of CE and segregating elements on shrinkage of ductile iron castings. CE = C + 1/3 Si, CEL = C + 0.25 Si. (From Kanno et al. [19].)

Compacted Graphite Cast Irons

Early production of compacted graphite iron involved the use of titanium containing magnesium ferrosilicon alloys. Titanium was used to reduce the nodularity in otherwise good ductile iron. Recent production practices use rare earth containing magnesium ferrosilicon alloys. A patented process called *SinterCast* uses thermal analysis techniques to control the nodularity below 20%, by "undertreating" the iron first and then adding just the right amount of active elements based on the test results. Inoculation requirement for CG iron is less than that for ductile iron. Too much inoculation will increase the nodularity. Figure 11.38 shows the relationship between Mg and sulfur for ductile and CG irons. Measured magnesium by spectrometers is total magnesium (MgS + MgO + Mg), hence, if final sulfur is high it will be combined with magnesium and thus the need for higher analyzed Mg residual to maintain the same nodularity [20, 21].

Process parameters to produce CG iron with low nodularity and free of carbides in thin sections down to 3 mm by both in-mold and in-ladle treatment processes have been established [22,23]. The best results in 6-mm-thick samples were obtained from base iron with 0.01% S and 0.028% Mg when insulating sand inserts were used. Fifty-seven percent compactness was obtained in 6-mm-thick sections with typical silica sand molds for 0.01% S in base iron and 0.028% residual Mg. The compactness could be increased to 66% in 6-mm-thick sections with relatively high S (0.02%) in base iron following in-mold treatment without sand inserts.

Key Elements and Their Effects on Cast Irons

All the elements normally present in the cast iron exert some influence on the microstructure of the iron. Carbon and silicon, of course, are fundamental and most important

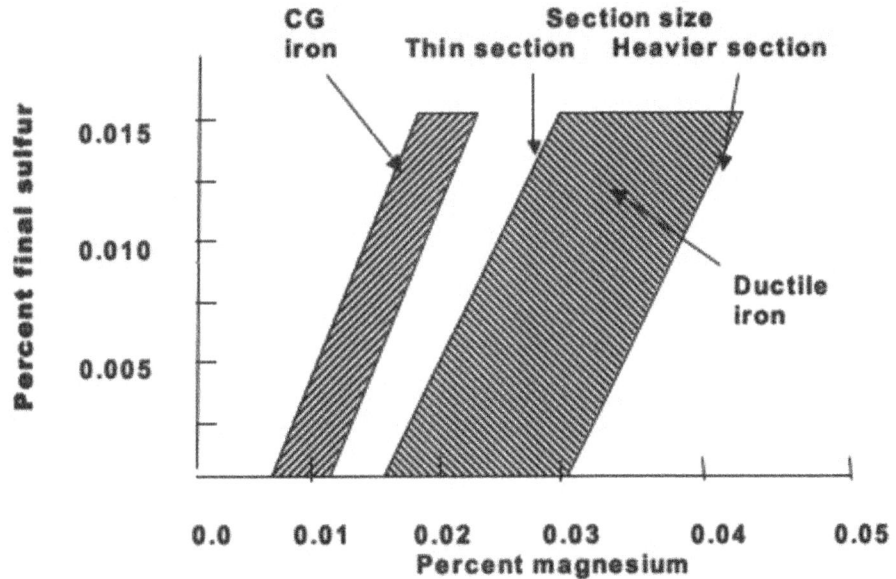

Figure 11.38   Relationship between final iron sulfur and magnesium residual on the graphite shape control. (From Heine [20], Alagarsamy [21].)

with respect to affecting the microstructure and mechanical properties of cast irons, and may be considered first.

Carbon

Carbon in graphitic cast iron is present from about 2.5 to 4.5% by weight. Two phases containing carbon occur: elemental carbon in the form of graphite and combined carbon as Fe_3C. Chemical analysis usually reports the total carbon percentage in the iron. Since the two forms may be determined separately by chemical analysis, the degree of graphitization may be assessed by the following relationships:

$$\% \text{ total carbon} = \% \text{ graphitic carbon} + \% \text{ combined carbon} \qquad (11.11)$$

and degree of saturation (Sc)

$$Sc = \% \text{ graphitic carbon}/\% \text{ total carbon} \qquad (11.12)$$

If graphitization is complete, there will not be any combined carbon, and the total carbon and graphitic carbon are the same. In pearlitic irons, the combined carbon content may vary from 0.5 to 0.8%.

Silicon

Silicon level varies in cast irons from 1.0 to 5.0%. The most important effect of silicon is on graphitization. It may be noted from Fig. 11.18 that increasing silicon shifts the carbon content at the eutectic point of the iron-carbon diagram to the left (from around 4.25 to 3.6%). The eutectic shift is often described by the following relationship:

$$\text{Eutectic carbon percentage (in Fe-C-Si alloy)} = 4.3 - \frac{1}{3}(\% \text{ Si}) \qquad (11.13)$$

The combined effect of carbon, silicon and phosphorous on the phase diagram can be expressed as carbon equivalent (CE). Several different formulae are used in different applications. To estimate the chemistry (C and Si) quickly in the melt shop by thermal analysis by determining the austenite- liquidus temperature (TAL), carbon equivalent liquidus(CEL) is used. CEL = C + 1/4Si + 1/2P is used generally for chemistry evaluation. It should be emphasized that melting variables affect the TAL for the same chemical composition. CE at the eutectic point is commonly given as

$$CE = C + 1/3Si + 1/2P \qquad (11.14)$$

In ductile and CG irons, the phosphorous term is omitted as its levels are very low.

If the carbon equivalent of an iron is calculated to be 4.3%, then the iron corresponds roughly to a eutectic alloy. Hypoeutectic and hypereutectic irons correspond to CE less than and higher than 4.3%, respectively. In very high hypereutectic irons, the freezing process may begin with the precipitation of graphite, and is said to form *kish in gray cast irons*. Because of its buoyancy, kish pops out of the melt into the air and can be observed as sparkly graphite flakes floating on the surface of the iron or in the air above the iron.

Not only does silicon shift the eutectic point to the left, but it also shifts the solubility limits of carbon in austenite to the left of equivalent points in the Fe-C system. At 2.0% Si solubility of carbon in austenite is only about 0.6% compared to 0.76% for 0% Si.

Microstructurally, silicon dissolves in ferrite and as a result, hardens and strengthens the ferrite. Ferrite in pure iron will have a hardness of 80 to 90 BHN, whereas with 2.0% Si, the hardness increases to 120 to 130 BHN.

Sulfur and Manganese

Sulfur, which may be present up to 0.25%, is an important modifying element in flake graphite cast irons. Sulfur content below 0.01% enables full graphitization. At higher sulfur levels (above 0.25%) pearlite will be the final microstructure and the iron is harder to machine.

The influence of sulfur needs to be considered relative to its reaction with manganese in the iron. Sulfur will form FeS and segregate to the grain boundaries during freezing. When manganese is present, MnS and complex manganese-iron sulfides are formed, depending on the manganese content. The manganese sulfides begin to precipitate early, and continue to do so during the entire freezing process and are therefore randomly distributed. When the sulfur is combined with Mn as MnS, the effect of sulfur on pearlite formation is lost. Manganese in excess of that which is combined with sulfur may promote pearlite. The following rules are advanced to express the Mn and S relationship:

1. $1.7 \times \%S = Mn$; theoretical value of Mn to react with S.
2. $1.7 \times S + 0.15 = \%Mn$; the manganese percentage for maximum ferrite and a minimum of pearlite.
3. $3 \times \%S + 0.35\%$; the manganese level to promote pearlitic microstructure.

Phosphorous

Formation of steadite (iron-iron phosphide eutectic) by phosphorous in cast iron has been mentioned earlier. Segregation of P may result in lowering of the temperature of final solidification to 1800°F (982°C). The percentage of steadite present in the final structure may amount to 10 times the P content in the iron. Because of segregation, the steadite is usually found in cell boundaries. Because P forms a eutectic as it segregates, P is often included in the carbon equivalent calculation as shown:

$$CE = \%C + \frac{1}{3}(\%Si + \%P) \qquad (11.15)$$

The iron phosphide is hard and brittle as is the carbide. Increasing P increases hardness and brittleness, especially above 0.3%. To a limited degree, increasing P contributes to increased fluidity and is recommended in very thin section castings to avoid misruns.

Influence of Melting and Melt Handling Procedures on Solidification

Solidification of cast irons are affected by many factors involved in melting and melt handling, including the presence of minor elements, superheating, holding time, and cleanliness of charge materials. This also affects the shrinkage development during solidification and risering techniques used to produce sound castings [22–25].

Superheating above 2730°F (1500°C) reduces nucleation potential and graphite precipitation is retarded. In gray irons undercooled graphite (type E and D) may form if the nucleation of graphite is not enhanced during melting and holding. One of the main reasons for inverse chill in ductile iron solidification is superheating the iron above this temperature. The term "Monday morning iron" is used to describe cast irons that have low nucleation potential and thus require special techniques such as addition of silicon carbide to the melt to avoid problems such as inverse chill, low nodule count, chill carbides, and shrinkage [26]. Graphite nucleation and growth should be able to keep up with the cooling rate of liquid iron. If the cooling rate is faster than the graphite precipitation rate, iron undercools below the carbide eutectic and solidifies as carbide eutectic.

Gray Iron Cell Counts with Superheating

When the iron is superheated, graphite nucleation and cell formation are delayed with increased undercooling. This will result in lower cell count and larger eutectic cells. When cell size is larger, graphite flake length is increased with lower hardness and tensile strength [4]. Figure 11.31 illustrates the effect of undercooling on cell count as a result of superheating.

High Carbon Equivalents

Gray iron is section sensitive; that is, when it is cooled slowly, graphite flakes grow longer with resultant low strength and hardness.

In ductile irons, when CE is higher than 4.6, and conditions are favorable for graphite precipitation and growth, such as higher pouring temperatures, larger casting sections, and well-inoculated iron, graphite will grow to a large size before austenite envelops the nodule. If the graphite size is large enough, the buoyancy forces will lift the nodules to the upper levels of castings. This movement of graphite nodules will result in what is called *graphite flotation* (Fig. 11.39). In extreme cases, nodules will not be compact and will look like they have exploded. When these events occur, the cope surface will be much higher in carbon and lower in hardness with inferior mechanical properties. Different forms of deformed graphite are discussed later in this chapter. Even when the cope surface exhibits exploded graphite (Fig. 11.40), lower levels (drag) of castings will be of typical microstructure similar to that of eutectic composition [9, 27].

Graphite Shape-Active Elements

Key elements that significantly advance the solidification of carbon as graphite in nodular form are magnesium, rare earths (cerium, lanthanum, neodymium, etc.), calcium, and barium. They combine with sulfur and oxygen in the iron and provide some nuclei for the precipitation of graphite in the spheroidal form. The amount of these elements needed to convert to ductile iron is very small. Calcium and barium are used as supplemental elements and are not used as main elements for control. Even though rare earths could precipitate carbon as nodular graphite, they are not used alone in ductile iron production. Magnesium is the element of choice in the production of ductile iron.

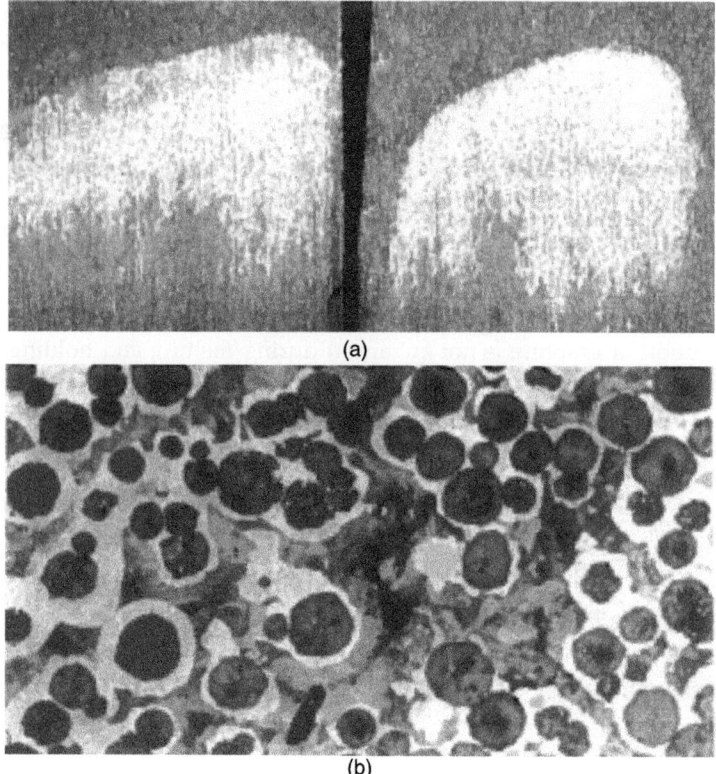

(a)

(b)

Figure 11.39   Graphite floatation zone (a) showing convection currents moving the graphite nodules to the upper levels of casting, and (b) increased graphite volume in the flotation zone. (From *The Sorrelmetal Book of Ductile Iron* [9].)

Atomic weights of some of the elements important in spheroidal graphite formation are listed below.

Magnesium = 24

Cerium = 140

Sulfur = 32

Oxygen = 16

It may be noted that to form spheroidal graphite, sulfur and oxygen need to be neutralized, and it would require quite a bit more rare earths to combine with sulfur and oxygen compared to magnesium. Magnesium and rare earths that are not combined with S and O are responsible for graphite shape control. Free magnesium that is necessary to make ductile iron is around 0.02% and depends on section size (cooling rate). Since the analyzed magnesium residual contains Mg combined with S and O, it is usually higher than 0.02%. If only rare earths are used, the requirement of rare earths will be much higher than Mg.

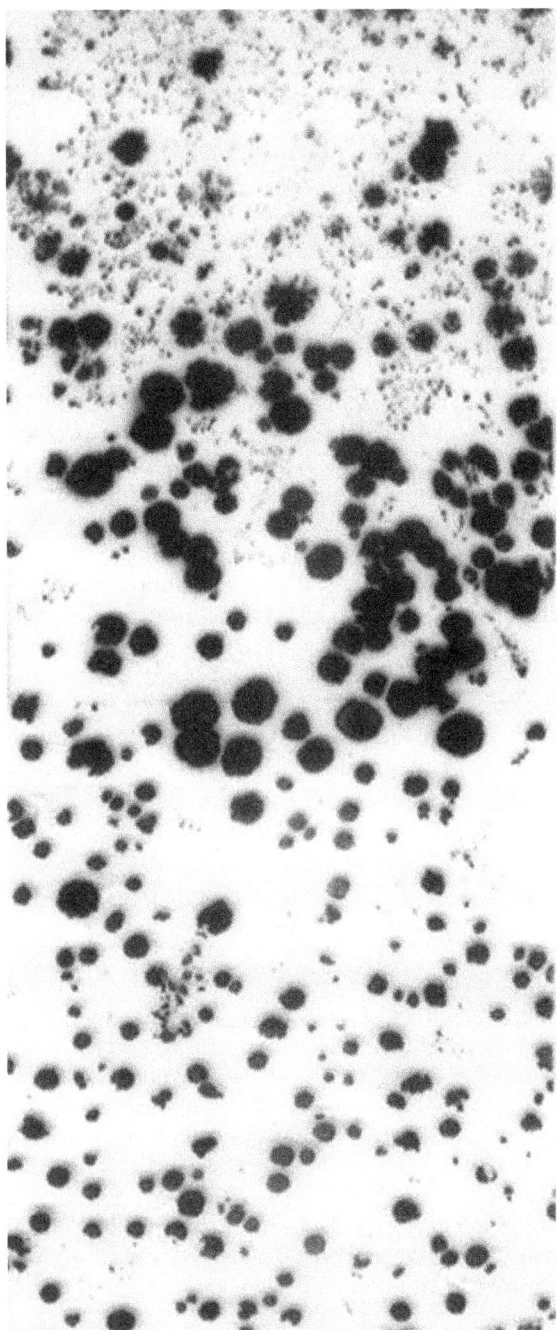

FIGURE 11.40 Various graphite shapes in single section of a casting. At the bottom of the section, nodular graphite shape of a typical ductile iron is seen. Above the typical ductile region, primary graphite nodules, which are larger in size, have floated up. Above the floatation zone, exploded graphite and chunk graphite can be seen. (From Alagarsamy [27].)

Chunk Graphite	Intercellular Flake Graphite	Deleterious Graphite
Ce, Ca, Si, Ni, C	Bi, Cu, Al, Pb, Sb, Sn, As, As, Cd	Zr, Zn, Se, Ti, N, S, O

Source: From AFS [8].

TABLE 11.8 Classification of Subversive Elements

When the residuals are in excess of that required for spheroidization, various negative effects are observed. Some of these effects are (1) shrinkage porosity, (2) carbide formation, (3) graphite flotation, and (4) exploded graphite.

Graphite morphology is also affected by trace elements and alloying elements that may be present from the raw materials and alloys used in melting and conditioning the iron. Elements which are present in small quantities but interfere with the normal spheroidal graphite growth are called subversive elements, and the resulting shape can seriously affect mechanical properties. These elements are very highly surface-active and exert significant effect on graphite morphology. Some of these elements are beneficial at certain levels and harmful at other levels. Table 11.8 shows three classifications of these elements as to their effect on graphite shape.

Exact mechanism of chunk graphite formation is not yet clear. Elements in the first column either increase the carbon equivalent or promote undercooling. Some of the elements in column 2 tend to segregate to cell boundaries, and achieve local concentrations high enough to form flake graphite. The undesirable effect of elements in column 2 can be neutralized by the addition of elements from column 1 and vice versa.

Elements in column 3 influence the graphite shape directly by interfering with the nodularizing effect of magnesium. They must be controlled to a low level either by controlling charge materials or by adding extra magnesium to neutralize their effect. Rare earths are very effective in neutralizing the effect of elements in column 2, that are prone to flake graphite formation, but Mg by itself is ineffective in doing so.

Deleterious Graphite Shapes

Graphite Flotation

When the CE is too high and iron is well inoculated, graphite precipitates at a higher temperature (above the graphite eutectic temperature), and has time to grow before the composition reaches eutectic. When the size of graphite nodules is large enough they tend to float in the liquid and will concentrate near the cope surface of the casting (Figs. 11.39 and 11.40). Convection currents may move the graphite nodules, as shown in Fig. 11.39a.

Exploded Graphite

When CE is too high (above 4.6%), and Mg and Ce levels are too high, graphite grows discontinuously and appears like a broken up nodule, as shown in Fig. 11.41. It is important to control Mg and rare earth levels with section thickness of casting produced.

Chunky Graphite

When the rare earth levels are too high for the section thickness, spheroidal graphite degenerates into chunky graphite (Fig. 11.42a). Addition of antimony and or bismuth may neutralize the effect of excess rare earths and minimize the formation of chunky graphite. A SEM micrograph of a chunk graphite cell is shown in Fig 11.42b.

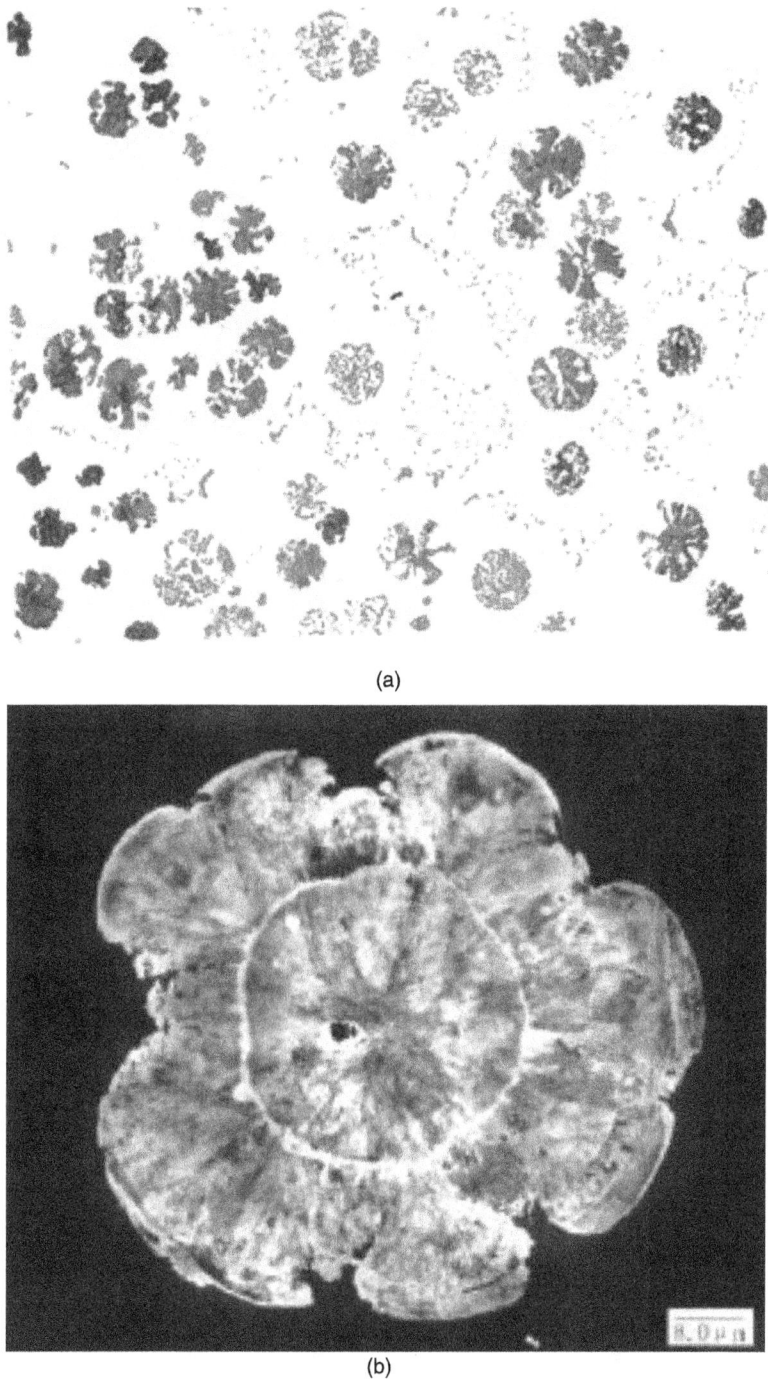

(a)

(b)

FIGURE 11.41   (*a*) Exploded graphite distribution in casting cope surface [9]. (*b*) Single nodule showing good spherical growth in the center and cauliflower-type growth on the outside. (From Loper and Fang [14].)

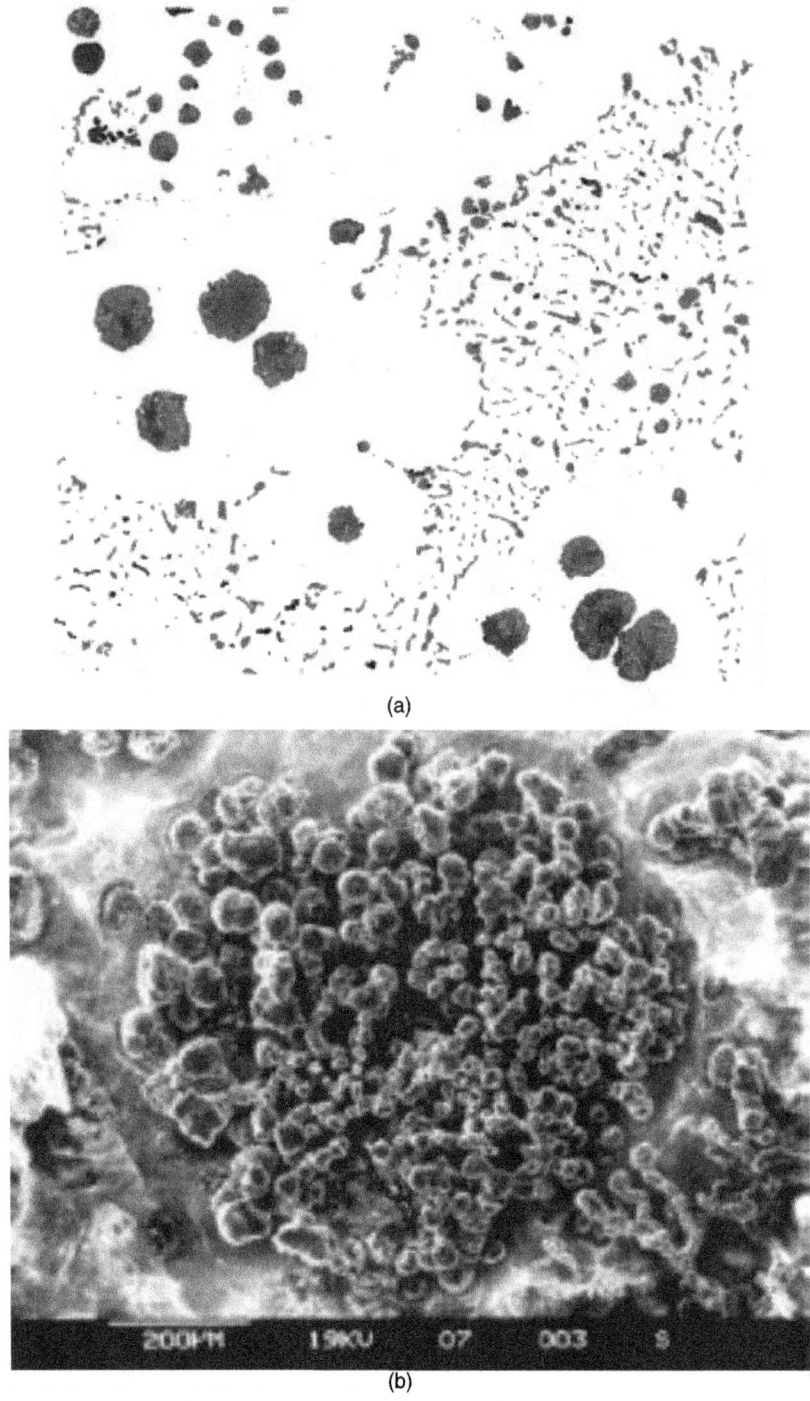

(a)

(b)

FIGURE 11.42   Chunk graphite (*a*) distribution [9] and (*b*) a SEM micrograph of a deep-etched chunk graphite cell. (From Loper and Fang [14].)

Vermicular Graphite

For the ductile iron, vermicular graphite is considered as faded ductile iron. As the magnesium residual is reduced, more and more vermicular graphite is formed.

A SEM micrograph and optical vermicular microstructures are shown in Fig. 11.43.

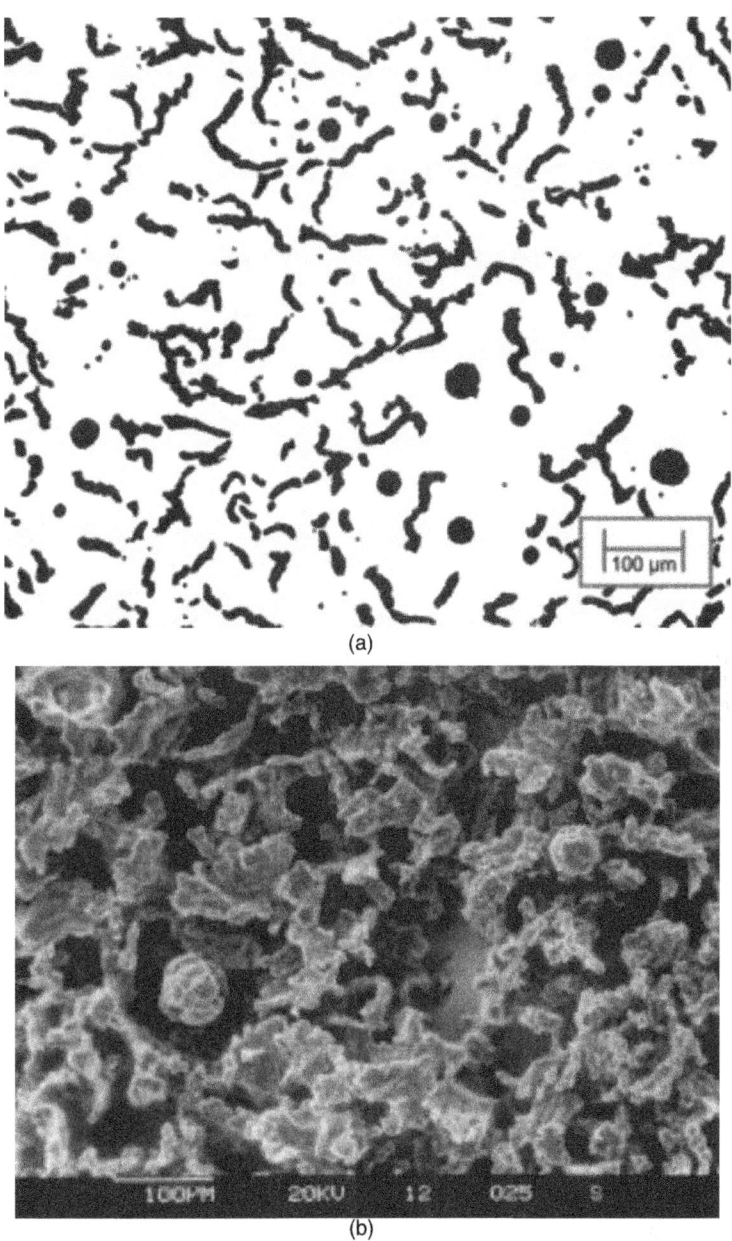

(a)

(b)

FIGURE 11.43   (a) Vermicular graphite distribution [3] and (b) a SEM micrograph of a deep-etched vermicular graphite cell. (Loper and Fang [14].)

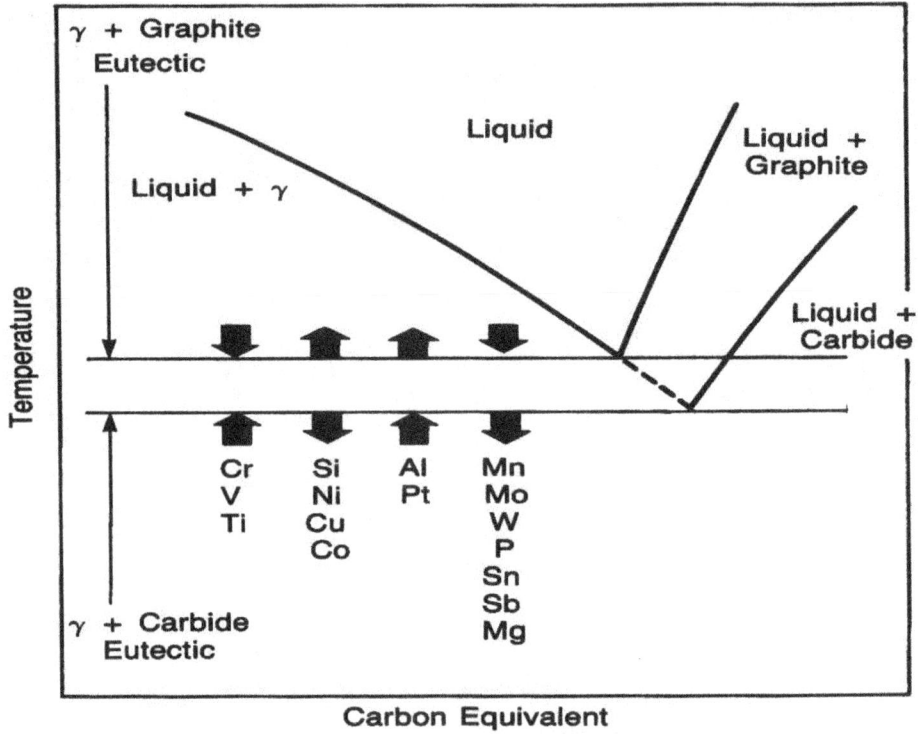

Figure 11.44 Effect of key elements on eutectic solidification temperatures. (From Janowak and Gundlach [28].)

Key Elements and Their Effects on Ductile Iron

Many of the elements in cast iron affect two areas of cast iron metallurgy: (1) the stable and metastable eutectic temperatures, and (2) the eutectoid transformation temperatures. Key elements affecting the eutectic temperatures are shown in Fig. 11.44. Some raise the stable eutectic and lower the metastable eutectic temperatures, and they are considered strong graphitizers. The ones that lower the graphite eutectic temperature and increase the carbide eutectic temperature are considered strong carbide formers.

Carbon

Next to iron, carbon is the most important element in ductile iron and affects many characteristics. The expansion resulting from precipitation of graphite as nodules counteracts the solidification shrinkage associated with the precipitation of austenite. This helps mold yield compared with steel castings and makes production of sound castings much easier. Carbon level varies from 3.3 to 4.0%, mainly depending on silicon level and casting section thickness. Carbon equivalent is controlled about the eutectic point, slightly higher than the eutectic. This ensures the least amount of shrinkage in the castings. Lower carbon is prone to shrinkage and carbide formation in thin sections. Higher carbons favor graphite flotation, exploded graphite, and even shrinkage. Carbon is considered a ferritizer. The presence of free graphite in the matrix causes the elastic modulus to decrease and thus, ultrasonic velocity. Increasing graphite content reduces dynamic properties.

Silicon

Silicon levels vary from 2.0 to 3.0% in regular ductile iron castings and 3.8 to 4.5% in high-silicon irons used at high temperatures. Silicon is a strong ferritizer and increases the hardness and yield strength of ferrite by solid-olution hardening. Increasing silicon increases shrinkage propensity in castings. Silicon increases oxidation resistance and is used up to 4.5% for castings such as exhaust manifolds. Silicon also increases the nil ductility temperature. Silicon is added in the melt as ferrosilicon alloys or silicon carbides. Silicon is also added through magnesium ferrosilicon during conversion to ductile iron and during postinoculation. A higher silicon level reduces chill carbide forming tendency.

Sulfur

Free sulfur is detrimental to graphite nodule shape. Base iron sulfur level is carefully controlled to a low level of 0.08 to 0.15%. High sulfur irons can be desulfurized prior to treatment, or special techniques need to be used to convert them to ductile iron. After magnesium treatment, sulfur in the final iron usually varies from 0.005 to 0.015%. Higher sulfur after treatment may lead to sulfur reversal and eventual deterioration of graphite nodules to vermicular and flake graphite. Measured magnesium residual in ductile iron depends on the sulfur residual as some of the magnesium is tied up with sulfur as sulfides.

Oxygen

There is a very low level of dissolved oxygen (<10 ppm) gas in ductile irons, especially after magnesium treatment. But oxygen is present as oxides of metals. Carbon and other elements combine with oxygen to form various oxides and gases that tend to exit through the molten iron. During melting, oxides of iron, manganese, and silicon are dispersed due to violent mixing and turbulence. These oxides can be neutralized by carbon and other elements like aluminum. These de-oxidation reactions are temperature and time dependent. In high production shops, oxides are present throughout the melt treatment and pouring operations to varying degrees depending on charge materials, melt practice, and other variables. Increased oxide content in the iron increases the chances for formation of defects such as pinholes, dross, and slag.

Alloying Elements

Special elements are added to ductile and other graphitic irons mainly to control the matrix during cooling through the eutectoid region as well as during various heat-treatment practices. Alloying elements enable the control of the matrix, independent of the cooling rate. Small amounts of alloys can be added to the treatment ladles to ensure good mixing. Larger additions of alloys are added in the melting furnaces to get better mixing. Elements that have high melting points should be added in the furnace. Many elements may have the same effect on the matrix and the cumulative effect should be considered when calculating additions. Some elements counteract the effect of others. When calculating the actual addition for a certain level of pearlite in the matrix, the cumulative effect of all pearlite-promoting elements should be considered together by giving a weighting factor for each of the elements [29]. Silicon effect is negative since it promotes ferrite.

Copper

This is the preferred element for stabilizing pearlite in the matrix. Copper does not segregate to grain boundaries and does not promote carbides. In many cases, copper is added in the treatment ladles as cover material for magnesium alloy, and thus improves magnesium recovery. Like silicon, copper also increases magnesium solubility. Many have observed that copper increases shrinkage tendency. Generally, copper is limited to 1.0% for pearlitic castings. If copper alone is inadequate to yield the pearlite content desired, either due to heavy section or when boron is present, then materials such as tin and manganese can be added to complement copper.

Manganese

Manganese affects cast irons during solidification as well as during eutectoid transformation. It segregates to the austenite grain boundaries increasing carbon solubility at the end of solidification, increasing the carbide-forming tendency, and increasing shrinkage porosity. Manganese strengthens the matrix along with copper to increase yield strength. Manganese also increases hardenability and austemperability. Due to segregation, it is usually limited to less than 0.5%, especially for austempered ductile irons. Manganese decreases machinability in pearlitic ductile iron.

Nickel

Nickel is a mild pearlite promoter. It increases yield strength even in ferritic irons. For low-temperature impact grade irons, where silicon level is low, nickel is added to increase yield strength. Addition of 2 to 4.0% nickel may produce as-cast acicular matrix. Higher levels of nickel, 18 to 35%, are used to get as-cast austenitic irons for high temperature and corrosion-resistant applications.

Molybdenum

Molybdenum is added in small quantities, less than 0.4%, to improve hardenability during austempering. It is also added (0.6 to 1.0%) for improving creep resistance for exhaust manifolds that operate at high temperatures. Molybdenum does not promote pearlite, but segregates to grain boundaries and stabilizes complex carbides.

Tin

Tin is a strong pearlite promoter and is considered to be 10 times stronger than copper. It is used only when necessary to stabilize pearlite if it cannot be done by copper alone, and the level of addition is limited to 0.025%. At higher levels of tin (>0.04%) ductile iron becomes increasingly brittle.

Chromium

Chromium is not added to ductile iron and is considered detrimental to ductility as it forms stable carbides in the grain boundary areas. Taking advantage of this fact, castings are produced with carbides to improve wear resistance. For carbidic austempered iron, castings are produced by a carbide stabilizer such as chromium, and by reducing inoculation and thus promoting stable carbides [30]. Stable carbides are not dissolved during the austenitizing step in the austempering process. Chromium is also added with nickel for different grades of austenitic gray and ductile irons.

Residual Elements (Trace Elements and Deleterious Elements)

Many elements are present in small quantities that are not added intentionally but are introduced from scrap steel, pig iron, and alloys. Some elements, even in small quantities, can be harmful to the graphite shape, and others may affect mechanical properties. Effects of many of these elements are known and can be compensated in most cases.

Aluminum—introduced mainly from ferrosilicon alloys. Aluminum easily oxidizes in the melting operation. About 0.03% aluminum has been shown to produce pinholes. Aluminum is used in steel as a deoxidizer and similar effects can be seen in irons also. With reduced alloy additions, aluminum has not been a significant element affecting ductile iron solidification and casting defects.

Antimony—is a strong pearlite promoter and also forms intercellular flake graphite above 0.004%. Sb is usually used in heavy sections to counteract the chunky graphite forming effect of rare earths and to improve nodule count.

Bismuth—promotes flake graphite even at levels around 0.003%. It is used in heavy section casting production to increase nodule count and is balanced by rare earths.

Boron—even in low levels (0.003%) affects pearlite formation. In normally pearlitic irons, excess boron will reduce the amount of pearlite, hardness, and tensile strength. Boron can come from steel scrap and from leaching out of silica lining bonded with boric acid in induction furnaces.

Lead—introduced from contaminated steel scrap with wheel weights, brass, and other leaded steels. Lead forms flake graphite even at 0.002% levels and magnesium itself is not adequate to neutralize the adverse effect of lead. Rare earths are, however, effective in neutralizing the effects of lead.

Phosphorous—forms grain boundary iron-phosphide eutectic carbides at higher levels and thus is limited to less than 0.05% in most specifications. For low-temperature impact requirements it is limited to less than 0.03%.

Titanium—is introduced from ferrous scrap and from alloying elements. Titanium forms flake graphite and is seen as having a poisoning effect in ductile iron. In heavy sections, Ti may segregate and may cause problems. It reduces the spheroidizing effect of magnesium and is added with magnesium to produce vermicular graphite irons. Titanium can be counteracted by rare earths to avoid flake graphite. Titanium can reduce machinability when present as titanium carbo-nitrides.

Zinc—comes from coating on ferrous metals and from contaminants in scrap. Zinc evaporates during melting.

Conversion from Flake Graphite to Nodular Graphite

In the beginning, ductile iron production was licensed by the International Nickel Company and the alloy used at that period was nickel magnesium alloy. Magnesium is easily soluble in nickel and was easy to introduce into molten iron without violent reactions.

Even though rare earths alone can be used to produce spheroidal graphite, magnesium- and magnesium-containing alloys are the preferred way to make ductile iron castings. Due to cost and the sensitivity to form carbides and other forms of graphite, rare earths are generally used to augment magnesium in designing treatment systems.

The high cost of nickel necessitated the development of magnesium ferrosilicon alloys for reducing the cost of ductile iron treatment. High volatility, reactivity, and low density of magnesium and differing base iron chemistries necessitated the development of various treatment methods to convert base iron to ductile iron. If pure magnesium is used for the conversion, special ladles and procedures need to be used for safety and efficiency of alloy usage [9].

Ladle Treatments Using Master Alloys

Ni-Mg and Simple Ladle Treatments

This is the simplest and was, for many years, a widely used method for Mg treatment. Generally, Mg alloy is placed in the bottom of the empty ladle and the base iron is tapped over the alloy. There is no need for fume collection since the reactions are mild and less turbulent as the Mg dissolves in the iron. This alloy does not provide nucleation sites and is necessary to have a robust postinoculation step. Ladles are usually tall and the height to diameter ratio is 2 and higher.

Sandwich Open Ladle

By providing a pocket at the bottom of the ladle, Mg alloy can be placed inside this pocket. Steel punchings, spill iron, or ferrosilicon alloys can be added on top of the Mg alloy as cover material for improving the Mg recovery. Use of sandwich ladles has led to the adoption of magnesium ferrosilicon alloys, thus reducing the cost of treatment. Due to the introduction of silicon with the alloy, base silicon levels need to be controlled lower than with nickel Mg alloys. Lower base silicon tends to reduce the lining life in silica-lined induction furnaces. Ferrosilicon-based alloys provide some nucleation from the treatment. Mg recovery depends on the ladle design, pocket design, alloy design (Ca and Mg levels), as well as operational practice such as tap temperature, ladle cleanliness, and tap time. This method has almost replaced open ladle treatment with nickel Mg alloys.

Covered Ladles or Tundish Ladles

To improve further the efficiency of treatment, ladles were designed with covers (called *tundish*) to reduce tap temperature, turbulence, and reaction with oxygen. There are many different designs of tundish ladles in use now with Mg recovery extending up to 80%.

Porous Plug Method

One or more porous plugs placed at the bottom of the ladles enable magnesium alloys to be added after tapping the metal into the ladle. Melt can be desulfurized, deslagged, and then magnesium alloys can be added on top of the melt and mixed by blowing nitrogen through the porous plugs. This method offers flexibility with varying base iron sulfur levels as well as large treatment sizes. Temperature loss and long cycle times are concerns with this process.

In-Mold Treatments and Other Chamber Methods

Iron can be converted to ductile iron inside a mold. As the base iron runs over alloy placed inside a chamber in the runner system, Mg reacts with sulfur and oxygen and is converted to ductile iron. Base iron sulfur needs to be low for this process as there is no removal of sulfur-rich slag prior to filling of the cavities. Higher sulfur irons may revert to flake graphite in slow cooling sections. There is no need for postinoculation with this process. Nodule counts are very high and thin sections can be produced without carbides. As each mold is a separate treatment all castings produced in this process need to be checked for nodularity, especially if they are used in safety-related applications.

Pure Magnesium Treatment Methods

When base iron sulfur is high, such as in irons melted in acid cupolas, treatment with master alloys are not preferred due to increased alloy usage, sulfur reversal, and silicon limitations. Processes using predominantly magnesium [31] have been developed to treat high sulfur irons. Due to very high vapor pressure of magnesium at the treatment temperatures and violence associated with the reaction, special equipment needs to be used for safety and efficiency.

Plunging Bells

Pure magnesium chunks placed inside a refractory bell is plunged into liquid iron in a deep ladle. The reactions are contained inside the bell and the ladle. Due to lower cost and high-volume capability, many ductile iron pipe producers adopted the plunging process. This process can be used for any level of sulfur in the iron. Other than pure magnesium chunks, magnesium-rich alloys, and refractory-coated ingots of magnesium could be used with this method.

Converters

George Fischer developed a converter vessel that has two chambers, one for placement of the magnesium and the other for reactive elements separated from the rest of the ladle by a perforated refractory plate [31]. Iron is tapped into the larger space in the ladle in a horizontal position and the alloy is placed in the chamber area and sealed with a cover. When the ladle is rotated to the vertical position, iron enters the alloy chamber in a controlled manner dissolving the magnesium. The main ladle is vented to let the magnesium vapors escape to prevent a possible explosion. Due to safety considerations, the treatment is restricted to a safe protected area. After the reaction, the ladle is tilted to the horizontal position for deslagging and transported to the pouring area. Sulfur-rich slag should be removed after each treatment to prevent sulfur reversal. This method is slower than other methods, hence, it is generally used for larger treatment sizes. After treatment, iron is usually transferred to holding or pressure-pour furnaces. Iron treated with pure magnesium is prone to carbide formation, due to lack of nucleation from the treatment. Efficient postinoculation is necessary to minimize carbide formation and shrinkage defects.

Wire Injectors

Magnesium-rich materials filled inside a hollow steel tube can be injected into a liquid bath in a ladle placed in a safe enclosure. Several different materials can be included in the tube (also called *wire*) for different situations. High- and low-sulfur irons can be treated by controlling the amount of addition (the length of wire injected). Speed of wire addition is also controlled to get maximum recovery. Large treatment sizes may require more than one wire feeder to reduce the time of treatment. Temperature loss is a factor with this method.

Postinoculation

After conversion to ductile iron another important step is inoculation of treated iron to nucleate graphite. It takes special nuclei for the carbon to precipitate as graphite nodules. Nuclei that are present from the treatment are shortlived due to treatment temperature and rapid dissolution of discrete particles. It is common practice to add nucleating agents to the iron during transfer to pouring ladles. Most of these inoculants are ferrosilicon based. Active elements are Ca, Ba, rare earths, and silicon. Increasingly, with automated molding and pouring systems, which use pressure-pour furnaces, inoculation is achieved by adding fine material in the pouring stream. Inoculation can also be in the mold by placing either solid inserts or fine materials in the runner system. The amount of inoculant used in-stream or in the mold is usually much less than that used in the ladles. Late inoculation is very effective in increasing nodule count and minimizing carbides.

Solid-State Transformation

Carbon Adjustment after Solidification

Right after eutectic solidification [at about 2100 to 2065°F (1150 to 1130°C)] carbon content of austenite will be around 1.7%. Solubility of carbon in austenite decreases as the iron cools down to eutectoid transformation region. Excess carbon in austenite, beyond the solubility limit for the temperature, deposits onto existing graphite nodules, and carbon in austenite drops down to around 0.6% at about 1300°F (723°C). If the iron is not alloyed with nickel and other austenite-stabilizing elements, austenite will not be stable below a temperature termed the critical temperature.

Further decrease in temperature is accompanied by rejection of carbon from the austenite as graphite and its precipitation on the graphite flakes in the eutectic. Carbon precipitation continues until the eutectoid temperature range is reached [about 1475 to 1400°F (802 to 760°C)] with 2% silicon. At the eutectoid temperature, the 2.0 Si iron contains about 0.6% C. Equilibrium cooling through this temperature range results in the transformation of austenite to ferrite and the precipitation of the remaining carbon on the graphite flakes. The final microstructure then consists of isolated areas of ferrite originating from the primary austenite dendrites and other areas of mixed ferrite and flake graphite having their origin in the austenite-graphite eutectic. The microstructural changes described above are those occurring in the ternary alloy of Fe-C-Si. Similar processes in commercial cast irons are much more complex since many other elements are present and a number of other factors are introduced. However, the simple alloy considered does point out the important stages of graphitization.

1. Graphitization during solidification
2. Graphitization by carbon precipitation from austenite (solid state)
3. Graphitization during eutectoid transformation (solid state)

The temperature at which the austenite transformation in slow cooling conditions to ferrite (upper critical temperature, UCT) occurs depends on chemistry, silicon being the most significant element. Silicon raises the upper critical temperature for the beginning of ferrite formation and lowers the pearlite formation temperature (lower critical temperature, LCT), as shown in Fig. 11.45. This increases the range of temperature (between upper and lower critical temperatures) where ferrite can form rejecting carbon, which diffuses to the existing graphite flakes, vermicular graphite, and nodules. Below the lower critical temperature, austenite cannot exist and it transforms into pearlite almost instantaneously. Pearlite phase is made up of two constituents: ferrite and iron carbide (Fe_3C). The amount of carbon dissolved in the austenite as well as cooling rate controls the hardness of pearlite and thickness of ferrite and carbide layers (fineness of pearlite).

Eutectoid transformation from austenite to either pearlite or ferrite is affected by alloying elements. Pearlite-stabilizing elements such as copper, manganese, chromium,

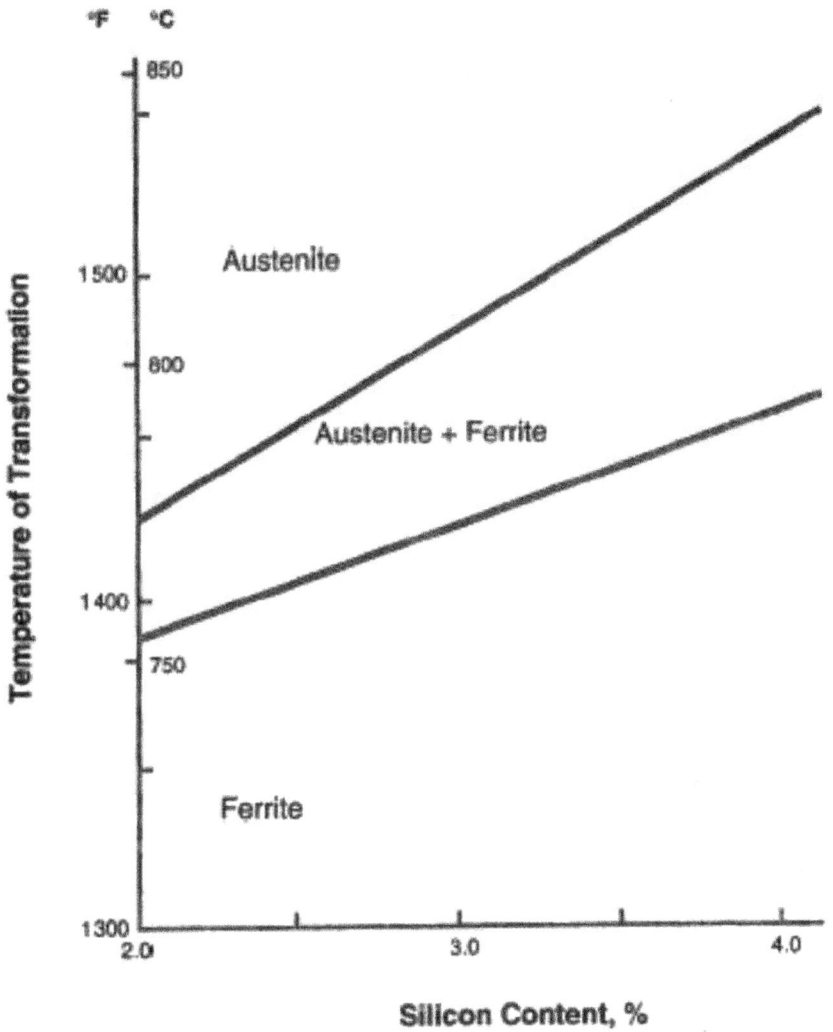

FIGURE 11.45 Effect of silicon on eutectoid transformation temperature. (From Davis [15].)

tin, and antimony retard carbon diffusion into graphite from austenite and thus retard ferrite formation. These elements have varying effect on the amount of pearlite formed at the end of the transformation. Eutectoid transformation temperatures were studied [32] for austempering controls and the effect of common elements on the upper and lower critical temperatures are given below. The equations given represent the typical alloys used in ductile iron castings for austempering, and the elemental concentrations are in weight percentages.

$$\text{UCT }°\text{F} = 1372 + 80\,(\text{Si}) - 63.3\,(\text{Mn}) - 19.7\,(\text{Ni}) - 50.3\,(\text{Cu}) + 6.4\,(\text{Mo}) + 169\,(\text{P}) \quad (11.16)$$

$$\text{LCT }°\text{F} = 1326.6 + 40.1\,(\text{Si}) - 88.4\,(\text{Mn}) - 9.1\,(\text{Ni}) - 12.6\,(\text{Cu}) + 3.4\,(\text{Mo}) + 138.1\,(\text{P}) \quad (11.17)$$

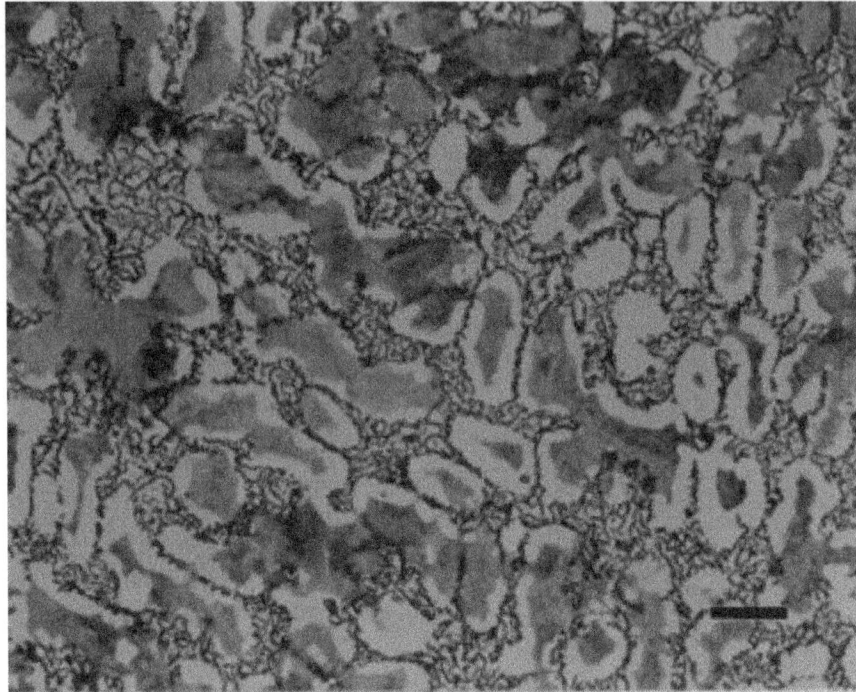

Figure 11.46 Graphite precipitation is suppressed in hypoeutectic gray iron by processing variables, resulting in undercooled graphite. Cooling through the transformation temperature range, ferrite is formed around graphite particles and on the periphery of proeutectic austenite. Pearlite forms at the center of large proeutectic austenite grains. (From Vander Voort [3].)

Graphite morphology has significant effect on the diffusion of carbon into graphite particles (Figs. 11.40 and 11.48). Carbon deposition onto graphite particles is affected by surface area and crystallographic structure of graphite. Carbon has been shown to deposit onto existing graphite particles to varying degrees with the change in graphite morphology. Ferrite easily forms with type D and E in flake graphite irons (Fig. 11.46). Ferrite can easily form around vermicular and nodular graphite (Fig. 11.47). Crystallographic structure plays a significant role in accepting the carbon deposed by austenite. In flake graphite with type A graphite, pearlite formation is favored compared to spheroidal or vermicular graphite. It is harder to get fully pearlitic matrix in compacted graphite irons for the same chemistry and cooling rates as gray irons. Graphite grows in contact with liquid, then from diffusion of carbon through austenite shell as solidification progresses, as well as during solid-state transformation of austenite; these stages can be clearly seen in Fig. 11.48.

Effect of Cooling Rate

Cooling rate through the eutectoid transformation range affects the amount of pearlite/ ferrite phases and lamellar spacing of pearlite. Faster cooling increases pearlite content, as there is less time above the lower critical temperature as the iron is cooling down. With faster cooling rate, pearlite transformation tends to be at lower temperature resulting in finer lamellae.

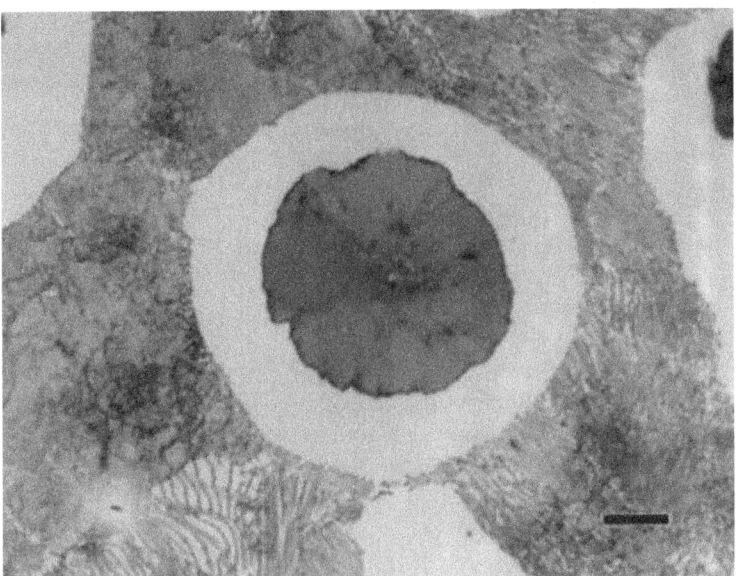

FIGURE 11.47 Annular ring of ferrite referred to as "bullseye" ferrite results from solid-state transformation of austenite surrounding the nodule. Surface of the graphite nodule readily accepts the carbon rejected by austenite. (From Vander Voort [3].)

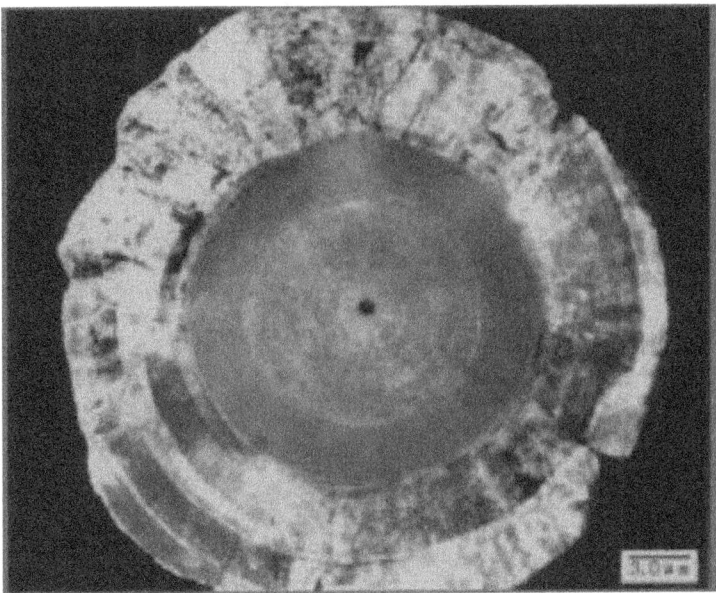

FIGURE 11.48 Graphite nodule growth. Three separate zones can be seen in this nodule separated by the deep-etching technique. In this picture, the nodule grew in contact with liquid, depicted by the solid inner sphere. The center annular ring is indicative of growth by carbon diffusion from liquid through austenite shell and the outermost ring shows the growth phase of the nodule during solid-state diffusion as austenite rejects carbon and transforms into ferrite. (From Loper and Fang [14].)

Foundry Practice

Molding

Many molding methods are used to make molds for iron castings. They can be classified in various ways:

- Resusable pattern processes
- Greensand, dry sand, chemically bonded sand, and permenant molds
- Expendable pattern processes
- Centrifugal castings

By far most tonnage of engineered castings are made by the greensand molding process. With the greensand system they can be divided into horizontally parted molds and vertically parted molds. Large tonnage of ductile iron pipes are made by the centrifugal casting process in metal molds coated with a refractory slurry.

Greensand Molding

A wide variety of sands are used for greensand molds. More production of iron castings has shifted to high-pressure molding of greensand. Base sand is generally silica based (silica, lake, and river sands) of varying grain distribution from 50 to 80 AFS grain fineness. Bonding clays are sodium and/or calcium ion bentonites (montmorillonite) and mixtures of both. Clay levels also vary from 6 to 12% depending on the molding equipment and sand system equipment such as mullers, cooling, and dust collection. Sea coal is added to improve casting peel during shakeout as well as to provide reducing atmosphere to minimize pinholes and graphite degeneration on the casting surface. Gray irons tend to have rougher surface finish than ductile irons and penetration defects are more common with gray iron castings. Exudation or expansion during final stages of solidification may cause metal penetration into molding sands or core sands. This expansion at the end is controlled by the balance between graphite precipitation and compensating shrinkage by austenite formation along with formation of grain boundary carbides aggravated by segregating elements like phosphorous, chromium, molybdenum, and manganese. With proper balance between expansion of graphite and shrinkage increasing factors, exudation penetration can be minimized. Greensands are not strong enough to contain the expansion associated with ductile iron and high CE gray iron solidification. Risering techniques are used to alleviate the pressure by backfeeding into the riser.

Chemically Bonded Sand Molds

Many different types of chemical binders, both organic and inorganic, are used to make molds and cores. For large castings chemically bonded sands are favored as the molds can be made in segments and assembled. One of the concerns with organic binder systems is the formation of lusturous carbon on the surface of castings, especially in ductile iron castings due to high carbon-equivalent irons. Pouring at higher temperatures, use of mold washes and oxidants added to the sand mix can control the lusturous carbon defects.

Gating and Risering

Gating systems are designed to control the mold filling with molten iron. Key requirements in a gating system include:

- Minimizing temperature loss to avoid casting defects such as misruns
- Minimizing metallurgical fade
- Minimizing oxidation and slag and dross generation during mold filling
- Slag separation and containment in the gating system

Apart from the factors mentioned, considerations are given to mold yield, easy separation and handling of gates during shakeout and charging back into melting furnaces.

Gating systems are designed for the combination of molding method and pouring equipment used in the foundry. Most pouring is done by gravity method, although there are instances of counter-gravity pouring of iron castings in special applications. See chapter 5 for more information on gating and risering.

Fluid Flow

Flow of molten metal through the runners and gates is controlled by physical laws applicable to fluid flow.

Shrinkage

There are three stages where shrinkage occurs: (1) during cooling of liquid to onset of solidification, (2) during solidification, and (3) after solidification to room temperature.

Liquid cast irons shrink at the rate of about 0.9% per 100°F (38°C). Actual liquid stage shrinkage amount of the liquid in the mold will depend on the average temperature of liquid after the mold is filled and the onset of solidification.

Volumetric shrinkage during solidification ranges from negative shrinkage in "soft" irons (high CE gray iron) to +1.94% in an iron containing about 0.90% combined carbon. As most ductile irons (high CE gray iron) are eutectic or hypereutectic, the solidification shrinkage is near zero to 0.5% depending on metallurgical quality of the ductile iron. The white irons undergo 4.0 to 5.5% contraction in volume within the same temperature range.

After the iron is completely solid at the solidification temperature there is shrinkage as the castings cool down to room temperature and is known as solid-state contraction. This shrinkage needs to be compensated in the pattern dimension (pattern size is larger by the amount of solid contraction). Patternmakers' rules (shrink rules) take care of this dimensional change. See Chapter 2 for shrinkage allowance.

Shrink Rules for Different Types of Irons

Gray iron: 1%

Annealed ferritic ductile iron: 0%

As-cast ductile iron: 0.7%

White irons: 1 to 2%

In practice, solid-state contraction depends also on factors other than metal such as: mass of casting, actual composition of metal, size and shape of casting, molding media, mold density, and cores [33]. Generally, contraction is less with an increase in mass and in gray irons, it increases with increasing tensile strength.

Risering Concepts

To compensate for the liquid shrinkage from mold-filling temperature to onset of eutectic solidification, a system of feeders or risers needs to be incorporated in the gating system. Unlike other alloys, graphitic cast irons need the feed metal during liquid cooling and not so much during eutectic solidification. There is expansion when graphite precipitates and shrinkage when autenite precipitates. The austenite contraction is compensated by graphite expansion. There can be net expansion or slight contraction during eutectic solidification. If there is adequate amount of carbon, as expressed in the equation below, there will be no shrinkage if the iron is processed correctly.

$$C + \frac{1}{7}Si > 3.9 \tag{11.18}$$

Calculating Feed Metal Requirements

To calculate the feed metal required for a casting, casting weight, pouring temperature, and the loss of temperature during mold filling should be known. Pouring temperature generally varies from 2450°F (1343°C) to 2600°F (1427°C) depending on the size and pouring time and number of castings in a mold. Lower temperature for heavier castings and higher temperature for thin-wall castings are used. As we have stated earlier, liquid shrinkage can be calculated by the following formula:

$$F = W \times (T - 2100) \times 0.9 \tag{11.19}$$

where F = liquid feed metal needed, lb
$\qquad W$ = weight of the casting, lb
$\qquad T$ = effective metal temperature right after mold filling

F can be converted to V (volume) by dividing it by density, d and hence, V, in cubic inches = F/d (d = 0.25 lb/in^3, density of moletn iron at the pouring temperature).

Total amount of metal needed to be fed to the casting will also include solidification shrinkage, which can vary from 0 to 5%, depending on the carbon equivalent and type and amount of graphite precipitation. In ductile irons, forward-segregating elements, superheated melt, low nucleation levels, and high Mg residuals increase the amount of feed metal required. In white irons, as there is no expansion from lack of graphite precipitation, total shrinkage volume will be at the high end (Table 11.9).

Material	Solidification Range,°F (°C)	Total shrinkage volume, %
White irons	2550–2065 (1400–1130)	4–8
Malleable irons	2550–2065 (1400–1130)	4–8
Gray irons	2400–2000 (1316–1093)	2–4
Ductile irons	2150–2000 (1177–1093)	2–5
Compacted graphite irons	2150–2000 (1177–1093)	2–4
Austenitic irons	2400–2000 (1316–1093)	3–7

Source: From *Metal Casters Reference and Guide* [33].

TABLE 11.9 Solidification Range for Cast Irons

Feeding from Riser

When the mold is full, depending on the thickness of gates, there could be some feeding from the pour cup until the gate freezes, isolating the riser and casting from the pouring cup and sprue. This is an important factor to remember not only for calculating the total feed metal required but it also affects the design and functioning of the riser. To keep the risers from freezing at the top, gates are made thin to solidify quickly, which will cause the riser to start feeding prior to skin formation, keeping the riser functioning well. If the riser top freezes, it may result in shrinkage at the riser connection to the casting. Generally, risers are located at the heavier section (larger modulus).

Geometric Design of Risers

Most solidification modeling programs use some form of this method: risers are attached to the significant section of the casting. Riser size is calculated based on the feed metal required and solidification time of the significant section. It is essential to ensure the last place to solidify is in the riser. Due to the metal flow through the riser connection, size of the connection may seem small, but it will be adequate to supply feed metal to the casting. Since most graphitic irons will expand during eutectic solidification, it is not necessary to keep the riser connection open until the end of solidification. Heine proposes to shape the riser tapered rather than cylindrical, for optimizing mold yield, and to make the shrinkage void (pipe) in the riser deep in order to keep it from freezing until the riser connection freezes. If there is considerable expansion after the connection freezes, the expanding liquid will exert significant pressure on the mold wall, which will increase the dimensions of the cavity resulting in internal shrinkage porosity [34].

Pressure Relief Risering

As the name suggests, in this method the riser connection is designed to stay open to relieve pressure in the mold cavity due to graphite expansion during eutectic solidification. In this method, the riser feeds liquid shrinkage to the casting cavity and acts as a sink to absorb expanding liquid [35]. The mold wall movement is minimized in this method.

Riserless Castings

There are many instances where no separate risers are used to ensure sound castings [35]. Liquid shrinkage is entirely compensated by feed metal from the pouring basin through runners and gates. Gates are thicker than with risers. This system does not work in all geometries. Molds have to be rigid, to contain the expansion pressures during eutectic solidification. Iron is generally poured at lower temperatures. Metallurgy needs to be controlled to minimize secondary shrinkage or end of solidification shrinkage.

Postcasting Operations

Shakeout

Castings are separated from the mold after they cool below the critical temperature. If castings are exposed to air by broken molds or early shakeout, hardness of graphitic iron castings and normally ferritic castings will increase. Shakeout time is controlled by the largest section or slowest cooling section of the casting. Minimum time in the mold cooling is necessary for consistent hardness control.

White iron castings and gray iron castings need to be handled gently, unlike ductile iron castings to prevent casting breakage and cracking. At the shakeout, molding and core sands are separated and any lumps are reduced for further processing of the sand. Castings will still be hot and need to be cooled before gates and risers are separated. After the runners and gates are removed, castings are cleaned.

Gate and Riser Removal

Gates and risers can be removed by the use of hammers and wedges on the casting cooling conveyor after the sand is separated. As cast irons are notch sensitive, it is easier to remove the risers when sharp notches are provided at the connection between the riser and the casting. In many instances risers may fall off in the shakeout equipment (vibratory shakers or rotary drums) if the connections are designed optimally. For larger castings and risers, hydraulic wedges are used for degating.

Shot Blasting

All ferrous castings can be cleaned in blast machines to remove adhering sand. There are continuous blast machines and batch-type machines. They may use steel shot, grit, and a mixture of both. Shots and grits may be heat treated to increase their life. Working mix, that is, avergae shot size and distribution, is maintained by adding new shots periodically. Steel shots and grits are thrown by impellers at high speed onto castings not only to remove sand and improve surface finish but also to improve the surface condition by imparting compressive stresses on the surface. This improves the fatigue strength of as-cast surfaces similar to shot peening. Sometimes when the castings are heat treated they may be blasted again to impart compressive stress on the surface.

Property Enhancement

Even as castings can be made to many different specifications in the as-cast condition by alloying and controlled cooling in the mold and outside, more varied and improved properties can be obtained by many different thermal treatments to the ferrous castings. Sometimes heat treatments are used to make properties more consistent and to improve machinability. Different heat-treatment processes are discussed below.

Heat Treatment of Iron Castings

Iron castings can be heat treated similar to steel castings. Due to the presence of higher carbon content there is no need to pack the castings with carbon, as would be the case for carburizing of low carbon steel castings. Annealing, stress relieving, normalizing, and quench and temper heat-treatment processes are similar to all irons to get similar effects on microstructures and mechanical properties. Except for stress relieving, which is heating to a temperature range of 1000 to 1200°F (550 to 670°C), all other heat treatments involve heating the castings to austenitizing temperature and then controlling the cooling rate to achieve different matrix microstructures. All these heat-treatment procedures do not change the graphite shape, except that some carbides may be dissolved and new graphite particles may precipitate in their place.

Subcritical Anneal

Castings can be annealed below the austenite transformation temperature to reduce hardness and pearlite content. Subcritical anneal will not break down carbides if they are present as massive carbides but only the carbides in pearlite. This is done to improve toughness and machinability of the castings and to relieve stresses. To achieve higher ferrite content when there are process problems such as early shakeout, improper chemistry, and thin sections with high pearlite content, subcritical heat treatment is used.

Full Anneal

When castings are heated above upper critical temperature, the matrix is transformed into austenite. If the original matrix is pearlite the transformation time is short, as there is very little carbon transport from nodules. If there are simple iron carbides, they may break down to form austenite and graphite particles. Castings are slowly cooled through the transformation range to result in a fully ferritic matrix.

Normalizing

This is a type of hardening treatment. Castings after austenitizing are cooled faster by cooling fans or air cooling in the open to get a fully pearlitic structure. This type of heat treatment is done to increase hardness, make the structure and hardness consistent throughout the casting, and have a minimum pearlite for further heat treatment such as induction or flame hardening. Normalizing improves machinability, but decreases the fatigue endurance limit for the same hardness.

Quench and Temper

To get martensitic structure, castings are cooled very fast, either in water or oil from above the upper critical temperature by avoiding the pearlite nose, as per the Time-Temperature-Transformation (TTT) diagram, as in Fig. 11.49. Special alloy additions to the irons are made to improve hardenabiliy. Castings are tempered immediately after quenching to avoid quench cracks. Tempering makes the martensite tougher and also reduces the hardness. End hardness is controlled by the tempering temperature and time. Martensite improves wear resistance and strength and is used in high-stress applications.

Austempering

Austempering of ductile irons is slightly different from austempering of steel. Due to the higher silicon level the resulting structure does not contain carbides. Depending on the austenitizing temperature and austempering temperature, varied properties can be obtained. Higher austenitizing temperature increases the carbon dissolved in austenite, which increases carbon-stabilized austenite in the final structure. By austempering, much higher strengths with good toughness can be achieved for various engineering applications. A typical austempering cycle is shown in Fig. 11.49.

Austempering temperatures vary from 500°F (260°C) for grade 5 to 725°F (385°C) for grade 1. At lower austempering temperatures, resulting strength is higher with lower elongation and at higher austempering temperatures strength is lower with higher elongation.

ASTM minimum properties are shown in Table 11.22. Grades 4 and 5 are generally used for wear-resistant applications and grades 1 and 2 for high strength and toughness applications.

Intercritical Austempering

By partially austenitizing the matrix and keeping the castings between upper and lower critical temperatures, a tougher grade (grade 750) of austempered ductile iron can be

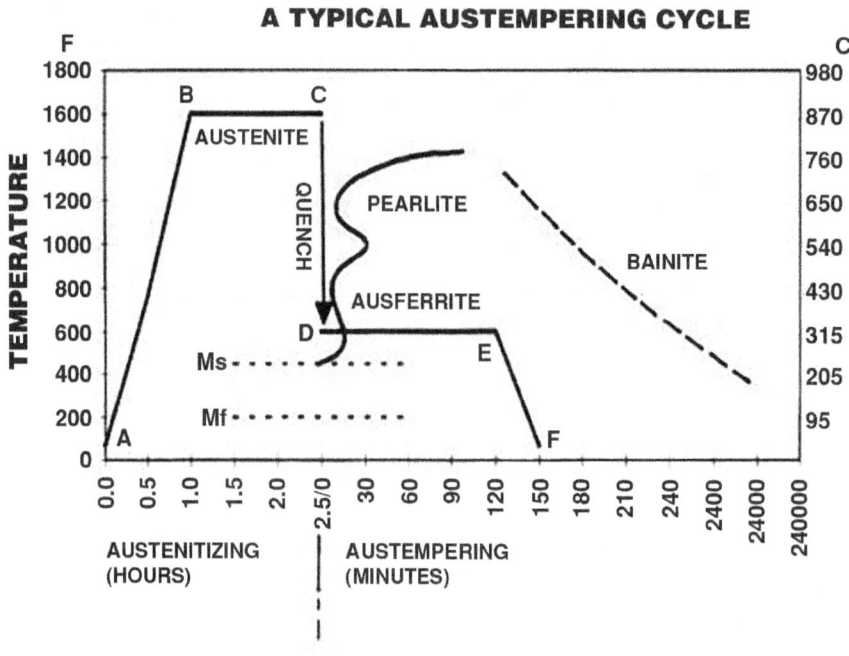

FIGURE 11.49   Diagram illustrating the different steps of austempering cycle superimposed on TTT diagram. (From Hayrynen et al. [32].)

produced. This process is ideally suited for load-bearing safety applications where a combination of high strength and ductility is required and cannot be met by as-cast pearlitic castings.

Carbidic Austempering

To increase wear resistance of austempered ductile iron castings, carbide stabilizing elements like chromium can be added to the iron during processing. Chromium carbides are stable under normal austeniting conditions. After austeniting when the castings are quenched in a salt bath, the original carbides are still present along with ausferrite, which increases wear resistance [30]. These types of castings are used in applications like ground-engaging tools (plough tips).

Surface Austempering or Selective Austempering

Castings such as camshafts and crankshafts can be heated on the surface by induction or gas flame above the austenitizing temperature and then quenched in a salt bath to obtain ausferrite structure on the surface of the casting. This treatment does not change the structure at the center of the casting. This process is economical for improving the surface wear resistance without affecting the rest of the casting microstructure.

Induction and Flame Hardening

These processes are used to change the matrix in local areas, by heating selectively some areas of the castings above austenitizing temperature and cooling that area fast to get martensite. These areas may be tempered to control the hardness.

Physical and Mechanical Properties of Cast Irons

Physical Properties

Important physical properties such as thermal conductivity and damping capacity of cast irons are given in Tables 11.10 to 11.12. Thermal conductivity is influenced by the microstructure of cast irons. Since graphite exhibits the highest thermal conductivity of all phases, gray iron has the better thermal conductivity in the family of cast irons. Again, in gray irons, thermal conductivity of graphite parallel to the basal plane is approximately four times that perpendicular to the basal plane. As shown in Table 11.10, thermal conductivity of gray iron increases with the carbon equivalent. The higher the thermal conductivity, the lower is the thermal gradients throughout the casting, and hence, lower thermal stresses in the component subjected to thermal stress in service. In general, thermal conductivity decreases with temperature.

Thermal conductivity of different cast irons are compared in Table 11.11; alloyed cast irons have lower thermal conductivity than ductile iron.

Gray iron possesses the highest damping capacity which is the ability to stop vibrations and ringing. This is shown in Table 11.12 for different ferrous and nonferrous common structural alloys. Because of its high damping capacity, gray iron is used for structures where vibration can cause stresses. Other examples are bases and supports, as well as moving parts.

			Thermal Conductivity at Different Test Temperatures, W/m·K		
C, %	Si, %	Carbon Equivalent, %	200°F (95°C)	400°F (205°C)	800°F (425°C)
3.50	2.25	4.25	57.1	51.9	
3.93	1.40	4.40	55.4	51.9	46.7
3.58	1.90	4.21	48.5		
3.16	1.54	3.67	46.7		41.5
2.92	1.75	3.50	36.3		
2.90	1.51	3.40 (Alloyed)	36.3	36.3	36.3

Source: From Davis [15].

TABLE 11.10 Thermal Conductivity of Gray Irons for Different C and Si Contents

	200°F (95°C)	400°F (205°C)	800°F (425°C)
Ductile irons			
Ferritic ductile iron	41.5	38.1	32.9
Pearlitic and alloyed	26	27.7	
High alloy irons			
36% Ni steel invar	10.4		
36% Ni cast irons	39.8		

Source: From Davis [15].

TABLE 11.11 Thermal Conductivity of Different Types of Cast Irons, W/m·K

Material	Relative Damping Capacity
Coarse flake gray iron	100–500
Fine flake gray iron	20–100
Malleable iron	8–15
Ductile iron	5–20
Pure iron	5
White iron	2–4
Eutectoid steel	4
Aluminum	0.4

Source: From Davis [15, p. 854].

TABLE 11.12   Relative Damping Capacities of Some Common Structural Alloys

Thermal Expansion

For the temperature range from room temperature to 1070°C (1960°F), the coefficient of thermal expansion varies from 9.2 to 16.9 μm/m·°C (5.1 to 9.4 μin/in·°F). At room temperature, the commonly used figure of 10 μm/m·°C (5.5 μin/in·°F) is accurate enough for moderate changes in temperature.

Density

Density of gray irons at room temperature varies from about 6.95 g/cm^3 for open-grained high-carbon irons to 7.35 g/cm^3 for close-grained low-carbon irons. The density of white iron is about 7.70 g/cm^3.

Shrink Rule

Patternmakers' rules (shrink rules) allow 1% linear contraction upon solidification and cooling of gray iron, 0% for annealed ductile iron, 0.7% for as-cast ductile iron, and 1 to 2% for white iron.

Mechanical Properties and Specifications

The single most important factor affecting the strength of graphitic iron castings is graphite shape, size, and distribution. The strength of irons similar in chemistry can vary from 20,000 to 120,000 psi in the as-cast condition just by modifying the graphite shape from flake to nodular. Strength can also be affected by manipulating the matrix for the same graphite structure. All of this can be seen in the various specifications as shown later in the chapter.

All cast irons are section sensitive (properties are dependent on section thickness), as shown in Fig. 11.50 [4]. Many properties and microstructures are affected by the cooling rate of the castings both during solidification and during solid-state transformation. Gray irons are more sensitive to variations in section size compared to malleable and ductile irons. During solidification thicker sections cool slower resulting in large cell size (large graphite flakes) in gray irons and lower nodule count in ductile irons. Low solidification rates increase segregation tendency of carbide-forming elements. At the eutectoid transformation region, heavier sections cool slowly resulting in more ferrite and less pearlite for the same chemistry. Even if the matrix stays pearlitic in gray irons, due to the coarseness of graphite flakes and pearlite lamellae, tensile strengths, and

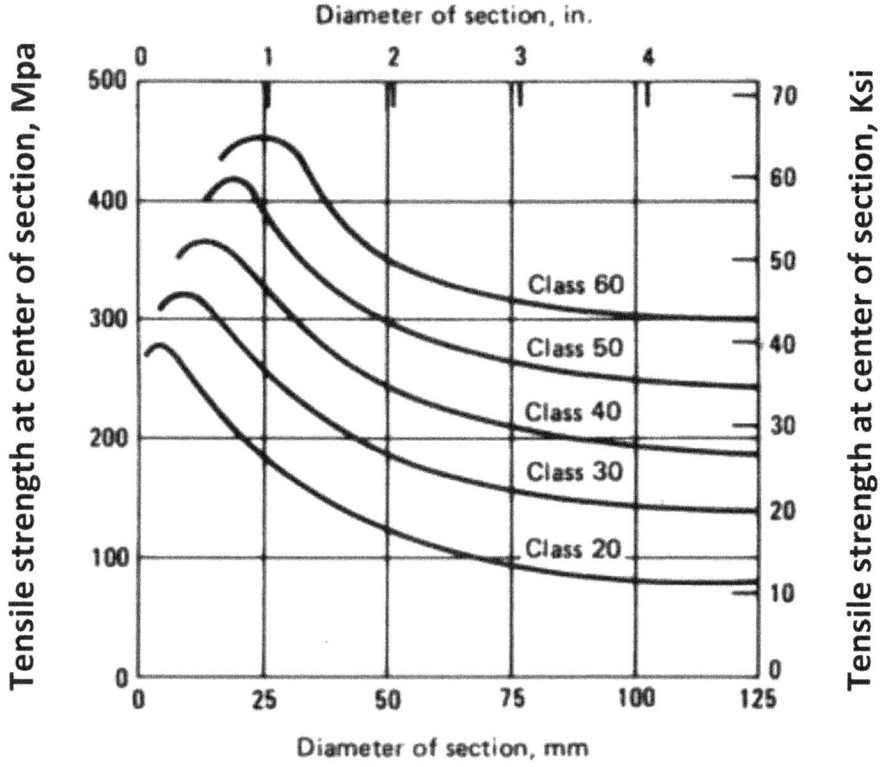

FIGURE 11.50   Decreasing tensile strength with increasing section thickness, illustrating the section sensitivity of gray irons. (From Walton [4].)

hardness are lower in heavier sections than in thinner sections. In ductile irons, heavier sections cooling slowly will have an increased amount of ferrite, unless castings are above the upper critical temperature during mold shakeout.

Different specifying bodies have issued property standards to be met by casting producers. It is not the intent of this book to provide the exact requirements for different grades of materials. Usually, casting producers and buyers agree upon the minimum properties expected from the castings, whether they are from separately cast test bars or from the casting themselves. The tables given here are for reference only and not for strict enforcement of the standards. There are many other details in standards that are not shown here and should be obtained from the specifying bodies to get the latest information. Mechanical properties of SAE J431 automotive gray irons and ISO 185 specifications for gray irons are given in Tables 11.13 and 11.14, respectively. Their graphite size chart, as in ISO 945, is shown in Table 11.15.

Malleable Iron

Mechanical properties of malleable irons are given in Table 11.16, where similar standards are grouped together with class or grade. Automotive grade malleable irons are also included in this table. (See *ASM Handbook*, p. 889.)Typical applications include

SAE Grade	Hardness, HB	Minimum Transverse Load		Minimum Deflection		Minimum Tensile Strength	
		kg	lb	mm	in	MPa	Ksi
G1800	187 max	780	1720	3.6	0.14	124	18
G2500	170–229	910	2000	4.3	0.17	173	25
G3000	187–241	1000	2200	5.1	0.20	207	30
G3500	207–255	1110	2450	6.1	0.24	241	35
G4000	217–269	1180	2600	6.9	0.27	276	40

Note: Properties determined from as-cast test bar (30.5 mm, or 1.2 in, diam).

TABLE 11.13 Mechanical Properties of SAE J431 Automotive Gray Cast Irons

Grade	Tensile Strength, Separately Cast Test Bars, MPa/Ksi
100	100/14.5
150	150/21.75
200	200/29
250	250/36.25
300	300/43.5
350	350/50.75

TABLE 11.14 ISO 185-2005 Gray Cast Iron Classification

Reference Number	Dimension of Graphite Particle Observed at 100 Magnification, mm
1	>100
2	50–100
3	25–50
4	12–25
5	6–12
6	3–6
7	1.5–3
8	<1.5

TABLE 11.15 ISO 945-2008 Graphite Size Chart 1 through 8

pistons, transmission gears, connecting rods, compressor crank shafts and hubs, flanges, pipe fittings, etc.

White Iron (Abrasion-Resistant Cast Iron)

To make the iron-iron carbide system stable or to increase the amount of carbides present, different elements such as Cr, Mo, V, etc., are added to the alloy during melting and treatment. The high-alloyed white cast irons fall into two major groups: (1) Nickel-chromium white irons, which are low-chromium alloys containing 3 to 5% Ni and 1 to 4% Cr, with

Specification No.	Class or Grade	Tensile Strength		Yield Strength		Hardness, HB	Elongation, %
		MPa	Ksi	MPa	Ksi		
Ferritic							
ASTM A 47 and A 338, ANSI G48.1, FED QQ-1-666c	32510	345	50	224	32	156 max	10
	35018	365	53	241	35	156 max	18
ASTM A 197	–	276	40	207	30	156 max	5
Pearlitic and Martensitic							
ASTM A 220, ANSI G48.2, MIL-11444B	40010	414	60	276	40	149–197	10
	45008	448	65	310	45	156–197	8
	45006	448	65	310	45	156–207	6
	50005	483	70	345	50	179–229	5
	60004	552	80	414	60	197–241	4
	70003	586	85	483	70	217–269	3
	80002	655	95	552	80	241–285	2
	90001	724	105	621	90	269–321	1
Automotive							
ASTM A 602, SAE J158	M3210	345	50	224	32	156 max	10
	M4504	448	65	310	45	163–217	4
	M5003	517	75	345	50	187–241	3
	M5503	517	75	379	55	187–241	3
	M7002	621	9	483	70	229–269	2
	M8501	724	105	586	85	269-302	1

Source: Courtesy of ASTM.

TABLE 11.16   Properties of Malleable Iron Castings

one alloy modification that contains 7 to 11% Cr. The nickel-chromium irons are also commonly identified as Ni-hard types 1 to 4. (2) The chromium-molybdenum irons contain 11 to 23% Cr, with up to 3% Mo and are often additionally alloyed with nickel or copper.

A third group comprises the 25 or 28% Cr white irons, which may contain other alloying additions of molybdenum and/or nickel up to 1.5%.

Carbidic cast irons such as Ni-hard, are used in wear-resistant or abrasion-resistant applications. Some Ni-hard microstructures are shown in Fig. 11.20. For further reading on abrasion-resistant irons, see Ref. 13.

Compositions and mechanical properties of abrasion-resistant cast irons or white irons are given in Tables 11.17 and 11.18, respectively, in accordance with ASTM A532/A532M

Ductile Iron

Ductile iron tensile properties are ensured by proper microstructure. Once the graphite shape is ensured to be at least 85% nodularity, there is very little change from the graphite

Class	Type	Designation	Composition, %				
			C	Si	Ni	Cr	Mo
I	A	Ni-Cr-HC	2.8–3.6	0.8 max	3.3–5.0	1.4–4.0	1.0 max
I	B	Ni-Cr-LC	2.4–3.0	0.8 max	3.3–5.0	1.4–4.4	1.0 max
I	C	Ni-Cr-GB	2.5–3.7	0.8 max	4.0 max	1.0–2.5	1.0 max
I	D	Ni-HiCr	2.5–3.6	2.0 max	4.5–7.0	7.0–11.0	1.5 max
II	A	121%Cr	2.0–3.3	1.5 max	2.5 max	11.0–14.0	3.0 max
II	B	15%Cr-Mo	2.0–3.3	1.5 max	2.5 max	14.0–18.0	3.0 max
II	D	20%Cr-Mo	2.0–3.3	1.0–2.2	2.5 max	18.0–23.0	3.0 max
III	A	25%Cr	2.0–3.3	1.5 max	2.5 max	23.0–30.0	3.0 max

Source: Courtesy of ASTM.

TABLE 11.17   Compositions of Abrasion-Resistant Cast Irons in Accordance with ASTM A 532/A 532M

Class	Type	Designation	Hardness, HB				Typical Section Thickness	
			Sand Cast, min.	Chill Cast, min.	Hardened, min.	Softened, max.	in	mm
I	A	Ni-Cr-HC	550	600	600	–	8	200
I	B	Ni-Cr-LC	550	600	600	–	8	200
I	C	Ni-Cr-GB	550	600	600	400		
I	D	Ni-HiCr	550	550	600	–	12	300
II	A	121%Cr	550	550	600	400		
II	B	15%Cr-Mo	450	–	600	400	4	100
II	D	20%Cr-Mo	450	–	600	400	8	200
III	A	25%Cr	450	–	600	400	8	200

Source: Courtesy of ASTM.

TABLE 11.18   Mechanical Property Requirements for Abrasion-Resistant Cast Irons in Accordance with ASTM A 532/A 532M

morphology, unlike the gray iron, where the tensile properties are greatly affected by graphite size and distribution. Strength and ductility are significantly affected by the matrix microstructure. When the matrix is predominantly pearlitic both yield and tensile strengths are higher and elongation is lower. On the other hand, when the matrix is ferritic, the strengths are lower and ductility is higher. This is shown clearly in the minimum requirements of these properties in published specifications (Tables 11.19 and 11.20).

Properties of all the regular ASTM A536 grades except 120-90-02 grade can be met in the as-cast condition when the chemistry is closely controlled. Casting users may require heat treatment for grades 60-40-18 and quench and temper heat treatment is normally used to get 120-90-02 grade. The ASTM standards were developed during the early production period of ductile iron, when it was common to use nickel-magnesium alloys. Current practice is without nickel except in special circumstances and hence, the mechanical properties may be different and customers may expect properties in the castings to meet their specifications, which may be different from the standards shown in Table 11.19.

ASTM A536-84 (2004) Ductile Iron			
	UTS, Ksi (MPa)	YS, Ksi (MPa)	E, %
60-40-18	60(414)	40(276)	18
65-45-12	65(448)	45(310)	12
80-55-06	80(552)	55(379)	6
100-70-03	100(689)	70(482)	3
120-90-02	120(827)	90(620)	2

Source: Courtesy of ASTM.

TABLE 11.19   ASTM Grades for Standard Ductile Irons

ISO1083-2004 Ductile Iron			
Grade/JS/S	UTS, MPa	YS, MPa	E, %
350-22	350	220	22
400-18	400	250	18
400-15	400	250	15
450-10	450	310	10
500-7	500	320	7
550-5	550	350	5
600-3	600	370	3
700-2	700	420	2
800-2	800	480	2
900-2	900	600	2
Impact Specifications, minimum Joules, Mean value of three tests (individual test)			
Temp °C	23	−20	−40
350-22-LT			12(9)
350-22-RT	17(14)		
400-18-RT		12(9)	
400-18-LT	14(11)		

TABLE 11.20   International Standard Grades for Ductile Iron

Table 11.19 shows the ASTM grades for as-cast and heat-treated ductile iron castings, properties measured in separately cast test bars. Table 11.20 lists the grades including low-temperature impact properties as per the International Standards Organization (ISO).

It may be necessary to keep the silicon and phosphorous content low to meet the impact specifications.

Austempered Ductile Iron

Mechanical properties of austempered ductile iron are given in Table 11.21 according to ASTM A897 and A897M-06 grades. It is important to ensure that austempered castings are free of carbides, inclusions, and nodule degenerations as they impair properties

Grade	Tensile Strength, MPa/Ksi	Yield Strength, MPa/Ksi	Elongation, %	Impact Energy,* J/ft·lb	Typical Hardness, BHN
750	750/110	500/70	11	110/80	241–302
1	900/130	550/90	9	100/75	269–341
2	1050/150	750/110	7	80/60	302–375
3	1200/175	850/125	4	60/45	341–444
4	1400/200	1100/155	2	35/25	388–477
5	1600/230	1300/185	1	20/15	402–512

*Note: Impact testing is at room temperature, unnotched.

Source: Courtesy of ASTM.

TABLE 11.21   ASTM A897 and A897M-06 Grades for Austempered Ductile Irons

significantly more than in as-cast condition. Nodule count may be maximized to reduce the effect of segregation.

Quality of ductile iron can be evaluated by looking at the relative elongation for tensile or yield strength [36]. One easy way is to plot the yield strength against elongation superimposed on the ASTM minimum properties curve, as shown in Fig. 11.51. When the test points fall below the ASTM minimum curve, as is the case with point A,

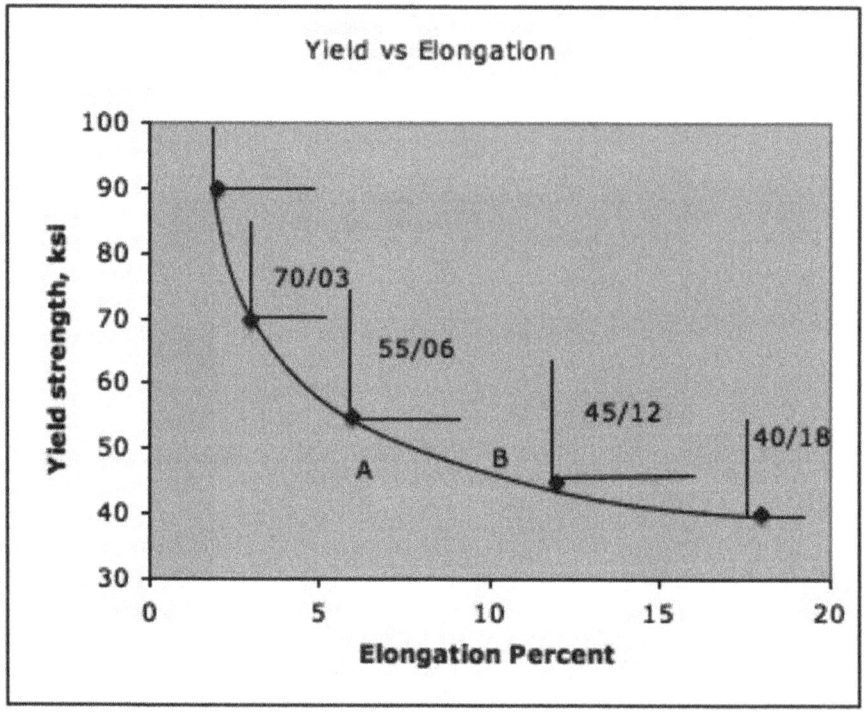

FIGURE 11.51   Yield strength versus elongation as per ASTM minimum properties for different grades. Test point A is inferior ductile, whereas point B is considered good-quality ductile iron even as it does not fall within one of the standard grades. (From Loper and Kotschi [36].)

there could be shortfall in the control of ductile iron production. When carbides, either chill or inverse chill, shrinkage porosity, nonmetallic inclusions and deformed graphite nodules are present in the microstructure, the test point may fall below the curve. When the test points are above the curve, ductile iron quality is above par even if it did not meet a given minimum yield or elongation, represented by point B in Fig. 11.51.

Compacted Graphite Iron

Factors influencing the mechanical properties at room and elevated temperatures of CG irons are composition, structure (nodularity and matrix), and section size. Structure is also affected by type of raw materials, preprocessing variables (superheat, holding time, desulfurization), melt processing (graphite compaction and postinoculation), and postsolidification processing (cooling rate, shakeout time, and heat treatment). Typical mechanical properties according to ASTM A842 are given in Table 11.22. Their strength properties are comparable to those of ductile irons. Elongation values are significantly higher than those of flake graphite irons.

Customers may specify upper limit on nodularity in the microstructure to ensure damping and other properties to meet their needs.

Tensile/Hardness Ratio

Tensile/hardness (T/H) ratio is a nondimensional number, which indicates the quality of irons. Brinnel hardness number is converted to compressive strength by multiplying it by 9.81, where the units are in MPa. In ductile iron, this ratio ranges from 0.27 to 0.33. This means the tensile strength is about 30% of compressive strength. A higher ratio indicates a better-quality ductile iron. In gray irons, ratio depends on the CE, and the strength of the irons. Ratio is lower with higher CE and low-strength irons with longer flakes.

Design Considerations

Castings are designed for easier production of molds, with minimal or no coring. Castings should be easily parted and placed in separate halves of the mold. Considerations should be given to ensure ease of placement of the cores in the molds and to remove the residual core sand after the castings are shaken out.

Generous draft (taper) should be provided for easy release of the pattern from the mold.

Plate castings of uniform thickness are harder to fill without misruns. Casting section could be tapered or ribs can be provided for easy and even filling of the mold cavity.

There are some special considerations apart from general design guidelines when designing iron castings. Cast irons have excellent fluidity during mold filling and thus thin sections can be cast. However, there is a minimum wall thickness for different grades

Grade	250	300	350	400	450
Min. tensile strength, MPa	250	300	350	400	450
Min. yield strength, MPa	175	210	245	280	315
Elongation, %	3	2	1	1	1

Source: Courtesy of ASTM.

TABLE 11.22 ASTM A842 CG Iron Specification Grades

depending on the chilling tendency of the irons. As gray and white cast irons are brittle, thin, unprotected sections could break during shakeout and processing in the foundry.

Gray Iron

Chill carbides tend to form at tips of thin sections, hence, wedge-type ends should be avoided. Unsupported thin sections like fins will be susceptible to breakage during normal shakeout operations unless special handling is available.

As gray irons are section sensitive, thin sections will be harder than thick sections, and tensile properties will vary according to cooling rate in the mold. Hence, it is preferable to have uniform sections in the castings.

Interrupted casting sections are susceptible to chipping during machining operations.

Even as gray iron has very minimal shrinkage during eutectic solidification, isolated sections may have shrinkage if intermediate sections are too thin.

Malleable Iron

Malleable iron castings are cast as white iron. White irons shrink more and require risering to eliminate shrinkage porosity. Directional solidification toward riser is beneficial to ensure solid castings. Thin wall castings can easily be made into white iron. If the casting sections are too big, iron can solidify as mottled or even gray, which will not meet the mechanical property requirements. Hence, the largest section that can be cast as white iron is limited.

Ductile and Compacted Graphite Iron

Ductile irons can be cast in thin as well as thick sections. Even though the solidification could be slower for thick sections, which will result in lower nodule count, strength is not reduced significantly as in gray iron castings with heavier sections. Similar to gray iron isolated heavy sections should be avoided to minimize shrinkage porosity. Reentrant corners with sharp radii can cause shrinkage porosity close to the mold surface, which might open up during machining. Mold density at these locations will be lower, which could reduce heat transfer. Special consideration is also given for inspection of ductile iron castings using ultrasonic method. Parallel surfaces are needed for effective testing of nodularity with ultrasonic systems. For the production of ferritic castings, shakeout time is important. If the shakeout time is short or the casting section is too big for the mold line, temperature at shakeout may be above the eutectoid transformation temperature. When castings are shaken out above the transformation temperature, hardness of castings will increase as pearlite formation is favored when the cooling rate is high. In compacted graphite iron castings, nodularity is dependent on cooling rate. Faster cooling rates increase nodularity, and slower cooling rates may favor flake graphite formation. Hence, uniform size is favorable to maintain vermicular graphite in all sections of the casting.

References

1. R. W. Heine, C. R. Loper, and P. C. Rosenthal, *Principles of Metal Casting*, McGraw-Hill Company, New York, 1967.
2. C. R. Loper Jr, "The Wonder of Cast Iron—From Art to an Engineering Material", *AFS Transactions*, Paper No. 07-190, 2007.

3. G. F. Vander Voort, Vander Voort Consulting, private communication.

4. C. F. Walton (ed.), *Gray and Ductile Iron Castings Handbook,* Gray and Ductile Founders' Society, Cleveland, OH, 1971. (It was revised in 1981 as *Iron Casting Handbook* by the Iron Casting Society, and was copyrighted to the American Foundry Society, Des Plaines, IL in 2003.)

5. G. M. Goodrich, "Introduction to Cast Irons," *ASM Handbook*, Vol. 15. ASM International, 2008, pp. 785–811.

6. *Susan P. Thomas Cupola Handbook*, 6th ed., American Foundry Society, Schaumburg, IL. 2008.

7. Staff Report, "Optimizing SiC Addition for Cupola Melting, Modern Casting," Jun. 2012, pp. 36–39.

8. AFS, courtesy of American Foundry Society.

9. *The Sorrelmetal Book of Ductile Iron*, Rio Tinto Iron & Titanium, 2004.

10. D. Stefanescu, *Ductile Iron Handbook*, AFS, 1993, Chapter 1, pp. 1–19.

11. C. R. Loper, Jr, and K. Fang, "Flake Graphite Structure in Cast Iron," *AFS Transactions*, 116: 665–672, 2008.

12. E. Fras, M. Gorny, and H. F. Lopez, "Eutectic Cell Count, Chilling Tendency and Chill in Flake Gaphite Cast Iron, Part I, Theoretical Analysis," *AFS Transactions*, Vol. 115, Paper No. 07-013, 2007.

13. R. B. Gundlach, "White Iron and High Alloyed Iron Castings," and "High-Alloy Graphitic Irons," *ASM Handbook*, Vol. 15, Castings, 2008, pp. 896–908.

14. C. R. Loper, Jr, and K. Fang, "Structure of Spheroidal Graphite in Cast Iron," *AFS Transactions*, 2008, pp. 673–682.

15. J. R. Davis (ed.), *ASM Specialty Handbook—Cast Irons*, ASM International, Materials, Park, OH, 1996, p. 17.

16. G. L. Rivera, R. E. Boeri, and J. A. Sikora, "Research Advances in Ductile Iron Solidification," *AFS Transactions*, Vol. 111, Paper No. 03-159, 2003.

17. T. Mizoguchi, J. H. Perepezko, and C. R. Loper, "Nucleation during the Solidification of Cast Irons," *AFS Transactions*, 105: 89–94, 1997.

18. L. Ruxanda, J. Beltran-Sanchez, J. Massone, and D. M. Stefanescu, "On the Eutectic Solidification of Spheroidal Graphite Iron: An Experimental and Mathematical Modeling Approach," *AFS Transactions*, Vol. 109, Paper No. 01-066, 2001.

19. T. Kanno, I. Kang, Y. Fukuda, T. Mizuki, and S. Kiguchi, "Effect of Pouring Temperature and Composition on Shrinkage Cavity in Spheroidal Graphite Cast Irons," *AFS Transactions*, Vol. 114, Paper No. 06-084, 2006.

20. R. W. Heine, "Major Aspects of Processing Cast Irons," *AFS Transactions*, 102: 985–1002, 1994.

21. A. Alagarsamy, "Ductile Iron Treatment Using Pure Mg in a Modified Tundish Ladle," *AFS Transactions*, 100: 235–238, 1992.

22. R. W. Heine, "Magnesium Requirements of Ductile Iron," *AFS Transactions*, 97: 485–488, 1989.

23. M. Popescu, R. Zavadil, and M. Sahoo, "Development of the Compacted Graphite Irons (CGI)—Potential Materials to Produce Lightweight Diesel Engines," CANMET-Materials Technology Laboratory, Report No. CANMET-MTL 2008-9 (TR), 32 pages.

24. C. Leon, U. Ekpoon, and R. W. Heine, "Relationship of Casting Defects to Solidification of Malleable Base White Cast Iron," *AFS Transactions,* 82: 323–344, 1974.

25. F. W. Jacobs, "Practical Application of Liquidus Control for Malleable Iron Melting," *AFS Transactions,* 64: 261–276, 1956.

26. M. Popescu, J. Thompson, R. Zavaid, and M. Sahoo, "Restoring Techniques for Monday Morning Iron," *AFS Transactions,* Vol. 110, Paper No. 02-158, 2002.

27. A. Alagarsamy, "Effect of Rare Earths on Nodularity," internal communication, Intermet Corporation, 1995.

28. J. F. Janowak and R. B. Gundlach, "Development of a Ductile Iron for Commercial Austempering," *AFS Transactions,* 91: 377–388, 1983.

29. D. Venugopalan and A. Alagarsamy, "Effects of Alloy Additions on the Microstructure and Mechanical Properties of Commercial Ductile Iron," *AFS Transactions,* 98: 395–400, 1990.

30. K. L. Hayrynen and K. R. Brandenberg, "Carbidic Austempered Ductile Iron (CADI)—The New Wear Material," *AFS Transactions,* Vol. 111, Paper No. 03–088, 2003.

31. I. Henych, "Some Metallurgical Aspects of Producing Ductile Iron and Holding Treated Metal," *AFS Transactions,* 100: 609–620, 1986.

32. K. L. Hayrynen, J. R. Keough, and B. V. Kovacs, "Effect of Alloying Elements on the Critical Temperature of Ductile Iron," DIS Research Report, No. 22, 1997.

33. *Metal Casters Reference and Guide,* 2d ed., AFS, 1989, p. 159.

34. R. W. Heine, "Design Method for Tapered Riser Feeding of Ductile Iron on Green Sand," *AFS Transactions,* 90: 147–158, 1982.

35. Stephen Karsay, *Gating and Risering, Ductile Iron-III,* QIT, 1981.

36. D. L. Crews, "Quality and Specifications of Ductile Iron," and discussion by C. R. Loper and R. M. Kotschi, *AFS Transactions,* 82: 223–228, 1974.

CHAPTER 12

Steel Castings

Raymond Monroe
Steel Founders' Society of America

Sudhari "Sam" Sahu
Creative Technical Solutions, Inc.

Introduction

Steel is iron-containing carbon (<2%) with small amounts of Mn and Si, and impurities like phosphorus and sulfur. Other elements are added in alloyed steels for achieving other desirable properties. These properties include not only the strength and other mechanical requirements but also improved corrosion resistance, wear resistance, heat resistance, or even change magnetic response or thermal expansion. Today all steel made is melted and cast. The melting of steel was not generally practiced until about 300 years ago. Then in England in 1730s, Benjamin Huntsman developed the crucible melting process that allowed high-quality steel to be cast. This was one of the enabling technologies for the Industrial Revolution. It also made possible the development of steel casting production. The first steel casting in the United States was produced in Buffalo, New York, in 1861.

In the U.S. economy, about 150 million tons of all steel products are required per year. The U.S. steelmaking industry only makes approximately 75% of this requirement and the balance is imported. The steel casting industry is small compared to total steel-making or iron casting industries, producing about 1.3 million tons a year. So steel casting is less than 1% of the steel industry and less than 10% of the iron casting industry.

Although cast steel accounts for only 10% of the total foundry industry sales, steel castings are used for vitally important components in the mining, railroad, truck, construction, military, and oil and gas industries. Some castings used in these industries are shown in Fig. 12.1 [1]. The largest market for steel castings is in the railroad industry. Currently there are approximately 230 steel foundries in the United States. They are specialized, each foundry serving a few industries and pouring a limited number of alloys. Even though the number of foundries is down compared to 40 years ago, the total production capacity is still the same. The surviving foundries have increased capacities and efficiencies. Due to the diversity of market requirements such as size, tolerances, chemistry, volume, etc., a single foundry cannot serve all of the markets and each company will tend to specialize in a portion of the total market.

(a)

(b)

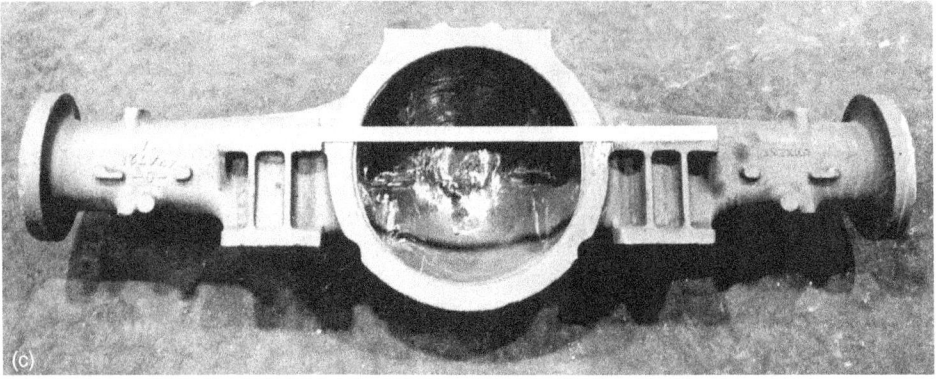

(c)

FIGURE 12.1 Castings used in railroad, truck, mining, petroleum, steel mill, marine, and defense industries. (a) Railroad side frame and boister; (b) railroad wheel; (c) truck axle housing

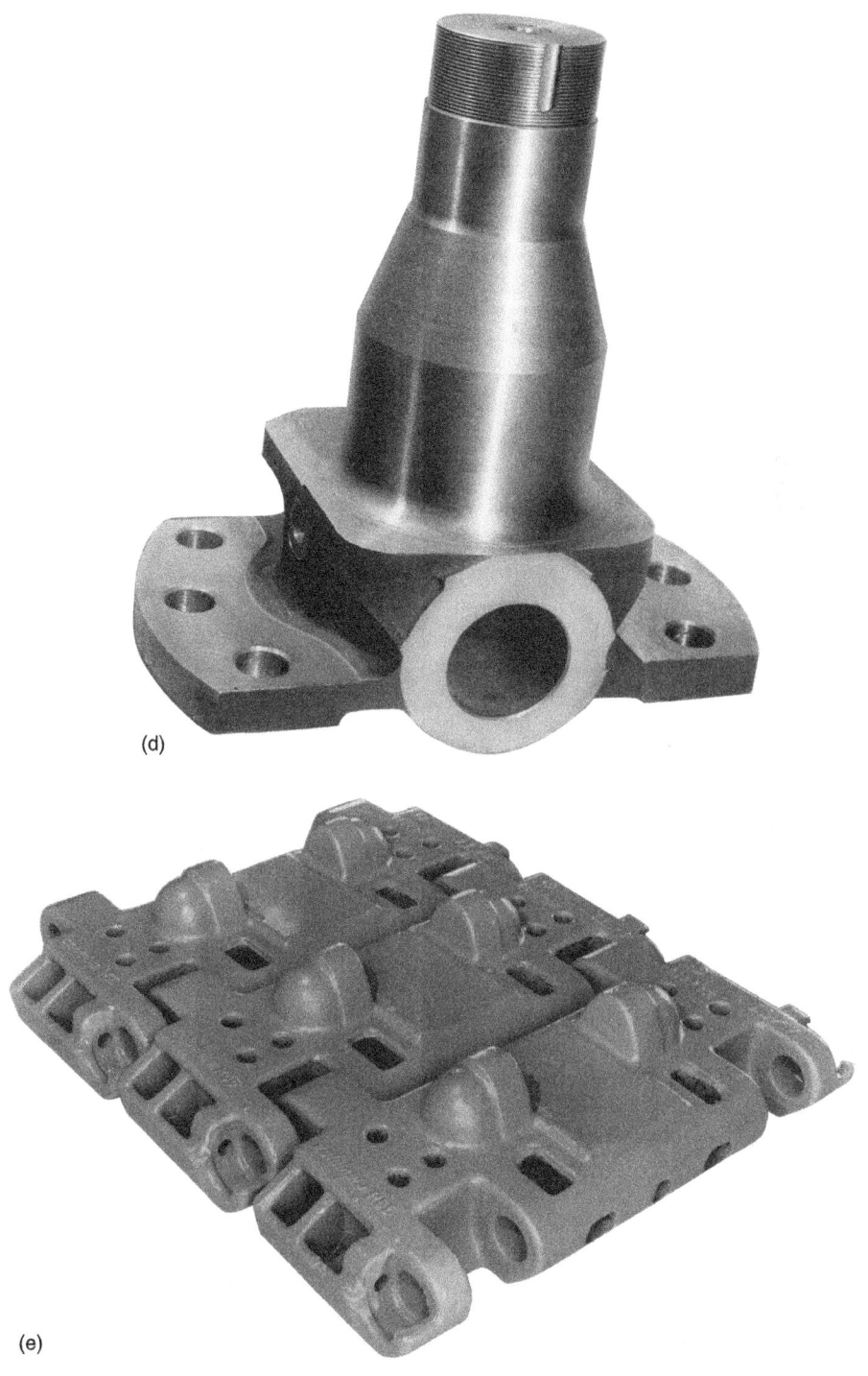

(d)

(e)

Figure 12.1 (*Continued*) (*d*) truck axle spindle; (*e*) mining crawler pad

FIGURE 12.1 (*Continued*) (*f*) mining jaw plate; (*g*) steel mill roll

(h)

Figure 12.1 (*Continued*) (*h*) oil field valve

(i)

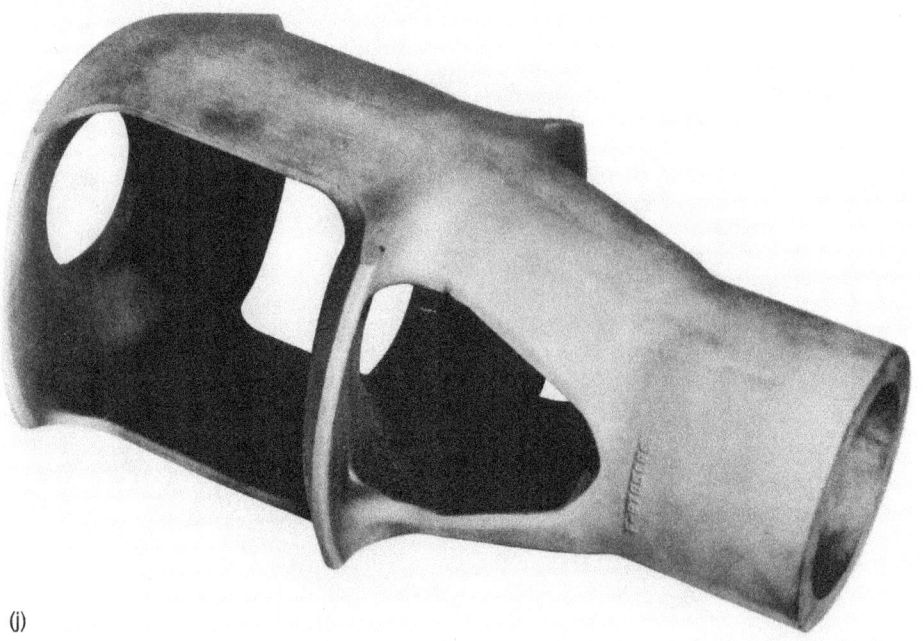

(j)

FIGURE 12.1 (Continued) (i) boat propeller; (j) muzzle brake for military gun. (From *Steel Castings Handbook* [1].)

Year	Shipment	Capacity
1975	1974	2726
1980	1980	2500
1990	1133	N/A
1992	986	1560
1995	1160	1410
2001	778	1260
2004	895	1135
2010	1190	1400
2011	1300	N/A

Source: From Monroe [2].

TABLE 12.1 Steel Casting Shipping and Total Capacity (in 1000 tons)

The total capacity of the steel casting industry in the United States is approximately 1.4 million tons as of 2010 [2], Table 12.1, with a sales tonnage of 1.3 million tons and a sales value of approximately US$3 billion. Steel castings are specified for applications which require weldability, abrasion resistance, high strength, impact toughness, creep strength, and corrosion resistance.

Many of the advantages that make steel long products like bars and plates, an outstanding material of construction, are also available in steel castings. Steel is strong with tensile strength ranging from 400 to 2000 MPa (60,000 to 300,000 psi). Steel is also ductile. The combination of strength and ductility gives steel great toughness and resistance to shock. These properties of steel can be controlled by its composition, specifically its carbon content. Steel's remarkable properties stem from the presence of carbon in iron. For example, when carbon is absent, iron is soft and weak. If carbon is added in as little as 0.2 to 0.3%, the strength is raised appreciably and the ductility, although reduced, is still considerable compared to other common materials. The result is that steel exhibits versatility found in no other metal.

Figure 12.2a shows this effect of carbon on tensile strength and percent reduction of area of plain carbon cast steel. The effect is similar on yield strength and percent elongation.

Properties of steel can be further controlled by heat treatment. Iron and steel change their crystal-lattice structure (i.e., the arrangements of atoms in the solid state) at high temperatures (above 750°C/1380°F) and that makes it possible to control their properties by controlling the cooling rate from that elevated temperature (850 to 950°C/1560 to 1740°F). Further control is also obtained by reheating (tempering or drawing) after rapid cooling (quenching), see Fig. 12.2b [3].

All steel starts as a cast product. Most steel is rolled into a bar or plate or other shape. The rolling operation causes the structure of the steel to have different properties in different directions. In contrast, steel castings have similar properties in every direction. This nondirectional character can be called "isotropic" and is absent in steel that has been rolled into structural shapes from ingots, billets, or other continuously cast shapes. Steel so worked is tough and strong when tested in the direction of rolling but is less ductile and more brittle if tested in a transverse direction. Cast steel does not possess this directionality and is therefore better suited to applications where this directionality might prove harmful.

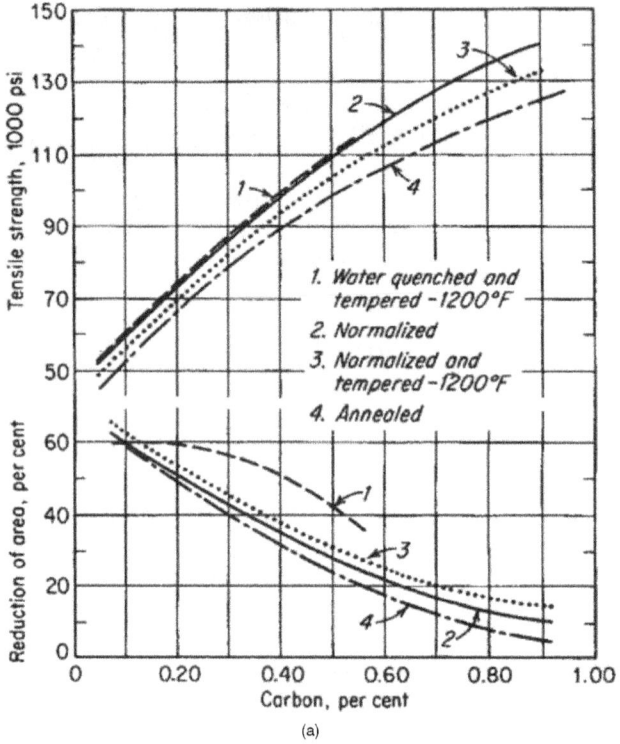

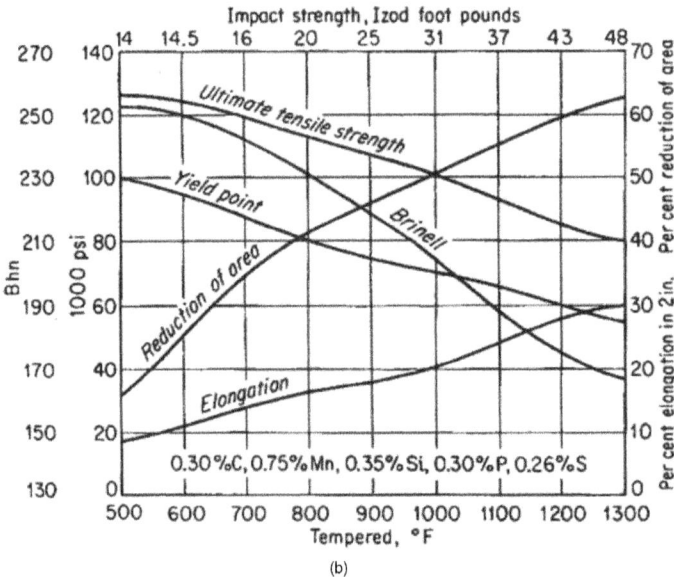

Figure 12.2 (a) Tensile strength and reduction of area versus carbon content for carbon cast steels after various heat treatments, and (b) the effect of tempering on the mechanical properties of a 0.30% carbon cast steel. (From Heine et al. [3].)

A distinct advantage of steel castings compared to iron castings is welding. The fact that steel can be readily welded with no serious loss of properties means welding can be used to integrate steel castings in larger steel structures by fabrication. Welding can also be used to meet all the customer requirements in the production of castings. Of perhaps greater potential importance is the opportunity to combine by welding steel castings with other steel shapes to produce a composite structure of castings and forged or rolled steel parts.

Rather ironically, steel's strength and ductility become a definite handicap in the foundry. Shaped castings including steel castings must have a plumbing system to fill the mold with liquid steel and this plumbing system is called the gating system. Solid steel occupies less volume and is more dense than the liquid. To compensate for this shrinkage, reservoirs designed to solidify last and provide liquid metal are added to the desired casting geometry. These reservoirs are called risers. After casting, removal of these gates and risers presents a problem since the ductility and strength of the steel preclude hammering them off, as is common in brittle alloys like cast iron. Saws, abrasive cut-off wheels, torches, etc., are required for this purpose, leading to higher finishing costs.

Steel has an excellent combination of properties making it the most common material for industrial equipment. For a foundry, it taxes the ingenuity of the designer and metallurgist to produce the best casting. The high pouring temperature of steel demands that attention be given to refractories, ladles, molding sands, metal transfer in the shop, filling the mold with no misruns, and related problems. The high solidification shrinkage of steel also introduces design and molding problems. Steel's reactivity with oxygen and other gases requires controlled melting and refining to produce good-quality castings.

Steel Melting in the Foundry

Cast steel is melted in two types of furnaces:

- Electric arc (acid and basic)
- Electric induction (acid, basic, or neutral)

Some open-hearth and crucible furnaces do exist in the worldwide industry but these are obsolete methods of steelmaking.

The choice of furnace and melting practice depends on many variables, including:

- The plant capacity or tonnage required
- The size of the castings
- The intricacy of the castings
- The type of steel to be produced, i.e., whether plain or alloyed, high or low carbon, etc.
- The raw materials available and prices thereof
- Power costs
- The amount of capital to be invested
- Previous experience

Electric furnaces for steelmaking either apply the current directly to the charge, as in arc melting, or indirectly to the charge as in induction melting. Induction melting can be done in small batches in sealed furnaces which allow steel to be vacuum melted. After melting, the steel can be further processed in a treatment station and an AOD vessel is the most common. The AOD (argon-oxygen decarburization) process is routinely applied for special grades and alloy requirements.

Most steel for castings is melted in an arc furnace. Arc furnaces can be either three-phase AC (alternating current) arc (vast majority) or single-phase DC (direct current) arc. DC furnaces are not common.

Electric Arc Furnaces

Electric arc furnaces are either lined with acid or basic refractories. Basic refractories are not a reference to the refractory's wet chemistry pH but are a chemical designation. Silica (SiO_2) found in quartz or beach sand is an acid refractory. Alumina (Al_2O_3) is neutral. Magnesia (MgO) or calcia (CaO) are basic refractories. Figure 12.3 is a schematic diagram of an arc furnace showing typical refractories used in acid and basic practices [4].

Acid refractories are inexpensive. They have limited temperature range of use and do not allow for reduction of sulfur or phosphorus in the steel. They are less aggressive in picking up hydrogen or nitrogen. They are "dirtier" since the steel made in an acid arc furnace has higher levels of sulfur and oxygen. Most large steel makers and European steel foundries do not use acid refractories. Their use in North America is declining. Since they are less capable of high-temperature operation, significant lining wear with each heat makes acid refractories more of a consumable addition to the slag than a lasting lining.

The acid process depends on having a good grade of scrap available, since neither phosphorus nor sulfur can be removed. With basic furnace practices, either P or S or both can be partially eliminated. Basic refractories are more expensive in their initial installation cost. The cost of melting a ton of steel can be the same or less than acid melting if the basic refractory lining is properly maintained. They hydrate (pickup water) when allowed to cool, so extending their life requires keeping the furnace lining hot.

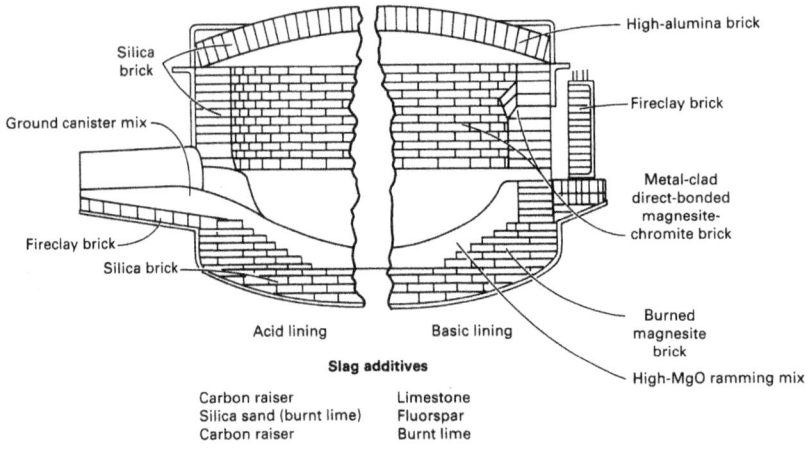

Figure 12.3 Cutaway view of electric arc furnace showing the typical refractories used. (From *Metals Handbook* [4].)

The lining loss during melting can be reduced by using water-cooled panels. If the steel shell of the furnace holding the refractory is cooled on the outside with water, the refractory operates at a lower temperature and experiences less loss of the lining during melting and holding. This reduces the energy efficiency of the furnace slightly but greatly extends the lining life. It is more important in basic melting to remove the sand and dirt from the charge materials, especially the returns, since this is silica that will become part of the slag and will erode the refractory and neutralize the basic slag.

In electric arc melting of steel, scrap is placed in a furnace, and the furnace uses electrodes to arc on the scrap and melt it. While it is possible to use a single electrode in a DC arc furnace, steel foundries use AC arc furnaces. The power company generates electricity in three opposing phases. AC arc furnaces use all three phases for melting. Each electrical phase is attached to a large graphite electrode that strikes an arc in the scrap. The arc furnace has a steel shell which is bowl shaped with a spout lined with a refractory. Scrap is placed on the refractory-lined shell. The furnace roof has three graphite electrodes that can move up and down. The electrodes are retracted and the roof swung into place. Then the electrodes are lowered with power and each electrode strikes an arc on the steel scrap below. Since there are three opposing electrodes, no ground is needed for the system since the electric potential is developed between the electrodes. The arcs are managed to maximize the power input into the scrap until it is melted. The three phases are managed through a large transformer that allows the melters to control the arc during the melting process.

The size of an arc furnace is measured by the diameter of the shell and the power available. The electrode size is another common measure of the furnace size. Furnaces as large as 300 tons are not uncommon in mini-mill steel production.

Electric arc melting includes charge, melt down, oxidation, and finishing. Charge materials are selected to meet the material requirements and are specified by chemistry, cleanliness, and size. If the scrap is rusty, then oxidation losses of alloys increase. If the scrap is oily, then the sulfur, carbon, and hydrogen can increase. It is also likely to create a black smoke that is objectionable and may clog the air collection system. If the scrap is dirty, recovery of alloys is lower and it increases the slag and refractory wear. If the scrap or any charge material is wet, it is dangerous and can result in explosions or eruptions of metal or slag, causing fire and injuring people. Lead-containing scrap can cause the steel to fail to meet mechanical properties and expose the melters and other foundry workers to unacceptable levels of lead in the air.

Scrap and charge materials need to have known chemistries. For alloys and other materials, this can be done through the purchase specification. For scrap, it is useful to check a reasonable sample (at least three pieces) to make sure the composition is correct. Most heats use returns (gates and risers that are removed from the castings). It is important to know the composition of the returns either with good records and storage or through analysis. Scrap should be at least 3 mm ($\frac{1}{8}$ in) thick and preferably over 12 mm ($\frac{1}{2}$ in) thick to provide adequate density of the charge and reduce oxidation in melting. The scrap should be less than one-quarter of the furnace diameter in the longest length. The scrap charge density should be over 1440 kg/m^3 (90 lb/ft^3). A layer of lighter material (like turnings) can be placed in the bottom of the furnace prior to dumping the bulk of the scrap charge to protect the bottom of the furnace from damage. Scrap that is too low in density will melt slowly and may take several charges to fill the furnace. Scrap that is too heavy or big may damage the furnace lining or shift and break an electrode during melting.

Some of the needed alloys may be present in the scrap or be added with the charge. Making alloyed steel grades requires adding alloying elements like chromium, molybdenum, and nickel. Carbon is included in the charge to meet the steel grade requirements and to supply the carbon for the oxygen blow. Since the elements in the original charge oxidize during melting and especially during the carbon blow, most of the alloying elements that oxidize must be added later to the heat.

After the furnace is charged with scrap and alloys, the roof is swung over the charge. The electrodes are lowered into the scrap with the power on and the melting begins. The electrodes melt steel underneath and then are lowered into the hole in the scrap allowing more efficient melting. The electrodes are continuously adjusted in power and height to ensure the maximum melting rate and energy efficiency. The arcs are continued until the scrap is all melted.

Arc furnaces are normally used in the most continuous manner feasible. Melting in a cold furnace takes more time and energy and so is more expensive. The refractories used for the shell and roof crack, spall, and degrade when they cool. They also hydrate (pick up moisture from the air) and this degrades their performance and life.

In addition to semicontinuous operation, energy can be saved and heat times reduced through preheating the charge. The practical problems of preheating scrap and other charge materials make this unusual in a steel foundry, although many foundries do a minimal preheat to make sure the charge materials are dry. The heat times can also be reduced and the energy efficiency improved through using jets of natural gas burners to heat the charge while melting under power.

Once the furnace is charged, the heat should be melted as fast as possible. This minimizes power required, alloy loss, gas pickup, and electrode and refractory consumption.

During meltdown, oxidation of iron, manganese, silicon, and carbon occurs. Part of the oxygen comes from the atmosphere and part from the solid oxides such as rust on charge material or sand (SiO_2) adhering to returns of risers and gates. As soon as a slag forms, oxidation may be thought of as occurring between the oxidizable elements dissolved in the metal and the FeO dissolved in the slag, although the original source of the oxygen in the slag could still be the atmosphere.

FeO, SiO_2, and MnO formed by oxidation become a part of the slag. In the case of the basic practices, the slag volume is further increased by the lime additions that have been made with the charge or during the meltdown.

The carbon content at meltdown is governed by the amount added with the charge and the amount lost during melting. In the arc furnace melting, enough carbon is included at meltdown to provide for a carbon boil during the refining period. Because of the lower temperature during meltdown, manganese and silicon oxidize more than carbon; so carbon is not removed as readily as these two elements during meltdown. Most of the carbon remains after melting. The silicon content is usually below 0.05% in basic furnaces and below 0.20% in acid furnaces at meltdown. The manganese losses are as great as those of silicon and the amount of Mn retained at meltdown will range between 0.05 and 0.40%, the low value coming from well-oxidized heats starting with a low-manganese charge, and the higher value from high-manganese content in the original charge.

Once the bath is melted, the carbon boil (the oxidation of carbon in the charge, $\underline{C} + \underline{O} \rightarrow CO$) is initiated. The carbon boil is instituted and controlled by the injection of oxygen. This reduces the heat time and removes some of the hydrogen and nitrogen. The heat charge contains excess carbon, at least 0.30% above the aim final, typically 0.40

to 0.60%. The oxygen is injected in the molten steel through a steel pipe when the molten bath reaches the desired temperature. The oxygen is added rapidly enough to reduce the carbon level in the bath quickly. This reaction resembles boiling and is called the *carbon boil*. To remove the hydrogen and nitrogen, it is necessary to remove the carbon quickly, 0.03 to 0.12% per minute. The CO bubbles forming are attractive to the nitrogen and hydrogen dissolved. This leads to a reduction in these normally undesirable gases. Nitrogen may be over 120 ppm in a dead melted (no carbon boil) steel but is reduced to below 80 ppm with a carbon boil. Similarly, hydrogen may be 4 to 6 ppm in a dead melted heat but can be reduced below 4 ppm with a carbon boil. Silicon essentially will be removed, half the manganese will be removed, some chromium will be lost into the slag, but the nickel and molybdenum will remain in the steel.

The carbon boil not only reduces the carbon to the desired level and removes nitrogen and hydrogen, it also raises the steel bath temperature. This addition of chemical energy improves the energy efficiency and reduces the heat time. It is important not to overshoot the desired carbon reduction during oxygen injection. Low-carbon contents may require re-adding the carbon which is difficult to dissolve and extends the heat time. It also allows nitrogen and hydrogen to be reabsorbed. If the carbon content is too low (<0.15%), then the heat will also see an increase in oxide inclusions.

After melting and the carbon blow, the slag is typically deoxidized. This is done by killing the heat with the addition of silicon and manganese. It has become a common practice to add some aluminum as well, to deoxidize both the steel and the slag. This allows the recovery of alloys and metallics in the slag. It is also associated with lower levels of inclusions.

A preliminary chemical sample is taken when the bath is molten to allow the melter to assemble any alloy additions needed to correct the heat composition to the desired chemistry. Alloys that oxidize must be added after the oxygen blow to avoid losing them in the boil. After the heat is killed and alloys are added, a heat chemistry is taken to make sure the alloy meets the requirements for the desired grade.

When the temperature of the heat and composition are correct, the heat can be tapped. This is done by tilting the furnace to pour the steel into a waiting ladle. Most arc furnaces in steel foundries are in pits and not on pedestals. The ladle is lowered next to the furnace in the pit and the heat with all the slag is tapped into the waiting ladle.

After blocking/deoxidation, the goal is to get the steel out of the furnace as quickly as possible. It may be necessary to wait for the chemistry to be confirmed, but any delay is costly and detrimental to quality. If the tapping is too delayed, then the arc may need to be restarted to reheat the steel. This is always detrimental. The arc ionizes the air and is a potent source of nitrogen and hydrogen. The arc also disrupts the slag covering allowing reoxidation of the heat. The goal is to make this less than 10 minutes. If the melting crew is capable, no additional electrical arcing is required after the beginning of oxygen injection. If more heat is needed then the arc can be resumed, but this is undesirable.

Basic refractories also allow the reduction of sulfur or phosphorus in the melting process. Maintaining a basic slag is critical to achieve these reductions. The basicity of a slag is related to the V ratio, the calcia plus magnesia divided by the silica, $V = (CaO + MgO)/SiO_2$. If the V ratio exceeds 1.2, the slag is basic. If it is less than 0.8, the slag is acid and in between the slag is considered neutral. While many steel foundries run basic lining in their arc furnace, few maintain a strongly basic slag. Most steel foundries with electric arc furnaces use basic refractories but do not manage the slag chemistry to have a basic slag.

Control of Phosphorus, Sulfur, and Oxygen in Arc Melting

Phosphorus

Removing phosphorus from steel heats requires a slag that is strongly basic and oxidizing. This is shown at a constant temperature of 1600°C in Fig. 12.4 [5].

Lower temperatures also increase dephosphorization. Other factors that influence dephosphorization are as follows:

- *Slag viscosity.* A viscous slag slows reaction and hinders dephosphorization. This happens if the slag becomes too basic. This is not the result of thermodynamic equilibrium, but of reaction rates or kinetics.

- *Active bath action.* A vigorous boil provides a stirring action which increases the contact area between metal and slag. Again, this is a question of kinetics rather than equilibrium.

- *Time.* Usually dephosphorization increases with time.

Phosphorus removal requires an oxidizing slag. Just as slag can be basic or acid, it can be oxidizing or reducing. When the slag contains higher levels of FeO or MnO, then it is oxidizing. Oxidizing slags are common at the end of the carbon blow. Oxidized slag is black in color when cooled; see Table 12.2 [6]. Removing phosphorus is favored by high

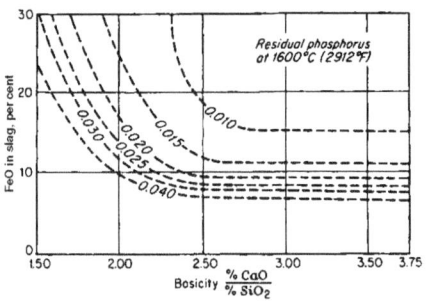

FIGURE 12.4   Relation between phosphorus removal and FeO content and basicity of slag. (From Heine et al. [5].)

Basic Slag			Acid Slag	
Color	FeO%	CaO/SiO$_2$	Color	FeO%
Black	14.85	3.33	Gray-black	30–40
Dark brown-green	11.50	2.19	Dark green	20–30
Light brown	1.05	1.91	Streaked dark green	–
Gray	1.05	1.88	Green	15–20
White	0.56	2.01	Blue-green	–
White	0.51	2.53	Jade-green	<15

Source: From Monroe [6].

TABLE 12.2   Slag Color for Acid and Basic Arc Melting

calcia and iron oxide and low silica. Since the phosphorus will revert to the steel from the slag when the slag and metal is deoxidized, the phosphorus-containing slag must be removed before deoxidation or killing the carbon blow. This is done by tilting the furnace backward and letting the slag fall into a slag pot. Most steel foundries with a basic practice do not routinely use a double slag practice to remove phosphorus because of increased cost. Phosphorus is routinely controlled in steelmaking through charge selection.

Sulfur

Conditions in the average oxidizing basic slag are not favorable for sulfur removal from the metal, and not possible in acid slags. Most steelmaking operations control the phosphorus content with charge materials and use a basic melting practice with a low-oxygen slag to reduce sulfur. The reaction to remove sulfur is

$$S + [CaO] \rightarrow [CaS] + O \tag{12.1}$$

The oxygen in the metal must be kept low by using a reducing agent, such as carbon, in the slag.

The essential features of a desulfurizing slag are as follows:

- High basicity
- High lime content
- Low iron oxide content
- Low temperature
- High carbon, silicon, and phosphorus in the metal since these all increase the activity of sulfur in the iron

The rate of the reaction is increased by use of a fluid slag with fluorspar additions which reduces slag viscosity and by stirring of the bath to improve slag-metal contact.

Since desulfurization requires both high basicity and low oxygen, the steel melted in basic practice is lower in oxygen as well as sulfur. The micro-cleanliness of basic-melted steel compared to acid-melted steel is obvious when looking at a polished unetched sample.

Oxygen

High oxygen content at the time of pouring tends to reduce the quality of steel. Oxygen is distributed between the metal and slag.

High oxygen in the metal is caused by

- Low carbon or silicon contents
- High FeO content in the slag
- Acid slag composition and low slag viscosity
- High temperature

The carbon boil is initiated in the melted steel when the temperature and carbon content are sufficiently high and the silicon is sufficiently low. During this period of refining, the oxygen in the melt is controlled primarily by the carbon content and is fairly low. It is critical not to reduce the carbon excessively, below 0.15%, or the castings produced will have higher levels of inclusions.

When the carbon is low enough (about 0.20% or less) it is no longer effective in controlling the oxygen content of the melt, and slag composition then becomes an important factor. Thus a high FeO content of the slag at this time will not only tend to further reduce the carbon, but will also tend to raise the oxygen content of the steel since at this stage more oxygen can go into solution because of the lowered carbon.

Not only a variation in FeO but other slag oxides affect the degree of oxidation. With either acid or base slags, control of the FeO content can be achieved, at least temporarily, through adjustment in slag composition. For instance, in an acid slag, lime may be added to achieve the transfer of oxygen to the bath because it replaces FeO in the slag, consequently making the FeO more available to oxidize the steel. In a basic heat, a lime addition would exert an opposite effect because it would thicken and chill the slag, thus tending to retain FeO and decrease its transfer to the metal. The slag oxygen can be decreased by the addition of aluminum or carbon to the slag.

If fluorspar is added to a basic slag, it will also affect the oxygen distribution by its effect on the fluidity. A thinner or fluid slag obtained by adding fluorspar will transfer oxygen more rapidly than a viscous one. In other words, the addition of fluorspar gives a more rapid approach to equilibrium through its effect on the rates of reaction. If a slag is too viscous early in the refining period, it will build up in oxygen content. If it is subsequently made more fluid by an increase in temperature or increase in FeO content, the retained oxygen may then be transferred to the metal.

A high temperature of the metal increases the oxygen content in the bath because of the increased solubility of oxygen with raised temperature. The increase in oxygen content at higher temperatures is seen in Fig. 12.5 [7].

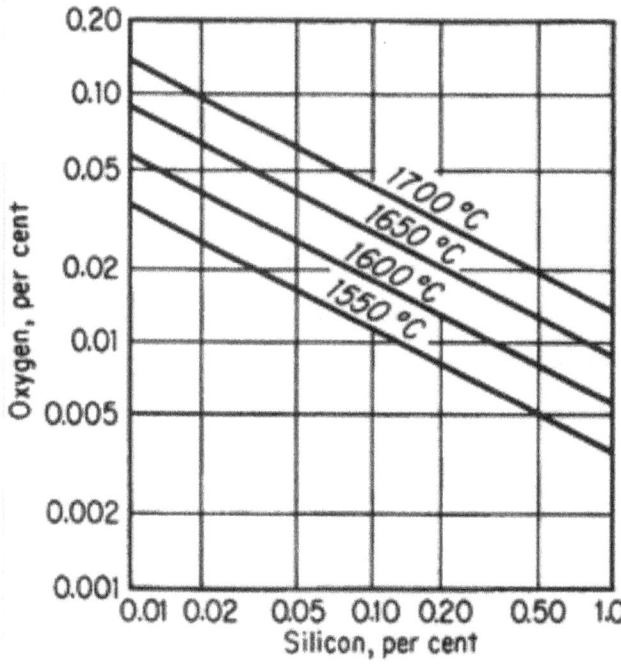

Figure 12.5   Effect of temperature on the silicon-oxygen equilibrium in steel. (From Heine et al. [7].)

Slag Control

The slag also controls the "condition" of the bath. This control can be exerted through an adjustment of the slag composition or of slag viscosity. For example, an increase in the FeO content of the slag will make it more oxidizing, but the actual rate at which reactions take place will be dependent on the slag viscosity. Of course, slag viscosity is dependent indirectly on slag composition but can be varied somewhat independently of the oxidizing power of the slag.

In the acid practice, the slag is siliceous, as indicated in Table 12.3 [8]. At meltdown, the iron oxide content is high, leading to a black color and good fluidity (low viscosity). After carbon boil is killed, the slag tends to be less fluid and has a lighter color. A slag low in FeO and MnO is green. This slag is desirable for retaining a higher manganese percentage in the steel. Although either slag appearance or slag viscosity serves as an indication of the FeO content of the slag, neither is an exact measure since changes in the lime or MnO content of the slag will also affect changes in color or viscosity. The fluidity of acid slag is closely associated with the SiO_2 content.

Adjustments in basic slags are made to aid in removing phosphorus and sulfur. Adjustments are also made to control the degree of oxidation at the end of a refining period. This degree of oxidation is affected both by the carbon content of the melt and by the basic-acid ratio of the slag. Figure 12.6, for example, shows that, for a given slag composition, the higher the iron oxide content of the slag (and also of the metal), the lower the carbon content [9]. It is also varied, however, by changing the base-acid ratio. An addition of lime to the slag should therefore result in a momentary stop in the transfer of FeO to the bath.

Changes in slag fluidity due to temperature or composition changes affect the oxidation rate by affecting the reaction rate at the metal-slag interface. Such changes due to composition are obtained by changing the base-acid ratio or by adding fluorspar.

Induction Melting

Induction melting practices are simpler than electric arc furnace practices. The furnaces are smaller and the furnace bath is deeper and well stirred by the induction currents. Induction melting practice consists of charging, melting, and finishing.

Induction furnaces can be acid, basic, or neutral. The vast majority of them are lined with high-alumina refractories which is considered neutral. Practically no refining takes place in the induction furnace, so the percentage of Al_2O_3 in the refractory depends upon the type of alloy being melted. Alumina content in the refractory is generally 70 to 90%, the balance being mainly silica. Figure 12.7 shows a cross section of an induction furnace [10].

Furnace	Composition, %				Slag Weight, % of Metal Charge
	CaO	SiO$_2$	MnO	FeO	
Basic electric	35–50	12–16	3–13	12–25	5–7
Acid electric	0–10	57–66	15–25	12–20	5+

Source: From Heine et al. [8].

TABLE 12.3   Approximate Composition and Weight of Slags Used in Cast-Steel Refining

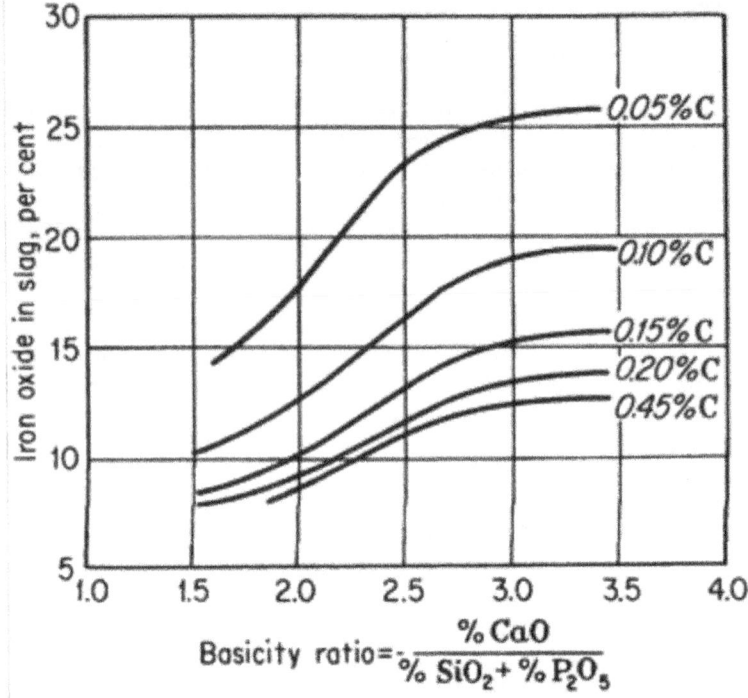

FIGURE 12.6 For a given slag composition in an open-hearth furnace, the higher the FeO content, the lower the carbon content. Also, for a given carbon content, the FeO content of the slag is higher, the larger the basicity ratio. (From Heine et al. [9].)

The charge of an induction furnace is formulated to directly achieve the desired steel composition. The charge materials are selected so that at meltdown, the desired chemistry is attained. There is no attempt to have the oxidation period to remove hydrogen and nitrogen. Since the furnace lining is neutral, sulfur and phosphorus cannot be

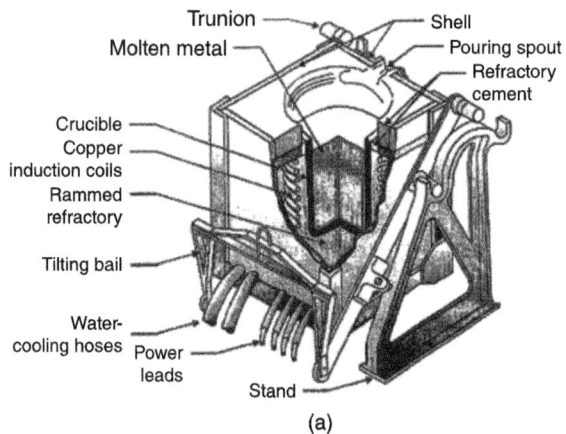

(a)

FIGURE 12.7 Components of a coreless-type induction furnace-: (a) operational elements

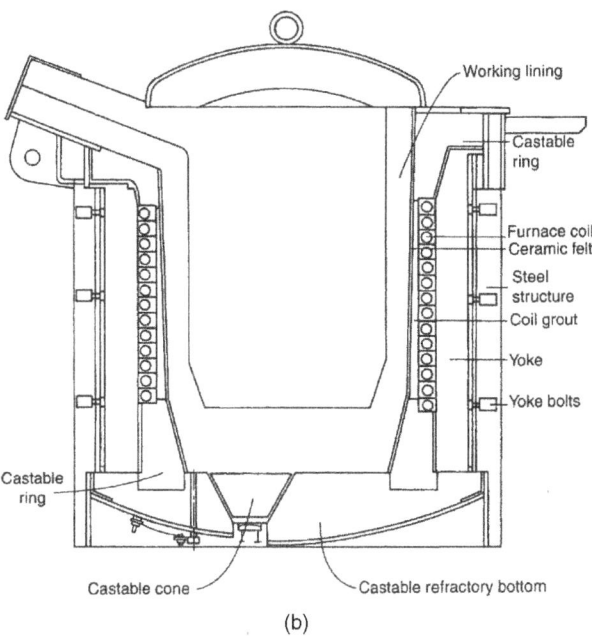

Working lining
Castable ring
Furnace coil
Ceramic felt
Steel structure
Coil grout
Yoke
Yoke bolts
Castable ring
Castable cone
Castable refractory bottom

(b)

Figure 12.7 (*Continued*) (*b*) cross section showing water-cooled copper induction coil and key structural components. The entire molten metal bath (which serves as the secondary) is surrounded by the coil (the primary) that encircles the working lining. (From *Metals Handbook* [10].)

removed. The final composition limits must be achieved through the selection of charge materials. No slag-making materials are added and often with clean charge materials, little slag is present in induction melting. Some refractories contain phosphorus in their binder and phosphorus pickup from the lining can be an issue in induction melting.

Some light steel scrap is normally added first just as in arc melting to provide a cushion for the dumping of heavier charge materials. It is common to add most of the alloys required as ferroalloys or alloying additions at the bottom of the charge or use returns or scrap that already contains the alloys required. The exception is ferrochromium or other high-chromium materials which absorb/react with nitrogen on heating. To avoid excess nitrogen, the chromium alloys are added to the molten bath. After charging the furnace, the heat is melted as quickly as possible.

During melting, it is necessary to be on guard against bridging. This happens when the charge at the top of the furnace is stuck and allows the molten pool to form in the bottom of the furnace. This molten pool can heat to extremely high temperatures, more than 3000°C (>5000°F), in a short period of time. This overheated pool can damage the furnace lining, put a hole in the water-cooled induction coils, and even lead to an explosion. Careful melting practice and attentive melters are keys to avoiding bridging.

Another concern after the charge is molten is a spontaneous carbon boil. While the melt practice does not include a carbon boil and no oxygen is applied, the dissolved oxygen and carbon can spontaneously start a carbon oxygen reaction. This can "boil" out of the furnace just like a pot of boiling water on a stove. It is a standard procedure to make sure that there are some deoxidizers like silicon or aluminum ready to kill the heat in case of a spontaneous carbon boil.

Similar to arc melting, many induction melters perform preliminary deoxidation in the furnace to reduce oxide inclusions in the castings. It is typical to add about 0.1% (2 lb/ton), of aluminum. With oxidation, the aluminum level drops to the desired 0.03 to 0.06% and this drop comes partially from the removal of oxygen and oxides from the heat.

When the heat is at the right temperature and composition, it is important to tap and pour. While reheating in an induction furnace is not as detrimental as in an arc furnace, it is undesirable. Every delay costs money, and steel quality degrades with increases in oxygen, nitrogen, and hydrogen.

To prevent the pickup of gases during melting and to try to improve the quality of induction melting, attempts have been made to protect or improve the metal during melting. One common technique is to cover the molten metal with inert gas, like argon. The furnace top is covered with a ceramic fiber blanket and argon gas is introduced over the melt either through a pipe or a frame underneath the blanket on top of the furnace. Another technique is to drop liquid argon through the furnace lid onto the heat during melting. Generally, the oxygen level is reduced to below 1% on top of the melt. Another innovation is the introduction of porous plugs into the bottom of the induction furnace.

Vacuum Melting

Some alloys processed by induction melting are melted under vacuum. Some furnaces apply a vacuum only during melting to control nitrogen and hydrogen. The more common arrangement is to melt and pour in vacuum. This is a requirement for some of the nickel- or cobalt-based superalloys. This class of materials relies on titanium and aluminum alloying which would oxidize if not processed in a vacuum.

AOD Refining

Argon oxygen decarburization (AOD) is a process for refining steels after melting. It was developed to permit the more economical production of stainless steel that requires low carbon contents with high chromium contents. In an electric arc furnace, getting low carbon contents below 0.03% with high chromium content over 18% is a problem. As the carbon contents fall during the carbon boil, the oxygen begins to oxidize and remove the chromium. The AOD process uses a mixture of oxygen and argon to preferentially remove the carbon without removing the chromium.

The AOD process, by using the flow of argon-oxygen bubbles through the steel with a strongly basic slag, not only allows the steelmaker to reduce carbon to the low levels required for the most advanced stainless steels, it also allows the steelmaker to get low sulfur levels. If the phosphorus is removed in an oxidizing basic slag in melting operations, the sulfur can be reduced to low levels with a reducing basic slag in the AOD vessel. The argon-oxygen bubbles also reduce the nitrogen and hydrogen levels.

Forging shops use the vacuum oxygen decarburization (VOD) process because high-strength forgings require hydrogen levels less than 1.5 ppm. Steel casting's even with high strength, can tolerate higher hydrogen contents and AOD equipment and processing is less expensive than VOD treatments. Steel foundries prefer AOD operations to make low-carbon stainless grades and low-sulfur high-strength alloy steel grades.

Deoxidation

When melting is complete, the steel is at the lowest carbon and at the highest dissolved oxygen content. The purpose of deoxidation is to reduce the oxygen content to a low value. Carbon, in reacting with oxygen, forms a gas and if this reaction occurs at solidification the resulting porosity will remain in the casting.

The relative effectiveness of a deoxidizer is given by equilibrium data showing the concentration of oxygen that can remain in solution in steel in equilibrium with a given concentration of deoxidizer (also in solution). Examples of such equilibrium lines are given by solid lines in Fig. 12.8 [11].

The dashed lines in this figure are for calculated values based on thermodynamic data. Actually, these lines represent the limits of solubility of oxygen in liquid steel in the presence of the added elements, and may be considered as lines on an isothermal section taken from ternary systems of iron, oxygen, and a deoxidizing element.

The lines in Fig. 12.8 are for a temperature of 1600°C (2912°F). If the temperature were lower, they would all shift to lower values, with the exception of that for carbon,

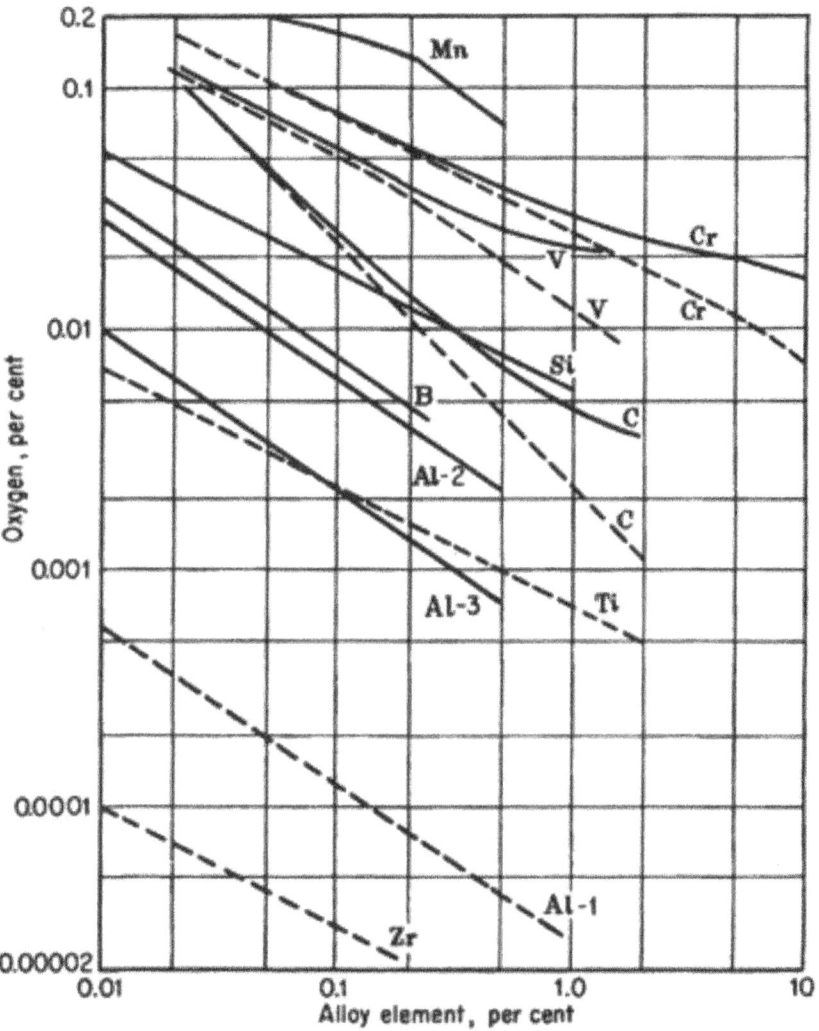

FIGURE 12.8 Equilibrium between various deoxidizers and the oxygen content in steel at a melt temperature of 1600°C (2912°F). (From Heine et al. [11].)

which would be shifted upward. An example of the shift is given in Fig. 12.5 for the silicon-oxygen equilibrium.

Deoxidizers like manganese and silicon are more effective when they are added together rather than separately. Entropy favors mixtures of deoxidizers to gain the maximum benefit. These result in the formation of complex deoxidation products such as manganese silicates. For that reason, some melters prefer to add ferromanganese and ferrosilicon together. Silicomanganese additions are preferred by some because in this ferroalloy the ratio of silicon to manganese is such that a relatively fluid deoxidation product is formed. Ferromanganese may be used after the silicomanganese to aid in coalescing these products. Not only is the oxygen content of the metal reduced to a low value by this procedure, but the deoxidation products are more fluid, and hence should coalesce and float out of the bath more readily. The addition of silicon and manganese to reduce the oxygen in the heat and prevent the formation of additional carbon monoxide in subsequent operations is referred to as "killing" the heat. Most steelmakers add aluminum or aluminum plus other elements like titanium or calcium as a final deoxidation to prevent porosity from carbon monoxide and try to improve the sulfide shape, avoid embrittling phase formation, and reduce the number of inclusions in the final product.

Ellingham Diagram for Oxides

Figure 12.9 plots the standard free energy ($\Delta G^0 = RTl_n Po_2$) of formation of various oxides (y-axis) as a function of temperature (x-axis) [12]. The more negative (lower) the value of ΔG^0 the more stable the oxide at a given temperature. It is seen from this figure that slopes for all oxidation reactions are positive except for the reaction $2C + O_2 = 2CO$. This is because most oxides are liquids or solids while CO is a gas. Entropy favors gas or CO formation at higher temperatures. This means that all oxidation reactions become less thermodynamically favorable as the temperature increases except $2C + O_2 = 2CO$, which becomes more favorable.

It is also seen from this diagram that during melting Mn and Si will oxidize preferentially to carbon below about 1520°C (2768°F). The oxygen blow is typically started around 1552°C (2825°F) and only carbon gets oxidized and the rate of carbon elimination accelerates as the temperature goes up.

The effectiveness of different deoxidizers can also be explained by the use of Fig. 12.9. At a given temperature, the lower the curve for a given metal, the more powerful is the metal as a deoxidizer. Thus Ti is more powerful than Mn and Si but less powerful than Al.

Inclusions

Many steel alloys are formulated to have good engineering properties with a large margin of safety for structural loads. One aspect of these steels is high ductility and toughness. Steels with high ductility and toughness are relatively insensitive to second phase microstructural features like microporosity or inclusions.

Inclusions in steel are both micro and macro features. Macro-inclusions (dirt) are the result of either entrained materials from molds or ladles or more commonly the result of reoxidation from reactions with air or other oxygen sources. Micro-inclusions are normally the result of solidification where the oxides and sulfides form. Micro-inclusion formation can occur prior to bulk solidification, during or at the end of solidification. The formation of inclusions during solidification may occur early resulting in geometrically shaped particles more or less randomly distributed (Fig. 12.10, left) [13], or

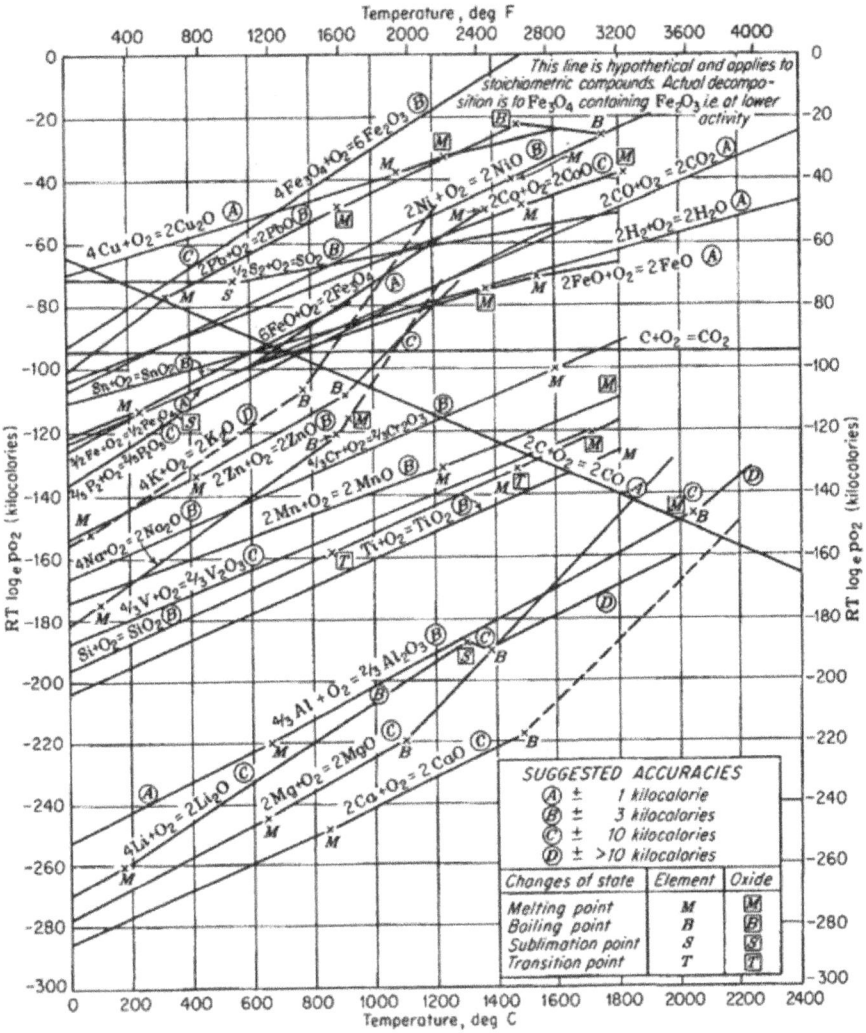

Figure 12.9 Ellingham diagram showing effects of temperature on the standard free energy of formation of a number of oxides. (From Heine et al. [12]).

the formation may not take place until some steel has already solidified, which will give rise to a segregation pattern for the inclusions.

Micro-inclusions differ as to composition, with oxides, sulfides, and occasionally nitrides or carbo-nitrides being found. The form and type of these inclusions are strongly influenced by the deoxidation practice used. Deoxidation is required to prevent the oxygen dissolved in the steel from reacting with the carbon to form carbon monoxide bubbles causing porosity. Not only is deoxidation of concern in producing castings free of pinholes and blowholes, but also it affects the microstructure and properties of the cast steel.

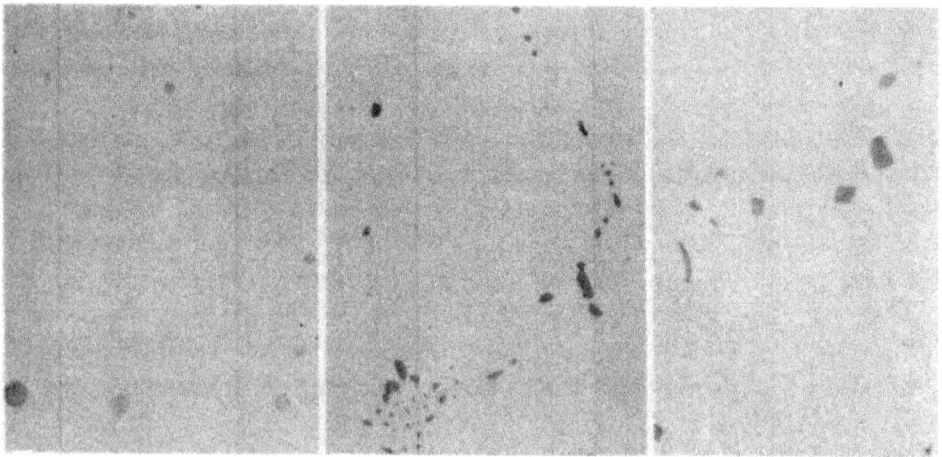

Figure 12.10 Typical appearance of sulfide inclusions classified on the basis of shape, ×500. (From Heine et al. [13].)

Steels deoxidized with manganese and silicon only will be found to contain two principal types of inclusions: (1) iron-manganese silicates and (2) sulfide inclusions. The silicates are glassy and tend to be globular and translucent and vary in color, depending on their composition. The sulfides, on the other hand, are opaque and do not reflect polarized light. They are also globular in shape, but are light gray in color under ordinary light and may be more irregularly shaped than the silicates. The actual quantity of these inclusions is controlled by the iron oxide content of the molten steel prior to deoxidation. The size and distribution are controlled by the cooling rate, i.e., the size of the casting section.

The combination of globular silicates and sulfides characteristic of manganese and silicon deoxidation suggests relatively low solubility of these constituents in the liquid steel, so that they begin to precipitate at a rather early stage during the solidification process, giving rise to more or less randomly distributed particles. These are essentially spherical in shape because they have grown freely in liquid steel. The low solubility is attributed to the relatively high oxygen content of the steel. These inclusions are called type I inclusions. An example of type I inclusions is given in Fig. 12.10, left. Manganese and silicon alone as deoxidizers is not normally employed since it can be inadequate to prevent porosity in the castings. The most common supplement to silicon and manganese is the addition of aluminum.

The addition of a small amount of aluminum (about 0.02 to 0.05%) drastically changes the appearance of the inclusions. There is a decrease in the quantity of the silicate inclusions, or they may not appear at all. Also, the reduced oxygen content of the liquid steel results in a sulfide which is appreciably more soluble than in the case of straight manganese and silicon deoxidation. This increased solubility of the sulfides in the liquid steel causes them to precipitate at a late stage during the freezing of the steel. They are on the austenite grain boundaries, where they appear as fine and elongated particles, suggestive of a eutectic mode of solidification (Fig. 12.10, center). This type of sulfide distribution is called type II. Clusters of fine, angular, whitish alumina particles may also be present as inclusions in these steels. The concentration of sulfide inclusions as fine dots or elongated particles on the grain boundaries reduces the ductility and toughness of the steel. This reduction in properties is of course dependent on the levels of sulfur in the steel and at low sulfur levels common in modern production, this effect is minimized.

If the aluminum content exceeds this critical amount around 0.03%, more alumina is present as inclusions. The sulfide, although still persisting on the austenite grain boundaries, are larger particles, irregular in outline, and farther apart than the type II sulfides. These sulfides are designated type III, and their shape and distribution are attributed to the presence of aluminum in the sulfides, which lowers their solubility in the iron, causing them to precipitate earlier in freezing. The sulfides can also be duplex in structure, being mixtures of Al_2S_3 and MnS (Fig. 12.10, right). Because of the absence of a film-type distribution of the sulfides, the ductility and toughness of steels containing these sulfides are almost on a par with those of steels containing type I inclusions. This is the normal goal of deoxidation.

The effect of aluminum deoxidation results in property changes of the degree indicated in Fig. 12.11 [14]. If aluminum additions are required to enhance casting characteristics, a poor sulfide distribution can be avoided by using sufficient aluminum to exceed the critical concentration range. This usually means an addition of 2 to 2.5 lb per ton (0.10 to 0.125%). Recovery of the aluminum results in a final aluminum around 0.03%. It should be noted that, if sulfur content is low, the changes in sulfide shape and properties are not so drastic as with higher sulfur percentages. These days because of the good quality of scrap, sulfur and phosphorus contents are low in steel. Most steels produced have lower than 0.025% sulfur and phosphorus each.

Other deoxidizers are used to improve the ductility of cast steel. A calcium-silicon addition preceding aluminum deoxidation has been found to give a ductility matching silicon-manganese deoxidized steel. Regular deoxidation with aluminum followed by an addition of calcium-manganese-silicon, has resulted in improved ductility, toughness, and resistance to hot tears. Inclusions were changed from type III to type I.

In an effort to maintain the minimum aluminum needed to avoid type II sulfide inclusions, it is tempting to add excess aluminum. Excess aluminum with heavier section steel castings can result in the formation of grain boundary aluminum nitride. This precipitates on the original as-cast austenite grain boundaries in castings that cool relatively slowly. It causes the steel to fracture in a brittle manner along the as-cast grain boundaries. The type of fracture surface produced is a "rock-candy."

Gating and Pouring Systems

After melting the steel, adjusting it to the final chemistry, and heating it to the desired temperature the steel is tapped into a ladle. Ladles are used to transport the steel to the mold and pour the casting. The ladles used in steel foundries are typically either bottom pour or lip pour ladles depending on whether the steel is poured from a nozzle in the bottom of the ladle or over the edge of the ladle. Larger castings and larger heats, greater than 5 tons, are almost always bottom poured. Smaller castings are typically lip poured. The lip poured ladle can either be open or have a spout with a dam. The spout with a dam or channel is like a teapot allowing the steel to be poured from the bottom of the ladle. These ladles are called *teapot ladles* and are the most common for small castings.

The designs of ladles for pouring castings are not based on the pouring requirements but on the logistical and mechanical requirements of production. For example, lip or teapot ladles rotate around their center of gravity. This requires constant adjustment of the position of the ladle lip by the operator and makes it impossible to deliver molten metal to the same location at the same rate as the level of metal in the ladle changes.

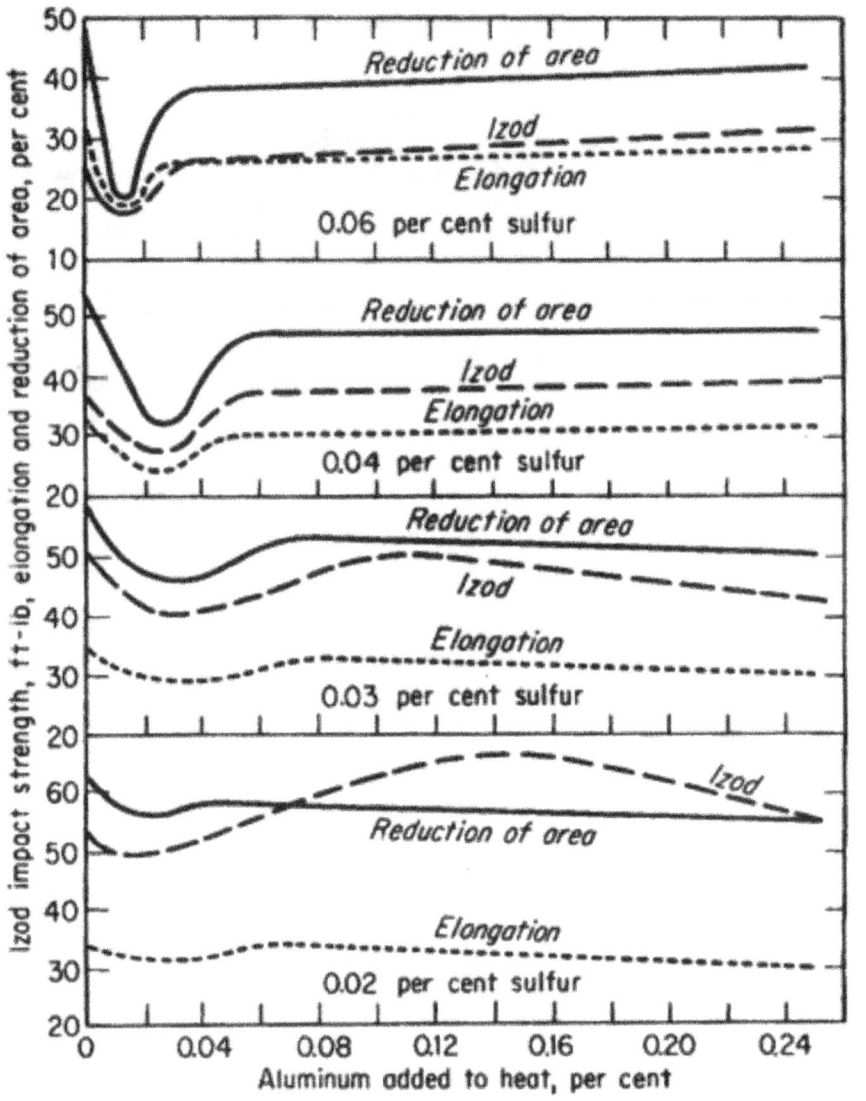

FigurE 12.11 Effect of aluminum on the properties of medium-carbon cast steels at various sulfur contents. (From Heine et al. [14].)

Bottom pour ladles are able to pour large quantities of metal. The compact design keeps the thermal losses low. The head height in the ladle varies dramatically during the pouring operation changing the pouring time from the first to final casting produced from the ladle.

An ideal gating system for pouring steel or any other metal into sand mold is one that fills the mold cavity in the shortest possible time without causing pronounced splashing. Slow pouring may cause deterioration of mold wall because of radiated heat.

Gating systems for steel casting production must consider both the size of casting and the type of ladle. Gating systems are discussed in detail in Chapter 5. The gating system for a large casting (>1000 lb) or from a bottom pour ladle is different than for smaller castings (<500 lb) poured with a teapot or lip pour ladle. The pouring system is coupled and the gating system must fill the casting from the ladle and cannot be designed without knowing the ladle pouring characteristics.

In the rigging of a casting, the calculation of riser sizes is an engineering task which may be carried out using empirical engineering approaches, such as modulus calculations, or through the simulation of solidification behavior using readily available computer software. This approach is not available for pouring and gating systems. While recommendations abound, simple empirical relationships or complex simulations do not give a systematic analysis of pouring and gating designs. Designs for gating and pouring systems are left to the art and experience of a rigging engineer who does not have agreed-on tools to evaluate his proposed design.

Our intuitions about flow and gating system design are typically ill founded because they are based on our personal observations of fluid flow. These observations are most useful in slow-flow steady-state systems. For example, gating systems are visualized as filling quickly and operating full for most of the pour. In reality, many designs only fill after most of the metal is poured and do not work the way they are visualized. Water modeling has been used to improve our understanding of pouring and gating system performance.

Entrapment of exogenous materials was considered to be the most common source of inclusions. Inclusions were thought to be slag, sand, mold coatings, or refractories. Inclusions were formed when these materials were poured into the casting or were eroded from the mold and then trapped during the solidification of the casting.

The initial work done by the Steel Founders' Society of America (SFSA) demonstrated that while the mold can erode and materials can be entrained, reactions with air during pouring were a major source of inclusions. In steel casting production, out of about 500 samples examined, reoxidation was the source of more than 80% of the inclusions in steel castings.

Most work to understand gating has been conducted on smaller lip pour. The job of the gating system is to deliver the liquid steel into the casting cavity to produce the highest quality casting possible. Traditionally it was thought that by providing a properly designed runner, the steel quality could be improved by allowing the flotation and removal of oxide inclusions in the gating system. More recent work suggests that pouring reduces the steel quality because it reacts with air and this forms inclusions during the pouring event. The amount of air entrainment appears proportional to the pouring time and metallostatic height in the pouring system. Increasing the pouring time increases the reoxidation of the steel. This seems to be a linear effect, twice the pouring time gives twice the inclusion formation. The metallostatic height or total head height in the system is not the sprue height but includes the ladle. It is measured from the free surface of liquid steel in the ladle to the lowest point where the steel flows in the mold. This is a much bigger effect than pouring time, which is why there is often no observed effect of pouring time on quality. Water modeling results identify the head height effect as the height to the 2.5 power. Inclusion measurements in steel foundries demonstrate repeatedly the detrimental effect of pouring system head height on inclusion content. This implies that gating and pouring systems should be designed to fill the casting as quickly as possible with the minimum height. One needs large cross-section gates with slow-moving streams. Minimizing the velocity with low head heights must not unduly extend the pouring times. Short pouring times should not be accomplished by increasing the velocity with added head height.

Since most of the detrimental effects on quality during pouring are the result of reactions, not entrapment, gating systems designed to allow inclusion flotation are not necessarily optimal. This is especially true for gating designs which do not consider both the casting geometry and the pouring system. A general approach to gating system design has not been established and an engineering criteria does not exist which has been useful in ranking gating systems. Large changes in casting quality are experienced with small changes in gating design.

For lip pour or teapot ladles, the gating system is a traditional foundry gating system with a pouring cup to pour into from the ladle, a down sprue, a runner, and then gates. The pouring cup should be large enough to hit easily with the stream from the ladle to allow accurate and rapid pouring. Avoiding splashing and air entrainment is improved if the ladle can drop the steel into the cup on the near wall at the shallowest angle possible. Some water modeling results suggested that a tilted sprue would be beneficial but no plant trials have confirmed that result. The sprue is recommended to be tapered to avoid additional air entrainment but there are no compelling production data supporting that requirement.

The sprue is normally the flow-controlling gating element. Water modeling shows that this traditional system does not function as designed or visualized. The gating system does not get full until the casting pour is almost complete. The pouring should be done as rapidly as possible. A typical pouring time is calculated from the pour weight of the casting; the pouring time in seconds is equal to the square root of the pouring weight in pounds. So a casting with a 100-lb pour weight would be poured in 10 seconds. A faster and preferred pour time would be two-thirds of that calculated typical time, or 6.7 seconds.

At the bottom of the sprue is a small basin. The basin was originally required in greensand to prevent erosion of the sand at the base of the sprue. It is useful to have a small basin to prevent the stream from "sheeting." If a liquid stream hits a flat surface, it is common for it to form a sheet of liquid at the impact point before it breaks into streams. This sheeting would cause larger amounts of surface and reaction. A basin prevents sheeting.

Runners take the liquid steel from the base of the sprue and deliver it to the gates for filling the mold cavity. Runners are placed in the drag and gates in the cope to ensure that the runners are full during pouring. The cross-sectional shape of the runner is normally rectangular. If it is cylindrical or square, the heat loss is a minimum and this would be desirable for thin section castings that are difficult to fill. As the aspect ratio (width/thickness) increases, the runner gets thinner in the vertical direction and extended in the horizontal direction, the runner fills more quickly, and has more surface area contact with the flowing steel. Since the inclusions will wet the wall and stick, this aids in the reduction of inclusions. It also allows the runner to fill more quickly, reducing exposure to air. It does result in greater heat loss as well. Runner extensions were used like the basin at the bottom of the sprue, to prevent erosion in greensand molds. Many designs for runner extensions that are clever have been developed but none has been demonstrated to improve casting quality. Minimizing the runner extension improves yield and reduces cost. Gates are generally designed to exit from the top of the runner in the cope and then enter the mold cavity. The entry point of the gates should be chosen carefully. It is desirable not to impinge on a surface and cause a fountain or splash zone that could result in increased reoxidation or mold erosion. It is not a good idea to have the gate create a waterfall effect while entering the mold cavity. It is helpful to have the gate enter at the base of a riser to allow it to be the hot section after filling.

Traditionally the runner system also used a gating ratio, the area of the sprue: runner area: gate area, as a design parameter. In steel gating systems, the general approach was to use a 1:2:2 ratio. Gating trials to determine the effect of gating design failed to show much variation regardless of the gating design chosen. Gating ratio was not shown to be important. The best casting results were obtained through the use of a filter at the bottom of a riser. No gating system other than pouring through the riser was used. The poorest results were obtained by designing a water park gating system with cascading waterfalls and fountains.

Pouring and gating larger castings is different. When the casting weight exceeds 5000 lb, the common practice is to pour with a bottom pour ladle. The gating system for bottom pour operations is fundamentally different from the lip pour/teapot system. The flow in bottom pouring should not be controlled by the operator or the gating system but by the nozzle on the ladle. The gating system must be able to accommodate the full flow from a full ladle. Typically, the sprue is an inch in diameter greater than the nozzle and goes to the lowest part of the mold. Then the gating system splits and is larger in cross section and enters the casting horizontally. Care is taken to minimize the fountain effect or a splashing or dropping stream. Cylindrical castings often have gating systems that create a torsional flow during filling.

Shrouded Pouring

For larger steel castings, protecting the stream with a ceramic shroud has shown dramatic improvements in casting quality. The shroud is attached to the nozzle of a bottom pour ladle. The casting is rigged to allow the shroud to be placed down an open riser into the bottom of the mold. Since the exit of the shroud is quickly submerged under the metal and the top is sealed, no reactions damage the metal. Reductions in inclusions from 60 to over 90% have been seen from shroud pouring larger castings Fig. 12.12 (a) and (b) [15].

(a)

Figure 12.12   (a) Photograph of shrouded pouring

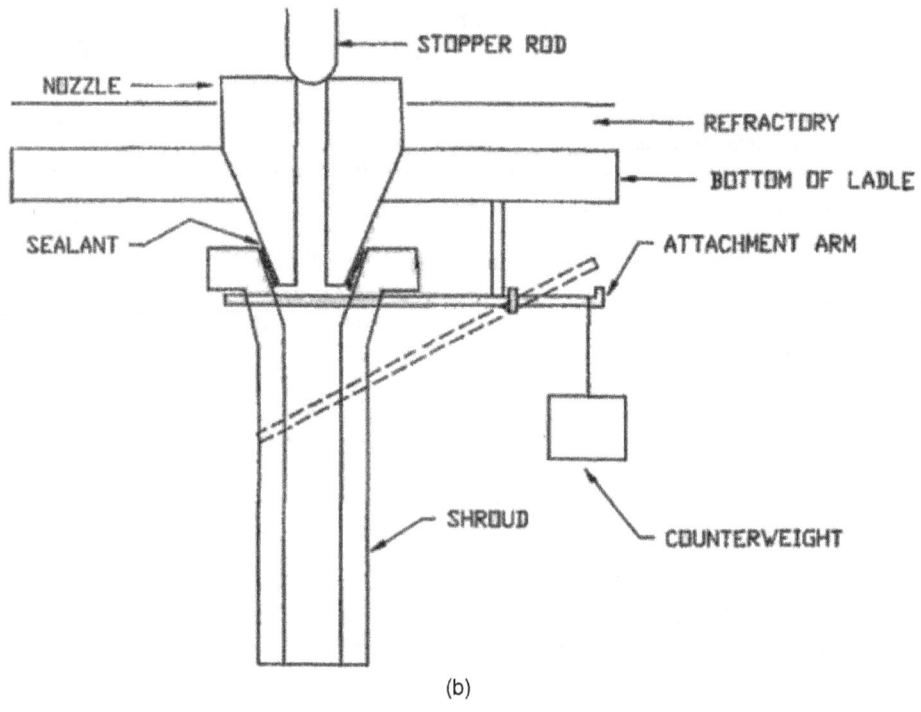

STOPPER ROD

NOZZLE

REFRACTORY

BOTTOM OF LADLE

SEALANT

ATTACHMENT ARM

SHROUD

COUNTERWEIGHT

(b)

FIGURE 12.12   *(Continued)* *(b)* sketch of typical shroud seated on a nozzle. (From Monroe [15].)

Risering Design

An understanding of the basics of risering technology is essential for the accurate risering of castings and maximization of casting yield [16]. It is particularly necessary for the successful application of feeding aids. It is necessary to riser steel casting because of volume contractions which take place during solidification of the casting. Typically, these consist of shrinkage of molten steel prior to the onset of freezing (~1% per 100°F) and contraction during freezing (~3%). Usually small steel castings (<500 lb) exhibit a total volume contraction of 5%; large castings made in nonrigid molds may have up to 8% or more total shrinkage.

The subject of risering castings has been covered in detail in Chapter 5 of this book. What follows below is one of the ways of risering steel castings [16, 17]. Some prior experience with gating and risering is needed to successfully use this method, which is called the modulus method. The freezing times of castings are approximately given by Chvorinov's rule

$$T = C \ (V/A)^2 \tag{12.2}$$

where

T = freezing time of casting
V = volume of the casting
A = cooling area of the casting
C = constant governed by metal and mold properties

For convenience, V/A ratio of the casting is generally termed the casting modulus, M_c

$$M_c = Vc/Ac \qquad (12.3)$$

It will be observed from Eq. (12.2) that *the freezing time of the casting (or a portion of the casting) in a given metal and mold type* is determined solely by the casting modulus, at least to a first approximation. The three rules of risering are

1. Risers must be *thermally adequate*
2. Riser *volume must be adequate*
3. For maximum soundness, *feeding distance rules must be obeyed*. Feeding distance dictates the number of risers to be used.

Procedure for risering castings:

1. Calculate the casting modulus M_c
2. Calculate the riser modulus $M_r = 1.2M_c$
3. Make sure the feeding distance is met
4. Make sure the riser volume is adequate

The rationale for $M_r = 1.2M_c$ is that the riser will stay liquid much longer than the casting. Since freezing time T is proportional to M^2, the riser freezing time will be 44% longer than the casting freezing time. The volume ratio of riser and casting, namely, V_r/V_c is dependent on freezing ratio, namely, M_r/M_c (Fig. 5.28) for both sand and sleeved risers. At a freezing ratio of 1.2 the riser volume is less than 20% of casting volume with adequate safety of freezing time.

For maximum soundness, the feeding distance given in Table 5.3 must be met. If some shrinkage is acceptable, the feeding distance is more [17]. The feeding distance is measured from the outside of the riser/riser sleeve to the farthest point the casting can be fed. The feeding distance is dependent on the casting width to thickness ratio, or W/T. Finally, the riser volume requirement is dependent on casting thickness:width:length [16]. For parts with this ratio of 1:15:30 the riser volume needed is only 8 to 14% of casting volume, the lower value being for an insulated riser and the higher value being for sand riser. For chunky parts with this ratio of 1:2:4, V_r/V_c will range from 26% for sleeved riser to 140% for sand riser. Hence, the importance of an insulated riser for chunky castings is obvious.

Solidification Process

An examination of the iron-carbon diagram, Fig. 12.13 [18] shows that iron-carbon alloys freeze by a peritectic reaction in the carbon range of 0.10 to 0.54% and as typical solid-solution alloys beyond these limits. At the peritectic temperature of 1496°C (2725°F), the first solid to freeze, delta ferrite (δ-iron), combines with liquid metal to form austenite. The amount of austenite formed and the amount of either δ-iron or liquid iron remaining after this interaction are dependent on the carbon content, δ-iron being in excess below 0.16% carbon and liquid in excess above this composition.

Austenite apparently forms from the δ-iron-liquid-iron combination by forming an encasement of austenite around each δ-dendrite. Subsequent diffusion during cooling should tend to promote chemical homogeneity. Just how much influence the peritectic reaction has on the microstructure and properties of cast steel is uncertain. Considering the high rate with which carbon can diffuse at the peritectic temperature, it would be expected that plain-iron-carbon alloys should approach an equilibrium structure quite

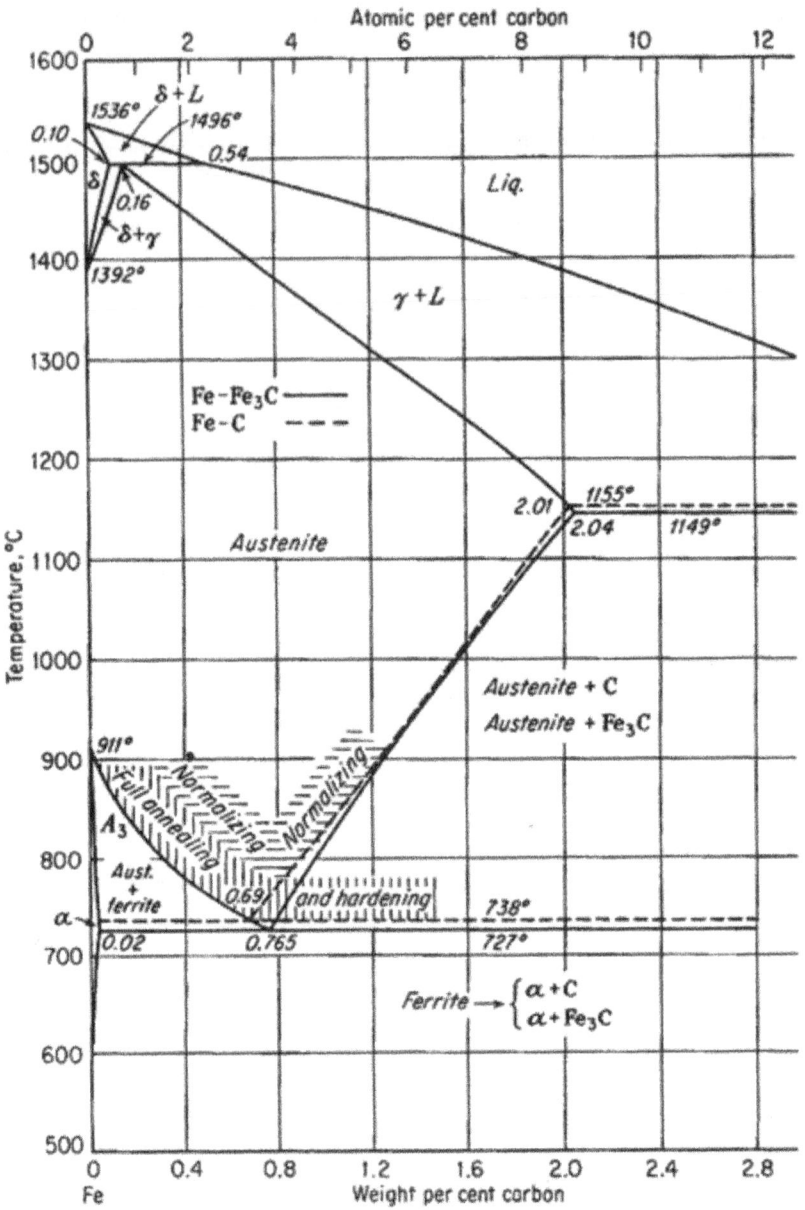

Figure 12.13 A portion of Fe-Fe₃C and Fe-C diagrams. (From Heine et al. [18].)

readily. This is not so certain, however, if the casting contains elements such as phosphorus, nickel, chromium, and molybdenum, which segregate along with carbon and which are not nearly so ready to diffuse as carbon.

Solidification of a steel casting may be visualized as the gradual thickening of a skin of solid metal that forms first at the mold-metal interface and grows toward the thermal center as heat is extracted from the surface through this skin. The skin, when first formed, is an array of interlocking dendrites. Because of heat-flow effects, the inward growth of the dendrites is usually more rapid than the lateral growth, with the result that a "mushy zone" of part-solid-part-liquid metal separates the completely solidified skin from the interior liquid. The last liquid to freeze in this mushy zone is found (1) between the several branches of a dendrite (interbranchial areas) and (2) between separate dendrites or interdendritic regions.

Segregation of carbon and of phosphorus and some other alloying elements follows the same pattern, with the last liquid to freeze highest in these elements and the first part of the dendrites highest in iron and lowest in other alloy elements. The size and shape of the dendrites depend upon the rate of cooling (section size of the casting) and the direction of heat flow. The dendritic structure formed during the freezing process affects the microstructure of the cast steel, and is related to the distribution of inclusions and pinhole porosity.

Molds and Cores for Steel Castings

The general topic of mold and core making has been covered in Chapter 3 of this book. Here some of the things are covered that apply to steel castings which are poured at high temperatures of up to 3100°F (1704°C).

The shape and dimensions of castings are determined by mold and core. The mold is made from a refractory material which resists the abrasive and chemical attack of liquid metal. The mold cavity must retain its shape until solidification is complete. The mold must have certain characteristics to meet the following requirements:

- Must be strong to support the weight of metal
- Must have good permeability
- Must have good erosion resistance
- Must be chemically inert
- Must have good shakeout properties
- Must be economical

Many different molding processes are used in steel casting production. The factors which govern the choice of a particular molding process are quantity, dimensional accuracy, type of pattern and core box, cost, etc. The molds are either permanent or are destroyed after one use.

Disposable Mold Processes

These could be categorized as follows:

- Greensand molding
- Dry sand molding
- Vacuum molding
- Expendable pattern process
- Shell process
- Nobake binders including organic and inorganic binders
- Investment process

Reusable Mold Processes

- Centrifugal process (horizontal and vertical)
- Permanent mold

Greensand Molding Processes

The traditional molding process for steel casting production is the greensand process. In this process, sand is coated with a mixture of bentonite clay, water, and other specialized ingredients. Silica sand is most commonly used in greensand molding because it is most abundant and the cheapest material available. It also has a range of desirable other properties. More expensive granular refractories can be used for special applications such as zircon, olivine, chromite, and mullite. Silica sand needed for steel castings must be of high purity, proper shape, grain size, and grain distribution.

Some advantages of greensand molding are as follows:

- Material costs are low.
- It lends itself to high-production automatic molding.
- Wood or plastic patterns can be used in hand molding, sand slinging, manual jolt, or squeeze machine molding.
- High-pressure molding produces a well-compacted mold giving better finish and dimensional tolerances.

The disadvantages are as follows:

- If the mold is not properly compacted, the dimensional accuracy of the casting can be impaired and the surface finish may be poor.
- High-pressure greensand molding requires metal pattern equipment, which adds to cost.

Special refractory coatings are brushed or sprayed onto the surface which will be in contact with the molten steel. The refractory used in this coating is usually finely crushed zircon, chromite, or mullite.

Nobake Molding Processes

More common in steel casting production is the nobake molding. The name nobake is a legacy of early chemically bonded systems that were baked in an oven prior to use. Modern chemically bonded molds use a catalyst to manage the chemical reaction for bonding but the industry still uses the term "nobake" for these systems. In chemically bonded molds a liquid polymeric resin is cured by a liquid or gas catalyst. In recent years, many job shop foundries have switched over to nobake molding which provides better dimensional control and surface finish. The cost of sand mold is more but reduction in finishing costs compensates for it.

Coremaking Processes

Many processes can be used to make either molds or cores. These include the shell process, oil sand, many different varieties of nobake systems, sodium silicate, gas-cured systems, and liquid catalyst systems. In the production of some steel castings, the mold may be made of cores which form both the external and internal surfaces of the casting.

Cores must be strong enough to maintain their shape as the molten metal fills the mold cavity. During solidification, the cores must collapse so that the casting can contract.

If the core does not completely collapse, high internal stresses can be set up in the casting causing it to tear in the mold.

As the temperature in the mold increases, the binders decompose and produce gases. These gases must be vented to the atmosphere to prevent formation of gas holes in the casting.

Investment Casting Process

This process is used for production of moderate to high-quantity parts especially for high-alloy steels and nickel-base alloys. This process has also been described as the lost wax and precision casting process.

Some advantages of investment casting are given below:

- Excellent surface finishes are produced.
- Very tight dimensional tolerances can be held.
- Machining of casting can be reduced or completely eliminated.
- It is suited for high-production small parts.

The disadvantages are given below:

- Cost of dies for patterns is expensive.
- Length of time required to make a shell mold is high.
- The size of casting produced is limited.

Postcasting Processes

After pouring, castings are allowed to solidify and cool. They are later removed from molds in the shakeout operation. A series of activities then follow which are generally referred to as cleaning, finishing, and heat treatment. These activities can be broadly categorized as

- Shakeout
- Abrasive blast cleaning
- Removal of ingates, risers, and unacceptable casting conditions
- Rough inspection
- Removal of remaining unacceptable conditions
- Finishing welding
- Heat treatment
- Final visual, dimensional, and NDT inspection

The type of alloy, desired quality level, and applicable specifications will dictate the order and extent of these activities. Casting inspection takes place on several occasions and dictates the extent of production operations such as welding, dimensional adjustment, and surface finishing.

Shakeout

The shakeout operation, i.e., the separation of casting from the mold and core sand, is the first step in finishing a casting after it has solidified and cooled in the mold. A considerable amount of energy is required to remove the adhering sand and oxide, and lumps of sand still bonded together. This energy may be supplied by a vibrating deck or grate, shaker pan conveyer, or any other type of vibratory mechanical action. Abrasive blasting may be used to remove individual sand grains and reduce the remaining lumps of bonded mold and core sand that adhere to the casting.

Abrasive Blast Cleaning

Blast cleaning is used at various stages in the finishing operation to remove adhering sand, oxide scale, and weld or air carbon-arc spatter, or to lightly peen the surface and produce a uniform matte finish which is pleasing to the eye. Blasting is usually done after shakeout, heat treatment, and processing steps such as torch cutting, air carbon-arc metal removal, or grinding.

Various types of equipment are employed for blast cleaning of steel castings, all of which utilize some means of exposing the surfaces of the casting to a stream of abrasives. Smaller castings are frequently blasted in a *tumble blast* equipment with an inside chamber composed of a belt of metal slats in the form of a continuous loop. The castings are cradled in the trough formed by the slats and continuously tumbled by the movement of the continuous slat belt. Throughout this process the abrasive, usually steel shot, is thrown at the castings from rotating wheels.

Casting may also be hung inside of cabinets on steel bars which rotate, or on a meat hook-type conveyer which passes continuously through the cabinet. This type of equipment may be used for castings of similar size to those processed in tumble equipment. The blast abrasive is also usually propelled by a wheel in this type of equipment.

Castings from a few hundred pounds up to a few thousand or many thousands of pounds may be blasted on table blast equipment. In this case, the castings rest on a circular table which rotates so that different parts of the castings are exposed to the blast emanating from rotating wheels.

Besides propelling the blast cleaning abrasive by centrifugal action from the vanes of a rapidly rotating wheel, abrasives may be propelled in a stream of compressed air. In these cases, the stream of abrasive is usually directed at the casting by means of a hand-held hose and nozzle.

Although cast steel shot, heat treated to a tough, hard structure is the most popular abrasive used in steel foundries, other abrasive materials that are used include steel grit, iron shot or grit, cut wire, or other metallic abrasives. Nonmetallic abrasives consisting of sand, glass beads, or alumina or zirconia grit can be used when the presence of iron residue is undesirable. Each abrasive material is selected on the basis of the intended work to be done and the type of surface desired after blasting.

Removal of Risers, Gates, and Metal Padding

The risers and the gating system must be cut to remove them from the casting. Various methods may be selected for this operation depending on the size and type of casting, type of metal, and the configuration of the contacts to be cut.

A popular method of cutting is oxy-gas torch cutting for burning, utilizing acetylene, natural gas, propane, or proprietary fuel gases. When cutting higher chromium or stainless steels, iron powder may be included in the stream of cutting oxygen. Castings from very small sizes up to those weighing several thousand pounds may have risers removed by oxy-gas cutting. Risers may also be removed by the air carbon-arc process, especially where various high-alloy grades of stainless steel or nickel-base alloys are involved. Plasma arc cutting is a process applicable to thinner section work. Other operations used for riser removal on small- to medium-size castings consist of high-speed friction bandsaws, mechanical shear, or abrasive cutoff wheel.

After removal of risers and gates, the contact areas are shaped to the casting contour by grinding, oxy-gas flame cutting, or more commonly by air carbon arc. Grinding equipment may include stand or swing grinders to abrasive belts or air-driven wheel grinders.

Some extraneous metal which has to be removed from casting is usually present, such as a fin at the parting line where the mold halves have come together or where the mold surface contacts a core or chill. These are usually removed by grinding, air-carbon arc washing, oxy-gas torch cutting, or chipping with a pneumatic chipping hammer and chisel.

Heat Treatment of Steel

An essential part of the processing of most steel castings includes one or more thermal cycle treatments collectively known as heat treatment. The mechanical properties, such as strength, hardness, and toughness are effectively controlled by these heat treatments in order that they comply with the specifications. In the case of corrosion-resistant castings, the heat treatment imparts improved corrosion resistance. In the case of castings which have been welded, the heat-treating cycle is called a stress relief because it reduces the thermal stresses present after welding, and provides uniform hardness in and around the welded area.

Heat treatment can be given to accomplish any number of objectives, such as

- To reduce segregation of carbon and other alloying elements
- To soften
- To improve machinability
- To harden
- To toughen
- To increase wear resistance
- To stress-relieve
- To remove hydrogen
- To improve corrosion resistance

So far as the heat treatments designed to alter the mechanical properties are concerned, cast steel can be considered in the same light as other steel products and is subject to the same treatments. There are additional features that should be mentioned when dealing with cast steel.

Carbon and low-alloy steel castings usually receive a cycle called normalizing treatment in which the casting is austenitized at 900 to 950°C (1650 to 1750°F) and then cooled in air to room temperature. If fully softened conditions are required for machining, the cooling rate is slowed by cooling in the furnace and the cycle is called *annealing*. When higher strength, hardness, and toughness are required of low-alloy steels, the castings are austenitized and quenched into an agitated liquid such as oil, synthetic quenchant or water depending on cooling rate required. This heat treatment is followed by a tempering heat treatment in which the casting is heated to within the range of 200 to 700°C (400 to 1300°F) as determined by the type of alloys present and the desired final mechanical properties and cooled to room temperature in air or quenchant. The heat treatment given to relieve a casting after welding is identical to the tempering operation, but may be performed at a slightly lower temperature. Another type of heat treatment given to high-alloy corrosion-resistant and to Hadfield 13% manganese steel castings is called solution annealing, which dissolves undesirable and harmful carbide particles. In this case, castings are heated to 1050°C (1950°F) or above and rapidly cooled by liquid quenching. Certain types of alloys may be given a subzero cryogenic treatment to improve their mechanical properties.

Homogenization

Homogenization is an effort in heat treatment to reduce the micro- and macrosegregation that may occur during solidification of the casting. Some elements are smaller than iron and are in the holes or interstices of the steel crystal structure like nitrogen or carbon. This allows these interstitial elements to diffuse freely. Most alloying elements are substitutional and replace an iron atom in the steel structure. This makes their diffusion slow and makes homogenization less effective. Since steel changes crystal structure when it is heated and forms austenite, this transformation allows for some grain refinement. Elevated temperatures also allow porosity and inclusions to become more rounded and less sharp by diffusion down the boundaries. This treatment at an elevated temperature achieves a better distribution of the segregated carbon and alloying elements. For heavy section castings, segregation is such that a long time at higher than average temperatures may be necessary. Current practices are to normalize a casting, especially a heavy section casting, prior to final heat treatment required by the specifications and properties. Additional cycles may be used to improve the mechanical properties slightly or to stress relieve the casting. Homogenization is most effective at higher temperature above 1050°C (1950°F) and longer times. Scaling of the casting by oxidation in the furnace is an issue at these long times and high temperatures.

Annealing

The annealing treatment is essentially the same in principle as that employed on other steels and involves heating to the austenitic state and cooling slowly. Its purpose is to

- Soften for machinability
- Relieve stresses

Stress-Relief Anneal

Stress relief can be accomplished by a full anneal, as already described or it may be done by using a subcritical temperature. Holding at 400°C (750°F) will reduce stresses about 50%, whereas a temperature of 550°C (1000°F) will reduce stresses more than 90%. Time at temperature is also a factor. If liquid quenching and tempering are employed, stress relief is accompanied by quenching.

Normalizing

Normalizing is common and is accomplished by heating the casting above the transformation temperature and then cooling in air. Normalizing gives higher strength and hardness than annealing, and is used as the final heat treatment, where strength requirements do not exceed 100,000 psi. The most common final heat treatment of a steel casting is normalizing. Normalizing is preferred as well when the product is used for elevated temperatures applications.

Tempering is not required after normalizing but has been shown to improve ductility in many steel grades. Tempering can also be employed as a stress relief, especially if the casting requires extensive welding in production.

The usual temperature range for annealing, normalizing, and quenching is shown in Fig. 12.13, and the effect of annealing, normalizing, and tempering on the properties of medium-carbon cast steels is illustrated in Fig. 12.14 [19].

Liquid Quench and Temper

Rapid cooling will avoid the transformation of steel to a ferrite-pearlite structure and the steel will form martensite instead. Martensite is the hardest product obtainable for a given carbon content. Reheating martensite to a subcritical temperature (tempering) will produce a dispersion of fine carbide in ferrite and will lead to a softening and toughening of the

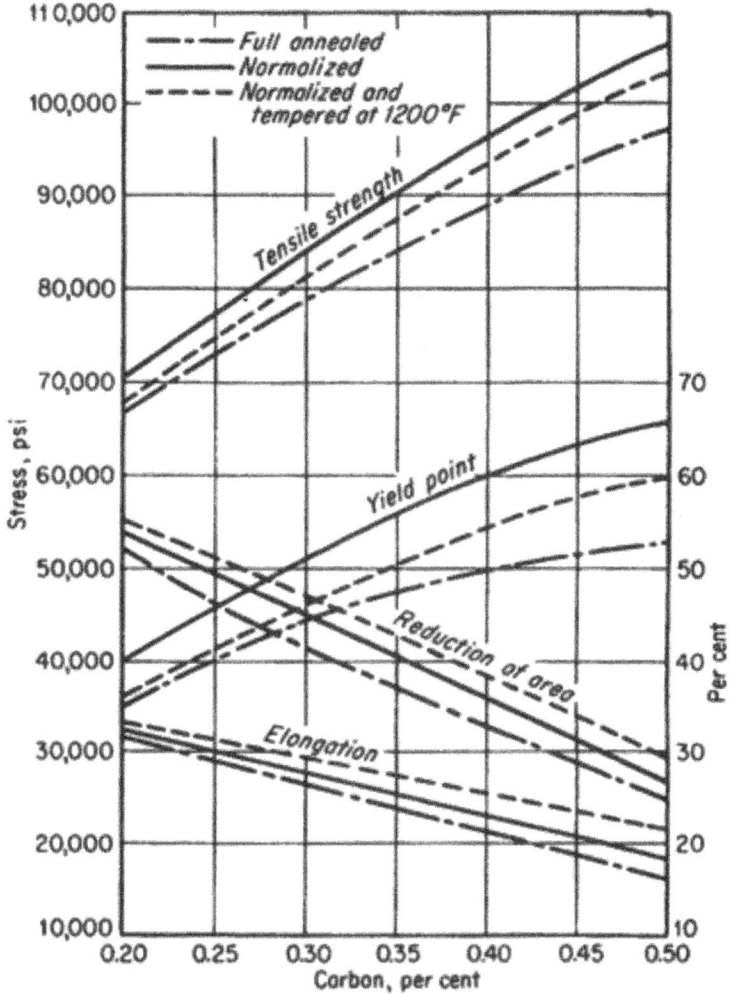

Figure 12.14 Effect of various heat treatments on properties of cast steel. (From Heine et al. [19].)

steel. Thus, the quenching and tempering treatment provides a means of controlling the properties of a given steel alloy within rather broad limits. Furthermore, this treatment gives the best combination of properties obtainable. In other words, a steel tempered to a given tensile strength (or hardness level) will have the highest ductility, toughness, or yield strength as compared with other methods of heat treating leading to the same hardness. Figure 12.15 correlates the effect of carbon content and tempering temperatures on the properties of medium-carbon cast steel [20]. There are embrittling microstructures that form in most steels and particular tempering temperatures. The highest strengths and hardnesses with an optimal toughness and ductility are formed with tempering from 180 to 220°C (350 to 425°F). Above this temperature, the strength declines and the ductility improves but the toughness is degraded. This is avoided by tempering above 550°C (1000°F). This gives the best combination of toughness and ductility at lower strengths.

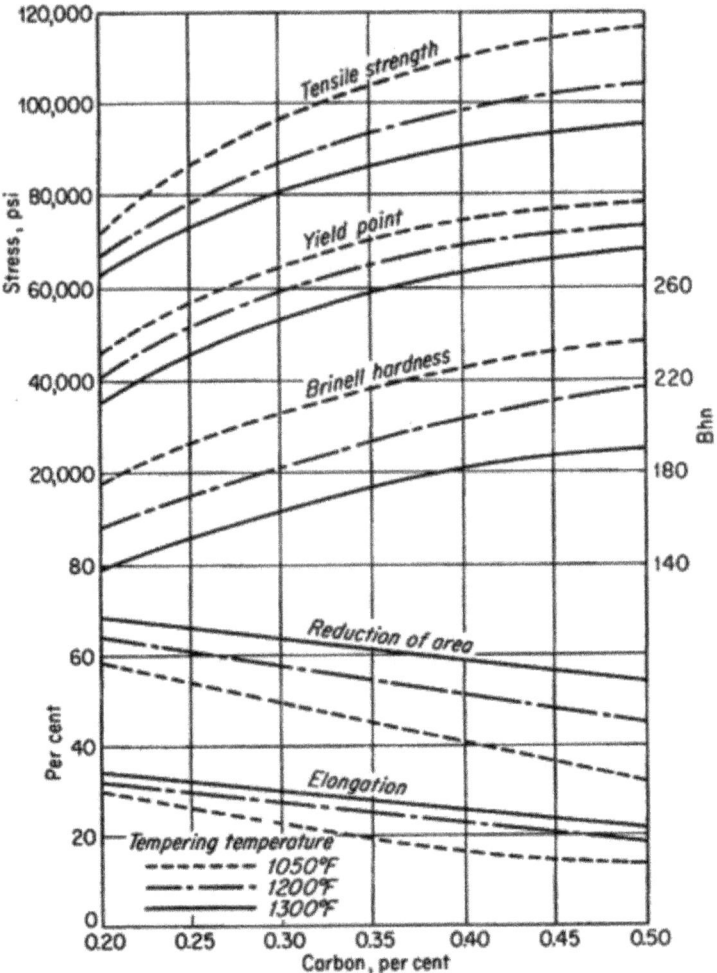

FIGURE 12.15   Effect of carbon content and tempering temperature on the properties of medium-carbon cast steel. (From Heine et al. [20].)

Quenching and tempering of cast steels follows the same principles and techniques used for all steel products. Therefore the problems of quenching media, steel selection, quench cracking, distortion, tempering treatment, hardenability requirements, etc., are the same as those encountered with other steel products.

Alloy Steels

The properties of low-alloy steels are controlled largely by the effects of the specific alloying elements on the heat-treatment characteristics and are not dependent on any intrinsic property that the particular alloy might confer. In other words, in this class of steels the alloying elements are used primarily to increase the ability to heat treat the steel in heavier sections.

On the other hand, the high-alloy steels are designed primarily for some specific property conferred by the alloy, such as corrosion resistance or heat resistance or some other property, rather than for their effect on heat treatment. In this group, for instance, would fall the austenitic manganese steels (14% manganese), which have outstanding wear resistance, and the 18-8 stainless steels for their corrosion resistance. The metallurgy of these steels also differs considerably from that of plain carbon and low-alloy steels, and reference should be made to later sections of this chapter.

The heat treatment of the low-alloy steels is basically the same as that of the plain carbon steels. The fact that alloy steels are designed to improve their heat-treatment response means that less severe quenching practice (oil or air quench rather than water quench) can be employed to obtain the desirable martensitic structure prior to tempering, or that heavier sections can be heat treated than would be the case for carbon steels.

Effect of Structure on Properties

A study of structure is important because the structure of the steel controls the properties of the casting. Figure 12.16 illustrates schematically how the mechanical properties of cast steel are influenced by the microstructure [21]. Heat treatment is an indispensable tool for the control of mechanical properties in the supplied steel casting.

The structure is also influenced by the mass effect. The mass, or size, of the casting not only affects the microstructure of the metal through its influence on the cooling rate, but also through its influence on inclusion size and distribution and segregation. Increasing mass of the casting results in the following effects near the center of the section:

- A decrease in strength, ductility, and toughness
- A decrease in density (an increase in microshrinkage or porosity)

Alloying Elements in Low-Alloy Steels

In addition to the manganese and silicon present in the carbon steel grades, low-alloy steel grades may contain varying amounts of Cr, Ni, Mo, etc. In the annealed condition, low-alloy steels have two microstructural features:

1. Ferrite: alpha iron with dissolved elements
2. Carbide: cementite with dissolved elements or alloy carbides

In ferrite, iron alloying elements increase its strength and hardness according to the general rule of solid solution; see Fig. 12.17 [22]. The same effect of alloying element is seen, in the annealed condition, even in low-carbon steel; see Fig. 12.18 [23].

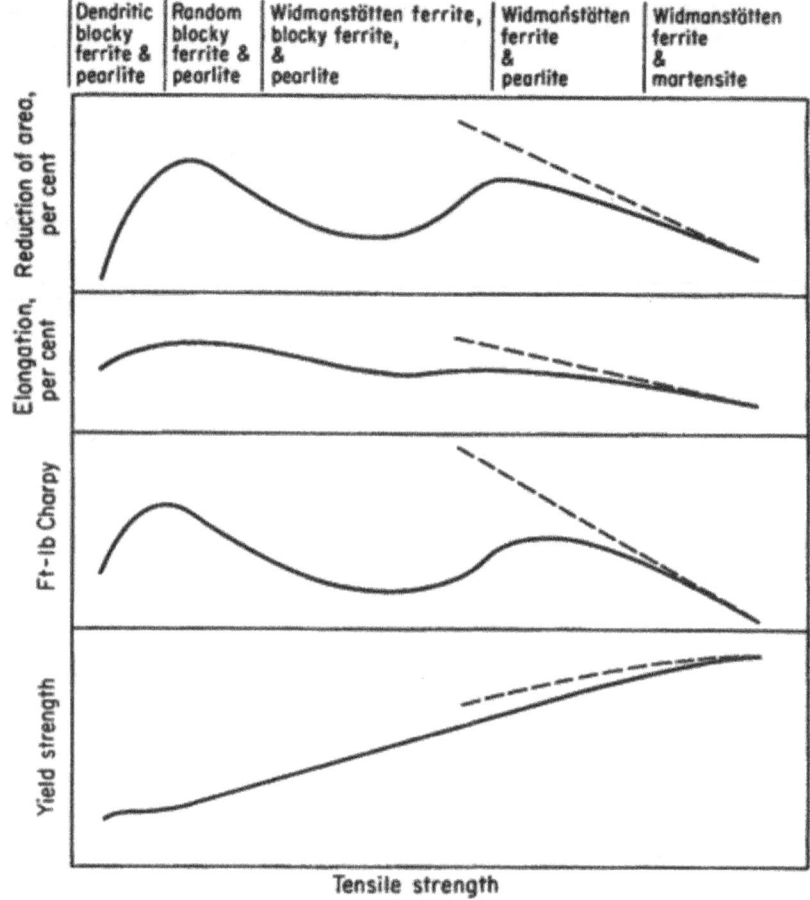

Figure 12.16 Idealized diagram showing the effect of microstructure on mechanical properties of cast steel. Cooling rate decreases from left to right. (From Heine et al. [21].)

Pearlite is a layered structure of cementite and ferrite and is characteristic of slower cooling rates. Steel that is rapidly cooled or quenched can form martensite instead of pearlite. Martensite is strong, hard, and brittle. When reheated to an intermediate temperature in a tempering operation, steel gains ductility and loses hardness and strength. By quenching and tempering a steel component a more desirable mix of properties is obtainable. The ability to form martensite by quenching is limited by section size in unalloyed steels. The ability to form martensite in different section sizes is referred to as hardenability.

Measuring of Hardenability and Its Significance

The main effect of alloying elements in steels is to increase the hardenability of steel. Hardenability is the property of steel that governs the depth to which hardening occurs in a section during quenching. It should not be confused with hardness, which is the resistance to penetration, as measured by Rockwell, Brinell, or other hardness tests. Hardenability is of considerable importance because it relates directly to the strength of steel deep inside the casting and to many other properties, notably, toughness and fatigue.

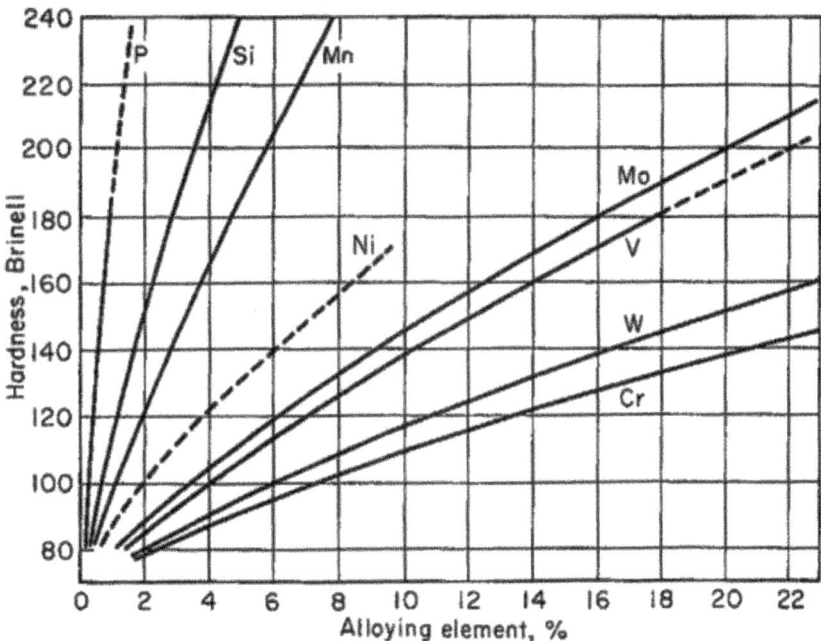

Figure 12.17 Probable hardening effect of the various elements as dissolved in iron. (From Bain and Paxton [22].)

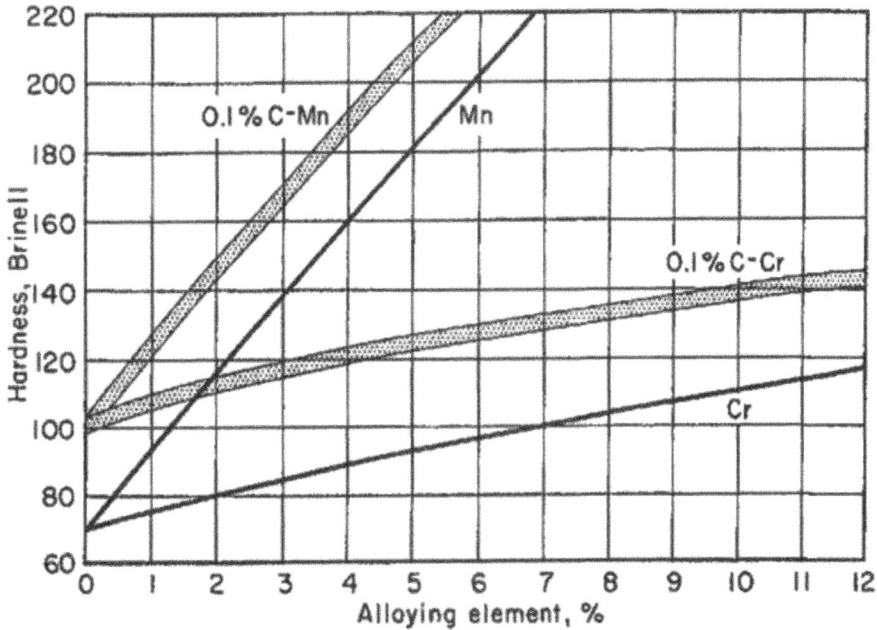

Figure 12.18 Hardness increase caused by chromium and manganese dissolved in ferrite. Bands indicate annealed 0.1% carbon steels; lines refer to substantially pure iron alloys. (From Bain and Paxton [23].)

The hardenability of alloy steels and carbon steels is measured in the Jominy end-quench test [24]. This procedure is the most commonly used test to evaluate hardenability. In the Jominy end-quench test, a 1-in (25.4-mm) diameter bar approximately 3½ in (88.9 mm) long is heated to the austenitizing temperature. After holding at temperature, the bar is removed from the furnace and placed over a water jet which quenches only the bottom of the specimen. The test parameters are fixed to standardized conditions [25]. This quenching produces a gradient in cooling rate along the length of the bar, with the highest rates at the quenched end. After quenching, the sides of the bar are carefully ground down about 0.015 in (0.38 mm), and Rockwell C hardness measurements are made at frequent intervals along the length. Results are illustrated in Fig. 12.19 [26].

It is apparent from Fig. 12.19 that the hardness is fairly constant for some distance along the bar and then drops off fairly rapidly, except for the HC steel. This distance of constant hardness is a measure of hardenability; the greater the distance, the higher the hardenability.

Note that the steels have essentially the same quenched hardness at the quenched end of the bar but their hardenabilities as measured by the distance of constant hardness are different. An alloy steel, such as HC in Fig. 12.19, could be used in heavier sections than the HA steel and still produce martensite through the section after quenching. Alternatively, in a given section where a drastic quench might be required for the HA steel, an oil or even an air quench might suffice for the HC steel.

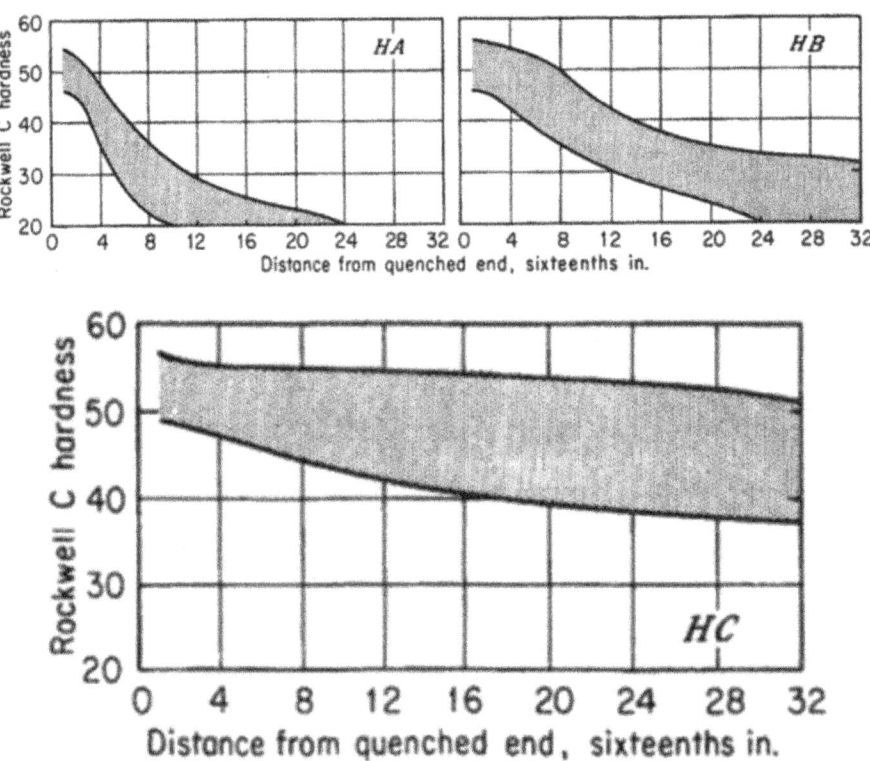

FIGURE 12.19   End-quench hardenability bands for three cast steels of 0.3% carbon content. (From Heine et al. [26].)

The common alloying elements used to increase hardenability include manganese, molybdenum, chromium, silicon, and nickel. They are listed in their approximate order of effectiveness. Other elements used are vanadium and boron. The latter element is used in percentages of less than 0.005% and is used to replace a portion of the more expensive elements. Boron, though, is hard to control in melting and is associated with cracking in processing.

A general idea of relative hardenability of steel can be gained by calculating its ideal critical diameter (DI). The DI is defined as the diameter of a round bar that can be hardened to 50% martensite in the ideal quench. A quench would be ideal if the surface of the quenched bar instantly reaches the temperature of the quench and does not change. The hardenability of alloy steel grades expressed as DI is dependent on grain size and carbon content of steel (Fig. 12.20a) [27] as well as amount of different alloying elements present (Fig. 12.20b) [27]. The grain size of steel castings depends

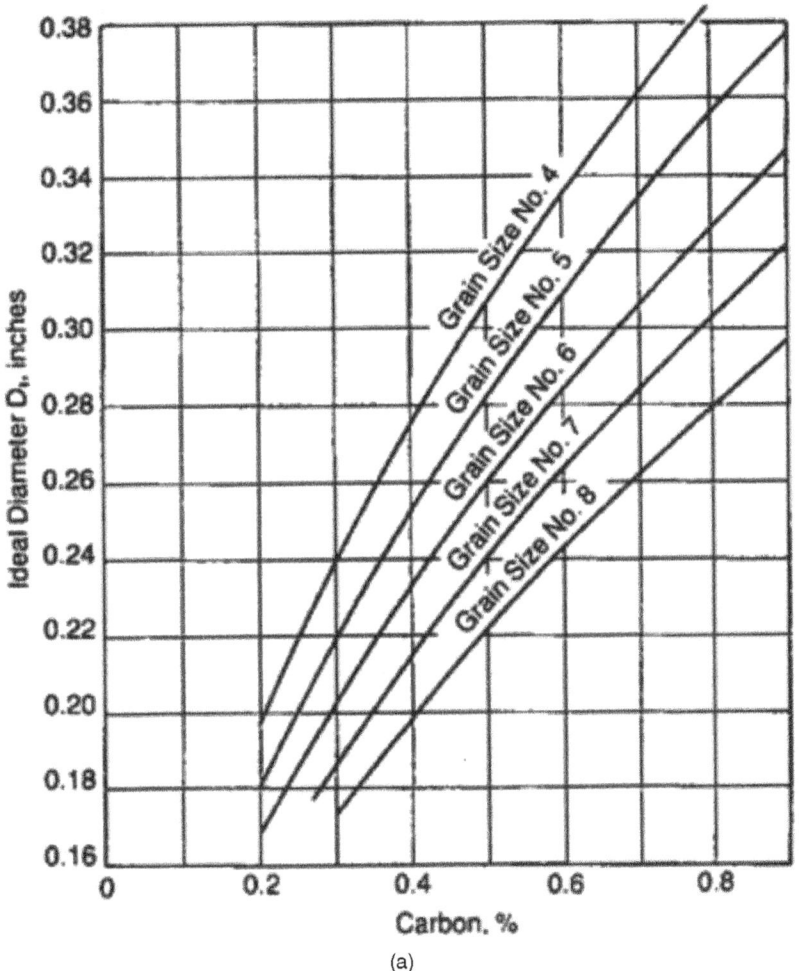

(a)

FIGURE 12.20 (a) Hardenability, expressed as ideal critical size, as a function of austenite grain size and carbon content of iron-carbon alloys. (*Continued*)

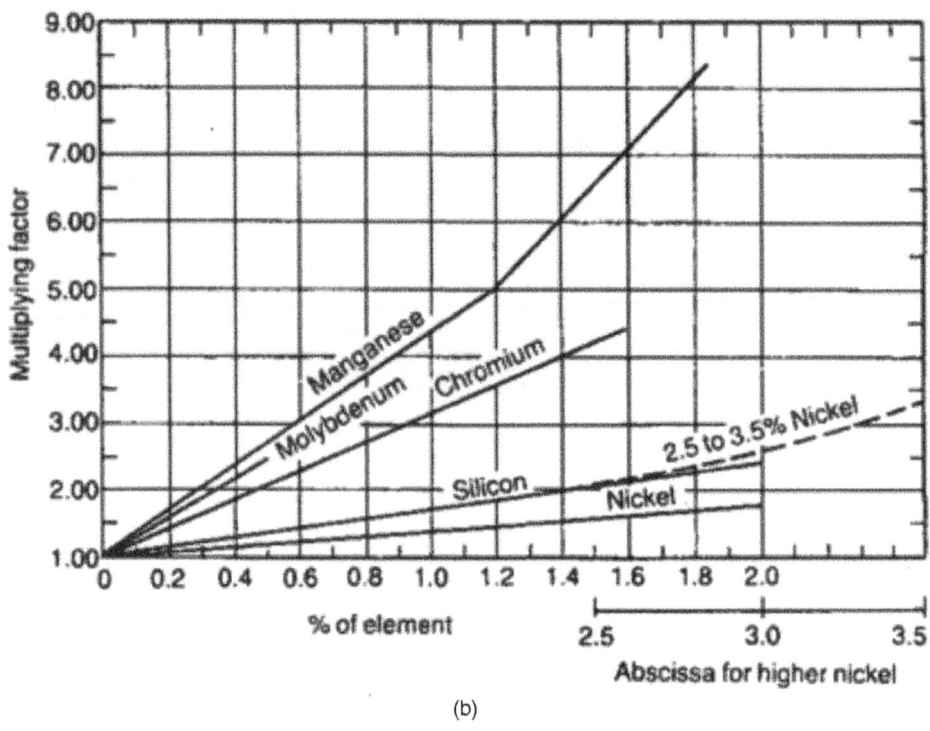

(b)

Figure 12.20 *(Continued)* *(b)* Multiplying factors as a function of the concentration of various common alloying elements in alloy steels. (From Krauss [27].)

mainly on section size but is also influenced by alloying. Smaller numbers are larger grain sizes, and smaller grain sizes generally give better mechanical properties. Steel castings generally have grain sizes of 4 or 5. The DI is calculated by taking the DI due to carbon and grain sizes from Fig. 12.20a and multiplying this with alloy factors picked up from Fig. 12.20b.

Isothermal Transformation Diagram

Diagrams that show the transformation of austenite as a function of time at constant temperature are referred to as isothermal transformation (IT) diagrams, or time-temperature-transformation (TTT) diagrams. Sometimes they are also referred to as S-curves. Figure 12.21 shows three IT curves for C-1080, C-1321, and C-4150 steel [28]. The distance of the nose of the IT curve from the temperature axis is another measure of the hardenability of the steel. The farther the nose is to the right from the temperature axis the higher the hardenability. Thus, the steel in Fig. 12.21b has a higher hardenability than the steel in Fig. 12.21a due to much higher Mn content; they have the same carbon content. The steel in Fig. 12.21c has a higher hardenability than the steel in Fig. 12.21b.

Liquid quenching followed by tempering is common practice. Alloy steels, when tempered or cooled through a tempering-temperature range of about 550 to 425°C (1050 to 800°F), exhibit a form of embrittlement referred to as *temper embrittlement*.

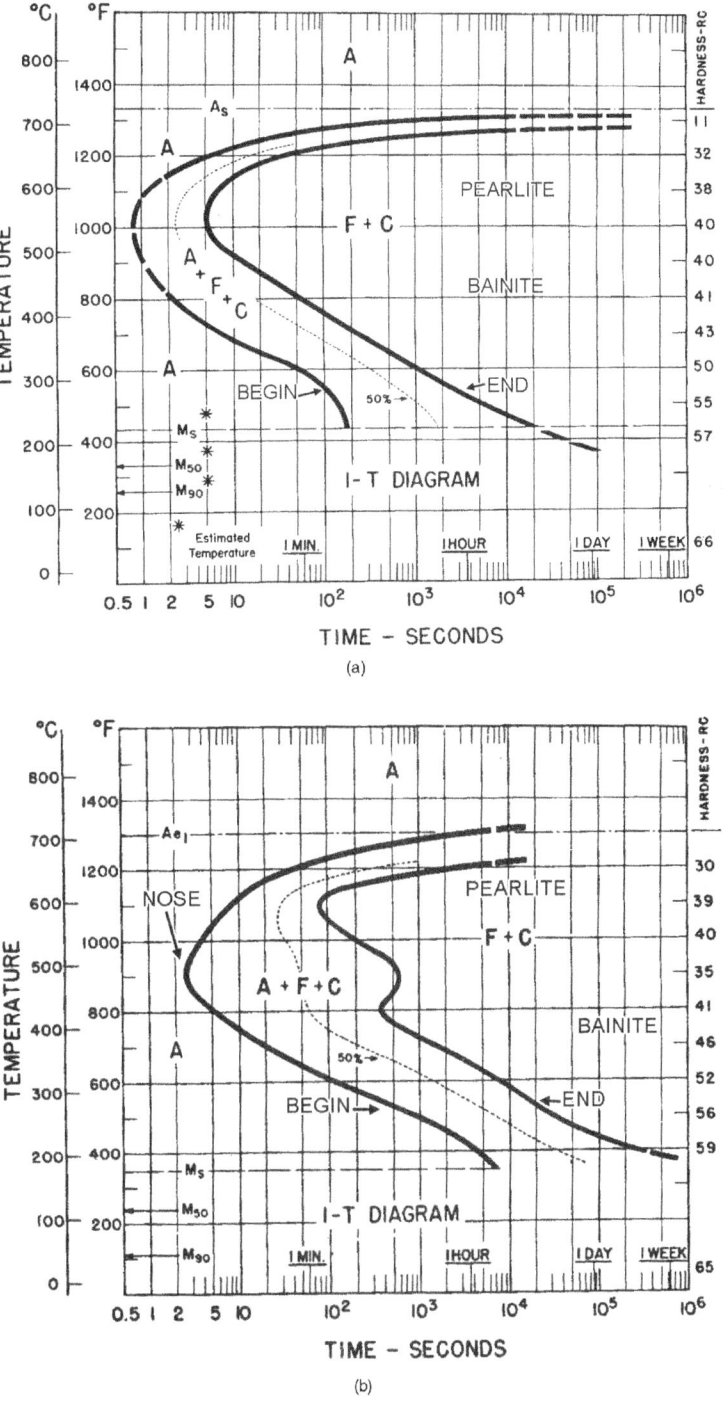

Figure 12.21 IT curves for C-1080, C-1321, and C-4150 steels. (*a*) C-1080 steel, C = 0.79, Mn = 0.76; (*b*) carburized 1321 (0.8C), C = 0.8, Mn = 1.88, austenitized at 1700°F, grain size: 5 to 8;

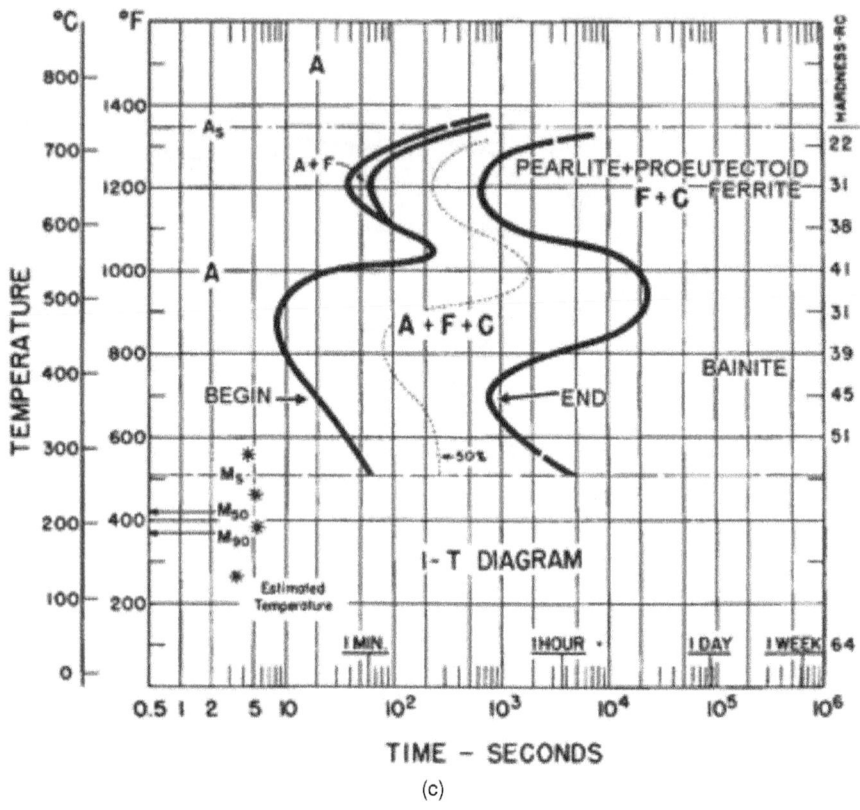

FIGURE 12.21 *(Continued)* (c) 4150 modified, C = 0.55, Mn = 0.60, Cr = 1.03, Mo = 0.19, Ni = 0.36, austenitized at 1550°F, grain size: 7 to 8. (From ASM [28].)

This embrittlement can be eliminated or reduced in severity by water quenching after tempering above the embrittlement range. It can be reduced by using molybdenum in the steel. The embrittlement is manifested by a reduction in toughness values at room temperature or at lower temperatures.

Despite differences in hardenability when low-alloy steels are heat treated, they have essentially identical properties when quenched and tempered to a given hardness or tensile-strength level. This is illustrated in Fig. 12.22 [29].

Production Heat Treating

The heat treating of a steel casting in most foundries is on a batch-lot basis in car-type or stationary furnaces. Where production volumes are large, automation of heat-treating operations can be achieved.

The heating rate for the castings is limited primarily by the furnace capacity and the need to avoid cracking and warping during heating. Ductile cast materials can be heated as quickly as possible. Less ductile materials must be heated more slowly to prevent the expanding heated exterior from causing an unsustainable tensile load inside the casting and having a casting crack in two. Large uniform-sectioned castings do

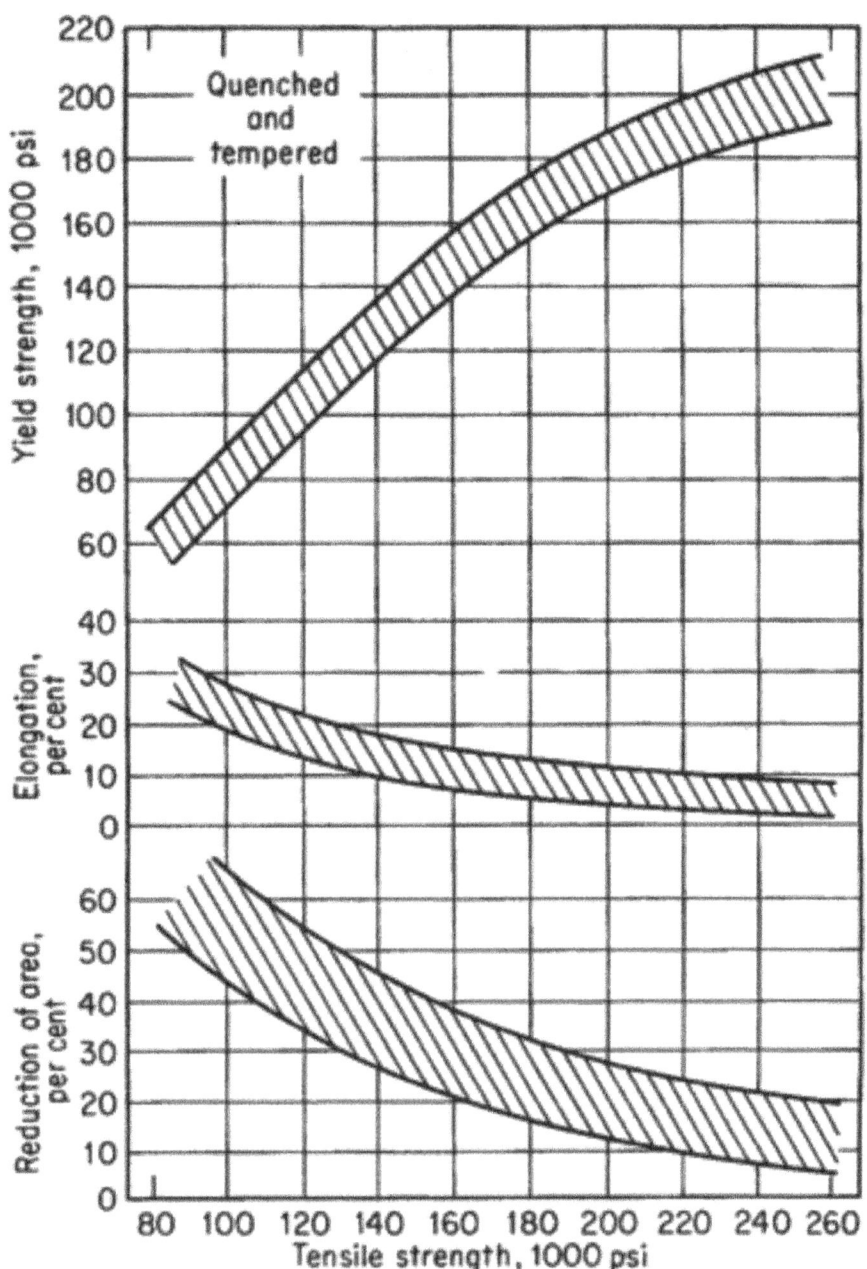

Figure 12.22 Tensile properties of low-alloy cast steels in quenched and tempered condition as a function of ultimate tensile strength. (From Heine et al. [29].)

not have a large temperature difference between the surface and center of the casting during heating. Because of the high conductivity of the steel, heat transfer inside is faster than the rate of heat transfer from the furnace to the steel surface. Even when the furnace is already at temperature before insertion of the steel, temperature gradients within a heavy section may not be over 200°F. For castings of nonuniform cross section, the thinner parts will heat up to furnace temperature much more rapidly than the heavier sections, and in so doing may develop sufficient stress to cause warping or cracking. In such cases, lower rates of heating are required. Since most of the cracking or distortion takes place only after the steel has reached its critical temperature, heating in the early stages can be quite rapid. In this connection, some tempering furnaces are now started at a higher temperature than required for the castings, and are so regulated that this temperature drops while the castings heat. Thus more rapid heating rates are attainable than could be realized by simply placing the castings in a furnace maintained at the desired final temperature.

Effective use can often be made of the various specialized heat-treating processes designed to avoid distorting or cracking during cooling. These processes, some of which are compared with the standard quench and temper (Fig. 12.23) [30], include

- *Martempering.* This is also called *marquenching*. This is a hardening treatment that consists of quenching the part from austenitizing temperature to a temperature above the M_s (martensite start temperature), usually by quenching into a salt bath, holding for a time sufficient for the temperature to become uniform, and then air cooling through the M_s to room temperature. Tempering is then performed as required (Fig. 12.23b). The equalization of temperature throughout a part prior to martensite formation ensures that the transformation stresses across the part will be minimal and therefore that cracking and distortion will also be minimized.

An important aspect of martempering is that no transformation product other than martensite should form. Therefore, steels that are suitable for martempering

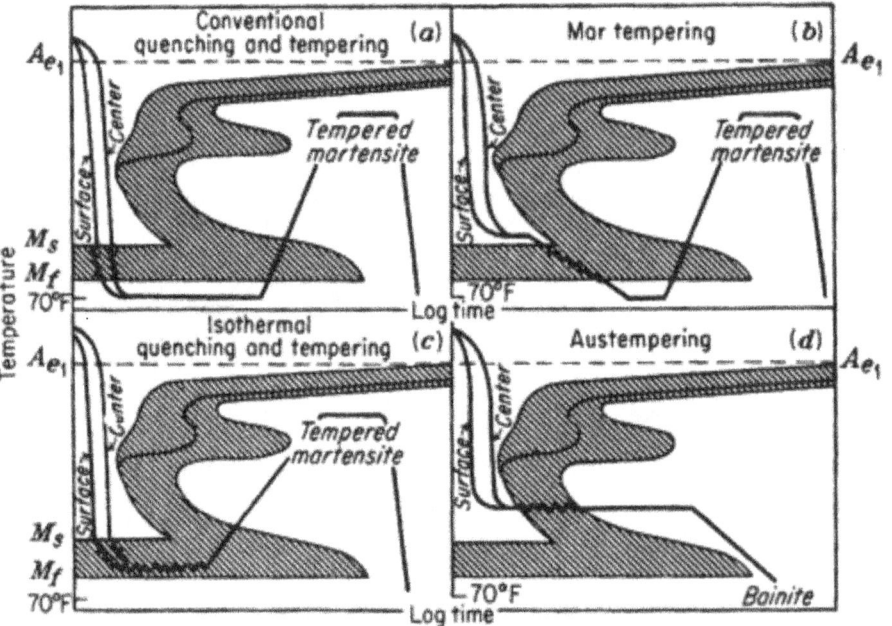

Figure 12.23   Various types of quenching processes. (From Heine et al. [30].)

must have sufficient hardenability not only with respect to higher temperature transformation products such as ferrite and pearlite, but also with respect to bainite that might form just above the M_s.

- *Austempering.* Austempering is another hardening treatment designed to reduce distortion and cracking in higher carbon steels. The object of austempering, however, is to form bainite rather than martensite (Fig. 12.23*d*). The steel is austenitized, quenched in molten salt held at a temperature above M_s and then allowed to transform to bainite at that temperature. No tempering is required. As shown, the temperature of the center and surface of an austempered part come together at the holding temperature, and the absence of thermal gradients in the subsequent transformation to bainite minimizes the stresses generated during austempering.

- *Isothermal quench.* This process is similar to martempering except the salt bath temperature used is much lower, between M_s and M_f (martensite finish temperature) (Fig. 12.23*c*).

- *Time quenching.* This is also called *interrupted quenching*. In this process, the part is quenched in oil or water for a length of time sufficient to bring the thickest section to below M_s and then pulled out of quench and subsequently tempered.

- *Differential hardening.* This is accomplished either by heating a part of the casting and quenching the whole or by heating the entire casting and quenching only a part of it.

Straightening

Stresses induced by contraction of the casting during solidification, cooling from shakeout, and sometimes heat treatment may cause a certain amount of warping of steel castings. Variation in metal section thickness, while desirable from the standpoint of design freedom, promotes nonuniform cooling, and resultant stresses and distortion. Rangy casting shapes are more susceptible to distortion than the more rigid designs, such as box sections.

Straightening methods involve the use of hydraulic presses for larger castings, with fixtures, templates, and gauges utilized to ensure dimensional tolerances. Smaller hydraulic presses or hand hammer and die-straightening methods may be utilized for smaller castings. Press straightening is generally done at normal room temperature, but higher strength castings may be heated to 100 to 200°C (200 to 400°F) for increased ductility during straightening.

Machining

Some steel foundries are equipped with machine shops capable of some or all of the required machining operations on castings to prepare them for the intended application. In other cases, preliminary or rough machining operations may be performed by steel foundries for their own purposes. Control of certain dimensions to closer than cast tolerances and removal of metal padding added to aid in feeding are examples of such purposes. Sometimes extra stock is added to large flat cope surfaces and removed during rough machining. This saves grinding and welding time since most of the inclusions are removed during rough machining.

Qualification of surfaces, before machining to final dimensions by the customer, is another reason why some machining is performed prior to shipment of castings to the customer.

Dimensional Layout

Many designs involve the use of castings in multicomponent assemblies demanding complex relationships of dimensional attributes between the cast components. In this case, the required dimensional attributes are specified on a casting drawing which increasingly involves the use of geometric tolerancing practices and CAD/CAM-generated reference points and surfaces. These complex dimensional situations require that both foundry and customer agree on the procedures and responsibilities for laying out the castings on the first article and production sampling basis. Many casting producers have facilities which include layout tables and layout machines to accomplish these tasks.

Many castings do not have critical dimensional requirements and may require only a few dimensional checks. These castings would be inspected using the more conventional layout and measurement tools, calipers, and gages. It should be remembered that when specifying critical dimensions and tolerancing, the type, accuracy, and condition of pattern equipment, and the type of molding system used to make the casting will have a large effect on casting dimensional repeatability.

Welding

During the finishing processes, the castings are inspected by visual and/or nondestructive methods to identify any areas which do not comply with specified quality standards. Based on economics a decision is made as to whether to finish by welding or scrap the part.

The finish welding of steel castings in the finishing process is very similar to weld fabrication of forged or rolled products. Preparation of the area to be welded may be necessary to provide a clean, properly shaped surface that will facilitate complete fusion of weld metal. Preparation by grinding or air carbon-arc includes removal of the discontinuity and shaping the resultant cavity to specific geometric features. Weld metal of the proper composition is chosen to approximate the base metal in composition and other desirable physical characteristics such as strength, toughness, or corrosion resistance.

Since the weldability of steel castings is very similar to that of other steel product forms, the welding processes and procedures are also similar. The manual shielded metal arc ("stick" electrode SMAW) and the gas shielded semiautomatic welding processes (GMAW/MIG) are commonly used. Other processes sometimes used include the gas tungsten-arc (GTAW/TIG), submerged arc, and electroslag welding processes.

Research has confirmed that most cast steel alloys have equal or better weldability than the equivalent wrought alloys. Furthermore, once cast steel castings are welded, the weld bead is finished to the contour of the casting, thus ensuring freedom from stress raising notches at the weld heat-affected zone (HAZ). The welded casting may be further processed through a stress-relieving heat treatment, with the outcome being that cast/welded structures have excellent properties in the weld metal, the HAZ, and the base metal. After corrosion-resistant castings are welded, they may be given a solution anneal heat treatment in order to restore the full corrosion-resistant properties of the alloy that may have been compromised by the heat of the welding process.

Nondestructive Testing

Magnetic Particle Inspection

Discontinuities which are located either on or close to the surface of a ferritic steel casting, can be revealed by magnetic particle testing. This test method is based upon the attraction and adherence of fine ferromagnetic particles to the component surface where localized leakage fields form as a result of a discontinuity at or just below the component surface. Nonmagnetic steel castings cannot be inspected in this manner.

During magnetic particle inspection, either a portion or all of the casting is magnetized by the application of a high-amperage current to the part either through prods or other direct contact methods, or through placing the casting within a magnetic field generated by passing current through a central conductor or cable coil. Magnetic yokes are another method of inducing magnetic fields within a casting. Discontinuities on or near the casting surface create highly localized magnetic fields that attract iron powder particles to the site of the discontinuity. The particles can either be applied in dry powder form (the dry method) or in a suspension in a liquid medium (the wet method). The particles can either be colored for easier visibility under normal lighting or can be coated with a florescent dye which renders the particle buildup visible in ultraviolet light.

Dye Penetrant Inspection

Discontinuities which extend to the surface of magnetic and nonmagnetic materials can be revealed by dyes that are applied by spraying, brushing, or dipping of the casting. Florescent dyes are available to improve the visibility of the discontinuities in ultraviolet light. The process of inspection involves application of the liquid penetrant, its removal by washing and wiping, and the subsequent application of the developer, frequently a white powder which is discolored by the dye seeping from the discontinuity.

Radiography

Radiography has become the major nondestructive test method for determining the presence of internal discontinuities. A low-power X-ray source (around 300 keV) is used for section thickness less than 1 in. Indium 192 is used as a radiation source primarily for casting sections under 1.5 in, and Cobalt 60 is generally employed for sections between 1 and 6 in. More powerful X-ray (>8 MeV) machines or linear accelerators are chosen by foundries producing heavy sections where exposure time with small isotope sources would be prohibitively long, and where the production program is specialized for castings requiring extensive documentation of internal integrity. Filmless radiography has also been introduced for examining up to 2-in-thick casting sections. The image is projected on a florescent screen, rather than being reproduced on film.

The ASTM (American Society for Testing and Materials) has prepared standards for radiographic testing and reference radiographs to rate the degree of soundness (E-446, E-136, E-230, E-192, and E-390). The reference radiographs represent graduated severity levels of discontinuities. Since the degree of casting integrity required varies with end use, the discontinuity severity levels must be specified according to their effect on the serviceability of the component.

Ultrasonic Testing

The ability of ultrasonic testing to determine the location of discontinuities below the component surface is the low cost of the inspection method, and the speed with which it can be performed has led to its increased use. Ultrasonic impulses are sent into the component from special probes. Attenuation and echoes of the sound impulses are monitored to determine the presence of discontinuities and their location.

Until recently the lack of permanent records, as obtained routinely in conventional radiography, has been the reason for customers' preferences for specifying radiography in nondestructive testing. The low cost of the ultrasonic technique, however, and the ability to determine the size and location of discontinuities, have been the reasons for foundries to adopt the technique in inspection for upgrading and repair of steel cast-ings prior to final radiography. These measures reduce the direct examination costs and expedite the movement of castings through the cleaning room.

The benefits of ultrasonic testing are greater for a casting with the section size exceeding 51 mm (2 in) although special techniques are available for thin sections. The application of ultrasonic testing is also largely restricted to ferritic steel because the large grain size of stainless steel castings severely reduces the sensitivity of the tech-nique and limits the method to such applications as wall thickness measurements and examination of machined weld ends for which special probes have been developed.

Pressure Testing

Pressure testing is employed for pressure-containing parts and flow control devices using primarily hydrostatic or air-pressure techniques, with visual leak detection re-maining the most widely used method. Other testing methods include the use of high-pressure air or nitrogen, Freon, helium, steam, and high vacuum.

Metallurgy of Cast Steel

This section defines the basics for classification of steel types and effect of individual elements on physical, mechanical, and thermodynamic properties of steel. Cast steel, as mentioned earlier, is basically an alloy of iron and carbon. In addition to carbon, which imparts basic properties to steel, the other elements which are normally present in rolled steel are also found in cast steels. These include Mn, Si, P, and S. These elements fall into the following typical ranges (by weight percentage):

Manganese	0.5–1.0
Silicon	0.2–0.8
Phosphorus, max	0.04
Sulfur, max	0.03

Small percentages of other residual metals such as Cu, Ni, Cr, Mo, V, etc., may also be present. Steel is produced in large part by recycling scrap where so many of these elements are present in the scrap supply. For plain-carbon steel castings, the carbon content determines one type of classification used for commercial steel.

- Low-carbon steel (C <0.20%)
- Medium-carbon steel (C = 0.20 to 0.50%)
- High-carbon steel (C >0.50%)

In addition to three classes of plain carbon steels listed above, two other classes are also defined as

1. Low-alloy steels (alloy content totaling <8%). Generally, alloying elements included are Mn, Si, Ni, Cr, Mo, and V.
2. High-alloy steels (alloy content totaling >8%). The alloying elements added include Mn, Si, Ni, Cr, Mo, Co, Nb, W, V, Al, etc. The total percentage of alloying elements may exceed 50.

Other methods of classification of steel, such as strength, microstructure, etc., are also practiced.

Manganese and Silicon

In carbon steels, both these elements are residuals resulting from the deoxidation practice. They occur in solution in the steel, i.e., they are dissolved in the iron (forms solid solution) and are not visible when a steel sample is examined under a microscope. Both these elements confer strength and hardness. They add strength by solid solution strengthening and through their influence on the transformation the steel undergoes when cooling from an elevated temperature. In other words, in heat treatment they increase the *hardenability* of the steel.

Sulfur

Manganese combines with the sulfur that is present in the steel to form nonmetallic sulfide inclusions in the steel. Sulfides, of course, are readily identified in a sample prepared for microexamination since they are not soluble in solid steel. Since all the sulfur appears as sulfide inclusions, the sulfur content is limited by specification to 0.06% maximum and is commonly below 0.03% to avoid harmful effect on ductility and toughness if excessive sulfur were present.

Phosphorus

Phosphorus is limited to 0.05% by specification and is typically below 0.04% because it tends to embrittle the steel at low temperatures. Like Si and Mn, it is soluble in iron and *does not* appear in the microstructure.

Effect of Carbon

At room temperature, iron has a body-centered cubic structure. This body-centered cubic structure is designated α-iron or ferrite. On heating, the iron changes crystal structure at 910°C (1670°F) to a face-centered cubic structure, designated γ-iron or austenite. At 1392°C (2538°F), iron returns to the body-centered cubic structure and this is designated δ-iron, or delta ferrite, although it is the same as the room temperature ferrite. It is not necessary to discuss these crystallographic structures here. The allotropic transformation occurring at 910°C (1670°F), modified by carbon determines the properties of the steel by controlling (1) carbon content and (2) the heat treatment.

The iron-carbon equilibrium diagram, shown in Fig 12.13 [18], depicts the changes effected by increasing the carbon content. It is seen that carbon dissolves in γ-iron to a maximum of about 2.0% at about 1149°C (2100°F). This solid solution is referred to hereafter as austenite. The solution of carbon in γ-iron also extends the maximum and minimum temperatures at which *austenite* is stable, respectively, 1496°C (2725°F) at about 0.16% carbon and 727°C (1340°F) at 0.76% carbon.

At room temperature, carbon is only slightly soluble (0.02%) in ferrite. When carbon is in excess of the solubility it forms iron carbide, called cementite. In steels cooled in air from temperatures where it is austenite, this cementite is present with ferrite in a lamellar structure referred to as pearlite (Fig 12.24) [31]. At 0.76% carbon this mixture of ferrite and carbide constitutes the entire microstructure of the steel, in other words, it is all pearlite. At carbon less than this amount, ferrite and pearlite appear in separate patches (Fig. 12.24), the relative proportions being determined by the carbon content. The steel changes from an entirely ferrite material at less than 0.02% carbon to an entirely pearlite material at 0.76% carbon. Ferrite is softer and more ductile than pearlite and it is the

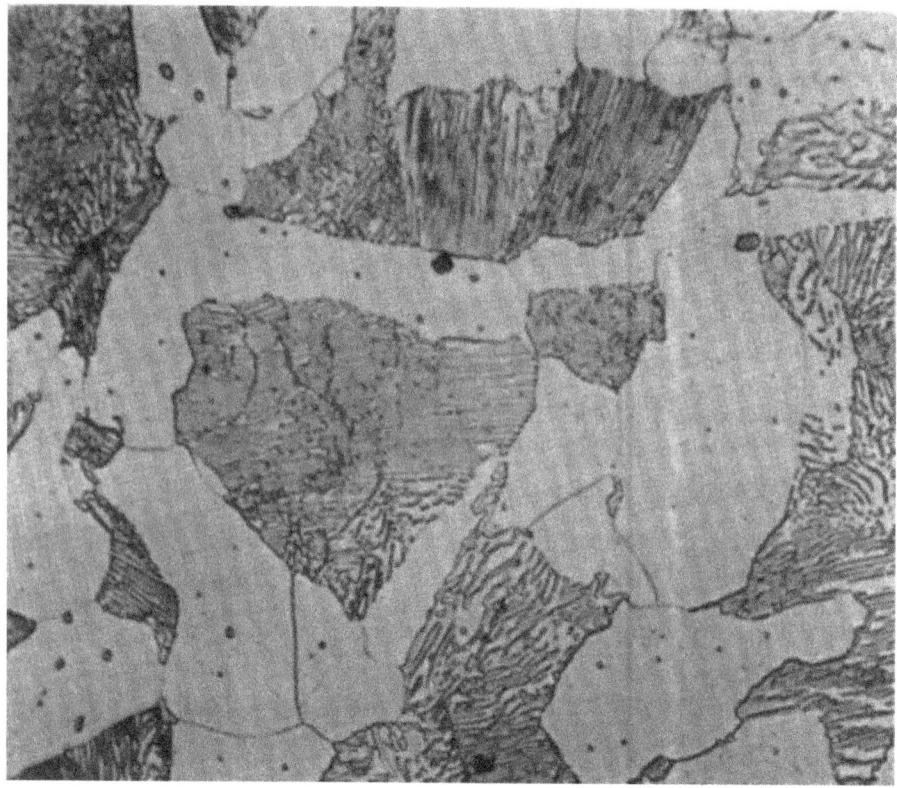

Figure 12.24   Microstructure of ferrite and pearlite in a 0.50% carbon steel, ×750. (From Heine et al. [31].)

increasing amount of pearlite as the carbon content increases that accounts for the effect of carbon on the properties of slow-cooled cast steel. Most steels have carbon contents below 0.76%.

Beyond 0.76% carbon, instead of a mix of pearlite and ferrite, cementite is present with the pearlite and amount of cementite increases as the carbon increases. Thus, in these higher-carbon steels, cementite is found not only in pearlite, but also in a more massive form as a network around the pearlite areas. Few steel castings are made, however, with carbon this high. In fact, higher carbon does increase the strength of the steel but at the cost of decreased ductility, toughness, and weldability. Because of this effect, steel castings are generally made with less than 0.30% carbon.

The contribution of carbon to the strength and hardening characteristics of cast steel is of the same degree and magnitude as in other steel products. The effect of carbon on the properties of cast steel was shown in Fig. 12.2*a* and *b*.

Control of Properties

The fineness of the pearlite structure in steels of less than 0.76% carbon is controlled by the rate of cooling from the austenite structure. This makes it possible to control the properties of the cast steel with heat treatment.

Material Grades

Steel can be divided into the following grades:

- Carbon and low alloy
- Wear resistant
- Corrosion resistant
- Heat resistant
- Specialty alloys

Carbon and Low-Alloy Steels

Carbon steel is steel that contains only carbon, manganese, and silicon as deliberate alloying elements. Cast steels containing more than the following amounts of a single alloying element are considered low-alloy cast steels [4]. They are utilized when different properties are required.

Element	Amount, %
Manganese	1.00
Silicon	0.80
Nickel	0.50
Copper	0.50
Chromium	0.25
Molybdenum	0.10
Vanadium	0.05
Tungsten	0.05

The compositions of low-alloy cast steels are characterized by carbon contents primarily under 0.50% and by small amounts of alloying elements.

Carbon-manganese steels. Cast steels containing 1.00 to 1.75% Mn and 0.20 to 0.50% C have received considerable attention from engineers in the past because of the excellent properties that can be developed with a single, relatively inexpensive alloying element and by a single normalizing or normalizing and tempering heat treatment. Carbon-manganese steels are also referred to as medium-manganese steels and are represented by the cast variation of 1300 series of wrought steels (1.60 to 1.90% Mn). Carbon levels are rarely used above 0.30% to prevent cracking and facilitate welding.

Manganese-molybdenum cast steels are similar to the medium-manganese steels with the added characteristics of higher strength at elevated temperatures, higher ratio of yield strength to tensile strength at room temperature, greater freedom from temper embrittlement, and greater hardenability. Therefore, these steels have replaced medium-manganese steel for certain applications. There are two general grades of manganese-molybdenum cast steels:

- 8000 series (1.0 to 1.35% Mn, 0.10 to 0.30% Mo)
- 8400 series (1.35 to 1.75% Mn, 0.25 to 0.55% Mo)

For both of these alloy types, the carbon content is frequently selected between 0.20 and 0.35%, depending on the heat treatment employed and the strength characteristics desired.

Manganese-nickel-chromium-molybdenum cast steels. The cast 9500 series low-alloy steels are primarily produced for their high hardenability. Sections exceeding 127 mm (5 in) in thickness can be quenched and tempered to obtain a fully tempered martensitic structure. The composition range employed for the 9500 series is

Element	Composition, %
Manganese	1.30–1.60
Nickel	0.40–0.70
Chromium	0.55–0.75
Molybdenum	0.30–0.40

Nickel cast steels. Among the oldest alloy cast steels are those containing nickel. These steels are characterized by high tensile strength and elastic limit, good ductility, and excellent resistance to impact. The cast steels of the 2300 series contain 2.0 to 4.0% Ni, depending on the grade required. Nickel alloy steels are commonly used to achieve high toughness values that require high impact toughness at low temperatures. These materials are quenched and tempered to gain the best toughness.

Nickel-chromium-molybdenum cast steels. The addition of molybdenum to nickel-chromium steel significantly improves hardenability and makes the steel less prone to temper embrittlement. Nickel-chromium-molybdenum cast steel is particularly well suited to the production of large castings because of its deep-hardening properties. In addition, the ability of these steels to retain strength at elevated temperatures extends their usefulness in many industrial applications.

Chromium-molybdenum cast steels. The cast 4100 series low-alloy steels contain chromium contents of about 1.00% or more with a smaller amount of molybdenum and provide an improvement in elevated-temperature properties. These chromium-molybdenum low-alloy cast steels are widely used. These alloys are commonly normalized and used at elevated temperatures to resist creep.

Copper-bearing cast steels. There are several types of copper-containing steels. Selection among these various types is primarily based on either their atmospheric corrosion resistance (weathering steels) or the age-hardening characteristics that copper adds to steel.

Wear-resistant steels. Cast wear-resistant steels are utilized in wear applications such as ground engaging equipment or scrap yard shredders. Wear-resistant alloys can be either austenitic Hadfield manganese steels or heat-treated low-alloy steels. Specialty alloys like chrome irons are also important. The mechanism causing the wear can be categorized as frictional, abrasive, corrosive, or deformation. Austenitic manganese steels are used in many impact-wear environments. The compositions of the austenitic manganese steels (Table 12.4) [32] can be varied to achieve differing combinations of strength, ductility, wear resistance, and machinability.

Mechanical and Physical Properties of Carbon and Low-Alloy Steels

ASTM specifications for carbon steels are A27, A148, A216, A352, A356, and A757, as given in Table 12.5 [33]. Low-alloy steels are covered under ASTM specifications A148, A217, A389, and A487 as given in Table 12.6 [34]. It should be noted that some of the specifications contain both carbon as well as low-alloy steels. These specifications contain both chemistry and minimum mechanical property requirements. Actual properties obtained during production must be over the minimum required. Table 12.7 lists typical properties obtained for various classes of carbon and low-alloy steels [35].

| Identification | | Chemistry | | | | | | | Typical Properties | |
| | | | | | | | | | HDN | CVN(ft·lb) |
Specification	Grade	C	Mn	Si	Ni	Cr	Mo	P	BHN	Room Temp.
ASTM A128	A	1.05–1.35	11.0 min	1.00 max				0.07 max	160	114
	B1	0.90–1.05	11.5–14.0	1.00 max				0.07 max		
	B2	1.05–1.20	11.5–14.0	1.00 max				0.07 max		
	B3	1.12–1.28	11.5–14.0	1.00 max				0.07 max		
	B4	1.20–1.35	11.5–14.0	1.00 max				0.07 max	190	31
	C	1.05–1.35	11.5–14.0	1.00 max		1.5–2.5		0.07 max		
	D	0.70–1.30	11.5–14.0	1.00 max	3.0–4.0			0.07 max		
	E1	0.70–1.30	11.5–14.0	1.00 max			0.9–1.2	0.07 max	185	74
	E2	1.05–1.45	11.5–14.0	1.00 max			1.8–2.1	0.07 max		
	F	1.05–1.35	6.0–8.0	1.00 max			0.9–1.2	0.07 max	160	38

Source: From *Steel Castings Handbook* [32].

TABLE **12.4** Austenitic Manganese Steel Castings

Class or Grade	(UNS) No.	Tensile Strength (a) MPa	Tensile Strength (a) Ksi	Yield Strength (a) MPa	Yield Strength (a) Ksi	Minimum Elongation in 50 mm (2 in), %	Minimum Reduction in area, %	Chemical Composition (b), % C	Chemical Composition (b), % Mn	Chemical Composition (b), % Si	Other Requirements	Condition or Specific Application
ASTM A27: Carbon Steel Castings for General Applications												
N-1	(J02500)	...	...	...	...	...	...	0.25(c)	0.75(c)	0.80	0.06% S, 0.05% P	Chemical analysis only
N-2	(J03500)	...	...	...	...	...	...	0.35(c)	0.60(c)	0.80	0.06% S, 0.05% P	Heat treated but not mechanically tested
U60-30	(J02500)	415	60	205	30	22	30	0.25(c)	0.75(c)	0.80	0.06% S, 0.05% P	Mechanically tested but not heat treated
60-30	(J03000)	415	60	205	30	24	35	0.30(c)	0.60(c)	0.80	0.06% S, 0.05% P	Heat treated and mechanically tested
65-35	(J03001)	450	65	240	35	24	35	0.30(c)	0.70(c)	0.80	0.06% S, 0.05% P	Heat treated and mechanically tested
70-36	(J03501)	485	70	250	36	22	30	0.35(c)	0.70(c)	0.80	0.06% S, 0.05% P	Heat treated and mechanically tested
70-40	(J02501)	485	70	275	40	22	30	0.25(c)	1.20(c)	0.80	0.06% S, 0.05% P	Heat treated and mechanically tested
ASTM A148: Carbon Steel Castings for Structural Applications (d)												
80-40	(D50400)	550	80	275	40	18	30	(e)	(e)	(e)	0.06% S, 0.05% P	Composition and heat treatment necessary to achieve specified mechanical properties
80-50	(D50500)	550	80	345	50	22	35	(e)	(e)	(e)	0.06% S, 0.05% P	Composition and heat treatment necessary to achieve specified mechanical properties
90-60	(D50600)	620	90	415	60	20	40	(e)	(e)	(e)	0.06% S, 0.05% P	Composition and heat treatment necessary to achieve specified mechanical properties
105-85	(D50850)	725	105	585	85	17	35	(e)	(e)	(e)	0.06% S, 0.05% P	Composition and heat treatment necessary to achieve specified mechanical properties
SAE J435 Oct. 2002: See Table 12.6 for Alloy Steel Castings Specified in SAE J435												
0000	...	...	...	...	...	...	...	0.12–0.22	0.50–0.90	0.60	187 HB max	Low-carbon steel suitable for carburizing
415	...	415	60	207	30	22	30	0.25(c)	0.75(c)	0.80	187 HB max	Carbon steel welding grade

450	...	450	65	241	35	24	35	0.30(c)	0.70(c)	0.80	131–187 HB	Carbon steel welding grade
585	...	585	85	310	45	16	24	0.40–0.50	0.50–0.90	0.80	170–229 HB	Carbon steel medium-strength grade
690	...	690	100	485	70	10	15	0.40–0.50	0.50–0.90	0.80	207–255 HB	Carbon steel medium-strength grade
550	...	550	80	345	50	22	35	...	...	...	163–207 HB	Medium-strength low-alloy steel
620	...	620	90	415	60	20	40	...	...	...	187–241 HB	Medium-strength low-alloy steel
ASTM A 216: Carbon Steel Castings Suitable for Fusion Welding and High-Temperature Service												
WCA	(J02502)	415–585	60–85	205	30	24	35	0.25	0.70(c)	0.60	(f)	Pressure-containing parts
WCB	(J03002)	485–655	70–95	250	36	22	35	0.30	1.00(c)	0.60	(f)	Pressure-containing parts
WCC	(J02503)	485–655	70–95	275	40	22	35	0.25	1.20(c)	0.50	(f)	Pressure-containing parts
Other ASTM: Cast Steel Specifications with Carbon Steel Grades (g)												
A 352-LCA	(J02504)	415–585	60–85	205	30	24	35	0.25	0.70(c)	0.60	(f) (i)	Low-temperature applications
A 352-LCB	(J03003)	450–620	65–90	240	35	24	35	0.30	1.00	0.60	(f) (i) (j)	Low-temperature applications
A 356-Grade 1	...	485	70	250	36	20	35	0.35	0.70(c)	0.60	0.035% P max, 0.030% S max	Castings for valve chests, throttle valves, and other heavy-walled components for steam turbines
A 757-AIQ	...	450	65	240	35	24	35	0.30	1.00	0.60	(i) (j) (k)	Castings for pressure-containing applications at low temperatures

(a) When a single value is shown, it is a minimum.
(b) When a single value is shown, it is a maximum.
(c) For each reduction of 0.01% C below the maximum specified, an increase of 0.04% Mn above the maximum given in the applicable ASTM specification.
(d) Grades may also include low-alloy steels: see Table 12.6 for the stronger grades of ASTM A 148.
(e) Unless specified by purchaser, the compositions of cast steels in ASTM A 148 are selected by the producer in order to achieve the specified mechanical properties.
(f) Specified residual elements include 0.30% Cu max, 0.50% Ni max, 0.50% Cr max, 0.20% Mo max, and 0.03% V max, with the total residual elements not exceeding 1.00%.
(g) These ASTM specifications also include alloy steel castings for the general type of applications listed in the table.
(h) Testing temperature of –32°C (–25°F).
(i) Charpy V-notch impact testing at the specified temperature with an energy value of 18 J (13 ft·lb) min for two specimens and an average of three.
(j) Testing temperature of –46°C (–50°F).
(k) Specified residual elements of 0.03% V, 0.50% Cu, 0.40% Cr, and 0.25% Mo, with total amount not exceeding 1.00%. Sulfur and phosphorus content, each 0.025% max.

Source: From *Metals Handbook* [33].

TABLE 12.5 Summary of Specification Requirements for Various Carbon Steel Castings. (Unless otherwise noted, all the grades listed in the table are restricted to a phosphorus content of 0.040% max and a sulfur content of 0.045%. max.)

Material Class (UN No.)	Tensile Strength (a) MPa	Ksi	Yield Strength (a) MPa	Ksi	Minimum Elongation in 50 mm (2 in), %	Minimum Reduction in area, %	Composition (b), % C	Mn	Si	Cr	Ni	Mo	Other
ASTM A 148: Steel Castings for Structural Applications (c)													
115-95	795	115	655	95	14	30	...	...	...	...	...	...	(d)
135-125	930	135	860	125	9	22	...	...	...	...	...	...	(d)
150-135	1035	150	930	135	7	18	...	...	...	...	...	...	(d)
160-145	1105	160	1000	145	6	12	...	...	...	...	...	...	(d)
165-150	1140	165	1035	150	5	20	...	...	...	...	...	...	(e)
165-150L	1140	165	1035	150	5	20	...	...	...	...	...	...	(e)
210-180	1450	210	1240	180	4	15	...	...	...	...	...	...	(e)
210-180L	1450	210	1240	180	4	15	...	...	...	...	...	...	(e)
260-210	1795	260	1450	210	3	6	...	...	...	...	...	...	(e)
260-210L	1795	260	1450	210	3	6	...	...	...	...	...	...	(e)
SAE J4350CT 2002: See Table 12.5 for the Carbon Steel Castings Specified in SAE J435 (f)													
725	725	105	585	85	17	35	...	...	...	...	...	...	(g)
830	830	120	655	95	14	30	...	...	...	...	...	...	(g)
1035	1035	150	860	125	9	22	...	...	...	...	...	...	(g)
1205	1205	175	1000	145	6	12	...	...	...	...	...	...	(g)
ASTM A 217: Alloy Steel Castings for Pressure-Containing Parts and High-Temperature Service													
WC1 (J12524)	450-620	65-90	240	35	24	35	0.25	0.50-0.80	0.60	0.35(h)	0.50(h)	0.45-0.65	(h)(i)
WC4 (J12082)	485-655	70-95	275	40	20	35	0.20	0.50-0.80	0.60	0.50-0.80	0.70-1.10	0.45-0.65	(i)(j)
WC5 (J22000)	485-655	70-95	275	40	20	35	0.20	0.40-0.70	0.60	0.50-0.90	0.60-1.00	0.90-1.20	(i)(j)
WC6 (J12072)	485-655	70-95	275	40	20	35	0.20	0.50-0.80	0.60	1.00-1.50	0.50(h)	0.45-0.65	(h) (i)
WC9 (J21890)	485-655	70-95	275	40	20	35	0.18	0.40-0.70	0.60	2.00-2.75	0.50(h)	0.90-1.20	(h) (i)
WC11 (J11972)	550-725	80-105	345	50	18	45	0.15-0.21	0.50-0.80	0.30-0.60	1.00-1.75	0.50(h)	0.45-0.65	(h)(k)
C5 (J42045)	620-795	90-115	415	60	18	35	0.20	0.40-0.70	0.75	4.00-6.50	0.50(h)	0.45-0.65	(h)(i)
C12 (J82090)	620-795	90-115	415	60	18	35	0.20	0.35-0.80	1.00	8.00-10.00	0.50(h)	0.90-1.20	(h)(i)

TABLE 12.6 Summary of Specification Requirements for Various Alloy Steel Castings with Chromium Contents up to 10% *(continued)*

Table 12.6 Summary of Specification Requirements for Various Alloy Steel Castings with Chromium Contents up to 10% (continued)

ASTM A 389: Alloy Steel Castings Suitable for Fusion Welding and Pressure-Containing Parts at High Temperatures

Grade													
C23 (J12080)	485	70	275	40	18	35	0.20	0.30–0.80	0.60	1.00–1.50	...	0.45–0.65	(g)(l)
C24 (J12092)	550	80	345	50	15	35	0.20	0.30–0.80	0.60	1.00–1.25	...	0.90–1.20	(g) (l)

ASTM A 487: Alloy Steel Castings for Pressure-Containing Parts at High Temperatures

Grade													
1A	585–760	85–110	380	55	22	40	0.30	1.00	0.80	0.35(m)	0.50(m)	0.25(m)(n)	0.5 Cu(g)(m)
1B	620–795	90–115	450	65	22	45	0.30	1.00	0.80	0.35(m)	0.50(m)	0.25(m)(n)	0.5 Cu(g)(m)
1C	620	90	450	65	22	45	0.30	1.00	0.80	0.35(m)	0.50(m)	0.25(m)(n)	0.5 Cu(g)(m)
2A	585–760	85–110	365	53	22	35	0.30	1.10–1.40	0.80	0.35(h)	0.50(h)	0.10–0.30	(h)(o)
2B	620–795	90–115	450	65	22	40	0.30	1.10–1.40	0.80	0.35(h)	0.50(h)	0.10–0.30	(h)(o)
2C	620	90	450	65	22	40	0.30	1.10–1.40	0.80	0.35(h)	0.50(h)	0.10–0.30	(h)(o)
4A	620–795	90–115	415	60	20	40	0.30	1.00	0.80	0.40–0.80	0.40–0.80	0.15–0.30	(j)(o)
4B	725–895	105–130	585	85	1	35	0.30	1.00	0.80	0.40–0.80	0.40–0.80	0.15–0.30	(j)(o)
4C	620	90	415	60	20	40	0.30	1.00	0.80	0.40–0.80	0.40–0.80	0.15–0.30	(j)(o)
4D	690	100	515	75	17	35	0.30	1.00	0.80	0.40–0.80	0.40–0.80	0.15–0.30	(j)(o)
4E	795	115	655	95	15	35	0.30	1.00	0.80	0.40–0.80	0.40–0.80	0.15–0.30	(j)(o)
6A	795	115	550	80	18	30	0.38	1.30–1.70	0.80	0.40–0.80	0.40–0.80	0.30–0.40	(j)(o)
6B	825	120	655	95	15	35	0.38	1.30–1.70	0.80	0.40–0.80	0.40–0.80	0.30–0.40	(j)(o)
7A(p)	795	115	690	100	15	30	0.20	0.60–1.00	0.80	0.40–0.80	0.70–1.00	0.40–0.60	(j)(o)(q)
8A	585–760	85–110	380	55	20	35	0.20	0.50–0.90	0.80	2.00–2.75	...	0.90–1.10	(j)(o)
8B	725	105	585	85	17	30	0.20	0.50–0.90	0.80	2.00–2.75	...	0.90–1.10	(j)(o)
8C	690	100	515	75	17	35	0.20	0.50–0.90	0.80	2.00–2.75	...	0.90–1.10	(j)(o)
9A	620	90	415	60	18	35	0.33	0.60–1.00	0.80	0.75–1.10	0.50(h)	0.15–0.30	(h)(o)
9B	725	105	585	85	16	35	0.33	0.60–1.00	0.80	0.75–1.10	0.50(h)	0.15–0.30	(h)(o)
9C	620	90	415	60	18	35	Composition same as 9A but with a slightly higher tempering temperature						
9D	690	100	515	75	17	35	0.33	0.60–1.00	0.80	0.75–1.10	0.50(h)	0.15–0.30	(h)(o)
10A	690	100	485	70	18	35	0.30	0.60–1.00	0.80	0.55–0.90	1.40–2.00	0.20–0.40	(j)(o)
10B	860	125	690	100	15	35	0.30	0.60–1.00	0.80	0.55–0.90	1.40–2.00	0.20–0.40	(j)(o)
11A	485–655	70–95	275	40	20	35	0.20	0.50–0.80	0.60	0.50–0.80	0.70–1.10	0.45–0.65	(o)(r)

Material Class (UN No.)	Tensile Strength (a) MPa	Tensile Strength (a) Ksi	Yield Strength (a) MPa	Yield Strength (a) Ksi	Minimum Elongation in 50 mm (2 in), %	Minimum Reduction in area, %	Composition (b), % C	Mn	Si	Cr	Ni	Mo	Other
11B	725–895	105–130	585	85	17	35	0.20	0.50–0.80	0.60	0.50–0.80	0.70–1.10	0.45–0.65	(o)(r)
12A	485–655	70–95	275	40	20	35	0.20	0.40–0.70	0.60	0.50–0.90	0.60–1.00	0.90–1.20	(o)(r)
12B	725–895	105–130	585	85	17	35	0.20	0.40–0.70	0.60	0.50–0.90	0.60–1.00	0.90–1.20	(o)(r)
13A	620–795	90–115	415	60	18	35	0.30	0.80–1.10	0.60	0.40(s)	1.40–1.75	0.20–0.30	(o)(s)
13B	725–895	105–130	585	85	17	35	0.30	0.80–1.10	0.60	0.40(s)	1.40–1.75	0.20–0.30	(o)(s)
14A	825–1000	120–145	655	95	14	30	0.55	0.80–1.10	0.60	0.40(s)	1.40–1.75	0.20–0.30	(o)(s)
16A (t)	485–655	70–95	275	40	22	35	0.12(u)	2.10(u)	0.50	0.20(r)	1.00–1.40	0.10(r)	(r)(v)

(a) When a single value is shown, it is a minimum.
(b) When a single value is shown, it is a maximum.
(c) Unless specified by the purchaser, the compositions of cast steels in ASTM A148 are selected by the producer and therefore may include either carbon or alloy steels: see Table 12.5 for the lower-grade steels specified in ASTM A148.
(d) 0.06% S (max), 0.05% P (max).
(e) 0.020% S (max), 0.020% P (max).
(f) Similar to the cast steel in ASTM A148.
(g) 0.045% S (max), 0.040% P (max).
(h) When residual maximums are specified for copper, nickel, chromium, tungsten, and vanadium, their total content shall not exceed 1.00%.
(i) 0.50% Cu (max), 0.10% W (max), 0.045% S (max), 0.04% P (max).
(j) When residual maximums are specified for copper, nickel, chromium, tungsten, and vanadium, their total residual content shall not exceed 0.60%.
(k) 0.35% Cu (max), 0.03% V (max), 0.015% S (max), 0.020% P (max).
(l) 0.15 to 0.25% V.
(m) The specified residuals of copper, nickel, chromium, and molybdenum (plus tungsten) shall not exceed a total content of 1.00%.
(n) Includes the residual content of tungsten.
(o) 0.50% Cu (max), 0.10% W (max), 0.03% V (max), 0.045% S (max), 0.04% P (max).
(p) Material class 7A is a proprietary steel and has a maximum thickness of 63.5 mm (2½ in).
(q) Specified elements include 0.15 to 0.50% Cu, 0.03 to 0.10% V, and 0.002 to 0.006% B.
(r) When residual maximums are specified for copper, nickel, chromium, tungsten, molybdenum, and vanadium, their total content shall not exceed 0.50%.
(s) When residual maximums are specified for copper, nickel, chromium, tungsten, and vanadium, their total content shall not exceed 0.75%.
(t) Low carbon grade with double austenitization.
(u) For each reduction of 0.01% C below the maximum, an increase of 0.04% Mn is permitted up to a maximum of 2.30%.
(v) 0.20% Cu (max), 0.10% W (max), 0.02% V (max), 0.02% S (max), 0.02% P (max).

Source: *Metals Handbook* [34].

TABLE 12.6 Summary of Specification Requirements for Various Alloy Steel Castings with Chromium Contents up to 10% (*continued*)

Class (a)	Heat Treatment (b)	Tensile Strength MPa	Tensile Strength Ksi	Yield Strength MPa	Yield Strength Ksi	Reduction in Area, %	Elongation, %	Hardness, HB	Fatigue Endurance Limit MPa	Fatigue Endurance Limit Ksi	Ratio of Endurance Limit to Tensile Strength
Carbon Steels											
60	A	434	63	241	35	54	30	131	207	30	0.48
65	N	469	68	262	38	48	28	131	207	30	0.44
70	N	517	75	290	42	45	27	143	241	35	0.47
80	NT	565	82	331	48	40	23	163	255	37	0.45
85	NT	621	90	379	55	38	20	179	269	39	0.43
100	QT	724	205	517	75	41	19	212	310	45	0.47
Low-Alloy Steels (c)											
65	NT	469	68	262	38	55	32	137	221	32	0.47
70	NT	510	74	303	44	50	28	143	241	35	0.47
80	NT	593	86	372	54	46	24	170	269	39	0.45
90	NT	655	95	441	64	44	20	192	290	42	0.44
105	NT	758	110	627	91	48	21	217	365	53	0.48
120	QT	883	128	772	112	38	16	262	427	62	0.48
150	QT	1089	158	979	142	30	13	311	510	74	0.47
175	QT	1234	179	1103	160	25	11	252	579	84	0.47
200	QT	1413	205	1172	170	21	8	401	607	88	0.43

(a) Class of steel based on tensile strength (Ksi).
(b) A = annealed; N = normalized; NT = normalized and tempered; QT = quenched and tempered.
(c) Below 8% total alloy content.
Source: From *Metals Handbook* [35].

TABLE 12.7 Properties of Various Classes of Cast Carbon and Low-Alloy Steels

For all alloy systems, the mechanical properties are controlled by the chemical composition and the microstructure of the alloy. With respect to carbon and alloy steels, the influence of microstructure is so great as to overshadow that of chemical composition and, for cast steels, the only practical method for changing the microstructure is by heat treatment (Figs. 12.25 and 12.26) [36, 37]. With few exceptions, the mechanical properties of cast carbon and low-alloy steels are controlled by heat treatment. Among the exceptions are the effect of carbon on increasing hardness and strength (Figs. 12.1a, 12.27, and 12.28) [3, 38, 39], the effect of nickel on increasing toughness, and the effect of combinations of chromium, molybdenum, vanadium, and tungsten on increasing elevated temperature strength.

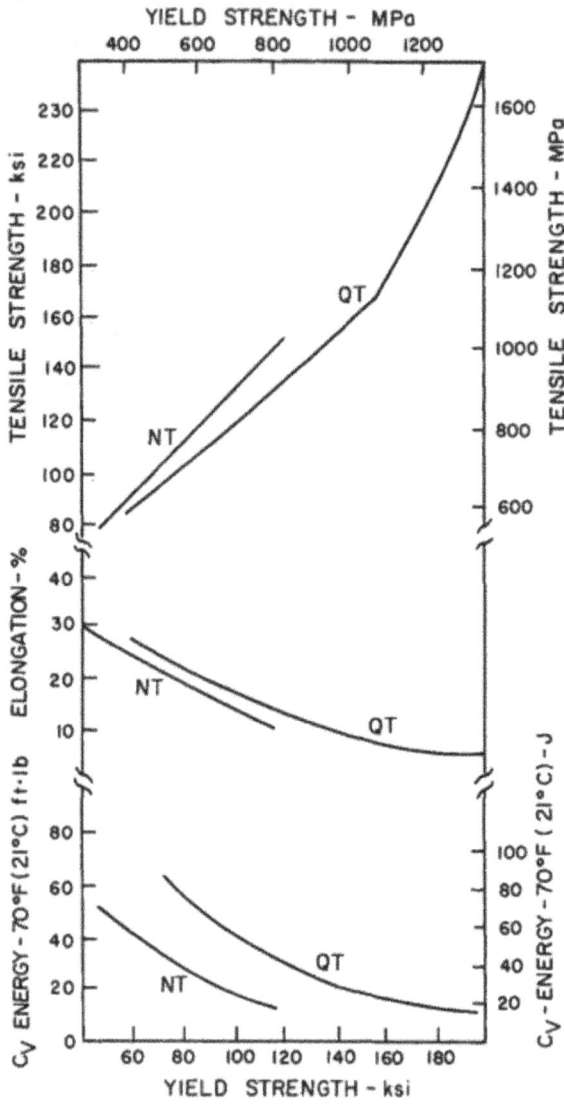

Figure 12.25   Room temperature properties of cast low-alloy steels. QT=Quench and temper, NT=Normalizing and temper (From *Steel Castings Handbook* [36].)

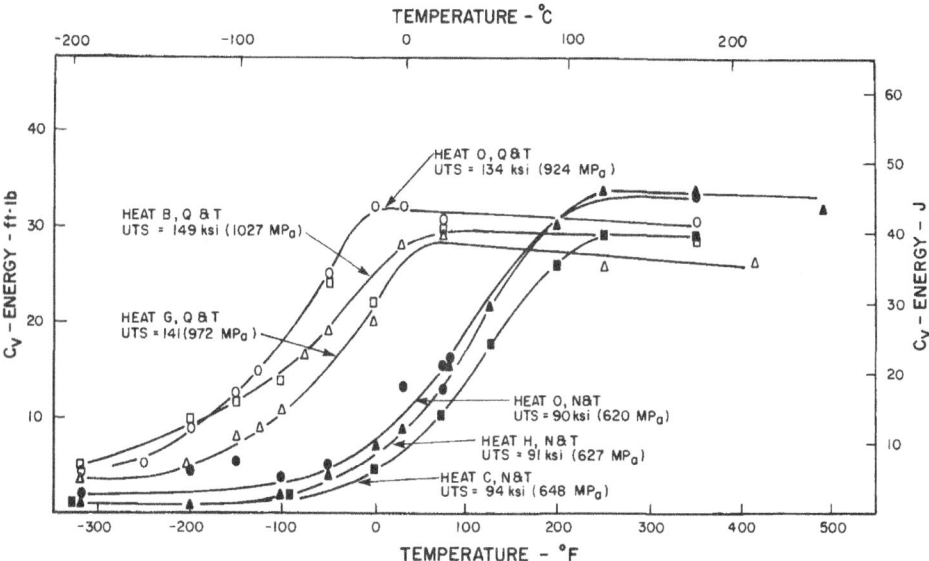

Figure 12.26 Charpy V-notch test results for various heats of 8630 steels in quenched and tempered and normalized and tempered condition. (From *Steel Castings Handbook* [37].)

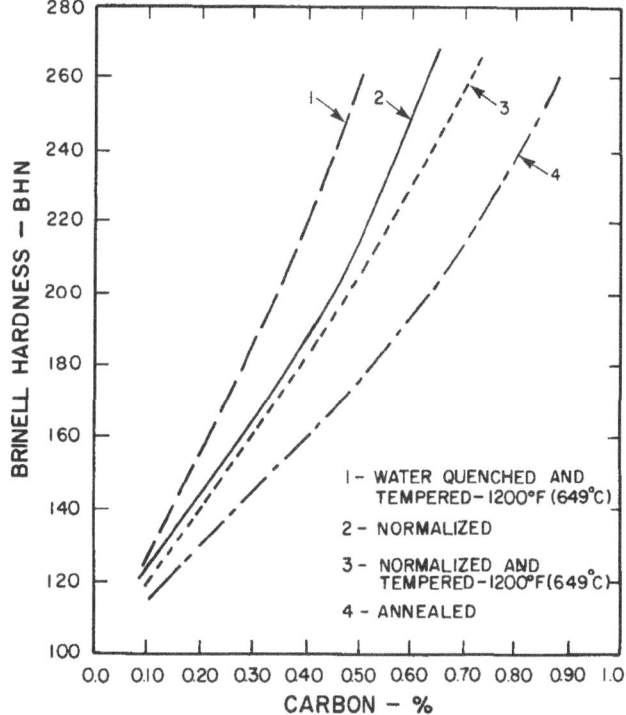

Figure 12.27 Hardness versus carbon content of cast carbon steels. (From *Steel Castings Handbook* [38].)

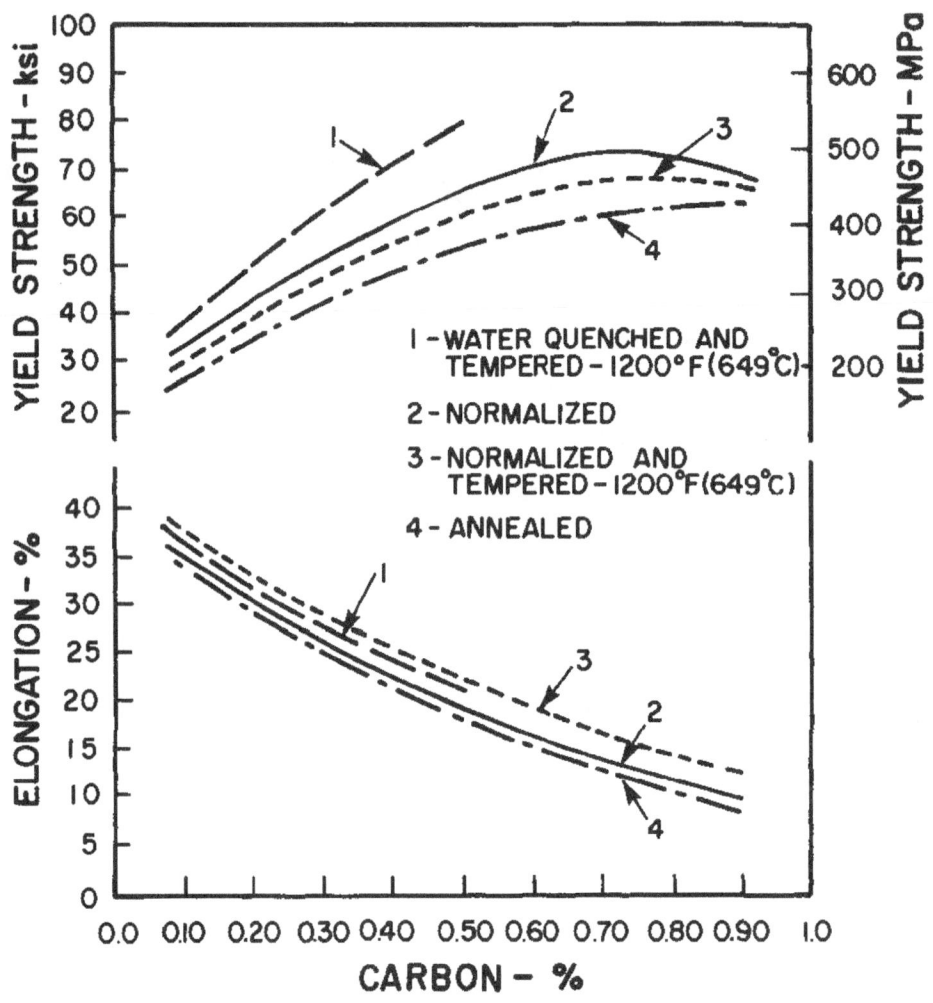

Figure 12.28 Yield strength and elongation versus carbon content of cast carbon steels. (From *Steel Castings Handbook* [39].)

Hardness and Strength

All elements dissolved in steel increase the hardness and strength; however, as compared with heat treatment, this effect is small. An exception is carbon, the effect of which is large as mentioned earlier, although not as large as the effect of heat treatment.

Because of the close relationship between hardness and tensile strength of alloy steels, as shown in Figs. 12.29 [40] and 12.30 [41], there would be a tendency to use carbon, an inexpensive element, to achieve a high-strength steel. The tendency has some validity but entails some penalties. For example, at a given strength level, the toughness (the resistance to brittle fracture) of a steel decreases with increasing carbon level (Fig. 12.31 [42]).

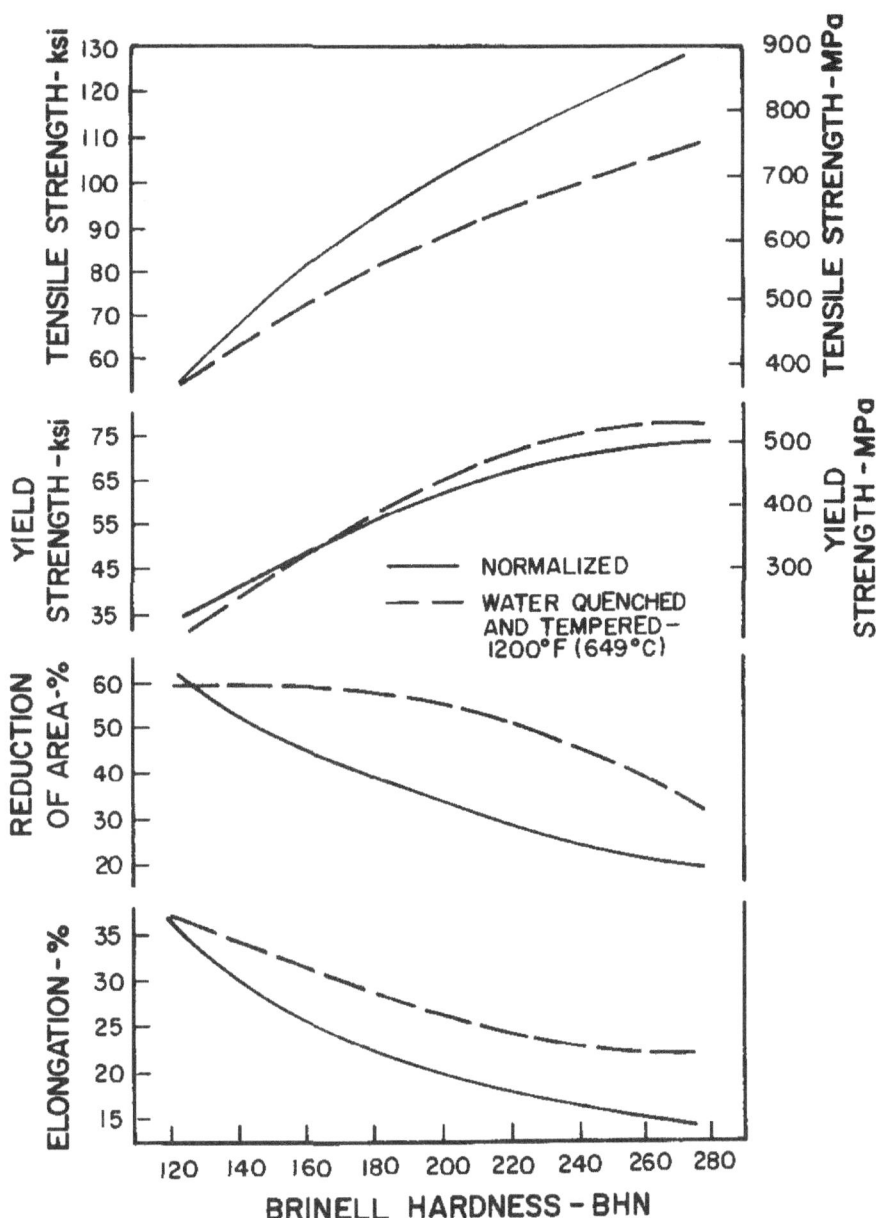

FIGURE 12.29 Tensile properties of cast carbon steels as a function of hardness. (From *Steel Castings Handbook* [40].)

Consequently, a preferred plan is to select a steel having a combination of the lowest carbon content and the required amount of alloying elements to achieve, in a tempered martensite structure, the desired strength. Of course, the alloy content selected must be that required to achieve the hardenability needed for the section size being considered.

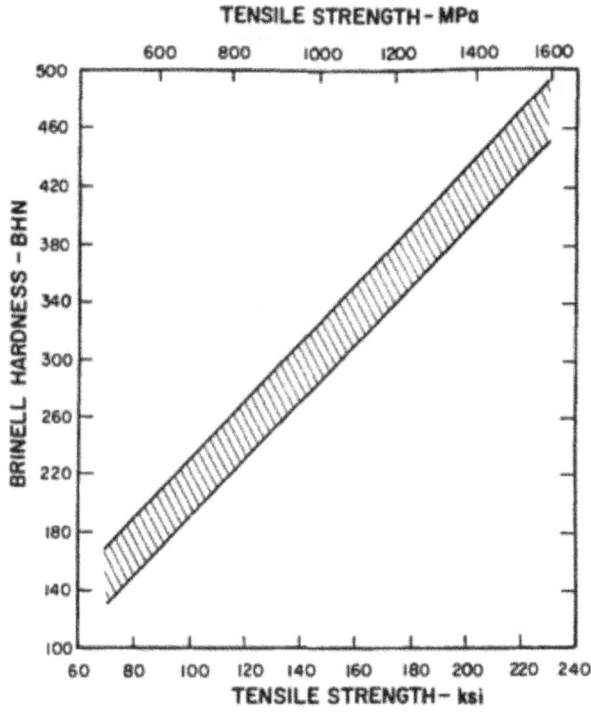

FIGURE 12.30   Hardness versus tensile strength of low-alloy steels regardless of treatment. (From *Steel Castings Handbook* [41].)

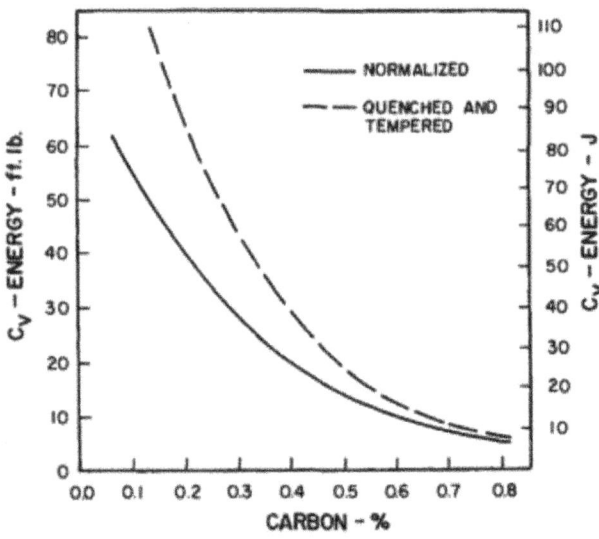

FIGURE 12.31   Room temperature Charpy V-notch values versus carbon content of cast carbon steel in normalized and tempered and quenched and tempered conditions (tempering temperature 1200°F or 650°C). (From *Steel Castings Handbook* [42].)

Toughness

Toughness, the ability of a steel to resist brittle, rapid fracture, is an essential property for structural components subject to high loading rates. All of the test methods used to measure toughness involve the use of samples containing a sharp notch or crack. The reason for employing such samples is that for every service failure caused by brittle fracture, the source of fracture has been found to be a discontinuity in the metal, and every engineering structure will contain some sort of one or more discontinuities (notches). Consequently, the test procedures for toughness evaluations are aimed at determining the resistance of the metal to the rapid propagation of a discontinuity. One such test is the Charpy V-notch impact test described in ASTM Specification E23, "Standard Test Methods for Notched Bar Impact Testing of Metallic Materials."

Fatigue Strength (Endurance Limit)

For cast steels, the fatigue strength, or endurance limit, as determined by tests on smooth bars, is generally in the range of 40 to 50% of the tensile strength. In nonpercentage terms, this relationship is expressed as 0.40 to 0.50 and is termed as endurance ratio. The endurance ratio is largely independent of the tensile strength, chemical composition, and heat treatment of steel.

Under conditions of rough surfaces, i.e., as-cast, machined (not polished), notches, and cracks, both the endurance limit and the endurance ratio decrease. Although the notch sensitivity generally appears to increase with increase in strength and the quenched and tempered condition appears to be less notch-sensitive than the normalized condition, the endurance ratio (in the notched condition) is in the narrow range of 0.27 to 0.32. Consequently, for the somewhat mild notch radius of about 0.015 in. the endurance limit increases with the tensile strength of the steel. For sharper notches, the endurance limit will decrease at high strengths. Cast steels suffer a smaller decrease in endurance limit due to notches than wrought steels. As shown in Figs. 12.32 and 12.33 [43, 44], when no notches are present (smooth surface) the endurance limit of the wrought steel is higher than that of cast steel, but when notches are present, the endurance limits of both the cast and wrought steels are essentially the same. A notched condition such as a rough surface, discontinuities, etc., represents a more realistic service condition so that in practical applications, where fatigue (alternating stresses) are expected, cast steel will perform at least as well as wrought steels.

Section Size, Mass Effects

Mass effects are common to all steels, whether rolled, forged, or cast, because the cooling rate during the heat-treating operation varies with section size, and because the microstructure components, grain size, and nonmetallic inclusions increase in size from surface to center. Mass effects are metallurgical in nature, distinct from the effects of discontinuities. An example of how the mass of component lowers strength properties for wrought AISI 8630 steel plate is shown in Fig. 12.34 [45]. Properties are plotted at 1/4T location, halfway between surface and center of the plate. The section size, or mass effect, is of particular importance to steel castings because the mechanical properties are typically assessed from test bars machined from standardized coupons which have fixed dimensions and are cast separately from or attached to the castings.

Hydrogen Effects

Hydrogen is an undesirable element in steel; none of the effects of hydrogen are good. When present in amounts as low as 3 parts per million, hydrogen in steel significantly

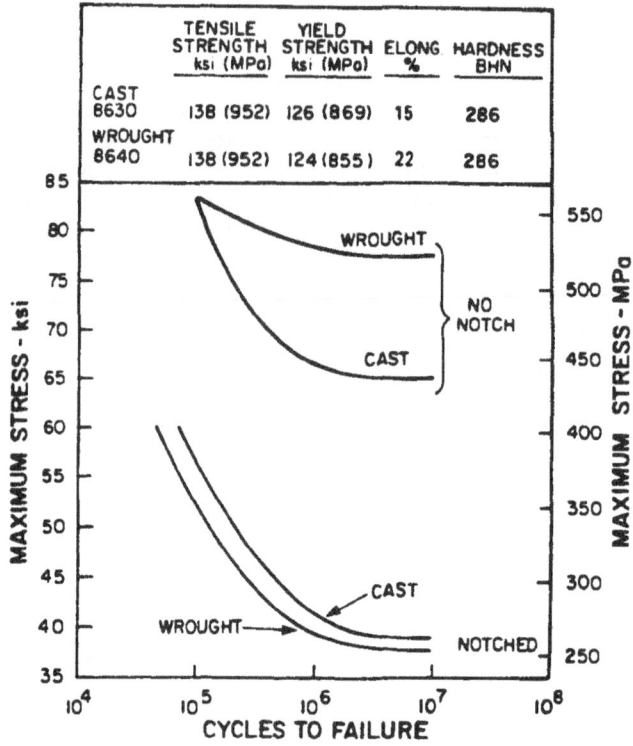

FIGURE 12.32   Fatigue characteristics (S-N curves) for cast and wrought 8600 series steels, quenched and tempered to the same hardness, both notched and unnotched R.R. Moore rotating beam tests, $K_t = 2.2$. (From *Steel Castings Handbook* [43].)

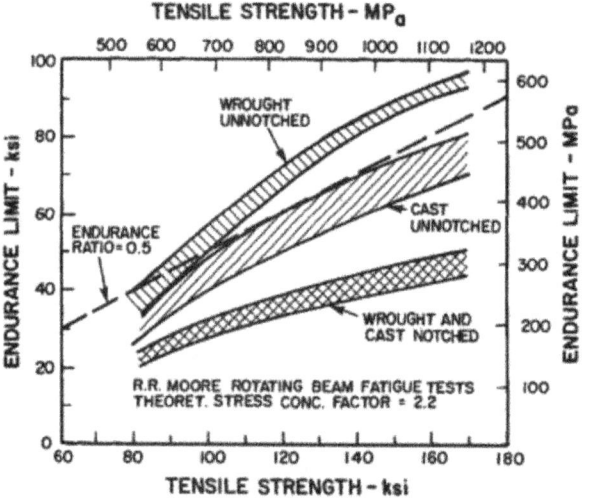

FIGURE 12.33   Relation between fatigue endurance limit (both notched and unnotched) and unnotched tensile strength for a number of cast and wrought steels with various heat treatments. (From *Steel Castings Handbook* [44].)

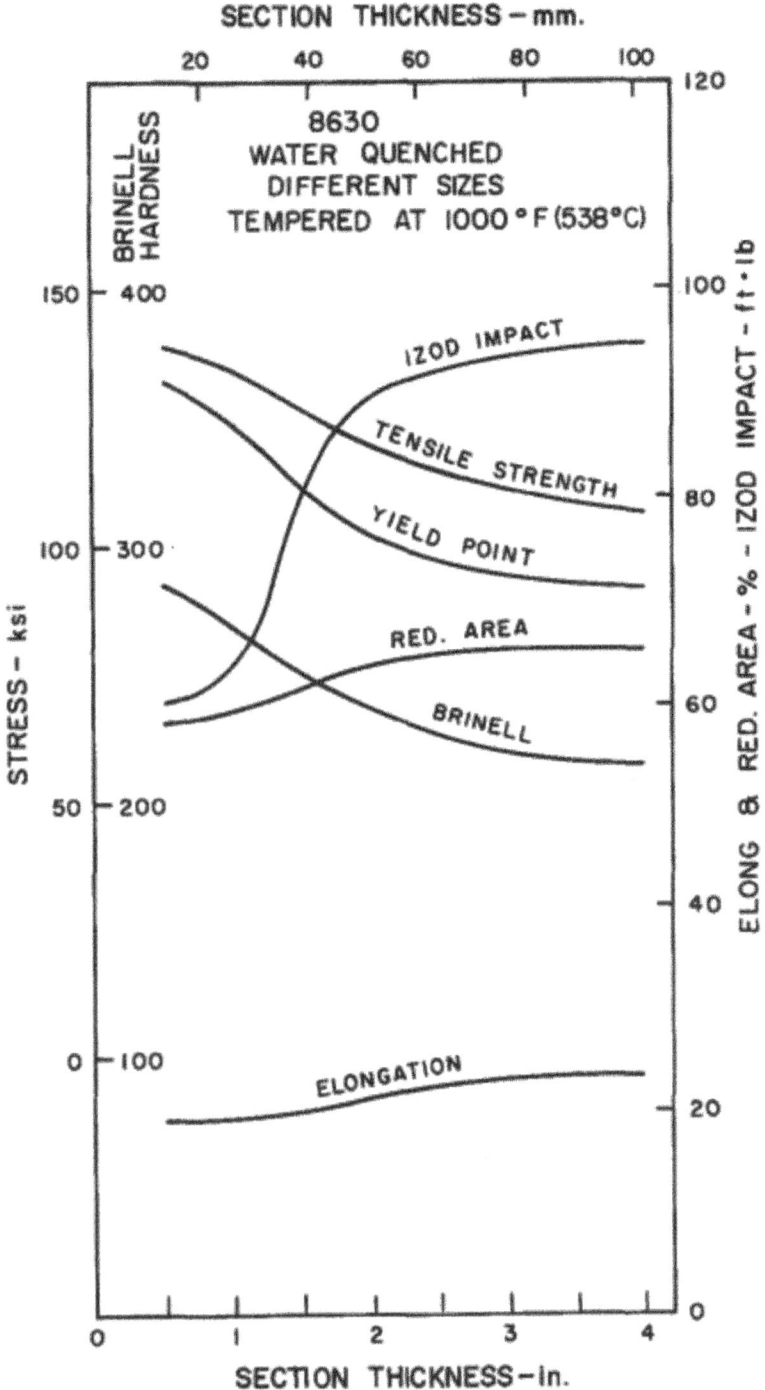

Figure 12.34 Section size effects on water-quenched and tempered wrought 8630 steel in sizes over 1 in (25 mm). The properties reported are those midway between surface and center. (From *Steel Castings Handbook* [45].)

reduces toughness and ductility. The damage gets worse as the casting thickness increases. This effect can be ameliorated by prolonged heating of the part at 204°C (400°F). Among the sources of hydrogen in steel, the most important are the raw materials used in the melting and refining process and traces of moisture in the molds used for casting. Care should be taken to dry all steel melting and refining additions as well as molds.

Corrosion Resistance

The cast carbon and low-alloy steels are not considered corrosion-resistant materials. As is true for wrought steels, minor additions of nickel, chromium, copper, and silicon will increase resistance of cast steels to atmospheric corrosion to the extent that for some applications no protection such as paint is required. At elevated temperatures, increasing amounts of chromium increase resistance to oxidation and, low-alloy steels for elevated temperature service contain higher amounts of chromium.

High-Alloy Steels

These are divided into two groups, namely, corrosion-resistant steels and heat-resistant steels.

Corrosion-Resistant Steels

Cast corrosion-resistant steels, more commonly called cast stainless steels, are alloyed to prevent corrosion from aqueous solutions or liquid vapor normally below 300°C (600°F). They are iron alloyed with chromium, nickel, and carbon. The principal applications for these steels are as materials of construction for food-processing, chemical-processing, and power-generating equipment involving corrosion service. Cast stainless steels are defined as ferrous alloys that contain a minimum of 12% Cr for corrosion resistance.

The most common formulation of cast stainless steel is 18% chromium and 8% nickel that gives a mainly austenitic structure with good toughness and ductility, good casting properties, and good corrosion resistance. Unlike the long product like pipe made from 18-8 or 304 stainless steel that are 100% austenitic, the cast grades like the weld metals add more chromium to get a microstructure that contains small amounts of ferrite to increase the strength and resist cracking in casting. This addition of chromium and mixed microstructure containing ferrite generally improves the corrosion resistance of the cast alloy compared to the common long products.

Corrosion-resistant cast steels are usually classified on the basis of composition or microstructure (Table 12.8) [46]. It should be recognized that these bases for classification are not completely independent in most cases; that is, classification by composition also often involves microstructural distinctions.

Table 12.8 lists the compositions of the commercial cast corrosion-resistant alloys. Alloys are grouped as chromium steels, chromium-nickel steels in which chromium is the predominant alloying element, and nickel-chromium steels in which nickel is the predominant alloying element. The serviceability of cast corrosion-resistant steels depends greatly on the absence of carbon, and especially precipitated carbides, in the alloy microstructure. Carbide formation on the grain boundaries reduces the corrosion resistance of these alloys by removing some of the chromium from resisting corrosion. Therefore, cast corrosion-resistant alloys are generally low in carbon (typically <0.08% C).

As shown in Table 12.8, high-alloy cast steels are also classified on the basis of microstructure. Structures may be austenitic, ferritic, martensitic, or duplex; the structure of a particular grade is primarily determined by composition. Chromium, nickel, and carbon

TABLE 12.8 Compositions and Microstructures of Corrosion-Resistant High-Alloy Cast Steels

Alloy	Wrought Alloy Type (a)	Most Common End-Use Microstructure	Composition(b), %								
			Cr	Ni	Mo	Si	Mn	P	S	C	Others
Chromium Steels											
CA15	410	Martensite	11.5–14.0	1.00	0.50	1.50	1.00	0.04	0.04	0.15	...
CA15M	...	Martensite	11.5–14.0	1.00	0.15–1.0	0.65	1.00	0.04	0.04	0.15	...
CA40	420	Martensite	11.5–14.0	1.00	0.5	1.50	1.00	0.04	0.04	0.20–0.40	...
CB30	431, 442	Ferrite + carbides	18.0–21.0	2.00	...	1.50	1.00	0.04	0.04	0.30	...
CC50	446	Ferrite + carbides	26.0–30.0	4.00	...	1.50	1.00	0.04	0.04	0.50	...
Chromium-Nickel Steels											
CA6NM	...	Martensite	11.5–14.0	3.5–4.5	0.40–1.0	1.00	1.00	0.04	0.03	0.06	...
CB7Cu	17-4PH	Martensite-age hardenable	15.5–17.0	3.6–4.6	...	1.50	1.00	0.04	0.04	0.07	2.3–3.3 Cu
CD4MCu	...	Austenite in ferrite-age hardenable	25.0–26.5	4.75–6.0	1.75–2.25	1.00	1.00	0.04	0.04	0.04	2.75–3.25 Cu
CE30	...	Ferrite in austenite	26.0–30.0	8.0–11.0	...	2.00	1.50	0.04	0.04	0.30	...
CF3	304L	Ferrite in austenite	17.0–21.0	8.0–12.0	...	2.00	1.50	0.04	0.04	0.03	...
CF8	304	Ferrite in austenite	18.0–21.0	8.0–11.0	...	2.00	1.50	0.04	0.04	0.08	...
CF20	302	Austenite	18.0–21.0	8.0–11.0	...	2.00	1.50	0.04	0.04	0.20	...
CF3M	316L	Ferrite in austenite	17.0–1.0	9.0–13.0	2.0–3.0	1.50	1.50	0.04	0.04	0.03	...
CF8M	316	Ferrite in austenite	17.0–21.0	9.0–13.0	2.0–3.0	1.50	1.50	0.04	0.04	0.08	...
CF12M	...	Ferrite in austenite or austenite	18.0–21.0	9.0–12.0	2.0–3.0	2.00	1.50	0.04	0.04	0.12	...
CF8C	347	Ferrite in austenite	18.0–21.0	9.0–12.0	...	2.00	1.50	0.04	0.04	0.08	Nb = 8 × C, 1.0 max
CF16F	303	Austenite	18.0–21.0	9.0–12.0	1.50	2.00	1.50	0.17	0.04	0.16	0.20–0.35 Se
CG8M	317	Ferrite in austenite	18.0–21.0	9.0–13.0	3.0–4.0	1.50	1.50	0.04	0.04	0.08	...
CH20	309	Austenite	22.0–26.0	12.0–15.5	...	2.00	1.50	0.04	0.04	0.20	...
CK20	310	Austenite	23.0–27.0	19.0–22.0	...	1.75	1.50	0.04	0.04	0.20	...
Nickel-Chromium Steel											
CN7M	320	Austenite	19.0–22.0	27.5–30.5	2.0–3.0	1.50	1.50	0.04	0.04	0.07	3.0–4.0 Cu

(a) Wrought alloy type numbers are of AISI designations for grades most closely corresponding to casting alloys. Wrought alloy type numbers are given only as a guide for determining corresponding cast and wrought grades. Buyers should use cast alloy designations when specifying castings.
(b) Maximum unless range is given. All compositions contain balance of iron.

Source: From *Metals Handbook* [46].

contents are particularly important in this regard. In general, straight chromium grades of high-alloy cast steel are either martensitic or ferritic, the chromium-nickel grades are either duplex or austenitic, and the nickel-chromium steels are fully austenitic.

The chemistry of stainless steel can be used to assess the microstructure of the alloy by use of the Schaeffler diagram (Fig. 12.35) [47]. One need only to calculate nickel equivalent, %Ni + 30(%C) + 0.5(%Mn) and chromium equivalent %Cr + %Mo + 1.5(%Si) + 0.5(%Nb) and refer to the chart.

Martensitic grades include alloys CA-15, CA-40, CA-15M, and CA-6NM. The CA-15 alloy contains the minimum amount of chromium necessary to make it essentially rust-proof. It has good resistance to atmospheric corrosion as well as many organic media in relatively mild service. A higher-carbon modification of CA-15, CA-40 can be heat treated to higher strength and hardness levels. Alloy CA-15M is a molybdenum-containing modification of CA-15 that provides improved elevated-temperature strength. Alloy CA-6NM is an iron-chromium-nickel-molybdenum alloy of low carbon content.

Austenitic grades include CH-20, CK-20, and CN-7M. The CH-20 and CK-20 alloys are high-chromium, high-carbon, wholly austenitic compositions in which the chromium exceeds the nickel content. The more highly alloyed CN-7M has excellent corrosion resistance in many environments and is often used in sulfuric acid service.

Ferritic grades are designated CB-30 and CC-50. Alloy CB-30 is practically nonhardenable by heat treatment. In these grades, the balance among the elements in the composition results in a wholly ferritic structure similar to wrought AISI type 442 stainless steel. Alloy

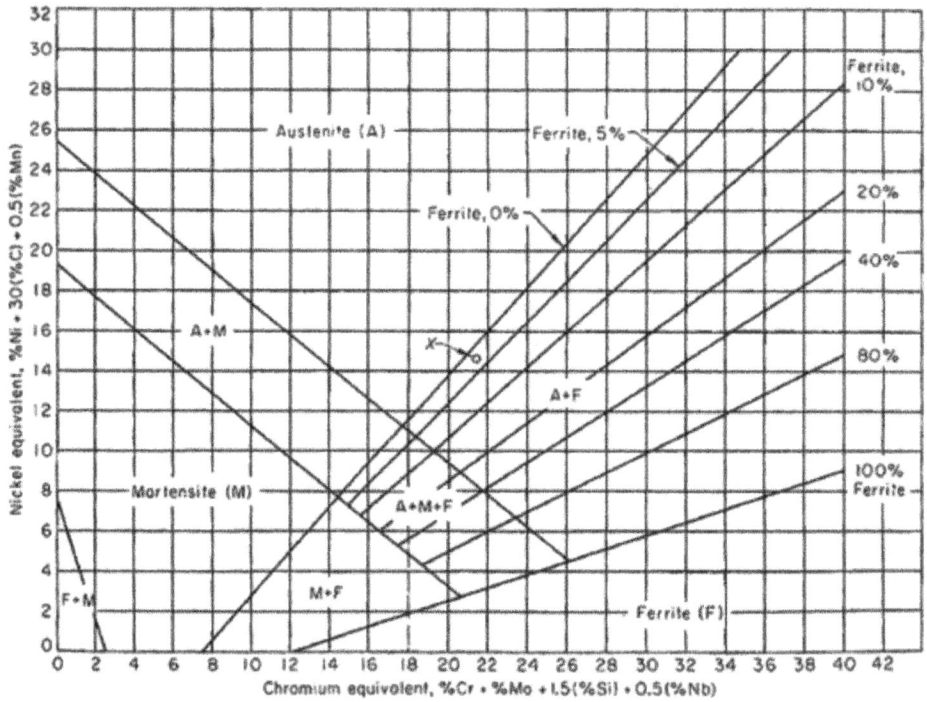

Figure 12.35 Schaeffler diagram showing the relationship between nickel and chromium equivalents and the microstructure. (From *Metals Handbook* [47].)

CC-50 has substantially more chromium than CB-30 and has relatively high resistance to localized corrosion in many environments. These grades are not commonly produced as castings. They are brittle as cast.

Austenitic-ferritic alloys include CE-30, CF-3, CF-3A, CF-8, CF-8A, CF-20, CF-3M, CF-3MA, CF-8M, CF-8C, CF-16F, and CG-8M. The microstructures of these alloys usually contain 5 to 40% ferrite, depending on the particular grade and the balance among the ferrite-promoting and austenite-promoting elements in the chemical composition.

Duplex alloys. Alloy CD-4MCu is the most highly alloyed duplex alloy. Wrought alloy Ferralium was developed by Langley Alloys and is essentially CD-4MCu with about 0.15% N added. With high levels of ferrite (about 40 to 50%) and low nickel, the duplex alloys have better resistance to stress-corrosion cracking (SCC) than CF-3M. Alloy CD-4MCu, which contains no nitrogen and has a relatively low molybdenum content, has only slightly better resistance to localized corrosion than CF-3M. Ferralium, which has nitrogen and slightly higher molybdenum than CD-4MCu, exhibits better localized corrosion resistance than either CF-3M or CD-4MCu.

First-generation duplex stainless steels, for example, AISI type 329 and CD-4MCu, have been in use for many years. The need for improvement in the weldability and corrosion resistance of these alloys resulted in the second-generation alloys, which are characterized by the addition of nitrogen as an alloying element.

Second-generation duplex stainless steels are usually about a 50-50 blend of ferrite and austenite. The new duplex alloys combine the near immunity to chloride stress corrosion cracking (SCC) of the ferritic grades with the toughness and ease of fabrication of the austenitics. Among the second-generation duplexes, wrought Alloy 2205 seems to have become the general-purpose stainless.

Precipitation-hardening grades. The alloys in this group are CB-7Cu and CD-4MCu. Alloy CB-7Cu is a low-carbon martensitic alloy that may contain minor amounts of retained austenite or ferrite. The copper precipitates in the martensite when the alloy is heat treated to the hardened (aged) condition.

Ferrite in Cast Stainless Steels

As noted, the CF alloys, or 18-8 stainless steels, constitute the most technologically important and highest-tonnage segment of corrosion-resistant casting production. These 19Cr-9Ni alloys are the cast counterparts of the AISI 300-series wrought stainless steels (Table 12.8). In general, the cast and wrought alloys possess equivalent resistance to corrosive media, and they are frequently used in conjunction with each other.

Important differences do exist, however, between the cast CF alloys and their wrought AISI counterparts. Most significant among these is the difference in alloy microstructure in the end-use condition. The CF-grade cast alloys have duplex structures (Table 12.8) and usually contain 5 to 40% ferrite, depending on the particular alloy. Their wrought counterparts are fully austenitic. The ferrite in cast stainless with duplex structures is magnetic, a point that is often confusing when cast stainless steels are compared to their wrought counterparts by checking their attraction to a magnet. This difference in microstructures is attributable to the fact that the chemical compositions of the cast and wrought alloys are not identical by intent. Ferrite is intentionally present in cast CF-grade stainless steels for three principal reasons: (1) to provide strength, (2) to improve weldability, and (3) to maximize resistance to corrosion in specific environments.

Ferrite control. From the preceding discussion, it is apparent that controlled ferrite contents in predominantly austenitic cast chromium-nickel steels, notably, the CF alloys, offer certain property advantages and that the amount of ferrite present will depend primarily on the compositional balance of the alloy.

The major elemental components of cast stainless steels are in competition to promote austenite or ferrite phases in the alloy microstructure. Chromium, silicon, molybdenum, and niobium promote the presence of ferrite in the alloy microstructure; nickel, carbon, nitrogen, and manganese promote the presence of austenite. By balancing the contents of ferrite-forming and austenite-forming elements within the specified ranges for the elements in a given alloy, it is possible to control the amount of ferrite present in the austenite matrix. The alloy can usually be made fully austenitic or with ferrite contents up to 30% or more in the austenite matrix.

The relationship between composition and microstructure in cast stainless steels permits the foundry engineer to predict and control the ferrite content of an alloy, as well as its resultant properties, by adjusting the composition of the alloy. This is accomplished with the Schoefer constitution diagram for cast chromium-nickel alloys (Fig. 12.36) [48]. This diagram was derived from an earlier diagram (Fig. 12.35) developed by Schaeffler for stainless steel weld metal. Use of Fig. 12.36 requires that all ferrite-stabilizing elements in the composition be converted into chromium equivalents and that all austenite-stabilizing

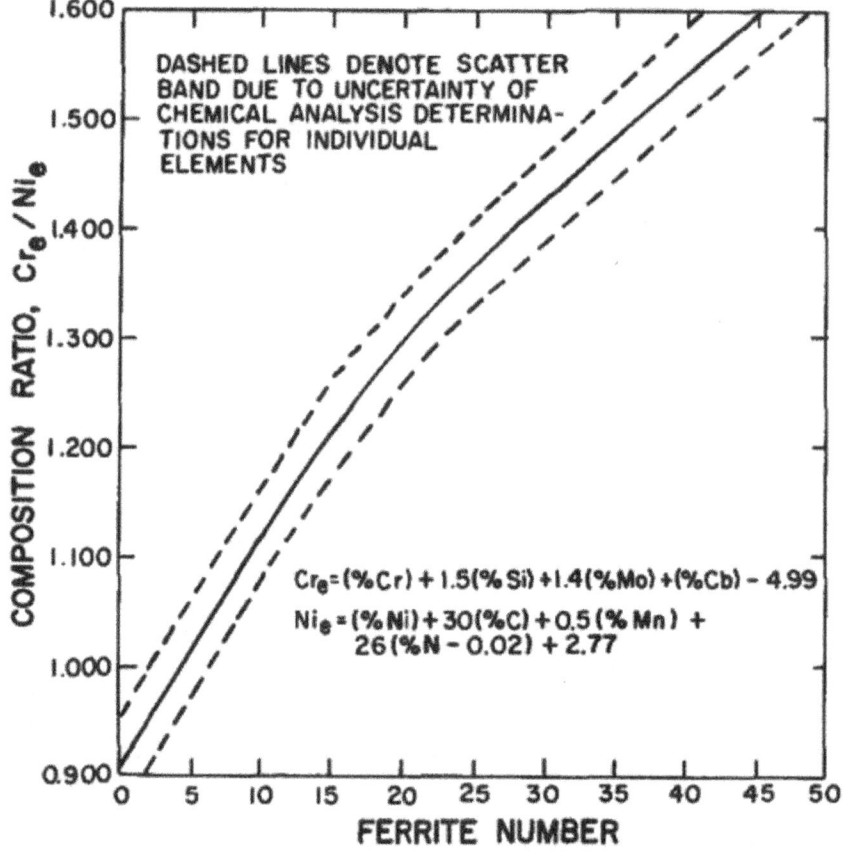

FIGURE 12.36 Schoefer diagram for estimation of ferrite content in stainless steel castings within the composition range of 16-26Cr, 6-14Ni, 4 max Mo, 1 max Cb, 0.2 max C, 0.19 max N, 2 max Mn, 2 max Si. (From *Steel Castings Handbook* [48].)

elements be converted into nickel equivalents by means of empirically derived coefficients representing the ferritizing or austenitizing power of each element. A composition ratio is then obtained from the total chromium equivalent Cr_e and nickel equivalent Ni_e, calculated for the alloy composition according to the following:

$$Cr_e = \%Cr + 1.5\,(\%Si) + 1.4\,(\%Mo) + \%Nb - 4.99 \qquad (12.4)$$

$$Ni_e = \%Ni + 30\,(\%C) + 0.5\,(\%Mn) + 26(\%N - 0.02) + 2.77 \qquad (12.5)$$

where the elemental concentrations are given in weight percent.

Heat-Resistant Steels

Cast heat-resistant steels are alloyed to prevent creep or other failure due to elevated operating temperatures or cyclic heating, from 650 to 1200°C (1200 to 2200°F). Strength at these elevated temperatures is only one of the criteria by which these materials are selected, because applications often involve aggressive environments to which the steel must be resistant. The atmospheres most commonly encountered are air, flue gases, or process gases; such atmospheres may be either oxidizing or reducing and may be sulfidizing or carburizing if sulfur or carbon are present. Only heat-resistant steels exhibit the required mechanical properties and corrosion resistance over long periods of time without excessive or unpredictable degradation. In addition to long-term strength and corrosion resistance, some cast heat-resistant steels exhibit special resistance to the effects of cyclic temperatures and changes in the nature of the operating environment.

A number of cast high-alloy grades have been developed and successfully used for a variety of service requirements. There are three principal categories, based on composition:

- Iron-chromium alloys
- Iron-chromium-nickel alloys
- Iron-nickel-chromium alloys

These alloy types resemble high-alloy corrosion-resistant steels except for their higher carbon contents, which impart greater strength at elevated temperature. The higher carbon content and, to a lesser extent, alloy composition ranges distinguish cast heat-resistant steel grades from their wrought counterparts. Unlike the corrosion-resistant grades, the heat-resistant grades are often supplied in the as-cast condition and are not heat treated. Table 12.9 summarizes the compositions of standard cast heat-resistant grades [49].

Iron-chromium alloys contain 8 to 30% Cr and little or no nickel. They are ferritic in structure and exhibit low ductility at ambient temperatures. Iron-chromium alloys are primarily used where resistance to gaseous corrosion is the predominant consideration because they possess relatively low strength at elevated temperatures. Examples of such alloys are the cast HA, HC, and HD grades listed in Table 12.9. HA is martensitic and commonly supplied as CA 15 or CA6NM. HC and HD are ferritic and difficult to cast, brittle, and not typically produced.

Iron-chromium-nickel alloys contain more than 18% Cr and more than 8% Ni, with the chromium content always exceeding the nickel. They exhibit an austenitic matrix, although several grades also contain some ferrite. These alloys exhibit greater strength and ductility at elevated temperatures than those in the iron-chromium group and withstand moderate thermal cycling. Examples of these alloys are the HE, HF, HH, HI, HK, and HL grades listed in Table 12.9.

Alloy Grade	ASTM Specs	Composition, Wt%								
		C	Mn	Si	P	S	Cr	Ni	Mo	Nb
HA	1,2,3	0.20	0.65(a)	1.0	0.04	0.04	8–10	...	0.90–1.20	...
HC	1,3	0.50	1.0	2.0	0.04	0.04	26–30	4.0	0.5	...
HD	1,3	0.50	1.5	2.0	0.04	0.04	26–30	4–7	0.5	...
HE	1,3	0.20–0.50	2.0	2.0	0.04	0.04	26–30	8–11	0.5	...
HF	1,3	0.20–0.40	2.0	2.0	0.04	0.04	18–23	8–12	0.5	...
HH	1,3,4	0.20–0.50	2.0	2.0	0.04	0.04	24–28	11–14	0.5	...
HI	1,3	0.20–0.50	2.0	2.0	0.04	0.04	26–30	14–18	0.5	...
HK	1,3,5	0.20–0.60	2.0	2.0	0.04	0.04	24–28	18–22	0.5	...
HL	1,3	0.20–0.60	2.0	2.0	0.04	0.04	28–32	18–22	0.5	...
HN	1,3	0.20–0.50	2.0	2.0	0.04	0.04	19–23	23–27	0.5	...
HP	1	0.35–0.75	2.0	2.5	0.04	0.04	24–28	33–37	0.5	...
HT	1,3,5	0.35–0.75	2.0	2.5	0.04	0.04	15–19	33–37	0.5	...
HU	1,3	0.35–0.75	2.0	2.5	0.04	0.04	17–21	37–41	0.5	...
HW	1,3	0.35–0.75	2.0	2.5	0.04	0.04	10–14	58–62	0.5	...
HX	1,3	0.35–0.75	2.0	2.0	0.04	0.04	15–19	64–68	0.5	...
CT15	5	0.35–0.75	1.5(b)	1.5(c)	0.03	0.03	19–21	31–34	0.5	0.5–1.5
50-50-Nb	6	0.10	0.3	0.5	0.02	0.02	47–52	Bal	...	1.4–1.7

Notes: Maximum values unless shown otherwise.
(a) Mn 0.35% min. (b) Mn 0.15% min. (c) Si 0.5% min.
Key to ASTM specifications: 1. A297, 2. A743, 3. A608, 4. A447, 5. A351, 6. A560.
The letter "H" indicates heat-resistant alloy. The second letter from "A" to "X" denotes increasing nickel from 0 to 68%. Numerals after the first two letters are used to designate the carbon content, i.e., HK40 is an HK alloy with 0.40%±0.05% carbon.
Source: From *Steel Castings Handbook* [49].

TABLE 12.9 Standard Designations and Compositions of Cast Heat-Resistant Steels

Iron-nickel-chromium alloys contain more than 10% Cr and more than 23% Ni, with the nickel content always exceeding that of chromium. These alloys are wholly austenitic and exhibit high strength at elevated temperatures. They can withstand considerable temperature cycling and severe thermal gradients and are well suited to many reducing, as well as oxidizing, environments. Examples of iron-nickel-chromium alloys are the HN, HP, HT, HU, HW, and HX grades listed in Table 12.9. Even though nickel is the major element (>50%) in the HW and HX grades, these grades are ordinarily referred to as high-alloy steels rather than nickel-base alloys.

Mechanical Properties of High-Alloy Steels

High-alloy steels fall into three groups: Austenitic Hadfield manganese steels, corrosion-resistant steels, and heat-resistant steels. High manganese steels are relatively soft and ductile in solution-annealed condition but develop very high hardness during impact. Table 12.2 lists typical BHN and impact values for them in annealed condition.

Chemistries of cast corrosion-resistant steels are given in Table 12.8 and are covered under ASTM specifications A217, A351, A743, A744, and A890 and the reader is advised to refer to them for minimum property requirements. Representative room-temperature mechanical properties of these alloys are given in Table 12.10 [50].

Heat-resistant cast steels are listed in Table 12.9. They are covered under various ASTM specifications as mentioned in the footnote of this table. Their representative short-term tensile properties in the cast condition are listed in Table 12.11 [51].

Microstructures of Steel Castings

Microstructure is defined as the structure of material seen under a microscope above 25× magnification [52]. Microstructure of a material can strongly influence physical and mechanical properties such as strength, toughness, ductility, hardness, corrosion resistance, high/low temperature behavior, wear resistance, and so on, which in turn govern the application of these materials in industrial practice. Microstructure of steel can be changed by various heat treatments as described earlier in this chapter.

In this section, some typical microstructures of the steel are shown for plain carbon, low-alloy, Hadfield manganese, and corrosion-resistant and heat-resistant steels. The steel grade and the heat treatment given are given in the respective captions.

The most common steel alloy cast is a carbon steel. The most typical heat treatment is normalized and this results in ferrite and pearlite in the microstructure as seen in Fig. 12.37 [53]. To improve the properties of a carbon steel, it may be quenched and tempered instead of normalized. In carbon steels of thicker sections, this results in a mixed microstructure of tempered martensite and ferrite and pearlite as seen in Fig. 12.38 [54].

Another way to improve the properties is to add alloy. If alloy is added then even in the normalized condition, the properties are higher. This can be seen in the alloy grade used to make ASTM A148 Grade 90-60 shown in Fig. 12.39 [55]. If the alloyed grade is quenched and tempered, the added alloy elements makes heat treatment more effective and the microstructure becomes predominantly martensite, as seen in Fig. 12.40 [56].

Nickel-alloyed steels are used for low-temperature service. ASTM A352 grade LC3 contains over 3% nickel and the microstructure of the alloy in normalized and tempered is shown in Fig. 12.41 [57]. To achieve the desired impact toughness these grades are quenched and tempered to obtain martensite in the structure, as seen in Fig. 12.42 [58].

Special steels like austenitic wear-resistant steel are formulated to have an austenite structure at room temperature. Austenite work hardens in impact service and is widely used in aggressive impact-wear environments. Austenite for these wear products is stabilized by the addition of manganese and the wear resistance is improved by the occurrence of carbides. The microstructure of ASTM A128 grade B-3 is shown in Fig. 12.43 [59].

Stainless steel gains much of its corrosion resistance from chromium. The strongest stainless alloys are limited to about 12% chrome to stabilize the formation of martensite. One common martensitic stainless steel is CA-15, shown in Fig. 12.44 [60].

Adding higher levels of chromium gives a mixture of phases but can achieve higher strengths with aging and precipitation. An alloy formulated to give high strength as a stainless grade is CB-7Cu, as shown in Fig. 12.45 [61].

Adding additional chromium and some nickel creates an alloy that has a mixture of ferrite and austenite or a duplex structure. Duplex stainless steels with approximately half ferrite and half austenite are increasingly common and one early duplex grade is CD-4MCu, shown in Fig. 12.46 [62].

The minimum amount of chromium and nickel to get a fully austenitic structure for the most common stainless grades is 18% chromium and 8% nickel. The chromium

Alloy	Heat Treatment Condition (1)	Tensile Strength Ksi	(MPa)	Yield Strength (0.2 % offset) Ksi	(MPa)	Elongation in 2 in (50 mm) %	Reduction in Area %	Brinell Hardness	Charpy Impact Energy ft-lb	(J)	Specimen
CA-6NM	1750 AC, 1100-1150 T	120	(827)	100	(689)	24	60	269	70	(94.9)	V-notch
CA-15	1800 AC, 1200 T	115	(793)	100	(689)	22	55	225	20	(27.1)	Keyhole notch
CA-40	1800 AC, 1100 T	150	(1034)	125	(862)	10	30	310	2	(2.7)	Keyhole notch
CB-7Cu	1900 OQ, 925 A	190	(1310)	170	(1172)	14	54	400	25	(33.9)	V-notch
CB-30	1450 AC	95	(655)	60	(414)	15		195	2	(2.7)	Keyhole notch
CC-50	1900 AC	97	(669)	65	(448)	18		210			
CD-4MCu	2050 FC to 1900 WQ	108	(745)	81	(558)	25		253	55	(74.6)	V-notch
A890-4A	2050 FC to 1850 WQ	95	(645)	66	(455)						
CE-30	2000 WQ	97	(669)	63	(434)	18		190	7	(9.5)	Keyhole notch
CF-3	1900 WQ	77	(531)	36	(248)	60		140	110	(149.2)	V-notch
CF-8	1900 WQ	77	(531)	37	(255)	55		140	74	(110.3)	Keyhole notch
CF-20	2000 WQ	77	(531)	36	(248)	50		163	60	(81.4)	Keyhole notch
CF-3M	1900 WQ	80	(552)	38	(262)	55		150	120	(162.7)	V-notch
CF3MN	1900 WQ	82	(565)	42	(289)	52		170			
CF-8M	1950 WQ	80	(552)	42	(290)	50		170	70	(94.9)	Keyhole notch
CF-8C	1950 WQ	77	(531)	38	(262)	39		149	30	(40.7)	Keyhole notch
CF-16F	2000 WQ	77	(531)	40	(276)	52		150	75	(101.7)	Keyhole notch
CG-8M	1900 WQ	82	(565)	44	(303)	45		176	80	(108.5)	V-notch
CH-20	2000 WQ	88	(607)	50	(345)	38		190	30	(40.7)	Keyhole notch
CK-20	2100 WQ	76	(524)	38	(262)	37		144	50	(67.8)	Izod V-notch
CN-7M	2050 WQ	69	(476)	31	(214)	48		130	70	(94.9)	Keyhole notch
CK3MCuN	2100 WQ	85	(586)	40	(276)	50					

(1) Numbers denote °F. AC = air cool, OQ = oil quench, T = temper, FC = furnace cool, WQ = water quench, A = age.

Source: From *Steel Castings Handbook* [50].

TABLE 12.10 Representative Room-Temperature Mechanical Properties of Cast Corrosion-Resistant Steels

Property at Indicated Temperature

Alloy	650°C (1200°F)					870°C (1600°F)					1095°C (2000°F)				
	Ultimate Tensile Strength		Yield Strength		Elong.,	Ultimate Tensile Strength		Yield Strength		Elong.,	Ultimate Tensile Strength		Yield Strength		Elong.,
	MPa	Ksi	MPa	Ksi	%	MPa	Ksi	MPa	Ksi	%	MPa	Ksi	MPa	Ksi	%
HD	...	...	...	...	...	159	23	...	...	18	...	...	...	...	...
HF	414	60	217	31	10	145	21	107	15.5	16	...	...	...	...	...
HH (type I)	...	...	...	...	...	127	18.5	93	13.5	30	38	5.5	...	...	...
HH (type II)	471	60.5	222	32	14	148	21.5	110	16	18	...	...	...	...	...
HI	...	...	...	...	...	179	26	...	...	12	...	...	...	...	...
HK	...	...	...	...	...	161	23	101	15	16	39	5.5	34	5	55
HL	...	...	...	...	...	210	30.5	...	...	...	...	...	...	...	...
HN	...	...	...	...	...	140	20	100	14.5	37	43	6	34	5	55
HP	...	...	...	...	...	179	26	121	17.5	27	52	7.5	43	6	69
HT	292	42.5	193	28	5	130	19	103	15	24	41	6	...	...	...
HU	...	...	...	...	...	135	19.5	...	...	20	...	...	...	...	...
HW	...	...	...	...	...	131	19	103	15	...	...	...	...	...	...
HX	303	45	138	20	8	141	20.5	121	17.5	48	...	...	...	...	...

Source: From *Metals Handbook* [51].

TABLE 12.11 Representative Short-Term Tensile Properties of Cast Heat-Resistant Alloys at Elevated Temperatures

665

FIGURE 12.37   As-cast structure of plain carbon (0.35% C) steel. The microstructure is ferrite (white) and pearlite (dark). (From *ASM Metals Handbook* [53].)

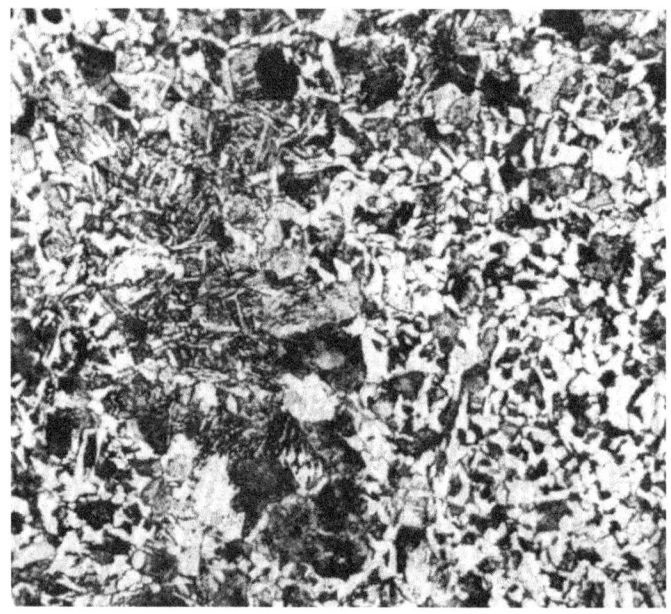

FIGURE 12.38   Same as Fig. 12.37, austenitized at 1650°F (881°C), quenched in water, and tempered at 1150°F (621°C). Structure: some tempered martensite, pearlite, and ferrite. (From *ASM Metals Handbook* [54].)

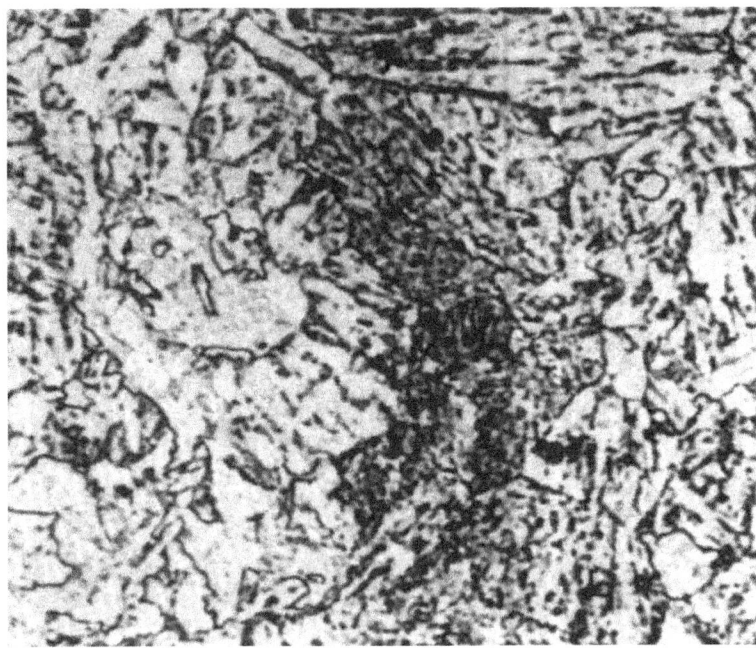

FIGURE 12.39   ASTM A148 steel, grade 90-60 (0.27% C, 0.80 Mn, 0.51 Si, 0.35 Mo) normalized at 1700°F (929°C) and tempered at 1300°F (704°C). Structure is ferrite and pearlite. (From *ASM Metals Handbook* [55].)

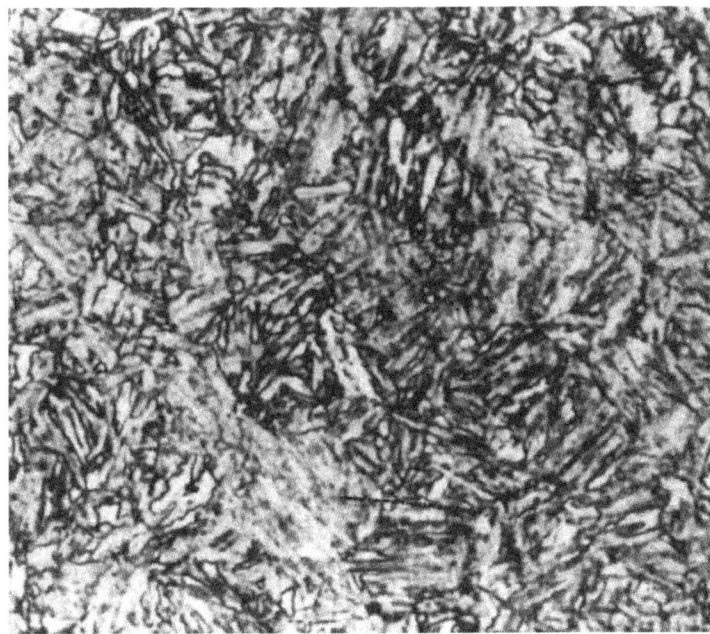

FIGURE 12.40   Same steel as in Fig.12.39, austenitized at 1700°F (927°C), water quenched and tempered at 1200°F (649°C). Mostly tempered martensite. (From *ASM Metals Handbook* [56].)

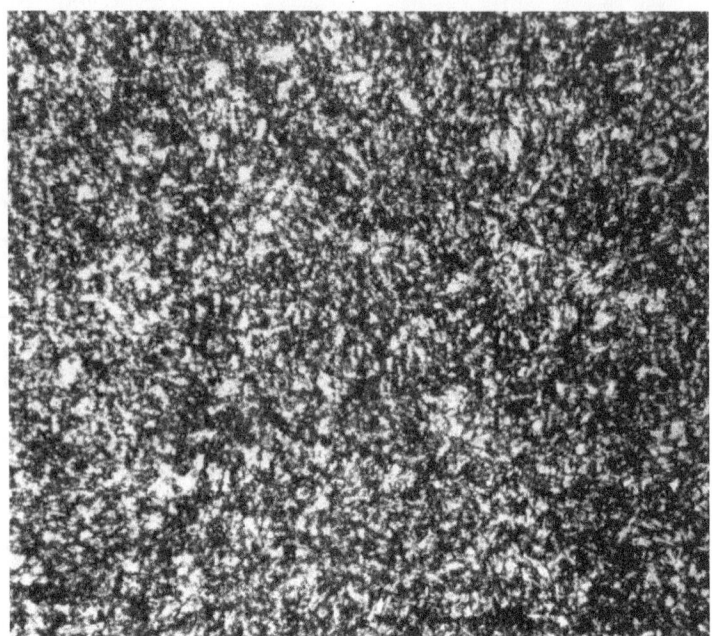

Figure 12.41   ASTM A352 steel, grade LC3, austenitized at 1650°F (899°C), air cooled to RT and tempered at 1150°F (621°C). Microstructure consists of ferrite (light) and pearlite (dark). (From *ASM Metals Handbook* [57].)

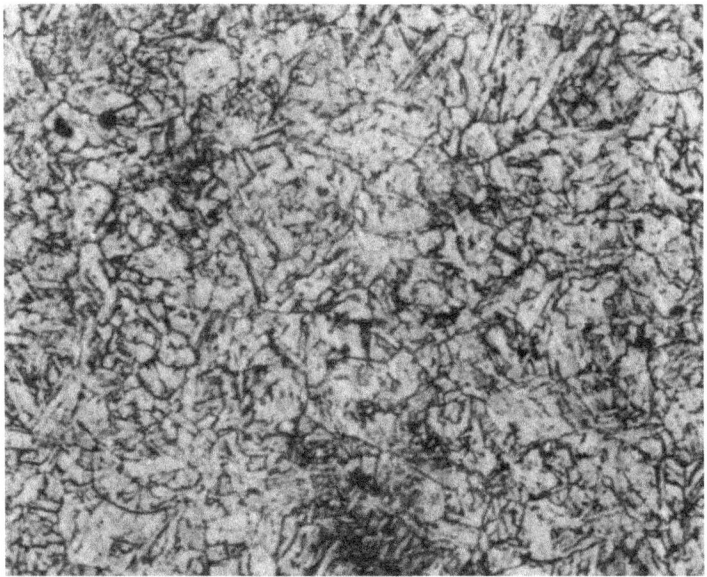

Figure 12.42   Same steel as in Fig.12.41, austenitized at 1600°F (871°C), water quenched and tempered at 1200°F (649°C). Microstructure is tempered martensite. (From *ASM Metals Handbook* [58].)

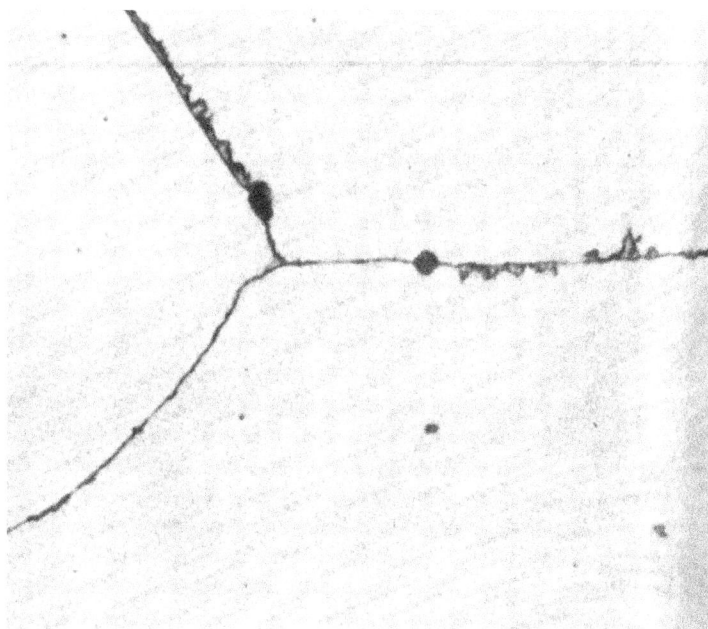

FIGURE 12.43 ASTM A128 austenitic manganese steel, grade B-3, as cast. Black lines are austenitic grain boundaries; black constituents are grain boundary carbides. (From *ASM Metals Handbook* [59].)

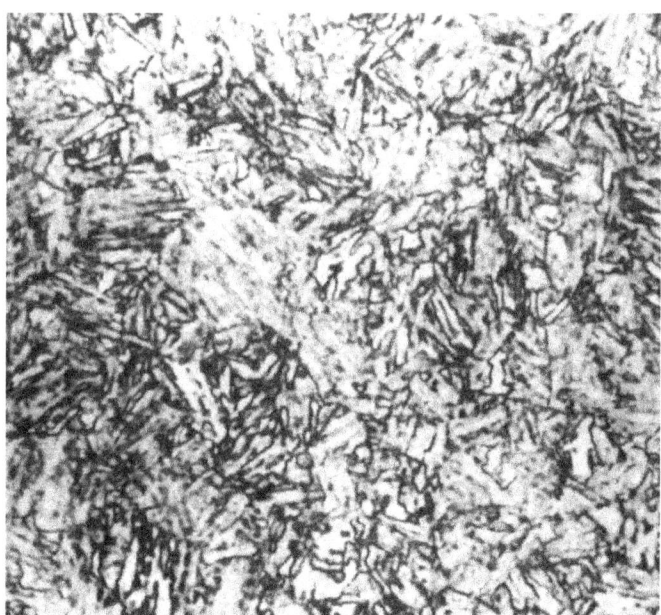

FIGURE 12.44 CA-15 alloy, austenitized at 1850°F (1010°C), air cooled, tempered at 1250°F (677°C). The structure shows traces of delta ferrite in a matrix of tempered martensite. (From *ASM Metals Handbook* [60].)

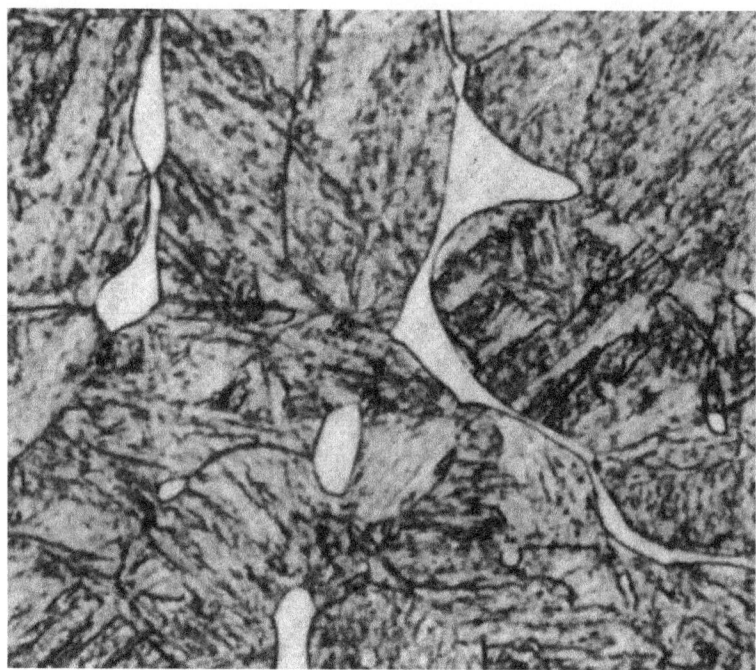

Figure 12.45 CB-7Cu alloy, austenitized at 1925°F (1052°C) and aged at 925°F (496°C). There are some ferrite pools (light) in a matrix of tempered martensite. (From *ASM Metals Handbook* [61].)

Figure 12.46 CD-4MCu alloy, as cast. Structure: jagged pools and particles of austenite in ferrite. Black specs are nonmetallic inclusions. (From *ASM Metals Handbook* [62].)

gives the corrosion resistance and the nickel stabilizes the structure as austenite. For welds and castings, the 18-8 grades are modified with added chromium to create a duplex microstructure but with only about 5 to 15% ferrite, not a 50-50 microstructure as seen in Fig. 12.46. Grade CF-3 is the low-carbon 19-9 cast grade from ASTM A743 that is predominantly austenite with some ferrite, as shown as-cast in Fig. 12.47 [63]. The higher-carbon grade CF-8 is shown in Fig. 12.48 [64]. These grades are normally heat treated to obtain the best corrosion resistance.

Higher alloy grades are used at elevated temperatures. While static room temperature properties of austenite are lower than ferritic or martensitic grades, at elevated temperatures austenite retains more strength. In creep-resistant service, higher nickel and chromium grades with higher carbon content to form carbides are supplied as cast. Two grades for elevated temperature service are shown in Fig. 12.49 [65] and 12.50 [66].

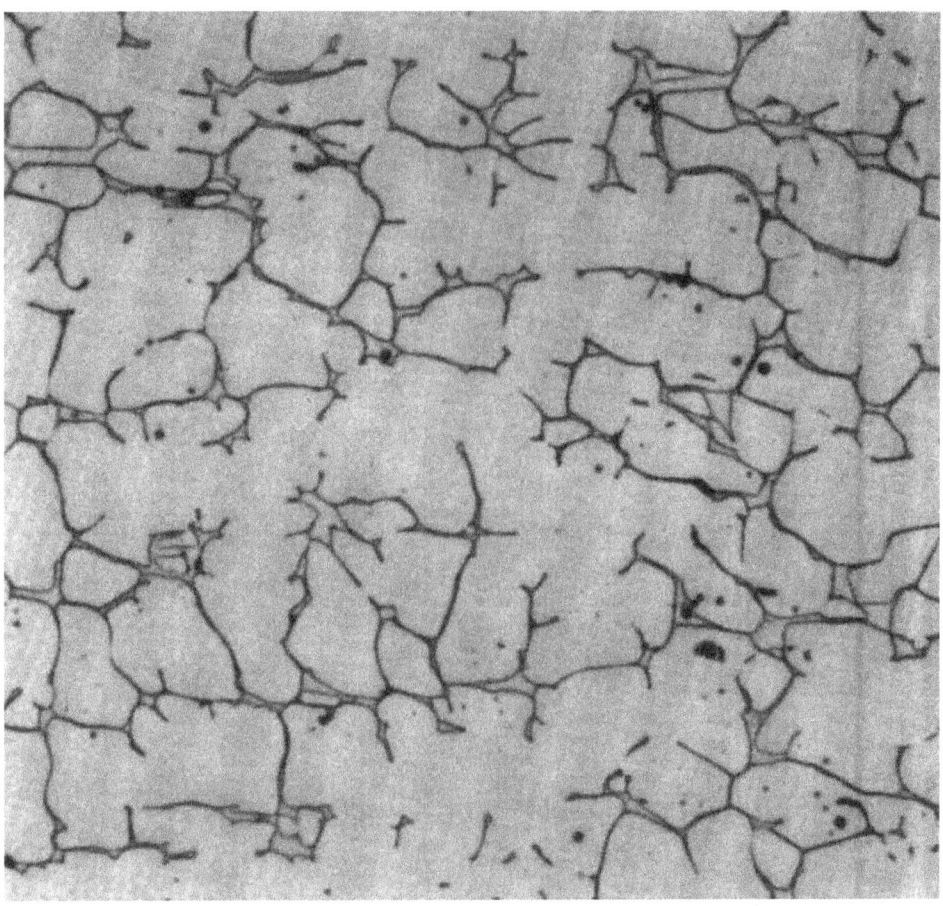

FIGURE 12.47 CF-3 alloy, as cast. The structure consists of a fine network of ferrite in a matrix of austenite. (From *ASM Metals Handbook* [63].)

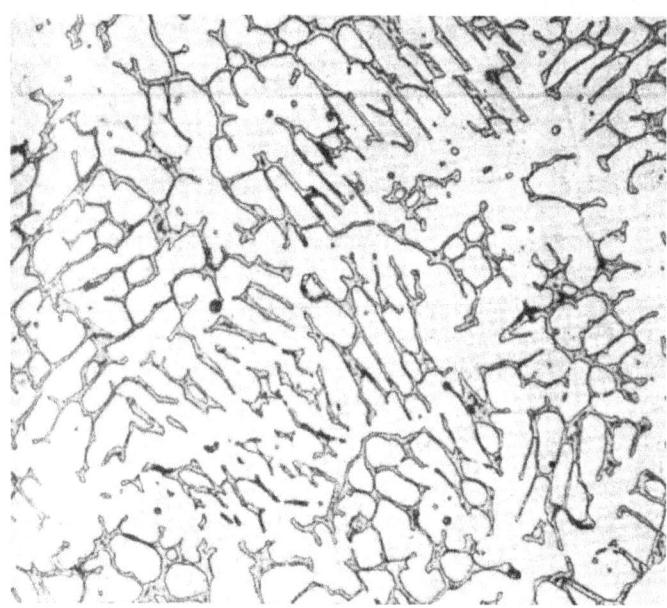

FIGURE 12.48 CF-8M alloy, as cast. Structure consists of ferrite networks containing some precipitated carbide particles in a matrix of austenite. (From *ASM Metals Handbook* [64].)

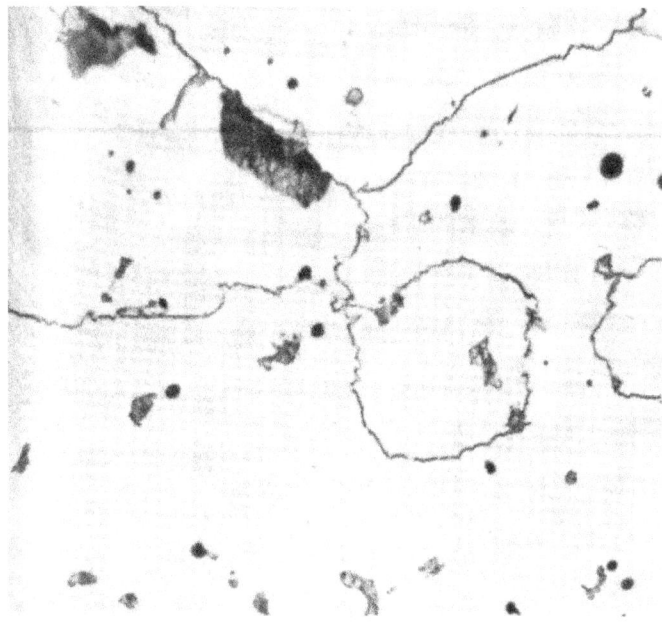

FIGURE 12.49 Alloy HK-35, as cast. Scattered eutectic carbide in austenitic matrix and at grain boundaries; patches of the lamellar constituents also are associated with grain boundaries. (From *ASM Metals Handbook* [65].)

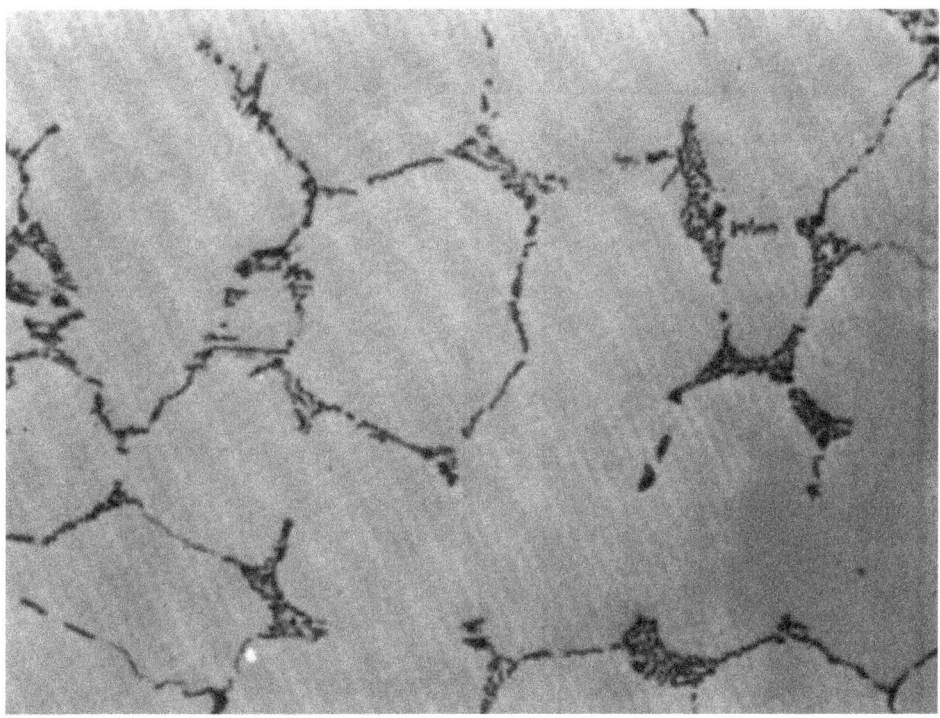

Figure 12.50   Alloy HT-44, as cast. The austenite matrix contains a complex network of eutectic carbide that outlines the boundaries of the original dendrites. Note that larger patches of primary carbide have a lamellar structure. (From *ASM Metals Handbook* [66].)

References

1. *Steel Castings Handbook*, 6th ed., Steel Founders Society of America (SFSA) and ASM International, Materials Park, OH, 1995.

2. R. Monroe, SFSA, 2013.

3. R. W. Heine, C. R. Loper, and P. C. Rosenthal, *Principles of Metal Casting*, McGraw-Hill, New York, 1967, p. 385.

4. *Metals Handbook*, Vol. 15, *Castings*, ASM International, Materials Park, OH 2008, p. 88.

5. Heine et al., p. 457.

6. R. Monroe, SFSA, 2013.

7. Heine et al., p. 465.

8. Heine et al., p. 462.

9. Heine et al., p. 463.

10. ASM International, p. 109.

11. Heine et al., p. 464.

12. Heine et al., pp. 444–452.

13. Heine et al., p. 478.

14. Heine et al., p. 480

15. R. Monroe, SFSA, 2013.

16. R. W. Ruddle, "Risering Steel Castings," Foseco, Cleveland, OH, 1979.

17. "Feeding and Risering Steel Castings," SFSA, Crystal Lake, IL, 2001.

18. Heine et al., p. 469.

19. Heine et al., p. 482.

20. Heine et al., p. 484.

21. Heine et al., p. 477.

22. E. C. Bain and H. W. Paxton, "Alloying Elements in Steel," ASM, 1966, Fig. 49, Metals Park, OH, p. 62.

23. Bain and Paxton, Fig. 50, p. 63.

24. W. E. Jominy, "Commercial Aspects of Hardenability Tests," *Metal Progress*, 38: 685–690, Nov. 1940.

25. American Society for Metals, *Metals Handbook*, 8th ed., Vol. 1, 1961, Metals Park, OH, p. 189.

26. Heine et al., p. 485.

27. G. Krauss, "Principles of Heat Treatments of Steel," 1980, ASM, Metals Park, OH, pp. 151–152.

28. "Atlas of IT and Cooling Transformation Diagrams," ASM, Metals Park, OH, 1977.

29. Heine et al., p. 486.

30. Heine et al., p. 489.

31. Heine et al., p. 470.

32. *Steel Castings Handbook*, Table 19.2, p. 19.4.

33. *Metals Handbook*, Table 5, Vol. 15, p. 954.

34. *Metals Handbook*, Table 6, Vol. 15, p. 955

35. *Metals Handbook* , Table 8, Vol. 15, p. 958.

36. *Steel Castings Handbook*, Fig. 18.1, p. 18.2.

37. *Steel Castings Handbook*, Fig. 18.2, p. 18.3.

38. *Steel Castings Handbook*, Fig. 18.3, p. 18.3.

39. *Steel Castings Handbook*, Fig. 18.10, p. 18.6.

40. *Steel Castings Handbook*, Fig. 18.4, p. 18.3.

41. *Steel Castings Handbook*, Fig. 18.12, p. 18.7.

42. *Steel Castings Handbook*, Fig. 18.13, p. 18.7.

43. *Steel Castings Handbook*, Fig. 18.16, p. 18.9.

44. *Steel Castings Handbook*, Fig. 18.17, p. 18.10.

45. *Steel Castings Handbook*, Fig. 18.18(a), p. 18.11.

46. *Metals Handbook*, Table 2, Vol. 15, p. 952.

47. *Metals Handbook*, Vol. 1, "Properties and Selection: Irons, Steels and High-Performance Alloys," 10th ed., ASM International, 1990, p. 899.

48. *Steel Castings Handbook*, Fig. 20.6, p. 20.6.

49. *Steel Castings Handbook*, Fig. 22.3, p. 22.2.

50. *Steel Castings Handbook*, Fig. 20.2, p. 20.4.

51. *Metals Handbook*, Table 15, Vol. 15, p. 968.

52. *ASM Metals Handbook*, "Atlas of Microstructures of Industrial Alloys," Vol. 7, 8th ed., Metals Park, OH, 1972.

53. *ASM Metals Handbook*, micrograph 556, Vol. 7, p. 71.

54. *ASM Metals Handbook*, micrograph 560, Vol. 7, p. 71.

55. *ASM Metals Handbook*, micrograph 568, Vol. 7, p. 72.

56. *ASM Metals Handbook*, micrograph 572, Vol. 7, p. 72.

57. *ASM Metals Handbook*, micrograph 632, Vol. 7, p. 79.

58. *ASM Metals Handbook*, micrograph 635, Vol. 7, p. 79.

59. *ASM Metals Handbook*, micrograph 638, Vol. 7, p. 80.

60. *ASM Metals Handbook*, micrograph 1226, Vol. 7, p. 153.

61. *ASM Metals Handbook*, micrograph 1231, Vol. 7, p. 153.

62. *ASM Metals Handbook*, micrograph 1232, Vol. 7, p. 153.

63. *ASM Metals Handbook*, micrograph 1234, Vol. 7, p. 154.

64. *ASM Metals Handbook*, micrograph 1248, Vol. 7, p. 155.

65. *ASM Metals Handbook*, micrograph 1462, Vol. 7, p. 158.

66. *ASM Metals Handbook*, micrograph 1501, Vol. 7, p. 186.

CHAPTER 13

Cleaning and Inspection

Gregory Miskinis
Waupaca Foundry, Inc.

The cleaning of castings generally refers to all the operations involved in the removal of adhering sand, the gating system, risers, parting line fins, wires, chaplets, or other metal not a part of the casting. Cleaning operations may also include a certain amount of metal finishing or machining to obtain the required casting dimensions, the salvage of castings having minor defects, and the inspection of the finished castings.

Cleaning Operations and Equipment

The series of operations performed in the cleaning department may be classified as follows:

1. Removal of gates and risers, rough cleaning
2. Surface cleaning, exterior and interior of casting
3. Trimming, the removal of fins, wires, and protuberances at gate and riser locations
4. Finishing, final surface cleaning, giving the casting its outward appearance
5. Inspection

Sometimes heat treatments are involved which necessitate cleaning the castings after the heat treatment. This might be done between steps 3 and 4 above. Often steps 1 to 5 are carried on simultaneously. Some of these, such as gate removal, may occur during shakeout operations.

Removal of Gates and Risers

The sprue, runners, and risers are firmly attached to the solidified casting. If the casting alloy is brittle, the gating system may be broken off by impact when the castings are dumped and vibrated in shakeout and mold knockout devices.

Flogging

Flogging with a hammer or sledge is a positive means of gate removal by impact. A worker may be stationed at the shakeout to flog the sprue and risers as the sand falls away from the casting. When the molds are set out on floors and dumped by hand, workers with hammers knock off the gates and toss the castings and gates into separate boxes for transfer to the cleaning room. To de-sprue the castings is to remove gates in this way. Gray and white-iron castings are especially amenable to gate removal by this method. An inherent danger of breaking off the gates is that the break may extend into the casting proper. This condition may be remedied by notching the ingate ahead of the casting. The protuberance remaining on the casting can then be ground flush with the casting wall.

Flogging may be applied advantageously to steel castings as long as the gates are of a size that can be knocked off by a person using about a 5 kg (12 lb.) maul. One author [1] estimates that the maximum size of gate which may be flogged from carbon-steel castings is one with 28 cm^2 or 4.4 in^2 cross-sectional area connection to the casting (6.7 cm or 2.625 in diameter round riser). The same author studied the impact in foot-pounds necessary to break steel knock-off risers and developed the graph shown in Fig. 13.1. It is important to note that the diameter referred to in Fig. 13.1 is the necked-down diameter of a Washburn riser (neck-down riser), and not the enlarged diameter. Notching, or necking down, is a common means of making flogging easier. Even certain brasses and bronzes, relatively ductile as-cast, may have risers flogged off if Washburn risers are used.

Mechanical Cutoff

Gates may be removed by band sawing, power sawing, abrasive cutoff wheels, or with a sprue cutter. A rapid method of removing a number of small castings from a central runner is provided by a sprue cutter or trim press. An example of a trim press is illustrated in Fig. 13.2, which cradles the casting and shears off the protruding gating. Ductile metal castings such as steel, ductile iron, brass, copper, and aluminum are conveniently handled by sprue cutters or trim presses, provided the gate and casting size is not too large to shear. Gates of ⅛ to ½ in in thickness may be readily sheared on the sprue cutter or trim press. In some cases, parting line fins may be removed and gates cut off simultaneously by using a punch press fitted with a die for performing these operations. Die castings are often treated in this way [3].

Abrasive cutoff wheels may be used for gate removal. The machine illustrated in Fig. 13.3 will cut hard or difficult-to-saw alloys, as well as the more common foundry alloys. Hand-held grinders using abrasive wheels are also used for removal of parting line fins, remnants of gating connections, and general surface finishing of castings as shown in Fig. 13.4.

Band sawing can be used for removing gates from many castings, both ferrous and nonferrous. Its most extensive use is on nonferrous castings and involves equipment such as shown in Fig. 13.5. Band sawing may be done by cutting or friction sawing. Actual sawing involves cutting, chip formation, and removal at speeds up to about 3500 fpm.

Friction sawing requires cutting speeds up to 15,000 fpm which will heat the metal to temperatures approaching its melting point. Friction sawing is used almost entirely on ferrous materials. Gate removal by sawing under the proper conditions requires a consideration of cutting speed, cutting pressure, section thickness, lubricant, saw-blade type, alloy type, and other conditions. Numerical data for a few of these factors are

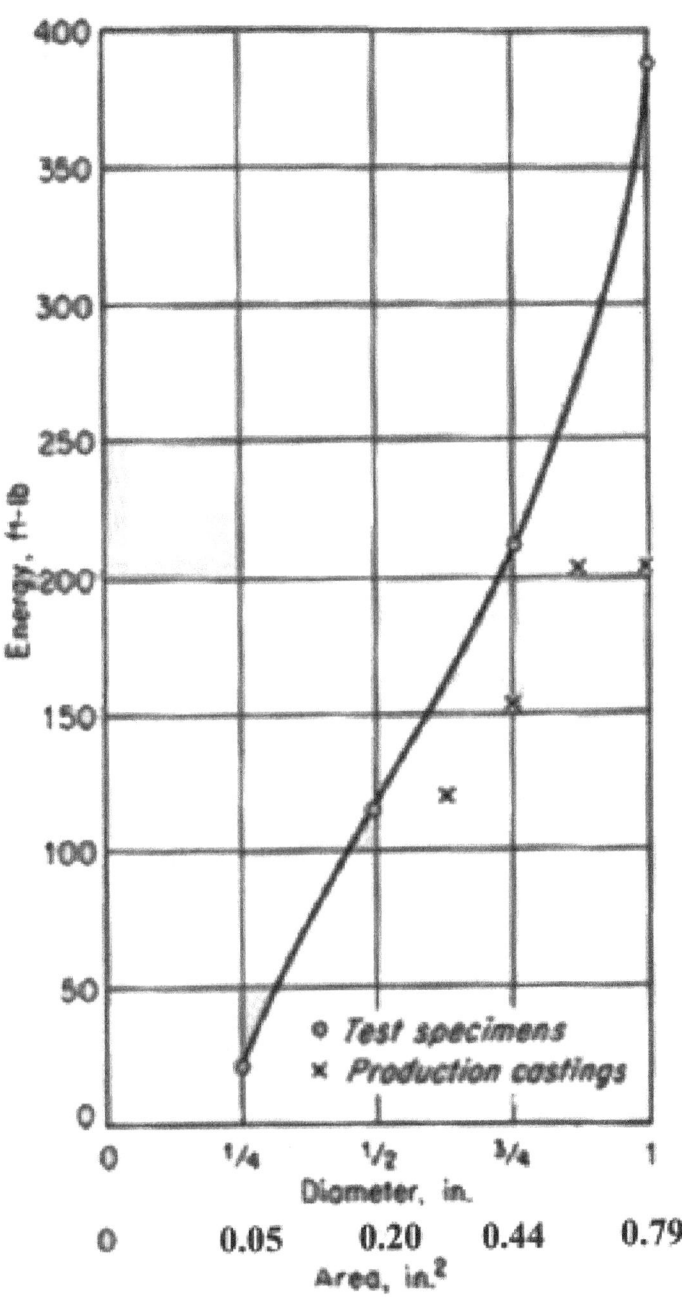

FIGURE 13.1 Impact load necessary to rupture necked-down steel risers. (From Brinson and Duma [1], Heine et al. [2].)

Figure 13.2 Hydraulic trim press used to remove excess gating and parting line flashing from a ductile iron caliper. (Courtesy of Waupaca Foundry, Inc.) (a color version of this figure is available at mhprofessional.com/pmc3)

Figure 13.3 Abrasive cutoff wheel used for cutting gates and risers. (Courtesy of De Pere Foundry, Inc.) (a color version of this figure is available at mhprofessional.com/pmc3)

(a)   (b)

FIGURE 13.4   (a) Parting line fin removal using a hand-held disc grinder (from De Pere Foundry Inc.); (b) grinding cooling fins of an aircraft cylinder heads. (Courtesy of Eck Industries, Inc.)

listed in Table 13.1. Low-speed sawing, ordinarily at less than 500 fpm of saw-blade velocity, is conventional band sawing. Sawing at higher speeds, however, is used whenever possible for gate removal. Band sawing makes it possible to a degree to follow the contour of the casting when removing gates and risers, something which cannot be done so easily with a sprue cutter or abrasive cutoff wheel.

Torch Cutting

Large risers and gates on steel castings are most conveniently removed by cutting torches. The sprue cutter is limited in metal thickness, whereas the cutting torch and oxygen lance may cut risers of practically any size. Torch cutting by oxy-fuel process relies on an oxygen/gas reaction [6]. The torch has preheating jets and a main oxygen orifice. The metal to be removed is preheated to ignition temperature, at which time it combusts with the excess oxygen being delivered by the torch. Cutting occurs as a result of melting and oxidation of metal into slag. Metal-section thickness determines the proper torch-nozzle

Figure 13.5   Gate removal using upright band saw. (Courtesy of Eck Industries, Inc.)

Alloy	Speed, fpm		Feeding Pressure	Lubricant	General Information
	Low Speed	**High Speed**			
Copper-base alloy	400 or less	2000 or less buttress	25 lb–1 in thickness	Used if chips weld to saw blade	Sawing speed decreases as hardness increases
Aluminum and magnesium	Up to 500 buttress blade	Up to 3500 on $\frac{1}{2}$ in section	Low	Ordinarily dry	4-pitch, 0.50-in, buttress blade, sawing speed decreases with section thickness
Ferrous (band sawing)	40–500	1500 or less	Moderate	May be used if chips weld to saw blade	
Ferrous (friction sawing)	–	3000–5000	20–40 lb	No	$\frac{1}{4}$–1 in width saw blade with 10–18 teeth per inch

Source: From Shippard [4], Chamberland [5], and Heine et al. [2].

Table 13.1   Band-Sawing Conditions for Gate Removal

Section Thickness, in	Diam. Cutting Orifice, in	Oxygen Pressure, psi	Cutting Speed, in/min	Oxygen, ft³/h	Acetylene, ft³/h
1.0	0.0465–0.0595	28–40	9–18	130–160	13–16
2.0	0.0670–0.0810	22–50	6–13	185–231	16–20
3.0	0.0670–0.0810	33–55	4–10	207–290	16–23
4.0	0.0810–0.0860	42–60	4–8	235–388	20–26
6.0	0.098–0.0995	36–80	3–5.4	400–567	25–32
10.0	0.0995–0.110	66–96	1.9–3.20	610–750	36–46
12.00	0.110–0.120	58–86	1.4–2.60	720–905	42–55
24.00	0.221–0.332	22–48	–	1600–3000	
36.00	0.290–0.500	12–38	–	3000–4600	

Source: From Heine et al. [2], adapted from American Welding Society [7].

TABLE 13.2 Data for Hand Cutting Carbon Steel

size and oxygen pressure necessary to keep the cutting reaction and kerf moving along. Table 13.2 lists some of the combinations required for effective cutting.

If castings of a given size and type are positioned in line, the speed and ease of gate removal are increased. The same type of setup is occasionally conveyorized to improve workflow. Torch cutting, as illustrated in Fig. 13.6, is most commonly employed in jobbing foundries on a variety of castings. Sometimes mechanized cutting may be used, employing a motorized cutting carriage, with a form template to guide the cutting torch around the casting surface. Extremely large risers require an oxygen lance as well as cutting torch to complete the cut. The torch maintains the cutting reaction zone on the near side of the riser, and the lance carries it through to the far side. Thus, cutting by hand or machine is seen to be an exceedingly versatile means of gate removal.

Some ferrous alloys, cast irons, and high-alloy steel are oxidation-resistant and do not cut well with the oxygen torch alone. Powder cutting has been developed to handle these materials. Iron powder is introduced at the oxygen stream after being preheated in the flame. The iron powder is picked up by an airstream and discharged through the preheat flame into the oxygen stream. The combination of the oxygen stream and burning iron attacks the metal by fluxing and oxidation. By this means, 18-8 stainless steel, high-chromium irons, high-temperature alloys, cast irons, and other oxidation-resistant alloys can be cut to remove gates rapidly.

FIGURE 13.6 Riser removal from large castings by torch cutting. (From *Cleaning Castings* [8].)

Gouging and powder washing are processes allied to torch and powder cutting which assist in cleaning castings. Gouging is used to clean out or remove surface defects on steel castings. Gouging is performed with the cutting torch, and involves preheating the defective area and cutting it or washing it out with a low-velocity, large-bore oxygen stream [9]. Slag inclusions, blowholes, cracks, fins, and wires may be cleaned out by gouging. Powder washing is a similar process but is used on the more oxidation-resistant alloys and in removing sand-metal encrustation due to metal penetration of molds or cores. Powder washing differs from gouging in that iron powder is required as in powder cutting. Much of the trimming of riser pads, heavy fins, chills, and other metal protuberances on medium and large steel castings may be done in this way.

Surface Cleaning

Surface-cleaning operations ordinarily follow the removal of gates and risers. However, surface cleaning may also facilitate gate removal. For instance, elimination of sand from the casting favors sawing and torch cutting, so that surface cleaning may then be done before gate removal in the case of nonferrous alloys and sometimes steel. When flogging is used on steel or cast iron, it is easier to perform the surface cleaning after gate removal.

Tumbling

Sand, scale, and some fins and wires may be removed by tumbling in a tumbling mill. The mill is filled with castings and some jack stars. Rotation of the mill causes the castings and stars to tumble and abrade each other. Twenty minutes to 1 hour of tumbling is used on gray -and malleable-iron castings. Tumbling has a burnishing action on the casting surfaces and causes the corners to be rounded. Excessive tumbling can cause overabrasion and deformation of the casting corners.

The tumbling action can be combined with sand, grit, or shot blasting in blast mills. Wet tumbling using water treated with caustic is employed for dust suppression. Continuous tumbling mills in which the castings and stars are charged at one end and discharged at the other are usually operated wet. Typical cleaning shot and grit sizes are given in Table 13.3.

Tumbling for deburring and brightening of copper-base castings is also practiced. Castings may be cleaned and deburred by using water and detergents in combination with sand, pumice stone, or other ceramic media in the tumbling barrels. Ball burnishing provides a means of imparting high luster to copper-base castings by tumbling. Steel balls and the castings in ratio of 2:1 are charged into the barrel along with dilute soap solution. After tumbling, the castings are dried in another mill by tumbling them in sawdust or wood shavings.

Blasting

Blasting the surface of castings is the most rapid means of removing sand and scale. Abrasives employed are sand, metal grit, and metal shot. Sand blasting may be performed using coarse sand, 6- to 30-mesh size, as the abrasive and air as the carrying medium. When air blasting is employed, the blasting must be done in cabinets or rooms as shown in Fig. 13.7, to provide a means of handling the dust arising from the disintegrating sand. Water blasting eliminates the dust problem.

| Shot Size | None on Screen | Min. on Screen | | Max. through Screen | |
No.	Opening, in	%	Opening, in	%	Opening, in
S1320	0.187	85	0.132	5	0.111
S1110	0.157	85	0.111	5	0.0937
S930	0.132	85	0.0937	5	0.0787
S780	0.111	80	0.0787	10	0.0661
S660	0.0937	80	0.0661	10	0.0555
S550	0.0787	80	0.0555	10	0.0459
S460	0.0661	75	0.0469	15	0.0394
S390	0.0555	75	0.0394	15	0.0331
S330	0.0469	75	0.0331	15	0.0232
S230	0.0394	70	0.0232	20	0.0165
S170	0.0331	70	0.0165	20	0.0117
S110	0.0232	65	0.0117	25	0.0070
S70	0.0165	65	0.0070	25	0.0049
G10	0.111	80	0.0787	10	0.0661
G12	0.0937	80	0.0661	10	0.0555
G14	0.787	80	0.0555	10	0.0469
G16	0.0661	676	0.0469	15	0.0394
G18	0.0555	75	0.0394	15	0.0280
G20	0.0469	70	0.0280	20	0.0165
G40	0.0394	70	0.0165	20	0.0117
G50	0.0280	65	0.0117	25	0.0070
G80	0.0165	65	0.0070	25	0.0049
G120	0.0117	60	0.0049	30	0.0029
G200	0.0070	55	0.0029		0.0017
G325	0.0049	20	0.0017		

Source: From Heine et al. [2], American Wheelabrator and Equipment Corp. [10].

TABLE 13.3 Typical Cleaning Shot and Grit Sizes (SAE shot numbers and screening tolerances)

Grit or shot blasting can be done by the airless-blast method. The metallic particles are thrown by centrifugal force from a rapidly rotating wheel such as that illustrated in Fig. 13.8.

The blasting wheel is incorporated in a variety of devices, which tumble castings or rotate them under the shot or grit stream so that all casting surfaces are exposed to its abrasive action. An example of equipment that incorporates the combination of tumbling and mechanical blasting is shown in Figs. 13.9, 13.10, and 13.11.

Another type of machine uses a series of rotating tables or moving belts that pass under the shot or grit stream. Rebound of the shot assists in cleaning surfaces not in direct

FIGURE 13.7   Air blasting of a machinery base. (Courtesy of De Pere Foundry Inc.)

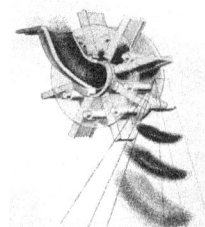

FIGURE 13.8   Grit blasting wheel. (From *Cleaning Castings* [8].)

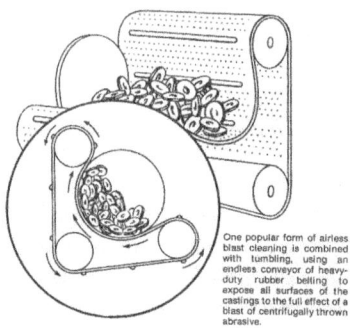

One popular form of airless blast cleaning is combined with tumbling, using an endless conveyor of heavy-duty rubber belting to expose all surfaces of the castings to the full effect of a blast of centrifugally thrown abrasive.

FIGURE 13.9   Schematic diagrams illustrating the mechanism for inducing the tumbling action of the machine shown in Fig. 13.10. (From *Cleaning Castings* [8].)

Figure 13.10 Tumbleblast barrel interior showing continuous metal drive belt. (Courtesy of Waupaca Foundry, Inc.)

Figure 13.11 Tumbleblast with typical load; slots at top of cabinet direct shot onto castings below. (Courtesy of Waupaca Foundry, Inc.)

line of the shot stream. On table-type models, the castings must be turned over for cleaning all surfaces. Conveyor belt-type machines, as shown in Figs. 13.12 and 13.13, utilize a mesh-type belt which allows the shot to pass through to contact all surfaces of the casting.

The mechanical-impact cleaning method may also be applied to conveyorized cleaning. Castings, suspended from conveyor hooks as illustrated in Fig. 13.14, enter a spinner cabinet or room where they are subjected to a shot or grit stream. The castings are rotated in the stream to expose all the surfaces to cleaning.

Metallic Abrasives

Shot or grit may be made of white iron, malleable iron, or steel. Shot is produced by allowing a stream of molten iron or steel to pass through a steam jet. The jet breaks up

FIGURE 13.12   Raw castings entering a belt blast machine. (Courtesy of Waupaca Foundry, Inc.)

FIGURE 13.13   Cleaned castings exit a belt blast machine. (Courtesy of Waupaca Foundry, Inc.)

the stream into fine particles which fall into a water quenching tank. Grit may be produced from brittle shot by crushing. Malleable iron or steel shot has fewer tendencies to break down into smaller particles than does white-iron shot. Typical shot and grit sizes are types in Table 13.4. Shot blasting has a battering effect on the metal surface and causes surface-metal flow as it is continued. Grit appears to have more of a gouging action and seems to be removing small particles of metal from the casting surface as blasting continues. Shot blasting therefore produces a rather shiny surface, whereas grit dulls the surface. Only very light shot blasting can be applied to nonferrous castings or the surface will be severely battered. Grit is not desirable because it will become embedded in the surface of copper or aluminum castings. Malleable iron, soft steel, copper, or bronze shot seems best for that application. In either shot or grit blasting, effectiveness is greatly reduced if the abrasive is contaminated with excessive amounts of sand or

FIGURE 13.14 Flywheels being loaded onto a rotating conveyor tree in a spinning cabinet blast machine. (Courtesy of Waupaca Foundry, Inc.)

Castings	SAE Size No.
Average gray or annealed malleable with pockets, burned-in sand, etc.	S390 shot
Light gray iron or annealed malleable	S330 shot
Hard malleable	S330 shot
Brass (all types)	G50 or G120 grit
Die castings	G50 grit
Aluminum	G50 grit
Steel	S390 shot

Source: From Heine et al. [2], American Wheelabrator Equipment Corp. [10].

TABLE 13.4 Recommended Sizes of Abrasive

fine metal particles. Hence, the cleaning units are equipped with a means of removing the fines as well as recycling the shot.

Other Types of Surface Cleaning

A number of methods of casting-surface cleaning other than those mentioned above are in use. Wire brushing, buffing, pickling, and various polishing procedures may be applied. Simple wire brushing can be an adequate means of taking off surface sand of nonferrous castings. Most of the other means of cleaning mentioned are used to impart an especially desirable surface finish for final cleaning of the casting.

Trimming and Sizing

Either after, before, or during the initial surface cleaning, the castings are trimmed to remove fins, gate and riser pads protruding beyond the casting surface, chaplets, wires, parting-line flash, or other appendages to the casting which are not part of its final dimensions. Trim dies may be utilized as single or multiple step stations as part of a robotic cell as illustrated in Fig. 13.15.

Shearing, punching, coining, and straightening are mechanical operations which may be employed to complete trimming and sizing castings. Coining or straightening is done with dies in presses under hydraulic pressures up to about 1000 tons, as described in Refs. 11, 12, and 13. In mild-steel and malleable-iron castings, holes may be punched out to eliminate drilling operations. Some castings may have surfaces which are milled, broached, or ground to a specified accuracy as required by the customer.

Chipping

Pneumatic chipping hammers may be used to remove fins, gate and riser pads, wires, etc., and to remove cores. A variety of hammer and chisel sizes are used for various casting alloys. A No. 2 hammer, with a 51 mm (2 in) piston stroke, will handle most iron and steel foundry work. Specifications for chipping on different types of casting are given in Table 13.5. Chisel types and the relationship of air operating pressure to air-hose setup and are illustrated in Fig. 13.16.

An air-hose arrangement that causes more than about 10% pressure drop from the sources to the pneumatic tool seriously lowers the efficiency of the tool. Air-hose combinations of length and diameter which are favorable and unfavorable at various flow rates are listed in Table 13.6. Pneumatic grinders or other tools are subject to the same conditions. Chipping operations may be expedited or improved by having chipping

FIGURE 13.15   Robotic differential carrier trim press cell. (Courtesy of Waupaca Foundry, Inc.) (a color version of this figure is available at mhprofessional.com/pmc3)

Piston Stroke, in	Piston diam., in	Length, in	Weight, lbs	Hose, in	Work Adapted For
1	0.625	8	3	½	Very light chipping
1.375	0.75	10	5.5	½	Light chipping and scaling
1	1.125	11.25	10	½	Aluminum casting and light cast-iron caulking
2	1.125	13.5	13.25	½	Heavy cast iron, light-steel casting, flue beading
3	1.125	14.5	14.5	½	Heavy-steel casting, billet chipping
4	1.125	15.75	15.75	½	Extra-heavy chipping

Source: From Ringer [9], Heine et al. [2].

TABLE 13.5 Specifications for Chipping Hammers

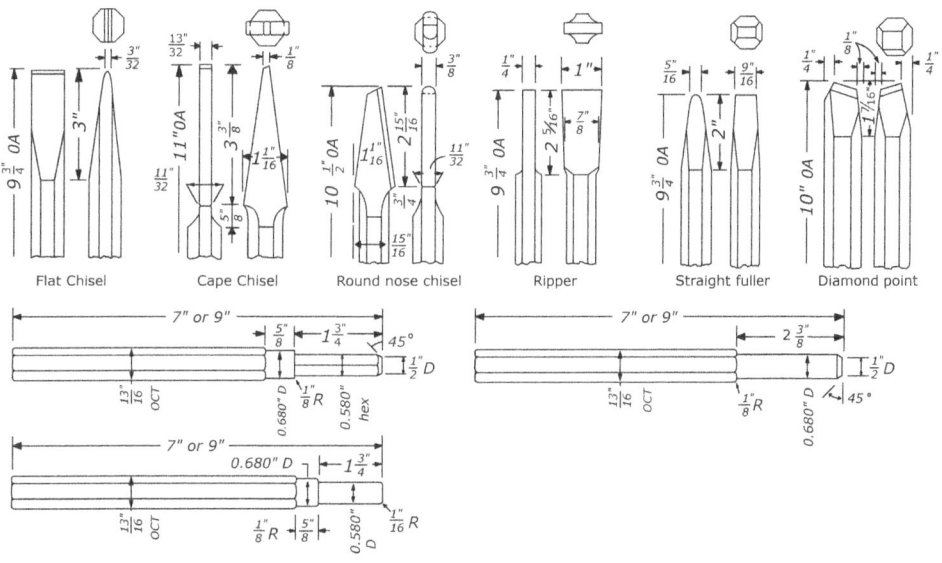

FIGURE 13.16 Chisel types for chipping castings. Air-hose arrangements for maintaining the efficiency of pneumatic tools are given in Table 13.6. (From Ringer [9], *Cleaning Castings* [8], Heine et al. [2].)

stations at conveyors used to transport castings in the cleaning room. Much chipping may be done most conveniently by hand with a hammer. Light gray and white-iron castings are especially adaptable to hand chipping. Pneumatic chipping is then done on the areas that are heavier and more difficult to trim.

Grinding

Grinding, or "snagging" of castings is practiced to remove excess metal. Four principal types of grinders are employed for this purpose: floor or bench grinders, portable grinders, swing-frame grinders, and programmed robotic grinders. In addition, specialized machines such as disk grinders, belts, and cutoffs may be used.

Air Flow, cfm	10 ft of ¼ in Hose	8 ft of ⁵⁄₁₆ in Hose	10 ft of ⅜ in Hose	12½ ft of ½ in Hose	25 ft of ½ in Hose	50 ft of ½ in Hose	12½ ft of ¾ in Hose	25 ft of ¾ in Hose	50 ft of ¾ in Hose	50 ft of ½ in Hose + 10 ft of ¼ in Hose	50 ft of ½ in Hose + 10 ft of ⅜ in Hose	50 ft of ½ in Hose + 8 ft of ⁵⁄₁₆ in Hose	50 ft of ½ in Hose + 12½ ft of ½ in Hose	50 ft of ¾ in Hose + 25 ft of ½ in Hose	50 ft of ¾ in Hose + 12½ ft of ¾ in Hose
						Pressure Drop, psi (based on 100 psi line pressure)									
10–11	5.0	0.9								5.3	0.7	1.4			
11–12	5.9	1.0								6.2	0.8	1.6			
12–13	6.8	1.2	0.4							7.2	0.9	1.9			
13–14	8.0	1.4	0.5							8.4	1.1	2.2			
14–15	9.3	1.6	0.6							9.8	1.3	2.5			
15–16	11.0	1.9	0.7							11.6	1.5	2.9			
16–18	14.0	2.4	0.8							15.0	1.9	3.5	1.7		
18–20	19.6	3.0	1.0							21.4	2.4	4.5	2.0		
20–25		4.3	1.4	0.7	1.0	1.3					3.5	6.4	2.6	1.3	
25–30		6.6	2.1	1.0	1.5	2.3					5.2	9.8	3.8	1.9	
30–35		9.5	3.1	1.3	2.1	3.6					7.3	13.7	5.3	2.6	
35–40		12.8	4.2	1.7	2.8	5.2					9.6	18.4	7.1	3.5	
40–50		19.3	6.3	2.4	4.1	8.0					14.0		10.4	5.2	1.8
50–60			9.6	3.7	6.3	12.2					21.8		16.0	7.8	2.3
60–70			13.5	5.3	9.0	17.4	0.9	1.4	1.9				22.8	11.1	3.0
70–80			18.7	7.1	12.4		1.1	1.7	2.5					15.0	3.7
80–90			25.0	9.0	16.1		1.4	2.2	3.2					19.8	4.6
90–100				11.1			1.7	2.7	4.0						5.8
100–120							2.3	3.5	5.6						7.9
120–140							3.2	4.8	8.0						11.2
140–160							4.3	6.6	11.0						15.5
160–180							5.6	8.7	15.2						20.4
180–200							7.2	11.0							
200–220							9.0								

Source: From Ringer [9], *Cleaning Castings* [8], Heine et al. [2].

TABLE 13.6   Recommended Hose Arrangements for Pneumatic Tools

Stand Grinding

Stand grinders are usually of the double-end type and may be of constant or adjustable-speed type, the latter compensating for wheel wear. Low-speed machines operate at speeds up to 6500 surface feet per minute (sfpm); high-speed machines at 9500 sfpm.

FIGURE 13.17   Grinding parting line fins using a stand grinder. (Courtesy of Waupaca Foundry, Inc.) (a color version of this figure is available at mhprofessional.com/pmc3)

Castings that can be handled manually are ground on machines of this type, as shown in Fig. 13.17. The operator presents the areas to be ground to the wheel face under suitable pressure. Traversing of the wheel face is desirable to prevent rounding or grooving of the wheel and the need for frequent dressing of the wheel.

Stand grinding is ordinarily hand work, depending upon the judgment of the operator for the selection of areas to be ground and amount of grinding required. When many castings of one type are ground, special guide fixtures or mechanized feeding and positioning fixtures may be used to speed up grinding. Figure 13.18 is an example where a casting rotational drive has been added to produce an automated grinding station.

FIGURE 13.18   Automated grinding station shown with a diamond impregnated grinding wheel. (Courtesy of Waupaca Foundry, Inc.)

Abrasive Wheels

Grinding wheels used for cleaning castings are made of either the aluminum oxide or silicon carbide types of abrasives. The aluminum oxide abrasives are bonded with vitrified clay or with a resinoid bond. The vitrified-clay-bonded wheels are limited to 6500 sfpm. The silicon carbide abrasives are resinoid-bonded and may operate up to 9500 sfpm. Newer materials such as natural and synthetic diamond or superabrasives are capable of operation at speeds in excess of 20,000 sfpm. Application of various common abrasives and bonds is given in Table 13.7. In general, the silicon carbide abrasives are used for casting materials of lower tensile strength. Coarser abrasive sizes are used for fast cutting, whereas fine grits produce a smoother finish. Grits may be graded by sieve number similar to foundry sands. High grain-size numbers in Table 13.7 indicate finer abrasives. Grain spacing and structure determine the number of cutting edges per unit area of wheel face. Snagging usually requires wide grain spacing to get rapid metal removal unless the loading pressure is high.

Wheel diameters of 36 to 91 cm (14 to 36 in) are common in foundries. The wheel velocity in sfpm varies as the wheel is worn. Of course, the maximum wheel diameter is limited by the safe upper limit of sfpm for the particular wheel and rpm of the grinding machines at hand. Worn wheels and stubs from heavy grinders may be transferred to smaller grinders for light work. Specific wheel applications are best determined by experience and testing. An advantage of the newer technology bonded wheels is that the loss of cutting effectiveness is coupled with very little dimensional change, less compensation for wear is necessary, and wheel life is greatly extended.

Swing Grinders

Swing grinders are employed when the castings are too heavy to be carried to the work. In this case, the grinder is mounted on a swing frame, the casting positioned under the

Casting	Grain Size	Wheel Type	Bond	Speed and Equipment
Gray iron	16–24	Silicon carbide	Vitreous	5000–6000 sfpm, floor stand and swing frame
		Silicon carbide	Resinous	7000–9500 sfpm, floor stand and swing frame
Brass	24–30	Silicon carbide	Vitreous	5000–6000 sfpm, floor stand
		Silicon carbide	Resinous	7000–9500 sfpm, floor stand
Steel	14–20	Aluminum oxide	Vitreous	5000–6000 sfpm, swing frame, floor stand, and portable
		Aluminum oxide	Resinous	7000–9500 sfpm, swing frame, floor stand and portable
Aluminum	24–30	Silicon oxide	Vitreous	5000–6000 sfpm, floor stand
		Aluminum oxide	Resinous or shellac	7100–9500 sfpm, floor stand
Malleable	20	Aluminum oxide or silicon carbide		

Source: From Sawtelle [14], Schuman [15], Work [16], Wagner [17], and Heine et al. [2].

TABLE 13.7   Abrasive Wheels Used for Grinding Castings

FIGURE 13.19   Swing frame grinder being used to remove parting line fins on a machinery base. (Courtesy of De Pere Foundry Inc.)

grinder, and the grinder is then worked over the casting surface, as illustrated in Fig. 13.19. Wheel sizes run from 30 to 61 cm (12 to 24 in) in diameter. When the wheel is worn to the point where an excessive loss in sfpm occurs, the rpm on the wheel may be increased by changing the position of belts on the motor and wheel pulleys. The work piece must be firmly positioned under the grinder. This may require supports or fixtures to hold pieces that do not have large stable flat surfaces.

Portable Grinders

Electric or air-operated portable grinders are employed for working over surfaces of castings that cannot be handled by swing or stand grinding. Cone, cup, disk, special shapes, and straight grinding wheels up to about 36 cm (14 in) in diameter may be mounted on portable machines, as shown in Fig. 13.20. Generally, only light grinding is intended.

FIGURE 13.20   Parting line fins being removed with a portable grinder. (Courtesy of Waupaca Foundry, Inc.)

Disk grinding is sometimes practiced in this way. Specially shaped wheels have safe operating speeds which are lower than those of regular wheels, and these must be observed to avoid the hazards of wheel disintegration.

Programmed Robotic Grinders

Robotic grinding is becoming more popular for intricate or highly irregular-shaped parts that have multiple surfaces to grind and especially where close tolerances are required. These systems are often coupled with diamond or superabrasive grinding discs, as industrial robots can apply higher pressures and speeds necessary for optimum performance from the abrasives, as illustrated in Fig. 13.21.

Rotary Tools

Cleaning of softer nonferrous alloys such as aluminum or magnesium alloys may be done by rotary filing tools or cutters. These may be electrically or air operated, and are

(a)

(b)

Figure 13.21 Robotic grinding. *(a)* diamond-impregenated wheel attached to an industrial robot arm *(b)* ductile iron crankshaft ground using a diamond-impregenated wheel (Courtesy of Waupaca Foundry, Inc.) (a color version of this figure is available at mhprofessional.com/pmc3)

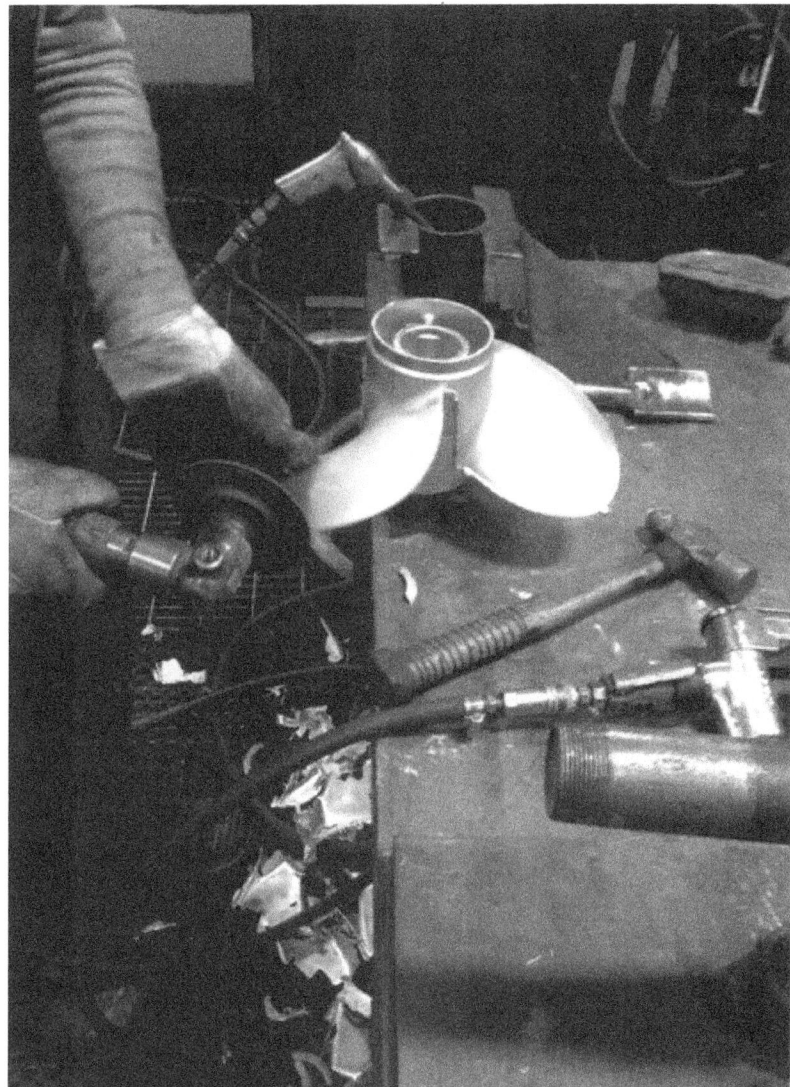

Figure 13.22   Finishing boat motor propeller blades using an air grinder. (Courtesy of Eck Industries, Inc.)

usually of the portable type, as shown in Fig. 13.22. The tool operator may then file off excess metal on any casting surface accessible to the tool.

Finishing

The latter stages of cleaning are often referred to as finishing. Many castings have received their final cleaning operations when grinding is completed. Others are given additional surface finishing such as machining, chemical treatment, polishing, buffing, blasting, and painting to put them into a suitable appearance for sale (Fig. 13.23). Heat treatments may come at various stages of cleaning. Since scaling, oxidation, or discoloration occurs during heat treatment, steel and malleable-iron castings receive their final surface cleaning after heat treatment. In nonferrous castings, chemical, electrolytic, and mechanical means of making the casting surface attractive may be utilized.

A special salt-bath cleaning operation may be used on ferrous castings to obtain maximum freedom from scale, sand, dirt, and grit. For example, gray-iron castings may be immersed into a molten salt bath of caustic, 95% sodium hydroxide, 5% sodium nitrates

Figure 13.23   Buffing the surface of a missile wing. (Courtesy of Eck Industries, Inc.)

and nitrites, at 425°C (800°F) for cleaning. The castings are then rinsed, dipped in acidified water, 15% hydrochloric acid, hot-water rinsed, and treated with soluble oil. The oil coating provides rust protection. Treatments of this type are utilized only when the additional cost is justified in a particular application or processing sequence [18].

Casting Inspection

Inspection comprises those operations which check the quality of the castings and result in their acceptance or rejection. Inspection procedures may be classed as follows:

- Visual, surface inspection for foundry defects
- Dimensional, requiring gauges for measurements
- Metallurgical, requiring chemical, physical, and other tests for metal quality

Complete inspection usually embraces all these types of procedures.

Visual Inspection

Certain types of casting defects are immediately obvious upon visual examination of the casting. Cracked castings, tears, dirt, blowholes, scabs, metal penetration, severe shifts, runouts, poured short, swell or strains, cracked mold or cores, and numerous other defects can be identified by the inspector. Casting defects of this type are usually associated with defective molds, cores, their materials of construction, flask equipment, the operations of molding and coremaking, and the rest of the factors in making and pouring the mold. Many of the causes of these defects have been discussed in previous chapters including chapter 14. Evidently, if excess amounts of these defects occur, faulty practices are being employed in the foundry. Their correction is a necessity and can be facilitated if the cause of the defect can be located. Inspectors identify the casting defects and assign their cause to some foundry operation or material so that corrective measures can be taken as have been discussed in chapter 14.

Visual inspection is the simplest method of inspection, and carried to its ultimate it will ensure a casting that "looks" good. This degree of inspection is satisfactory for some castings such as sash weights, manhole covers, drains, and counterbalance weights. However, countless numbers of castings for manufacturing require more exacting inspection for dimensional accuracy and metallurgical standards.

Optical or Vision System Inspection

Visual inspection by human eyes still accounts for a large percentage of casting inspection, however, it is easily understood that it is not 100% reliable. Modern vision systems couple high-resolution digital cameras with computer software to perform both simple color matching type inspection to more detailed contour and dimensional compliance [19]. In the former case, the system may be "taught" to recognize differences in surface color or tone due to surface contaminants or irregularities (dents, holes, protrusions). Prominent diameters, lengths, and heights may be programmed by controlling the camera's position or in combination with three-dimensional surface inspections (Fig. 13.24).

Dimensional Inspection

Dimensional inspection of castings involves the principles of gauging as it is applied to any machined elements. Surface plates, height and depth gauges, layout tables, dividing heads, go and no-go gauges, snap and plug gauges, templates, dies, contour gauges, etc.,

FIGURE **13.24** Brake rotor inspection using an automated vision system. (Courtesy of Waupaca Foundry, Inc.) (a color version of this figure is available at mhprofessional.com/pmc3)

as used in standard layout and inspection procedures, may be applied to castings (Fig. 13.25). Agreement between the machine shop and foundry, or purchaser and vendor, is necessary so that gauging and dimensional checking may be carried out in a mutually acceptable way. Locating points used as starting points for machining and dimensional

FIGURE **13.25** Dimensional inspection using a standard layout table. (Courtesy of Waupaca Foundry, Inc.)

inspection should be selected by common consent. Castings may be sectioned to check metal-wall thickness. The area of dimensional inspection and accuracy involves the entire field of castings' utilization by the machine shop. There is no intent in this textbook to discuss this mechanical-engineering field. It needs to be pointed out, however, that the closest cooperation between the foundry, pattern shop, and machine shop utilizing the casting will result in the most efficient use of good gauging and machining practice.

Three-Dimensional Surface Inspections

Three-dimensional inspection data can be created outright or overlaid to its original digital nominal CAD format to track deviations in the casting process. Software can calculate precise deviations and illustrate them in a color-coded format to visually show inconsistencies in the casting relative to its designed intent. Positive or negative surface areas can be used to recalculate where the pattern tooling requires correction or track the life of tooling by repeatedly scanning it over time to calculate wear patterns. The commonly employed techniques used for three-dimensional inspection are laser and white light scanning [20].

Laser Scanning

Laser scanning is a noncontact, relatively fast method of surface contour measurement. The most common method of data collection uses triangulation. A laser line is reflected off the casting surface and a receiving camera determines its depth in the field of view. The angle between the laser emitter and the camera is known as the fixed angle. This allows the software to calculate the three-dimensional profile of the object based on where the laser line is in the field of view. To track its relative position in space requires specific hardware to mount the laser scanner to, as illustrated in Fig. 13.26. The most

Figure 13.26   Dimensional analysis of an engine bedplate using a laser scanner. (Courtesy of Waupaca Foundry, Inc.) (a color version of this figure is available at mhprofessional.com/pmc3)

common method employed is a user-driven portable CMM (coordinate measuring machine) articulated arm. These articulated arms contain a series of joints that contain encoders to track the end of the arm position relative to its base. The encoders are sensor-equipped bearings that relay precise coordinates back to a home or zero position. The intelligent joints, in conjunction with the laser scanning head, allow software to merge multiple scans into the same coordinate position with very high accuracy. The laser scanning process is best described analogously as "spray painting." The laser emitter is passed over the casting surface using a series of overlapping logical segments, or strokes. This free-form movement allows the scanner to see the part from a broad range of different angles. User-driven CMM arms offer an advantage when scanning complicated parts and there is a significant time savings because no programming is required to determine the part shape. Portable base fixtures allow the inspection of large or difficult to transport castings.

The laser scanning heads can also be mounted to traditional CMM machines. This system uses a more permanent structure. Standard CMM systems are comprised of a thick granite base and a bridge system that contains mechanical drives that allow the laser head to move in three directions, commonly known as XYZ positions. Some systems have adapted the mechanics to incorporate additional axis of rotation to make the scanning process easier. Because this system is mechanically driven, programming is required by the operator to tell the scanner which location to move to. Bridge CMM systems are more accurate because there is less mechanical movement, reducing the amount of error in the positional changes. Because they are typically a permanent structure, part size and capabilities can be limited. If a part is too big for the table, the machine may not be able to measure it. Also, parts have to be brought to it; it can't be carried out to the field for onsite measurements.

White Light Technology

White light scanning, also known as structured light, is currently one of the cleanest forms of three-dimensional scanning, typically used as a benchmark for quality three-dimensional data. Its technology is comprised of some very simple hardware coupled with a series of complex mathematics. Multiple cameras are symmetrically and sometimes asymmetrically mounted around a central light source, or projector, as shown in Fig. 13.27.

Behind the light source is a slide mechanism. The mechanism contains a series of varied thickness vertical stripes, known as fringes. When the scan is initialized the projector displays a sheet of light onto the object being scanned. The different fringe patterns are cycled onto the part and the cameras record the distortion in the fringe pattern, as shown in Fig 13.27. Software converts the deviations into three-dimensional coordinates. These projections can contain up to 12 million data points per scan volume (the size of the rectangle of projected light). By varying the optics, similar to a conventional camera, the resolution of the scans can be changed. This allows you to scan areas the size of a postage stamp, up to the size of a small car. The scanning process is repeated to capture all different angles of the object.

In order to align the scans together to form the final part shape, two different methods are used. The most common method uses small circular black and white targets, or index markers. These are randomly placed onto the surface of the part. The software uses the contrast between the black and white to determine the target center. This allows the software to generate a reference frame. Each successive scan must contain three targets

Figure 13.27 White light projector and fringe pattern projected onto the subject casting surface. (Courtesy of Capture 3d.)

from the previous scan in order to calculate the relative position on the part. The software places successive scans in the correct coordinate position with an extreme amount of accuracy. The other method employs a user-driven, contour-based alignment. Instead of index marks, the operator visually picks three common geometry areas on two scans. Once identified, the software uses an algorithm to fit the two data sets together based on the contour of the geometry. This process is repeated until the entire surface is scanned and the data sets are stitched together to make the final three-dimensional part shape.

There are limitations to white light scanners. As most systems use two cameras or a stereo configuration, they may not both be able to see parts of the casting at the same time. For example, both cameras may not see the bottom of a deep hole or pocket at the same time, resulting in no data being captured in this area. Another limitation sometimes is surface finish and color. Some highly reflective surfaces, like die-cast nonferrous metals are hard to scan due to reflectivity of the projector. This can be combated by coating or painting the object a white or light gray color (if the casting tolerances allow the additional thickness of the coating). Dark colors also can be an issue as the scanners use the contrast of colors in their calculations. If the surface is too dark, it may not be seen by the cameras. Projecting different light spectrums onto the part, such as blue or green light systems, and incorporating multiple scan heads, as shown in Fig. 13.28 capturing different finishes, complex geometries that present greater color variation or contrast are possible.

Metallurgical Inspection

Metallurgical inspection includes chemical analysis, mechanical-property test, evaluation of casting soundness, and product testing of special properties such as electrical conductivity, resistivity, magnetic effects, corrosion resistance, response to heat treatment, strength in assemblies, surface conditions, coatings and surface treatments, and others.

Chemical Analysis

The methods of chemical analysis for many casting alloys, ferrous and nonferrous, have been developed and adopted as standard through the work of the ASTM. The ASTM "Methods of Chemical Analysis of Metals" sets forth the standard and tentative standard procedures adopted for ferrous and nonferrous metals by American industry. Many short-cut methods of analysis have been developed for specific casting alloys. Since this textbook does not deal with chemical analytical methods, there will be no further discussion of this subject.

Figure 13.28 Blue light three-head scanner shown with color-coded dimensional model created. (Courtesy of Capture 3d.) (A color version of this figure is available at mhprofessional.com/pmc3.)

Casting Soundness

Shrinkage cavities, blowholes, gas holes, porosity, hot tears, cracks, entrained slag, lapped or cold-shut surfaces, etc., are all considered as contributing to lack of casting soundness when they are present. These defects are of greater or lesser importance, depending on the casting application. Many castings with internal shrinkage, porosity, or other defects that do not interfere with the functioning of the casting are quite acceptable to the user. Where the requirements are high and factors of safety low, however, the very highest degree of metallurgical quality is required. In castings for aircraft, ordnance, and other highly precision-engineered applications, absolute soundness and optimum properties are needed. These objectives will be met only when the casting inspection includes methods which check the casting for soundness defects not visibly detectable. Diverse methods of discovering soundness defects have been devised, and are exhaustively treated by various authors.

Pressure Testing

Pressure testing is used to locate leaks in a casting or to check the overall strength of a casting in resistance to bursting under hydraulic pressure. Equipment for sealing off castings and finding leaks is discussed in Ref. 13. Proof loading by hydraulic pressure involves introducing a fluid, oil, or water into a casting as shown in Fig. 13.29. The casting is subjected to a pressure which is in excess of the maximum stress that the casting is supposed to encounter in service. Cast pipe or tubes are often proof-tested in this way.

Sectioning

Castings may be sawed up, and the sections examined for soundness. This procedure is desirable since the interior of the casting, section thickness, as well as its soundness, may be studied. Macroetching may be used to locate suspected shrinkage, porosity, or cracks.

Radiographic Examination

Radiographic examination, or as generically referred to as X-ray, is a nondestructive testing (NDT) method that examines the entire casting for soundness. X-rays were first discovered in 1895 by Wilhelm Conrad Roentgen, a professor at the Wuerzburg University in Germany while working with cathode ray tubes. The technique was readily adopted in the medical and dental fields to examine low-density tissue and bones. Industrial use was initially limited by the short service life of the first X-ray vacuum tubes which required high voltage and long exposure times to penetrate higher-density metals. Refinements in the form of tubes capable of handling 100 kV in 1913, provided sufficient penetrating power for industrial purposes. X-ray tubes capable of handling 200 kV in 1922 are widely responsible for permitting thick steel castings to be inspected and X-ray generators producing 1000 kV manufactured by General Electric in 1931 led to the technique's acceptance by the American Society of Mechanical Engineers (ASME) for approval of fusion-welded pressure vessels.

The technique was further advanced by the discovery of gamma rays by Henri Becquerel and subsequent discovery of other radioactive elements by Marie and Pierre Curie. The shorter wavelength gamma ray (one angstrom X-ray versus 0.0001 angstrom gamma ray) allowed thicker steel castings to be examined. In general, X-ray source machines are a better choice for casting radiography than radioactive isotopes in that the latter can never be turned off and safely managing the source is a constant responsibility. While isotopes provide shorter exposure times, the fixed power is less flexible and performance decays over time [21–23].

Figure 13.29 Leak testing an aircraft cylinder head. (Courtesy of Eck Industries, Inc.)

X-rays are generated by directing a stream of electrons at a target material such as tungsten. A focusing cup concentrates the flow of electrons into a small area called the focal spot. The general layout of an X-ray apparatus is shown in Fig. 13.30. The important parameters governing the quality of the X-ray image are

- Focal spot size of the X-ray tube head.
- Geometric distances between the X-ray tube head and the imaging device or substrate (film) and the casting and the imaging device.
- X-ray source energy level used (kV and mA).

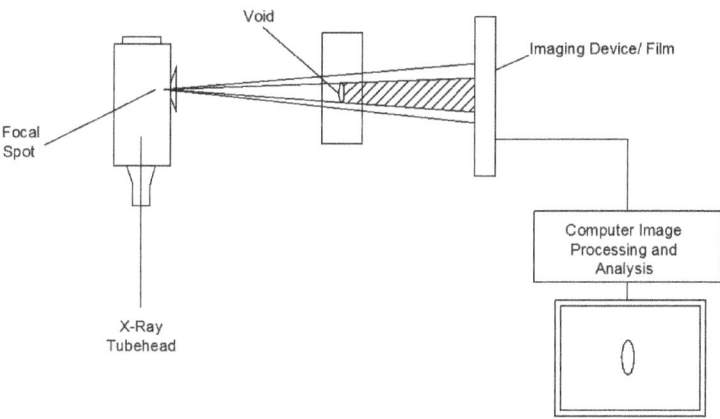

FIGURE 13.30 Arrangement of a typical X-ray examination system.

The size of the focal spot is a very important parameter, as illustrated in Fig. 13.31. Too large a focal spot creates a greater area of unsharpness (Penumbra effect), but allows greater contrast by allowing a larger photon flux density. A smaller focal spot will reduce the unsharpness, but limits the available contrast. The highest-quality image is not created solely by the selection of a specific parameter, but the balance of all. The balance of the major parameters is very important when inspecting light alloy castings, where the density contrast between the base metal and inclusions may be smaller.

There are no specific rules governing the geometric distances as this is largely controlled by the physical size of the casting and the resulting field of view available with the equipment used. Placing the casting as close to the imaging device as possible is the general rule.

The energy input is normally governed by first selecting sufficient kV to penetrate the casting and then maximize the available mA. Image quality indicators (IQIs) or penetrameters provide a means to confirm the contrast sensitivity and definition of the radiograph, as illustrated in Fig. 13.32. Proper selection and use create a reference point to ensure consistency and quality between radiographs and prevent defects from going undetected. The shapes and form of IQIs are dictated by the individual casting standards or codes. Placard or hole-type requirements are found in ASTM Standard E1025 and wire-type IQIs are covered in ASTM Standard E747.

The three commonly used radiographic methods are film radiography (FR), computed (or digital) radiography (CR), and real-time radiography (RTR). The basic principles of radiography remain the same, regardless of the method used. All describe techniques where collimated X ray or gamma ray beams are passed through a subject casting and are impinged onto film or an imaging substrate or device. As illustrated in Fig. 13.30, the beam passes through the casting, and the energy is attenuated in proportion to the material thickness and the presence of any voids, inclusions, or discontinuities within the casting. As a result, a shadowgraph of the object is created with a depiction of the internal structure defined by changes in density. The presence of any void (shrinkage porosity or gas hole) would appear as darkened areas. Conversely, high-density inclusions would appear as lightened areas.

Film radiography (FR) is the most commonly used radiographic NDT method. Procedures governing the method and interpretation for film radiography are offered by both ATSM and ISO standards [24, 25]. FR offers the best sensitivity and contrast when

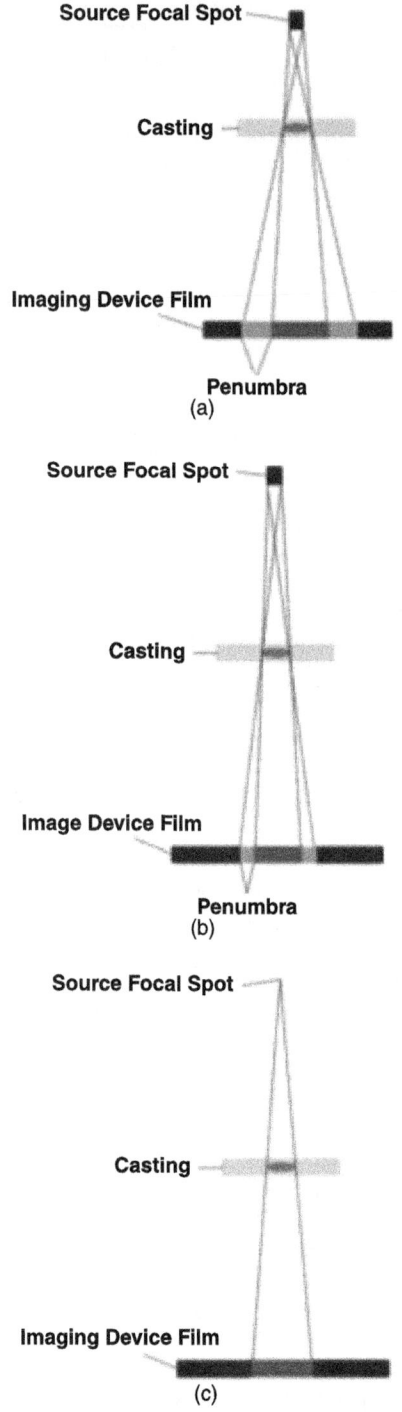

FIGURE 13.31   Illustration of the influence of the X-ray source focal spot size and resulting geometric factors on image quality unsharpness, or Penumbra. (Courtesy of Waupaca Foundry.)

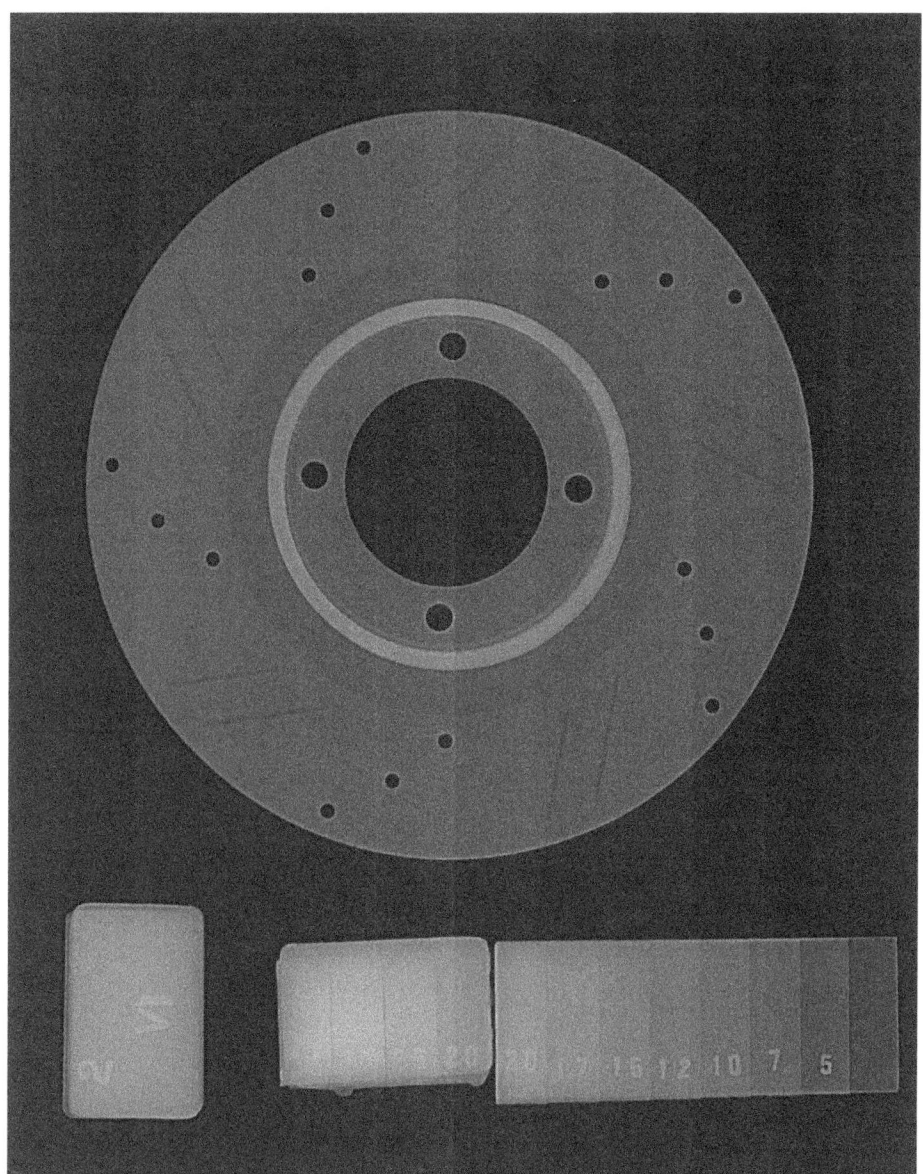

FIGURE 13.32 Computed X-ray image of a machined brake disc shown with image quality indicators. (Courtesy of InspecTech Corp.)

interpreting castings with very small discontinuities as the film substrate is highly refined. Developing costs and cumbersome handling and storage of the film are the major drawbacks of the technique. As interpretation of the X-ray results is largely a manual task, the skill of the reader becomes a factor in the selection of a testing laboratory [26].

Computed radiography (CR), also known as digital radiography, is becoming the industry choice for casting radiography. CR is a film-less radiographic method which uses a photosensitive phosphor imaging plate housed in a special cassette placed under the object to be examined. The resulting image is processed through a special laser scanner or CR reader that digitizes the image. The digital image can then be viewed and enhanced using a special imaging software with functions that can adjust contrast, brightness, magnification, and filters [27]. CR can be used with higher energies on thicker material, but sensitivity and contrast will be sacrificed. The rising cost of film, ease at which digital files can be transmitted to the customer, and the ease of record storage are largely responsible for the growing popularity of the technique. As the possibility that the digital images can be manipulated, some industries may not accept the technique for some applications.

Real-time radiography (RTR), also known as fluoroscopy or direct radiography, is used primarily in high-volume situations or in field examinations. The image is produced electronically on a fluorescent screen or flat panel detector (Fig. 13.33) as opposed to on film or cassette, so the image appears shortly after the subject has been radiated. Unlike FR and CR, the image formed is a "positive" with lighter areas indicating greater level of transmitted energy. In other words, brighter areas would represent voids, lower density, or thinner sections. This is opposite of the negative image produced in film radiography. Sensitivity and contrast are reduced with increased system speed. Higher energy input also lowers contrast and sensitivity. Due to these limitations, the technique is best used for large discontinuities (voids) or castings with few changes in section thickness. RTR is gaining greater acceptance due to the lower cost of the equipment and ease of protecting and storing digital images. When coupled to an axial rotation system, a three-dimensional X-ray image may be created [28].

Ultrasonic Velocity

Ultrasonic velocity inspection uses the speed of sound through a metal to determine cast wall thickness, detect internal flaws, or to characterize the material against a

Figure 13.33   Direct radiography detector plate used in place of film or image cassette. (Courtesy of InspecTech Corp.)

standard of known quality. A short-duration sound pulse is transmitted through the test piece and the time of flight of the sound's pulse from the transmitter to a reflective surface and back are measured and the velocity calculated. Deviations from the known velocity of an acceptable material can be used to determine the compliance to a quality standard, as in the case of nodular cast iron, or indicate the potential for internal defects [29].

Magnetic-Particle Inspection

This method of inspection is used on magnetic ferrous castings for detecting invisible surface or slightly subsurface defects. When a current or magnetizing force is passed through the metal, fields are set up as illustrated in Fig. 13.34. The process imparts a magnetic field into the casting by direct (contact) or indirect magnetization (induction coil) using AC or DC current depending on the part geometry, material, and the type of discontinuity sought. Polar effects exist at the defects, which cause magnetic particles to be aligned around the defect. The magnaflux indication is obtained when magnetic particles align themselves preferentially in the flux field. Magnaflux equipment may be of the portable type which has electrodes that can be positioned anywhere desired on the casting. Larger units are stationary and have fixed electrodes, which are clamped to the casting and provide low-voltage high-current magnetization. The magnetic particles are applied dry, by an airstream, with the portable machine, or wet on the larger machines. The presence of a discontinuity causes the magnetic flux to dissipate and the particles will be attracted to this area giving the indication of an anomaly. Accumulations of particles are normally viewed under ultraviolet light. The indication is then further evaluated to determine its type and what additional action should be taken, if any [30]. Magnetic indications are not obtained from defects alone. Certain shapes, sharp corners, fillets, and welds give indications that might look like defects. Hence, an experienced inspector is required to use this tool effectively. After magnafluxing, the parts must be demagnetized. Reference 31 is suggested for review of the principle of this inspection method.

Fluorescent or Dye Penetrant Inspection

Invisible surface defects of nonmagnetic alloys cannot be located by magnaflux inspection. However, a similar inspection tool is available. A penetrating oil, mixed with whiting powder, may be applied to the casting surface. After the casting has been wiped

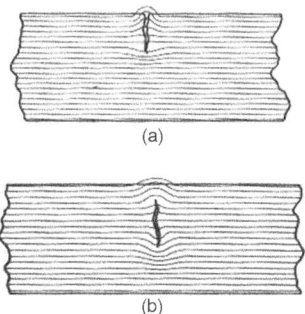

(a)

(b)

FIGURE 13.34 Magnetic flux field in a magnetized bar containing (a) surface and (b) subsurface defect. (From Thomas [31], Heine et al. [2].)

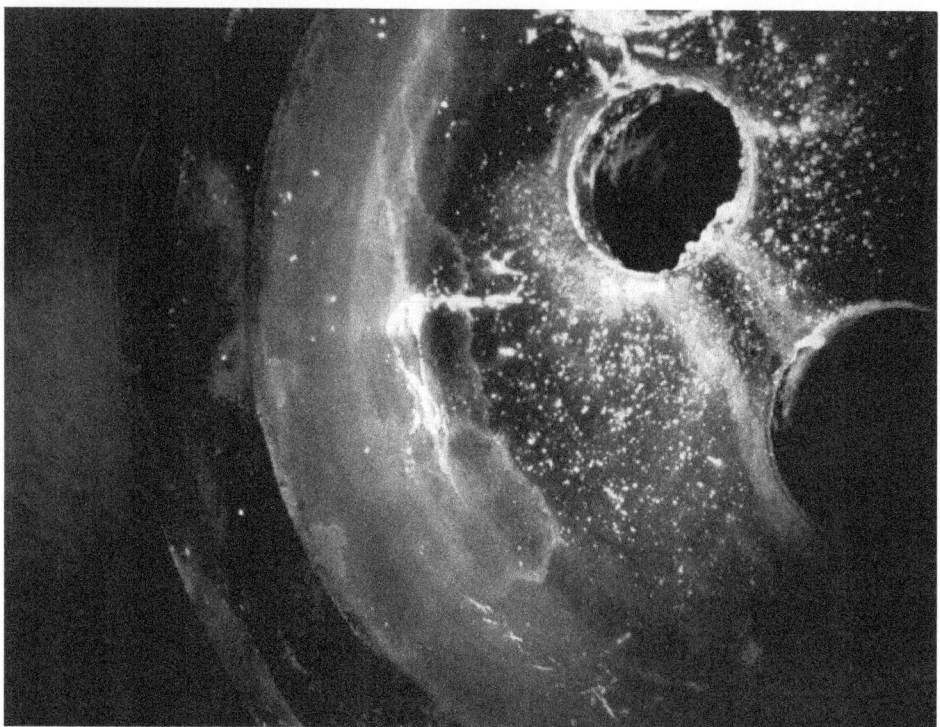

Figure 13.35 Lap defect revealed using fluorescent dye penetrant. (Courtesy of Eck Industries, Inc.) (a color version of this figure is available at mhprofessional.com/pmc3)

dry, the oil will creep out of cracks or other defects and become visible at those places. Recent developments of this technique have employed fluorescent liquids. When these are wiped or cleaned off the casting, the defects filled with the fluorescent oil may be readily observed under black light, as shown in Fig. 13.35. These tests are limited to surface defects [32].

Dye penetrant inspection is commonly used as a low-cost inspection method to detect surface rupture type defects (hairline cracks) in all types of castings. The method uses a low surface tension oil or fluid to which a fluorescent or nonfluorescent dye has been added to penetrate into a clean and dry surface by capillary action (Fig. 13.36). After a suitable time period, the excess fluid or oil is removed from the surface and a developer (often chalk-based) is applied (Fig. 13.37). The developer draws additional penetrant out from the flaw (Fig. 13.38). Inspection is performed under white or ultraviolet light depending upon the type of dye used in the penetrant [33].

Super or Ultrasonic Testing

Sound waves above the audible frequency (16,000 cycles per second) may be used to locate defects. If supersonic vibrations are initiated at one casting surface, they will be less than that of the opposite casting surface. The sound vibrations and reflection measurements are made with equipment known as the "supersonic reflectoscope." A vibrating quartz crystal applies the waves, and an oscilloscope can be used to detect the

FIGURE 13.36 Dye penetrant applied to surface. (Courtesy of InspecTech Corp.)

FIGURE 13.37 Crack developer has been applied to subject part. (Courtesy of InspecTech Corp.)

reflections from casting surfaces and defects. Of course, the problem of locating the defects is great since the entire casting must be laboriously surveyed by hand with the equipment. References 34 to 36 may be studied for further details on these inspection tools.

Figure 13.38 Crack indication revealed after penetrant seeps to surface. (Courtesy of InspecTech Corp.)

Resonance Acoustic Method

Traditional NDT techniques, ultrasonic velocity, magnetic particle or dye penetrant testing, focus on detecting and/or diagnosing specific defects or conditions. A castings natural frequency is governed by its mass (material density), stiffness (modulus), and damping as determined by its material properties and geometry. Resonance acoustic method (RAM) inspection is a nondestructive inspection method that uses changes of the natural vibrational frequency of the whole casting to detect internal and external structural flaws such as shrinkage porosity, cracks, hot tears, cold laps, and oxide inclusions [36]. While not suitable for detecting specific defects, RAM can provide reliable inspection if the primary objective is to identify potential nonconforming castings rather than specific identification of the nonconforming feature, known as the "if" versus "why" versus a definitive go/no-go argument. This technique measures a casting's resonances by striking the part with an impact device and analyzing the acoustic ringing produced. Software compares the acoustic response of a part and evaluates it against a statistical library of values generated from a control set of known good parts. Traditional NDT methods can be used for further subjective analysis on the smaller quantity of "rejected" parts. Traditionally, this technique is used for small- to medium-sized components, typically less than 0.5 m^2 in footprint [36].

Mechanical Property Testing

Castings and test bars must be tested to see that mechanical property specifications are met. Tensile, hardness, impact, fatigue, and other properties are tested in accordance with standard procedures adopted by ASTM and other specifying groups. Since these tests may be studied in the ASTM standards books and other technical work, they will not be considered in this book. Some special test bars peculiar to certain casting alloys can be considered. Special proof tests are used on certain types of castings. Pipe, for example, may be subjected to a pressure test to prove its reliability. Assemblies of cast chain links are often subjected to a maximum tensile proof load to ensure quality. Thus, mechanical tests may be used to qualify castings for their service requirements, as well as for determining the normal strength properties.

Salvage

Castings which have been rejected because of failure to meet inspection requirements are not necessarily scrapped. If the defects are not too serious, salvage is possible in many cases by welding or other treatment and refinishing. Whether salvage is permissible or economical depends greatly on factors such as the casting alloy, casting size and shape, relative cost of new casting versus repairing the defects, difficulty of salvage, availability of repair equipment and methods, quality requirements, and any agreements between the foundry and casting user relative to salvage castings. The salvaging process ordinarily consists of welding the defective areas in an effort to re-cast the defective or missing metal in subject areas. Castings that are defective because they are leakers under pressure tests may be reclaimed by sealing processes.

Welding

Castings reclaimed by welding defects are treated as follows:

- The defective areas are prepared for welding by chipping, grinding, gouging, or powder washing in the case of ferrous alloys. Nonferrous alloys require defect removal by mechanical means such as filing, grinding, or other tooling. Cracks should be completely removed before welding.

- By welding the actual repair area. The welding of castings is discussed in references listed at the end of this chapter including chapters 7, 8 and 12.

- By cleaning. The welded areas may be cleaned by the methods described earlier, and then the castings must be reinspected to pass the required standards.

References

1. S. W. Brinson and J. A. Duma, "Observations on Knock-off Risers as Applied to Steel Castings," *Transactions of AFS*, 56: 586, 1948.

2. R. W. Heine, C. R. Loper, Jr., and P. C. Rosenthal, "Principles of Metal Casting," McGraw-Hill Book Company, New York, 1967.

3. D. Hannah, "Die Casting Trim Die Procedure," *Transactions of AFS*, 60: 784, 1961.

4. G. H. Shippard, "Band Sawing in Foundries," *Transactions of AFS*, Vol. 58, p. 621, 1950.

5. H. J. Chamberland, "Band Sawing Nonferrous Castings," *Foundry*, Vol. 60, September 1952.

6. R. S. Babcock, Oxygen Cutting Processes in Steel Foundries, *Transactions of AFS*, 58, 346–357, 1950.

7. American Welding Society, *Welding Handbook*, Vol. 1, 1962.

8. "Cleaning Castings," *AFS*, 1992, Des Plaines, IL.

9. A. G. Ringer, "Use of Portable Air Tools in Foundry Cleaning Rooms," *Transactions of AFS*, 58: 510, 1950.

10. American Wheelabrator and Equipment Corp., *Wheelabrator Operators Manual*, 1965.

11. American Foundrymen's Society, "Malleable Straightening Dies," *Transactions of AFS*, 68: 801, 1960.

12. D. T. Martin, "Outline of Inspection for Pearlitic Malleable Castings," *Transactions of AFS*, 58: 692, 1950.

13. K. M. Smith, "Dimensional Checking and Pressure Testing of Gray Iron Castings," *Transactions of AFS*, 59: 304, 1951.

14. D. E. Sawtelle, "Mechanized Malleable Foundry Finishing and Inspection," *Transactions of AFS*, 55: 388, 1947.

15. L. N. Schuman, "Equipment and Methods of Straightening and Dimensional Inspection of Malleable Iron Castings," *Transactions of AFS*, 59: 418, 1951.

16. B. H. Work, "Foundry Cleaning Room Abrasive Operation," *Transactions of AFS*, 58: 685, 1950.

17. H. W. Wagner, "Snagging Operations," *Foundry*, Vol. 81, p. 112, January, 1953.

18. R. H. Herrman, "Salt Bath Cleaning of Gray Iron Castings," *Foundry*, Vol. 78, August, 1950.

19. P. Gamage and S. Q. Xie, "A Real Time Vision System for Defect Inspection in Cast Extrusion Manufacturing Process," *The International Journal of Advanced Manufacturing Technology*, 40(1–2): 144, 2009.

20. Capture3D, "Optical Measuring Techniques for Industrial Casting and Forging Processes," 2013.

21. www.ndt-ed.org., "Introduction to Radiographic Testing," 2013.

22. W. Harara, "Digital Radiography of Aluminum Castings by Fluoroscopy," *Russian Journal of Nondestructive Testing*, 48(6): 384, 2012.

23. F. Brant, http://www.ndt.net/article/v05n05/brant/brant.htm. "The Use of X-ray Inspection Techniques to Improve Quality and Reduce Costs," NTD net5(05): May 2000.

24. ASTM E1742-12, Standard Practice for Radiographic Examination.

25. ISO 4993:2009, Steel and Iron Castings-Radiographic Inspection.

26. K. Carlson, S. Ou, R. Hardin, and C. Beckermann, "Analysis of ASTM X-ray Shrinkage Rating for Steel Castings," *Proceedings of the 54th SFSA Technical and Operating Conference*, Paper No. 1.6, Steel Founders' Society of America, Chicago, IL, 2000.

27. GE Inspection Technologies, "Radiography (X-ray)-Non-Destructive Testing," 2013.

28. H. Boerner and H. Strecker, "Automated X-ray Inspection of Aluminum Casting," *IEEE Transactions of Pattern Analysis and Machine Intelligence*, 10(1): 79, 1988.

29. ASTM E494-10, Standard Practice for Measuring Ultrasonic Velocity in Materials.

30. ISO 4986:2010, Steel Castings-Magnetic Particle Inspection.

31. W. E. Thomas, "Casting Industry Application of Magnetic Particle Inspection," *Transactions of AFS*, 55: 482, 1947.

32. ASTM E16512, Standard Practice for Liquid Penetrant Examination for General Industry.

33. A. Lindgren, "Fluorescent Magnetic Particle Method for Casting Inspection," *Transactions of AFS*, 67: 635, 1959.

34. G. L. Kehl, *Principles of Metallographic Laboratory Practice*, 3d ed., McGraw-Hill Book Company, New York, 1949.

35. B. V. Kovacs and G. S. Cole, "On the Interaction of Acoustic Waves with SG Iron Castings," *Transactions of AFS*, 83: 497, 1975.

36. R. W. Bono, "Resonant Acoustic Nondestructive Testing," *Advanced Material and Processes*, 164(9): 35, Sep. 2006.

CHAPTER 14

Casting Defects

Laurence V. Whiting
Formerly with CANMET Materials Technology Laboratory
Natural Resources Canada

Mahi Sahoo
Suraja Consulting Inc.

Introduction

Production of castings involves many processes and many variables within each process. When either one or more variables fall outside control limits, castings may not meet the specifications in their entirety. Even when the individual variables are within the range, some combination of variables may result in rejection of castings. Defective castings occur in the normal day-to-day operation of any foundry. These defects result in scrap or extensive rework, which unless steps are made to reduce their frequency, will seriously impact the foundry's bottom line and competitive position.

Defect analysis and control should have the following goals:

- Identify defects which result in rejection of castings or unnecessary rework.
- Eliminate the causes of the defects.
- Improve quality continuously and reduce scrap and production costs.

The term "defect" describes not only a nonconformity in a casting that results in its rejection, either by the foundry's quality inspector or the customer, but any nonconformity that increases the processing cost of the casting, such as surface roughness, hardness, or additional padding. Identifying defects is more important and more difficult than it might initially appear. One foundry industry study resulting from the mobile foundry visits to over 76 Canadian foundries indicated that almost 80% of the casting defects were caused by sand molding [1]. The areas found to affect casting quality were sand control, metal composition, metal temperature control, and gating design.

Statistical methods are important in the identification of conditions that lead to casting defects. While it has been often shown that foundries have the appropriate average sand properties or metal conditions, it is frequently observed that the scatter is unacceptable. If, for example, a foundry's greensand has a moisture content of 3.5% plus or minus 1%, see Fig. 14.1, this implies that moisture is between 2.5 and 4.5%, 95% of the time. As the compactibility of a given sand changes by 3 to 4 points for every 0.1%

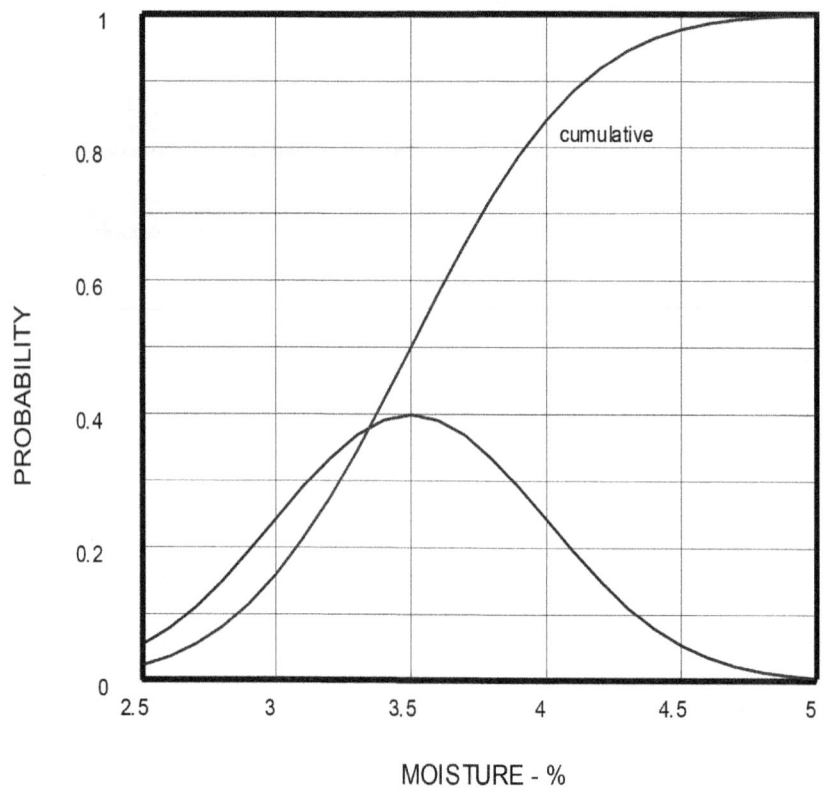

FIGURE 14.1 Understanding the relationship of probability and moisture content can help reduce casting defects. (From *Casting Copper-Base Alloys* [2].)

change in moisture, one could expect the compactibility to vary by plus or minus 25 to 30 points, and green strengths to vary from 7 to 21 kPa (10 to 30 psi).

Suppose a defect, such as blowholes, can occur at moisture contents above 4%. This doesn't mean that blowholes occur with 100% probability above 4% moisture and 0% probability at moisture levels below 4%, but the probability rises quickly as the moisture levels approach and exceed 4%. In this example, the moisture exceeds 4%, 16% of the time. If the compactibility were reduced by 6 to 8 points, and the moisture by 0.2% without changing the scatter, then this is equivalent to shifting the distribution centered at 3.5 to one centered around 3.3% moisture. Moisture levels would still exceed 4%, 10% of the time. If testing was conducted by making molds from only a few batches of sand, the defect might still occur and one could infer that decreasing the moisture is not the solution to the problem. Elimination of causes requires that during normal production both defects and processing variables have been measured correctly and defined statistically. When both defects and causes are known, cause and effect relationships can be established and critical causes eliminated.

Figure 14.2 shows the effect compactibility can have on many castings prone to certain defects. Defects induced by high compactibility and high moisture are mold wall movement, shrinkage, blows, pinholes, surface roughness, gas tracks, scabbing,

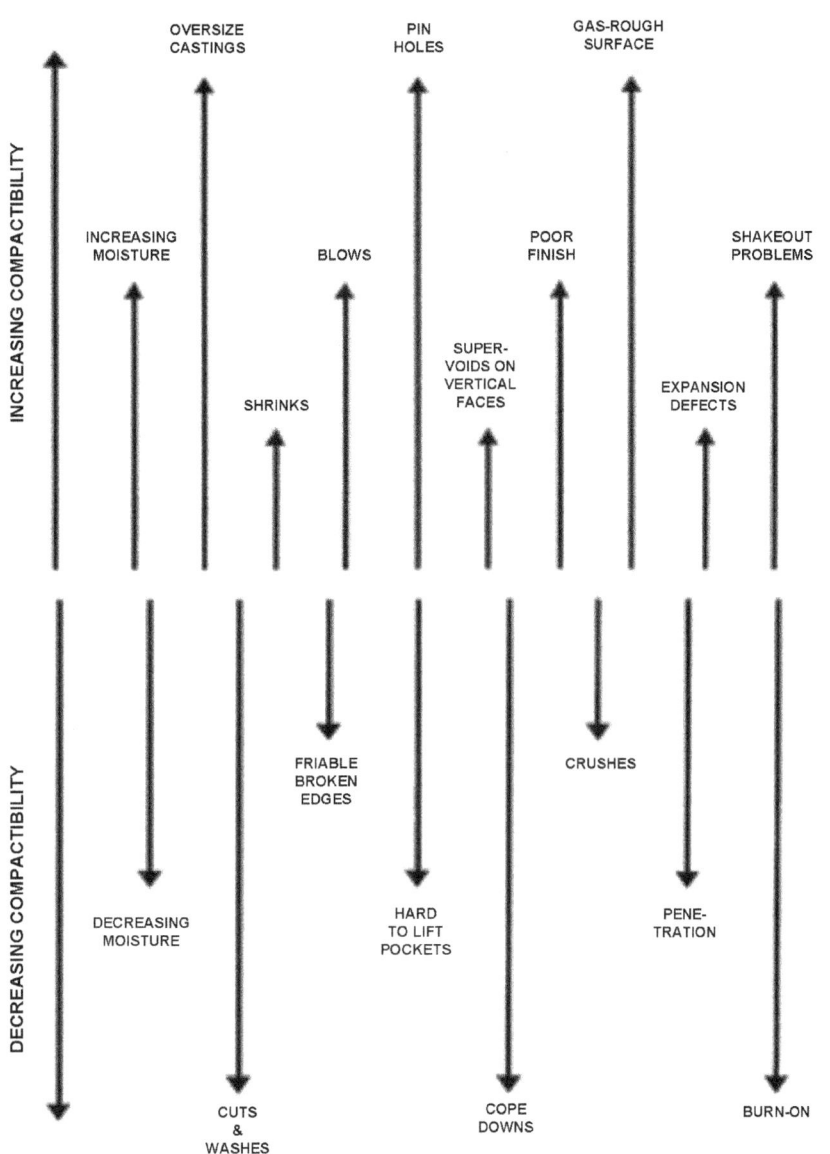

FIGURE 14.2 Effect of moisture content on compactibility and the type of resultant defects. (From *Casting Copper-Base Alloys* [2].)

rat tails, and shakeout problems. Defects caused by sand with low compactibility and low moisture are cuts and washes, broken edges, dropped copes, crushes, penetration, burn-on, and difficult draws. Hot sand further complicates the issue. Thus moving the average sand moisture will move the range of compactibilities and change the frequency of the various defects and their relative costs. The solution should not be to change the average compactibility, but to tighten the process and thus the range, hopefully

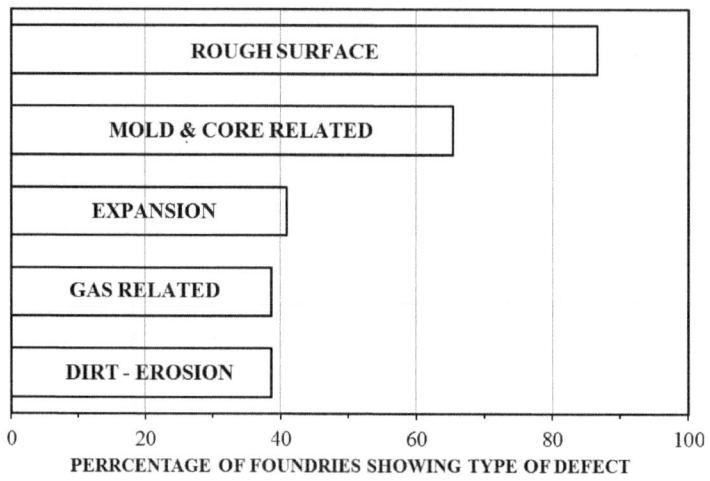

FIGURE 14.3   Histogram depicting the more common casting defects (From Buhr et al. [1].)

reducing the occurrence of all defects. Management must know the control limits on various properties of the system sand, and have some idea of the process specification limits. The range is as important as the average.

Tools such as Pareto analysis help deal with defects that occur concurrently and help select the most common or most costly defect. Figure 14.2 shows how compactibility and moisture can alter the frequency of many defects in castings that are prone to some, but perhaps not all, defects shown here. In the past, foundries assumed that working on one defect at a time would solve the problem. They should work toward minimizing all the major defects and not just blindly exchange one defect for another. Pareto analysis ensures you work on the important 80% and ignore the trivial many. Figure 14.3 shows a histogram of important defects identified by CANMET's first Mobile Foundry Laboratory Project [1, 2]. In this case, the trivial many are not shown.

Defect Identification

Before any casting defects can be solved one should know what defect is being solved. To determine the particular defect type there are many steps that can be taken to ensure it is not misidentified. There are simple and more sophisticated tools available for identification.

- Visual examination
- Comparison with known defect standard castings
- Stereo microscope
- Scanning electron microscope (SEM)
- Defect handbooks or manuals (in-house, and other published information as in Refs. 3 to 8)

It is important to gather all the relevant facts, before any decision is made as to the identity of the defect. It is also important to preserve the evidence within the defect, as shown below [9].

- Castings should be examined before shot blasting or after light sand blasting. This may preserve pockets with material inside that will help to identify the defect and possibly its cause.
- Collecting the castings with the gating system still attached will yield clues where and how the defect may have originated. It will also help to examine if the risers fed the shrinkage.
- The shakeout area is a good place to gather evidence.

There are some defects that are not easy to identify correctly until more information is known. For example, until one is certain of the defect, it is advisable to keep the defect generic such as inclusions rather than specific such as dross, slag or porosity caused by shrink or gas. When identifying the defect, the following are helpful:

- Where and how a particular type of defect occurs and knowledge of the production process and the conditions at the time.
- The surface condition of the defect—Is the surface smooth, rough, or irregular?
- What is the material inside the defect—molding sand, core sand, dross, slag, or other materials used in the foundry?
- Examine the gating system if the flow is impinging in certain areas.
- Is the defect on the surface or in the interior of the casting?
- Is it in the cope or drag or vertical walls?

Information from the examination of many defective castings or by answering the above questions will help to narrow down the type and nature of defect. Defects do not necessarily originate during mold filling. They may have occurred prior to mold filling or after mold filling, as shown below [9].

Prior to Mold Filling

1. Sand inclusions: sand that is present in the mold cavity as a result of core setting and not removed prior to pouring. These are usually at the bottom section of the cavity.
2. Shift: due to misalignment of cope to drag or from side to side.
3. Crush: improper fit between mold halves and/or molds and cores.
4. Dimensional: wrong dimensions, missing details, cores, etc.
5. Core setting: wrong core, misplaced core, no core, wrong chaplets, etc.

During Mold Filling

- Misrun
- Short pour
- Laps (interrupted pour)

- Runout
- Pinholes
- Dross, slag
- Sand inclusions
- Core blow
- Erosion scabs
- Expansion scabs
- Leakers (inclusions, laps, cracks, holes, oxide films, chaplets not fused)
- Explosive penetration
- Metal: out of specification, improper metallurgical condition, not properly degassed, etc.

After Mold Filling (in the mold)

- Shrinkage, suck-in, porosity
- Incorrect structure
- Metal penetration, rough surface, burn-on
- Swell
- Hot tears
- Warped castings
- Broken, cracked, and other damage occurring from rough handling
- Hammer marks, dents
- Overblasted castings
- Excess or insufficient grinding

Tools for Defect Analysis

Many analytical tools have been developed to analyze production variables, identify critical casting problems and potential solutions, and ultimately control costs, improve quality, and lead to continuous improvement. Among those that foundrymen find useful are

- Histograms
- Scatter plots
- Scrap analysis charts
- Pareto charts
- Statistical process control (SPC)/control charts
- Failure mode and effect analysis (FMEA)
- Design of experiments (DOE)
- Cause and effect diagrams (Ishikawa diagram or fishbone diagram)

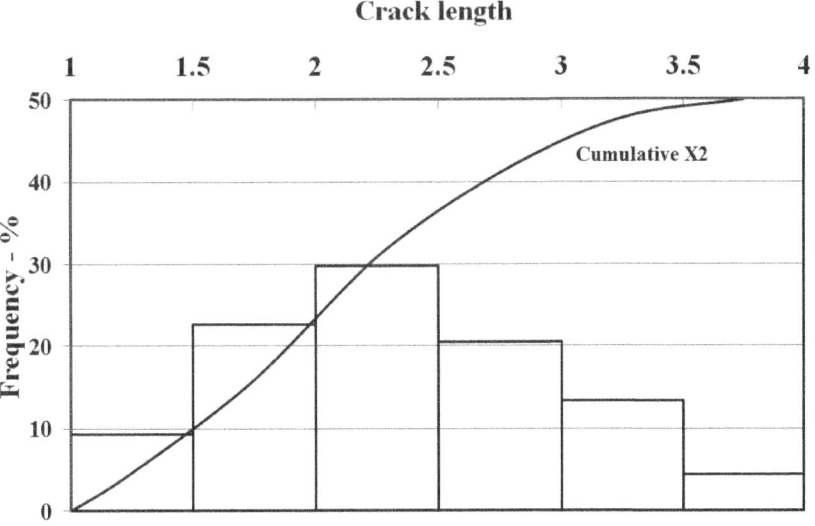

FIGURE **14.4** Example of a histogram.

Histograms

Histograms display a single measurement, for example, crack size, mold temperature, or sand screen size as a bar graph of the distribution. Rectangles are erected on the class intervals that are marked off on one of the axes, with the other axis representing the frequency. The size of the rectangles is proportional to the class frequencies. One could also show the cumulative frequency on a second axis parallel to the frequency axis. If the class size is decreased and number of observations increased, the histogram smooths and eventually becomes a frequency distribution. Figure 14.4 is a histogram showing how many cracks of different sizes occurred in a particular production casting. Histograms differ from Pareto charts, which may deal with the characteristics of a defective product, e.g., types of defects. Histograms can also show the variation in processes in which repetitive events may have different outcomes that vary over time. They may also indicate whether the defect is normally distributed, skewed, or divided. Skewed distributions can result in all individuals falling within the limits, but the population failing to perform because the mean is too low.

Scatter Diagrams

Scatter diagrams are used to determine whether a relationship exists between variables, i.e., they test possible cause and effect relationships. Although they cannot prove that one variable causes the other, but may indicate whether a relationship exists and the strength of that relationship. Figure 14.5 is an example of a scatter graph showing the temperature of a permanent mold just prior to closing and before pouring another casting. Although the points are scattered, one gets the feeling that there is a relationship between mold temperature and time from the start of the experiment. Regression analysis could be used to calculate a relationship with an intercept and slope. A test can be used to determine whether a relationship exists (count the number of points in the opposite quadrants divided by the two sample means, and check that the count is less than the maximum allowed for the sample size).

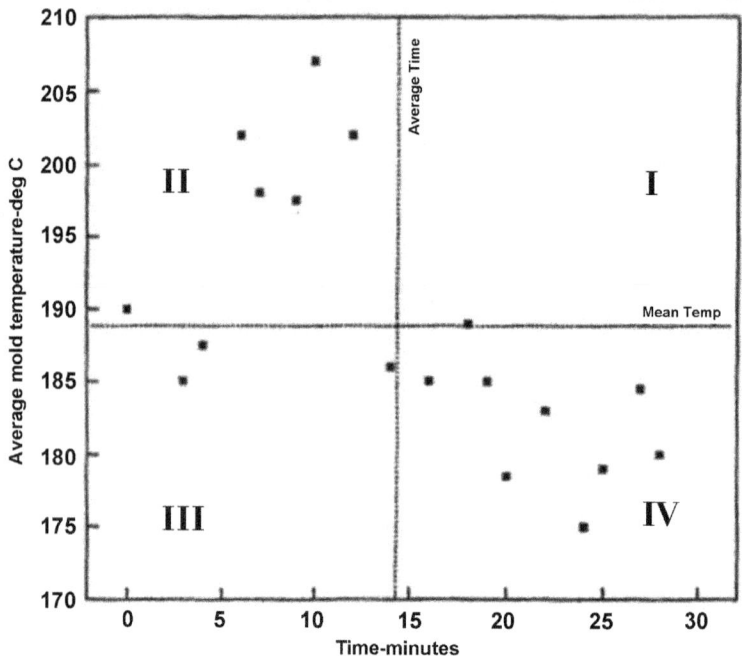

FIGURE 14.5 Example of a scatter diagram for mold temperature.

Scrap Analysis Charts

Scrap analysis charts collect information on frequency, occurrence of defects with time, equipment, operator, mold cavity, part number, etc. These charts take many forms depending on the function, and a typical foundry will use several types to track scrap problems. The most common is a scrap analysis by part number, defect type, and time. The usual time scale is a single day's production, although automotive foundries might use shorter times. These charts show the scrap rate, a ranking of scrap, and an identification of trends in scrap production by part and type of defect. Scrap analysis charts for individual casting machines, operators, or mold location, encourage the collection of data for each source defect type, operator, and other operational variables, and note changes in these variables. Figure 14.6 is an example of a daily scrap analysis chart for one day, showing the number and types of defect produced for each job.

Pareto Charts

Pareto charts are a form of histogram or vertical bar graph that separate the vital few from the trivial many and determine the priority of the problems to be tracked. Pareto charts focus attention and efforts on the truly important problems. The impact of the factors (number of defects, cost of defects, etc.) are plotted to scale, as vertical bars in decreasing order of impact or priority. A cumulative line can be plotted on the same chart to emphasize that most of the cost or defects are caused by only a few defects (the "vital few"). Generally, only a few factors are worth correcting, as the cost of correcting

Attribute Control Chart for Castings with Many Characteristics

Part No.	TOTAL SCRAP	Process	GREENSAND	Dept.	INSPECTION
Metal	CAST IRON GRADE 30	Plant	BOOTH STREET	Date	JULY 1990
Frequency	JULY PRODUCTION	Line	MMC	Inspector	WHITING

Rank	Defect	Scrap Frequency																			#	%
1	SAND	5	2	5	0	6	1	1	6	1	3	6	2		2	2	1				45	23.0
2	SLAG	2	5	3	4	3	4	3	1		3	6	4		1	1		1			41	20.5
3	SHRINKAGE	1		1			2	2	4		2		4	1							17	8.5
5	MISRUN		1			2	2	6	2			2									15	7.5
7	DIMENSIONS	1			4				1							1	2				9	4.5
9	HARD SPOTS	1		1	2									1		3					8	4.0
7	BLOW	2	2		4		1														9	4.5
9	SCABS			1			3			1		1		1		1					8	4.0
4	BROKEN CORE	1				1		3	2			6		1	2						16	8.0
13	RAN OUT			1		1	1									2					5	2.5
11	POURED SHORT			1	1					4											6	3.0
11	INCORRECT LIFT	2	1		2	1															6	3.0
6	MISC	1	1			3		2	1		1	2	1	1		1					14	7.0

| Total 200 8948 | Scrap No. | 13 | 16 | 9 | 21 | 8 | 21 | 16 | 8 | 12 | 9 | 16 | 10 | 12 | 3 | 7 | 6 | 6 | 6 | | | |
| | Castings prod. | 370 | 636 | 178 | 1073 | 153 | 811 | 941 | 481 | 255 | 1000 | 400 | 588 | 608 | 185 | 397 | 250 | 462 | 160 | | | Significant Defect |

$UCL = \bar{p} + 3\sqrt{\bar{p}(100 - \bar{p})/\bar{n}}$

4.22

$LCL = \bar{p} - 3\sqrt{\bar{p}(100 - \bar{p})/\bar{n}}$

0.25

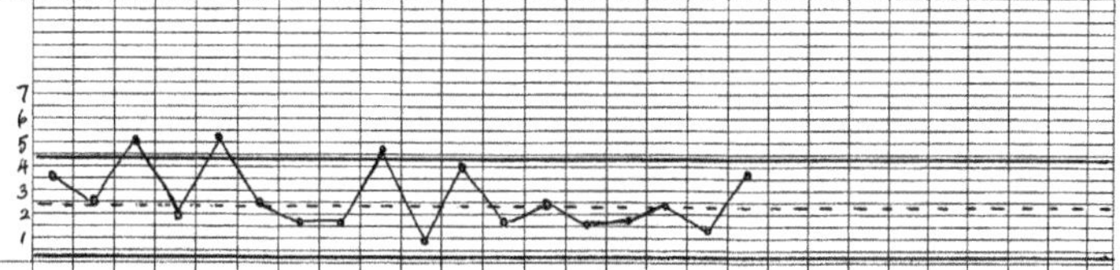

$\bar{p}$ 2.23 $\bar{n}$ 497

(plotted values): 3.51 2.52 5.06 1.96 52.3 2.47 1.70 1.66 4.71 0.9 4.0 1.70 2.30 1.62 1.76 2.40 1.30 3.75

Significant Defect: SAND and SLAG

FIGURE 14.6 Scrap analysis chart with p-chart.

trivial problems (trivial many) may not only be prohibitive but counterproductive. Figure 14.7 shows the information in Fig. 14.6 displayed as a Pareto chart.

The steps in Pareto analysis which will prioritize the actions are:

1. List factors of interest and decide on how they should be classified: by defect type, cause, casting type, process (machine, man, or location in the plant), by time (day, shift, etc.).

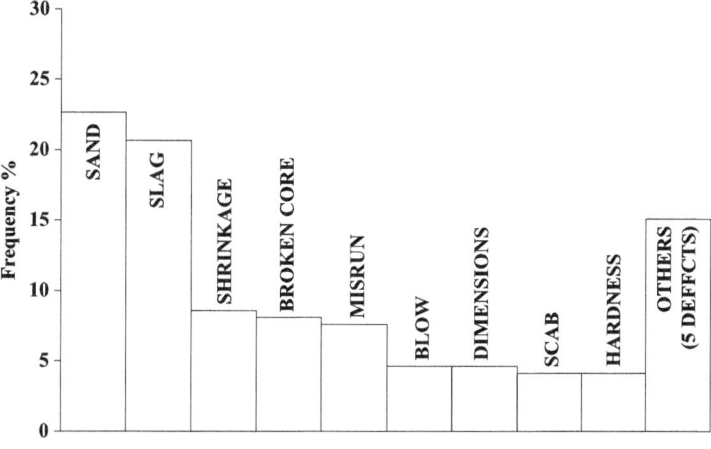

FIGURE 14.7 Pareto analysis data from Fig. 14.6.

2. The impact of each factor (defect, cost, time for correction) should be selected to optimize the impact of each Pareto chart.

3. The various factors (number of defects, cost, etc.) are plotted in decreasing order of impact.

4. A cumulative line combining all factors will emphasize the impact of the critical few.

5. A critical level needs to be determined above which the foundry will commit resources.

Control Charts

Control charts show the variation of a response variable in a sequence (in time) of sample statistics. They consist of a central line (CL) denoting the sample mean and two statistically determined upper (upper control limit, UCL) and lower (lower control limit, LCL) lines drawn on either side of the process average [10]. Generally, the upper and lower limits are three standard deviations from the sample mean. Only three samples per thousand should fall outside these limits. Although there is still a small chance of this happening, any deviation should be cause for action. Some people even use two sigma lines as warnings. Foundries get into problems when the process control limits conflict with the customer specifications. However, the problems of statistical process control, sample size, frequency of testing, etc., are also important. Figure 14.8 shows a casting weight control chart for a piston.

Statistical Process Control

Either one variable outside a control range or combination of variables farther away from the target values but still within the specification range may cause a defect. It is necessary

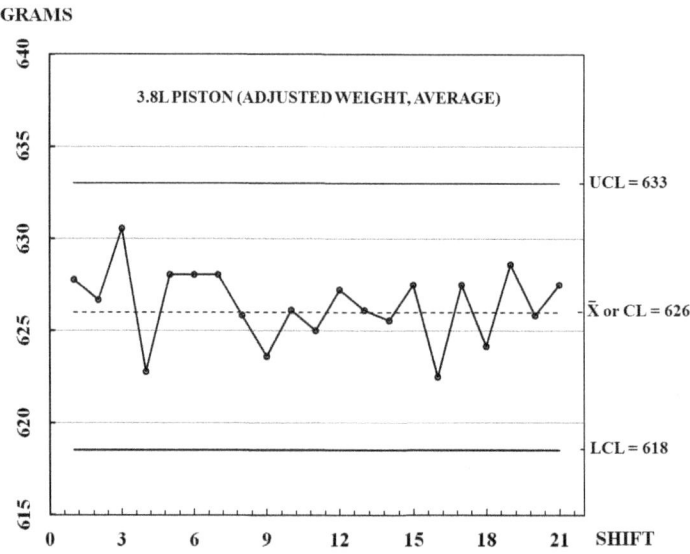

FIGURE 14.8 Control chart for weight of a permanent mold-cast piston. (From Enright [10].)

to determine the main reason for the defect in order to solve the problem. There are several different methods used to find out the process variable responsible for the defect. To this end, statistical process control (SPC) should be considered. It uses appropriate and reliable measurement of casting process variables and resultant casting properties (defects, physical properties, and downstream problem costs). The defect should be capable of being defined as a deviation from a measureable target or range. Customers' concerns may also need to be addressed. Measurements must be reliable and repeatable. In addition, critical production variables must be measured not only reliably but appropriately before any meaningful correlation between cause and effect can be proven. It requires that the variable be measured at the appropriate time and location so that it truly represents the process. Take pouring temperature as an example. It should be measured at the pouring station. If only one casting is poured, no problem, but if 15 are poured, it is not uncommon for temperature of cast iron to fall 150°F degrees (80°C) from the first to the last. Radiation losses in the ladle depend on the exposed metal surface as the ladle is tilted and is proportional to AT^4, where A is the area and T is the temperature. It is not unusual for first castings to suffer from an interrupted pour (overly full ladle) and the last to misrun. It is therefore critical that measurements be accurate, reliable, and meaningful. For this reason, individuals doing the measurements need to be committed to the process, aware of process flow charts, sampling procedures, and standard operating practices. Errors occur because (1) devices are not calibrated and checked on regular schedules, (2) operator bias and error, (3) incorrect calculations, (4) incorrect time or location of measurement, or (5) entry of false data.

Statistical methods were developed in the 1960s and 1970s. One of the early pioneers was W. Edwards Deming in the United States. However, his reputation was made when Japanese industry adopted him and his methods, and by the 1980s his name became synonymous with quality factory production on both sides of the Pacific Ocean. Control charts (X-bar-R charts) can be used for any variable that is normally distributed (i.e., follows a "bell curve," Fig. 14.9). If true, then control limits, upper UCL and lower LCL, can be calculated using statistical tables for the particular sample size. Figure 14.10 shows examples of X-bar-R charts for processes out of control. Many foundry variables (e.g., greensand properties, metal temperature, composition) should not be measured in "batches." A moving range chart should be used for these.

Other statistical charts are employed when properties cannot be measured, and are judged by the presence or absence of defects, number, or some attribute that is ranked

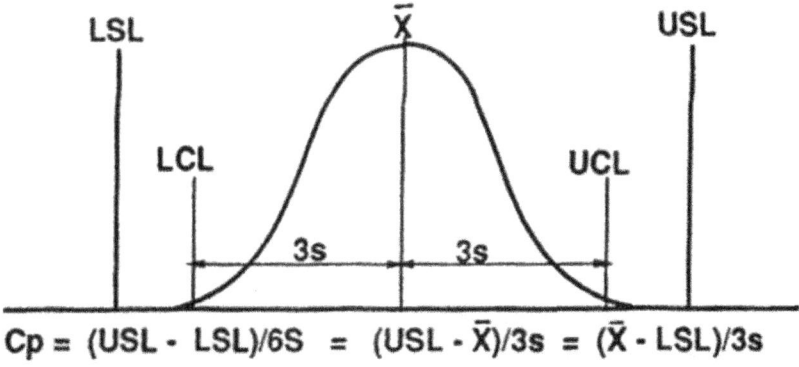

$$Cp = (USL - LSL)/6S = (USL - \bar{X})/3s = (\bar{X} - LSL)/3s$$

FIGURE 14.9 Bell curve.

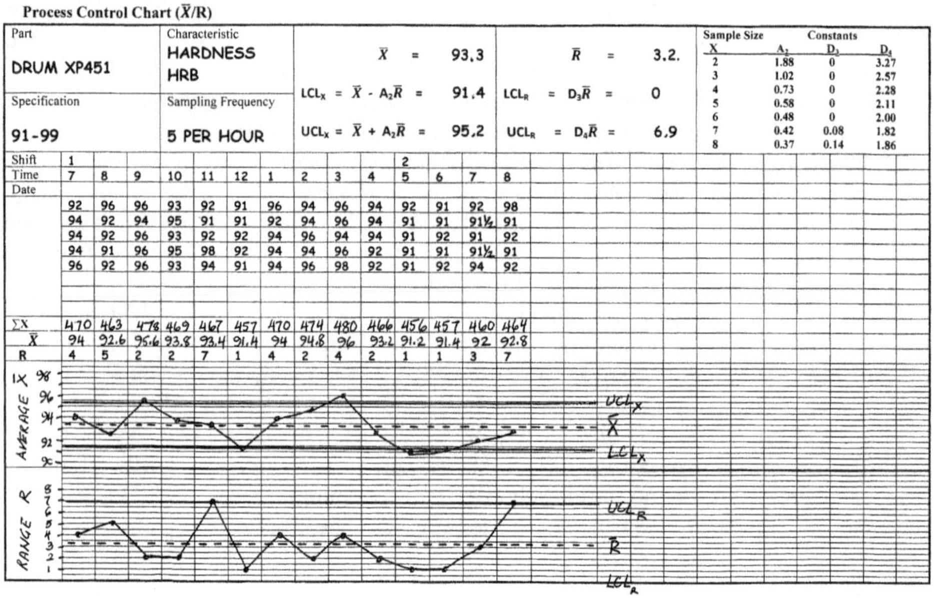

Process Control Chart ($\bar{X}$/R)

Part	Characteristic													Sample Size	Constants		
DRUM XP451	HARDNESS HRB	$\bar{X}$ = 93.3				$\bar{R}$ = 3.2.								X	A₂	D₃	D₄
														2	1.88	0	3.27
Specification	Sampling Frequency	$LCL_x = \bar{X} - A_2\bar{R}$ = 91.4				$LCL_R = D_3\bar{R}$ = 0								3	1.02	0	2.57
														4	0.73	0	2.28
91–99	5 PER HOUR	$UCL_x = \bar{X} + A_2\bar{R}$ = 95.2				$UCL_R = D_4\bar{R}$ = 6.9								5	0.58	0	2.11
														6	0.48	0	2.00
														7	0.42	0.08	1.82
														8	0.37	0.14	1.86

Shift	1							2						
Time	7	8	9	10	11	12	1	2	3	4	5	6	7	8
Date														
	92	96	96	93	92	91	96	94	96	94	92	91	92	98
	94	92	94	95	91	91	92	94	96	94	91	91	91½	91
	94	92	96	93	92	92	94	96	94	94	91	92	91	92
	94	91	96	95	98	92	94	94	96	92	91	91	91½	91
	96	92	96	93	94	91	94	96	98	92	91	92	94	92
ΣX	470	463	478	469	467	457	470	474	480	466	456	457	460	464
$\bar{X}$	94	92.6	95.6	93.8	93.4	91.4	94	94.8	96	93.2	91.2	91.4	92	92.8
R	4	5	2	2	7	1	4	2	4	2	1	1	3	7

Figure 14.10 Example of $\bar{X}$-R chart.

on a scale. Such attributes must be evaluated using attribute control charts, such as the p-chart shown in Fig. 14.6 for a gray iron foundry. In this example, multiple attributes (defects) are described by the fraction nonconforming or total percentage of scrapped castings. Figure 14.11 shows a decision tree which indicates which chart to use.

Process control specifications are not the same as upper and lower control limits, which describe the actual behavior of the variable. Process control specifications are the

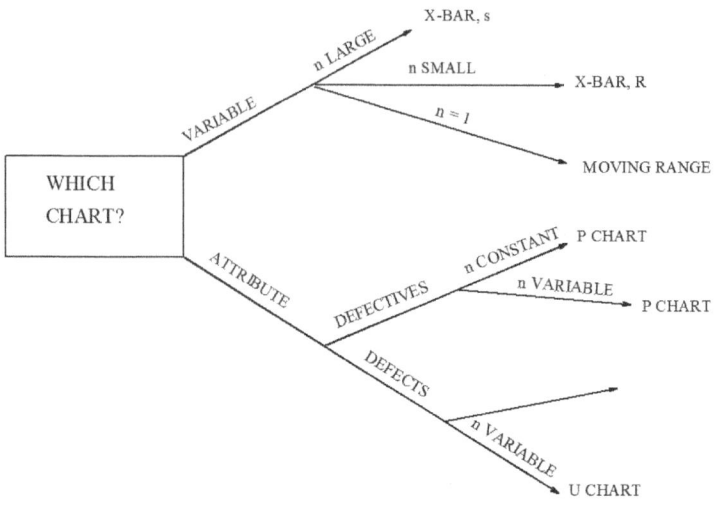

Figure 14.11 Decision tree to decide the chart to be used.

desired limits for the variable either defined by the customer, or determined by experimental design as an optimum, etc. Process capability compares the control limits with the specification limits. If the control limits are well inside the specified limits, then the process is unlikely to produce defective castings. The statistical capability of a process can be calculated from statistical data. Consider the case of an automobile company making piston cylinders at one plant, opens a second plant. Six control charts had previously monitored and controlled production, yet another showed parts had variable weight—not good for an engine. The aluminum alloy, although in specification, contained light silicon and heavy copper ingredients that affected the alloy's density. The mold cavity dictated a constant volume. The density by the chemistry of each melt had to be controlled [11].

SPC reduces the quality and productivity costs due to assignable causes. However, as Taguchi pointed out, SPC may not be enough, as process capability might have to be improved, or more costly procedures adopted. These might be avoided by making the design less sensitive to process variability. Taguchi wanted robust designs that withstood real-world wear and tear. Demming was only concerned that population not exceed the limit and the only rejects were those that fell outside, whereas Taguchi was concerned with costs and assumed that cost built up exponentially about the mean, and outside the limits customers were lost.

Failure Mode and Effects Analysis

Failure mode and effects analysis (FMEA) is a method of improving processes and reducing defects by prevention rather than detection. FMEA offers two types of analysis, one for design and the other for process. Design FMEA is concerned with how the design of a casting (shape, tolerances, materials, machining, assembly, etc.) can potentially lead to failure. Process FMEA considers the effect on defects of the entire foundry process (materials, machines, procedures, operators, process measurements, and controls).

FMEA identifies and prioritizes actions required to prevent defects by a methodical examination of each part of the process and how it can lead to failure, the mode of the failure, and the effect of failure on the product. It then considers corrective actions, probabilities of occurrence and detection, severity, priority for taking action, and verification of that action. FMEA uses a team approach to problem solving and requires its members to be familiar with defect analysis tools and use flow charts and an SPC program to characterize and control processes.

FMEA uses forms to guide the team to:

- Identify the failure (defect) and all possible modes of failure.
- Identify the causes (why) the part failed and how it failed.
- Determine priorities for future action by calculating a risk priority number (RPN) from assigned points for various levels of probability of occurrence, probability of detection, and severity of the effects of failure. The RPN and Pareto analysis both indicate priorities for action.
- Prepare a list of corrective actions that minimize failure and subsequent effects.
- Implement a plan to determine whether corrective actions are effective under production conditions.
- Modify standard operating practices to include the successful corrective actions and inform clients and management of the results.

Design of Experiments

Design of experiments (DOE) is used when changing one variable at a time while holding others constant and then another to see what corrects the problem is either usually ineffective or problematic. Statistical design of experiments was developed by R.A. Fisher in England during the 1920s to determine the best way to grow crops in the real world rather than in a controlled environment (farm vs. greenhouse). He understood that most systems are complex and interactions can be important. These interactions will not be observed, unless more than one variable is changed simultaneously. Fisher introduced the concept of factorial design to vary all the factors simultaneously. If three variables are considered relevant then it would require 2^3 or 8 experiments, if eight variables then 2^8 or 256 experiments in a full factorial design. He and his collaborators soon realized that fractional factorials could impart enough information to run confirming experiments with only the critical variables. The confirming experiments should not be highly fractionated. In 1945, L.H.C. Tippet used a factorial design to solve a problem in textile manufacturing. He employed just 25 runs to examine 5^5 or 3125 experiments required by a full factorial. Because not all experiments can be run simultaneously, the method usually randomizes the order of the fraction of experiments to be run to eliminate the many environmental variables not considered controllable (i.e., the humidity). While early experimenters analyzed the results manually, computer programs are now available to not only select the fraction to be examined, but later to analyze the relationships. Use of computer software is the only way to analyze DOE data.

Ishikawa (Fishbone or Cause and Effect) Diagrams

The foundry process can be regarded as an organized series of causes that produce a desired defect-free casting. The prevention of defects involves several steps, identifying the defect, identifying the sources of variation that cause them, and eliminating the controllable cause of variation. One effective way of identifying potential causes of defects is the use of cause and effect diagrams. These are also known by several names: Ishikawa diagram, fishbone diagram, or brainstorming diagram. This diagnostic tool provides a geometric correlation between an effect and all possible causes. As such the diagram helps aid a brainstorming session enumerate and consider most of the probable causes of a specific defect. While a cause and effect diagram is effective in correlating effects with potential causes, it does not prove cause. Proof of cause, or multiple causes, requires further investigation of the process or even designed experiments conducted on- or offline.

A cause and effect diagram has the following structure, as shown in Fig. 14.12:

- The name of the effect (defect) is put in a box on the right-hand side of a sheet of paper.
- An horizontal line is drawn across the paper with an arrow ending at the box.
- The names of the main family of causes: methods, equipment, materials, measurements, data and information systems, environment, money, management, and people, are placed in boxes and joined by slanting arrows to form a fishbone geometrical form.
- Branches and subbranches are placed on each main branch to identify all possible causes and subcauses that belong to each family.

The cause and effect diagrams for some defects common to both ferrous and nonferrous alloys are shown in Figs. 14.13 to 14.18 [2].

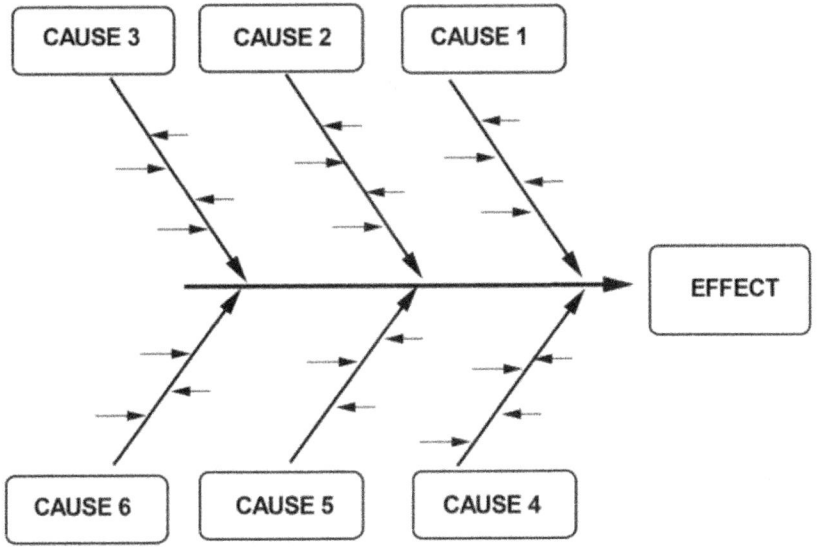

FIGURE 14.12   Structure of a cause and effect diagram showing major and minor causes.

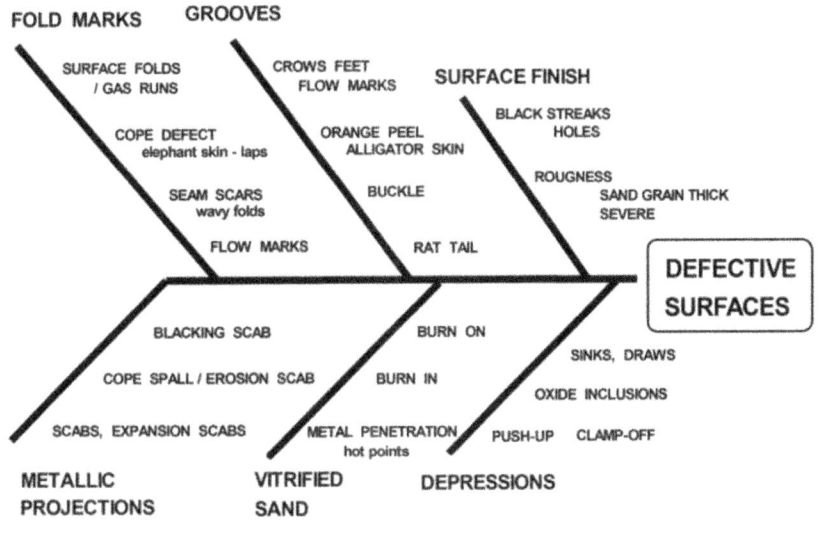

FIGURE 14.13   Defect classification chart for defective casting surfaces. (From *Casting Copper-Base Alloys* [2].)

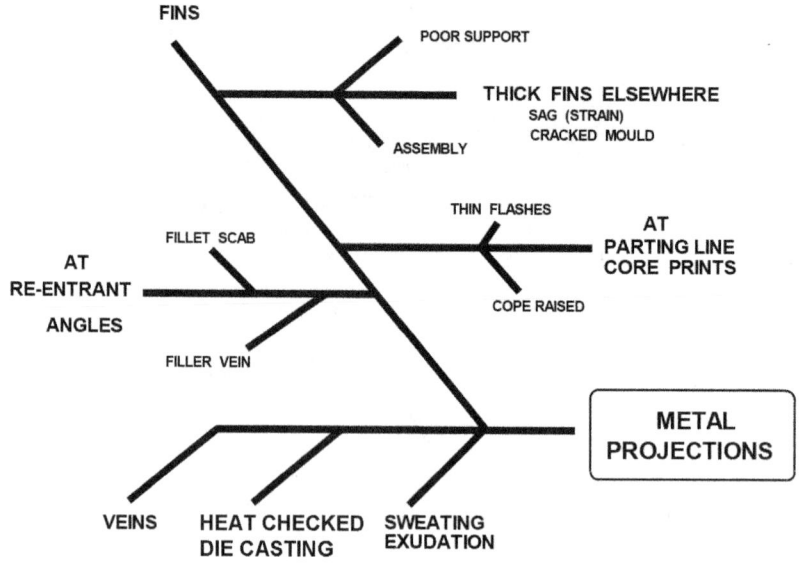

FIGURE 14.14   Defect classification chart for castings with fins, etc. (From *Casting Copper-Base Alloys* [2].)

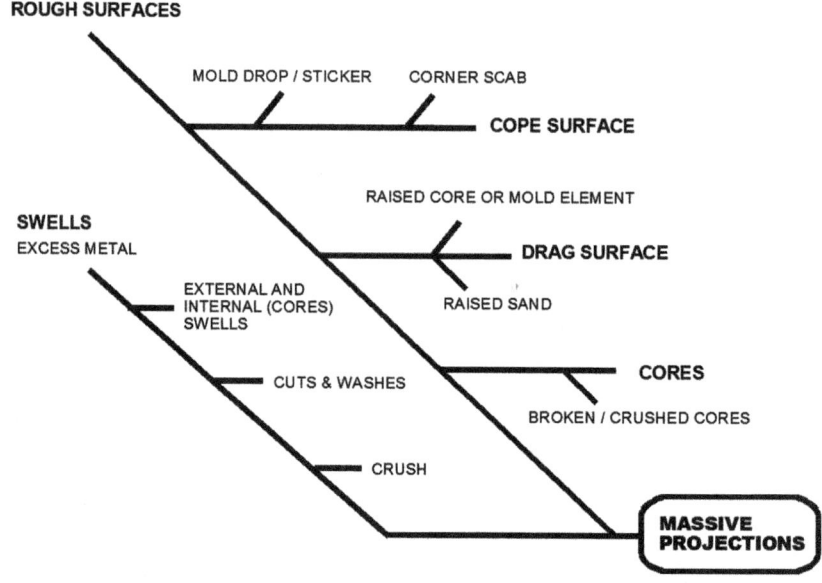

FIGURE 14.15   Defect classification chart for castings with massive projections. (From *Casting Copper-Base Alloys* [2].)

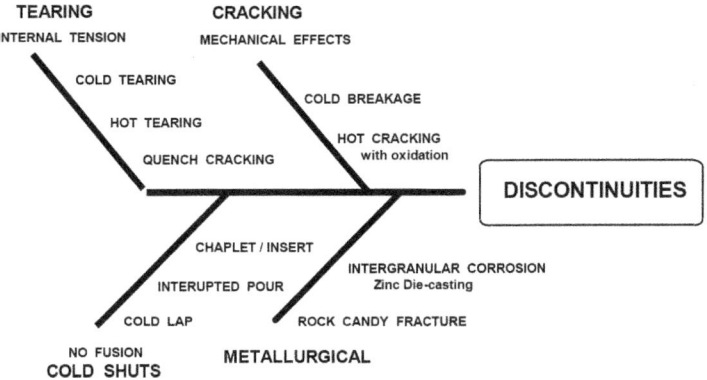

FIGURE 14.16 Defect classification chart for castings with discontinuities. (From *Casting Copper-Base Alloys* [2].)

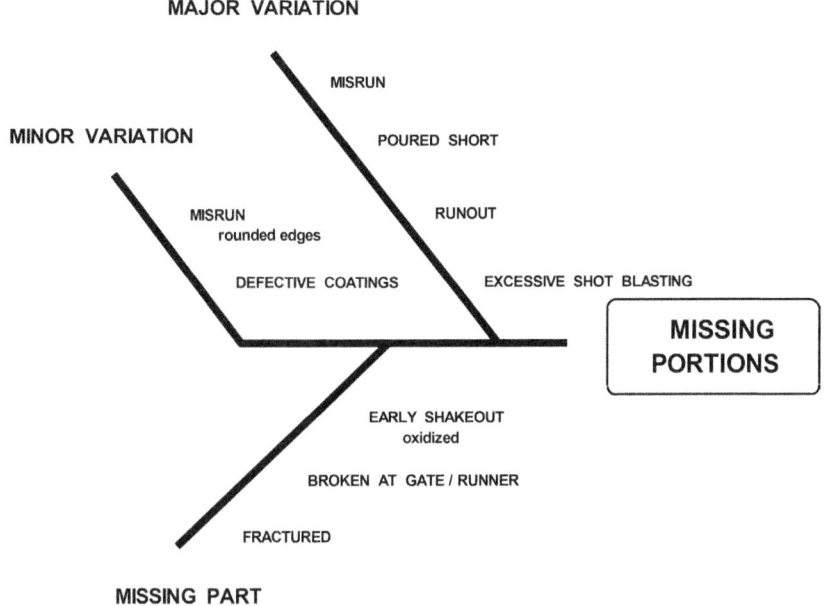

FIGURE 14.17 Defect classification chart for castings with parts missing. (From *Casting Copper-Base Alloys* [2].)

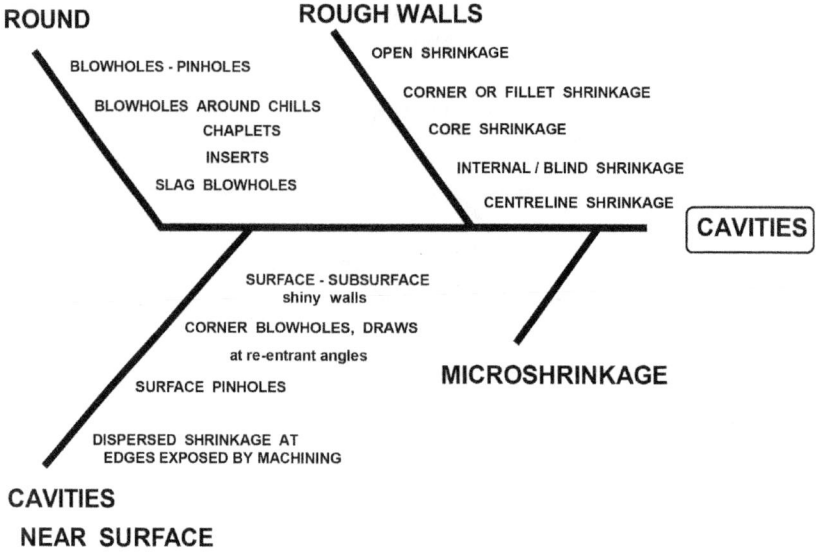

FIGURE 14.18   Defect classification chart for castings with cavities. (From *Casting Copper-Base Alloys* [2].)

These diagrams are designed to assist participants in a brainstorming session to quickly identify many issues that need to be examined in the session. They are not exhaustive, but are provided to expedite the process and indicate the various statistical studies that might be done in advance of the brainstorming session and subsequent cause elimination session(s). The brainstorming session should encourage problem solving by pursuing the following rules:

- Encourage free thinking and discourage criticism of any suggestion.
- Create a list of ideas first, discuss and organize them later.
- Put ideas on the diagram.
- Use further brainstorming to revise the diagram until all causes and subcauses have been identified.
- Later, as more information is gathered, further refinements can be made to the cause and effect diagram.

Upon completion of the cause and effect diagram for the defect in question, statistical methods such as control charts on the various processes should be used to encourage open discussion on the various process parameters and access measurements of identified causes and show the statistical distribution of the cause. Process upper and lower control limits need to be determined from the statistical data. The customer might also specify an acceptable range. Process capabilities need to be determined. If the actual control limits are well within the process capabilities, then the variable can be adequately controlled to avoid producing defects. While this might not be the actual cause of the defect, it might need to be further explored. If a cause and effect relationship is proven, the foundry might need to change processes or equipment if the existing process cannot meet the required limits.

At the end of the brainstorming sessions, a consensus must be reached on which process variables need to be investigated. Ideally, the number should be small, between 3 and 5. The least effective way to solve the problem is to change one variable at a time while holding the others fixed. A full factorial investigation reveals all interrelations. However, depending on the number, it might be cheaper to use design of experiments to investigate a partial factorial design.

Key Casting Defects

The defect numbers are those used in the *International Atlas of Casting Defects* [3]. The letter refers to direct observation of the defective casting or a precise description of those involving only the criteria of shape, appearance, location and dimensions. The seven categories are

1. Metallic projection
2. Cavities
3. Discontinuities
4. Defective surface
5. Incomplete casting
6. Incorrect dimensions or shape
7. Inclusions or structural anomalies

Each is then divided into groups or subgroups designated by numerals. Within each subgroup a third numeral is assigned, so that each defect is described by a letter followed by three numbers. Some defects could fall into more than a single category.

It has been decided to only give a few illustrative examples of casting defects that are representative of defects caused by sand, ferrous, copper, aluminum, and magnesium metals.

Sand

A211 Swells

Description

Swells show as massive and irregular projections on the surfaces of castings, generally rough and extended, and usually with the defect's rough surface and penetration. A schematic of swell formation is given in Fig. 14.19 and typical swell defects are shown in Figs. 14.20 and 14.21.

Causes

When metal pressure exceeds the strength of the sand, the casting swells and is oversized. It can be caused by excessive metallostatic head or pressure of liquid metal or insufficient sand density caused by lack of compaction. Metal pressure can also be excessive due to the expansion of the metal matrix such as the formation of graphite during cooling of gray and ductile iron. In this case, the sand mold

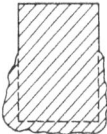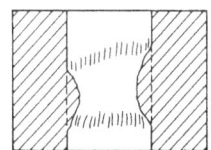

FIGURE 14.19 Schematic to show swell formation. (From *International Atlas of Casting Defects* [3].)

Figure 14.20 Dry sand steel casting with external swells on a cylinder on a part rammed up by hand. (From *International Atlas of Casting Defects* [3].)

Figure 14.21   Casting with mold wall movement. (From *Analysis of Casting Defects* [4].)

lacks the rigidity, and may be accompanied by other defects: internal shrinkage and porosity.

Remedy

Swells may be avoided by either (1) increasing the ramming density in areas of the mold subject to the defect, increasing the binder content of the sand; (2) modifying the location of the casting in the mold together with the gating and risering system to reduce the pressure of liquid metal; or (3) substituting dried or chemically bonded sand for greensand.

G131 Sand Inclusions

Description

Sand inclusions are irregular-shaped masses of sand, usually near the cope surface. Although sometimes visible on the surface, they will appear on machining. This defect occurs with scabs and other expansion defects. Figure 14.22 is a schematic of the formation of sand inclusions on the cope surface. Different sand inclusions are shown in Figs. 14.23 to 14.28.

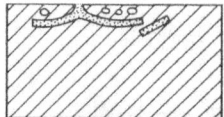

Figure 14.22 Schematic of formation of sand inclusions on the cope surface. (From *International Atlas of Casting Defects* [3].)

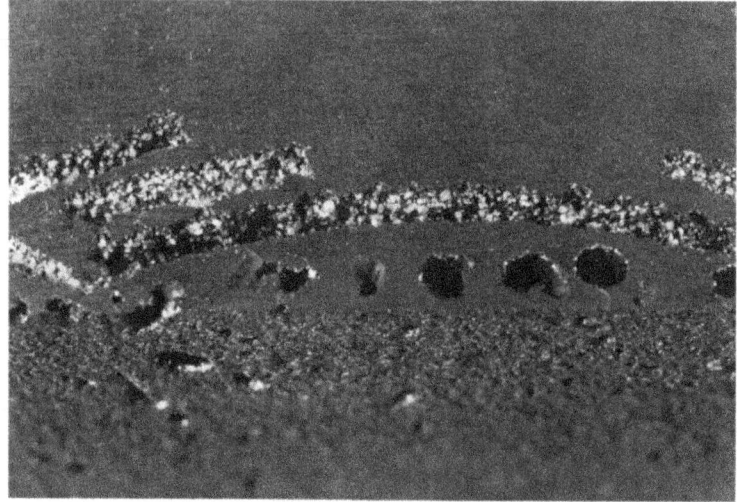

Figure 14.23 Gray iron casting with crusted sand inclusions and residual cavities. (From *International Atlas of Casting Defects* [3].)

Figure 14.24 Bronze casting in greensand; poor density of the core. (From *International Atlas of Casting Defects* [3].)

Figure 14.25 Steel casting in dry sand with sand inclusions caused by erosion during pouring. (From *International Atlas of Casting Defects* [3].)

Figure 14.26 Gay iron in greensand; sand inclusions at upper right. (From *International Atlas of Casting Defects* [3].)

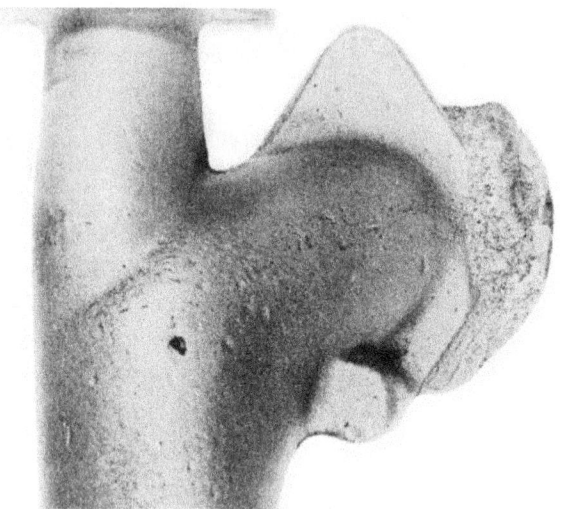

FIGURE 14.27 Broken edge washed to another part of the casting. (From *Analysis of Casting Defects* [4].)

FIGURE 14.28 Loose sand left by molder. (From *Analysis of Casting Defects* [4].)

Causes

Expansion scabs occur when pieces of sand bulge then detach from the surface of the mold due to silica expansion or weak sand detaching at the condensation zone.

Sand inclusions can also occur from poor molding habits, sand falling from above, crush, failure to clean cores, etc.

Sand washed from in or around the gating system can also contribute to this defect.

Remedy

As western bentonite has higher hot strength than other clays, its proportion in the system sand should be increased.

- Harder, more uniform ramming should improve the situation.
- The gating system should be examined to see if the metal flow can be more evenly distributed, with particular emphasis on not hitting unsupported corners.
- Mold walls can be protected with a refractory coating.
- The gating systems for steel castings are often made with ceramic inserts to protect the mold from direct contact with hot metal.

D231 Expansion Scabs

Description

Expansion scabs are irregular rough layers of metal several millimeters thick, often with sharp edges connected to the casting. They can be detached with a chisel.

- Cope scab on the upper mold surface
- Drag scab on the lower mold surface
- Buckle when the metal has not penetrated under the sand

A schematic of the expansion scab is shown in Fig. 14.29.

Causes
Cope Scabs

The heat from the molten metal drives moisture back into the mold, forming a condensation zone in the cooler sand [Fig. 14.30(1)]. This weakens this layer of sand, while the layer near the metal is dried and whose expansion is restrained by compression in the mold wall. The expanding sand buckles [Fig. 14.30(2)], and eventually cracks just as the metal arrives, letting metal into the cavity [Fig. 14.30(3)]. When the raised sand does not crack, the buckle causes a ridged depression in the casting. Photographs of scabs are shown in Figs. 14.31 and 14.32.

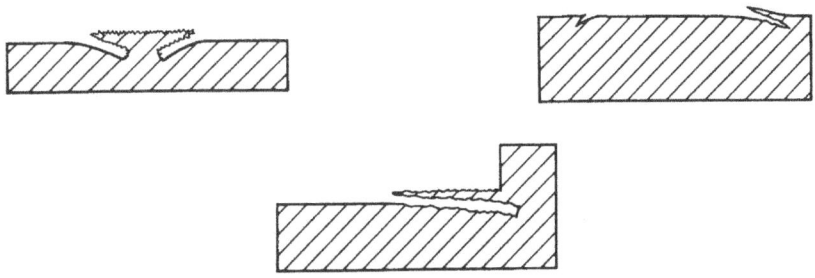

Figure 14.29 Schematics of appearance of three types of scabs. (From *International Atlas of Casting Defects* [3].)

Condensation zone

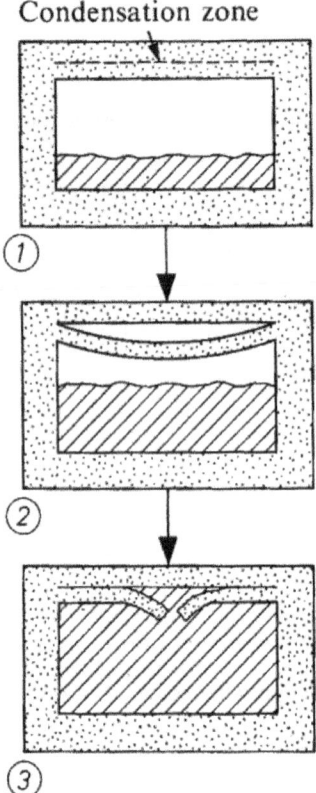

Figure 14.30 Schematic representation of formation of a scab. (From *International Atlas of Casting Defects* [3].)

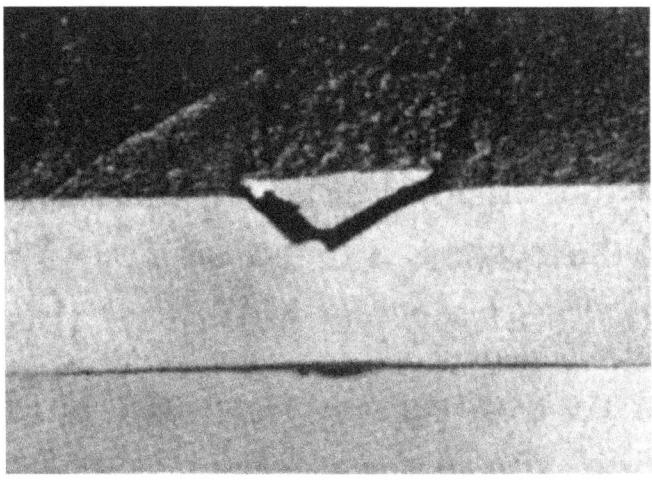

Figure 14.31 Cross section of a scab. (From *International Atlas of Casting Defects* [3].)

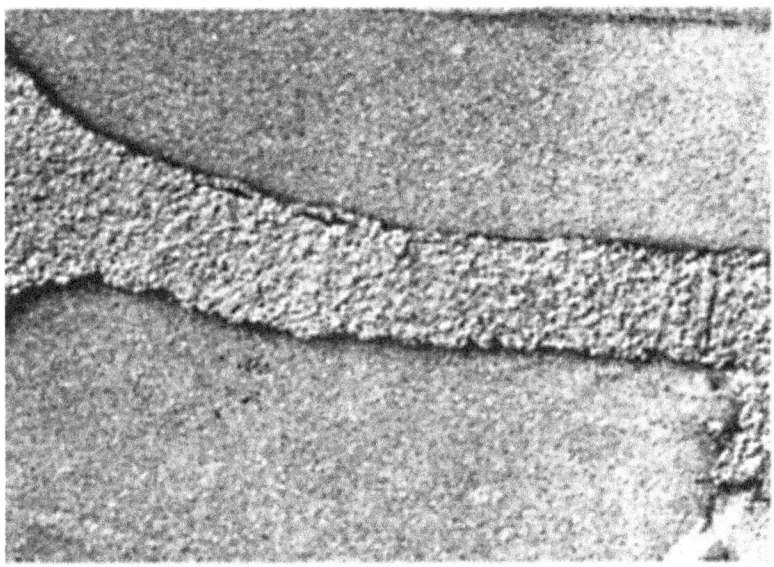

Figure 14.32 Scab. (From *International Atlas of Casting Defects* [3].)

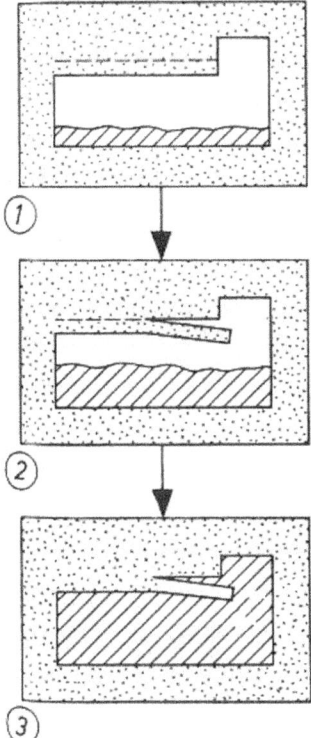

Figure 14.33 Formation of a fillet scab. (From *International Atlas of Casting Defects* [3].)

When the sand is unsupported or unrestrained by the compressive strength of the mold wall, the drying sand can detach, allowing metal in behind to form a fillet when solidified [Fig. 14.33(3)]. Photographs of fillet scabs are shown in Figs. 14.34 and 14.35.

Drag Scabs

The stream of metal causes an underlying section of sand to separate, which when it cracks leaves a wavy line across the drag surface of the casting—these lines are known as rat tails. If two parallel streams of metal cause the sand between the streams to buckle, then a combination of drag scabs and rat tails result.

Remedies

The sand strength needs to be increased by increasing the clay level, particularly western bentonite, by increasing the mulling efficiency and time duration of the mulling

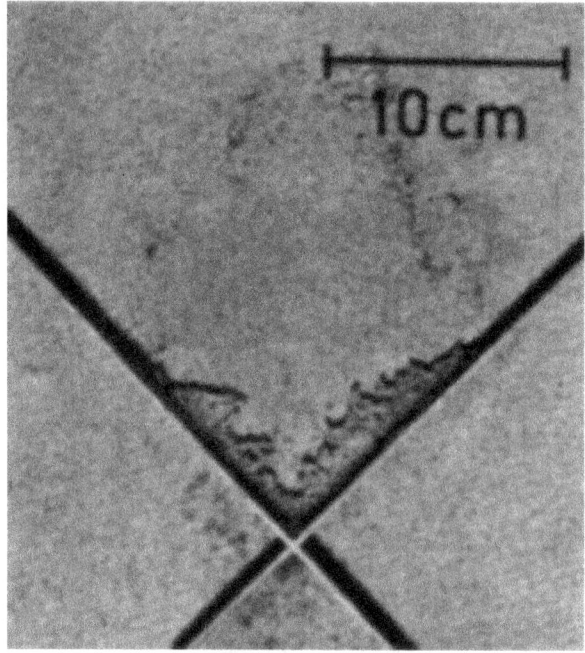

FIGURE 14.34   An example of fillet scab. (From *International Atlas of Casting Defects* [3].)

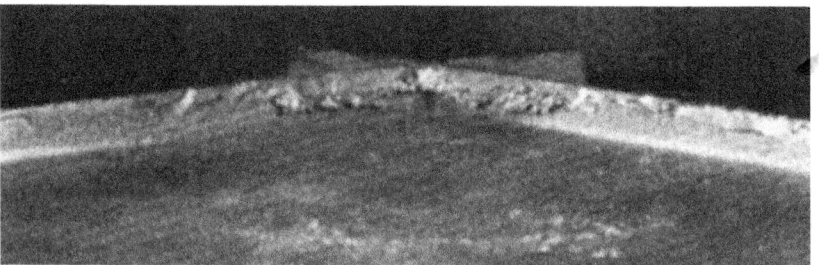

FIGURE 14.35   Another example of fillet scab. (From *International Atlas of Casting Defects* [3].)

cycle, checking the sand temperature, or using coarser sand and generally lowering the moisture content of the sand.

- Reduce the stresses on the mold by rapid, even filling of the mold, or increasing the carbonaceous material content of the sand by addition of seacoal, wood flour, or cellulose.
- Venting the mold to increase permeability and speed filling.

Defects Typical on Ferrous Castings
D221 Burn-On and D222 Burn-In
Description
Burn-on is a strongly adherent crust of sand on the casting surface that is not removed by normal shot-blast cleaning, but requires grinding.

Burn-in is a thin layer of fused sand which has a vitreous appearance. Both occur on the hottest part of the casting. A schematic of burn-on is given in Fig. 14.36 and photographs of the defects are shown in Figs. 14.37 to 14.40.

FIGURE 14.36   Schematic of burn-on. (From *International Atlas of Casting Defects* [3].)

FIGURE 14.37   Malleable iron in greensand showing burn-on. (From *International Atlas of Casting Defects* [3].)

FIGURE **14.38** Bronze castings with burn-on. (From *International Atlas of Casting Defects* [3].)

FIGURE **14.39** Cast iron in greensand showing burn-in. Defect was eliminated by reducing the pouring temperature below 2550°F (1400°C). (From *International Atlas of Casting Defects* [3].)

Causes

Burn-on is a chemical reaction of metal with sand (formation of fayalite), which can be assisted by metal penetration from lack of ramming. Metal wets the sand.

- Sand reacts with slags (metal oxides or sulfides) in the liquid metal.
- Both are caused by lack of refractoriness or too low a sintering point of the sand, and excessive pouring temperatures.

Remedies

Burn-on. Improve sand reclamation and reduce fines. Burn-on may be prevented by (1) obtaining good rammed density, particularly in hollows, e.g., between risers and casting

Fusion of Silica on Steel and Ductile

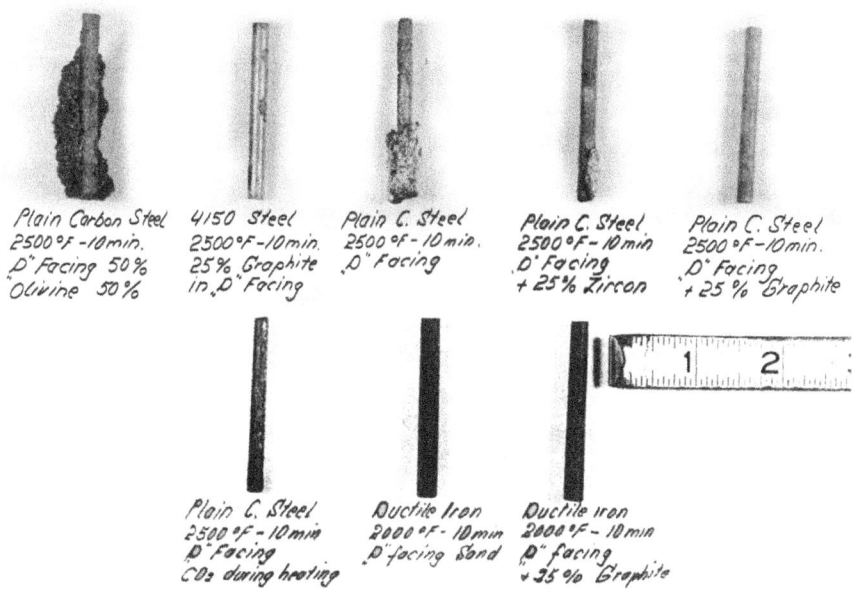

Plain Carbon Steel
2500°F - 10 min.
D° Facing 50%
Olivine 50%

4150 Steel
2500°F - 10 min.
25% Graphite
in D° Facing

Plain C. Steel
2500°F - 10 min.
D° Facing

Plain C. Steel
2500°F - 10 min.
D° Facing
+ 25% Zircon

Plain C. Steel
2500°F - 10 min.
D° Facing
+ 25% Graphite

Plain C. Steel
2500°F - 10 min
D° Facing
CO₂ during heating

Ductile Iron
2000°F - 10 min
D° facing Sand

Ductile Iron
2000°F - 10 min
D° facing
+ 25% Graphite

FIGURE 14.40   Effect of atmosphere, sand and temperature on sand fusion. (From *Analysis of Casting Defects* [4].)

walls; (2) using mold and core washes when excessive head pressure or sand coarseness exists; (3) increasing hydrocarbon additives; or (4) decreasing the pouring temperature. In some cases, the metallostatic head on fixed-height molding machines can be decreased by increasing the size of the pouring cup and pouring short.

Burn-in. Use an aggregate with higher refractoriness. Use a gating system which avoids hot spots and lower the pouring temperature.

G211 Chills Spots

Description
White iron structure, particularly in thin sections, corners, or edges. Gradual transition from gray to white iron structure. This is depicted in Fig. 14.41. Examples are given in Figs. 14.42 and 14.43.

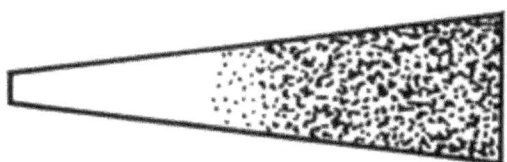

FIGURE 14.41   Schematic of chill spots in white iron. (From *International Atlas of Casting Defects* [3].)

FIGURE 14.42 The large cross section at the hub caused the outer edges of the vanes to be chilled as they were the last to fill. (From *Analysis of Casting Defects* [4].)

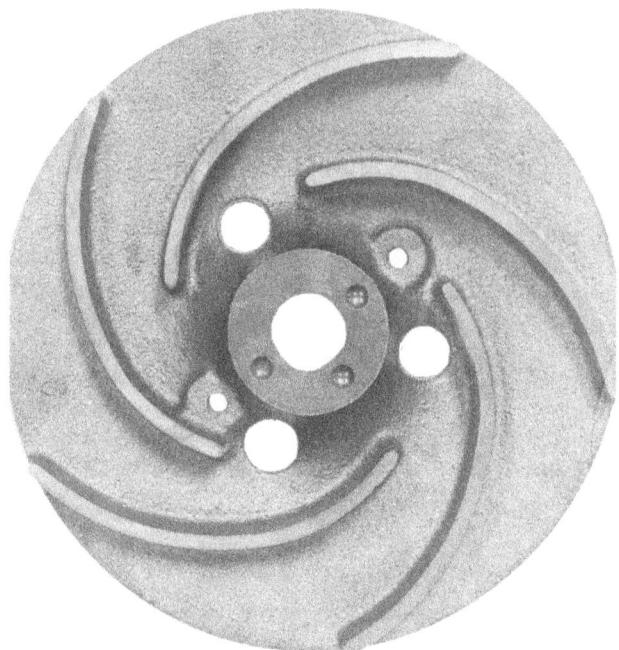

FIGURE 14.43 The casting has nonuniform sections. The thick hub caused the thin veins to chill. (From *Analysis of Casting Defects* [4].)

Cause

The carbon equivalent and/or the carbon-silicon ratio are not correct for the thickness, or for the cooling rate of the casting.

Remedy

- Ensure inoculation procedure is correct and adequate.
- Ensure mold-parting line and core prints are tight, so that fins do not affect the cooling rate.
- Control carbide-forming elements such as chromium in the charge.
- Avoid prolonged holding or overheating of the metal as it can increase the mottled area.

D112 Elephant Skin

Description

This is rough-grained or a folded surface with a shriveled appearance, like a network of groves. It is usually on the upper surface of the thick section of magnesium-treated ductile iron. Figure 14.44 gives a schematic of elephant skin with photographs in Fig. 14.45.

Causes

Magnesium oxides, sulfides, and silicates are usually dispersed throughout the liquid after treatment. In extreme cases, once in the casting cavity, when two streams meet, the silicate films do not allow complete fusion.

The defect might be confused with lustrous carbon which can appear as wrinkles. Lustrous carbon is caused by too much carbonaceous material in the molding sand, and is encouraged by slow mold filling. Iron oxide is usually the solution.

Remedy

- The gating system produces too much turbulence; check whether the sprue size is correct.
- Use good quality charge materials and the minimum amount of magnesium ladle treatment.
- Control carbon content. Do not hold the metal in the ladle longer than necessary.
- Use a teapot ladle to minimize turbulence.

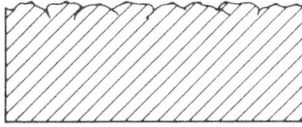

Figure 14.44 Schematic of formation of elephant skin. (From *International Atlas of Casting Defects* [3].)

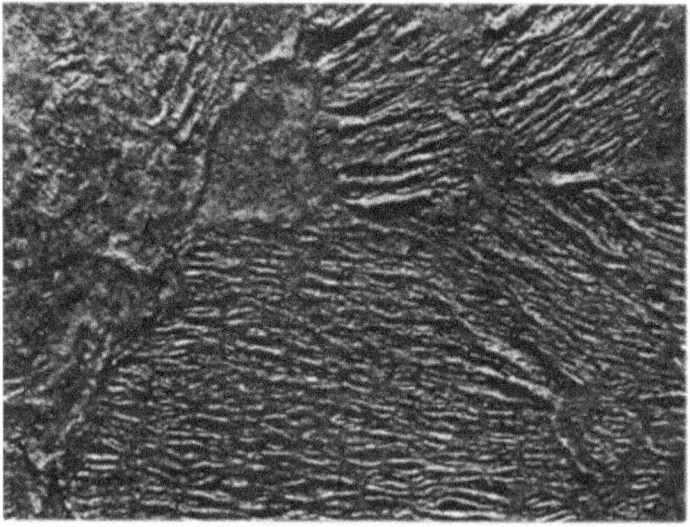

Narrow Folds

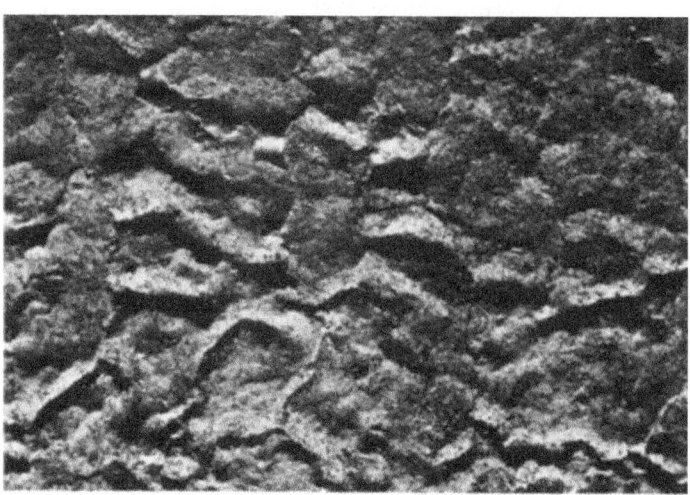

Coarse Folds

FIGURE 14.45   Interior surface of ductile iron pipe. (From *International Atlas of Casting Defects* [3].)

Defects Typical on Copper Castings

A311 Sweating

Description

These are smooth-surfaced metallic projections which are nearly spherical in shape and found on casting surfaces which have not been in contact with the mold. In copper alloys, these can be phosphide sweat, tin sweat, lead sweat, etc. The beads are generally

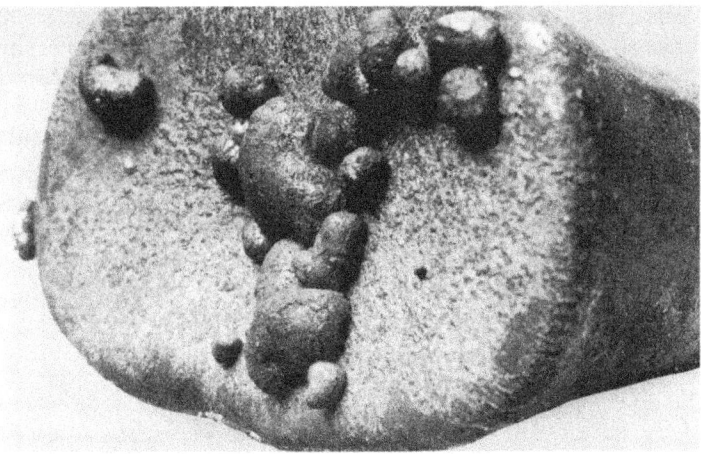

FIGURE 14.46   An example of tin sweat in a copper alloy containing approximately 85% Cu, 14% Sn, and 1% Zn. (From *International Atlas of Casting Defects* [3].)

of different composition than the base metal. An example of a tin sweat is given in Fig. 14.46, where the base metal composition is about 85 Cu, 14 Sn, and 1 Zn. However, the exuded materials had about 74 Cu, 23 Sn, and 1 Zn. In precision castings, the defect occurs most often in reentrant angles, holes, and undercuts. Sweating also occurs in light metals and cast irons. In cast iron, the droplets are generally richer in phosphorus.

Causes

In copper alloys, it is the pressure caused by the release of dissolved gases which gives rise to this defect. Hence, sweating may be associated with macroporosity. Contraction of the solidified portions of the casting can also play a role in the formation of the defect.

- In light metals, heat treatment where the temperature exceeds the solidus temperature can cause this defect.
- In cast iron, the pressure that may develop due to eutectic graphitization, evolution of dissolved gases, or contraction of the solidified portion of the casting would cause the eutectic to exude toward the surface.
- In precision investment casting with expendable patterns, air bubbles can lodge in the dip coat if not properly applied.

Remedy

- Avoid gas contamination during melting.
- Lower the temperature of the heat-treating furnace to its normal value.
- Use proper precautions during dipping and coating.

B100 Gas

Description

Gas defects arise from the presence of gas in the liquid metal which forms porosity in the casting as the metal solidifies. The porosity can appear as a cavity, visual blowhole, pinholes, or blisters. The largest cavities are most often isolated, of variable dimensions, and have an external (exogenous) origin. The smallest (pinholes) appear in groups of varying dimensions, can be uniform in size, and distributed throughout the casting (endogenous). Figure 14.47 is a sketch of the endogenous and exogenous blowholes and pinholes. Some examples are shown in Figs. 14.48 to 14.51. Sources of the gas—which may include hydrogen, air, or volatile organics—include gas soluble in liquid metal, gas

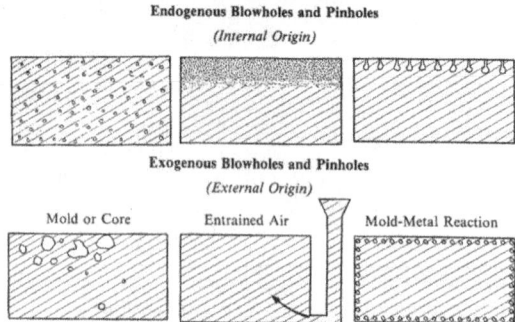

FIGURE 14.47   Drawing to show presence of endogenous and exogenous blowholes and pinholes. (From *International Atlas of Casting Defects* [3].)

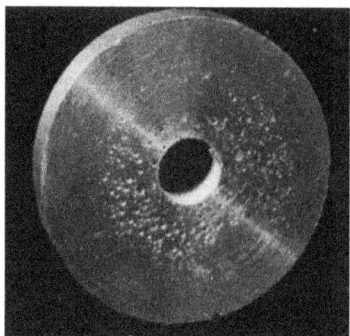

FIGURE 14.48   Blowholes in a tin bronze cast in dry sand due to gassy melt. (From *International Atlas of Casting Defects* [3].)

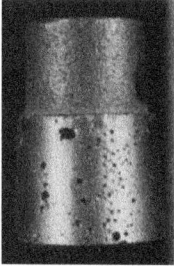

FIGURE 14.49   Blowholes of different sizes in bronze stopper cast in greensand. Gassy melt and/ or mold/metal reaction with excessive moisture in the sand. (From *International Atlas of Casting Defects* [3].)

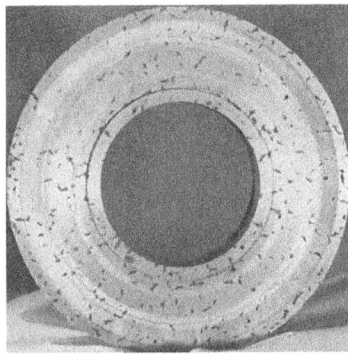

Figure 14.50 Gas defect in gray iron casting containing 0.035% Al. (From *Analysis of Casting Defects* [4].)

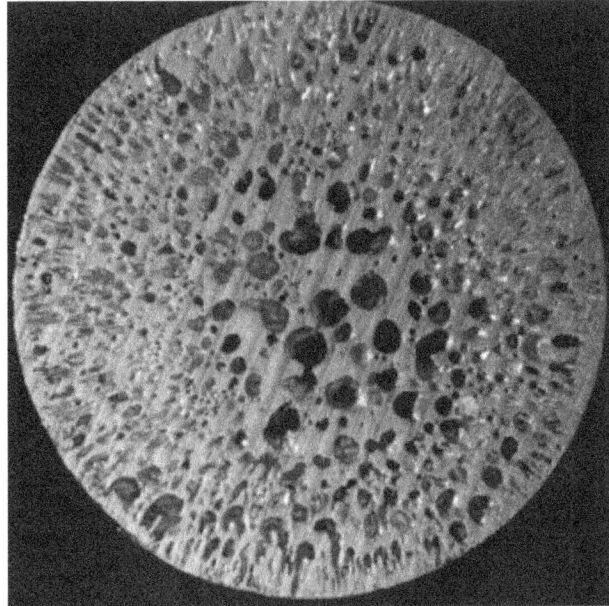

Figure 14.51 Gas porosity in a high-copper casting. (From *Casting Defects Handbook* [7].)

arising from the moisture in the sand, core outgassing, entrapped air, and mold and core wash outgassing. Gas defects have adverse effects on machinability, surface finishing, and mechanical properties.

Causes

Possible causes can be

- Metallurgical (melt too gassy or inadequately degassed and/or deoxidized; excessive melt and/or pouring temperature; metallic impurities in the melt with affinity for hydrogen)

- Sand, core, molding practice (high moisture content in sand; poorly mixed sand; high clay content; excessive binder; undercured cores; inadequate venting; inadequate permeability; wet coatings; etc.)

- Pouring, gating (turbulent or interrupted pouring; too fast or too slow pouring; short pours unable to fill mold cavity; sprue and/or gating system inadequately sized; cold and damp ladles)

Remedy

Gas-related defects can be controlled and avoided by controlling the three factors mentioned above (metallurgical, sand quality, and molding practice and proper pouring) and designing a good gating system to avoid turbulent pouring and turbulent flow of liquid metal into the mold cavity.

B311 Macroshrinkage, Microshrinkage, Shrinkage Porosity

Description

This defect can be revealed by a pressure-leakage test, examination by the naked eye (macroshrinkage), or by the aid of a magnifying glass (microshrinkage) in sections exposed by machining or destructive testing. It may also show up on casting surfaces in the form of secondary defects in coatings (enameling, galvanizing, etc.), or at surface depressions (sinks).

The defect has a spongy appearance, sometimes dendritic or in the form of small, superimposed cavities. It is generally localized in the last sections to solidify (heavy walls, intersections, reentrant angles, cores, or adjacent gates and risers). Drawings of the appearance of such defects at L sections and T sections are shown in Figs. 14.52 and 14.53, respectively. Alloys of long freezing ranges are invariably prone to this type of defect. These should not be confused with gas defects which are, in general, round in shape without any jagged appearance. An example of a shrinkage defect is shown in Fig. 14.54.

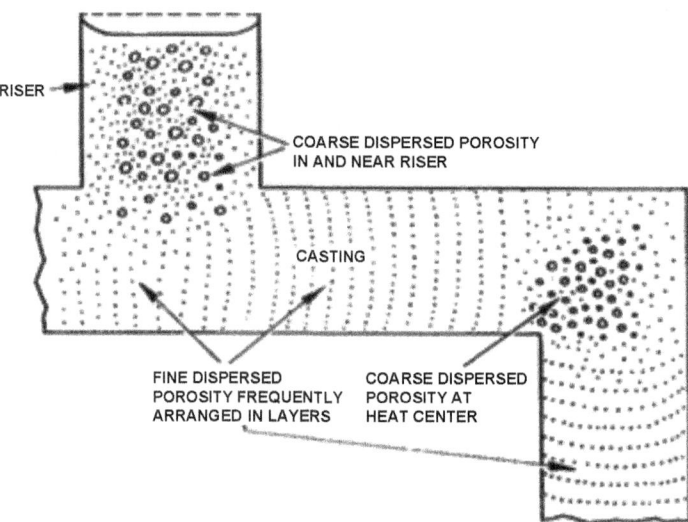

RISER

COARSE DISPERSED POROSITY
IN AND NEAR RISER

CASTING

FINE DISPERSED
POROSITY FREQUENTLY
ARRANGED IN LAYERS

COARSE DISPERSED
POROSITY AT
HEAT CENTER

Figure 14.52 Diagrammatic representation of shrinkage found in castings of wide freezing range. (From *Casting Copper-Base Alloys* [2].)

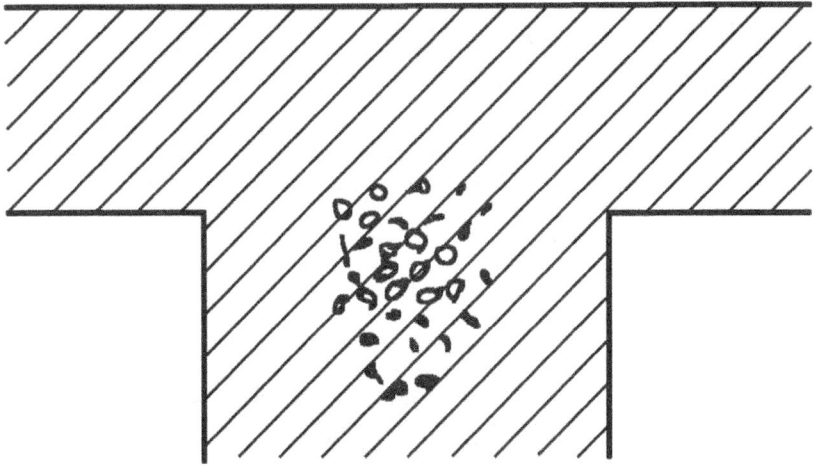

FIGURE 14.53   Drawing of a T section to show presence of shrinkage porosity. (From *International Atlas of Casting Defects* [3].)

FIGURE 14.54   Typical shrinkage defect in a thick section of a casting. (From *Analysis of Casting Defects* [4].)

Causes

This defect is more intense with abrupt changes in sections, isolated heavy sections that cannot be fed, and alloys of wide freezing ranges.

Remedies

These defects can be eliminated by (1) improved part design where reentrant angles and heavy sections are avoided; (2) incorporating better gating, feeding, and chilling to provide directional solidification; (3) selecting alloys of narrow freezing ranges; and (4) using grain refinement to produce fine equiaxed grain structures.

Defects Typical on Aluminum Castings

D114 Flow Marks

Description

A flow mark is a visual surface defect, encountered in metal molds where there is non-uniform metal flow during filling, and is often associated with a thin oxide film. The defect appears as lines which trace the flow of the liquid metal. Typical examples of flow marks are given in Figs. 14.55 and 14.56.

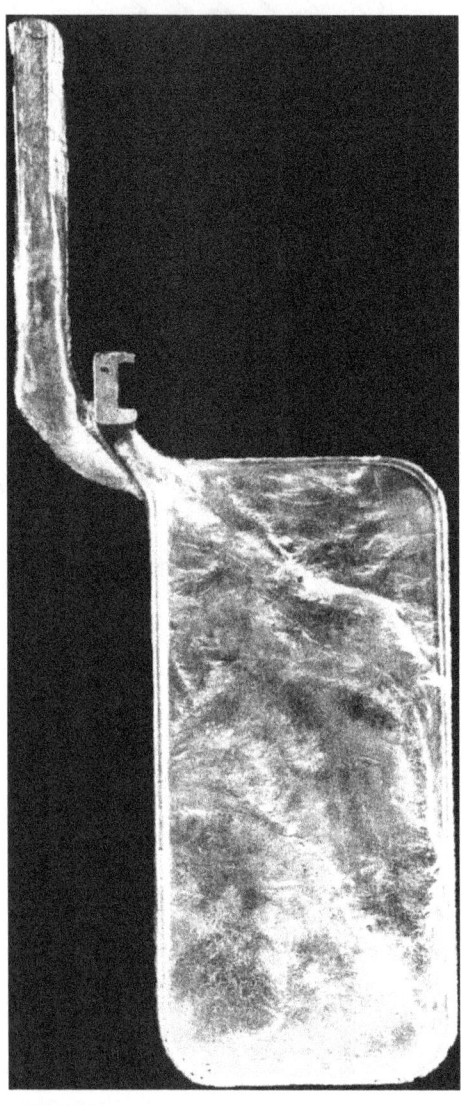

FIGURE 14.55 Flow marks in an aluminum permanent mold casting. (From *International Atlas of Casting Defects* [3].)

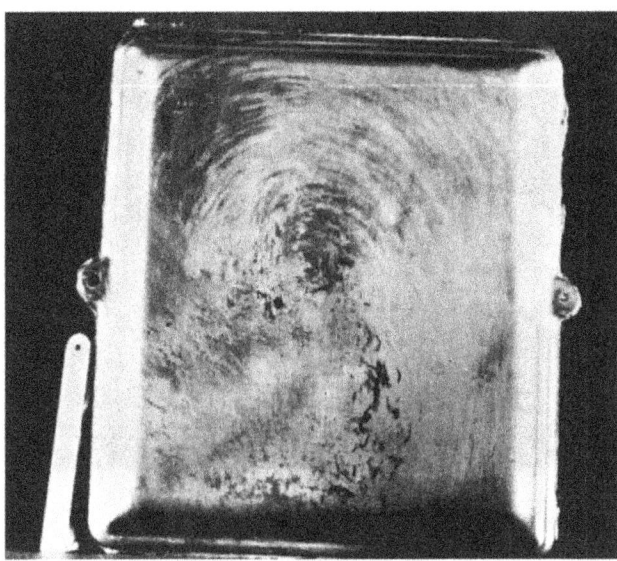

FIGURE **14.56**   Flow marks in an aluminum die casting. (From *International Atlas of Casting Defects* [3].)

Causes

Possible causes for this defect are oxide films which lodge at the surface, partially marking the paths of metal flow through the mold, interrupted or irregular pouring rate, gassy metal, low pouring temperature allowing premature solidification in thin sections, and rough handling producing turbulent metal flow.

Remedy

Flow marks can be avoided by increasing the mold temperature, modifying gate size and location, using tilt pouring to minimize turbulent flow, etc.

C221 Hot Tearing

Description

Hot tearing is a defect that occurs in nonferrous castings (sand as well as permanent mold) during the solidification process before the casting becomes completely solid, and in which physical constraints restrict the casting from normal contraction. These are more or less intercrystalline fissures of irregular outline. The cracks often show a fine dendritic structure with an oxidized surface. Typical hot-tear cracks are shown in Figs. 14.57 and 14.58.

Causes

As the solidifying alloy is subjected to constraint or deformation in the semisolid state, there is hindered contraction due to (1) faulty design (large differences in section thickness, abrupt transition from one thickness to another, too many branching/connected sections); (2) sand (excessively rigid mold and cores, vitrification of sand); (3) permanent molding/die casting (insufficient taper, extraction of cores which are improperly aligned or guided, mold opened too soon, excessive pouring temperature, low mold temperature); or (4) metallurgical causes (long freezing range nonferrous alloy such as tin bronzes, Al-Cu based alloys; insufficient P content in cast iron for permanent mold casting, high S and Al in cast steel; large grain structures due to lack of grain refinement, or presence of tramp elements such as Bi in monels).

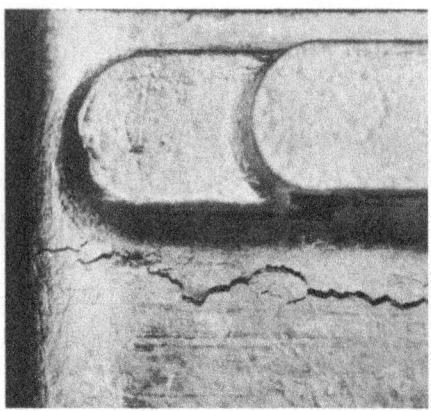

FIGURE 14.57 Hot-tear cracks in an aluminum alloy casting due to restrained contraction during solidification. (From *Casting Defects Handbook* [6].)

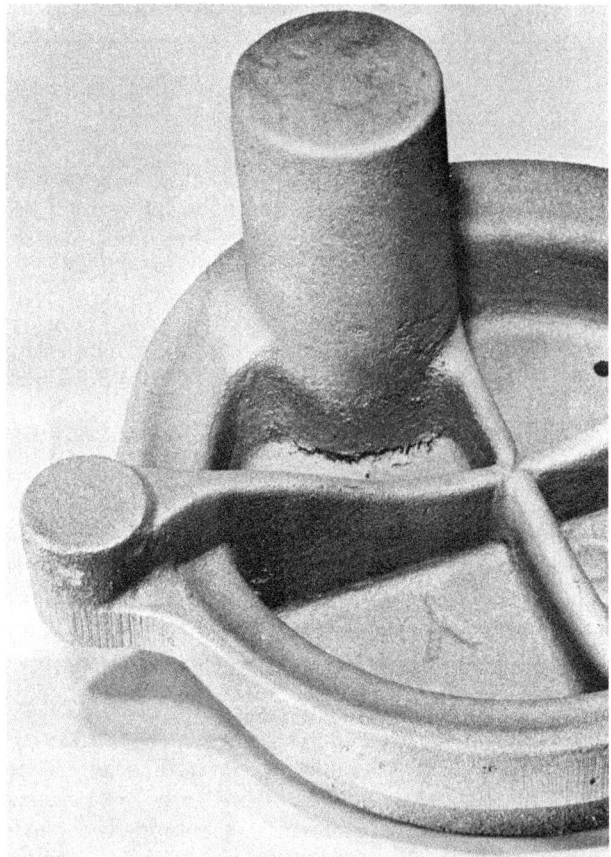

FIGURE 14.58 Hot-tear cracks in a copper alloy casting at the riser-casting interface. (From *International Atlas of Casting Defects* [3].)

Remedy

General causes listed above should be corrected to eliminate hot tearing. Chapters 4 (solidification), 7 (aluminum alloys), and 9 (magnesium alloys) have addressed issues relating to grain refinement and proper mold temperature for permanent mold casting to control hot tearing.

Defects Typical on Magnesium Castings

Oxidation and Fold Formation

Casting defects observed during sand and permanent mold casting of nonferrous alloys such as aluminum are applicable to magnesium. However, fold formation is the major defect during lost-foam casting of magnesium alloys. Oxidation and fold formation can be more severe for Mg alloys in comparison with Al alloys as the former is more reactive to oxygen. Typical fold defects are shown in Fig. 14.59.

Figure 14.59   Fold and inclusion defects in AM50 magnesium alloy. (From Han et al. [12].)

Causes

Fold and inclusion defects can be due to merging of two flow fronts during mold filling. Turbulent mold filling can also create fresh melt/air surfaces and oxides on these fresh surfaces. Oxide defects can be due to transfer of oxides created in the melting furnace or residual metal oxide left in the pouring ladle during pouring.

Remedy

Avoid abrupt changes in the section thickness and sharp corners to minimize turbulence during mold filling. Steel filters could be incorporated into the gating system to trap the oxides during pouring.

References

1. R. K. Buhr, R. D. Warda, L. V. Whiting, K. G. Davis, and M. Sahoo, "The State of Foundry Technology as Measured by a Mobile Foundry," *Transactions of American Foundrymen's Society*, 96: 171–182, 1988.

2. *Casting Copper-Base Alloys*, edited by S. Ducharme, M. Sahoo and K. Sadayappan 2d ed., Schaumburg, IL, 2007.

3. *International Atlas of Casting Defects*, AFS, Des Plaines, IL, 1999.

4. *Analysis of Casting Defects*, AFS, Des Plaines, IL, 1974.

5. *Manual of Casting Defects*, 2d ed., S&B Industrial Minerals, GMBH, Marl, Germany, 2006.

6. *Casting Defects Handbook: Aluminum & Aluminum Alloys*, AFS, Schaumburg, IL, 2011.

7. *Casting Defects Handbook: Copper and Copper-Base Alloys*, AFS, Schaumburg, IL, 2010.

8. *Casting Defects Handbook: Iron and Steel*, AFS, Schaumburg, IL, 2008.

9. Al Algarsamy, Alagarsamy Consulting, private communication.

10. T. P. Enright, "Statistical Methods Reduce Casting Defects at Ford," *Modern Casting*, 78: 51–52, Nov. 1988.

11. G. N. Booth, "Defining Quality through SPC: Foundry Applications," *Modern Casting*, 75: 27–32, May 1985.

12. "Lost Foam Casting," chapter by Q. Han, C. Song, and M. Marlatt, *Technology for Magnesium Production: Design, Products and Applications*, M. Sahoo (as principal author), American Foundry Society, 2011, pp. 161–182.

Index

The page numbers containing italic letters f, t, and n refer to figures, tables, and notes on those pages.

S